Applied Mathematical Sciences

Volume 213

The mathematization of all sciences, the fading of traditional scientific boundaries, the impact of computer technology, the growing importance of computer modeling and the necessity of scientific planning all create the need both in education and research for books that are introductory to and abreast of these developments. The purpose of this series is to provide such books, suitable for the user of mathematics, the mathematician interested in applications, and the student scientist. In particular, this series will provide an outlet for topics of immediate interest because of the novelty of its treatment of an application or of mathematics being applied or lying close to applications. These books should be accessible to readers versed in mathematics or science and engineering, and will feature a lively tutorial style, a focus on topics of current interest, and present clear exposition of broad appeal. A compliment to the Applied Mathematical Sciences series is the Texts in Applied Mathematics series, which publishes textbooks suitable for advanced undergraduate and beginning graduate courses.

Muhammad Sahimi

Applications of Percolation Theory

Second Edition

Springer

Muhammad Sahimi
Department of Chemical Engineering
and Materials Science
University of Southern California
Los Angeles, CA, USA

ISSN 0066-5452 ISSN 2196-968X (electronic)
Applied Mathematical Sciences
ISBN 978-3-031-20385-5 ISBN 978-3-031-20386-2 (eBook)
https://doi.org/10.1007/978-3-031-20386-2

Mathematics Subject Classification: 60K35, 82B43, 82B44, 60K37, 60K40, 60-02

This Springer imprint is published by the registered company Springer Nature Switzerland AG
The registered company address is: Gewerbestrasse 11, 6330 Cham, Switzerland

Dedicated to
the people of my native land, my beloved Iran,
may they live in freedom, peace, and
prosperity;
and to
the memory of my brother, Ali (1958–1981),
who lost his life for his ideals

Preface to the Second Edition

Shortly after the first edition of this book was published in 1994, a European physicist remarked that "percolation as a research field is dead." Since then, however, percolation theory and its applications have advanced greatly. On the experimental side, new instruments have made it possible to measure various properties of heterogeneous systems. As a result, characterization of heterogeneous materials and media, which have provided fertile grounds for applying percolation theory to model their properties, can now be done in great details and have demonstrated the crucial role of connectivity on their properties. Many new concepts and ideas, such as complex graphs and networks, as well as extensions of the classical percolation model, such as explosive and optimal percolation, have also been developed that have made it possible to greatly extend the range of possible applications of the theory. New methods of analysis have also made it possible to analyze precise data for large-scale industrial and societal problems, which have demonstrated the crucial effect of the connectivity on the properties of such systems, hence motivating further development and applications of percolation theory. As a result, not only are percolation theory and its applications not dead, but they are in fact thriving.

Such new developments motivated the preparation of the second edition of this book. But this new edition does not represent merely an updated version of the first edition. All the chapters have been rewritten completely, updated, and expanded. Multiple new chapters have been added that demonstrate clearly the depth and breadth of the role of connectivity and percolation in an enormous number of phenomena.

Throughout my life, I have been blessed by great mentors. My mentors for life, my mother, Fatemeh Fakour Rashid (1928–2006), and father, Habibollah Sahimi (1916–1997), instilled in me my love for reading and science. Over 48 years after taking the first of many courses with him, I am still influenced by Dr. Hassan Dabiri, my first academic mentor when I was attending the University of Tehran in Iran, and his outstanding qualities, both as an academic mentor and as a wonderful human being. My advisors for my Ph.D. degree at the University of Minnesota, the late Profs. H. Ted Davis (1937–2009) and L. E. Skip Scriven (1931–2007), introduced me to percolation theory.

As the famous song by John Lennon and Paul McCartney goes, *I get by with a little help from my friends*, except that in my case my students and collaborators have given me *a lot* of help, and have contributed to my understanding of the topics described in this book. First and foremost, I have been blessed by many outstanding doctoral students and postdoctoral fellows with whom I have worked throughout my academic career on some of the problems described in this book. They include Drs. Sepehr Arbabi, Fatemeh Ebrahimi, Hossein Hamzehpour, Mehrdad Hashemi, Abdossalam Imdakm, Ehsan Nedaaee, and Sumit Mukhopadhyay. Over the years, I have also been most fortunate to have fruitful collaborations with many friends and colleagues on research problems related to what is studied in this book, including Profs. Behzad Ghanbarian, Joe Goddard, Barry Hughes, Allen Hunt, Mark Knackstedt, Reza Rahimi Tabar, Charles Sammis, Nima Shokri, the late Dietrich Stauffer, and Theodorte Tsotsis. I am extremely grateful to them.

Michael C. Poulson was the publisher of the first edition of this book with Taylor & Francis, as well as a close friend. He passed away on 31 December 1996 at the age of 50. In preparing this edition, I greatly missed his wise advices, great humor, and cheerful personality.

My wife Mahnoush, son Ali, and daughter Niloofar are the sunshine of my life. They put up with my long absence from family life, and my spending countless number of hours at home in front of the computer to write this book. This edition would not have been completed without their love, patience, and understanding.

I dedicate this book to the people of my native land, my beloved Iran, and the memory of my younger brother Ali (1958–1981). He was a university student when he lost his life on 19 September 1981 during his struggle for his ideals for a better Iran. I will miss him until I meet him again.

Los Angeles, CA, USA Muhammad Sahimi
August 2022

Preface to the First Edition

Disorder plays a fundamental role in many processes of industrial and scientific interest. Of all the disordered systems, porous media are perhaps the best-known example, but other types of disordered systems, such as polymers and composite materials, are also important and have been studied for a long time. With the advent of new experimental techniques, it has become possible to study the structure of such systems and gain a much deeper understanding of their properties. New techniques have also allowed us to design the structure of many disordered systems in such a way that they possess the properties that we desire.

During the past two decades, the development of a class of powerful theoretical methods has enabled us to interpret the experimental observations, and predict many properties of disordered systems. Included in this class are renormalization group theory, modern versions of the effective-medium approximation, and percolation theory. Concepts of percolation theory have, however, played one of the most important roles in our current understanding of disordered systems and their properties. This book attempts to summarize and discuss some of the most important applications of percolation theory to the modeling of various phenomena in disordered systems. Among such applications phenomena that occur in porous media have perhaps received the widest attention. Thus, we now have a fairly good understanding of two-phase flow problems in porous media, and recognize that, at least in certain limits, such problems represent percolation phenomena. Oil recovery processes have thus benefited from the application of percolation to two-phase flow problems. Reaction and diffusion in porous materials, such as catalysts and coal particles, have also benefited from the insights that percolation has provided us. Such applications are discussed in details in this book. However, other well-known applications of percolation, such as those to polymers and gels, composite materials, and rock masses are also discussed.

In this book, we consider applications of percolation theory to those phenomena for which there are well-defined percolation models, *and* a direct comparison between the predictions of the models and experimental data is possible. Percolation has been applied to many phenomena; it has, however, not always been possible to compare the predictions with the data and, therefore, it has not always been possible to check the quantitative accuracy of the predictions. Although a theoretician may justifiably argue that, "an application is an application is an application," the scope of the book is limited to such "practical" applications, since it is impossible to discuss in one book all applications of percolation theory, and this is a book written by an application-oriented chemical engineer. New applications of percolation are still being developed, and, in the coming years, such applications will find widespread use in many branches of science and technology.

Over the past decade, Dietrich Stauffer has greatly contributed to my understanding of percolation theory, disordered systems, and critical phenomena. He has done this through his "referee's reports," e-mail messages, letters, and our collaborations on various problems. Without his constant encouragement and support, this book would not have been written. He also read most of the book and offered constructive criticisms and very useful suggestions. I am deeply grateful to him. I would also like to thank Ted Davis and Skip Scriven who introduced me to percolation, and Barry Hughes for his many stimulating discussions and fruitful collaboration in my graduate school years. Many other people have contributed to my understanding of percolation theory, a list of whom is too long to be given here. I would like to thank all of them.

Most of this book was written while I was visiting the HLRZ Supercomputer Center at KFA Jülich, Germany, as an Alexander von Humboldt Foundation Research Fellow. I would like to thank Hans Herrmann and the Center for their warm hospitality, and the Foundation for financial support.

Los Angeles, CA, USA
April 1993

Muhammad Sahimi

Contents

1 Macroscopic Connectivity as the Essential Characteristic of Heterogeneous Systems ... 1
- 1.1 Introduction ... 1
- 1.2 What is Percolation? ... 3
- 1.3 The Scope of the Book ... 5
- 1.4 Further Reading ... 6

2 Percolation Theory: Classical and Poor Man's ... 7
- 2.1 Introduction ... 7
- 2.2 Historical Origin of Percolation Theory ... 8
- 2.3 Random Bond and Site Percolation ... 10
- 2.4 Percolation Thresholds ... 11
- 2.5 Bicontinuous Materials, Phase-Inversion Symmetry, and Percolation ... 13
- 2.6 Generating a Single Percolation Cluster: The Leath–Alexandrowicz Algorithm ... 13
- 2.7 The Newman–Ziff Algorithm ... 14
- 2.8 Percolation Properties ... 15
  - 2.8.1 Connectivity Properties ... 15
  - 2.8.2 Flow and Transport Properties ... 17
  - 2.8.3 The Structure of the Sample-Spanning Cluster ... 18
- 2.9 Universal Power Laws for Percolation Properties ... 18
  - 2.9.1 Connectivity Properties ... 19
  - 2.9.2 Flow and Transport Properties ... 19
- 2.10 Scale Dependence of Percolation Properties ... 21
- 2.11 Finite-Size Scaling ... 23
- 2.12 Experimental Measurement of Percolation Properties ... 24
- 2.13 Poor Man's Percolation Theory: Effective-Medium Approximation ... 29

3 Extensions of the Classical Percolation Model 33
3.1 Introduction 33
3.2 Continuum Percolation 33
3.2.1 Percolation Threshold of Continua 34
3.2.2 The Ornstein–Zernike Formulation 37
3.2.3 Percolation Thresholds: Materials with Very Low or High Thresholds 38
3.2.4 Differences Between Lattice and Continuum Percolation 38
3.2.5 Boiling: An Application of Continuum Percolation 42
3.3 Correlated Percolation 43
3.3.1 Percolation with Short-Range Correlations 43
3.3.2 Percolation with Long-Range Correlations 44
3.3.3 Correlated-Site Percolation and Unusual Properties of Low-Temperature Water 45
3.3.4 Site-Bond Correlated Percolation and Gelation 47
3.4 Bootstrap Percolation 48
3.4.1 k-Core Percolation 50
3.5 Invasion Percolation 50
3.6 Explosive Percolation 51
3.7 Directed Percolation 51
3.8 Dynamic Percolation 52
3.9 Percolation Model of Galactic Structures 52
3.10 Vector Percolation 53

4 Characterization of Porous Media and Materials 55
4.1 Introduction 55
4.2 Diagenetic Processes and Formation of Rock 57
4.3 Geometrical-Percolation Models of Diagenetic Processes 58
4.4 Percolation Model of the Tortuosity 59
4.4.1 Tortuosity and Continuum Percolation 62
4.4.2 Percolation Models for Tortuosity of Saturated and Unsaturated Porous Media 63
4.5 Characterization of Pore-Space Geometry 65
4.5.1 The Young–Laplace Equation 65
4.5.2 Contact Angle, the Washburn Equation, and Capillary Pressure 66
4.5.3 Mercury Porosimetry 68
4.5.4 Invasion and Random Percolation Models 70
4.6 Sorption in Porous Media 77
4.6.1 Percolation Models of Sorption 80
4.6.2 Determining Connectivity of Porous Materials Using Percolation 82

5 Connectivity of Fracture and Fault Networks 85
5.1 Introduction 85
5.2 Why Should Percolation be Relevant to Fractures? 86
5.3 Self-Similar Fractal Structures 87
5.4 Experimental Evidence for Self-Similarity of Fracture Networks and Relevance of Percolation 88
5.5 Excluded Area and Excluded Volume 89
5.6 Models of Fracture Networks and Their Percolation Threshold 91
5.7 Two-Dimensional Models 91
5.8 Three-Dimensional Models 94
5.9 Percolation Threshold of Fracture Networks 96

6 Earthquakes, Critical Phenomena, and Percolation 101
6.1 Introduction 101
6.2 Earthquakes as a Critical Phenomenon 103
6.3 Spatial Distribution of Earthquakes Hypocenters and its Relation with Percolation 106
6.4 Microseismicity and Percolation 108
6.5 The Gutenberg–Richter Law and its Relation with Percolation 110
6.6 Percolation Model of Regional Seismicity 112

7 Flow and Transport Properties of Porous Materials 117
7.1 Introduction 117
7.2 Percolation, Poor Man's Percolation, or Critical-Path Analysis? 117
7.3 Selecting the Correct Percolation Model of the Unsaturated Zone in Soil 120
7.4 Diffusivity 121
7.4.1 Percolation Approach 121
7.4.2 Poor Man's Percolation: Effective-Medium Approximation 124
7.4.3 Effective-Medium Approximation and Percolation 124
7.5 Permeability 125
7.5.1 Poor Man's Percolation: Effective-Medium Approximation 126
7.5.2 Critical-Path Analysis 128
7.5.3 Percolation-CPA Models of Hydraulic Conductivity and Air Permeability 133
7.6 Electrical Conductivity 134
7.6.1 Effective-Medium Approximation 135
7.6.2 Combining Effective-Medium Approximation and Percolation 136
7.6.3 Critical-Path Analysis 137

7.6.4 Renormalization Group Method 139
7.6.5 Renormalized Effective-Medium Approximation 144
7.6.6 Percolation and CPA models of Electrical Conductivity and Electrokinetic Coupling 146
7.7 Thermal Conductivity 149
7.7.1 Percolation Model 150
7.7.2 Effective-Medium Approximation 150
7.7.3 Percolation-Modified Effective-Medium Approximation 150
7.7.4 Universality Across Flow and Transport Properties 151

8 Mixing and Dispersion in Flow Through Porous Media 155
8.1 Introduction 155
8.1.1 The Phenomenon of Dispersion 155
8.2 Mechanisms of Dispersion 156
8.3 The Convective–Diffusion Equation 157
8.4 Taylor–Aris Dispersion in a Capillary Tube 159
8.5 Classification of Dispersion Regimes in Porous Media 160
8.6 Percolation and Pore-Network Models 162
8.7 Effect of Percolation and Connectivity 163
8.7.1 Scaling Theory 164
8.8 Critical-Path Analysis 169
8.9 Comparison with Experiments 172

9 Multiphase Fluid Flow in Porous Media 175
9.1 Introduction 175
9.2 Random Percolation Model 176
9.3 Computing the Relative Permeabilities with Random Percolation Model 179
9.3.1 Use of the Bethe Lattices 179
9.3.2 Poor Man's Percolation: Effective-Medium Approximation 180
9.4 Random Site-Correlated Bond Percolation Models 185
9.5 Invasion Percolation 185
9.5.1 Comparison with Experimental Data 186
9.6 Flow of Thin Wetting Films in Pores 186
9.7 Invasion Percolation with Two Invaders and Two Defenders 189
9.8 Random Percolation with Trapping 193
9.9 Roughening and Pinning of a Fluid Interface 193
9.10 Comparison with Experimental Data and Relation with Directed Percolation 194
9.11 Finite-Size Effects and Devil's Staircase 195

9.12 Displacement Under the Influence of Gravity: Gradient Percolation ... 197
9.12.1 Comparison with Experiments ... 198
9.13 Immiscible Displacements at Finite Capillary Numbers: Site-Bond Invasion Percolation ... 199
9.14 Evaporation and Drying in Porous Media ... 202
9.14.1 Pore-Network Simulation ... 204
9.14.2 Scaling Theory of Drying ... 206

10 Percolation in Evolving Media ... 209
10.1 Introduction ... 209
10.2 Noncatalytic Gas–Solid Reactions with Fragmentation ... 209
10.2.1 The Reaction-Limited Regime ... 210
10.2.2 Diffusion-Limited Regime ... 211
10.3 Percolation Models of Coal Gasification with Fragmentation ... 213
10.3.1 The Reaction-Limited Regime ... 213
10.3.1.1 Statistics of Percolation ... 214
10.3.1.2 Chemical Reactions and Kinetics ... 215
10.3.1.3 Comparison with Experimental Data ... 218
10.3.2 Hybrid Percolation-Continuum Models of Diffusion-Limited Regime ... 220
10.3.3 Pore-Network Models ... 221
10.3.3.1 Fragmentation During Gasification ... 223
10.4 Noncatalytic Gas–Solid Reactions with Pore Blocking ... 224
10.5 Percolation Models of Catalyst Deactivation ... 227
10.5.1 Reaction-Limited Deactivation ... 228
10.5.2 Diffusion-Limited Deactivation ... 230
10.5.2.1 Hybrid Continuum Models ... 231
10.5.2.2 Pore-Network Models ... 232
10.5.3 Catalyst Deactivation at the Reactor Scale ... 234
10.6 Reaction Kinetics and Diffusion-Controlled Annihilation ... 235
10.7 Clogging of Porous Media by Precipitation of Particles Suspended in a Flowing Fluid ... 238

11 Vector Percolation and Rigidity of Materials ... 245
11.1 Introduction ... 245
11.2 Derivation of Elasstic Networks From Continuum Elasticity ... 245
11.3 The Born Model ... 247
11.4 The Central-Force Networks ... 249
11.4.1 Elastic Networks in Biological Materials ... 250
11.5 Static and Dynamic Rigidity and Floppiness of Networks ... 251
11.6 Rigidity Percolation ... 253
11.7 Elastic and Superelastic Percolation Models ... 253
11.7.1 Numerical Simulation and Finite-Size Scaling ... 255

11.7.2 The Correlation Length of Rigidity Percolation 256
11.7.3 Finite-Size Scaling Analysis 256
11.8 The Force Distribution 257
11.9 Determination of the Elastic Percolation Threshold 258
11.9.1 Estimating the Percolation Threshold by Moments of the Force Distribution 258
11.9.2 Constraint-Counting Method 259
11.9.3 The Pebble Game 260
11.10 Elastic Percolation Networks with Bond-Bending Forces 261
11.10.1 The Percolation Threshold 262
11.10.2 The Force Distribution 264
11.11 Nature of Phase Transition in Elastic Percolation Netwoks 265
11.12 Scaling Properties of the Elastic Moduli 267
11.13 Poor Man's Percolation: Effective-Medium Approximation 269
11.13.1 The Born Model 270
11.13.2 Central-Force Percolation 272
11.13.3 The Bond-Bending Models 273
11.13.4 Filamentous Networks 275
11.14 Critical-Path Analysis 276
11.15 Fixed Points of Elastic Percolation and University of the Poisson's Ratio 277

12 Transport Properties of Composite Materials 279
12.1 Introduction 279
12.2 Electrical Conductivity of Powders and Polymer Composites 281
12.2.1 Powders 281
12.2.2 Polymer Composites 286
12.2.3 Polymer–Graphene Composites 289
12.2.4 Polymer–Carbon Nanotube Composites 290
12.2.5 Effect of Thickness and Particle-Size Distribution 292
12.3 Metal–Insulators Composites 293
12.4 Tunneling Versus Percolation 297
12.5 AC Conductivity and Dielectric Properties of Heterogeneous Composites 298
12.5.1 AC Conductivity 299
12.5.2 Dielectric Constant 299
12.5.3 Experimental Verification 301
12.6 Hall Conductivity 307
12.6.1 Poor Man's Percolation: Effective-Medium Approximation 308
12.6.2 Scaling Properties 311
12.6.3 Comparison with Experimental Data 312

12.7 Mechanical Properties of Heterogeneous Materials 313
12.7.1 Foams 313
12.7.2 Porous Materials 316
12.7.3 Superrigid Materials 318
12.8 Percolation Aspects of Classical Superconductivity 318
12.8.1 Magnetoconductivity 320
12.8.2 Magnetic Properties 321
12.8.3 Comparison with Experimental Data 322
12.8.3.1 The London Penetration Depth 322
12.8.3.2 The Specific Heat 323
12.8.3.3 The Critical Current 324
12.8.3.4 The Critical Fields 325
12.8.3.5 Differential Diamagnetic Susceptibility 326
12.9 Thermal Percolation 327

13 Rheology and Elastic Properties of Network Glasses, Branched Polymers, and Gels 331
13.1 Introduction 331
13.2 Connectivity of Network Glasses and Rigidity Transition 335
13.2.1 Experimental Confirmation of the Phillips–Thorpe Theory 336
13.2.2 Network Glasses with High Coordination Numbers 339
13.2.3 Network Glasses with Dangling Atoms 341
13.2.4 Stress Transition in Network Glasses 342
13.3 Branched Polymers and Gels 343
13.3.1 Percolation Model of Polymerization and Gelation 344
13.3.2 Morphology of Branched Polymers and Gels 345
13.3.2.1 Gel Polymers 345
13.3.2.2 Experimental Confirmation of the Percolation Model 346
13.3.2.3 Lattice Animal Model of Branched Polymers 348
13.3.2.4 Experimental Confirmation of Lattice Animal Model of Branched Polymers 351
13.4 Elastic Moduli of Critical Gels 352
13.4.1 The Spectrum of the Relaxation Times 355
13.4.2 Experimental Confirmation 357
13.4.3 Chemical Gels 357
13.4.4 Enthalpic Versus Entropic Elasticity 359
13.4.5 Physical Gels 361
13.5 Viscosity of Near Critical Gelling Solutions 361

14 Vibrational Density of States of Heterogeneous Materials 365
14.1 Introduction 365
14.2 Scalar Percolation Approximation for the Density of States 367
14.2.1 Poor Man's Percolation: Effective-Medium Approximation 368
14.2.2 Scaling Theory of Phonons and Fractons 370
14.2.3 Crossover from Phonons to Fractons 372
14.2.4 Experimental and Numerical Confirmation of the Scalar Approximation 373
14.2.4.1 Large-Scale Numerical Simulations 373
14.2.4.2 Experimental Data 376
14.3 Vector Percolation and Vibrational Density of States 381
14.3.1 Scaling Theory 381
14.3.2 Crossover Between Scalar and Vector Models of Density of States 383
14.3.3 Experimental and Numerical Confirmation of Vector Percolation Model of Density of States 384
14.3.3.1 Large-Scale Computer Simulation 384
14.3.3.2 Experimental Data 386

15 Hopping Conductivity of Heterogeneous Materials 389
15.1 Introduction 389
15.2 The Miller-Abrahams Network 390
15.3 The Symmetric Hopping Model 393
15.3.1 Exact Solution for One-Dimensional Materials 394
15.3.2 Exact Solution for Bethe Lattices 396
15.3.3 Poor Man's Percolation: Effective-Medium Approximation 397
15.4 Poor Man's Percolation: EMA Derivation of Mott's Formula 401
15.5 Critical-Path Analysis 404
15.5.1 Effect of a Variable Density of States 406
15.5.2 Effect of Coulomb Interactions 407
15.5.3 Comparison with Experimental Data 408
15.6 Effect of Fractal Morphology on Hopping Conductivity 413
15.7 Universality of Frequency-Dependent Hopping Conductivity 415

16 Applications of Invasion Percolation 419
16.1 Introduction 419
16.2 Invasion Percolation 420
16.3 Extensions of Invasion Percolation 422
16.3.1 Self-Avoiding and Loopless Compressible Invasion Percolation 422
16.3.2 Gradient Invasion Percolation 422
16.3.3 Thermal Invasion Percolation 423

16.4 Dynamics of Invasion Percolation: The Relation with Self-Organized Criticality 423
16.5 Applications 425
16.5.1 Modeling of Stream Networks 425
16.5.2 Healthy and Cancerous Vascular Networks 428
16.5.3 Simulating Spin Systems Near and Away from Criticality 429
16.5.4 Optimization Problems 430
16.5.5 The Queuing Problem 432
16.5.6 Social Dynamics 433

17 Percolation in Random Graphs and Complex Networks 435
17.1 Introduction 435
17.2 Erdös-Rényi Graph 436
17.2.1 Properties of Erdös-Réyni Graph 436
17.2.2 Percolation in Erdös-Réyni Graph 437
17.3 Small-World and Scale-Free Networks 437
17.4 Generation of Scale-Free Networks 440
17.5 Percolation Properties 441

18 Percolation in Biological Systems 443
18.1 Introduction 443
18.2 Antigen–Antibody Reactions and Aggregation 444
18.3 Network Formation on Lymphocyte Membranes 446
18.4 Percolation in Immunological Systems 447
18.5 Percolation Conductivity in Biological Materials 449
18.6 Neuromorphic Computing 452
18.7 Sensory Transmission and Loss of Consciousness 459
18.8 Actomyosin Networks 460
18.9 Percolation Transition in the Mutation-Propagating Module Cluster in Colorectal Tumorigenesis 462
18.10 Essential Nodes in Brain: Optimal Percolation 466
18.11 Flexibility of Thought in High Creative Individuals 468
18.12 Cardiac Fibrosis and Arrhythmia 470
18.13 Connectivity of Temporal Hierarchical Mobility Networks During COVID-19 475
18.14 Protein Structure 476
18.15 Molecular Motors and Mechanical Properties of Cells and Active Gels 478
18.16 Percolation and Evolution 484
18.17 Microbial Communications 485

19 Percolation Theory, Ecology, Hydrology, and Geochemistry 489
19.1 Introduction 489
19.2 Chemical Weathering 490
19.3 The Water Cycle 490

19.4 Scaling Theory of Low-Temperature Geochemistry 492
19.5 Weathering Rinds 498
19.6 Vegetation Growth 498
19.7 River Drainage 501
19.8 Water Balance 504
19.9 Variability and Climate 506
19.10 Plants 507
19.11 Improving Production of Plants with Plague Susceptibility 507
19.12 Spread of Fungi in Soil 511
19.13 Pattern of Tropical Deforestation 513

20 Explosive Percolation and Its Applications 517
20.1 Introduction 517
20.2 Explosive Percolation 518
20.2.1 Explosive Percolation Through the Achlioptas Process 518
20.2.2 Explosive Percolation Through Achlioptas Process and the Product Rule 519
20.2.3 Explosive Percolation Through the BFW Model 522
20.2.4 Explosive Percolation by the n-Edge Model 522
20.2.5 Explosive Percolation by the Spanning Cluster—Avoiding Rule 522
20.3 Continuous Versus Discontinuous Phase Transition: Powder Keg and Unusual Scaling 523
20.3.1 The Importance of Selecting the Correct Order Parameter 526
20.3.2 Reversible Versus Irreversible Phase Transition 527
20.4 Applications of Explosive Percolation 528
20.4.1 Electric Breakdown 528
20.4.2 Formation of Bundles of Nanotubes with Uniform Size 531
20.4.3 Crackling Noise in Fractional Percolation 533
20.4.4 Diffusion-Limited Cluster Aggregation 534
20.4.5 Emergence of Molecular Life 535
20.4.6 Evolutionary Process in Protein Networks 537
20.4.7 Global Trade Network and Transportation 539
20.4.8 Social Networks 542
20.4.9 Explosive Immunization 542
20.4.10 Optimal Dismantling and Explosive "Death" 546
20.4.11 Spreading of Information 547

21 Directed Percolation and Its Applications 549
21.1 Introduction 549
21.2 Statistics of Vortex Filaments and Percolation 550
21.3 Directed Percolation 555

21.4 General Model of Directed Percolation and Random Resistor–Diode Networks ... 558
21.5 Directed Percolation and Transition Between Laminar and Turbulent Flows ... 559
21.5.1 Numerical Simulation ... 561
21.5.2 Experimental Confirmation ... 568
21.6 Turbulent Flow and Ecological Collapse ... 573
21.7 Catalytic Reactions ... 574
21.8 Depinning of Interface in Two-Phase Flow in Porous Media ... 575
21.9 Flowing Sand ... 578
21.10 Neuronal Avalanche in Brain ... 579
21.11 Satellite Networks ... 582

22 Percolation in Large-Scale Systems ... 585
22.1 Introduction ... 585
22.2 Mobile Ad Hoc Networks ... 585
22.3 Topography of Earth and Moon ... 590
22.4 Ancient Sea Level on Mars ... 595
22.5 Forest Fires ... 597
22.5.1 Self-Organized Critical Model of Forrest Fires ... 600
22.6 Percolation in Sea Ice ... 601
22.7 Lifecycle of Industrial Products and Consumer Demand ... 604
22.8 Spreading of Technological Innovations ... 607
22.9 Social Percolation ... 608

References ... 611

Index ... 671

Chapter 1
Macroscopic Connectivity as the Essential Characteristic of Heterogeneous Systems

1.1 Introduction

Let us begin this book with the introduction of a paper of the author published nearly four decades ago (Hughes et al. 1984), which has been modified slightly:

It is a fact of life, which is as challenging to the mind of the scientists as it is frustrating to their aspirations, that Nature is disordered. In nowhere but the theoretician's supermarket can we buy clean, pure, perfectly characterized and geometrically immaculate materials. Engineers work in a world of composites and mixtures, and biologists must grapple with even more complex systems. Even the experimentalist who focusses on the purest of substances—carefully grown crystals—can seldom escape the effects of defects, trace impurities, and finite boundaries. There are few concepts in science more elegant to contemplate than an infinite, perfectly periodic crystal lattice (which physicist have been using for decades), and few systems as remote from experimental reality. We are, therefore, obliged to come to terms with disordered media: variations in shape and constitution often so ill-characterized that we must deem the media's morphology to be random, if we are to describe it. The morphology of a medium has two major aspects: topology—the interconnectiveness of individual microscopic elements of the medium—and geometry—the shape and size of the individual elements.

As if this were not enough bad news, we know that however random the stage upon which the drama of Nature is played out, it is also at times very difficult to follow the drama's script. We believe, at least above the quantum mechanical level, in the doctrine of determinism, yet important *continuous* systems exist in which deterministic descriptions are beyond hope. A well-known example is diffusion for which, over certain length scales, we observe an apparent random process, implying that diffusion is governed by *disordered dynamics*.

The statement remains as valid as ever and, therefore, we still must address the same type of phenomena occurring around us. Nature is disordered, both in her structure and the processes that she supports, with the two types of disorder often being

M. Sahimi, *Applications of Percolation Theory*, Applied Mathematical Sciences 213,
https://doi.org/10.1007/978-3-031-20386-2_1

concurrent and coupled. An example is fluid flow through a porous medium where the interplay between the disordered morphology of the pore space and the dynamics of fluid motion gives rise to a rich variety of phenomena, some of which will be described later in this book. Despite considerable progress in understanding flow phenomena in porous materials (Sahimi 2011), many problems remain unsolved, precisely due to the disordered morphology of the pore space and disordered dynamics of fluid flow.

Despite the rather obvious disorder in Nature, many phenomena that occur in heterogeneous systems were, for several decades, analyzed by physicists, engineers, and others only by statistical mechanics, or by the application of such models as Boltzmann's equation. Remarkable progress was made by representing the systems of interest as periodic structures or lattices. As one always must confront the real world, however, it became clear that a statistical physics of disordered media must be devised that can provide methods for predicting the macroscopic properties of such systems from laws governing the microscopic world or, alternatively, for deducing microscopic properties of such systems from the macroscopic information that can be measured by experimental techniques. Such a statistical physics of disordered systems must take into account the effect of *both* the system's connectivity and geometry. But, whereas the role of geometry had already been appreciated in the early years of twentieth century, the effect of topology was ignored for many decades, or was treated in an unrealistic manner, simply because it was thought to be too difficult to be taken into account.

Science never stops its progress. A study of its history indicates, however, that its progress is not usually made with a constant rate, but rather in a sporadic manner. There are periods of time when a phenomenon appears so difficult that we do not even know where to start analyzing it, but then there are also periods when a quantum-leap discovery removes an obstacle to scientific advances and enables great progress. An example is the discovery of a new class of superconducting materials. Superconductivity was first discovered by the Dutch physicist Heike Kamerlingh-Onnes[1] in 1911 in metallic mercury below 4 K. It took 75 years to fabricate materials that were superconducting at temperatures much higher than 4 K. IBM researchers Bednorz and Müller (1986) showed that it is possible to have superconductivity in (La, Ba)CuO alloys at temperatures $T_c > 30$ K. Subsequently, it was shown by Takagi et al. (1987) that the phase $La_{2-x}Ba_xCuO_4$ with $x \sim 0.15$ is responsible for bulk superconductivity with $T_c \sim 35$ K. Since then thousands of papers have been published on the subject of high-temperature superconductivity, and materials have been fabricated that become superconducting at room temperature (Snider et al. 2020), albeit at very high pressures. The discovery had such a great impact on the subject

[1] **Heike Kamerlingh-Onnes** (1853–1926) was a Dutch physicist who pioneered refrigeration techniques, and utilized them to study how materials behave when cooled to nearly absolute zero. He succeeded in liquifying helium, for which he was awarded the Nobel Prize in physics in 1913, discovered superconductivity in 1911, and coined the word "enthalpy" in thermodynamics. The Onnes effect that refers to the creeping of superfluid helium is named in his honor, as is the crater Kamerlingh-Onnes on the Moon.

that Bednorz and Müller were honored with the Noble Prize in physics in 1987, only a year after publishing their work.

Over the past five decades, statistical physics of disordered systems has been in a rapidly moving stage of progress, partly because standard methods for calculating the average properties of disordered media have been established by the theoreticians, while, at the same time, more and more experimental data have been accumulated due to the advances in instrumentation and development of many novel experimental techniques. But, perhaps, one of most important reasons for the rapid development of the statistical physics of disordered media is that the role of the connectivity of the microscopic elements of a disordered system and its effect on its macroscopic properties has been appreciated, and taken into account. This has become possible through the development of percolation theory, the applications of which are the subject of this book.

1.2 What is Percolation?

In early September of 1978, I arrived in Minneapolis, Minnesota, to attend graduate school and study for Ph.D. degree at the University of Minnesota. For nearly 6 years I lived in Minneapolis, a wonderful city of many lakes and, naturally, I wanted to live near a lake to enjoy the beautiful (but humid) summers, and to watch little kids skate on the lake during the long winters, and people do ice fishing. The magnificent Mississippi River runs through Minneapolis, dividing the campus of the University into the East and West Bank sections, and the chemical engineering department where I was studying was in the East Bank. For nearly 2 years, I lived in southeast Minneapolis near Lake Calhoun, about 6 miles from the campus (via Interstate I94). Every winter Minnesota experiences many snow storms that force temporary closure of many streets in one or both directions for sweeping the snow. Even during summers some streets are closed for repairing the damage caused by the long winters, to the point that there is a saying among people, "There are two seasons in Minnesota: winter and road construction." There are also many other types of "defects" or disorder in the structure of the streets of Minneapolis. For example, at that time there was a large K-Mart store that blocked Nicollet Avenue, one of the most important routes in the city that starts in the downtown area and ends in the suburb in the south. Many railways also cut the streets, and numerous lakes have created natural blockage for them. It often seems as if the streets are closed at random.

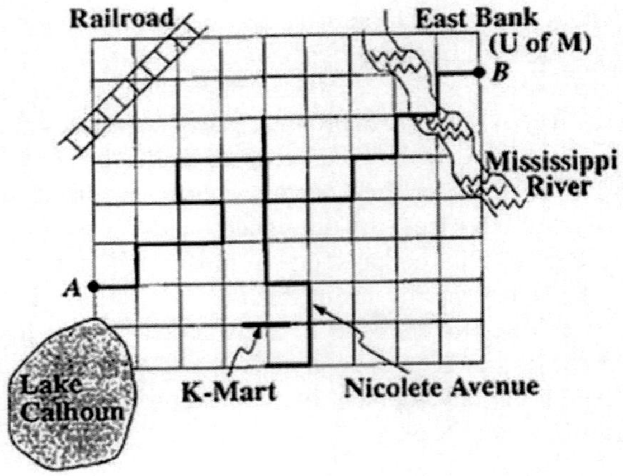

Fig. 1.1 Idealization of streets of Minneapolis, Minnesota, and determining a percolating path between points *A* and *B*

For my doctoral studies I had two great mentors: Skip Scriven[2] and Ted Davis,[3] in the then top-ranked chemical engineering department in the United States. I was expected to work on my research projects 6 days a week, from Monday through Saturday (working on Sundays was "voluntary," but strongly "encouraged!"). The question was: what fraction of the streets of Minneapolis (i.e., the number of open streets divided by the total number of streets) between my apartment and the campus had to be open to traffic in order for me to reach the university on time (particularly on Saturdays, when either Skip or Ted "randomly" stopped by their students' office to see how they were doing)?

To answer the question, suppose that we idealize the streets of Minneapolis as the bonds of a very large square network. Some of the streets were always blocked, with the blockage caused by heavy snow, a rail track, a lake, or a store like K-Mart; see Fig. 1.1. So, suppose that the fraction of the streets open to traffic was p. Clearly, if too many streets were closed, I could not reach the university on time. If, on the other hand, most streets were open, almost any route would take me to the campus. Naturally, then, one is led to thinking that there must be a critical fraction p_c of open streets, such that for $p < p_c$ I could not reach the University, but for $p > p_c$ I could always get to work on time. As we will see later on, $p = p_c$ is a special point.

[2] **Lawrence Edward "Skip" Scriven** (1931–2007) was Regents Professor of Chemical Engineering and Materials Science at the University of Minnesota. A member of the National Academy of Engineering, he made seminal contributions to coating flows (for which an award is named in his honor), flow through porous media, interfacial phenomena, and complex fluids. He received numerous awards from the American Institute of Chemical Engineering, American Chemical Society, American Physical Society, and American Mathematical Society.

[3] **Howard Theodore "Ted" Davis** (1937–2009) was Regents Professor of Chemical Engineering and Materials Science, former Chairman of his Department, and former Dean of the Institute of Technology at the University of Minnesota. A member of the National Academy of Engineering, he made fundamental contributions to statistical mechanics of surfaces and interfaces, flow through porous media, and complex fluids.

Let us consider another example. Suppose that, instead of representing the streets of Minneapolis, the bonds of the square network of Fig. 1.1 represent conductors, some of which have a unit conductance, while the rest are insulators with zero conductance. Set the voltage at point A in Fig. 1.1 to unity, and at B at zero. The question is: what is the minimum fraction of the bonds with a unit (or any other non-zero value) conductance, such that electrical current flows from A to B? The answer to this seemingly simple question is relevant to the technologically important question of the conductivity of composite materials, such as carbon black composites that are widely used in many applications, which consist of a mixture of conducting and insulating phases. As in the first example, if too many bonds—resistors—are insulating, no current will flow from A to B, whereas for sufficiently large number of conducting bonds electrical current flows between the two points with ease, and the system as a whole is a conductor.

Such questions are answered by percolation theory. Percolation tells us when a system is *macroscopically connected.* The critical point at which the transition between a macroscopically connected system and a disconnected one occurs for the first time is called the *percolation threshold* of the system. Because in its simplest form a percolation network is generated by simply blocking bonds at random, it is useful as a simple model of disordered media. Moreover, since the main concepts of percolation theory are simple, writing a computer program for simulating a percolation process is straightforward and, therefore, percolation can also serve as a simple tool for introducing students to computer simulations. Stauffer[4] and Aharony (1994) give a simple introduction to essential concepts of percolation. In Chap. 2, we will summarize the essential concepts and ideas of percolation theory.

1.3 The Scope of the Book

Over the past five decades percolation has been applied to modeling of a wide variety of phenomena in disordered media. It is very difficult, if not impossible, to describe all such applications in one book. In selecting those applications that are described in this book, three criteria were used:

(i) The application is quantitative, in the sense that there is a quantitative comparison between the predictions of the percolation and experimental data, or with credible computer simulations.

(ii) The problem is interesting and has scientific, societal, or technological importance.

(iii) This author has a clear understanding of the problem, and how the application of percolation has been developed.

---

[4] **Dietrich Stauffer** (1943–2019) was a Professor of Physics at Cologne University in Germany, who made seminal contributions to percolation theory, polymerization and gelation, phase transition, and applications of statistical physics to social and biological problems.

Based on the three criteria, I have selected the classes of problems to which percolation has been applied. Chapter 2 contains a summary of the main properties that will be used in the rest of this book.

Every effort has been made to explain the percolation approach in simple terms. In all cases, the predictions are compared with the experimental data, as well as precise numerical simulations, in order to establish the relevance of percolation concepts and their application to the problem. We also give what we believe are the most relevant references to each subject, or provide references to recent reviews on the subjects.

1.4 Further Reading

The focus of this book is on applications of percolation theory. There are several other books that describe theoretical foundations of percolation. The books by Kesten[5] (1982) and Grimmett (1999) focus on mathematical aspects of percolation. The two-volume book by Hughes (1995) has excellent discussions of theoretical aspects of percolation, as well as numerical results. The book by Meester and Roy (1996) focuses on continuum percolation, whose main features will be described in Chap. 3. The book by Bollobás and Riordan (2006) is limited in scope, but presents good discussions of the classical results, as well as application of conformal invariance to two-dimensional percolation systems, and site percolation in Voronoi structures (see Chap. 3). The book edited by Sahimi and Hunt (2021) represents a collection of reviews, written by leading experts, on various aspects of percolation, including some of the issues not discussed in this book.

The books by Sahimi (2011), Hunt et al. (2014a), and King and Masihi (2018) develop application of percolation theory to problems in porous media, from pore to field scales. The two-volume book by Sahimi (2003a, b) describes in great detail the applications of percolation theory to predicting and estimating properties of heterogeneous materials, while the book by Torquato (2002) also covers, but by a much more limited scope, some aspects of the same. Moss de Oliveira et al. (1999) describe some applications of percolation to social problems, from war and money, to computers.

A relatively recent and very good review of percolation is that of Saberi (2015).

[5] **Harry Kesten** (1931–2019) was an American mathematician. A member of the National Academy of Sciences, Kesten made important contributions to theory of random walks on groups and graphs, random matrices, branching processes, and percolation theory, for which he received numerous awards.

Chapter 2
Percolation Theory: Classical and Poor Man's

2.1 Introduction

As pointed out in Chap. 1, both Nature and people produce a wide variety of materials with enormous variations in their morphology. The morphology of a heterogeneous medium consists of its geometry, topology, and surface structure. The geometry describes the shapes and sizes of the micro- and mesoscale elements of a material or medium. The shapes can vary anywhere from completely ordered and Euclidean, to complex and seemingly chaotic. In general, Euclidean shapes are formed under close-to-equilibrium conditions, although even in such cases equilibrium thermodynamics is often inadequate for describing the process that gives rise to such shapes. The topology describes how the micro- or mesoscale elements are connected to one another. In addition to geometry and topology, many systems, especially those that are formed under far-from-equilibrium conditions, have very complex surface structure. In particular, many materials have very rough surface with the roughness seemingly following a very complex pattern. Despite the complexity, however, the rough surface is often amenable to precise characterization. Percolation theory quantifies the effect of the topology or connectedness, although, as we will see later in this book, in certain cases it can also be used for explaining certain properties of materials and systems associated with the other two aspects of the morphology. Note, however, that when we speak of surface roughness, the length scale over which the roughness is relevant must be specified, because even the most rugged landscapes look perfectly smooth when viewed from the outer space.

Examples of materials and media with disordered morphology are abundant. A well-known class of such systems consists of microporous materials, such as membranes and catalysts. The geometry of such materials is disordered, as the pores have a variety of shapes and sizes. Their topology is also disordered, since the connectivity of their pores varies in the pore space, with some pores being connected to a larger number of pores than others. The surface of the pores, over certain length scales

M. Sahimi, *Applications of Percolation Theory*, Applied Mathematical Sciences 213,
https://doi.org/10.1007/978-3-031-20386-2_2

that depend on the type of the material, is also rough. Many other examples will be described throughout this book.

The enormous variations in the morphology of the natural or man-made materials are such that, up until a few decades ago, the task of describing and characterizing such morphologies seemed hopeless. Several important developments changed considerably the bleak outlook for characterizing the morphology of disordered materials and media, and brightened the prospects for gaining deeper understanding of microstructure of disordered systems. Among such developments are the advent of powerful computers and novel experimental techniques, as well as advances in instrumentation, which allow highly sophisticated computations and modeling of disordered media, and measuring their properties.

Another important development for characterizing the effect of the topology, i.e., the connectivity of disordered media, on their macroscopic properties has been transferring percolation theory from its original mathematically abstract form to a powerful tool of studying disordered materials and media. Moreover, as we discuss in the subsequent chapters, percolation concepts are essential to the correct interpretation of experimental data for various properties of disordered materials and media.

The purpose of this chapter is to describe and discuss the essential concepts and ideas of percolation theory. In its original form, percolation theory was developed for completely random systems, i.e., those in which there are no correlations in the structure of a system. Over the past four decades, however, a variety of other percolation models have been developed in order to study a broad class of disordered materials and media. In this chapter, we discuss the classical random percolation. Chapter 3 will describe several variations of the classical random percolation.

2.2 Historical Origin of Percolation Theory

Percolation theory was first developed by Flory (1941)[1] and Stockmayer (1943)[2] for describing polymerization by which small branching molecules react and form macromolecules. It also appeared in the work of Good (1949) on the theory of branching processes. However, the trio did not use the terminology of percolation, or even referred to it as percolation theory. Moreover, Flory and Stockmayer developed their theory of polymerization using a special lattice, namely, the Bethe lattice, an endlessly branching structure without any closed loops, an example of which is shown in Fig. 2.1. We will come back to their theory later in this book.

[1] **Paul John Flory** (1910–1985) made fundamental contributions to understanding of macromolecules and polymers in a solution, for which he received the Nobel Prize in Chemistry in 1974. He taught at Stanford and Carnegie Mellon Universities, and also worked in the industry.

[2] **Walter Hugo Stockmayer** (1914–2004) was a renowned chemist who made seminal contributions to polymer science. He taught at Dartmouth College and founded *Macromolecules*, the premier journal dedicated to polymers.

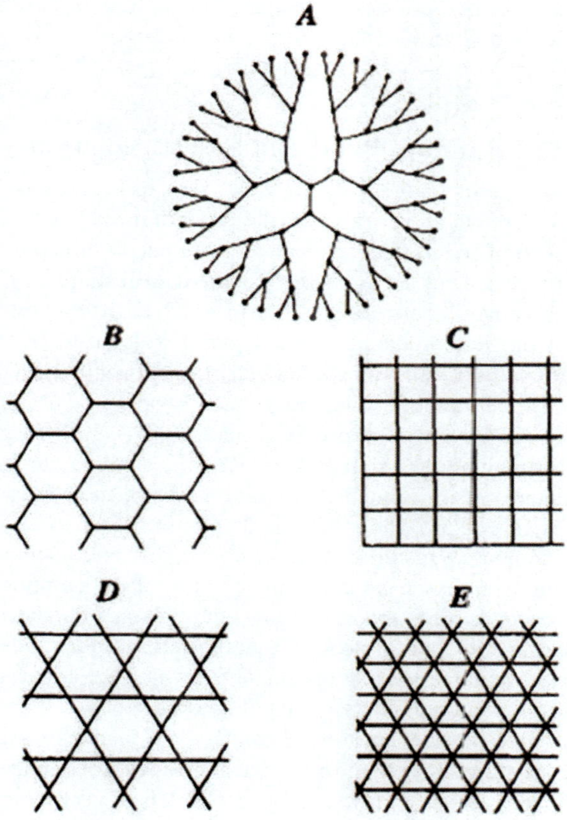

Fig. 2.1 **A** The Bethe lattice with $Z = 3$. **B** The honeycomb lattice with $Z = 3$. **C** The square lattice with $Z = 4$. **D** The Kagomé lattice with $Z = 4$, and **E** the triangular lattice with $Z = 6$

In its present form, percolation theory first appeared in the mathematics literature, and was introduced by Broadbent and Hammersley[3] (1957), who dealt with the problem of spreading of a hypothetical fluid through a random medium. The word "fluid" is, however, quite generic: a liquid or a gas or, more generally, any phenomenon or "information" that spreads throughout a system that contains some stochastic elements.

The underlying randomness might, however, be of two very different types. In one type, the randomness is imposed by the *fluid*, in the sense that it is the fluid that decides which path to take in the medium, in which case the motion of the fluid is described by the classical diffusion process. On the other hand, the randomness may also be imposed by the *medium* itself; this was the new situation that was studied by Broadbent and Hammersley (1957). They dubbed the new phenomenon,

---

[3] **John Michael Hammersley** (1920–2004) was a British mathematician who made important contributions to the theory of self-avoiding walks (a model of linear polymers), percolation theory, and development of Monte Carlo methods.

a percolation process, since they thought that the spread of a fluid through a random medium resembled flow of coffee in a percolator.

2.3 Random Bond and Site Percolation

Since percolation deals with the effect of the connectivity of elements of a disordered medium on its effective properties, we need a model in order to make our discussions concrete. Thus, let us assume, for the sake of simplicity, that we represent a disordered medium by a network or lattice in which each bond represents a microscopic piece of the medium. In their original paper, Broadbent and Hammersley studied two distinct problems. One was the *bond percolation problem* in which the bonds of the network are either randomly and independently *occupied* or intact with probability p, or they are *vacant* or removed with probability $1 - p$. In a large network, this is equivalent to removing a fraction $(1 - p)$ of all the bonds at random. Figure 2.2 shows a random square network in which a fraction 0.45 of the bonds has been removed at random, i.e., $p = 0.55$. The intact bonds represent, for example, the conducting regions of a composite material made of conductors and insulators, in which case the vacant bonds represent the insulating portion of the composite. In a heterogeneous solid, the conducting and insulating materials are not necessarily distributed randomly throughout, but for now we ignore such complications. In Chap. 3, we will consider more general percolation models in which the correlations between various regions of a disordered medium are not ignored.

Two sites are considered *connected* if there exists at least one path between them consisting solely of the occupied bonds. A set of connected sites surrounded by vacant bonds is called a *cluster*. If the network is of very large extent, and if p is sufficiently small, the size of any connected cluster is likely to be small. On the other hand, if $p \approx 1$, the network should be almost entirely connected, except for occasional small

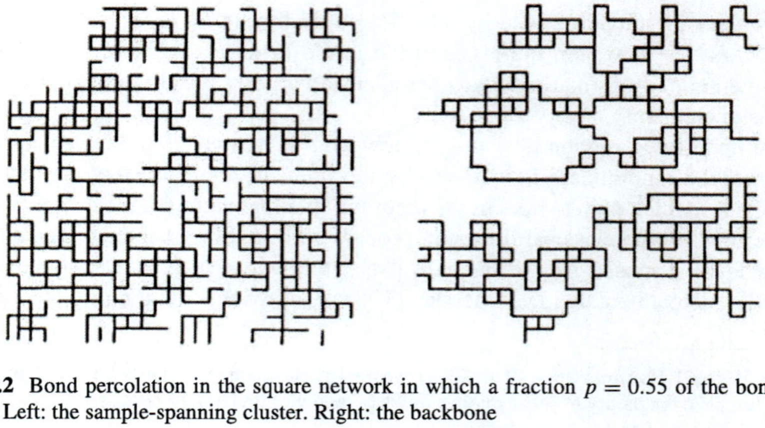

Fig. 2.2 Bond percolation in the square network in which a fraction $p = 0.55$ of the bonds are intact. Left: the sample-spanning cluster. Right: the backbone

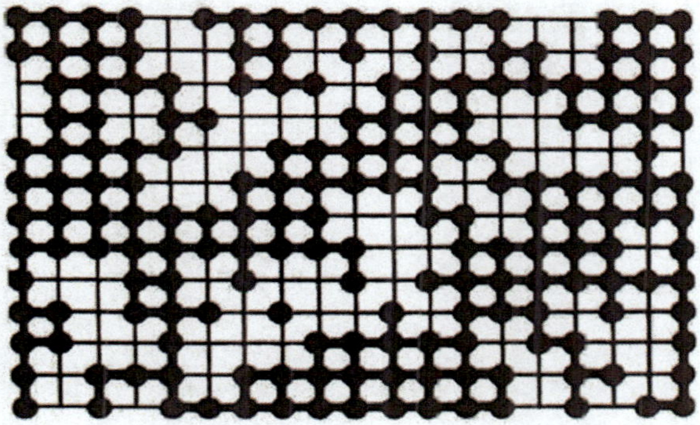

Fig. 2.3 Site percolation on the square network in which a fraction $p = 0.5927$ of the sites are intact

holes due to the vacant bonds. Therefore, at some well-defined value of p, there should be a transition in the topological (connectivity) structure of the network; this value is called the *bond percolation threshold* p_{cb}, which is the largest fraction of occupied bonds such that for $p < p_{cb}$ there is no sample-spanning cluster (SSC) of occupied bonds, whereas for $p > p_{cb}$ an SSC of occupied bonds does form.

The second problem studied by Broadbent and Hammersley was *site percolation process* in which the sites of a network are occupied with probability p and vacant with probability $1 - p$. Two nearest neighbor sites are considered connected if they are both occupied, and connected clusters on the network are defined in a manner similar to that for bond percolation. There is also a *site percolation threshold* p_{cs} such that for $p > p_{cs}$ there is an SSC of occupied sites, whereas for $p < p_{cs}$ the network is macroscopically disconnected. Figure 2.3 shows site percolation in a square network at the percolation threshold $p_{cs} \simeq 0.5927$.

2.4 Percolation Thresholds

The derivation of the exact values of p_{cb} and p_{cs} is an extremely difficult problem. In fact, such a derivation has been possible to date only for certain lattices related to the Bethe lattice and for some two-dimensional (2D) networks. For the Bethe lattice, one has

$$p_{cb} = p_{cs} = \frac{1}{Z - 1}, \tag{2.1}$$

where Z is the coordination number of the lattice, i.e., the number of bonds connected to the same site. We compile the current estimates of p_{cb} and p_{cs} (or their exact values if they are known) for three common 2D lattices in Table 2.1, while the most accurate numerical values of p_{cb} and p_{cs} for four common 3D lattices are listed in Table 2.2. Wierman (2021) provides a list of several other exact percolation thresholds of

Table 2.1 Percolation threshold, and $B_c = Zp_{cb}$, for three common 2D networks

Network	Z	p_{cb}	B_c	p_{cs}
Honeycomb	3	$1 - 2\sin(\pi/18) \simeq 0.6527$*	1.96	0.6962
Square	4	1/2*	2	0.5927
Triangular	6	$2\sin(\pi/18) \simeq 0.3473$*	2.08	1/2*

*Exact result

Table 2.2 Numerical estimates of the percolation thresholds, and $B_c = Zp_{cb}$, for four common 3D networks

Network	Z	p_{cb}	B_c	p_{cs}
Diamond	4	0.3886	1.55	0.4299
Simple cubic	6	0.2488	1.49	0.3116
BCC	8	0.1795	1.44	0.2464
FCC	12	0.119	1.43	0.199

2D lattices.[4] The bond percolation thresholds of d−dimensional lattices are well approximated by

$$p_{cb} \simeq \frac{d}{Z(d-1)} \,. \tag{2.2}$$

Also shown in Tables 2.1 and 2.2 is the product $B_c = Zp_{cb}$, which is an almost invariant of percolation networks with

$$B_c \simeq \frac{d}{d-1} \,. \tag{2.3}$$

The significance of B_c is discussed in Chap. 3. Generally speaking, since in any lattice a bond has more nearest neighbor bonds than a site has nearest neighbor sites, we have $p_{cs} \geq p_{cb}$. Moreover, because lattices in higher dimensions are better connected than those in lower dimensions, percolation thresholds decrease with increasing d.

Let us point out that Coupette and Schilling (2022) introduced a new criterion for percolation problems on arbitrary graphs and lattices. They defined a metric that characterizes the distance between two sites, taken to be the number of edges or bonds visited on the shortest path between the two sites, decomposed the graph into a hierarchy of neighborhoods, and expressed the percolation problem as a branching process. When they implemented their criterion for percolation, i.e., formation of the SSC at the percolation threshold, it predicted the exact percolation thresholds for a large number of exactly solved percolation problems, including the Bethe lattice, random graphs, and small-world networks (see Chap. 17), as well as bond percolation on 2D lattices with a triangular hypergraph, and site percolation on 2D lattices with a

[4] The exact percolation thresholds of a variety of other lattices are listed at: http://en.wikipedia.com/wiki/Percolation_threshold.

generalized triangular hypergraph, and certain continuum percolation problems (see Chap. 3).

2.5 Bicontinuous Materials, Phase-Inversion Symmetry, and Percolation

A two-phase medium in which both phases simultaneously form sample-spanning clusters is called *bicontinuous*. The question, which is important from a practical view point, is, under what conditions can both phases of a two-phase system simultaneously form sample-spanning clusters? If the occupied and vacant bonds or sites of a lattice are viewed as representing the two phases, then percolation theory provides a clear answer to the question of bicontinuity. Random percolation possesses *phase-inversion symmetry*. This means that the morphology of phase 1 at volume fraction p is statistically identical to that of phase 2 with the same volume fraction. Clearly, then, no random 1D medium can be bicontinuous, as its (bond or site) percolation threshold is 1. For $d \geq 2$, any system possessing phase-inversion symmetry is bicontinuous for $p_c < p < 1 - p_c$ (where p_c is either the site or bond percolation threshold), provided that $p_c < 1/2$. Thus, all the 3D lattices listed in Table 2.2 can be bicontinuous in either bond or site percolation. For 3D systems, however, neither $p_c < 1/2$ nor phase-inversion symmetry is a necessary condition for bicontinuity.

Two-dimensional media are much more difficult to be made bicontinuous. In fact, in many respects, randomly disordered two-phase materials correspond to site percolation systems. If so, then since there is no 2D network with $p_{cs} < 1/2$ (see Table 2.1), we conclude that no 2D randomly disordered medium can be bicontinuous.

2.6 Generating a Single Percolation Cluster: The Leath–Alexandrowicz Algorithm

For many applications, removal of sites or bonds may not be a suitable way of generating a model of a two-phase medium because, in addition to the sample-spanning cluster, isolated finite clusters are generated as well. For example, in some applications, only the sample-spanning cluster is of interest or, at least, the process of interest begins with a single cluster and, therefore, we must first delete all the isolated clusters from the system. A method that was originally developed by Leath (1976) and Alexandrowicz (1980) generates only the sample-spanning cluster for $p > p_c$, or the largest cluster for $p \leq p_c$.

In their method, one begins with a single occupied site at the center of the network. One then identifies the nearest neighbor sites of the occupied site and considers them occupied and adds them to the cluster, if random numbers R, uniformly distributed in (0, 1) and assigned to the sites, are less than the given p. The perimeter—the nearest neighbor empty sites—of these sites are then identified and the process of occupying the sites continues in the same way. If a selected perimeter site is not designated as occupied, then it remains unoccupied forever. The generalization of this method for generating a cluster of occupied bonds is obvious.

Counting the number of percolation clusters in computer simulation of percolation systems is a difficult task. An algorithm due to Hoshen and Kopelman (1976) (see Al-Futaisi and Patzek 2003, for non-lattice percolation systems) performs this task very efficiently. The execution time of the algorithm scales with the number of bond or sites N as $O(N^2 \ln N)$, which is large. It has, however, very low memory requirements of order of $N^{1-1/d}$, which makes it very useful for studies of particularly large, but low-dimensional systems.

2.7 The Newman–Ziff Algorithm

What if, as is often the case, one wishes to calculate the values of some observable quantity Q, such as the average cluster size, over a range of p? One straightforward method is by "brute force": one carries out repeated simulations for many closely spaced values of p in the range of interest, which makes the calculation much slower, and if one wishes a continuous curve $Q(p)$ in the range of interest, then theoretically one must compute Q for an "infinite" number of values of p. In practice, however, one can compute Q for a finite number of p-values and then interpolate between them, but this will introduce some error into the results. Newman and Ziff (2001) addressed this problem by developing a most efficient algorithm.

In the algorithm developed by Newman and Ziff (2001), one adds sites or bonds one at a time to an initially empty system, and as each new site or bond is added, recalculates the cluster distribution or other property, such as the spanning probability—the probability of forming a sample-spanning cluster—by an efficient tree-based updating process. The clusters are represented as a tree with a root, and other sites on the cluster are linked to the root either directly or indirectly by means of pointers. When a new bond links two distinct clusters together, the root of the smaller cluster is turned into a pointer to the root of the larger cluster. This is carried out efficiently by the union-find algorithm, a fundamental computer science procedure that also automatically allows paths to the root to be compressed, whenever a union is carried out, which keeps the dynamics fast.

The algorithm uses recursion and requires just a small number of lines of computer code. It allows a quantity, such as the size of the largest cluster, to be determined as a function of the number of bonds added to the system, from zero to fully occupied. To convert to the dependence on the occupation probability p, one has to carry out a convolution of the results with a binomial distribution of the expected number of occupied bonds for a given value of p; this convolution can also be carried out relatively easily. This procedure results in determining a quantity of interest for all values of p in a single sweep across the lattice. Repeating and averaging yields results to arbitrarily high precision. The algorithm can also be applied efficiently to other models, such as continuum percolation (see Chap. 3) and explosive percolation (see Chap. 20). Its execution time scales linearly with the size of the system, which is most efficient.

Note that the Newman–Ziff algorithm shares some similarities with the Hoshen–Kopelman algorithm, which can also be implemented in terms of trees, but in that case bonds are added with a fixed probability row by row, while in the Newman–Ziff algorithm bonds are added throughout the system, and with an effective probability that varies from zero to one.

2.8 Percolation Properties

In addition to the percolation thresholds p_{cb} and p_{cs}, several other important quantities characterize any percolation system. Some of them describe the connectivity of the percolation clusters, while others characterize their effective flow and transport properties. We now describe these properties.

2.8.1 Connectivity Properties

For simplicity, we use p_c to denote p_{cs} or p_{cb}.

1. The *percolation probability* $P(p)$ is the probability that, when the fraction of occupied bonds is p, a given site belongs to the infinite, or the SSC of the occupied bonds.
2. The *accessible fraction* $X^A(p)$ is that fraction of occupied bonds (or sites) that belong to the infinite cluster.
3. The *backbone fraction* $X^B(p)$ is the fraction of occupied bonds in the infinite cluster that actually participate in a transport process, such as conduction, as some of the bonds in the infinite cluster are dead end and do not carry any current. Therefore, $X^A(p) \geq X^B(p)$.
4. The *correlation length* $\xi_p(p)$ is the typical radius of percolation clusters for $p < p_c$, and the typical radius of the "holes" above p_c that consists of the vacant bonds or sites. For $p > p_c$, ξ_p is also the length scale over which the system is macroscopically homogeneous. Thus, in any Monte Carlo simulation of a percolation model, the linear size L of the network must be larger than ξ_p, in order for the results to be independent of L.
5. The *average number of clusters of size s* (per lattice site) $n_s(p)$ is an important quantity in many of the problems of interest, because it corresponds to, for example, the number of conducting or insulating clusters of a given size in a conductor–insulator composite solid.

6. The probability that two sites, one at the origin and another one at a distance $\mathbf{r}$, are both occupied and belong to the same cluster of occupied sites is $p^2 P_2(\mathbf{r})$, where $P_2(\mathbf{r})$ is called the *pair-connectedness function*.
7. The *mean cluster size* S (also called the *site-averaged* cluster number) is the average number of sites in the cluster that contains a randomly selected site, and is given by

$$S = \frac{\sum_s s^2 n_s(p)}{\sum_s s n_s(p)} \ . \tag{2.4}$$

Essam (1972) showed that S and the pair-connectedness function $P_2(\mathbf{r})$ are related through a simple relation:

$$S = 1 + p \sum_{\mathbf{r}} P_2(\mathbf{r}) \ . \tag{2.5}$$

Figure 2.4 shows the typical $p-$dependence of some of these quantities in site percolation on the simple-cubic lattice. In this figure, X^I is the fraction of the isolated sites, i.e., $X^I = p - X^A$.

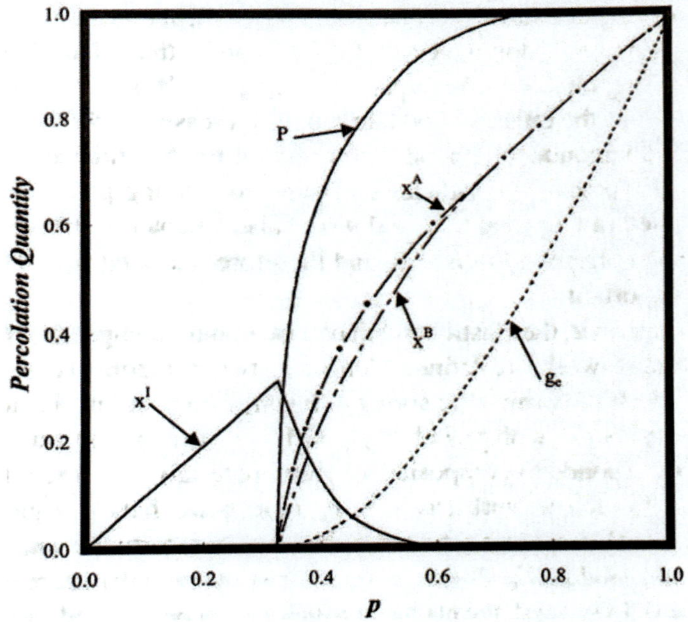

Fig. 2.4 Typical dependence of several percolation quantities on the fraction p of the active sites in a simple-cubic lattice. X^I is the fraction of the isolated active sites, that is, $X^I = p - X^A$

2.8.2 *Flow and Transport Properties*

Because a major application of percolation theory has been to modeling of flow and transport in disordered materials, and, in particular, composite solids and porous materials, we must also consider their effective flow and transport properties, namely, the effective conductivity, diffusivity, elastic moduli, dielectric constant, and permeability. We first consider the conductivity of a two-phase composite material modeled as a two-component network in which each randomly selected bond has a conductance g_1 with probability p or g_2 with probability $q = 1 - p$. It is straightforward to show that the effective electrical conductivity g_e of the network is a *homogeneous function* that takes on the following form:

$$g_e(p, g_1, g_2) = g_1 F(p, h) , \tag{2.6}$$

where $h = g_2/g_1$. Due to the assumption of randomness of the material's morphology, g_e is invariant under the interchange of g_1 and g_2 (phase-inversion symmetry; see above) and, therefore, we must have

$$g_e(p, g_1, g_2) = g_e(q, g_2, g_1) , \quad \text{and} \quad F(p, h) = hF(q, 1/h) . \tag{2.7}$$

The limit in which $g_2 = 0$ and g_1 is finite corresponds to a conductor–insulator mixture. In this case, as $p \to p_c$, more and more bonds are insulating, the conduction paths become very tortuous and, therefore, g_e decreases. Eventually, at $p = p_c$ one has $g_e(p_c) = 0$ since no sample-spanning conduction path exists any more. More generally, the conductance g_1 can be selected from a statistical distribution, which is in fact the case in practice, such as porous materials and composite solids.

The limit in which $g_1 = \infty$ and g_2 is finite represents a conductor–superconductor mixture. All quantum-mechanical aspects of real superconductors are ignored in this definition, as we are concerned only with the effect of the local connectivity of the material on the conductivity. It is clear that the effective conductivity g_e of this system is dominated by the superconducting bonds. If $p < p_c$, then a sample-spanning cluster of the superconducting bonds does not exist, and g_e is finite. As $p \to p_c^-$, the effective conductivity g_e increases until an SSC of superconducting bonds is formed for the first time at $p = p_c$, where g_e *diverges*. The practical significance of superconducting percolation networks will be discussed later in this book. Note that both limits correspond to $h = 0$ and, therefore, the point $h = 0$ at $p = p_c$ is particularly important.

In a similar manner, the elastic moduli of a two-phase composite solid, modeled by a percolation network, are defined. Consider a two-component network in which each bond is an elastic element, such as a spring or beam, with an elastic constant e_1 with probability p or e_2 with probability $q = 1 - p$. The limit in which $e_2 = 0$ and e_1 is finite corresponds to composites of rigid materials and holes, such as porous solids. In such networks, as $p \to p_c$, more bonds have no rigidity, the paths for transmission of stress or elastic forces become very tortuous and, therefore, the effective elastic moduli E_e—Young's, bulk, or shear modulus—decrease. At $p = p_c$ one has $E_e(p_c) = 0$. In general, the elastic constant e_1 can be selected from a statistical distribution.

The limit in which $e_1 = \infty$ and e_2 is finite represents mixtures of rigid–superrigid materials. If $p < p_c$, then a sample-spanning cluster of the superrigid bonds cannot form, and E_e is finite. As $p \to p_c^-$, the effective elastic moduli increase until the percolation threshold p_c of the rigid phase is reached at which an SSC of the superrigid bonds is formed for the first time, and E_e *diverge*. The significance of superrigid percolation networks will be discussed later in this book.

Similarly, the effective dielectric constant ϵ_e of a two-phase insulating material, modeled by a percolation network, may be defined and, in fact, as shown later on, ϵ_e is closely related to the conductor–superconductor model described above. Finally, the effective diffusivity D_e of a porous material, as well as its effective permeability K_e, is also defined in a similar manner. We will come back to them later in this book.

2.8.3 *The Structure of the Sample-Spanning Cluster*

The SSC in which transport of current or stress occurs may be divided into two parts: the dead-end part that carries no current or stress, and the aforementioned backbone, the multiply connected part of the cluster. Near p_c the bonds in the backbone are also divided into two groups: those that are in the *blobs* that are multiply connected and make the transport paths very tortuous, and the *red bonds* that are those that, if cut, they would split the backbone into two pieces. Such bonds are called red because in transport of heat or electrical currents, they carry the largest fluxes and, therefore, in analogy with a real electrical network become very hot and, hence, red. Figure 2.2 shows the backbone of the SSC in the square network. Currently, the most efficient algorithm for identifying the backbone of a percolation network is due to Sheppard et al. (1999); see also Knackstedt et al. (2000).

2.9 Universal Power Laws for Percolation Properties

One of the most important characteristics of percolation systems is their *universal* properties. The behavior of many percolation quantities near p_c is insensitive to the microstructure, such as the coordination number of the network, and to whether the percolation problem is a site or a bond process, or a variant of either one. The quantitative statement of this universality is that many percolation properties follow power laws near p_c, and the *critical exponents* that characterize such power laws are universal and depend only on the Euclidean dimensionality d of the system.

2.9.1 Connectivity Properties

In general, the following power laws hold near p_c:

$$P(p) \sim (p - p_c)^{\beta} , \tag{2.8}$$

$$X^A(p) \sim (p - p_c)^{\beta} , \tag{2.9}$$

$$X^B(p) \sim (p - p_c)^{\beta_{\mathrm{bb}}} , \tag{2.10}$$

$$\xi_p(p) \sim |p - p_c|^{-\nu} , \tag{2.11}$$

$$S(p) \sim |p - p_c|^{-\gamma} , \tag{2.12}$$

$$P_2(\mathbf{r}) \sim \begin{cases} r^{2-d-\eta}, & p = p_c , \\ \exp(-r/\xi_p), & \text{otherwise} , \end{cases} \tag{2.13}$$

where $r = |\mathbf{r}|$. For large clusters near p_c, the cluster size distribution $n_s(p)$ is described by the following scaling law:

$$n_s \sim s^{-\tau} f[(p - p_c)s^{\sigma}] , \tag{2.14}$$

where τ and σ are two more universal critical exponents, and $f(x)$ is a scaling function such that $f(0)$ is not singular.

These exponents are not all independent. For example, one has $\tau = 2 + \beta\sigma$, $\nu d = \beta + 1/\sigma = 2\beta + \gamma$, and $\gamma = (2 - \eta)\nu$. In fact, knowledge of ν and another exponent suffices for determining all of the above percolation exponents. For random percolation models, or for correlated percolation in which the correlations have a finite extent, all the above critical exponents are completely universal.

2.9.2 Flow and Transport Properties

Similar power laws are also followed by flow and transport properties of percolation systems. In particular,

$$g_e(p) \sim (p - p_c)^{\mu} , \quad \text{conductor–insulator composites} \tag{2.15}$$

$$g_e(p) \sim (p_c - p)^{-s} , \quad \text{conductor–superconductor composites} \tag{2.16}$$

$$E_e(p) \sim (p - p_c)^{f} , \quad \text{rigid–soft composites} \tag{2.17}$$

$$E_e(p) \sim (p_c - p)^{-\chi} , \quad \text{rigid–superrigid composites} , \tag{2.18}$$

$$K_e(p) \sim (p - p_c)^{e} . \tag{2.19}$$

In many, but not all, cases, one has $\mu = e$. We will come back to this point in Chap. 3.

The power law that characterizes the behavior of the effective diffusivity $D_e(p)$ near p_c is derived from that of $g_e(p)$. According to Einstein's relation, $g_e \propto \rho D_e$, where ρ is the density of the carriers. Although a diffusing species can move on all the clusters, above p_c only diffusion on the sample-spanning cluster contributes to D_e, so that $\rho \sim X^A(p)$. Hence, $g_e(p) \sim X^A(p)D_e(p)$ and, therefore,

$$D_e(p) \sim (p - p_c)^{\mu-\beta} . \tag{2.20}$$

For length scales $L < \xi_p$, the resistance R between two end points of a box of linear size L scales with L as $R \sim L^{\tilde{\zeta}}$. It is not difficult to show that

$$\mu = (d - 2)\nu + \zeta , \tag{2.21}$$

where $\zeta = \tilde{\zeta}\nu$. It was shown by Straley (1977) that in 2D, $\mu = s$.

Equations (2.15) and (2.16) can be unified by using the two-component resistor network described earlier. In the critical region, i.e., near p_c where both $|p - p_c|$ and $h = g_2/g_1$ are small, the effective conductivity g_e follows the following scaling law (Efros and Shklovskii 1976; Straley 1976):

$$g_e \sim g_1|p - p_c|^\mu \Phi_\pm(h|p - p_c|^{-\mu-s}) , \tag{2.22}$$

where Φ_+ and Φ_- are two homogeneous functions corresponding, respectively, to the regions above and below p_c, and are similar to μ and s, universal. For any fixed and non-zero h, g_e has a smooth dependence on $p - p_c$. This becomes clearer if we rewrite Eq. (2.22) as

$$g_e \sim g_1 h^{\mu/(\mu+s)} \Psi\left[|p - p_c|h^{-1/(\mu+s)}\right] , \tag{2.23}$$

where $\Psi(x) = x^\mu \Phi_+(x^{-\mu-s}) = (-x)^\mu \Phi_-[(-x)^{-\mu-s}]$.

Since the function $\Psi(x)$ is universal, the implication of Eq. (2.23) is that if one plots $g_e/[g_1 h^{\mu/(\mu+s)}]$ versus $|p - p_c|h^{-1/(\mu+s)}$ for all networks, or more generally randomly disordered materials, with the same Euclidean dimensionality, all the results, or measurements, should collapse onto a single universal curve. The same type of scaling laws also holds for the permeability of percolating and pore-network models of porous media (Ghanbarian et al. 2021).

This provides a powerful tool for estimating the conductivity of a composite for any value of h, given the conductivities for two other values of h, so that the universal curve can be constructed. Somewhat similar, but more complex, scaling equations can be developed for the elastic moduli, dielectric constant, and other properties of percolation systems, but we defer their description to future chapters. The implied prefactors in all the above power laws depend on the type of the network and are not universal.

No exact relation is known between the transport and connectivity exponents. This is perhaps because the transport exponents describe *dynamic* properties of percolation

Table 2.3 The critical exponents and fractal dimensions of 2D and 3D percolation. The geometrical exponents and the fractal dimension D_f for 2D systems, as well as all the quantities for the Bethe lattices, are exact. Values of μ, s, T, f, and χ represent the exponents for disordered materials and media with a statistical distribution that does not violate inequality (24)

Exponent	$d = 2$	$d = 3$	Bethe lattice
β	5/36	0.41	1
$\beta_{\rm bb}$	0.48	1.05	2
τ	187/91	2.18	5/2
gamma	43/18	1.85	1
σ	36/91	0.45	1/2
ν	4/3	0.88	1/2
η	5/24	−0.068	0
D_f	91/48	2.53	4
$D_{\rm bb}$	1.64	1.87	2
D_r	3/4	1.36	2
$D_{\rm min}$	1.13	1.34	2
μ	1.3	2.0	3
s	1.3	0.73	0
e	1.3	2.0	3
f	3.96	3.78	4
χ	1.3	0.73	0

systems, whereas the connectivity exponents describe the *static* properties, and in general there is no reason to expect a direct relation between the two.

The exponents μ, s, f, χ, and e are universal if two conditions are satisfied. (a) There is no any long-range correlations in the system, and (b) the distribution $\psi(x)$ of the conductance or elastic constant of the bonds should be such that (Kogut and Straley 1979; Straley 1982; Sahimi et al. 1983b)

$$\psi_{-1} = \int_0^\infty \frac{1}{x}\psi(x)dx < \infty\,. \tag{2.24}$$

Some correlated or continuous systems violate one or both of these conditions. In Chap. 3, we will discuss the implications of this violation. In Table 2.3, the current estimates of the critical exponents in 2D and 3D are compiled.

2.10 Scale Dependence of Percolation Properties

As mentioned earlier, the correlation length ξ_p has the physical significance that for length scales $L > \xi_p$ the percolation system is macroscopically homogeneous. For $L < \xi_p$, however, the system is *not* homogeneous and its macroscopic properties depend on L. In this case, the SSC is statistically self-similar at all the length scales less than ξ_p, and its mass M—the total number of occupied bonds or sites that it contains—scales with L as

$$M \propto L^{D_f} , \quad L < \xi_p, \tag{2.25}$$

where D_f is called the fractal dimension of the cluster. For $L > \xi_p$ one has $M \propto L^d$ and, thus, $D_f = d$, where d is the Euclidean dimension of the system. The crossover between the fractal and Euclidean regimes takes place at $L \simeq \xi_p$. Thus, we unify the two cases by writing a scaling equation for $M(L, \xi_p)$:

$$M(L, \xi_p) = L^{D_f} h(L/\xi_p) , \tag{2.26}$$

where $h(x)$ is a scaling function. For $L \gg \xi_p$ ($x \gg 1$), we expect to have $M \sim L^d$, and hence $h(x) \sim x^{D_f - d}$, whereas for $x \ll 1$ we must have $h(x) \simeq$ constant.

The fractal dimension D_f of the SSC is related to the percolation exponents defined earlier. The mass M of the cluster is proportional to $X^A(p)\xi_p^d$, since only a fraction $X^A(p)$ of all the occupied bonds or sites are in the cluster. Therefore, we must have $M \propto \xi_p^{d-\beta/\nu}$. For $L < \xi_p$ we must replace ξ_p with L, since in this regime L is the only relevant length scale. Thus, $M \propto L^{d-\beta/\nu}$, which, when compared with Eq. (2.26), yields

$$D_f = d - \beta/\nu , \tag{2.27}$$

so that $D_f(d = 2) = 91/48 \simeq 1.9$ and $D_f(d = 3) \simeq 2.53$. Similarly, for $L < \xi_p$ the backbone is also a fractal object and its fractal dimension D_{bb} is given by

$$D_{\mathrm{bb}} = d - \beta_{bb}/\nu . \tag{2.28}$$

Note that at $p = p_c$ the correlation length is divergent, so that at this point the SSC and its backbone are fractal objects at *any* length scale.

The number M_r of the red bonds of the backbone that are in a box of linear size $L < \xi_p$ scales with L as $M_r \propto L^{D_r}$, where D_r is the fractal dimension of the set of the red bonds. It has been shown (Coniglio, 1981) that

$$D_r = \nu^{-1} . \tag{2.29}$$

Another important concept is the *minimum* or *chemical* path between two points of a percolation cluster, which is the shortest path between the two points. For $L < \xi_p$, the length L_{min} of the path scales with L is given by

$$L_{\mathrm{min}} \sim L^{D_{\mathrm{min}}} , \tag{2.30}$$

so that D_{min} is the fractal dimension of the minimal path.

If $L < \xi_p$, i.e., if the percolation system is in the fractal regime, all the percolation properties become scale dependent. Since $|p - p_c| \sim \xi_p^{-1/\nu}$, one can rewrite all the above power laws in terms of ξ_p. For example, Eq. (2.29) is rewritten as $P(p) \sim \xi_p^{-\beta/\nu}$. For $L < \xi_p$ one replaces ξ_p by L, since in this regime L is the only relevant length scale in the system and, therefore,

$$P(L) \sim L^{-\beta/\nu} , \tag{2.31}$$

$$X^A(L) \sim L^{-\beta/\nu} , \tag{2.32}$$

$$X^B(L) \sim L^{-\beta_{bb}/\nu} , \tag{2.33}$$

$$g_e(L) \sim L^{-\mu/\nu} , \quad \text{conductor–insulator composites} \tag{2.34}$$

$$g_e(L) \sim L^{s/\nu} , \quad \text{conductor–superconductor composites} \tag{2.35}$$

$$E_e(L) \sim L^{-f/\nu} , \quad \text{rigid–soft composites} \tag{2.36}$$

$$E_e(L) \sim L^{\chi/\nu} , \quad \text{rigid–superrigid composites} \tag{2.37}$$

$$D_e(L) \sim L^{-\theta} , \tag{2.38}$$

$$K_e(L) \sim L^{-e/\nu} , \tag{2.39}$$

where $\theta = (\mu - \beta)/\nu$. Thus, *scale-dependent properties are a signature of a fractal morphology*. Similar to the mass of the sample-spanning cluster for which we could write down scaling Eq. (2.46), we can also write down scaling equations for length scale dependence of all other percolation properties. For example, Eqs. (2.15) and (2.34) are combined to yield

$$g_e(L, \xi_p) = \xi_p^{-\mu/\nu} G(L/\xi_p) , \tag{2.40}$$

where $G(x)$ is a (universal) scaling function. Once it is established that a material possesses a fractal morphology, many classical laws of physics must be significantly modified. For example, Fick's law of diffusion with a constant diffusivity can no longer describe diffusion processes in fractal materials, a subject that will be described and discussed later in this book.

2.11 Finite-Size Scaling

So far, we have described percolation in disordered media that are of infinite extent. In practical applications and in computer simulations, however, one deals with systems of finite extents, in which as p_c is approached, ξ_p eventually exceeds the linear size L of the material, and L becomes the dominant length scale of the system. Fisher (1971)[5] developed a theory for the scaling properties of a finite system near a critical

---

[5] **Michael Ellis Fisher** (1931–2021) was an English physicist, as well as chemist and mathematician, who made many seminal contributions to statistical physics, including, but not limited to, the theory of phase transitions and critical phenomena. A member of the National Academy of Sciences and

point, such as a percolation threshold, which is usually called *finite-size scaling* theory, according to which the variation of *any* property P_L of a system of linear size L is written as

$$P_L \sim L^{-\zeta} f(u) \tag{2.41}$$

with $u = L^{1/\nu}(p - p_c) \sim (L/\xi_p)^{1/\nu}$, where $f(u)$ is a non-singular function. If, in the limit $L \to \infty$, P_∞ follows a power law, such as $P_\infty \sim (p - p_c)^\delta$, then one must have $\zeta = \delta/\nu$. Therefore, the variations with L of $P_L(p)$ in a *finite network* at p_c of the *infinite network* can be used to obtain information about the quantities of interest for an *infinite network near* p_c.

Finite-size scaling theory has been used successfully for computing accurate estimates of the critical exponents, and even the percolation thresholds, from simulation of finite systems. The finite size of a network causes a shift in its percolation threshold (Levinshtein et al. 1976; Reynolds et al. 1980):

$$p_c - p_c(L) \sim L^{-1/\nu} . \tag{2.42}$$

In this equation, p_c is the percolation threshold of the infinite system, and $p_c(L)$ is an *effective percolation threshold* for a finite system of linear dimension L. Note that Eqs. (2.42) and (2.43) are valid for very large network size L, whereas in practice very large systems cannot easily be simulated. To remedy this situation, Eq. (2.42) is modified to

$$P_L \sim L^{-\zeta}[a_1 + a_2 h_1(L) + a_2 h_2(L) + \cdots] , \tag{2.43}$$

where a_1, a_2, and a_3 are three fitting parameters, and h_1 and h_2 are called the *correction-to-scaling* functions. These functions are particularly important when small and moderate system sizes L are used in the simulations. For *flow and transport properties*, $h_1 = (\ln L)^{-1}$ and $h_2 = L^{-1}$ often provide accurate estimates of ζ (Sahimi and Arbabi 1991). Equation (2.43) provides us with a means of estimating a critical exponent: calculate P_L at $p = p_c$ for several system sizes L and fit the results to Eq. (2.43) to estimate ζ and thus $\delta = \zeta\nu$.

2.12 Experimental Measurement of Percolation Properties

Let us now discuss experimental measurements of percolation quantities in some heterogeneous systems. To do so, we first consider conductor–insulator composites that have wide applications, and describe measurement of their morphological properties. Measurement of their transport properties will be described and analyzed in the subsequent chapters.

American Philosophical Society, and a fellow of Academy of Arts and Sciences, Fisher won many awards, including Irving Langmuir Prize and Lars Onsager Prize of American Physical Society, and Boltzmann Medal of the International Union of Pure and Applied Physics.

Thin metal films are two-phase mixtures of metal and non-metal components that have interesting morphological and transport properties. Their electrical conductivity varies continuously with the volume fraction ϕ of the metallic phase. If ϕ is large enough, the metal phase forms an SSC, while the non-metallic phase is dispersed in the composite in the form of isolated islands. In this regime, the electrical conductivity is large and its temperature coefficient of resistance is positive. If the volume fraction of the metallic phase is close to its percolation threshold ϕ_c, one has a complex composite of isolated metallic islands, and a tortuous and SSC of the metallic component, as well as islands of the non-metallic component. If the non-metallic phase forms a SSC, the system is in the dielectric regime and its electrical conductivity is very small.

Transmission electron microscopy (TEM) indicates that thin metal films have a discontinuous structure in which the metallic phase constitutes only a fraction of the total volume of the system. Such films have a nearly 2D structure and are formed in the early stages of film growth by a variety of techniques, such as evaporation or sputtering. The metallic phase first forms isolated islands that at later stages of the process join together and form a continuous film. Generally speaking, a metal film is considered thin if its thickness ℓ is less than the percolation correlation length ξ_p of the 3D composite.

Important studies of the morphological properties of thin metallic films were carried out by Voss et al. (1982) and Kapitulnik and Deutscher (1982), the results of which were published in back-to-back papers in the *Physical Review Letters.* Voss et al. prepared Au films by electron-beam evaporation onto 30-nm-thick amorphous Si_3N_4 windows, grown on a Si wafer frame. A series of samples were prepared simultaneously that ranged from 6 to 10 nm thick, and varied from electrically insulating to conducting. The transmission electron micrographs of the material were digitized and, with the use of threshold detection and a connectivity-checking algorithm, the individual Au clusters were isolated and their statistical properties were analyzed.

Voss et al. found that the Au clusters were irregularly shaped and ramified. At large scales, most of the film properties were uniform, but at small scales they were not. Figure 2.5 shows their results. Since the deposited Au atoms had some initial mobility, the Au clusters were not totally random, as a result of which the percolation threshold of the system, $\phi_c \simeq 0.74$, was larger than 0.5, the expected value for 2D continuum percolation (see Chap. 3). Voss et al. found, however, that at ϕ_c the largest Au cluster is a fractal object with a fractal dimension, $D_f \simeq 1.9$, in excellent agreement with that of 2D percolation, $D_f = 91/48 \simeq 1.896$.

Kapitulnik and Deutscher (1982) prepared samples by successive deposition of amorphous Ge as the substrate, and of thin Pb films as the metal. This allowed them to obtain deposition with only short-range correlations. The size of the Pb crystallites, defined as the average Pb thickness, was about 200 Å. Since the continuity of the metallic cluster is controlled by its thickness, varying the sample thickness is equivalent to generating percolation networks with varying fractions of conducting components. Kapitulnik and Deutscher (1982) deposited the thin films on the TEM grids and photographed it with very large magnification. The pictures were then

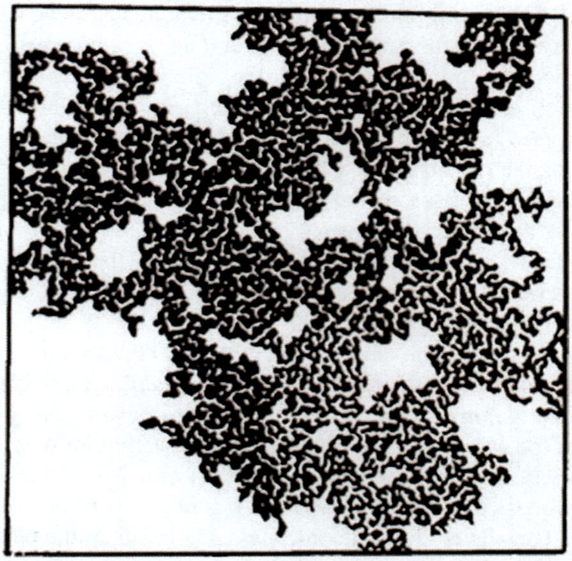

Fig. 2.5 The Au clusters at $p = 0.71$ (after Voss et al. 1982)

analyzed for various percolation properties. Figure 2.6 presents part of the SSC of Pb and its backbone.

Figure 2.7 presents their results for the density $\rho = M/L^2$ of the SSC and its backbone. Over about one order of magnitude variations in L, the SSC and its backbone exhibit fractal behavior, with $D_f \simeq 1.9$ for the SSC and $D_{\text{bb}} \simeq 1.65$ for the backbone, in excellent agreement with the fractal dimensions for 2D percolation (see Table 2.3). Kapitulnik and Deutscher (1982) also measured n_s, the number of metallic clusters of mass s below the percolation threshold. According to Eq. (2.14), one must have $n_s \sim s^{-\tau}$, and measurements of Kapitulnik and Deutscher, shown in Fig. 2.8, yielded $\tau \simeq 2.1 \pm 0.2$, completely consistent with $\tau = 187/91 \simeq 2.054$ for 2D percolation.

Viswanathan and Heaney (1995) succeeded in direct imaging of a percolation network in a 3D conductor–insulator composite, and obtaining accurate estimates of its percolation properties. They used electric force microscopy (EFM), a type of scanning probe microscopy that measures electric field gradients near the surface of a sample, using a sharp conductive tip. The gradients are measured by the grounded tip which is oscillated near its resonant frequency. By altering the effective spring constant of the tip as it encounters a force gradient from the electric field, the cantilever's resonant frequency is varied. The resulting change is then monitored, producing a map of the strength of the electric field gradients. The EFM mode generates two images, one of topography and the other of electric field gradients. The maps were generated by the "lift mode" technique by which a line is scanned to give topography; a second pass is made over the same line in a prescribed distance above the topography to image the electric field gradients. A commercial Digital

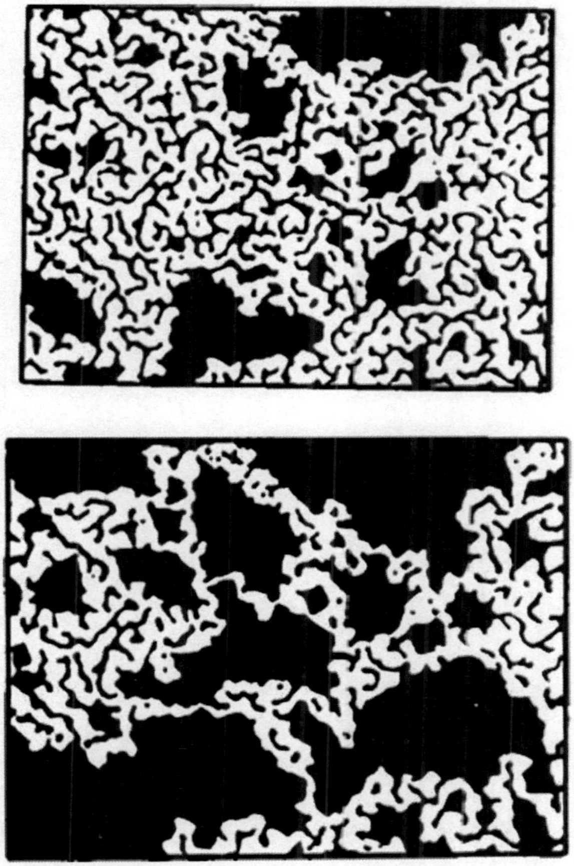

Fig. 2.6 The sample-spanning cluster (top) and its backbone (bottom) of thin Pb films (after Kapitulnik and Deutscher 1982)

Instruments Nanoscope Multimode Atomic Force Microscopy for EFM imaging was used for imaging the composites. Tips were metal coated and made of single-crystal silicon with a 5 nm radius curvature. The composite samples were attached on top of the metal substrates using conductive silver paint. The samples were composed of commercial carbon-black and polymer. The carbon-black resistivity was of order of 10^{-2} Ωcm. It consisted of 200 nm mean diameter aggregates composed of smaller fused semi-spherical particles of 80 nm mean diameter. The polymer was high-density polyethylene with a resistivity of 10^{18} Ωcm.

To quantify the morphology of the sample, the EFM image was binarized in order to highlight the conductive features. The resulting image was then analyzed in terms of the scaling theory of percolation described above. A point that was part of the conductive area or "mass" M in squares centered on this origin was counted as a function of the square linear size L. The density $\rho(L) = M(L)/L^2$ of the conductive area was then computed. The procedure was repeated for 25

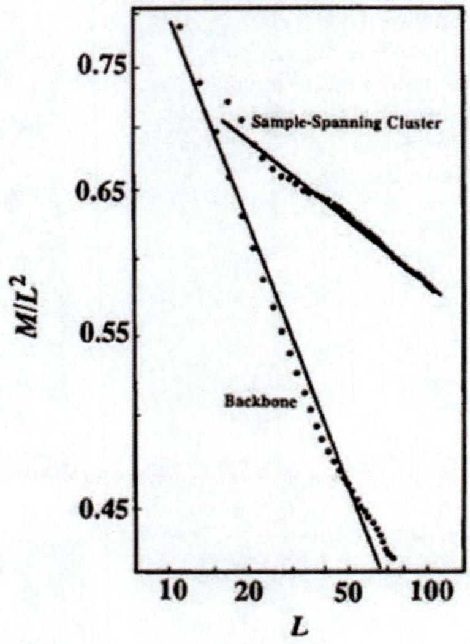

Fig. 2.7 The mass M of the sample-spanning cluster and its backbone of Pb films versus the linear size L of the material (after Kapitulnik and Deutscher 1982)

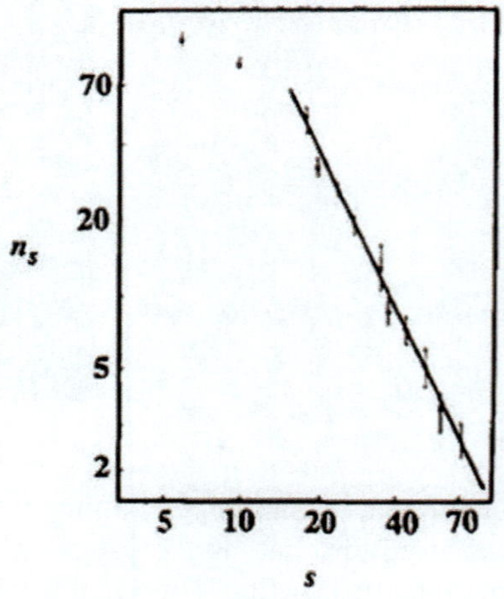

Fig. 2.8 The number n_s of Pb clusters of size s (after Kapitulnik and Deutscher 1982)

non-overlapping areas on the binarized image, and the results were averaged. Figure 2.9 shows their results, which indicate the existence of three regimes. For $L < 0.6$ μm, the density satisfied a power law corresponding to $M(L) \propto L^{D_{\rm eff}}$, where $D_{\rm eff}$ is the effective fractal dimension of the conductive areas. The data yielded $D_{\rm eff} \simeq 0.9 \pm 0.1$. Since the fractal dimension $D_{\rm eff}$ of an object embedded in a $d-$dimensional space and cut by a surface of dimensionality D_s is given by

$$D_{\rm eff} = D_f - (d - D_s) , \tag{2.44}$$

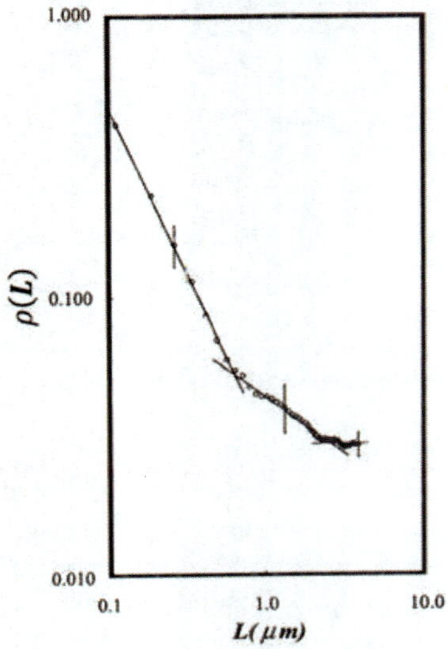

Fig. 2.9 Logarithmic plot of the average density $\rho(L)$ of the conducting regions versus the linear size L of the conductor–insulator composite sample. The minimum conducting unit is a square of size $L = 0.037$ μm (after Viswanathan and Heaney 1995)

the fractal dimension D_f of the carbon-black aggregate is estimated to be $D_f \simeq 1.9 \pm 0.1$, in excellent agreement with that of 2D percolation. Note also that $L \approx 0.6$ μm is an estimate of the percolation correlation length ξ_p for the 2D system. Had the material been strictly 2D, the curve in Fig. 2.9 for $L > 0, 6$ μm would have become a constant, signifying the crossover from the fractal to compact morphology.

For 0.8 μm $< L <$ 2 μm, the density followed the same type of power law that corresponds to $D_f \simeq 2.6 \pm 0.1$, in good agreement with the fractal dimension of 3D percolation clusters at the percolation threshold, $D_f \simeq 2.53$ (see Table 2.3), while for $L > 3$ μm the density was roughly constant, corresponding to $D_f = 3$ for a homogeneous material. The crossover from the power-law behavior occurred at $L \simeq \xi_p$, where ξ_p is the correlation length of percolation, and hence for these materials, $\xi_p \simeq 3$ μm. These measurements all agree with the predictions of percolation theory, hence firmly establishing the quantitative relevance of percolation to describing the morphology (and transport properties) of heterogeneous materials.

2.13 Poor Man's Percolation Theory: Effective-Medium Approximation

Consider a regular network of coordination number Z in which the conductance g of the bonds between the sites is randomly and independently distributed according to a normalized conductance distribution $\chi(g)$. The effective-medium approximation (EMA) replaces such a network with an *effective* network in which the conductance of all the bonds is g_e. To estimate g_e, the conductance of one bond in the effective

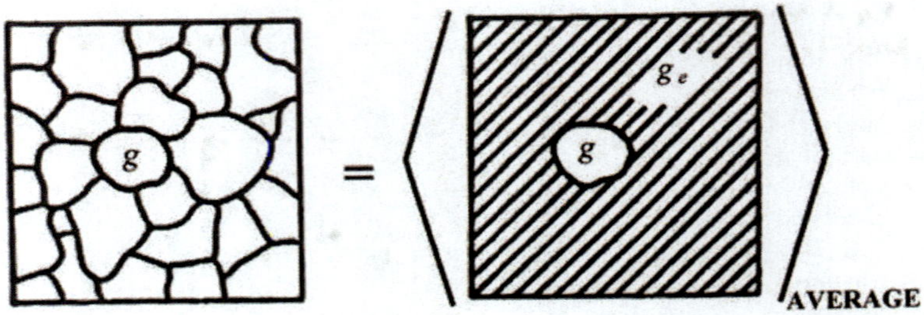

Fig. 2.10 Schematic implementation of the EMA, transforming a disordered medium (left) to a homogeneous effective one (right)

network is replaced by its true value in the original system. This induces a perturbation in the distribution of the potentials throughout the effective network. The perturbation is determined, and g_e is computed by requiring that the average perturbation, where the average is taken with respect to the conductance distribution, should vanish. This is shown schematically in Fig. 2.10.

Since the effective network is uniform, g_e is also the overall conductance of the network (or proportional to it). The EMA then predicts that g_e is the solution of the following equation, first derived by Landauer (1952)[6] for a continuum system, and then by Kirkpatrick (1971, 1973) for a network,

$$\int_0^\infty \frac{g - g_e}{g + (Z/2 - 1)g_e} \psi(g)dg = 0 \,. \tag{2.45}$$

The EMA may be viewed as *poor man's percolation theory*[7] because not only it provides a simple description of flow and transport in pore or conductance networks with percolation disorder, i.e., those in which a fraction $1 - p$ of the pores or bonds do not allow flow or conduction, while the remaining fraction p do, but also its prediction for the percolation thresholds is *nontrivial*, i.e., it is neither 0 nor 1, but something in between. The EMA also does predict power-law dependence of the permeability and conductivity of the networks on $(p - p_c)$ near p_c, except that the predicted exponents are incorrect. The EMA has also been extended to frequency-dependent conductivity (Odagaki and Lax 1981; Sahimi et al. 1983b). Surprisingly,

[6] **Rolf William Landauer** (1927–1999) was a German-American physicist who made important contributions in diverse areas of the thermodynamics of information processing, condensed matter physics, and the conductivity of disordered media. In 1961, he discovered Landauer's principle, which states that in any logically irreversible operation that manipulates information, such as erasing a bit of memory, entropy increases and an associated amount of energy is dissipated as heat, which is relevant to reversible computing, quantum information, and quantum computing. Landauer was awarded the Stuart Ballantine Medal of the Franklin Institute, the Oliver Buckley Prize of the American Physical Society, and the IEEE Edison Medal, among many other honors.

[7] To my knowledge, Rolf Landauer coined this phrase; see Kirkpatrick (1973); footnote 4.

the EMA even provides (Sahimi 1984) an estimate of D_f, the fractal dimension of the SSC, defined earlier.

As mentioned above, despite its simplicity and lack of detailed information about the network's microstructure, the EMA predicts a nontrivial p_c. Suppose that the bond conductance distribution is given by

$$\psi(g) = (1-p)\delta_+(0) + pw(g) \; . \tag{2.46}$$

That is, the conductances of a fraction p of the bonds are distributed according to a normalized distribution $w(g)$, while the rest of the bonds with a fraction $1-p$ do not conduct at all (or do not allow fluid flow). Then, Eq. (2.45) becomes

$$\frac{1-p}{1-Z/2} + p\int_0^\infty \frac{g-g_e}{g+(Z/2-1)g_e} w(g)dg = 0 \; , \tag{2.47}$$

the solution of which yields the effective conductance g_e of a random network of conducting and insulating bonds in bond percolation. If we set $g_e = 0$ and $p = p_{cb}$ in Eq. (2.47), since g_e must vanish at $p = p_{cb}$, we obtain

$$p_{cb} = 2/Z \; , \tag{2.48}$$

Since $p_{cb}Z = \langle Z\rangle$ represents the *average* coordination number $\langle Z\rangle$ of the network at the percolation threshold, Eq. (2.48) implies that for a network to be conducting, one must have $\langle Z\rangle \geq 2$. Recall from the earlier discussions that for $d-$dimensional networks, $\langle Z\rangle = B_c = p_{cb}Z \simeq d/(d-1)$ represents a very accurate approximation for the average coordination number at p_{cb}. Therefore, the EMA prediction for p_{cb} of 2D networks is very accurate, but not so for 3D networks [compare the predictions of Eq. (2.48) with those listed in Table 2.2]. Given the EMA prediction for p_{cb}, the basic EMA for computing the effective conductance g_e may be rewritten as

$$\int_0^\infty \frac{g-g_e}{g+(1/p_{cb}-1)g_e} \psi(g)dg = 0 \; , \tag{2.49}$$

which allows one to treat p_{cb} as an adjustable parameter, or use its numerical estimate, instead of $2/Z$ as predicted by the EMA, and, thus, have accurate predictions for g_e under a wide variety of conditions. Doing so is, however, semi-empirical.

In addition, if we set $w(g) = \delta(g - g_0)$ in Eq. (2.49), we obtain

$$\frac{g_e}{g_0} = \frac{p-2/Z}{1-2/Z} = \frac{p-p_{cb}}{1-p_{cb}} \propto (p-p_{cb}) \; . \tag{2.50}$$

Thus, according to the EMA, in all dimensions d ($d = 2, 3$) the exponent μ of the power-law behavior of g_e near p_{cb} is $\mu = 1$. This result is independent of $w(g)$, unless $w(g) = (1-\alpha)g^{-\alpha}$, where $0 \leq \alpha < 1$, in which case the EMA predicts (Kogut and Straley 1979) that

$$g_e \sim (p - p_{cb})^{1+\alpha/(1-\alpha)} , \tag{2.51}$$

so that $\mu = 1 + \alpha/(1 - \alpha)$. Thus, in this particular case, the exponent μ is no longer universal and depends on the details of the conductance distribution. The EMA prediction of a distribution-dependent non-universality of μ is also correct, although its precise prediction for the non-universal μ is not.

In general, the EMA is very accurate if the network is not close to its percolation threshold, *regardless* of the structure of $w(g)$. Its predictions are also more accurate for 2D networks than for 3D ones. Improvement of the EMA for predicting the effective conductivity near p_c was proposed by Sahimi et al. (1983c); Sahimi (1984) and Sahimi (1988). We will invoke the EMA—poor man's percolation theory—throughout this book to determine various flow and transport properties of disordered materials and media.

Before beginning to describe various applications of percolation in the following chapters, one crucial point must be emphasized. When we speak of application of percolation theory to any physical phenomenon, we do not necessarily mean random percolation. As Chap. 3 indicates, one can build on the classical random percolation and develop more complex variations of the model that better describe various applications. Thus, when we speak of a percolation problem, we mean one in which the connectivity of the microscopic elements of a system has a profound effect on its macroscopic properties, such that if the system is not in a state above its connectivity threshold, macroscopic properties, such as permeability and diffusivity, vanish. If the microscopic elements are randomly connected or disconnected, one has the classical random percolation. If there are correlations between the microscopic elements, then one has a variant of the percolation model.

Chapter 3
Extensions of the Classical Percolation Model

3.1 Introduction

The classical percolation model described in Chap. 2 is appropriate when there are no correlations between the bonds, sites, or both. Moreover, the percolation models described were defined in networks or lattices. In practice, however, almost any system of significant interest whose morphology may be described by percolation concept contains some correlations. For example, natural porous media often have very small percolation thresholds; that is, their pore space remains connected down to very small values of the porosity, the volume fraction of the pores. Such low percolation thresholds can be explained at least partly based on the existence of extended positive correlations in the pore space, which usually reduce the percolation threshold.

There are also many disordered materials and media that cannot be represented by a lattice or network model. Examples include packed beds of particles, and porous media that contain a spatial distribution of fractures and faults. The question then is, how should such non-lattice systems be modeled, if percolation is essential to describing their macroscopic properties?

To address such issues many variants of the classical random percolation model have been developed. The purpose of this chapter is to describe such models. Some of such models will be relevant to the applications of percolation theory that we will begin to describe in Chap. 4, while others have been developed for describing the systems that are beyond the scope of this book.

3.2 Continuum Percolation

Percolation in continua is of great interest, since in many practical applications one must deal with systems in which real objects, such as particles that have various shapes and sizes are distributed spatially. Such materials and media are better and

M. Sahimi, *Applications of Percolation Theory*, Applied Mathematical Sciences 213,
https://doi.org/10.1007/978-3-031-20386-2_3

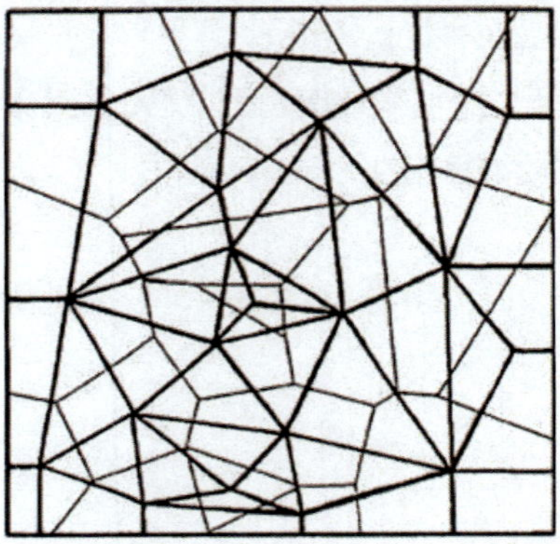

Fig. 3.1 Two-dimensional Voronoi tessellation of space (thin lines) and the corresponding Voronoi network (thick lines)

more accurately represented by a continuum percolation, which is in fact directly applicable to characterization and modeling of the morphology and effective transport properties of microemulsions, polymer blends, sintered materials, sol–gel transitions, and many more. An excellent review of the subject is given by Balberg (2021).

There are at least three types of continuum percolation systems:

(i) One has a random or correlated distribution of inclusions, such as circles, spheres, or ellipses, in an otherwise uniform background. The correlations arise when one imposes some constraints on the system. For example, if the particles are not allowed to overlap, or if the extent of their penetrability or overlap is fixed, then the resulting morphology contains correlations whose extent depends on the type of the constraints imposed. In such continua percolation is defined either as the formation of a sample-spanning cluster (SSC) of the channels *between* the inclusions, or as the formation of a SSC of touching or overlapping inclusions.

(ii) The percolating continuum is modeled by tessellating the space into regular or random polyhedra (Winterfeld et al. 1981). A (volume) fraction of the polygons or polyhedra is designated as one phase of the material, while the rest constitutes the second phase, with each phase having its own distinct spatial properties. Figure 3.1 presents an example of such a model.

(iii) One constructs a distribution of conducting sticks of a given aspect ratio, or plates of a given extent, inserted in a uniform background. Such models have been proposed for representing fibrous materials, as well as fractured porous media. An example is shown in Fig. 3.2 (see also Chap. 5).

3.2.1 Percolation Threshold of Continua

Several techniques may be used to estimate the percolation threshold of a continuous percolation system. Except in one special case and somewhat similar to lattice percolation, the percolation thresholds of continuous d-dimensional systems with

Fig. 3.2 A two-dimensional model of fibrous materials, or fractured porous media

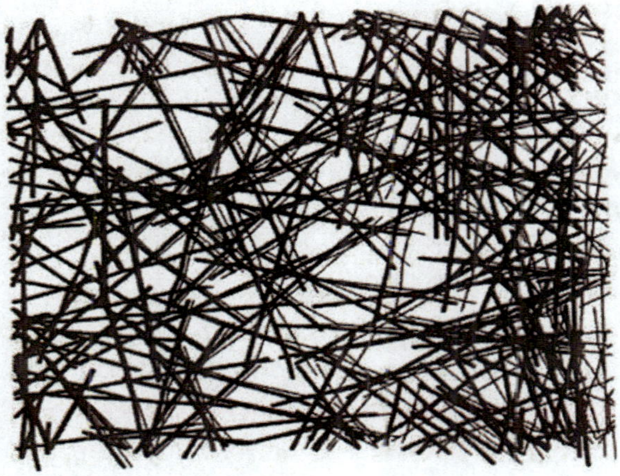

$d \geq 2$ are not known exactly. The exceptions are for certain two-dimensional (2D) symmetric-cell models. Two-phase symmetric-cell models are constructed by partitioning space into cells of arbitrary shapes and sizes, with the cells being randomly designated as phases 1 and 2 with probabilities ϕ_1 and ϕ_2. A simple example is a 2D system that is tessellated into square cells, where each cell belongs either to phase 1 or phase 2 with probabilities ϕ_1 and ϕ_2. The Voronoi tessellation shown in Fig. 3.1 provides another example of such models. Unlike distributions of particles in a matrix, the symmetric-cell models possess phase-inversion symmetry defined in Chap. 2.

It is generally believed that for any 2D symmetric-cell model in which the centers of the cells are the sites of a fully triangulated lattice, one has, $\phi_{1c} = \phi_{2c} = 1/2$, although we are not aware of any rigorous proof of this theorem. In this case, the continuum percolation threshold and the site percolation threshold of the underlying lattice are identical. Examples include hexagonal and Voronoi tessellations, both of which have exact site percolation thresholds of 1/2.

Consider the case in which particles are distributed in a uniform matrix. Let us refer to the particles as phase 2 of the system, with phase 1 being the background matrix. Pike and Seager (1974) and Haan and Zwanzig (1977) were among the first to obtain accurate estimates of the percolation threshold ϕ_{2c} for randomly-distributed overlapping disks and spheres. The most accurate current estimate of ϕ_{2c} for 2D (disk) systems is due to Quintanilla et al. (2000) who reported, $\phi_{2c} \simeq 0.67637 \pm 0.00005$, while for 3D spherical particles systems, Rintoul and Torquato (1997) provided the estimate, $\phi_{2c} \simeq 0.2895 \pm 0.00005$. The corresponding values for overlapping oriented squares and cubes are (Pike and Seager 1974; Haan and Zwanzig 1977), $\phi_{2c} \simeq 0.67 \pm 0.01$ and 0.28 ± 0.01, respectively. In these models, all the particles had the same size.

In general, if the particles do not have the same size, then ϕ_{2c} depends continuously on the size distribution of the particles. Numerical simulations indicated (see, for example, Pike and Seager (1974); Lorenz et al. (1993)), however, that ϕ_{2c} depends only weakly on the particle size distribution, although certain pathological exceptions to this general result can also be constructed. Meester et al. (1994) conjectured that the monodisperse distribution *minimizes* the percolation threshold.

One of the most important discoveries for continuum percolation was made by Scher and Zallen (1970). They found that for hard-core spheres placed on the sites of a lattice, the onset of percolation is at a volume fraction that is nearly a constant, irrespective of the lattice. Thus, a critical occupied volume fraction ϕ_c, or percolation threshold, was defined by

$$\phi_{2c} = p_{cs} f_l \ , \tag{3.1}$$

where f_l is the filling factor of a lattice—the maximum volume fraction of a lattice that can be filled by hard spheres of equal volume—when each site of the lattice is occupied by a sphere in such a way that two nearest-neighbor hard spheres touch one another at the midpoint. ϕ_{2c} appears to be an almost invariant of the system with a value of about 0.17 for 3D systems, and 0.45 for 2D ones. Shante and Kirkpatrick (1971) generalized Scher and Zallen's idea to permeable spheres and showed that the average number B_c of bonds per sites at p_c is related to ϕ_c by

$$\phi_{2c} = 1 - \exp(-\frac{1}{8}B_c) \ , \tag{3.2}$$

and that for continuous systems, $B_c \to p_{cs}Z$, as Z becomes large. It is clear from Table 2.3 that in 3D, $B_c \simeq 1.5$.

It has been shown (see, for example, Balberg and Binenbaum, 1985, and references therein) that the morphological exponents of percolation, defined by Eqs. (2.8)–(2.14), are equal for lattice and continuous systems and, thus, at least in certain cases the former may be used for studying the latter.

Similar to lattice systems described in Chap. 2, one may also speak of two-phase composites that are bicontinuous. Hence, extending the criteria of bicontinuity for lattice systems described above to continuum systems, we say that a $d-$dimensional two-phase continuum system that possesses phase-inversion symmetry is bicontinuous for $\phi_{2c} < \phi_2 < 1 - \phi_{2c}$ for $d \geq 2$. However, these are not necessary conditions. For example, a bicontinuous structure without phase-inversion symmetry but with $\phi_{2c} < 1/2$ is a 3D distribution of identical overlapping spheres, where for $0.29 < \phi_2 < 0.97$ both the particle phase and the space between the particles are percolating. An example of a bicontinuous composite without phase-inversion symmetry and with ϕ_{2c} is a close-packed face-centered cubic lattice. At the percolation threshold of the particle (sphere) phase, $\phi_{2c} = \pi/\sqrt{18}$, the space between the particles also percolates.

3.2.2 The Ornstein–Zernike Formulation

Continuum percolation can be described by the Ornstein–Zernike (OZ) formulation, according to which the total correlation between two particles at $\mathbf{r}_1$ and $\mathbf{r}_2$ can be written as the sum of two contributions: (a) A *direct* effect of the particle at $\mathbf{r}_1$ on the particle at $\mathbf{r}_2$, which is generally short-ranged and is characterized by a type of correlation function $D_2(\mathbf{r})$, and (b) an *indirect* effect in which a particle at $\mathbf{r}_1$ affects a particle at $\mathbf{r}_3$, which in turn affects a particle at $\mathbf{r}_2$, directly or indirectly through other particles. The indirect effect is weighted by the density and averaged over all possible $\mathbf{r}_3$. It should, therefore, be clear to the reader that the indirect effect has a longer range than $D_2(\mathbf{r})$ does. We do not give the details of the OZ formulation, but note that Coniglio et al. (1977) utilized the OZ formulation to derive the following equation for the pair-connectedness function $P_2(\mathbf{r})$ defined in Chap. 2:

$$P_2(\mathbf{r}) = D_2(\mathbf{r}) + \rho D_2(\mathbf{r}) \otimes P_2(\mathbf{r}) \,, \tag{3.3}$$

where $D_2(\mathbf{r})$ is the *direct connectedness function* [$D_2(\mathbf{r})$ is the analog of the usual two-point correlation function $C(r)$], ρ is the number density of the particles, and $\otimes$ denotes a convolution integral. Taking the Fourier transform of both sides of Eq. (3.3) yields

$$\tilde{P}_2(\omega) = \frac{\tilde{D}_2(\omega)}{1 - \rho \tilde{D}_2(\omega)} \,. \tag{3.4}$$

On the other hand, the mean cluster size is related to $P_2(\mathbf{r})$ through the relation

$$S = 1 + \rho \int P_2(\mathbf{r}) d\mathbf{r} \tag{3.5}$$

which is in fact the continuum analog of Eq. (2.5). Equations (3.3)–(3.5) are the basis for studying continuum percolation systems using a variety of models and approximations.

Note that Eq. (3.5) implies that

$$S = 1 + \rho \tilde{P}_2(0) = [1 - \rho \tilde{D}_2(0)]^{-1} \,. \tag{3.6}$$

Since at the percolation threshold the mean cluster size S diverges, Eq. (3.6) immediately yields an estimate for the critical value ρ_c of the number density ρ:

$$\rho_c = [\tilde{D}_2(0)]^{-1} = \left[\int D_2(\mathbf{r}) d\mathbf{r} \right]^{-1} \,. \tag{3.7}$$

In the OZ formulation, the average coordination number $\langle Z \rangle$ of the system is defined by

$$\langle Z \rangle = \rho \Omega_s(d) \int_0^D r^{d-1} P_2(r) dr \ , \tag{3.8}$$

where $\Omega_s(d)$ is the total solid angle contained in a $d-$dimensional sphere, $\Omega_s(d) = 2\pi^{d/2}/\Gamma(d/2)$, with Γ being the gamma function, and D is the diameter of the (possibly overlapping) spheres.

3.2.3 *Percolation Thresholds: Materials with Very Low or High Thresholds*

One can fabricate materials that have exceedingly small percolation thresholds. A well-known example is *aerogel*, an ultra-light material formed by a sort of a gelation process (see Chap. 13 for description and discussion of gelation). Aerogels have a fractal structure and their solid volume fraction is typically 0.5%, indicating that the percolation threshold of the solid phase is even smaller than 0.5%. One may also fabricate two-phase composite materials by compacting binary mixtures of particles of widely disparate sizes in such a way that the smaller particles form an SSC that resides essentially on the surfaces of the larger particles (see, for example, Malliaris and Turner 1971; Kusy 1977).

In the opposite limit, one may also have solid composites with percolation thresholds that approach unity. This can be achieved if the particle size distribution is very broad, and each particle possesses a "soft" repulsive interparticle potential with a range larger than the size of the particle. Unlike monodisperse particles, polydispersity causes the particles to fill the space, but the repulsive interactions prevent the formation of an SSC until the system is essentially completely filled by the particles. An example of this type of material is colloidal dispersions.

3.2.4 *Differences Between Lattice and Continuum Percolation*

Flow and transport in percolating continua can be quite different from those in lattice percolation models. Consider, for example, a model in which one inserts at random circular or spherical inclusions in an otherwise homogeneous medium. For the obvious reason, this is called the *Swiss-cheese model*; see Fig. 3.3. If flow and transport occur through the matrix (through the channels between the non-overlapping spheres), then the problem is mapped onto an equivalent one on the *edges* of the Voronoi polygons or polyhedra (Kerstein 1983). Such a network has an exact coordination number of 4 in 3D. To generate a Voronoi tessellation of space, one distributes Poisson (random) points in space, each of which is the basis for a Voronoi polygon or polyhedron, which is that part of the space that is closer to its Poisson point than

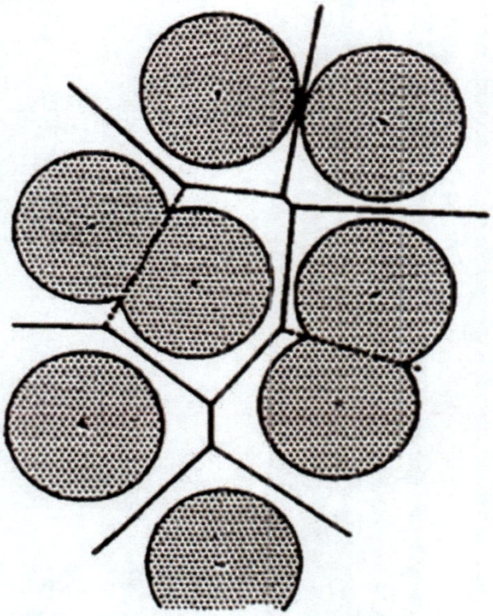

Fig. 3.3 Two-dimensional Swiss-cheese model

to any other Poisson point. A Voronoi *network* is constructed by connecting to each other the Poisson points in the neighboring polygons or polyhedra; see Fig. 3.3. One may also construct the *inverted Swiss-cheese model* in which the roles of the two phases in the Swiss-cheese model are switched, i.e., flow and transport occur through the circular or spherical inclusions, in which case the transport problem can be mapped onto an equivalent one in the Voronoi network.

It is assumed in both models that all the inclusions have the same size. van der Marck (1996) proposed a more general method for mapping transport in continua, which consists of a distribution of inclusions of various sizes in a uniform background, onto an equivalent problem in an equivalent network model. We should, however, point out that if one utilizes the equivalent network of a given continuum, one can no longer assign the transport properties of the network's bonds from an arbitrary distribution, because there is a *natural* distribution of the conductances (or other transport properties) of the transport channels in the continuum that must be constructed based on the shapes and sizes of the channels (see below).

The Voronoi network was utilized by Jerauld et al. (1984a, b) and Sahimi and Tsotsis (1997) as a prototype of irregular networks to study transport in disordered composites. The *average* coordination number of the Voronoi network is about 6 and 15.5 in 2D and 3D, respectively. Jerauld et al. (1984a, b) showed that the geometrical critical exponents of percolation in such networks are the same as those for regular networks. Moreover, they established that as long as the average coordination number of a regular network and a topologically-random one (for example, the 2D Voronoi and triangular networks) are about the same, many transport properties of the two systems are, for all practical purposes, identical.

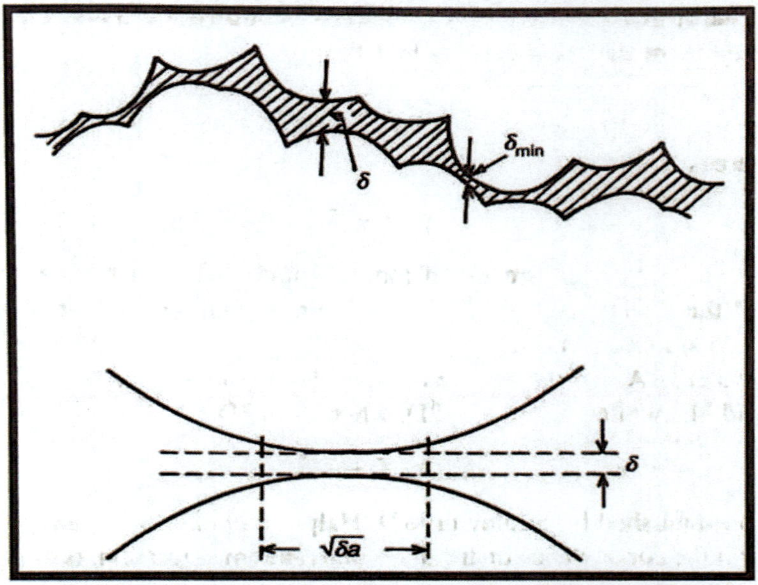

Fig. 3.4 The conducting necks (upper part of the figure) in the Swiss-cheese model, and determining the approximate channels (the dashed lines) that represent them

Certain continuum percolation models violate, however, the condition expressed by inequality (2.24) for the universality of the transport exponents (Halperin et al. 1985; Feng et al. 1987), in which case the flow and transport exponents for the continuum models are not necessarily the same as those in random networks and, thus, they must be estimated separately. The differences between the critical exponents of lattice and continuum percolation are caused by the aforementioned natural distributions of the transport properties that give rise to new scaling laws which cannot be predicted by the network models, unless the same natural distributions are also utilized in the network models as well.

Consider, as an example, the Swiss-cheese model. Figure 3.4 shows a typical channel between two disks (spheres in 3D) of radius a. We imagine that a rectangular strip of width δ is formed, in which case the length $2l$ of the strip is on the order of $\sqrt{\delta a}$. This is because the length l forms the base of a right-angled triangle with the apex at the center of a sphere, the sphere's center being the adjacent site, and the hypotenuse being of length $a+\delta$, so that $a^2+l^2=(a+\delta)^2$, hence yielding $l\sim\sqrt{\delta a}$. The conductivity g of any such bond in d dimensions is then, $g\sim\delta^{d-1}/l\sim\delta^{d-3/2}/\sqrt{l}$. The distribution function $\psi(g)$ of such conductivities is then finite as $\delta\to\delta_{\min}$. The node-link model described in Chap. 2 can then be utilized for estimating the effective conductivity of such a model (Halperin et al. 1985), which is then given by

$$g_e\sim\xi_p^{-(d-1)}\xi_p\Sigma_L\sim\xi_p^{2-d}\Sigma_L\ , \tag{3.9}$$

where

$$\Sigma_L^{-1} = \sum_i g_i^{-1} \sim \int_{\delta_{\min}}^{\infty} \psi(g) g^{-x} dg \ , \tag{3.10}$$

so that the effective conductivity depends on the structure of $\psi(g)$.

Based on such considerations, Halperin et al. (1985) and Feng et al. (1987) showed that the critical exponents that characterize the power laws for flow and transport properties of the Swiss-cheese and the inverted Swiss-cheese models near their percolation thresholds can be very different from those of a random network. In particular, if $\mu_{\rm sc}$ and $\mu_{\rm isc}$ denote, respectively, the critical exponents of the conductivity of the Swiss-cheese and the inverted Swiss-cheese models, then in 2D, $\mu_{\rm sc} = \mu_{\rm isc} = \mu$, while in 3D, $\mu_{\rm isc} = \mu$, and

$$\mu_{\rm sc} \simeq \mu + 1/2 \ . \tag{3.11}$$

Moreover, the critical exponent $f_{\rm sc}$ of the elastic moduli of the Swiss-cheese model (see Chap. 2) is related to that of the lattice models f through

$$f_{\rm sc} \simeq f + 3/2 \ , \tag{3.12}$$

in 2D, whereas in 3D

$$f_{\rm sc} \simeq f + 5/2 \ , \tag{3.13}$$

so that the elastic moduli are much more strongly affected by the continuous structure of the system. Equations (3.11)–(3.13) imply that near p_c, the conductivity and elastic moduli curves of the Swiss-cheese model are much flatter than those of lattice systems. As for the inverted Swiss-cheese model, one has, $\mu_{\rm isc} = \mu$, in both 2D and 3D, while $f_{\rm isc} = f$ in 2D, whereas in 3D

$$f_{\rm isc} \simeq f + 1/2 \ . \tag{3.14}$$

It was also established by Straley (1982); Halperin et al. (1985), and Feng et al. (1987) that if the conductance distribution of a random continuum is of power-law form, $\psi(g) = (1-p)\delta(g) + p(1-\alpha)g^{-\alpha}$, where $0 \le \alpha \le 1$, then, the conductivity critical exponent $\mu(\alpha)$ of the continuum is *not* universal, but is given by

$$\mu_l(\alpha) = 1 + (d-2)\nu + \frac{\alpha}{1-\alpha} \le \mu(\alpha) \le \mu_u(\alpha) = \mu + \frac{1}{1-\alpha} \ . \tag{3.15}$$

Such power-law conductance distributions are actually very common among disordered continua. For example, for the Swiss-cheese models in d-dimensions one has, $\alpha = (2d-5)/(2d-3)$. In 2D, $\alpha = -1$, and the conductance distribution is not singular at $g = 0$ and, therefore, $\mu_{\rm sc} = \mu$. In 3D, $\alpha = 1/3$, and the conductance distribution is singular at $g = 0$, hence giving rise to Eq. (3.11). Moreover, Straley (1982) argued that

$$\mu(\alpha) = \min[\mu, \mu_l(\alpha)] \ . \tag{3.16}$$

Not only have these predictions been confirmed by computer simulations, but they have also received experimental verification (see, for example, Garfunkel and Weissman 1985; Koch et al. 1985; Chen and Chou 1985; Rudman et al. 1986). A more detailed discussion of continuum percolation is given by Balberg (2021).

3.2.5 Boiling: An Application of Continuum Percolation

Although continuum percolation has been used in a variety of contexts to explain experimental data for flow and transport properties of heterogeneous materials and media, let us describe briefly a distinct application of it, which is boiling, a phenomenon that we observe in everyday life.

The dynamics of boiling is controlled by the heat flux that is transferred from a heated surface. As the heat flux increases, it produces a rise in the surface temperature, which in turn increases the site density of bubble nucleation, and departure frequency. Boiling is susceptible to an instability phenomenon known as the *boiling crisis*, which is triggered when the heat flux reaches a critical heat flux (CHF). When that happens, one observes a sudden transition from a nucleate boiling regime in which one has discrete bubbles on the surface, to a film boiling regime in which a stable vapor layer blankets the entire heating surface. Clearly, when the discrete bubbles join and form the vapor layer is a connectivity phenomenon, i.e., a percolation process. The percolation transition and formation of the vapor layer cause degrade strongly the heat removal process, which could lead to a catastrophic increase in the heater temperature.

Zhang et al. (2019) carried out a precise experiment in order to study boiling. Their experimental data for the bubble surface distributions during the pool and flow boiling of water indicated that, at the CHF, the distributions of the footprint surface A follow a power law, $1/A^{\zeta}$, with $\zeta < 3$, demonstrating the scale-free nature of the boiling crisis. To explain the data, Zhang et al. (2019) developed a continuum percolation model. Their experiments provided data for the nucleation site density N_n, the bubble growth time t_g, and the bubble departure frequency f for each heat flux. Thus, given a surface of area A equal to the experimental heater active area, they generated at random AN_n nucleation sites. The probability of a bubble growing out of a certain nucleation site is ft_g. Thus, if a random number $R \in [0, 1]$ is smaller than ft_g, a bubble is generated; otherwise, the next nucleation site is examined and the procedure is repeated. The radius of each bubble was also determined according to the experimental distribution of the individual bubbles. In agreement with their experiment, the transition from the aforementioned first regime to the second one was signaled by the formation of an SSC of the bubble.

3.3 Correlated Percolation

In the percolation models described in Chap. 2, no correlations between various segments of the system were assumed. Disorder in many important heterogeneous media is not, however, completely random. There usually are correlations whose extent may be finite but large. For example, in the packing of solid particles, there are short-range correlations. The universal power laws of percolation systems with the finite extent of correlations are the same as those of random percolation, if the length scale of interest is larger than the correlation length. Moreover, if the correlation function $C(r)$ in a $d-$dimensional system decays as r^{-d} or faster, then the power laws that described the properties of the system near the percolation threshold are identical with those of random percolation. This is not totally unexpected. Recall from Chap. 2 that for length scales $L < \xi_p$, where ξ_p is the percolation correlation length, the SSC is a fractal object. Thus, even in random percolation, as p, the fraction of occupied bonds or sites, approaches p_c, correlations begin to build up and, hence, the introduction of any type of correlations with an extent shorter than ξ_p cannot change its scaling properties. In many other cases, e.g., in some disordered elastic materials, there are long-range correlations. Therefore, the question of correlations in percolation models is important. An excellent review of the subject was given by Coniglio and Fierro (2021). Here, we only briefly describe some properties of correlated percolation and provide a few applications.

3.3.1 *Percolation with Short-Range Correlations*

The earliest correlated percolation model that we are aware of is a correlated bond percolation model developed by Kirkpatrick (1973). In his model, which was intended for describing hopping conduction in disordered solids (see Chap. 15), a random number s_i, uniformly distributed in $(-1, +1)$, was attributed to each site of a lattice. Each bond ij was assigned a number b_{ij}, given by,

$$b_{ij} = \frac{1}{2}(|s_i| + |s_j| + |s_i - s_j|) , \tag{3.17}$$

and all bonds with $b_{ij} > \epsilon$ were removed, where ϵ is some selected value. Thus, a bond remains intact only if b_{ij} is sufficiently small, implying that $|s_i|$ and $|s_j|$ must both be small, implying that if a bond is present or conducting, its neighbors are also likely to be conducting and, hence, the conducting bonds are clustered together. Arbabi and Sahimi (1988a) used a similar model to study the effect of correlations on the elastic properties of percolation networks. Many other percolation models with short-range correlations have been developed, most of which were described elsewhere (Sahimi, 1998).

Positive correlations—those that cluster the conducting bonds together – as opposed to negative correlations that make it more likely that a conducting bond is next to an insulating one, regardless of being short- or long ranged, *usually* reduce the percolation threshold of the system. For example, in Kirkpatrick's model, $p_{cb} \simeq 0.1$ for the simple-cubic lattice, sharply lower than $p_{cb} \simeq 0.249$ for random bond percolation (see Table 2.2). This is due to the clustering of the conducting bonds caused by the positive correlations, as a result of which the formation of an SSC of transport paths is possible, even when the fraction of the conducting phase is low. On the other hand, Bug et al. (1985) presented evidence based on computer simulations that indicate that positive correlations may sometimes increase the percolation threshold, while at other times decrease it, depending on the distance between two occupied sites that are considered connected, the the dimensionality of the system, and its temperature.

3.3.2 *Percolation with Long-Range Correlations*

Interesting percolation models with long-range correlations in which the correlations decay with the distance between 2 points on the SSC were developed by Weinrib (1984); see also Weinrib and Halperin (1983) and Prakash et al. (1992). In the latter case, the correlation function $C(r)$ in a $d-$dimensional system was assumed to be given by

$$C(r) \sim r^{-(d-\lambda)} , \tag{3.18}$$

where $-2 \leq \lambda \leq 2$ is a parameter of the model, such that $0 \leq \lambda \leq 2$ represents positive correlations, while $-2 \leq \lambda \leq 0$ corresponds to negative correlations.

Percolation models in which the correlations increase with the distance between two sites or bonds on the SSC were first studied by Sahimi (1994), Sahimi (1995) and Sahimi and Mukhopadhyay (1996). Their model was motivated by its applications to flow in geological formations. Although their model in a finite-size system exhibited well-defined scaling properties near the percolation threshold (which were non-universal), it turned out that in the thermodynamic limit, i.e., in the limit of an infinite system, there is no percolation transition, and the sample-spanning cluster remains compact. Knackstedt et al. (2000) studied the invasion percolation model (see below, and Chaps. 9 and 16). It turned out that, depending on whether the long-range correlations were positive (negative), the percolation transition was first-order (second-order).

Let us now describe two applications of correlated percolation. More applications are described by Coniglio and Fierro (2021).

3.3.3 Correlated-Site Percolation and Unusual Properties of Low-Temperature Water

Liquid water and its isotope D_2O have some very unusual properties that have been known for a long time. For example, as the temperature is lowered below the melting point, several static response functions, such as the isothermal compressibility, constant-pressure specific heat, and thermal expansivity increase at an increasingly rapid rate, whereas the mass density decreases rapidly. In addition, under conditions of supercooling, liquid H_2 and D_2 exhibit additional anomalous properties. Supercooling is a phenomenon in which the temperature of a liquid is lowered below the freezing point without forming a solid. In the case of water, for example, ice formation requires a seed crystal or a nucleation site (see, for example, Sastry et al. 1996; Debenedetti and Stanley 2003).

To explain some of the unusual properties of liquid water, Stanley (1979) proposed a correlated-site percolation model. Recall from Chap. 2 that in a lattice of N sites, each of which has Z nearest neighbors (Z is the connectivity of coordination number), one has $NZ/2$ pairs of nearest-neighbor sites. In random bond percolation one randomly places bonds between a fraction p_B of the pairs. Then, each site may be viewed as a member of one of $(Z+1)$ different species, depending on whether $0,\ 1,\ 2,\cdots,Z$ bonds emanate from that site. Due to the randomness of bond distribution, the "mole fraction" f_j of species j is simply

$$f_j = \binom{Z}{j} p_B^j (1-p_B)^{Z-j} \ . \tag{3.19}$$

Note that although the total number of sites belonging to each species is determined solely by p_B, the connectivity properties are very different from those of pure percolation since, for example, it is impossible for a species-Z site to appear as a nearest neighbor of a species-0 site.

Next, Stanley (1979) considered a bichromatic percolation model in which the sites were colored or partitioned into only two classes. If, for example, the members of class 1 (black sites) are the Nf_z species-Z sites, so that class 2 (white sites) are the remaining $N(1-f_z)$ sites, then, computer simulations with the square lattice produce what is shown in Fig. 3.5. One can also consider the full connectivity problem that would involve all $(Z+1)$ species, which is an example of *polychromatic percolation*, first suggested by Zallen (1977) and Halley and Holcomb (1978). To make connection to the properties of water, Stanley (1979) argued as follows.

Assume that the sites represent oxygen atoms of a 3D network with coordination number $Z=4$, while the bonds correspond to hydrogen bonds between neighboring pairs of O atoms. Geiger et al. (1979) demonstrated, from that for any reasonable definition of a hydrogen bond, H_2O at 10 °C is well above the bond percolation threshold and, therefore, the liquid consists of a single connected (hydrogen-bonded) network that is infinite in extent (the "gel" network), as well as numerous finite networks, similar to the correlated-site model suggested by Stanley (1979). The

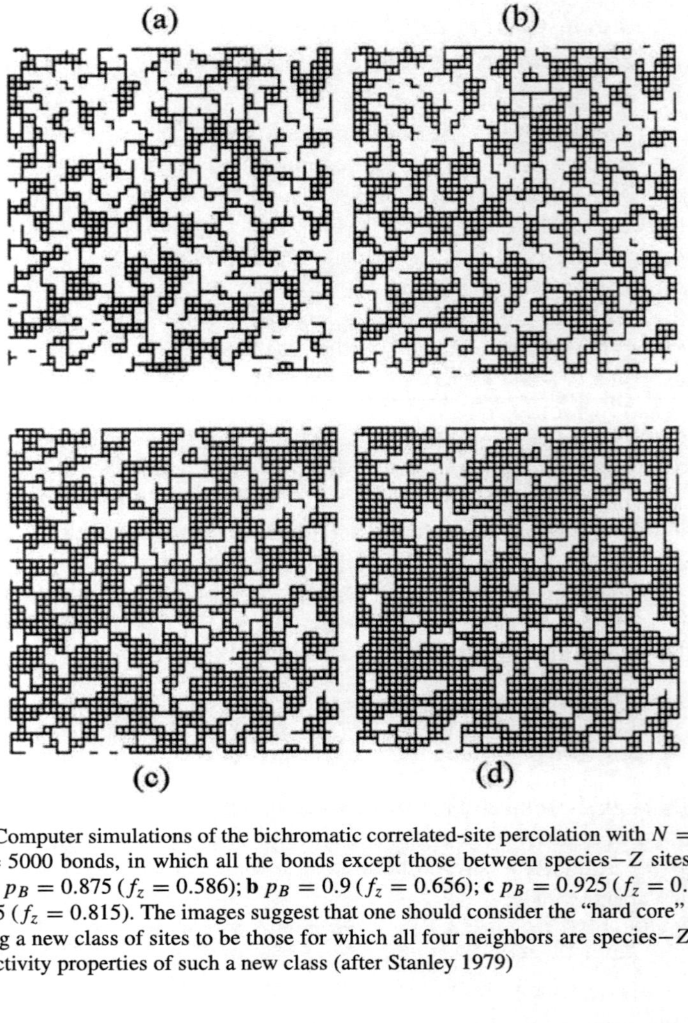

Fig. 3.5 Computer simulations of the bichromatic correlated-site percolation with $N = 2500$ sites and $2N = 5000$ bonds, in which all the bonds except those between species$-Z$ sites have been omitted. **a** $p_B = 0.875$ ($f_z = 0.586$); **b** $p_B = 0.9$ ($f_z = 0.656$); **c** $p_B = 0.925$ ($f_z = 0.732$), and **d** $p_B = 0.95$ ($f_z = 0.815$). The images suggest that one should consider the "hard core" of a cluster by defining a new class of sites to be those for which all four neighbors are species$-Z$, and study the connectivity properties of such a new class (after Stanley 1979)

volume V_j per oxygen atom depends on the number of bonds j emanating from the atom, with $V_0 < V_1 < V_2 < V_3 < V_4$. Then, if we partition the system into cells of characteristic dimension L, with which a "local" density ρ_L is associated. The quantity to study is then the fluctuations of the local density from cell to cell. The positions of each species are correlated, and the density is related to the site species and, thus, the density fluctuations are correlated. The "isothermal compressibility" of the model is enhanced.

Next, partition the oxygen atoms into five species, those that are bonded to 0, 1, 2,$\cdots$, and 4 other oxygens, and consider the "patches" (or clusters) of the gel defined by connected regions of the four-bonded species. As temperature T decreases, the probability $p_B(T)$ increases (there are more bonds), as does the fraction $f_4(T)$ of species−4 molecules. The mean size $\langle s \rangle$ of a patch is only of the order $10 - 10^2$ water molecules per patch in the supercooled region. Thus, one might expect the local density of the patches to be less than that of the surrounding gel network, and the patches give rise to spatial density fluctuations whose magnitude also increases as T decreases. Therefore, as T decreases, the size of the correlated patches increases, the overall mean density $\langle \rho \rangle$ decreases, and the density fluctuations associated with the patches, namely, $\langle (\rho - \langle \rho \rangle)^2 \rangle / \langle \rho \rangle$, which is proportional to isothermal compressibility, increases, which is similar to actual water.

Stanley (1979) argued that other anomalous properties of water can be explained qualitatively by his correlated-site percolation problem. Stanley and Texeira (1980) presented a detailed comparison between the predictions of the model and the experimental data on H_20 and D_2O at low temperatures and found general agreement with all existing data. See Bianco and Franzese (2019) for good discussions and references to the works that have been done since the introduction of the original model.

3.3.4 *Site-Bond Correlated Percolation and Gelation*

Coniglio et al. (1979, 1982) proposed a site-bond correlated percolation model for polymer gelation. We follow Coniglio and Fierro (2021) in order to describe the model. Consider a system of monomers in a solvent, and assume that the monomers interact with each other in two ways. One is the usual Van der Waals interaction, while the second way is a directional interaction that leads to the formation of a chemical bond. For simplicity, we may set the monomer–solvent interaction and the solvent–solvent interaction to be zero and, instead, include them in an effective monomer–monomer interaction ε_{ij}, which can be reasonably approximated by a nearest-neighbor interaction,

$$\varepsilon_{ij} = \begin{cases} -W, & \text{van der Waals interaction} \\ -E, & \text{bonding energy} \end{cases} \tag{3.20}$$

where we expect $E \gg W$. The second interaction occurs only when the monomers are in particular configurations, which, for simplicity, we may assume that there is one configuration that corresponds to the interaction of strength E and $\Omega \gg 1$ configurations that correspond to the interaction of strength W. Coniglio et al. (1979, 1982) showed that such a system is equivalent to a lattice gas model with an effective nearest-neighbor interaction $-\varepsilon$ given by,

$$\exp(\beta\varepsilon) = \exp(\beta E) + \Omega \exp(\beta W) , \tag{3.21}$$

where $\beta = 1/(k_B T)$, with k_B being the Boltzmann's constant. Microscopically, however, the system behaves very differently from a lattice gas, since two nearest neighbors monomer in a standard lattice gas feel one interaction, whereas in the present gelation model they feel with probability p_B a strong chemical interaction $-E$, and a much weaker interaction $-W$ with probability $(1 - p_B)$, where p_B is given by

$$p_B = \frac{\exp(\beta E)}{\exp(\beta E) + \Omega \exp(\beta W)} . \tag{3.22}$$

Given the model, one may study the percolation line of the clusters made by monomers that are connected by chemical bonds by introducing bonds between nearest-neighbor particles in the lattice gas with nearest-neighbor interactions in which the bonds are present with probability p_B given by Eq. (3.22). Changing the solvent results in changing the effective interaction W. The lifetimes of the bonds are on the order of $e^{\beta E}$, and because E is very large, the lifetime could also be very large. If the bond lifetime is infinite, the bonded clusters are permanent and the viscosity diverges due to the divergence of the mean cluster size (see Chap. 13), and the percolation line is easily detected. In fact, the line has been detected experimentally in at least three physical systems, namely, (a) microemulsions of water in oil (Chen et al. 1994), triblock copolymers in unicellular systems (Mallamace et al. 2000), and gelatin water methanol systems (Tanaka et al. 1979). More details are given by Coniglio and Fierro (2021).

3.4 Bootstrap Percolation

Pollak and Riess (1975) and Chalupa et al. (1979) proposed a percolation model known as the *bootstrap percolation.* In this problem sites of a lattice are initially randomly occupied. Then, those sites that do not have at least Z_c nearest-neighbor occupied sites are removed (note that $Z_c = 0$ is the usual random percolation). The *interactions* between the sites are short ranged, but the correlations between them increase as the distance between two occupied sites also increases. Figure 3.6 shows a 2D example of a bootstrap percolation cluster. An excellent review of the subject is given by De Gregoria et al. (2021).

Extensive numerical studies of bootstrap percolation were begun by Kogut and Leath (1981). Since then, bootstrap percolation has been studied extensively (for a review see Adler 1991; see also Stauffer and de Arcangelis 1996), and has proven to contain a rich variety of behavior that is a strong function of the parameter Z_c. For example, an important question is the nature of the percolation transition. It now appears that for sufficiently high values of $Z_c \leq Z$ (where Z is the coordination number of the lattice), the percolation transition is discontinuous and first-order, whereas for low values of Z_c the transition is continuous and second-order. A first-order phase transition is characterized by sharp discontinuities at p_c, namely, percolation quantities vary smoothly with p anywhere from $p = 1$ or $p = 0$ to just above p_c, but at p_c abruptly (discontinuously) change their values to zero. If the bootstrap phase transition is first-order, then the percolation threshold of the system is $p_c = 1$, the SSC is compact (i.e., its fractal dimension $D_f = d$), and power laws of percolation described in Chap. 2 are no longer valid. For example, bootstrap percolation tran-

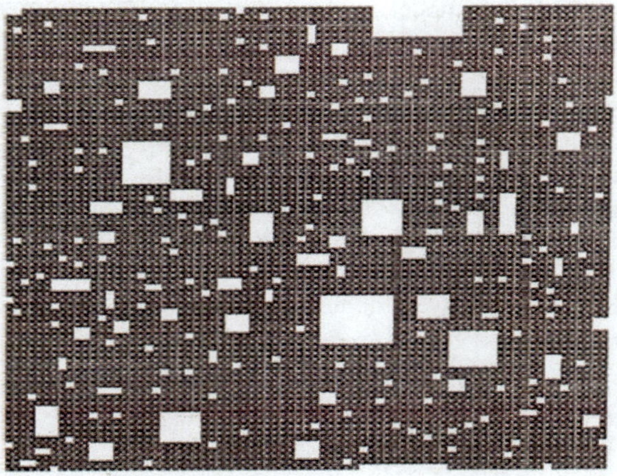

Fig. 3.6 Two-dimensional bootstrap percolation on the square network. White areas indicate the absent or cut bonds

sition is first order in the simple-cubic lattice with $Z_c \geq 4$, in the triangular lattice with $Z_c \geq 4$, and in the square lattice with $Z_c \geq 3$.

We must, however, point out that, although when bootstrap percolation transition is first-order, the percolation threshold is $p_c = 1$, extensive simulations have shown that sample-size effects are very large and strong, so much so that the approach to the asymptotic value, $p_c = 1$, is very slow, and that in small or even moderately large lattices one may obtain an *effective percolation threshold* p_c less than 1. Sample-size effect also smear the first-order phase transition. The reason for the peculiar behavior of this model seems to be the long-range nature of the correlations between two occupied sites, and the fact that for large enough Z_c these correlations do not decay with the distance between the two sites, rather they *increase* with the distance.

The richness of bootstrap percolation is not, however, limited to the nature of the percolation transition at p_c. In the first simulation of this model, Kogut and Leath (1981) found that in the triangular and simple-cubic lattices with $Z_c = 3$ the value of the critical exponent β, defined by Eq. (2.8), seems to be different from that of random percolation. Adler and Stauffer (1990) carried out an extensive study of this model in the simple-cubic lattice and found that, $\beta(Z_c = 0) = \beta(Z_c = 1) = \beta(Z_c = 2) \simeq 0.41$, the same as that of random percolation (see Table 2.3), but that, $\beta(Z_c = 3) \simeq 0.6$, whereas the corresponding value of ν, the exponent for the percolation correlation length (see Chap. 2), appeared to be the same as in random percolation for $0 \leq Z_c \leq 3$.

Conduction in bootstrap percolation systems was studied by Sahimi and Ray (1991) in both 2D and 3D, who showed that in the simple-cubic lattice the critical exponent μ of the conductivity for $Z_c = 0, 1, 2$, and 3 is the same as that of random percolation listed in Table 2.3. Sahimi and Ray also speculated that bootstrap percolation with large enough Z_c (i.e., when the percolation transition is first order) may be relevant to modeling of mechanical and fracture properties of solids since, similar to bootstrap percolation with large enough Z_c, one may also have first-order phase transitions in such phenomena.

The original motivation for developing this model was to explain the behavior of some disordered materials in which magnetic impurities are randomly distributed in a host of non-magnetic metals. It is believed that in some of such materials an impurity atom cannot sustain a localized magnetic moment unless it is surrounded by a minimum number of magnetic neighbors (see, for example, Jaccarino and Walker 1965). Bootstrap percolation has also been used for studying orientational order in orthoparahydrogen mixtures (Adler et al. 1987) and dynamics of glass transitions (Ertel et al. 1988).

3.4.1 *k-Core Percolation*

A percolation model closely related to bootstrap percolation is the so-called $k-$core percolation. The $k-$core is the largest connected cluster of a network with occupation probability p in which all nodes have at least k occupied neighbors, each of which is also in the subgraph. The structure is formed by removing all sites with less than k neighbors, which is continued until all the remaining nodes have at least k neighbors. The exact solution of the problem on the Bethe lattice was derived by Chalupa et al. (1979). The problem received renewed attention when $k-$core structures emerged in the structure of complex networks (see Chap. 17), and was studied by Pittel et al. (1996); Fernholz and Ramachandran (2003), and Balogh and Pittel (2007). Dorogovtsev et al. (2006) derived the exact solution for uncorrelated, damaged, or undamaged complex networks.

3.5 Invasion Percolation

Invasion percolation (IP) was first proposed by Lenormand and Bories (1980), Chandler et al. (1982), and Wilkinson and Willemsen (1983). In the IP model the network is initially filled with a fluid called the *defender*—the fluid to be displaced. To each site of the network is assigned a random number uniformly distributed in [0, 1]. Then, the displacing fluid—the *invader*—is injected into the medium to displace the defender. It does so by choosing at each time step the site next to the interface that has the smallest random number. If the random numbers are interpreted as the resistance that the sites offer to the invading fluid, then choosing the site with the smallest random number is equivalent to selecting a pore with the largest size and, hence, the IP model simulates the drainage process, i.e., the displacement of a wetting fluid by a non-wetting one. A slightly more tedious procedure can be used for working with bonds instead of sites.

A similar IP model can be devised for imbibition, during which a wetting fluid is drawn spontaneously into a porous medium and into the smallest constrictions for which the capillary pressure is large and negative, whereas it enters last into the widest pores. Displacement events are, therefore, ranked in terms of the largest opening that the invading fluid must travel through, since it is from the larger capillaries or bonds that it is most difficult to displace the defender. Imbibition is, therefore, a *site* IP, whereas drainage in which the invader has the most difficulty with the smallest constrictions, is a *bond* IP.

In addition to simulating slow two-phase flow in porous media, the IP has found many other applications. They will be described in detail in Chaps. 9, 16, and 19.

3.6 Explosive Percolation

An important new percolation model that leads not only to an unusual phase transition, but also to many novel phenomena not seen in the classical percolation and its variants described in this chapter, is *explosive percolation* (EP), first introduced by Achlioptas et al. (2009). It is now known that EP can be the result of many types of evolutionary processes that delay the formation of large clusters and break the multiplicative coalescence rule of standard models of percolation described in Chap. 2. Because of this, many unique phenomena can occur during the emergence of large-scale connectivity. In Chap. 20 we will describe the EP and many of its potential and realized applications.

3.7 Directed Percolation

Directed percolation (DP) is a generalization of isotropic percolation in which the bonds of a lattice are given a preferred direction, so that flow of a fluid, electrical current, traffic, or information can only occur between lattice sites that are connected in the correct (along the assigned direction) orientation. Therefore, whereas in isotropic percolation spreading happens in all directions, in the DP it can only occur along the assigned direction. Note that the cluster of connected sites in DP is a subset of the cluster in the isotropic case.

Interestingly, Janssen (1981) and Grassberger (1982) conjectured that any model that features a fluctuating phase and a unique absorbing state, a positive one-component order parameter, short-range interactions, and no additional symmetries or quenched disorder, should belong to the DP universality class, which has proven to be correct. In Chap. 21 we will describe DP in detail, and discuss several of its important applications.

3.8 Dynamic Percolation

In the classical percolation problem the configuration of the system does not change with time. That is, the probability that a bond is open or closed is independent of time. In many problems, such a description of a disordered medium is inadequate, since the configuration of the system changes with time. For example, in polymeric ionic conductors above the glass transition temperature, the electrical conductivity is dominated by the microscopic motion of the medium. Another example is the problem of oxygen binding to hemoglobin and myoglobin. It is known that the entrance to the haeme pocket is blocked by a number of side chains and, thus, oxygen could not bind if the side chains were fixed at their equilibrium positions. Because of the dynamic nature of proteins, one may expect that a 'gate' would fluctuate between open and closed positions. Thus, one has a diffusion process in a disordered medium whose structure varies with time.

To model such phenomena, various dynamic percolation models in which the probability that a bond or site of a network is open varies with time were proposed by Druger et al. (1983, 1985); Harrison and Zwanzig (1985); Sahimi (1986a), and Bunde et al. (1991); see also Grassberger (1989); Granek and Nitzan (1990), and Loring (1991). Stauffer and Sahimi (2005) studied diffusion in complex networks (see Chap. 17) with annealed disordered. Quite different dynamic percolation models have been suggested for gelation phenomena, and microemulsion systems will be described later in this book.

3.9 Percolation Model of Galactic Structures

It has been argued that the process of star formation may be a percolation process, and that the percolation transition from a disconnected to a connected state plays an important role in the stabilization and control of star formation. Schulman and Seiden (1986) proposed a theory according to which the structure of spiral galaxies arises from a percolation transition that underlies the phenomenon of propagating star formation. According to their theory, the appearance of spiral arms is a consequence of the differential rotation of the galaxy, as well as the characteristic divergence of correlation lengths for continuous phase transitions. Several other properties of spiral galaxies, such as the distribution of the gaseous components and the luminosity, arise directly from a feedback mechanism that pins the star formation rate close to the critical point of the phase transition. See also Seiden and Schulman (1990) for a review of this interesting application of percolation.

3.10 Vector Percolation

Every percolation model described so far is a scalar model. But if each bond is an elastic element, instead of a pore, or a resistor, or some other scalar entity, then, one has *vector percolation* that gives rise to elastic percolation networks (EPNs). The EPNs are discrete models of disordered two-phase materials in which the contrast between the elastic properties of the two phases is large. As described in Chap. 11, in at least one type of vector percolation, namely, the rigidity percolation model, the deformation of the material represents a linear, vector, but non-local transport process that possesses complex properties that are absent in many other discrete, or continuum, models of scalar transport in disordered materials. It will be shown that in rigidity percolation the mere fact that there is an SSC is not sufficient for supporting the transmission of stress and strain across the material, because there are long-range correlations in the system, giving rise to stress transport and the aforementioned non-locality in the EPNs.

Chapter 4
Characterization of Porous Media and Materials

4.1 Introduction

One of the earliest and most fruitful applications of percolation theory has been to the problems involving porous materials and media. This is not really surprising. After all, in their original paper on the foundation of percolation theory, Broadbent and Hammersley (1957) had expressed the hope that their theory would someday be used for solving some practical problems involving porous media. But, despite their hope, explicit use of the concepts of percolation theory for describing various phenomena in porous media began only in the 1970s.

Torelli and Scheidegger (1972) were probably the first who recognized the relevance of percolation theory to modeling fluid flow and transport in porous media. They were studying hydrodynamic dispersion in porous media (to be studied later in this book), and pointed out that percolation theory, if appropriately modified and applied, may provide useful insights into its properties. They did not, however, use percolation to model the phenomenon and, in fact, they did not even report any results in their original paper.

Melrose and Brandner (1974) suggested that the entrapment of oil in an oil reservoir is similar to a percolation process. They proposed that an approach based on percolation may yield deeper insights into the problem, but they did not actually calculate any quantity linked with percolation. In their paper on conduction in composite materials Davis et al. (1975) remarked that, "Although, to our knowledge, no quantitative work has been done on the subject, we believe that 2-phase oil-water flow in oil fields is a percolation process in which the connectivity of each phase determines the relative permeability of that phase." But, they did not also report any result for the relative permeability.

Larson et al. (1977) suggested that percolation theory may be useful for describing entrapment of one fluid phase by another in porous media, and used the theory to actually compute some properties of two-phase flow in porous media by making an

M. Sahimi, *Applications of Percolation Theory*, Applied Mathematical Sciences 213,
https://doi.org/10.1007/978-3-031-20386-2_4

analogy between the percolation cluster-size distribution n_s (see Chap. 2) and the distribution of isolated blobs of oil in a porous medium.

Almost simultaneously, Chatzis and Dullien (1977) calculated several percolation properties of various two-dimensional (2D) and 3D networks, pointed out how they may be used for simulating two-phase flows in porous media, and compared their predictions with the measured capillary pressure curves (see below). Shortly after the papers by Larson et al. and Chatzis and Dullien, de Gennes[1] and Guyon (1978) also suggested that two-phase flow in porous media may belong to the class of percolation processes, used visualization of mercury porosimetry (see below) as an example, and suggested ideas for how to use percolation concepts to model the phenomenon that they were describing. It was only after these pioneering works that the utility of percolation theory to describing the morphology of porous media and the various phenomena that occur there gained recognition. In fact, there is hardly any phenomenon in porous media to which percolation theory is *not* relevant.

This chapter focuses on characterization of the morphology of porous media based on percolation theory. Later in this book, we will describe the applications of the theory to fluid flow and transport in porous media. There are many advanced experimental techniques for characterizing the morphology of porous media, which are extensively discussed by Dullien (1992) and Sahimi (2011). What we are interested in here are, however, those experimental methods that benefit from the application of percolation theory, not only for better and more accurate interpretation of the data but also for modeling of the phenomena for which the data have been measured. The porous media considered in this chapter are quite general. They may vary anywhere from porous catalysts, to membranes, adsorbents, and packed beds of particles and, at the largest length scales, oil, gas, and geothermal reservoirs, groundwater aquifers, soil, and many other types of porous media.

The morphology of a porous medium depends, however, on its heterogeneity and the length scale over which the medium is studied. Therefore, we restrict our attention in this chapter to porous media that are *macroscopically homogeneous*—those whose properties, for large enough samples, are *independent of their size*—although they may be microscopically heterogeneous. Later in this book, we will describe the application of percolation theory to *macroscopically heterogeneous porous media*—those that contain heterogeneities over multiple and disparate length scales.

One critical aspect of the application of percolation theory to problems in porous media is the ability to measure critical percolation properties for a given porous medium, either as a laboratory sample or as 3D digitized image. Liu and Regenauer-Lieb (2021) provide an excellent review of the available techniques for making such measurements.

Before we begin describing characterization of porous media based on percolation theory, it is instructive to explain the processes that give rise to porous rock. These

[1] **Pierre-Gilles de Gennes** (1932–2007) was a French physicist who made fundamental contributions to superconductivity, liquid crystals, polymers physics, and wetting. A professor at Orsay and College de France, he was director of École Supérieure de Physique et de Chimie Industrielles de la Ville de Parios (ESPCI). Among other awards, he received the Nobel Prize for Physics in 1991.

are called *diagenetic processes*, and percolation theory has proven useful even to their modeling.

4.2 Diagenetic Processes and Formation of Rock

Formation of rock begins with deposition of sediments, followed by compaction and alteration processes that cause drastic changes in the morphology of porous formations. Consider, for example, sandstones that are assemblages of discrete grains with a wide variety of chemical compounds and mixtures. If the environment around sandstones changes, the grains begin to react chemically and produce new compounds that deposit on the surface of sand grains. As a result, the mechanical properties of the grains, such as their resistance to fracturing, also change. The chemical and physical changes in the sand after the deposition constitute what is referred to as the diagenetic processes, whose main features are, (a) mechanical deformation of grains; (b) solution of grain minerals; (c) alteration of grains; and (d) precipitation of pore-filling minerals, cements, and other materials. These features strongly influence the volume and the content of porous formations, because they control their *porosity* ϕ, the volume fraction of its pores.

Immediately after deposition, diagenesis starts and continues during the burial and uplifting of rock, until outcrop weathering reduces it again to sediment. The changes produce an end product with specific diagenetic features, the nature of which depends on the initial mineralogical composition of the formation, as well as the composition of the surrounding basin-fill sediments. The diagenetic history of a porous medium depends on several time-dependent factors, as the formation varies with temperature, pressure, and chemistry of the pore fluid.

Porosity of porous formations has either a *primary* or *secondary* origin. The former is due to the original pore space of the sediment, whereas the latter is caused by unstable grains or cements undergoing chemical and physical changes through reaction with the formation of water that have partially or entirely passed into solution. Therefore, if the pore space is somehow restored, the original porosity that had been protected from precipitation by the deposition of minerals is converted into secondary porosity.

In Chap. 2, we distinguished p, the fraction of the open bonds or sites of a percolation network, regardless of whether they can be reached from two opposing external surface of the network, from the accessible fraction $X^A(p)$ of such bonds or sites, the fraction of the bonds or sites that are in the sample-spanning cluster (SSC) that connects two opposing surfaces. Thus, we always have $X^A(p) \le p$. Likewise, we must distinguish the *total* porosity ϕ of a porous medium from the *accessible* porosity ϕ_A, the volume fraction of that part of the pore space that can be reached from two opposite external surfaces. In this sense, p and $X^A(p)$ are the analogs of ϕ and $\phi_A(\phi)$. Therefore, we may also define a *critical porosity* ϕ_c—the analog of the percolation threshold p_c—such that, for $\phi \le \phi_c$, there would be no SSC of the voids that connect two opposing surfaces of a porous medium, whereas, for $\phi > \phi_c$, there is an SSC through which flow and transport occur.

Generally speaking, solution pores constitute more than half of the entire pore space of many sedimentary rocks. The significance of secondary porosity in sandstones was only relatively recently recognized. It is now well established that five kinds of pores with various shapes and sizes may contain secondary porosity. Four of them are grouped together and referred to as "pores." The fifth type, open fractures, is different from the other four types of pores and is considered separately later on.

The diagenetic processes lead to a porosity smaller than the initial value, and pores that take on essentially any shape or size. Perhaps the most important consequence of the diagenetic processes is that the pores remain interconnected even when the porosity is very low. For example, it has been found by scanning electron microscopy that some pores can be connected to up to 20 others, implying that the critical porosity ϕ_c is very low. Thus, random percolation cannot quantitatively model the formation of pore space of natural porous media and its porosity.

4.3 Geometrical-Percolation Models of Diagenetic Processes

Since diagenetic processes for all types of natural porous media seem to be similar, and because there appear to be many similarities between the geometries of various types of rock, one may hope that many fundamental aspects of pore formation are *universal*, independent of many microscopic properties of rock. If so, one may be able to develop a general model of pore formation and growth that can explain many features of various rocks. A study of the literature indicates that there are essentially two types of modeling approaches to this problem. The first approach, which we call *chemical modeling*, relies on the continuum equations of transport (diffusion and convection) and reaction (see Sahimi et al. (1990), for a review), but ignores the effect of the morphology, and in particular, the connectivity of the pores. The second approach is what we refer to as *geometrical modeling* in which the details of the reaction kinetics and transport are ignored. Instead, starting from a model of unconsolidated pore space, the approach takes into account the effect of the connectivity and percolation of pores and grains in order to model the diagenetic processes. Two of such models are those of Wong et al. (1984), usually called the *shrinking tube model*, and of Roberts and Schwartz (1985), known as the *grain-consolidation model*.

In the model developed by Wong et al. (1984) one begins with a network of interconnected bonds in which each bond is a resistor whose resistance R_i is selected from a probability distribution, and represents cylindrical fluid-filled tubes (pores) with random radius r_i. To mimic the consolidation process and reduction of the porosity during the diagenesis, a tube is selected at random and its radius is reduced by a fixed factor x, $r_i \to xr_i$, where $0 < x < 1$. The model is, of course, too simple to mimic the actual deposition of irregularly shaped particles in an irregularly shaped pore, but has two attractive features: (a) it preserves for *any* $x > 0$ the network's macroscopic connectivity, even when its porosity has almost vanished, implying that the critical porosity is zero, and (b) the change in the pore radius r_i at any step of the simulation depends on r_i in that step. Both features are also true for the actual diagenetic processes. Wong et al. (1984) used the model to provide qualitative explanations for several empirical laws that relate the electrical conductivity and permeability of a porous medium to its porosity, including Archie's law (Archie 1942) that has been found to be accurate for sedimentary rocks, and is given by

$$g_e = g_f \phi^m , \tag{4.1}$$

where g_f is the electrical conductivity of a fluid (such as brine) that has saturated the pore space, and g_e is the effective conductivity of the medium. The Archie exponent m has been found to vary anywhere from 1.3 to 4 (Dashtian et al. 2015). Note that Archie's law implies that the critical porosity or percolation threshold of a porous medium for which the law is valid is zero.

In the grain-consolidation model, one begins with a dense packing of spherical grains of random radii R; see Fig. 4.1a. The radii of the particles are then increase in unison, reducing the porosity. If the spheres overlap, the grains are truncated. The expansion and truncation may continue until they yield a porous medium with any desired porosity; see Fig. 4.1b. For the porous medium shown in Fig. 4.1c, the critical porosity is, $\phi_c = 0.03 \pm 0.004$.

The initial (primary) and final porosities of the porous medium depend on how the particles were originally distributed. For example, in the packing shown in Fig. 4.1, the particles are initially distributed randomly, but without overlap. The initial porosity of the packing is 0.364. On the other hand, if one begins with a regular packing in which spherical grains of unit radius are placed at the nodes of a simple-cubic lattice and follow the same algorithm, one finds that, $\phi_c \simeq 0.349$, which is close to critical porosity of a random packing of spheres. But, if a body-centered cubic lattice is used as the starting packing, one obtains, $\phi_c \simeq 0.0055$. Therefore, the model is very flexible, and because the porosity of sandstones and similar porous formations is usually less than 0.4, it provides a reasonable model of the diagenetic process. Moreover, the resulting porous media resemble closely natural sandstones. Schwartz et al. (1989) studied an extension of the model in which the initial grains were ellipsoidal, in order to model anisotropic porous media. Note that the diagenetic processes tend to favor distribution of grains that are roughly equal in size, in which case the algorithm of Roberts and Schwartz (1985) is very efficient and accurate. We will come back to this model later in this book.

4.4 Percolation Model of the Tortuosity

Let us consider the tortuosity of a porous medium, which is due to the irregular morphology of its pore space. It is clear that fluid flow and transport in porous media are directly influenced by the tortuous nature of the fluids' paths. As described

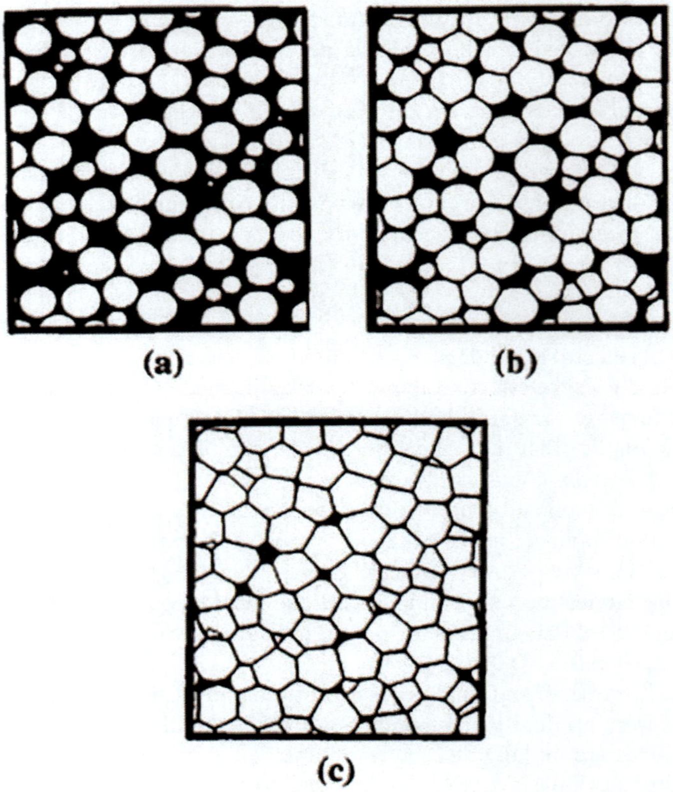

Fig. 4.1 Cross sections of three states of the grain consolidation. The porosity ϕ is, **a** 0.364; **b** 0.2, and **c** 0.030, which is also the critical porosity or percolation threshold (after Roberts and Schwartz 1985)

by Ghanbarian et al. (2013a) in their comprehensive review, one may define the tortuosity in four distinct ways that are based on (a) morphological; (b) hydraulic; (c) electrical; and (d) diffusion properties of a pore space. Among these, we focus on one definition, namely, one that is based solely on the geometry of the pore space, which is also the most popular definition of tortuosity.

Consider, first, a slab of area A and thickness L_s with N nonintersecting winding pores going through it. Suppose that each pore, filled with a fluid of conductivity g_f, has an effective length L_p and uniform cross-sectional area $A_p = \pi r_p^2$. Then, the pore's conductance is $g_f A_p/L_p$, and the formation factor F of the medium is defined by $F = g_f/g_e$. Thus, for the slab that we are considering, $F = (L_p/NA_p)/(L_s/A)$. The porosity is given by $\phi = NL_pA_p/L_s$, and the tortuosity is defined by

$$\tau = F\phi = \left(\frac{L_p}{L}\right)^2 . \tag{4.2}$$

Since we assumed that the pores are nonintersecting, Eq. (4.2) may be interpreted as the square of the ratio of the actual distance that a fluid's molecules travel in the pore space between two opposite faces of the medium, and the straight-line distance (the medium's thickness) between the two faces.

Tortuosity has also been defined in terms of flow and transport properties of porous media (see Ghanbarian et al. 2013a, for a comprehensive review of the subject). For example, if D_e and D_0 are, respectively, the effective diffusivities of diffusing probes in a porous medium and in the bulk (outside the porous medium), then τ may also be defined by

$$\tau = \phi_A \frac{D_0}{D_e} . \tag{4.3}$$

Often ϕ, the total porosity, is used in Eq. (4.3), instead of the accessible porosity ϕ_A. By definition, τ is always greater than one, and depends, of course, on the porosity. For $\phi \approx \phi_c$, τ is very large, and is expected to *diverge* at ϕ_c. One may also define τ in terms of the percolation properties. For example, Larson and Davis (1982) proposed, $\tau = P(p)/g_e(p)$, where $P(p)$ and $g_e(p)$ are, respectively, the percolation probability and effective conductivity of a percolation system (see Chap. 2). Since near p_c, one has (see Chap. 2), $P(p) \sim (p - p_c)^\beta$ and $g_2 \sim (p - p_c)^\mu$, then $\tau \propto (p - p_c)^{\beta - \mu}$. As $\mu > \beta$, then, according to this definition, τ diverges at $p = p_c$.

It is, therefore, clear that tortuosity may be defined by a variety of ways, based on various properties of a porous medium. Thus, considering the *geometrical* tortuosity τ_g, the ratio of the average length of the geometrical flow paths through the medium, $\langle L_g \rangle$, and the length L_s of a straight line across the medium,

$$\tau_g = \frac{\langle L_g \rangle}{L_s} . \tag{4.4}$$

The geometrical tortuosity *coefficient* is then defined by $T_g = 1/\tau_g$; clearly, $\tau_g > 1$. In a medium with a percolation threshold p_c in which a fraction p of the bonds (throats in a pore space) are conducting (are open to fluid flow and transport), the geometrical tortuosity is given by Ghanbarian et al. (2013b)

$$\tau_g \propto \left| p - p_c + \left(\frac{C}{L_s}\right)^{1/\nu} \right|^{\nu(1-D_x)} . \tag{4.5}$$

Here, ν is the exponent of the power law that describes the percolation correlation length ξ_p near p_c (see Chap. 2), and D_x is either $D_{\rm bb}$, the fractal dimension of the percolation backbone (Chap. 2), the multiply-connected part of the SSC, or $D_{\rm op}$, the fractal dimension of the *optimal path*, the most "energetically-favorable" path in the pore space. According to Lee et al. (1999), whereas the $D_{\rm op}$ describes the scaling of the *most probable traveling length*, D_{bb} describes the scaling of the *most probable traveling time*. Therefore, depending on whether we use D_{op} or D_{bb}, the models presented here describe tortuosity based on either the pore space or the dynamics of fluid flow. In Eq. (4.5), C is a constant on the order of unity.

Note that the sample-spanning percolation cluster is a fractal object over any length scale $L \ll \xi_p$. Thus, if $L_s \ll \xi_p$, it implies that it is impossible to eliminate finite-size effects in such a porous medium. In other words, if finite-size scaling is not incorporated into the interpretation of results of computer simulations when the porosity approaches its critical threshold, the effect of percolation, which is clear in physical systems of much greater size, will not be understood.

4.4.1 *Tortuosity and Continuum Percolation*

If percolation theory is to be invoked for describing phenomena in soil and other types of porous media that are partially saturated by water (or another wetting fluid), the relevant question is whether wetting of the pore space by water is modeled more accurately by bond or site percolation (Chap. 2). A similar question arises in the reverse process of drying, i.e., when water evaporates and leaves the pore space. Sahimi (1993b) proposed that wetting is better modeled by the site percolation model, whereas drying is akin to bond percolation. On the other hand, since soils and other types of porous media represent continuous, rather than discrete media, it may perhaps be better, at least for some specific applications, if the phenomenon of interest is modeled by the continuum percolation model described in Chap. 3 (Hunt 2001). In that case, the moisture content S of the pore space would play the role of the percolation probability p. Thus, to use Eq. (4.5) in the context of the continuum percolation model, we replace, respectively, p and p_c by S and S_c, where S_c represents the critical water content (saturation) at the percolation threshold of the water phase. We, therefore, obtain (Ghanbarian et al. 2013b),

$$\tau_g(S) \propto \left[S - S_c + \left(\frac{C}{L_s}\right)^{1/\nu}\right]^{\nu(1-D_x)} . \tag{4.6}$$

Aside from the constant C, the two parameters of Eq. (4.6) that must be specified are the critical moisture content S_c and the fractal dimension D_x. The former is the minimum moisture content required for the existence of an SSC of interconnected, water-filled pores. Note that, similar to the percolation threshold p_c, the critical water content of porous media is not a universal quantity, and depends on the pore space structure, and in particular its connectivity. Note also that S_c is the same as what is usually referred to as the irreducible or residual water saturation, which is used in multiphase flows in oil and gas reservoirs. Liu and Regenauer-Lieb (2011) proposed a morphological technique to estimate the percolation threshold from the microtomography and analysis of 3D images of porous media, which can also presumably be used for estimating S_c. But, in the absence of such a technique, S_c may be approximated by the water content retained at the permanent wilting point (1500 kPa), which is the residual water content S_r. Moldrup et al. (2001) presented a correlation that relates S_c to the vanishing of diffusion coefficient.

4.4.2 Percolation Models for Tortuosity of Saturated and Unsaturated Porous Media

Equation (4.6) may be rewritten in terms of the porosity ϕ of the pore space by setting $S = \phi$, which yields $\tau_g^{(s)}$ for a saturated porous medium (Ghanbarian et al. 2013b),

$$\tau_g^{(s)} \propto \left[\phi - S_c + \left(\frac{C}{L_s}\right)^{1/\nu}\right]^{\nu(1-D_x)}. \tag{4.7}$$

The implied prefactor can be calculated by noting that tortuosity should approach 1 as $\phi \to 1$. Doing so leads, however, to two approximations in the model:

(i) In order to derive Eqs. (4.5)–(4.7), we utilized the power laws of percolation theory, which are, in principle, applicable and accurate only near the percolation threshold. Therefore, for $p \gg p_c$ ($\phi \gg S_c$), the predictions of the percolation model, Eq. (4.7), are only approximations, and normalizing $\tau_g^{(s)}$ by using a value at $p = 1$ ($\phi = 1$), which is far from p_c (S_c), should introduce some error. Nonetheless, comparison with experimental data (Ghanbarian et al. 2013b) indicated that the approach is still accurate.

(ii) Normalizing Eq. (4.7) also produces a prefactor that depends on C/L_s, implying that it is scale-dependent. Matyka et al. (2008) and Duda et al. (2011) reported extensive numerical results, and in order to be consistent with their data, we eliminate C/L_s from the expression for the prefactor. Analysis of Ghanbarian et al. (2013b) indicated that doing so produces only up to 5% discrepancy in the final expression for $\tau_g^{(s)}$. We also note that if $L_c \gg C$, then the approximation has no discernable effect on the predictions.

Thus, normalizing Eq. (4.7) by making the two aforementioned approximations yields

$$\tau_g^{(s)} = \left[\frac{\phi - S_c + (C/L_s)^{1/\nu}}{1 - \theta_c}\right]^{\nu(1-D_x)}. \tag{4.8}$$

The relative tortuosity τ_r is then given by

$$\tau_r \equiv \frac{\tau_g(S)}{\tau_g^{(s)}} = \left[\frac{S - S_c + (C/L_s)^{1/\nu}}{\phi - S_c + (C/L_s)^{1/\nu}}\right]^{\nu(1-D_x)}. \tag{4.9}$$

Combining Eqs. (4.8) and (4.9) yields the geometrical tortuosity for an unsaturated porous medium (Ghanbarian et al. 2013b)

$$\tau_g(S) = \left[\frac{S - S_c + (C/L_s)^{1/\nu}}{1 - \theta_c}\right]^{\nu(1-D_x)} . \tag{4.10}$$

The typical porous media used in experiments are much larger than their models utilized in computer simulations. Thus, the range of porosities and saturations for which finite-size effects are important is small.

Extensive comparisons between the percolation predictions and the experimental data, as well as with the existing phenomenological theories, such as that of Koponen et al. (1997) were made by Ghanbarian et al. (2013b). Koponen et al. (1997) computed numerically the hydraulic tortuosity for flow in a 2D saturated porous medium with a percolation threshold of 0.33 ($S_c = 0.33$). To compare the predictions of Eq. (4.6) with their results, Ghanbarian et al. (2013b) set $C/L_s \approx 0.0085$ and 0.0043 for, respectively, porosities larger and smaller than 0.55, the two ranges simulated by Koponen et al. Figure 4.2a indicates good agreement with the results of Matyka et al. (2008), if $D_x = D_{\text{opt}} \approx 1.21$ (spatial tortuosity), $\nu = 4/3$ (universal

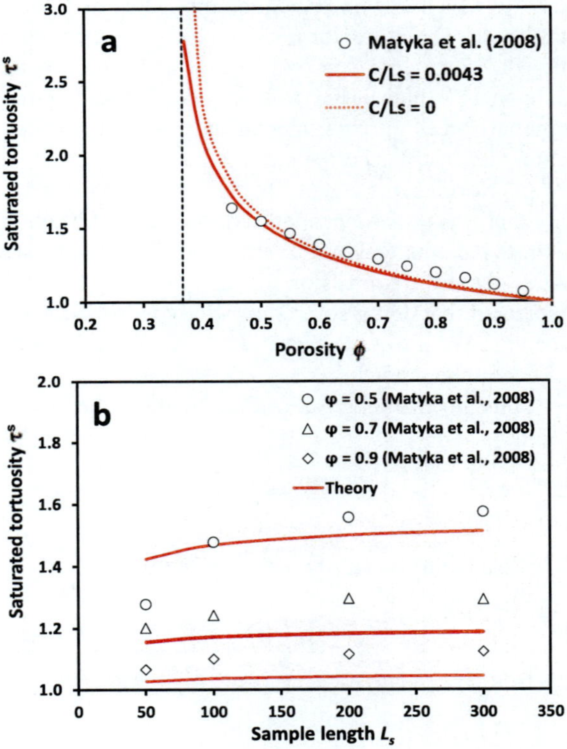

Fig. 4.2 Comparison of the predictions of percolation model of tortuosity for saturated porous media with the experimental data (after Ghanbarian et al. 2013b)

value of ν in 2D), and $S_t = 0.33$ are used in the model. Figure 4.2b presents the effect of finite size of a porous medium on the tortuosity. If the finite-size effect, $(C/L_s)^{1/\nu}$, is eliminated from the equation, one obtains $\tau_g^{(s)} \approx 0.988$, instead of 1.0, only about 1 percent error.

4.5 Characterization of Pore-Space Geometry

Two widely used methods of measuring the pore-size distribution (PSD) of porous materials are mercury porosimetry and measurements of sorption isotherms. Despite the relative simplicity of the measurements, interpretation of the data requires careful modeling of the pore space. Here, we describe the application of percolation to modeling and interpreting mercury porosimetry, as well as sorption data for porous materials. But, before doing so, some fundamental concepts must be carefully described.

4.5.1 *The Young–Laplace Equation*

A bubble collapses in order to minimize its free surface energy. As the bubble shrinks, however, the gas pressure inside the bubble increases to counter the reduction in the bubble's radius. If the bubble's radius decreases from r to $r - \Delta r$, the free energy decreases by, $-\Delta G = 8\pi r \Delta r \sigma$, and the corresponding volume change is $4\pi r^2 \Delta r$. The shrinking compresses the gas inside the bubble, while the air outside the bubble expands. The network ΔW associated with the compression–expansion is, $-\Delta W = (P_i - P_o)(4\pi r^2 \Delta r)$, where P_i and P_o are, respectively, the pressure inside and outside the bubble. Because the network is the change in the free energy, one obtains

$$P_i - P_o = \Delta P = \frac{2\sigma}{r} \; . \tag{4.11}$$

ΔP is sometimes referred to as the Laplace[2] pressure.

Equation (4.11) is a special limit of a more general equation, known as the *Young*[3]*-Laplace equation* (Laplace 1806; Young 1855), which generalizes Eq. (4.11) to nonspherical surfaces. If we consider an area element on a nonspherical curved surface, then there are two *principal radii of curvature*, r_1 and r_2, and the Young–Laplace equation is given by

---

[2] **Pierre-Simon, marquis de Laplace** (1749–1827) made pivotal contributions to mathematical astronomy, statistics, and other branches of science. He formulated the Laplace equation and Laplace transform, and was one of the first scientists who contemplated the existence of black holes, publishing a paper about it in 1799. John Michell had independently suggested the same in 1784.

[3] **Thomas Young** (1773–1829) was a British scientist who made important contributions to wave theory of light, solid mechanics (Young's modulus is named after him), and energy, and was also a prominent Egyptologist.

$$P_i - P_o = \Delta P = \left(\frac{1}{r_1} + \frac{1}{r_2}\right)\sigma \,. \tag{4.12}$$

For a sphere, $r_1 = r_2$ and Eq. (4.12) reduces to (4.11).

4.5.2 *Contact Angle, the Washburn Equation, and Capillary Pressure*

If a circular liquid drop is at rest on a solid surface, its circumference of the contact area is drawn toward the drop's center by σ_{sl}, the solid–liquid interfacial tension. The (equilibrium) vapor pressure of the liquid generates an adsorbed layer on the solid surface that causes the circumference to move away from the drop's center, generating a solid–vapor surface (or interfacial) tension, σ_{sv}. Simultaneously, the interfacial tension σ_{lv} between the liquid and vapor is equivalent to the liquid's own surface tension σ, and acts tangentially to the contact angle θ—the angle at which the interface between the two fluids intersects the solid surface—and draws the liquid toward the center. A force balance yields

$$\sigma_{\mathrm{sv}} = \sigma_{\mathrm{sl}} + \sigma_{\mathrm{lv}} \cos\theta \,. \tag{4.13}$$

Equation (4.13) is known as the Young–Dupré equation, and is used to determine the contact angle:

$$\cos\theta = \frac{\sigma_{\mathrm{sv}} - \sigma_{\mathrm{sl}}}{\sigma_{\mathrm{lv}}} \,. \tag{4.14}$$

If $\theta < 65°$, then the surface is said to be *wetted* by the fluid, because the adhesive force, $F_A = \sigma_{\mathrm{sv}}/\sigma_{\mathrm{lv}}$, exceeds the cohesive force, $F_C = \sigma_{\mathrm{sl}}/\sigma_{\mathrm{lv}}$, and the liquid spreads spontaneously on the surface. For $105° < \theta < 180°$, the surface is not wetted by the liquid, because, in this case, $F_C > F_A$ and the liquid in the form of a drop remains stationary on the surface and takes on a pseudo-spherical shape. In between, $65° < \theta < 105°$, the surface is said to be *intermediately wet*.

Suppose that a capillary tube of radius r is immersed in a liquid. Then, the liquid will rise in the tube if the contact angle $\theta > 90°$, or will be depressed below the surface of the liquid, if $\theta < 90°$. Assume that the net height of capillary rise or depression is h. Consider, first, the case in which the liquid rises in the tube, and assume that point A is in the gas phase next to the interface in the tube, and that point B is in the liquid in the tube at the same level as the free surface of the liquid outside the tube. Thus, A and B are separated by a distance h. Clearly, the pressure P_A at A is given by $P_A = P_B + hg(\rho_l - \rho_g)$, where ρ_l and ρ_g are the densities of the liquid and the gas, and g is the gravitational constant. Therefore, $\Delta P = P_A - P_B = hg(\rho_l - \rho_g)$, which, in view of Eq. (4.11), implies that, $hg(\rho_l - \rho_g) = 2\sigma/r$ and is valid when the liquid completely wets the surface, $\theta = 0°$. If $\theta > 0$, the work required for moving up the capillary tube is, $W = (\sigma_{\mathrm{sl}} - \sigma_{\mathrm{sv}})\Delta A = -(\sigma_{\mathrm{lv}} \cos\theta)\Delta A$, where ΔA is the area of the tube's wall covered by the liquid, and we used Eq. (4.13). The same work is required to force out a column of height h out of the tube. Suppose that a volume V of the liquid is forced out of the tube with gas at a constant pressure ΔP_g above ambient. Then, $W = V\Delta P_g$. For a capillary tube of radius r and length ℓ, $V = \pi r^2\ell$ and $A = 2\pi\ell$. Thus,

$$\Delta P = -\frac{2\sigma}{r}\cos\theta \tag{4.15}$$

which is known as the Washburn equation (Washburn 1921).

An important quantity for characterizing a porous medium is the *capillary pressure* P_c. Consider two immiscible fluids, such as water and oil, which are in contact in a confined system, such as a cylindrical tube, separated by an interface. The pressure field is discontinuous across the interface between the two fluid phases, implying that the pressures at two points, one in the water phase at pressure P_w and one in the oil phase at pressure P_o, with both being in the immediate vicinity of the interface, are not equal. The difference $P_c = P_o - P_w$ is called the capillary pressure. Note that the pressure is usually higher in the nonwetting fluid, oil in this case. P_c is given by the Young–Laplace equation:

$$P_c = \sigma_{\rm ow}\left(\frac{1}{r_1}+\frac{1}{r_2}\right)\cos\theta\ . \tag{4.16}$$

In a cylindrical tube $r_1 = r_2 = r$, where r is the radius of the tube, and Eq. (4.16) reduces to the Washburn equation. The principal radii of curvature are measured from the nonwetting fluid side.

P_c depends uniquely on the amount of the fluid in the system and, hence, on its saturation, the fluid's volume fraction. The situation is different, however, in a porous medium with many irregularly shaped pores, because different saturations may yield the same P_c, and vice versa, since there are several interfaces in the porous medium at different locations. If, for example, we initially fill a porous medium with water and then expel it gradually by injecting oil into the pore space, assuming that water is the wetting fluid, then, injecting oil into the medium requires applying a pressure that must be increased as more oil enters the pore space. At each stage of the experiment and for water saturation S_w, there is a corresponding capillary pressure P_c. If the experiment is continued for a long enough time, a point will be reached at which no more water is produced, even though there is still some water left in the pore space. At this point—at the highest P_c—S_w is called the *irreducible water saturation*, denoted by $S_{\rm iw}$, and the process of expelling water by oil is called *drainage*.

Next, consider the same porous medium at the end of drainage. We begin expelling the oil gradually by injecting water into the medium—a process called *imbibition*. We obtain, however, *another* capillary pressure curve very different from what was measured during drainage. Hence, there is *hysteresis* in the capillary pressure–saturation curves. Figure 4.3 presents a typical P_c diagram. Note that even when $P_c = 0$, there is still some oil left in the medium, with its saturation referred to as the *residual* oil saturation, $S_{\rm or}$. Clearly, drainage or imbibition does not have to begin at $S_w = 1$ or

Fig. 4.3 Typical capillary pressure curve for a water-wet porous medium

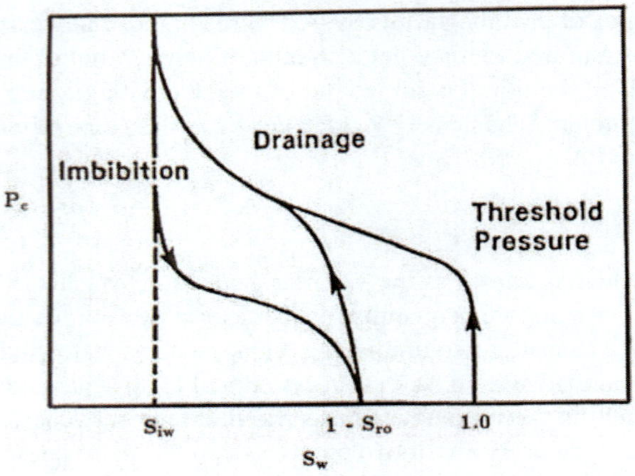

at $S_w = S_{\text{iw}}$, but at any other saturation in between the two, giving rise to *secondary drainage*, secondary imbibition, and so on, each having its own distinct capillary pressure curve, also shown in Fig. 4.3. The resulting set of curves is usually called the *scanning loops*.

4.5.3 Mercury Porosimetry

As a probe of a porous material mercury porosimetry (MP) is usually used for pores between 3 nm and 100 μm. In the MP, the porous medium is first evacuated and immersed in mercury. Since mercury does not wet usually the pores' surface, it enters the pore space only if a pressure is applied and increased, either incrementally or continuously. The volume of the injected mercury is measured as a function of the applied pressure. Clearly, injecting mercury corresponds to drainage, although it is usually called *mercury intrusion*. Because smaller pores require higher pressures, one must apply increasingly larger pressures in order for the mercury to penetrate them. Given a particular porous medium, there is a *unique* characteristic maximum pressure associated with it, which is the pressure required to completely saturate the sample with mercury.

The pressure is then lowered back to ambient, expelling mercury from the pores. With vacuum acting as the "wetting phase," *mercury extrusion* or retraction represents an imbibition process. The shapes of the intrusion–extrusion curves widely vary, as they depend on the morphology of porous media. Complete intrusion–extrusion experiments also form the aforementioned scanning loops, representing the complete experiments that bring the system back to ambient pressure, but where the highest pressure in the experiments is less than the characteristic (highest) pressure of the sample (see above). Thus, some of the pore spaces contain no mercury.

Wardlaw and co-workers (Wardlaw and Taylor 1976; Wardlaw and McKellar 1981) constructed a plot of cumulative residual saturation (the amount of mercury entrapped) versus initial saturation (corresponding to the highest pressure in the cycle) from the scanning loop data, and found that by adding the cumulative residual saturation curve to the initial intrusion curve, a curve equivalent to a final re-intrusion (re-injection) curve is produced. The resulting re-injection curve is one that results from increasing the pressure back up to the characteristic maximum pressure, starting from ambient pressure, with the sample being in the state at the end of the primary retraction curve. Wardlaw and McKellar (1981) suggested that by subtracting the initial intrusion curve from the respective re-injection curve, one would obtain a plot of residual saturation versus initial saturation without needing to carry out a complete set of scanning-loop experiments. Their empirical results were then proven to be true for any porous model (Androutsopoulos and Salmas 2000).

There is a characteristic hysteresis between the intrusion and extrusion curves because the path that the intrusion curve follows is not the same as that during intrusion. In addition, some mercury stays in the medium even after the pressure is lowered back to atmospheric pressure that often, hours after the pressure is back to atmospheric, continues to slowly leave the sample. The hysteresis may depend on the *history* of the system. That is, the way the experiment is carried out. In some cases, hysteresis can be eliminated by performing the experiment very slowly, but it cannot be eliminated in other case, in which case it is called *permanent hysteresis*.

The contact angle of mercury with a wide variety of surfaces is between 135° and 142°. As a fluid moves or flows over a solid surface, however, its contact angle may vary because it advances on a dry surface but recedes on a wetted one. Thus, one may have a *contact-angle hysteresis*. Usually, the advancing contact angle θ_A (associated with intrusion or drainage) is larger than the receding one, θ_R. For mercury, a contact angle hysteresis between 10° and 20° has been reported for a variety of surfaces.

All mercury porosimetry curves have three common features:

(i) Hysteresis, so that, at a given pressure, the volume indicated on the extrusion curve is larger than the corresponding volume on the intrusion curve, whereas, at a fixed volume of injected mercury, the pressure on the extrusion curve is smaller than the corresponding pressure on the intrusion curve. Hysteresis occurs because the *pore potential* prevents the extrusion of mercury from a pore until a pressure less than the nominal extrusion pressure is reached, hence causing the hysteresis.

(ii) Lack of formation of a closed loop, because some mercury remains in the porous sample at the end of the first intrusion–extrusion experiment.

(iii) Hysteresis in curves for a second intrusion–extrusion experiment, carried out at the end of the first one. The loops will, however, close if the cycles continue, implying that no further entrapment of mercury in the porous medium is possible.

Leverett (1941) defined a reduced capillary pressure function, usually called the Leverett J−function,

$$J = \frac{P_c}{\sigma \cos\theta}\sqrt{\frac{K_e}{\phi}}, \tag{4.17}$$

where K_e is the effective permeability of the pore space, which has been found to successfully correlate capillary pressure data originating from a specific lithologic type within the same formation. It is not, however, of general applicability, with the reason being inadequacy of $\sqrt{K_e/\phi}$ for taking into account the individual differences between pore structures of various porous media.

4.5.4 Invasion and Random Percolation Models

When mercury invades a porous medium, the sequence of the pores and throats that it fills depends not only on the PSD, but also on the coordination number of the pore space. Thus, mercury porosimetry is a percolation process. Any porous medium can, in principle, be mapped onto an equivalent network of pores and throats (sites and bonds), and Chatzis and Dullien (1977) and Wall and Brown (1981) developed network models of mercury porosimetry, and mentioned specifically their connection with percolation. Larson and Morrow (1981) used percolation concepts to derive analytical formulae for mercury porosimetry curves. Sahimi (2011) provides a comprehensive review of pore-network models of mercury porosimetry.

A representative example of a pore-network model of mercury porosimetry is that of Tsakiroglou and Payatakes (1990; see also Mata et al. 2001). In their work, the porous medium was represented by a simple cubic network in which the throats, assumed to be cylindrical, had converging–diverging ends, as the pores were spherical. The diameters of the pores and throats were selected from representative statistical distributions. The throats' length was adjusted such that the network's porosity matched that of the porous medium.

Pore-network simulation of mercury intrusion is an invasion percolation (IP) process, already described briefly in Chap. 3, and will be taken up again later in this book. The IP is typically simulated based on either the sites (pores) or the bonds (throats), whereas a complete simulation of mercury intrusion involves both. Sahimi et al. (1998) developed a site–bond IP algorithm that can simulate mercury intrusion completely. The simulation algorithm proceeds as follows.

(i) Mercury menisci are placed at the entrance to all the throats on the pore-network's external surface.

(ii) Small enough pressure is applied at the beginning such that no meniscus enters a throat yet, and then it is increased by a small amount. Then, all the menisci at the throats' entrances are examined. If the applied pressure exceeds the capillary pressure of any throat, it is filled by mercury and the meniscus is placed at the entrance to the downstream pore connected to that throat.

(iii) All the menisci at the entrances to the downstream pores are examined, after all of them at the pore entrances have been moved, if necessary. If the applied pressure exceeds the capillary pressure for entering a pore, the pore is filled with mercury and the new menisci are placed at the entrance to the empty throats that are connected to that pore. If a pore contains several menisci at its entrances, the smallest P_c determines whether it is filled with mercury.

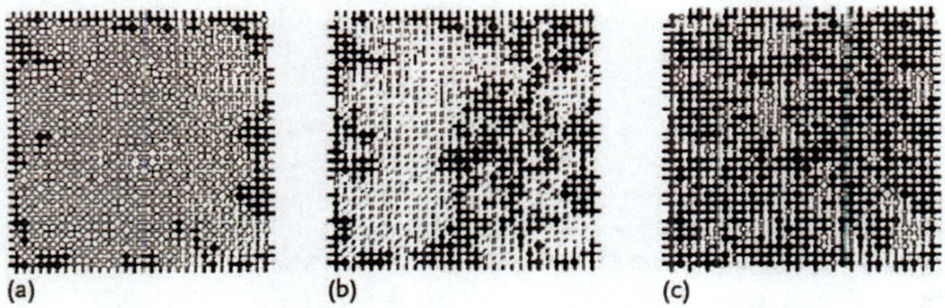

Fig. 4.4 Configurations of the pore network during mercury porosimetry. The applied external pressure increases from (**a**) to (**b**) to (**c**) (after Tsakiroglou and Payatakes 1990)

(iv) The new menisci are examined to decide whether any new throat can be filled with mercury. If the applied P_c is not large enough to do so, it is increased by a small amount. Then, all the menisci are examined again, and so on. The simulation continues until mercury-filled throat and pores form an SSC percolating the network. This is precisely how an IP process is simulated. Figure 4.4 shows typical configurations of a two-layer network of coordination number 5 at various stages of mercury injection and intrusion, demonstrating the striking similarity of the configurations with percolation clusters, and more generally the IP clusters.

Simulation of extrusion, the reverse of the IP algorithm, begins at the end of the intrusion, and proceeds as follows.

(i) The applied pressure at the end of intrusion is lowered by a small amount, and pores with menisci that are connected to the external mercury sink through continuous mercury paths and can move under the present P_c, as well as throats that contain mercury that must snap-off under the same conditions, are searched for. Snap-off happens (Mohanty et al. 1980) when the pressure is lowered and the curvature of the collars decreases until some interfaces become unstable and rupture. The phenomenon may leave pockets of isolated mercury in some throats or pores, which are then ignored for the rest of the simulation.

(ii) The applied pressure is decreased and the search described in (i) continues.

While the IP algorithm describes mercury porosimetry comprehensively, its simulation is time-consuming and difficult. Larson and Morrow (1981) assumed that mercury porosimetry can be approximated by random percolation, and developed a complete formulation for it, which we now described. But, we must first specify the conditions under which random percolation, as well as the IP, can be used to properly describe any two-phase flow phenomenon in porous media, including mercury porosimetry.

(i) The capillary pressure across a meniscus separating the two fluids must be greater than any other pressure difference in the process.

(ii) Frictional losses due to viscosity must be small compared to the capillary work.

To quantify the two conditions, a dimensionless group called the *capillary number* Ca is defined by

$$\mathrm{Ca} = \frac{\eta v}{\sigma} , \tag{4.18}$$

where v is the average fluid velocity and η the average viscosity. Then, one must have $\mathrm{Ca} \ll 1$ in order to fulfill the two criteria. For now, we ignore the size of the pores (if the pores are large enough, pressure drop in them will be small). Consider, then, a size distribution $f(r)$ for the throats. Larson and Morrow (1981) wrote down the capillary pressure as, $P_c = \sigma C$, where C represents the curvature of a meniscus, $C \propto 1/r$. During mercury injection, the network is exposed to mercury under an applied pressure that, if σ is constant, is equivalent to being exposed to some meniscus curvature C. As C (P_c) increases, it exceeds the entry curvature C_i (P_i) for a larger and larger fraction of the throats. In fact, C exceeds C_i for a fraction $Y_i(C)$ of the void volume given by

$$Y_i(C) = \int_0^C y_i(C_i)dC_i , \tag{4.19}$$

with $y_i(C_i)$ being the *pore entry-curvature* distribution given by

$$y_i(C_i) = \int_0^\infty y(C_i, C_w)dC_w , \tag{4.20}$$

where $y(C_i, C_w)$ is the *joint* probability distribution for throat entry and withdrawal curvatures, C_i and C_w, with the properties that, $y(C_i, C_w) \geq 0$, for $C_i \geq 0$ and, $C_w \geq 0$ and $y(C_i, C_w) = 0$, for $C_i < C_w$. Moreover, since $y(C_i, C_w)$ is a probability distribution, it must be normalized,

$$\int_0^\infty \int_0^{C_i} y(C_i, C_w)dC_i dC_w = 1 . \tag{4.21}$$

Given the physical meaning of $y_i(C_i)$, $Y_i(C)$ is the fraction of the throats with entry curvatures less than C. Not all the throats with an entry curvature C are, however, accessible from the external surface of the network (pore space), where accessibility is defined in the percolation sense. Thus, Larson and Morrow (1981) assumed that the saturation of mercury (the nonwetting fluid) is given by

$$S_{nw} = X^A[Y_i(C)] . \tag{4.22}$$

As explained below, Eq. (4.22) is only a rough estimate of S_{nw}, because one must take into account the effect of the size distribution of the throats. Therefore, for any given capillary pressure, $P_c = \sigma C$, the fraction $Y_i(C)$ is calculated, based on which the saturation S_{nw} is determined.

Mercury retraction (extrusion) begins at the end of intrusion. As the applied pressure at that stage is lowered, so also is the curvature C that can *potentially* eject mercury (the nonwetting fluid) from those throats for which the withdrawal curvature is between C_w and $C_w - dC_w$, and the entry curvature is C_i or smaller. The fraction of such throats is

$$y_r(C_i, C) = \int_0^{C_i} y(C_i', C)dC_i' . \tag{4.23}$$

But all such throats cannot actually expel mercury, because some of them were not invaded during the injection in the first place, as they were not accessible, while others, although containing mercury, cannot expel it because they are not connected to the external surface of the network via a path of mercury-filled throats. Thus, during retraction, the fraction of the throats with injection curvature smaller than or equal to C_i and withdrawal curvature smaller than or equal to C_w is given by

$$Y_r(C_i, C) = \int_0^{C} y_r(C_i, C_w)dC_w . \tag{4.24}$$

Only a portion $X^A(Y_r)$ of such throats still contain mercury and are also accessible, implying that their fraction is $X^A(Y_r)/Y_r$. Thus, as the curvature (applied pressure) C_w is reduced by dC_w, saturation of mercury also decreases by $dS_{\rm nw}$ given by

$$-dS_{\rm nw} = \frac{X^A[Y_r(C_i, C_w)]}{Y_r(C_i, C_w)} y_r(C_i, C_w)dC_w , \tag{4.25}$$

which, after integration, yields

$$S_{\rm nw} = S_0 - \int_C^{C_i} \frac{X^A[Y_r(C_i, C_w)]}{Y_r(C_i, C_w)} y_r(C_i, C_w)dC_w , \tag{4.26}$$

where S_0 is the initial saturation at which mercury retraction began. Therefore, given a capillary pressure P_c, one determines the corresponding saturation $S_{\rm nw}$. If P_c is zero, so also is the corresponding curvature, which is the point at which the saturation of mercury (the nonwetting fluid) is at its residual value, $S_{\rm rnw}$, determined from Eq. (4.26) by setting $C = 0$. This completes formulation of the retraction curve, at the end of which one may begin a second injection process, for which it is not difficult to see that one must have

$$y_i(C_i, C) = \int_0^{C} \frac{X^A[Y_r(C_i, C_w)]}{Y_r(C_i, C_w)} y(C, C_w)dC_w , \tag{4.27}$$

using the fact that, for $C < C_w$, one has, $y(C, C_w) = 0$. The corresponding saturation is given by

$$S_{\rm nw} = S_{\rm rnw} + \int_0^{C} y_i(C_i, C')dC' . \tag{4.28}$$

In a similar manner, one may also consider a secondary retraction process.

A given curvature distribution corresponds to a throat-size distribution since $r \propto 1/C$. Therefore, for given capillary pressure curves, one may assume a functional form for the throat-size distribution $f(r)$ with a few adjustable parameters, which are estimated such that the predicted capillary pressure curves agree with the data. The advantage of the model is that, if the accessibility function X^A is available, or can be computed, calculating the P_c curves will be straightforward. For 3D networks, however, no closed-form formula is known for X^A; only its numerical values have been computed. Thus, the integrals in Eqs. (4.26)–(4.28) must be evaluated numerically, a task that is still simpler than the computations associated with the IP algorithm.

Larson and Morrow (1981) assumed that the accessibility function can be approximated by that of a Bethe lattice of coordination number 4 (Chap. 2) for which $X^A(p)$ is known analytically. For a Bethe lattice of coordination number Z and $p \geq p_c$ (p_c is the percolation threshold), one has (Larson and Davis 1982)

$$X^A(p) = p\left[1 - R(p)^{2Z-2}\right] , \tag{4.29}$$

where $R(p)$ is the root of the equation, $p\sum_{j=2}^{Z-1} R^{Z-j} + p - 1 = 0$. Furthermore, Larson and Morrow (1981) assumed that

$$y(C_i, C_w) = 6g(C_i)g(C_w)\int_{C_w}^{C_i} g(C)dC , \quad C_i \geq C_w \tag{4.30}$$

which possesses all the properties that were described above with, $g(x) = 2x \exp(-x^2)$ and $x = \log C$, which results in a broad distribution of the curvatures and, therefore, a broad $f(r)$. Figure 4.5 compares the predictions of the model of Larson and Morrow with mercury porosimetry data for a Becher dolomite with porosity, $\phi = 0.174$. The qualitative agreement between the predictions and the data is striking. Also shown is the effect of sample size (thickness) on the capillary pressure curves.

The most important reason for the success of the model is the fact that the physical concept of percolation accessibility that determines which pores and throats can be invaded by mercury, and from which pores and throats it can be withdrawn, has been explicitly utilized.

An improved random percolation model was developed by Heiba et al.[4] (1982, 1992). They noted that during injection and retraction the statistical distributions of the throat accessible to and occupied by the nonwetting fluid (mercury), which we refer to as the *sub-distributions*, are not identical. Consequently, the throat-size distribution of the subset of pore space occupied by the nonwetting fluid differs from

[4] The original model was developed in early 1982 and presented in September of that year at the annual meeting of the Society of Petroleum Engineers. The paper was accepted for publication in 1982, subject to some minor clarifications. It took, however, *10* years to make the minor revisions, since the first two authors had moved on, and the last two were preoccupied with other things. In the meantime, an entire "industry" had risen based on that SPE preprint! When the paper was eventually published in 1992, it probably had set a world record for procrastination!

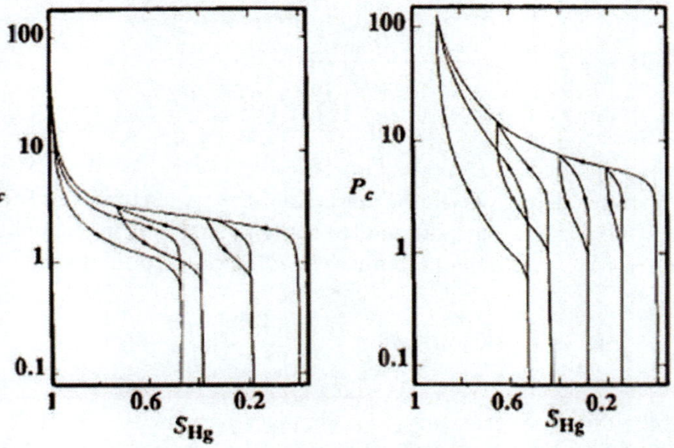

Fig. 4.5 Comparison of capillary pressure data (**a**) with the predictions of the percolation model (**b**), in which dashed curves represent the pressure in a pore network that is 10 pore throat thick, while the solid curves show the results for an infinitely large network (after Larson and Morrow 1981)

the *overall* throat-size distribution. Heiba et al. derived analytical formulae for such sub-distributions, which we now present.

Consider injection of the nonwetting phase. The fraction of the throats that are *allowed* to the nonwetting fluid—i.e., the throats with the right capillary pressure that can *potentially* be filled by the fluid—is

$$Y_i(r_{\min}) = \int_{r_{\min}}^{\infty} f(r)dr \,, \tag{4.31}$$

where $r_{\min}$ is the minimum throat radius that the nonwetting fluid can invade. The fraction of the throats that are accessible to and, thus, occupied by the fluid is $X^A(Y_i)$. Therefore, during injection, the distribution $f_i(r)$ of the throat radii that are occupied by the nonwetting fluid is

$$f_i(r) = \frac{f(r)}{Y_i(r_{\min})} \,, \quad r \geq r_{\min} \,, \tag{4.32}$$

and, clearly, $f_i(r) = 0$ for $r < r_{\min}$. The basis for Eqs. (4.31) and (4.32) is that during injection the *largest* throats, i.e., those with the *smallest* entry curvature or capillary pressure, are filled. Thus, the saturation of the nonwetting fluid (mercury) during injection is given by

$$S_{\mathrm{nw}} = \frac{X^A[Y_i(r)]}{Y_i(r_{\min})} \frac{\int_{r_{\min}}^{\infty} f(r)V_p(r)dr}{\int_0^{\infty} f(r)V_p(r)dr}, \tag{4.33}$$

where $V_p(r)$ is the volume of a throat of radius r.

As the pressure decreases during retraction, the nonwetting fluid (mercury) is expelled, first from the *smallest* throats. In reality, mercury is first expelled from the smallest pores, which the model ignores. The *allowed* fraction of such throats is

$$Y_r = \int_0^{r_0} f(r)dr + \left[1 - \frac{X^A(Y_{i,t})}{Y_{i,t}}\right] \int_{r_0}^{\infty} f(r)dr\ , \tag{4.34}$$

where r_0 is the throat's radius at a given capillary pressure P_c such that mercury is expelled from all the throats for which $r \le r_0$. Here, $Y_{i,t} = Y_i(r_{\min,t})$, and $r_{\min,t}$ is the throat radius at the end of injection. The first term of the right side of Eq. (4.35) is simply the fraction of the throats from which the nonwetting fluid is expelled, if there were no inaccessible throats at the end of injection. At that point, however, a fraction $1 - X^A(Y_{i,t})/Y_{i,t}$ of the throats could not be reached by the nonwetting fluid. Consequently, the second term of the right side of Eq. (4.35) is the fraction of the throats that were not invaded by nonwetting fluid at the end of injection. Hence, the size distribution of the throats from which mercury is expelled is given by

$$f_r(r) = \begin{cases} \dfrac{f(r)}{Y_r}\left[1 - \dfrac{X^A(Y_{i,t})}{Y_{i,t}}\right], & r > r_0 \\[2ex] \dfrac{f(r)}{Y_r}\left\{1 - \dfrac{X^A(Y_{i,t})}{Y_{i,t}}\left[1 - \dfrac{X^A(Y_r)}{Y_r}\right]\right\}, & r_{\min,t} < r < r_0 \end{cases} \tag{4.35}$$

Clearly, $f_r(r) = f(r)/Y_r$ for $r < r_{\min,t}$. Therefore, the model calculates the saturation correctly, as it takes into account the effect of the size distribution of the throats and the dependence of their volume on their effective radii.

As already pointed out, the thickness of a porous medium does influence the capillary pressure curves, since it increases the accessibility of pore space, which reduces the sharpness of the injection-curve knee. Injection curves for unconsolidated packings of particles indicate a rather strong dependence on the sample thickness for porous media up to about 10 particle diameters or about 30 throat diameters. The size effect for thicker porous media is relatively weak, and if the thickness exceeds 20 particle diameters, no appreciable sample size can be detected. As mentioned earlier, Larson and Morrow (1981) carried out an extensive study of the effect of sample size on the capillary pressure curves; see also Thompson et al. (1987a, b).

Although mercury porosimetry is a relatively straightforward experiment, the interpretation of the data is not simple, which is usually done using Eq. (4.18). But, the correct interpretation of the data requires the use of pore-network models of the type described earlier. Note that, as long as P_c is less than a critical value, $P_c = 2\sigma\cos\theta/r_{\min}$, where $r_{\min}$ is a minimum pore throat radius, no significant, sample-spanning invasion of the pore space by mercury occurs. $r_{\min}$ is calculated by Eq. (4.32) in which $Y_i(r_{\min})$ is set to be equal to the bond percolation threshold of the network:

$$p_c = \int_{r_{\min}}^{\infty} f(r)dr \ . \tag{4.36}$$

Clearly, higher connectivity implies lower percolation thresholds and, hence, larger $r_{\min}$.

Portsmouth and Gladden (1991) developed a method by which the connectivity of a porous material can be determined by mercury porosimetry. Their simulations using a pore network with a Gaussian PSD indicated that the mercury remaining in the network after the applied pressure was reduced to the ambient value, following a complete primary intrusion and retraction cycle, was a function of the reduced standard deviations (SD)—the SD divided by the mean—of the pore-size distribution and the connectivity. The intersection of a measured quantity of entrapped mercury with the characteristic entrapment function yields two estimates of pore connectivity. To differentiate between the two estimates, Portsmouth and Gladden (1991) introduced a new pressure sequence for mercury porosimetry, and showed that the mini-hysteresis loops produced by the new pressure sequence provide the information that can distinguish between the two possible values of connectivity for the network structure, and provide a more accurate determination of the characteristics of the SD and the mean of the PSD.

4.6 Sorption in Porous Media

An important method of determining the PSD and pore surface area of porous media is based on the sorption of a gas, usually nitrogen, in the pore space, although one can, in principle, use other gases, such as CO_2. Extensive discussions of gas sorption in porous media were given by Dabrowski (2001) and Sahimi (2011). The method allows one to determine the PSD with pore sizes ranging from about 0.35 nm to over 100 μm. Although sorprion of a gas in a porous medium can occur by chemisorption in which the gas forms chemical bonds with the pore surface, we describe and study here *physisorption* or *reversible* sorption that occurs when a gas is brought into contact with a solid surface. The important characteristics of the phenomenon are that, it (a) is in most cases accompanied by low heat of adsorption; (b) is fully reversible; (c) reaches equilibrium rather quickly because it requires no activation energy; (d) can lead to multilayer adsorption; and (e) can fill the pores completely.

According to the International Union of Pure and Applied Chemistry (IUPAC), all sorption isotherms fall into one of the six classes shown in Fig. 4.6, two of which manifest hysteresis between adsorption and desorption, while the other four represent completely reversible adsorption. The sorption types are

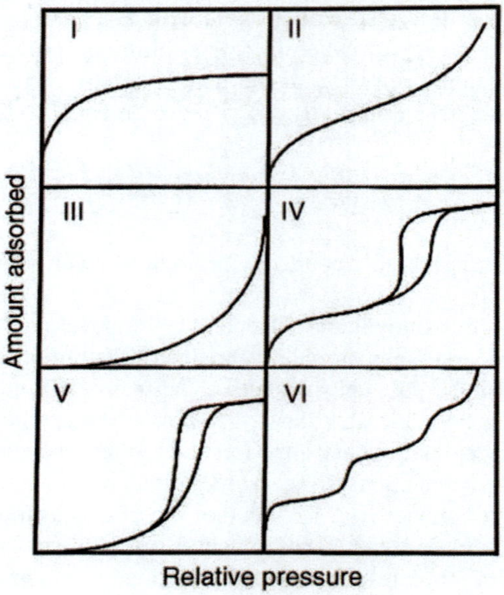

Fig. 4.6 Six types of sorption isotherms

Type-I, which forms when adsorption is limited to at most a few molecular layers, and the adsorbed amount reaches a limiting value when $P/P_0 \to 1$, with P and P_0 being, respectively, the condensation and saturation pressures. Chemisorption forms only a monolayer, and adsorption in nanoporous materials exhibit Type-I isotherms.

Type-II, which occurs in porous materials with macropores, so that the adsorbed amount may continue to increase. The inflection point of the curve is the pressure at which monolayer adsorption has ended and multilayer sorption has begun.

Type-III, which are rare because they occur if the attractive adsorbate–adsorbent interactions are weak, while the interactions between the adsorbates themselves are strong.

Type-IV, which occurs in mesoporous materials, with their inflection point indicating again the end of monolayer and the beginning of multilayer formation.

Type-V, which are similar to Type-IV, except that the initial part of the isotherm is similar to Type-III.

Type-VI, which are unusual because they exhibit a stepwise structure associated with uniform, nonporous surface, with the sharpness of the steps depending on the homogeneity of the surface, the adsorbate type, and the temperature.

The hysteresis loops in sorption isotherms have also been classified by the IUPAC, and they are referred to as the H1, H2, H3, and H4 loops, shown in Fig. 4.7. According to Sing et al. (1985), at least for the H1, H2, and H3 type the connectivity of the pore space plays an important role.

Consider nitrogen adsorption in a single pore. As the pressure increases, an adsorbed film of nitrogen forms on the pore's wall with a thickness that increases with increasing pressure. At condensation pressure P, the pore is filled with a (liq-

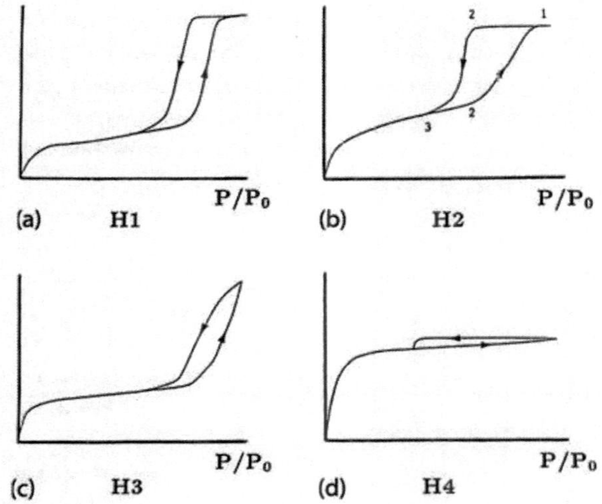

Fig. 4.7 Hysteresis loops in sorption isotherms

uidlike) condensed phase, resulting in a step increase in the adsorption isotherm. The condensation pressure is given by the Kelvin[5] equation that, for a pore of radius r is given by

$$\ln\left(\frac{P}{P_0}\right) = -\frac{r_c}{r}, \quad r_c = \frac{2s\sigma V_L}{RT}, \tag{4.37}$$

in which σ is the liquid–vapor surface tension, R the gas constant, T the temperature, V_L the molar volume of the liquid, and s is a geometrical factor, such that for spherical pores $s = 1$ for both adsorption and desorption, whereas, for cylindrical pores, $s = 1/2$ for adsorption, but $s = 1$ for desorption. At any reduced pressure, P/P_0 sorption is uniquely parameterized by a critical radius r_c such that adsorption or desorption corresponds to an increase or decrease in r.

Connectivity of the pores does not play a role in adsorption, during which the entire pore space is accessible to the adsorbate, the vapor condenses in all pores of size $r > r_c$, and liquid nitrogen fills the pores. For $r < r_c$, pores fill up rapidly and continuously. *Primary* adorption refers to the first cycle during which all that matters is the pores' size. Primary desorption begins at the end of primary adsorption. Similar to mercury porosimetry, as P is reduced, the desorption isotherm does not retrace that of adsorption, but forms a hysteresis loop before rejoining the adsorption isotherm. Unlike the primary adsorption, however, the pores' connectivity does matter, because only those with a radius $r > r_c$ can desorb—vaporize their condensate—*if* they have access to either the bulk vapor in primary desorption, or to the isolated vapor pock-

[5] **William Thomson, 1st Baron Kelvin**, or **Lord Kelvin** (1824–1907) was an Irish mathematical physicist and engineer who made many seminal contributions to electricity and development of the first and second laws of thermodynamics. His name is associated with many phenomena, including Joule–Thomson effect, Kelvin waves, Kelvin–Helmholtz instability and mechanism, and Kelvin equation.

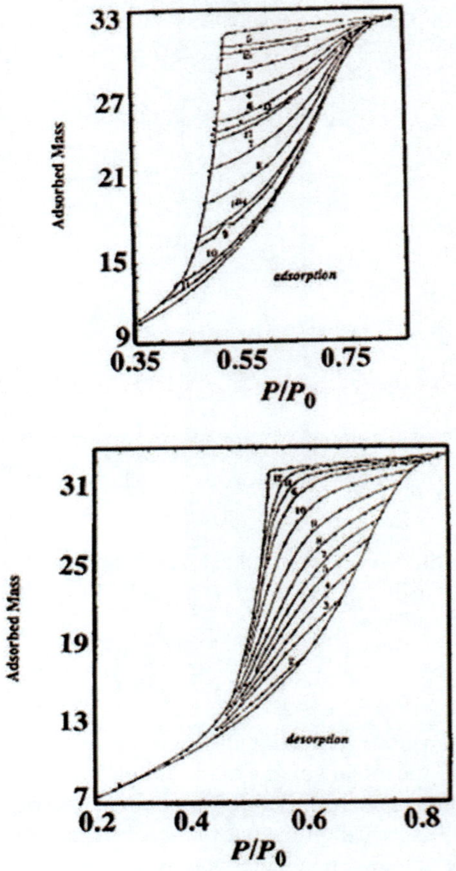

Fig. 4.8 Typical adsorption–desorption isotherms, where the adsorbed mass is in mg (after Mason 1988)

ets in the secondary desorption that occurs after the secondary adsorption. Typical adsorption–desorption isotherms are presented in Fig. 4.8.

4.6.1 Percolation Models of Sorption

Sorption in porous materials can be simulated using pore-network models that take into account the effect of the pore connectivity, i.e., the percolation effect. One must, in principle, use a pore network with converging–diverging throats, in order to take proper account of pore-scale physics of the phenomena. Certain assumptions must also be made in order to make the simulations tractable. For example, one may assume that a pore is filled with the adsorbate when its radius is equal to the radius of curvature of the meniscus. Indeed, Zhdanov et al. (1987) assumed that filling of the pores is controlled only by their radii.

Mason (1982, 1983, 1988) and Parlar and Yortsos (1988, 1989) modeled adsorption assuming that a pore is filled when its largest adjacent throat is filled, implying that the fraction of filled pores in a pore-network model is given by

$$p_s = 1 - (1 - p)^Z , \tag{4.38}$$

where p is the fraction of the throats with a size larger than a certain radius r, which is related to the pressure through Kelvin's equation, Eq. (4.38),

$$p = \int_r^{\infty} f_t(r)dr , \tag{4.39}$$

with $f_t(r)$ being the distribution of the throats' sizes, and Z the pore space coordination number or connectivity.

Adsorption ends when all or a certain fraction of the pores is filled with the condensate, after which desorption begins, but only if the desorbed molecules in a given pore have access to other pores that are filled with vapor. This requirement means that desorption is similar to mercury extrusion (Wall and Brown 1981), as well as the IP and, therefore, should be modeled as a site–bond percolation (Sahimi et al. 1998), because the fraction of the vapor-filled pores (the unoccupied sites in random percolation) is controlled by the throats (the bonds in random percolation). Ball and Evans (1989) demonstrated the accuracy of a percolation model to match experimental data of Burgess et al. (1989).

Consider desorption, which is controlled by the throats, and begins at the percolation threshold at which an SSC of the throats that contain vapor has been formed. The onset of primary desorption is defined in terms of a radius r_i such that

$$\int_{r_i}^{\infty} f_t(r)dr = p_{cb} , \tag{4.40}$$

where p_{cb} is the bond percolation threshold. Adsorption, on the other hand, is controlled by the pores. Thus, if $V_p(r)$ is the volume of a pore and $f_p(r)$ its size distribution, then, because percolation plays no role in primary adsorption, the saturation S_L of the liquid during adsorption is given by

$$S_L = 1 - \frac{\int_{r_c}^{\infty} f_p(r)V_p(r)dr}{\int_0^{\infty} f_p(r)V_p(r)dr} . \tag{4.41}$$

During primary desorption, the adsorbates in a pore vaporize if $r > r_c$, and if that pore is accessible to other vapor-filled pores. Thus, the fraction of pore or throats that are actually occupied by the vapor is given by

$$X_j = X_j^A \ , \quad j = \text{pores or throats} \ , \tag{4.42}$$

where X_j^A is the percolation accessibility function. The size distribution of the liquid-filled pores is given by

$$f_{L,j}(r) = \begin{cases} f_j(r)/(1 - X_j) & , r < r_c \\ f_j(r)(1 - X_j/p_j)/(1 - X_j) & , r > r_c \end{cases} \quad j = \text{pores or throats} \tag{4.43}$$

with p_j given by

$$p_j = \int_{r_c}^{\infty} f_j(r)dr \ , \quad j = \text{pores or throats} \ , \tag{4.44}$$

representing the fraction of pores or throats that have a radius greater than r_a. The liquid saturation during desorption is given by

$$S_L = (1 - X_p)\frac{\int_0^\infty f_{L,p}(r)V_p(r)dr}{\int_0^\infty f_p(r)V_p(r)dr} \ . \tag{4.45}$$

Similar equations are derived for the secondary adsorption and desorption. Following Heiba et al. (1982), and using the analogy between mercury injection and sorption, Parlar and Yortsos (1988) used many formulas derived by the former to study sorption.

Given desorption data, one may determine the distribution $f_t(r)$. To do so, assuming that there are no correlations in the morphology of the pore space, one proceeds as follows:

(i) Given an average connectivity Z and pore space thickness L, one first computes numerically the accessibility $X^A(p, L)$ for the fraction p of the occupied pores (sites).

(ii) The desorption isotherm is then converted into an equivalent curve that expresses the liquid saturation $S_L(r)$ as a function of the throat radius r_t, using an estimate of the porosity, utilizing Kelvin's equation, and assuming a functional form for $V_t(r)$.

(iii) The function $p(r)$ is then computed. Given that, $1 - p = \int_0^r f_t(r)dr$, numerical differentiation of the curve $p(r)$ yields $f_t(r)$.

4.6.2 *Determining Connectivity of Porous Materials Using Percolation*

Estimating coordination number Z, or the connectivity, of an irregular pore space, one in which Z varies spatially, is difficult. One must define an average coordination number $\langle Z \rangle$, with the averaging taken over a large enough sample. For microscopically disordered, macroscopically homogeneous media, $\langle Z \rangle$ is independent of the sample size. Several methods have been developed for estimating $\langle Z \rangle$ that were described by Sahimi (2011). What we are interested in here is a method that utilizes percolation to interpret sorption data in order to estimate $\langle Z\rangle$.

Seaton (1991) and Liu et al. (1992) developed such a method based on percolation theory in order to estimate the mean connectivity of a porous material. The method is based on finite-size scaling analysis described in Chap. 2. Recall that for a network of linear size L—a pore space of thickness L—one has

$$X^A(p) = L^{-\beta/\nu} f[(p - p_c)L^{1/\nu}] , \tag{4.46}$$

which is rewritten as

$$\langle Z\rangle X^A(p) = L^{-\beta/\nu} f[(\bar{Z}p - B_c)L^{1/\nu}] , \tag{4.47}$$

exploiting the fact that, $B_c = \langle Z\rangle p_{cb}$ is an almost invariant of the system (see Chap. 2). A desorption isotherm has three typical segments:

(i) A linear segment, which is due to the decompression of the liquid nitrogen (or another gas) in the pores. In the corresponding percolation network, p, the fraction of open pores—those in which the nitrogen pressure is below the condensation pressure—increases, but the accessible fraction $X^A(p)$ of the pores is still zero because an SSC of open pores has not been formed yet.

(ii) At end of the first interval, the network reaches its percolation threshold p_c, an SSC of the open pores is formed, and the metastable liquid nitrogen in the cluster's pores vaporizes. If one decreases the pressure further, the number of pores containing metastable nitrogen, and the number of pores whose nitrogen has vaporized, both increase.

(iii) Around the knee of the isotherm almost all the pores in which the nitrogen pressure is below their condensation pressure can also vaporize and, therefore, $X^A \simeq p$.

Therefore, determining $\langle Z\rangle$ based on sorption data and percolation consists of two steps:

(i) $X^A(p)$ is determined from the sorption data, and

(ii) assuming a relation between the radius and length of a throat, $\langle Z\rangle$ and L are determined by fitting Eq. (4.47) to $X^A(p)$.

Note that $X^A(p)/p$, the ratio of the fraction of the pores in the SSC and the fraction of pores below their condensation pressures, is simply N_p/N_b, with N_b being the number of moles of nitrogen that would desorb, if all the pores containing nitrogen below their condensation pressures had access to the vapor phase, and N_p is the number of moles of nitrogen that have actually desorbed at that pressure. Let N_A be the number of moles of nitrogen that are present in the pore space at a given pressure during adsorption; N_D be the number of moles of nitrogen that are present in the pores at that pressure during desorption; and N_F, the number of moles of nitrogen that would have been present in the pores at that pressure during desorption, had no nitrogen vaporized from the pores that contain nitrogen below its condensation pressure. Then, $N_p = N_F - N_D$, $N_b = N_F - N_A$, and

$$\frac{X^A(p)}{p} = \frac{N_F - N_D}{N_F - N_A} , \tag{4.48}$$

so that $X^A(p)/p$ is given in terms of measurable quantities. One must also determine p, so that $X^A(p)$ can be calculated based on Eq. (4.48). p is given by an equation similar to Eq. (4.34) in which r is the throat radius in which the adsorbate condenses at the specified pressure, and $f(r)$ is its statistical distribution. Therefore, given $f(r)$, p, and $X^A(p)$ are determined, and Eq. (4.47) is used to estimate $\langle Z \rangle$.

Measurement of adsorption/desorption isotherms is typically time-consuming, and can be very problematical very close to the bulk condensation pressure of nitrogen. Murray et al. (1999) demonstrated that mercury intrusion and nitrogen desorption share the same mechanism, implying that the mercury intrusion curve can be used in place of the nitrogen desorption isotherm in the above connectivity analysis.

Chapter 5
Connectivity of Fracture and Fault Networks

5.1 Introduction

The discussions in Chap. 4 that were concerned with the characterization of porous media were limited to those that are microscopically disordered, but macroscopically homogeneous. If measured on length scales larger than the so-called representative elementary volume, which sets the scale for macroscopic homogeneity, the properties of such porous media are independent of their size.

Macroscopically heterogeneous porous media—those in which the probability density functions of the various properties vary spatially—are also highly important, as they are ubiquitous in nature. A complete description of all properties of such porous media is well beyond the scope of this book. The interested reader is referred to Sahimi (2011) for a full exposition of this important subject. In this chapter, we restrict our attention to morphological properties at the largest scale of heterogeneities in rocks that influence and interfere with fluid flow, transport, and other phenomena occurring in them, such as earthquakes and wave propagation, namely, fractures and faults, and describe the relevance of percolation to their description.

Fractures, natural or man-made, are critical to the economics of oil and natural gas production from underground reservoirs, extraction of vapor from geothermal reservoirs for use in power plants, and development of groundwater resources. More recently, hydraulic fracturing, or fracking, has played a fundamental role in the production of oil and gas from unconventional reservoirs, such as shale formations. In all such cases, fractures provide high-permeability paths for fluid flow and transport in porous formations that without which they would not produce at high rates.

At the same time, with the increasing importance of groundwater pollution, storage of nuclear wastes in rock, and sequestration of CO_2 in large-scale porous formations, there is an even greater need to better understand the structure of the fracture and fault networks. Fault patterns are closely related to fracture networks, and play a fundamental role in seismic events that lead to earthquakes. Despite their clear significance, however, the field of characterization of fracture and fault patterns is not as

M. Sahimi, *Applications of Percolation Theory*, Applied Mathematical Sciences 213,
https://doi.org/10.1007/978-3-031-20386-2_5

well developed as unfractured porous media. Thus, the goal of this chapter is describing the application of percolation concepts to the characterization of connectivity of fracture and fault patterns, and other properties.

5.2 Why Should Percolation be Relevant to Fractures?

Before embarking on the discussion of the main topic of this chapter, we address a basic question: Why should percolation concepts be relevant to fracture networks? To answer this question we appeal to what is usually referred to as the *critical-path analysis* (CPA). Later in this book, we will describe the CPA in more detail and discuss its applications to a variety of heterogeneous materials, including porous media. For now, it suffices to provide a qualitative description of the CPA and how invoking it justifies the relevance of percolation to characterization of fracture networks.

Ambegaokar et al. (1971) argued, and many sets of precise simulations confirmed, that transport processes, and in particular conduction, in a highly heterogeneous material can be reduced to one in a percolation system at or very near the percolation threshold. The idea is that a finite fraction of a strongly disordered material has very small conductivity (or permeability, diffusivity, etc.) and, hence, makes a negligible contribution to the overall conductivity or other flow or transport properties. Therefore, the low-conductivity regions can be safely eliminated from a model of such materials, which reduces the connectivity and transforms the model to one of a percolation medium. In other words, the effective conductivity, or other properties of the materials can be approximated by the conductivity of a very small part of them, which is usually composed of a small number of paths that contribute overwhelmingly to the effective property of interest.

If we adopt the CPA for description of fractured porous media, we arrive at the conclusion that a sample-spanning fracture network must have the morphology of the sample-spanning percolation cluster of fractures, because the permeability of the fractures is much larger than those of the pores, and represents the main conduits for fluid flow and transport.

Another way of understanding the relevance of percolation to fracture and fault networks is by appealing to the physics of fracture nucleation and propagation. Suppose, first, that a porous formation is only weakly heterogeneous. The initial stages of nucleation and propagation of fractures do not occur randomly, because once a fracture is nucleated, stress enhancement at its tip is larger than at any other point of the medium and, therefore, the next microcrack almost surely develops at its tip. As a result, the fracture looks almost like a straight line. But, large-scale porous media are characterized by a broad spatial distribution of the elastic moduli (see, e.g., Sahimi and Tajer, 2005). In that case, fracture growth in one or a few locations continues until a much stronger region is encountered, at which time fracture growth slows down, or stops altogether, in that region and another fracture nucleates in a weaker block in another part of the medium. The cracks also interact with each other, and eventually intersect. The new crack also stops growing when it encounters another strong block, and so on. Thus, it was proposed (Chelidze 1982; Madden 1983; Guéguen et al. 1991; Sahimi et al. 1993; Robertson et al. 1995; Renshaw 1998; Tafti et al. 2013) that, at large length scales nucleation and propagation of fractures in a highly heterogeneous media may be well approximated by a percolation process.

There is, however, no consensus about this issue. Barton (1995a, b), for example, argued, based on his extensive experimental work, that fracture networks are well connected. This implies that such networks are well above their connectivity threshold. This does not, however, imply that percolation concepts are not relevant to characterization of the connectivity of fracture networks, but that the universal power laws of percolation near the connectivity threshold, described in Chap. 2, may not be applicable. But, if we accept the premise that stress build-up is the driving force for the growth of the nucleated fractures, then, as soon as a sample-spanning cluster (SSC) of the connected fractures is formed, the built-up stress should be relieved. In that case, the driving force for continuing the growth of the fractures will be much weaker (or vanishes altogether), implying that the connectivity of fracture networks should be similar to that of a percolation network at, or at least very near the percolation threshold. It is, of course, clear that such a SSC of fractures does not necessarily have the structure of *random* percolation, but it does resemble some sort of an SSC at or near the critical threshold.

We will return to the relationship between percolation and rock fracture in Chap. 6, where we will describe the connection between earthquakes and fracture and fault distribution in rock. We will see that the connection provides even more convicning evidence of the relevance of percolation to describing fracture network of rock.

To characterize and quantify percolation properties of fracture networks, two of their fundamental properties should be defined:

(i) *Fracture density*: The areal fracture density is defined as the sum of fracture trace lengths per unit area. For an isotropic network, this is the same as fracture area per unit volume.

(ii) *Fracture connectivity*, which is similar to the coordination number or connectivity of a porous medium. The connectivity of fractures is characterized in terms of the ratio of three types of fracture termination: (a) a *blind termination* in which a fracture ends in the rock matrix, and is similar to a dead-end pore in a porous medium; (b) an *abutting termination* in which a fracture ends when it intersects another fracture; and (c) a crossing in which two fractures intersect, but neither is terminated.

5.3 Self-Similar Fractal Structures

As described in Chap. 2, the SSC at the percolation threshold p_c is a self-similar fractal object with well-defined fractal dimension D_f. Thus, one was of demonstrating the relevance of percolation to fracture networks is by showing that (a) a large class of fracture networks of rock consists of self-similar fractals and (b) their fractal dimension D_f is consistent with that of the SSC at p_c. This is, of course, a necessary, but not sufficient conditions, because two fractal structures can have more or less the same D_f, but have very different structures.

Here, we describe the fractal dimension in a more general and systematic way. A simple and quite general way of defining the fractal dimension D_f is by covering it by non-overlapping d-dimensional spheres of Euclidean radius r, or boxes of linear size r. The number $N(r)$ of such spheres that completely cover the system is then counted, and D_f is defined by

$$D_f = \lim_{r \to 0} \frac{\ln N}{\ln(1/r)}. \tag{5.1}$$

Equation (5.1) constitutes what is usually called the box-counting method. For non-fractal structures, $D_f = d$, with d being the Euclidean dimensionality of the space in which the structures are embedded. To use Eq. (5.1) for estimating the fractal dimension of a fracture network, $N(r)$ should be taken as the number of fractures of length r. Moreover, we should keep in mind that Eq. (5.1) assumes implicitly the existence of lower and upper cutoff length scales for the fractality of the structures. They are, respectively, the radius r of the spheres and the linear size L of the system. Recall also that, as described in Chap. 2, D_f may also be defined through the relationship between a system's mass M and its characteristic length scale L. Thus, if the system is composed of elementary objects of radius r and mass m, then,

$$M = cm(L/r)^{D_f}, \tag{5.2}$$

where c is a geometrical constant of order 1. Since the dependence of M on m and r is fixed, Eq. (5.2) implies that

$$M(L) \propto L^{D_f}. \tag{5.3}$$

In practice, one uses the average mass of an ensemble of samples with similar structures, rather than a single sample, in which case,

$$\langle M \rangle = cm(L/r)^{D_f}, \tag{5.4}$$

where $\langle \cdot \rangle$ indicates the ensemble average. If natural fracture networks are self-similar, it is only in a statistical sense, as their self-similarity is only in an average sense.

5.4 Experimental Evidence for Self-Similarity of Fracture Networks and Relevance of Percolation

The possible self-similarity of fracture networks has been studied extensively. The earliest of such studies is perhaps an investigation that began in 1985 by the United States Geological Survey, with the goal of characterizing the geologic and hydrologic framework at Yucca Mountain in Nevada (Barton and Larsen 1985; Barton et al. 1987; summarized in Barton and Hsieh 1989). The site was proposed by the United States Department of Energy as a potential underground storage area for high-level radioactive wastes. Barton and Larsen (1985) developed what is referred to as the *pavement method* in which one clears a subplanar surface and maps the fracture surface, in order to measure its various properties. Barton and co-workers demonstrated that fractured pavements are scale-invariant and fractal, with the distribution of the fractures's length, ranging from 20 cm to 20 m, being characterized by a single fractal dimension $D_f \approx 1.6-1.7$, estimated by the box-counting method. A more careful analysis of the same data by LaPointe (1988), including three fracture-trace maps, indicated that the 3D fracture networks have $D_f \approx 2.37, 2.52$, and 2.68, all of which differ from $D_f \approx 2.53$ for the SSC in random percolation at p_c by less than 6 percent.

Velde et al. (1991) analyzed fractal patterns of fractures in granites, while Barton (1992) studied 2D maps of fracture traces spanning nearly ten orders of magnitude, from microfractures in Archean Albites to large fractures in South Atlantic seafloors. Barton reported that, $D_f \approx 1.3-1.7$. Watanabe and Takahashi (1995) and Babadagli (2001) characterized the structure of geothermal reservoirs and their self-similar structure.

Fracture patterns in Stockbridge (near Canamn, Connecticut) dolomite marble were analyzed by Nolen-Hoeksema and Gordon (1987) and Chelidze and Guéguen (1990). They are highly branched, forming an interconnected network, whose fractal dimension reported by Chelidze and Guéguen (1990) was, $D_f \simeq 2.5$, practically identical with that of the percolation SSC, $D_f \approx 2.53$. Nakaya et al. (2003) analyzed 2D fracture patterns whose sizes were anywhere from 6.5 cm × 6.5 cm to 100 km × 100 km, and estimated the fractal dimension of the patterns. They also constructed the distribution of the fractures' lengths. Let $N_\ell(\ell)$ be the number of fractures with a length larger than ℓ. Then, the distribution is of power-law type:

$$N_\ell \propto \ell^{-a}. \tag{5.5}$$

Nakaya et al. looked for the conditions under which the networks are hydraulically connected, i.e., an SSC of connected fractures is formed, and found that only when $1.62 \leq D_f \leq 1.91$, the networks are connected. Note that $D_f \approx 1.62$ is the fractal dimension of the backbone (flow-carrying) part of the SSC in random percolation, whereas $D_f \approx 1.91$ is essentially the fractal dimension of the SSC itself, hence providing further evidence for the relevance of percolation concepts to characterizing the fracture networks.

5.5 Excluded Area and Excluded Volume

The connectivity of fracture networks is closely linked with another important concept, namely, the excluded volume (EV) (Balberg et al. 1984a). The EV is the volume that surrounds an object in which the center of another object must be in order for the two objects to be connected. A similar definition may be used for 2D objects, with the volume replaced by surface area. Thus, the excluded area $\mathcal{A}_e$ of an object is defined as the area around that object into which another similar object's center is not allowed if overlapping of the two objects is to be avoided. We must, however, point out that the EV is meaningful only if the objects' locations are uniformly distributed in space.

Consider, for example, identical objects. The following is a summary of $\mathcal{V}_e$, the EV for some simple cases:

$$\mathcal{V}_e = \begin{cases} \frac{2}{\pi}\ell^2 & \text{line segments in a plane,} \\ 4S & \text{disks in a plane,} \\ 2b\ell(1+4/\pi^2)+(2/\pi)(\ell^2+b^2) & \text{rectangles of length } \ell \text{ and width } b \\ \pi^2 R^3 & \text{disks in 3D space,} \\ 8V & \text{spheres,} \\ \frac{1}{2}\pi^2(e_1 l_1^2 l_2 + e_2 l_1 l_2^2) & \text{ellipses in 3D space,} \end{cases} \tag{5.6}$$

where ℓ is the line segment's length, R and S are the disk's radius and area, respectively, and V is the sphere's volume. The first equation is for the case in which the line segments, such as thin fractures, are randomly and uniformly oriented, whereas the second equation is applicable to the case in which the disks have identical orientations. The third equation was derived by Charlaix et al. (1984) for a random spatial distribution of disks. The last expression is due to de Dreuzy et al. (2000) for ellipses of eccentricities e_1 and e_2 and of major axis lengths l_1 and l_2. If the line segments' length or the radius of the disks is distributed according to some statistical distributions, then

$$\mathcal{V}_e = \begin{cases} \frac{2}{\pi}\langle\ell\rangle^2 & \text{line segments in a plane,} \\ \frac{\pi^2}{8}\langle R^2\rangle\langle R\rangle & \text{disks in 3D space,} \end{cases} \tag{5.7}$$

Balberg et al. (1984a) suggested another type of averaging given by

$$\mathcal{V}_e = \begin{cases} \frac{2}{\pi}\langle\ell^2\rangle & \text{line segments in a plane,} \\ \frac{\pi^2}{8}\langle R^3\rangle & \text{disks in 3D space.} \end{cases} \tag{5.8}$$

For elongated objects, the excluded area depends on the relative orientation of the objects. Hence, for a statistical distribution of objects, one should define an average excluded area that, for a network of fractures with length l and width b, is given by (Balberg et al. 1984b; Yazdi et al. 2011)

$$\mathcal{A}_e = 2bl(1+\frac{4}{\pi^2}) + \frac{2}{\pi}(l^2+b^2). \tag{5.9}$$

Ishihara (1950) developed a general method for randomly oriented objects. Suppose that 3D convex objects A and B have surface areas S_A and S_B, volumes V_A and V_B, and surface-averaged mean radius of curvatures $\mathcal{R}_A$ and $\mathcal{R}_B$. Then, the *mutual excluded volume* $\mathcal{V}_{AB}$ is given by

$$\mathcal{V}_{AB} = V_A + V_B + (S_A\mathcal{R}_A + S_B\mathcal{R}_B). \tag{5.10}$$

If Eq. (5.10) is averaged over the distributions of the objects' shapes and sizes, it will reproduce the result for spheres given by Eq. (5.6). If the objects are flat and convex, one obtains

$$\mathcal{V}_{AB} = \frac{1}{4}(S_A P_A + S_B P_B), \tag{5.11}$$

with P_A being the perimeter of object A. When Eq. (5.10) is averaged over the size distribution of the objects with identical shapes, one obtains

$$\mathcal{V}_{AB} = \frac{1}{2}\langle S\rangle\langle P\rangle, \tag{5.12}$$

which is a general result. A different expression, $\mathcal{V}_e = \frac{1}{2}\langle SP\rangle$, has also been suggested, and used sometimes.

5.6 Models of Fracture Networks and Their Percolation Threshold

Percolation threshold of disordered media is not a universal property and depends on their morphology. Therefore, to describe the connectivity of fracture networks, one must first discuss the various models of such networks. Although natural fractures are not, generally speaking, planar, in many models of fracture networks, the fractures have been represented by planar objects, which has proven to be useful.

5.7 Two-Dimensional Models

A large number of 2D models of fracture networks in which the fractures have finite extents have been developed (Dienes 1980; Long et al. 1982; Englman et al. 1983; Schwartz et al. 1983; Charlaix et al. 1984, 1987a; Endo et al. 1984; Robinson 1984a, b; Long and Witherspoon 1985; Long and Billaux 1987; Shimo and Long 1987; Guéguen and Dienes 1989; Hestir and Long 1990; Robinson and Gale 1990; Bour and Davy 1997, 1998; Darcel et al. 2003; Yazdi et al. 2011), in which fractures were represented by 1D finite line segments.

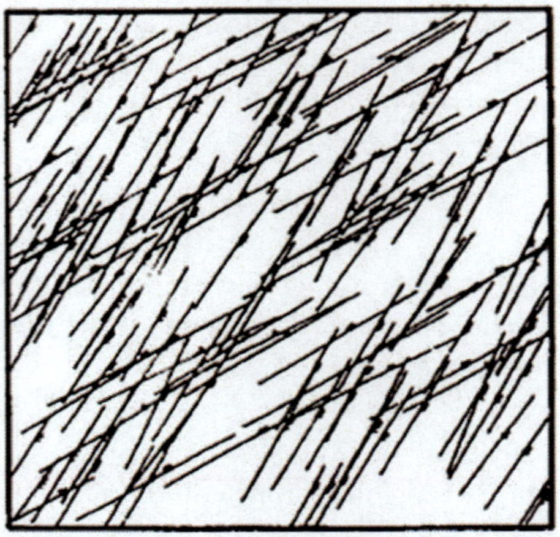

Fig. 5.1 A typical 2D Poisson fracture network in which the fractures are represented by lines of various lengths

In the earliest 2D models, the fractures (or their centers) were distributed at random in a plane. An example is the Poisson model (Long et al. 1982) in which one selects a square block of size $L \times L$ the (x, y) coordinates of the centers of a specified number of lines or fracture density from a uniform distribution in $(0, L)$. The orientations of the lines are also selected from a given distribution. Then, the lines are assigned randomly selected lengths. An example is shown in Fig. 5.1. Such models of 2D fracture networks are essentially the same as the percolation networks of sticks studied by Balberg et al. (1983, 1984b) and Balberg and Binenbaum (1983). The latter authors studied 2D anisotropic networks of conducting sticks that are presumably better suited for modeling of 2D naturally fractured porous media. Moreover, Balberg et al. (1984b) considered a 3D fracture network in which the fractures were finite cylinders of length L and radius r, and studied the dependence of the network's percolation threshold on the aspect ratio L/r and the macroscopic anisotropy.

The main difference between such networks of fractures and the usual percolation networks is that, in the latter case, an upper bound to all the properties is obtained when p, the fraction of conducting or open bonds, is unity. One also normalizes all the quantities of interest with their values at $p = 1$ because, otherwise, they would depend on the network's linear size. But, there is no theoretical upper bound to the degree of fracturing, or its density, implying that one cannot determine how "filled" a fracture network can be. Hestir and Long (1990) pointed out that one should hold λ_ℓ, the average frequency of fractures intersecting a sample line, constant, and study the properties of the network. Since any fracture network can be rescaled to a given constant λ_ℓ, the maximum network permeability is obtained when all the fractures are infinitely long.

To establish a precise one-to-one correspondence between a fracture network and a percolation one, one must relate the parameters of the former to those of the latter, such as p and $\langle Z\rangle$, the average coordination number or connectivity. Robinson (1984a, b) and Charlaix et al. (1987a) used ζ, the average number of intersections per fracture, as the measure of the connectivity. Suppose that the average length of the fractures is $\langle\ell\rangle$, their orientation distribution is $g(\theta)$, and the density of the fractures' center is ρ. Then, $\lambda_\ell = \langle\ell\rangle\rho$, and

$$\zeta = \lambda_\ell\langle\ell\rangle\Theta, \tag{5.13}$$

with

$$\Theta = \int_0^\pi \int_0^\pi \sin|\theta - \theta_0| g(\theta) g(\theta_0) d\theta d\theta_0. \tag{5.14}$$

Thus, if the fractures' orientations are uniformly distributed, $g(\theta) = 1/\pi$, then $\Theta = 2/\pi$. For every ζ, there is a corresponding $\rho(\zeta)$, the analog of p in the percolation networks. Therefore, a critical value of ζ, ζ_c, may also be defined, enabling one to rewrite all the results of percolation theory in terms of ζ, if we replace p everywhere in the percolation results with $\rho(\zeta)$, given a relation between the two.

As an example, consider Robinson (1984a, b) model that was analyzed by Hestir and Long (1990). In his model, $\rho(\zeta)$ was the average fraction of a fracture that is available for flow. Consider a fracture of length ℓ with $n_i(\ell)$ intersections. If ℓ is constant, then $n_i(\ell)$ will be a Poisson process. The average fraction of a fracture available for flow, i.e., the fraction between the two end sites separated by the distance ℓ, is $[n_i(\ell) - 1]/[n_i(\ell) + 1]$. Therefore, P_n, the probability that $n_i(\ell) = n$ is given by $P_n = (n!)^{-1}\zeta^n \exp(-\zeta)$, and

$$\rho(\zeta) = \sum_{n=2}^{\infty} \frac{n-1}{n+1} P_n = \left(1 + \frac{2}{\zeta}\right)[1 + \exp(-\zeta)] - \frac{4}{\zeta}. \tag{5.15}$$

Given Eq. (5.15), all the results of percolation theory described in Chap. 2 may be copied for Robinson's model by replacing p everywhere in percolation theory with ρ given by Eq. (5.15).

Experimental data for the distribution of fractures' length have been obtained using 1D, 2D, and 3D samples. For all the 1D data reported, which consisted of the trace lengths of fractures that intersect a scanline (Cruden 1977; Priest 1981), the length distribution was exponential. Since it is difficult to measure directly the fracture properties of 3D samples, most of the reported data are from 2D maps that consist of fracture traces in a plane intersecting a 3D fractured sample. In most cases, the distribution of fractures' length was reported to be of power-law type, Eq. (5.5). Hatton et al. (1993) analyzed extensive experimental data on fractured surfaces and found a power-law distribution for the length of the microcracks, Eq. (5.5), with $2 \le a \le 2.7$, where the numerical value of a was dependent upon the fluid content of sample, with its lower values corresponding to dry samples. Bonnet et al. (2001) compiled the existing data and the corresponding exponents. Most of the data indicated that $a \ge 2$, although smaller values have also been reported. Given the power-law distribution (5.5), it is clear that the percolation threshold of such fracture networks should depend on the distribution, since even a single long fracture can span the fracture network. Indeed, Berkowitz (1995) and Watanabe and Takahashi (1995), who studied 2D networks, found the percolation threshold to depend on the second moment of the fracture length distribution.

Since the fracture network of rock is often a self-similar fractal, given that the length of the fractures is not only not a constant but also often follows a power-law distribution, Eq. (5.5), the 2D models that have been described so far have been generalized (Darcel et al. 2003), in order to generate models of fractal fracture networks with a power-law distribution of the fractures' length. In such models, the fractal fracture network is generated by a multifractal process (Schertzer and Lovejoy 1987; Meakin 1991; Mukhopadhyay and Sahimi 2000), characterized by a fractal dimension D_f. Three connectivity regimes were then identified by Darcel et al. (2003).

(i) If $a < D_f + 1$, then the distribution of the fractures' length is dominant, and the fractal spatial distribution of the fracture has little effect. The network's macroscopic connectivity is determined by the largest fractures with lengths that are on the order of the system size.

(ii) If For $a > D_f + 1$, the network's connectivity is dominated by fractures that are much smaller than the system size, and the effect of the correlation is strong.

(iii) If $a = D_f + 1$, which corresponds to the transition between (i) and (ii), connectivity properties are scale-invariant. That is, the network's percolation threshold represents a critical density, with the average number of intersections per fracture at the threshold being scale invariant.

All the 2D fracture networks that described so far were represented by straight line segments with zero thickness. Yazdi et al. (2011) and Hamzehpour et al. (2021) generalized the model by representing the fractures by rectangles of length ℓ and width b. Though the generalization seems to be simple, it has a profound effect on the percolation, and flow and transport properties of the fracture network; see also Li and Östling (2013).

5.8 Three-Dimensional Models

Bour and Davy (1997, 1998) extended the 2D networks of sticks or line segments to 3D networks. Other than their work, 3D fractures have been represented mostly by either discs of finite radius (Warburton 1980; Long et al. 1985; Charlaix et al. 1987a; Billaux et al. 1989; Piggot 1997) or by flat planes of finite dimensions (Robinson 1984b; Wilke et al. 1985; Berkowitz 1995). In some cases, the fractures are very long so that they can be considered as essentially infinite, in which case a simple stochastic model based on the Poisson distribution, which is an extension of the 2D networks, may suffice.

A third class of models in which the fractures are represented by regular or irregular polygons has been developed by Adler, Thovert, and co-workers see (Adler et al. 2013, for a comprehensive discussion of the work of this group). The motivation for such models is that some experimental evidence indicates (Pollard 1976) that at least

Fig. 5.2 A 3D network of fracture, represented by intersection polygons (after Koudina et al. 1998)

some 3D fractures are either roughly elliptical or discs-shaped. Thus, Huseby et al. (1997, 2001) and Koudina et al. (1998) developed a model of fracture networks in which fractures are represented by convex polygons; see Fig. 5.2. They considered a model in which plane polygonal fractures are inserted in a simulation cell of linear dimension L. The normal vectors of the fractures are randomly and isotropically distributed, with the location of their center following a Poisson distribution. The contour of a fracture is a convex polygon, inscribed in a circle of radius R. The spatial location of a fracture is determined by the center of the circle that surrounds it. The vertices of the polygon are evenly (for a regular polygon) or randomly (for irregular polygons) distributed on the perimeter of the circle. The number of vertices of each fracture is selected randomly and uniformly in the interval $[3, N_{\max}]$. They are distributed on the disk's contour with angles that are distributed uniformly in $[0, 2\pi]$. The orientation of the polygon within its plane is distributed uniformly. Thus, the fracture network is completely characterized by the fractures' shape, the ratio L/R, and the number of fractures per unit volume.

One must also determine the fractures' intersection lines that are characterized by their end-point coordinates. In addition, the triple points, where three fractures intersect one another, must also be identified. Two other important aspects are checking whether the fracture network is sample-spanning to ensure the existence of macroscopic flow and transport, and identifying the connected part of the fracture network, because it simplifies the computations of flow and transport properties. If the matrix also contributes significantly to the flow or transport processes, identifying both the connected and disconnected parts of the fracture network is important.

5.9 Percolation Threshold of Fracture Networks

Consider, first, estimates of the percolation thresholds without considering the effect of the excluded volume. Assuming that a 3D fracture network consists of orthogonal families of equal squares of unit length, Robinson (1984b) defined a dimensionless density ρ of the fracture by

$$\rho = \frac{Nr^3}{L^3}, \tag{5.16}$$

where N is the number of fractures in the network of linear size L, and r is the half-width of the fractures (squares). Thus, the critical density at the connectivity threshold is defined by

$$\rho_c = \frac{N_c r^3}{L^3}, \tag{5.17}$$

where N_c is the number of fractures at the threshold. Robinson estimated that $\rho_c \approx 0.19$. If the orientations of the fractures are uniformly distributed, then Robinson (1984b) estimated that, $0.15 < \rho_c < 0.3$. Computer simulations of Robinson (1984b) also indicated that in 3D networks of planar fractures at their connectivity threshold, the number of intersections per plane is about 2, whereas, in 2D networks of fractures with constant length, the average number of fractures intersecting a given fracture at the percolation threshold is about 3.1.

If the fractures are polydispersed, and one takes into account the effect of the excluded volume, then (Charlaix et al. 1984) defined the critical density by

$$\rho_c = \frac{N_c \langle r^2 \rangle \langle r \rangle}{L^3}. \tag{5.18}$$

Balberg et al. (1984b), who first discussed the concept of excluded volume in the context of percolation networks, suggested that one must define ρ_c as

$$\rho_c = \frac{N_c \langle r^3 \rangle}{L^3}, \tag{5.19}$$

which is an intuitive generalization of Eq. (5.18). Charlaix et al. (1984) suggested that at the percolation threshold of a 3D fracture network that consists of equal flat disks of radius r with a density of ρ_c of disks per unit volume, one must have

$$\rho_c r^3 \sim 0.15 - 0.3. \tag{5.20}$$

We note that r^3 is proportional to the volume of a disk. Thus, as Eq. (5.18) suggests, for a fracture network consisting of polydispersed disks with a distribution of disks' radii, r^3 must be replaced by $\pi^2 \langle r^2 \rangle \langle r \rangle$ in Eq. (5.16), which is $1/2\langle\text{surface}\rangle\langle\text{perimeter}\rangle$, with $\langle \cdots \rangle$ denoting an average with respect to the distribution of the radii.

Table 5.1 Critical fracture density ρ_c for some 2D and 3D networks of objects, obtained by numerical simulations

Shape of the object	Spatial dimension	ρ_c
Orthogonal sticks	2	3.2
Randomly oriented sticks	2	3.6
Disks or parallel objects	2	4.5
Orthogonal elongated rods	3	0.7
Randomly oriented elongated rods	3	1.4
Orthogonal squares	3	2
Randomly oriented squares	2	2.46
Spheres or parallel objects	3	2.80

Using the excluded volume (or area), we define a (dimensionless) density ρ_d (Adler and Thovert 1999) by

$$\rho_d = \rho \mathcal{V}_e, \tag{5.21}$$

ρ_d is the number of objects per volume $\mathcal{V}_e$, with ρ being the volumetric density of the objects per unit volume. Given the definition of the excluded volume $\mathcal{V}_e$, Eq. (5.21) implies that ρ_d is the average number of intersections per object or fracture. Equation (5.21) is a clear definition of the fracture density, because it relates ρ_d to the shape of the fractures and their orientation distribution that specify $\mathcal{V}_e$, which contains information about both volumetric and connectivity properties.

At the same time, fracture networks at their connectivity threshold represent the real-world examples of continuum percolation described in Chap. 3. Therefore, determination of the critical fracture density is tantamount to estimating the percolation threshold of a continuum system. In Table 5.1, we list the percolation thresholds of various continuum systems that have been estimated by several research groups. Note that the estimates are based on Eq. (5.21).

For regular polygons with 3–20 vertices, as well as for rectangles with an aspect ratio of 2, Huseby et al. (1997) obtained a nearly constant percolation threshold:

$$\rho_c \simeq 2.26 \pm 0.04, \tag{5.22}$$

although extensive simulations of Yazdi et al. (2011), who studied 2D fracture networks of rectangular fractures of finite width b indicated that ρ_c depends on b. Note that

$$\mathcal{V}_e = \begin{cases} R^3 N_s^2 \cos(\pi/N_s)\sin^2(\pi/N_s) & \text{regular } N_s\text{-polygons,} \\ \dfrac{8R^3 A(A+1)}{(A^2+1)^{3/2}} & \text{rectangles with aspect ratio } A, \end{cases} \tag{5.23}$$

where R is the radius of the fracture (the radius of the circles in which the polygons are embedded), and N_s is the number of the sites or vertices of the polygons. If the rectangles do not have the same size, R^3 is replaced by $\langle R\rangle\langle R^2\rangle$. The percolation properties of polydispersed fracture networks of convex polygons, distributed randomly and isotropically, were studied by Mourzenko et al. (2005). We define a volumetric number density of fracture per fracture size, $F(R)$, by

$$F(R) = \rho n(R). \tag{5.24}$$

Mourzenko et al. (2005) studied the case with,

$$n(R) = \alpha R^{-\alpha}. \tag{5.25}$$

The polygons were assumed to have identical shapes, but different sizes. Then, their excluded volume is given by

$$\mathcal{V}_e = \frac{1}{2} v_e \left(R_1 R_2^2 + R_1^2 R_2\right). \tag{5.26}$$

Here, s_e is a dimensionless shape factor with, $s_e = \pi^2,\ 9\sqrt{3}/2$, and $4\sqrt{2}$ for, respectively, disks, hexagons, and squares. Then, for polydispersed fracture networks, the most relevant critical density is given by

$$\rho_c = \rho_c v_e \langle R^3\rangle. \tag{5.27}$$

If we consider a cubic domain of linear size L, together with, $L' = L/R_M$, where R_M is the maximum radius of the fractures. Then, large-scale simulations for several values of L', the parameter α of the distribution (5.25), and various polygon shapes, including hexagons, triangles, and squares, indicated that all the results fall in a narrow range. For example, for $L' = 4$, one has

$$\rho_c \simeq 2.95 \pm 0.12. \tag{5.28}$$

Note that ρ_c is defined by Eq. (5.27), not (5.21). Néda et al. (1999) and Sangare and Adler (2009) studied the percolation properties of isotropically oriented circular cylinders of various aspect ratios. Such cylinders are used as a model of thick fractures.

Berkowitz et al. (2000) proposed a general relation for the fracture connectivity, suggesting that, in order to include the effect of the length scale of the measurements, the resolution, and the distribution of the fracture lengths, one should write $n(\ell, L)$, the number density of fractures having a length in $[\ell, \ell + d\ell]$, as

$$n(\ell, L) = AL^{D_f}\ell^{-a}, \tag{5.29}$$

where D_f is the fractal dimension of the set of intersecting fractures. The length scale L (the linear size of the sample) is included because, as described earlier, large-scale fracture networks often possess fractal morphology. Berkowitz et al. (2000) used Eq. (5.17) to define the percolation threshold or critical fracture density ρ_c, and identified it as the point at which the fracture network is connected across the sample with probability of 1/2. For 3D random fractures in the form of ellipses of average eccentricity $\langle e \rangle$, de Dreuzy et al. (2000) suggested a slightly different expression,

$$\rho_c = \pi^3 N_c \langle e \rangle \frac{\langle \ell^3 \rangle}{L^3}, \tag{5.30}$$

where $\langle \ell^3 \rangle$ is the third moment of the distribution of fractures' length ℓ.

If the fractures' lengths are all fixed and much smaller than the linear size L of the network, then Eq. (5.17) is reasonable. If, however, the fractures can be as long as the system itself, then ρ_c would be the sum of the two terms that describe the contributions of small and large fractures (Bour and Davy 1997,1998). Assuming that the probability that the small-scale domains are connected by the presence of large fractures is small, $\rho(L)$, the fracture connectivity, is given by

$$\rho(L) = \int_{\ell_{\min}}^{L} \frac{n(\ell, L)\ell^{D_f}}{L^{D_f}} d\ell + \int_{L}^{\ell_{\max}} n(\ell, L)\, d\ell. \tag{5.31}$$

Note that L^{D_f} is the volume of the studied area. Here, $\ell_{\min}$ and $\ell_{\max}$ are, respectively, the minimum and maximum lengths of the fractures in the domain. Assuming that $\ell_{\max} \ll L$ and $p(L) = \rho_c$, and using Eqs. (5.21) and (5.31) is solved for the critical length L_c:

$$L_c = \left[(D_f - a + 1)(a - 1)\frac{\rho_c}{D_f A} + \ell_{\min}^{D_f - a + 1} \frac{(a-1)}{D_f} \right]^{1/(D_f - a + 1)}. \tag{5.32}$$

Equation (5.32) is valid for $a > 1$ and $a \neq D_f + 1$, and indicates that if $\ell_{\min}$ is sufficiently small, then, for $1 < a < D_f + 1$, the connectivity threshold is reached at system sizes that depend on the prefactor A that is related to the density of the fractures. Moreover, since $L_c > 0$, for $a > D_f + 1$, it follows from Eq. (5.32) that the connectivity threshold is reached when

$$A\ell_{\min}^{D_f - a + 1} > (a - D_f - 1) rho_c. \tag{5.33}$$

As pointed out earlier, if $a = D_f + 1$, then the fracture network is self-similar in the sense that the number of fractures having lengths that are of the order of the system size is constant. Hence, Eq. (5.32) yields

$$\ln\left(\frac{L}{\ell_{\min}}\right) \geq \frac{\rho_c}{A} - D_f^{-1}. \tag{5.34}$$

Because $L > \ell_{\min}$, inequality (5.34) is satisfied if $A > \rho_c D_f$. If, however, $A < \rho_c D_f$, then the connectivity is reached only for very large systems

$$L \geq \ell_{\min} \exp\left(\frac{\rho_c}{D_f} - D_f^{-1}\right). \tag{5.35}$$

Note that inequalities (5.34) and (5.35) indicate that the connectivity of a fracture network is controlled by the *smallest* fractures and, therefore, it can be estimated only through fracture maps with high enough resolutions that allow one to inspect the smallest fractures. In practice, $\ell_{\min}$ is the resolution length of the map and, therefore, varies with the observation length scale and, hence, with L. In other words, the connectivity depends on $L/\ell_{\min}$.

Thovert et al. (2017) studied percolation properties of 3D fracture networks with a wide range of fracture shapes, including disks and ellipses with an aspect ratio $f = 1 - 16$ (denoted by $f-$ellipses), regular polygons with $n = 3, 4, 6$, and 20 vertices, and elongated polygons with aspect ratios $f = 1 - 16$ The $f-$lozenges, $f-$ellipses, and $f-$rectangles belong to the class of the Lamé curves, or superellipses (Lamé 1818), described by

$$\left|\frac{x}{a}\right|^q + \left|\frac{y}{b}\right|^q = 1, \tag{5.36}$$

with $q = 1, 2$ and ∞, respectively, and aspect ration, $f = a/b$.

Pervago et al. (2018) studied the percolation threshold of models of fractured porous media consisting of vugs and fractures represented by, respectively, elliptical and spheroidal elements. They approximated the pore shapes by ellipses (in 2D) and spheroids (in 3D) and varied their aspect ratios, and developed different types of models, varying from vugs (spheres), to fractures (oblate spheroids) and channels (prolate spheroids). They calculated the critical percolation porosity for a network consisting of elements with constant shapes; a network composed of elements with the uniform logarithmic distribution of aspect ratios, and a network containing elements of two different shapes.

Xu et al. (2019) studied percolation of randomly orientated congruent polyhedral objects that are used to model pores, particles, and fractures, including regular tetrahedral, cubic, octahedral, dodecahedral, and icosahedral pores, usually referred to as the Platonic objects. Using numerical simulations, they obtained precise estimates of the percolation critical fraction ρ_c and percolation transition width Δ of Platonic objects. Their results indicated that ρ_c and Δ ascend in the order of tetrahedra < cubes < octahedra < dodecahedra < icosahedra, and that ρ_c and Δ increase with an increasing number of faces of polyhedral objects close to spheres.

Chapter 6
Earthquakes, Critical Phenomena, and Percolation

6.1 Introduction

Earthquakes are one of the most complex phenomena in nature. Their occurrence, especially when their magnitude is large, has important economical and societal consequences. Thus, understanding factors that contribute to an earthquake and predicting when and where it may occur have been problems of fundamental importance for hundreds of years. But, although the first documented earthquake occurred in 1177 B.C. in China, it was only in the nineteenth century that an understanding of how earthquakes develop and occur began to emerge.

The question that is asked most frequently is, are earthquakes predictable? If by "predicting" we mean describing how an eathquake comes about, then we already have a predictive theory, the *plate tectonics paradigm*, or hypothesis, according to which the Earth's outer crust, the lithosphere, consists of many plates that slide over its asthenosphere, a hot lubricating layer. Sometimes, the plates slide by and brush against each other. In the region between the two plates, which is usually referred to as a *transform boundary*, confined energy is accumulated in the rock. A fault line in the Earth's crust, where blocks of crust move in various directions, also forms. Most, if not all, earthquakes happen along transform boundary fault lines within the plates' interior. The paleoseismological dating of large prehistoric earthquakes has confirmed the hypothesis.

If, however, what we mean by "predicting" is the time at which a large earthquake may happen, then, the problem is still unsolved, despite the fact that many theories and "predictive" methods have been proposed. Two of such approaches are mentioned here briefly, because they are completely different from what we describe in the rest of this chapter, and they exhibit some predictive power.

In one approach, developed by Rahimi Tabar et al. (2006), precursory seismic data before an earthquake are treated as a Markov process, and the background noise is distinguished from the actual fluctuations due to an impending earthquake. Analyzing the time series for the vertical velocity of the Earth, Rahimi Tabar et

M. Sahimi, *Applications of Percolation Theory*, Applied Mathematical Sciences 213,
https://doi.org/10.1007/978-3-031-20386-2_6

al. showed that a short time before an earthquake the Markov time scale t_M of the series increases sharply, hence providing an alarm for an impending earthquake. To distinguish a false alarm from a reliable one, Rahimi Tabar *et al.* computed a second quantity, T_1, for the time series $V(t)$ of Earth's vertical velocity. They showed that T_1 also changes strongly before an earthquake occurs. To compute T_1, a time-dependent function $S_p(\tau)$, which represents a structure function of order p and is defined by

$$S_p(\tau) = \langle |V(t+\tau) - V(t)|^p \rangle \sim \langle |V(t+\tau) - V(t)|^3 \rangle^{\zeta_p} , \tag{6.1}$$

is first computed based on the extended self-similarity (ESS) (Benzi et al. 1996; Bergman 2002) of the seismic time series, where τ is the lag (in units of the data points). Since the first nontrivial moment (beyond the average and variance) of a distribution is S_3, and because for a Gaussian process, $\zeta_p = \frac{1}{3}p$, while deviations from this relation represent non-Gaussian behavior, S_3 plays a prominent role in the EES method. It is also well known that the moments S_p with $p < 1$ contain information on frequent events in a time series.

Prior to an earthquake the number of frequent events (development of cracks that join up) suddenly rises, indicated by a sudden change in S_p with $p < 1$. Therefore, close to an earthquake the quantity, $T_1 = [S_p(\tau = 1)^2 + S_3(\tau = 1)^2]^{1/2}$, suddenly changes and provides a second alert. The utility of T_1 is due to the fact that, it can be accurately estimated with very few data points, say 50, hence enabling online analysis of the data collected over intervals of about one second. Both the Markov time scale t_M and $T_1(t)$ are computed online as the data come in, and an alarm for an earthquake is accepted if *both* t_M and T_1 indicate it *simultaneously*. Rahimi et al. checked the accuracy of the method against hundreds of past earthquakes and demonstrated that its failure rate is very low.

In the second approach, proposed by Manshour et al. (2009, 2010) the Earth's vertical velocity time series was also analyzed by using methods that had been originally developed for complex hierarchical systems, and in particular for turbulent flows. Analysis of the fluctuations of the detrended increments of the time series by Manshour et al. revealed a pronounced change of the shapes of their probability density functions (PDF). Before and close to an earthquake the shape of the PDF and the correlations in the increments both manifested significant changes. For a moderate or large earthquake, the typical time at which the PDF underwent the transition from a Gaussian to a non-Gaussian was about 5–10 hours, hence representing a new precursor for detecting such earthquakes.

Aside from the two aforementioned approaches that are completely different from all other theories for predicting earthquakes, another idea was based on the hope that the intervals between large earthquakes are quasi-periodic, so that knowledge of the time of some of the past large earthquakes in a particular area may lead to prediction for the time of the upcoming ones. For example, analyzing data for over a century, (Sieh et al. 1989) reported that the average recurrence interval for the last ten large earthquakes on the San Andreas Fault north of Los Angeles is about 132 years. But, the average time interval is misleading, because the actual intervals between the earthquakes range from 44 to 332 years, so that the average recurrence time has no predictive power. Thus, the hope for quasi-periodicity of the times of large earthquakes has been more or less dashed.

There was also hope that one can identify physical observable precursors very close to large earthquakes. Although still subject to some debate, precursory anomalies, such as changes in the seismic recordings, and anomalous variations in the chemical, hydrological, mechanical, and electromagnetic properties of the area in which earthquakes occur, usually precede a large earthquake (Scherer 1990; Keilis-Borok and Soloviev 2002; Sornette 2004). One school of thought believes that the anomalies occur within days to weeks before the main event, but probably not much earlier (Jones and Molnar 1979), and that the spatial precursory patterns develop at short distances from impending large earthquakes. For example, an attempt was made to utilize the fact that, in laboratory experiments on rock samples under compression, microfracture damage and the associated dilatancy occur before the macroscopic shear failure of the sample happens. But, although the Haicheng earthquake of magnitude of $M = 7.3$ in China, which occurred on 4 February 1975, was successfully predicted, the same method failed to predict other large earthquakes around the world and, thus, that hope was also dashed (Turcotte 1991).

6.2 Earthquakes as a Critical Phenomenon

As mentioned above, over the past three decades attention has turned to analyzing seismicity data with the hope that the results can identify precursors to large earthquakes. Such approaches attempt to use statistical physics of disordered media, and are based on interpreting regional seismicity as a characteristic of a system at or near a statistically stationary dynamical critical point. The approach is called *self-organized criticality* by Bak[1] et al. (Bak et al. 1987; see also Bak and Tang 1989; for simple and illuminating discussions see, Bak 1996).

A self-organized critical state is characterized by power-law distributions of event sizes, as well as long-range spatial correlations of fluctuations around the stationary state. The link between self-organized criticality and earthquakes was motivated by the fact that the latter is also characterized by several power laws, such as the Gutenberg–Richter frequency–magnitude relation for earthquakes, and the fractal distribution of hypocenters (see below). We do not, however, discuss this approach in this book, not only because the subject is still controversial but also because it is not directly related to percolation theory that we describe and discuss here. We refer the interested readers to the paper by Bak et al. (2002). Note also that there is a link between self-organized criticality and directed percolation (Vaźquez and Costa 1999; Basu and Mohanty 2014), described briefly in Chap. 3, which will be studied in detail in Chap. 21.

[1] **Per Bak** (1948–2002) was a Danish theoretical physicist who made important contributions to the physics of complex systems, including theory of self-organized criticality.

The second school of thought proposes that large system-wide events give rise to large perturbations that move the system away from the statistically stationary dynamical critical point. In this picture, the precursory anomalies may occur up to *decades* before large earthquakes, at distances much larger than the length of the main shock rupture. Thus, due to their strong memory of past events, predicting large earthquakes may become more likely. The concept is closely linked to the classical theory of critical points (Scherer 1990; Knopoff et al. 1996; Sobolev and Tyupkin 2000; Keilis-Borok and Soloviev 2002; Sornette 2004), which was originally advocated by (Keilis-Borok and Malinovskaya 1964), with a report documenting the existence of long-range correlations in the precursors.

Sornette and Sammis (1995) proposed that large earthquakes represent critical points, as they are the culmination of a nonstationary precursory activity that accelerates up to the critical point. This is completely different from the self-organized criticality in which all events belong to the same global population and contribute to the shaping of the self-organized critical state. In the critical point view, on the other hand, a large earthquake is a special point—just as percolation threshold is a special point—which plays a special role, representing the end of a cycle on the associated fault network. Because of this, the critical-point model of earthquakes has a much better chance of being predictive, and can be implemented by monitoring the approach of the fault network toward a critical state. This is also similar to the approach of propagating fractures to the critical point at which the fracture network becomes sample-spanning and percolating. Sahimi and Arbabi (1992) showed that regardless of the initial state, at the critical point of the sample-spanning fracture network one has a *universal* Poisson's ratio.

As described by Sammis and Sornette (2002), the basic tenets of the critical-point model for regional seismicity are as follows:

(i) A large earthquake can occur only when the crust has reached a critical state. After an event is nucleated, it can grow by jumping geometrical and rheological barriers, if highly stressed patches are correlated at the scale of the fault network.

(ii) By destroying stress correlations, a large earthquake moves portions of the network out of the critical state. Destruction of the correlation leads to the observed shadow zones for intermediate-size events. Seismic shadow zones are areas of the Earth's surface in which seismographs cannot detect an earthquake after the waves have passed through the earth. More specifically, the seismic shadows represent the effect of seismic waves that strike the core–mantle boundary. The P (primary) and S (secondary or shear) waves radiate spherically away from an earthquake's focus in all directions and return to the surface by many paths. The S waves do not, however, reappear beyond an angular distance of about 103°, because they are stopped by the liquid, while due to refraction at the mantle–core boundary, the P waves do not arrive between 103° and 140°.

(iii) Long-range stress correlations are developed again by tectonic loading combined with the stress transfer from smaller events. This makes the next large event possible, and the approach to criticality can be monitored by using regional seismicity.

Two points are worth emphasizing (Sammis and Sornette 2002): (a) As described, the critical-point approach does not predict that the seismic cycle, as defined, is periodic. In fact, the variations in recurrence intervals of large earthquakes are due to details of the stress transfer from smaller events, as well as those from the nucleation process. They both depend strongly on the heterogeneity of rock. (b) The critical-point approach does not also claim that a large earthquake *must* occur when the system reaches the critical state, but only that a large earthquake is *possible*, when that state has been reached.

Various methods for monitoring the approach to criticality have been proposed. As the critical point is approached, the correlation length increases, and Zöller et al. (2001) estimated it directly by using the earthquake catalog and single-link cluster analysis. In this type of analysis (see, for example, Everitt (2011)), also called agglomerative clustering, one groups the clusters (in our case, the clusters of earthquakes) in a bottom-up fashion, and at each step combines two clusters that contain the closest pair of elements that do not yet belong to the same clusters in the pair. The method has the advantage of working with smaller events, thus improving the statistical significance of the results. The second approach exploits the fact that growing correlation is associated with increasingly larger events and, thus, monitoring the sizes is tantamount to observing the growing correlation length (Sykes and Jaumé 1990).

To better quantify the effect of the growing size of the intermediate earthquakes on the path to large ones at the critical point, Bufe and Varnes (1993) proposed the following equation for the so-called "cumulative Benioff strain,"

$$S = \sum_{i=1}^{N(t)} E_i^{1/2} = A + B(t - t_c)^m \ , \tag{6.2}$$

where E_i is the energy of the ith event, t_c is the time that the system reaches the critical state in which a large event becomes possible, $N(t)$ is the number of events at time t, and A, $B < 0$, and $m > 0$ are constant. Using arguments from the renormalization group theory of critical phenomena, Sornette and Sammis (1995) suggested that there are significant correction-to-scaling terms to power law (2) (see Chap. 2), and that the corrections are *log-periodic*, so that one has the following scaling law:

$$S = A + B(t_c - t)^m \left\{ 1 + C \cos\left[2\pi \frac{\log(t_c - t)}{\log D} + E \right] \right\} \ , \tag{6.3}$$

where C, D, and E are also constants. Such correction terms arise if the critical exponents are complex numbers. In Eqs. (6.2) and (6.3), the time t is made dimensionless with some suitable variable or time scale. In practice, one should fit the data to Eq. (6.3) to estimate the various parameters and, in particular, t_c. Sornette and Sammis (1995) demonstrated that Eq. (6.3) does have some predictive power for some large earthquakes.

If large earthquakes do represent a critical-point phenomenon, then one may be led to the idea that criticality is caused by failure in the Earth's crust that can be thought of as a scaling-up process in which failure at one scale is part of damage accumulation and creation of fractures at a larger scale. Hence, Eq. (6.3) should also be observed for any measure that characterizes nucleation and propagation of fractures in heterogeneous rock. Using extensive simulation of fracture propagation in a model of heterogeneous rock, Sahimi and Arbabi (1996) demonstrated that this is indeed the case and that, near the macroscopic failure point, the cumulative elastic energy released during formation of fractures in heterogeneous rock does follow an equation similar to Eq. (6.3). There are also some experimental evidence for Eq. (6.3), not only for earthquakes (Mastrogiannis et al., 2018) but also for icequakes (Faillettaz et al. 2017).

If we accept the notion of occurrence of large earthquakes as a critical-point phenomenon, and given that, as described in Chap. 5, the fracture network of heterogeneous rock has the structure of a percolation cluster, then, the connection between percolation and earthquakes becomes plausible. This is the subject of this chapter. Let us also note that Newman and Turcotte (2002) proposed a model that combined elements of critical-point view with the approach based on self-organized criticality.

6.3 Spatial Distribution of Earthquakes Hypocenters and its Relation with Percolation

A quantitative connection between the spatial distribution of earthquakes' epicenters or hypocenters and percolation was first proposed by Sahimi et al. (1993). They computed the fractal dimension of the cluster that consisted of the spatial distribution of the earthquakes' epi- and hypocenters by analyzing four seismic datasets from four distinct regions in California. For each set, the fractal dimension D_f of the set of the centers was estimated by counting the minimum number of cubes N_c of edge length L to completely cover the three-dimensional dataset at various length scales L, where L was taken to be the location error of the set. Figure 6.1 presents their results, from which one obtains $D_f \approx 1.8$ for *all* four cases.

What is the interpretation of the estimated value of D_f? One interpretation is that since D_f is close to 2, it implies that most tectonic events occur on a subset of near-vertical faults, because they have lower normal stress. There is, however, an alternative interpretation, first proposed by Sahimi et al. (1993). They noted that geologic features are rarely simple planes, and in fact, most faults are comprised of many shear fractures that are visible at a number of scales. Therefore, given that fracture network of rock at large scales appears to be similar to the largest, or the sample-spanning percolation cluster at the percolation threshold p_c (see Chap. 5), the fault patterns should also be similar to the same percolation clusters, which have a fractal dimension $D_f \approx 2.53$ in three dimensions. The vast majority of earthquakes are distributed on the regional fault networks. However, the hypocenters must belong

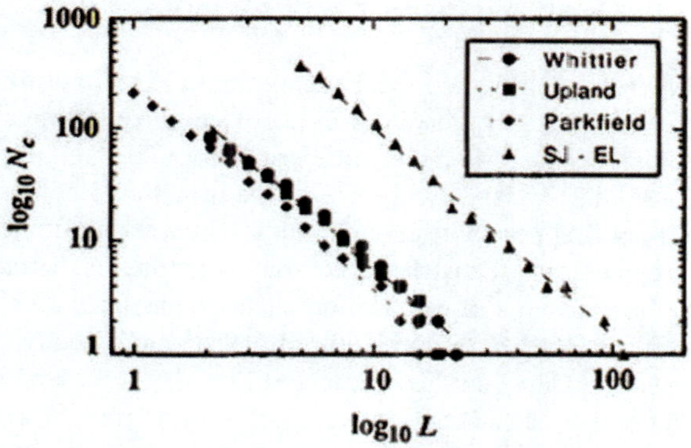

Fig. 6.1 Dependence of the minimum number of occupied cube N_c on the edge length L (in kilometers) to completely cover the dataset for four sets of data for the spatial distribution of earthquakes epicenters in California. The slope of all the plots is ≈ 1.8 (after Sahimi et al. 1993)

to the *active* part of the network, i.e., the part where finite strain deformation can occur. The most straightforward way for this to happen is if the hypocenters are on the *backbone* of the fault network, i.e., the multiply-connected part of the percolation cluster that allows stress transport to occur (Chap. 2). Indeed, the estimate, $D_f \approx 1.8$ is in agreement with the fractal dimension $D_{\rm bb} \approx 1.87$ of three-dimensional percolation backbone (see Chap. 2). Moreover, the fact that the four datasets yielded the same estimate of D_f is an indication that it is a universal property.

Following up on the work reported by Sahimi et al. (1993), Robertson et al. (1995) estimated the fractal dimension of the spatial distribution of the hypocenters of several aftershock sequences in south and central California, and reported D_f to be between 1.82 and 2.07, with an average of about 1.95. This is again very close to the fractal dimension $D_{\rm bb}$ of the percolation backbone.

As a further check of the hypothesis proposed by Sahimi et al. (1993), Nakanishi et al. (1993) carried out extensive simulation of diffusion in the clusters formed by the earthquakes' hypocenters, using random walk. Suppose that $\langle R^2(t)\rangle$ is the mean-squared displacement of the random walkers in the cluster at time t. Since the cluster of the earthquakes hypocenters is a fractal object, then one expects to

$$\langle R^2(t)\rangle \propto t^{2/D_w} , \tag{6.4}$$

where D_w is the fractal dimension of the random walk in the cluster, and is related to the percolation exponents given in Chap. 2 by

$$D_w = 2 + \frac{\mu - \beta_{\rm bb}}{\nu} , \tag{6.5}$$

with ν, $\beta_{\rm bb}$, and μ being, respectively, the exponents that characterize the power laws for the percolation correlation length, the fraction of sites in the backbone, and the conductivity near the percolation threshold. Using estimates of the exponents for 3D percolation networks, $\nu \approx 0.88$, $\beta_{\rm bb} \approx 1.05$ and $\mu \approx 2$, we obtain $D_w \approx 3.08$, so that

$$\langle R^2(t)\rangle \propto t^{0.65} , \tag{6.6}$$

which is in agreement with the numerical simulation of Nakanishi et al. (1993).

More recently, Pastén et al. (2011) analyzed the spatial distributions of hypo- and epicenters of earthquakes in central Chile. They reported that, $D_f \approx 2.02 \pm 0.05$ and 1.73 ± 0.02, respectively, which are to within 5 percent of the fractal dimensions of 3D and 2D percolation backbones. Thus, the notion that earthquakes' hypocenters or epicenters are distributed on the backbone of the fault networks, which is similar to the backbone of percolation clusters, has been receiving more support.

Note that Kagan and Knopoff (1981) had already studied the spatial distribution of earthquakes, but had estimated that, $D_f \approx 2$, which they interpreted as meaning that earthquakes occur on *planes*. They reported, however, that if they explore larger depths and areas, the estimate of D_f would decrease. On the other hand, if the hypocenters analyzed by (Sahimi et al. 1993) are plotted in three-dimensional space and rotated around all three axes, there does not seem to be any one plane about which the earthquakes cluster. The San Andreas–Elsinore (SA-EL) earthquakes are located between two major faults and over a very wide zone of deformation, whereas the Parkfield earthquakes are distributed over a relatively thin zone of deformation along the SA faults, yet they both have $D_f \approx 1.8$. There usually exists a large length scale cutoff in the three-dimensional distribution of earthquakes hypocenters, namely, the brittle crust length beyond which we may expect a crossover to a quasi-two-dimensional distribution. However, even in the SA-EL case, where the earthquakes are distributed over a very wide region, such a crossover is not evident.

6.4 Microseismicity and Percolation

An important field in geophysics that has received increasing attention is microseismicity. During injection of a fluid into porous formations, such as when a fluid is injected into shales in order to create fracture and allow the formation to produce, or when water is injected into a geothermal reservoir to extract vapor for use in power generation, small-magnitude seismic events, called microseismic events, often occur and are recorded. The magnitude M of such events is small, $M \leq 3$. The nature of the induced microseismic events that result from fluid injection into porous formations is, of course, different from the tectonic events described so far. It has been used for characterization of fracture network of rock (Aminzadeh et al. 2013) and the growth of hydraulic fractures during injection.

Tafti et al. (2013) were the first to suggest a connection between the spatial distribution of microseismic events and percolation clusters. They used the catalogs provided by the Lawrence Berkeley National Laboratory and online datasets from the Northern California Earthquake Data Center for an area in the northwest region of the Geyser geothermal field (GGF), one of the largest of its kind in the world. Injection of cold water into the GGF induces microseismic events, with the locations of the hypocenters recoded. Figure 6.2 present the distribution of the event sizes M.

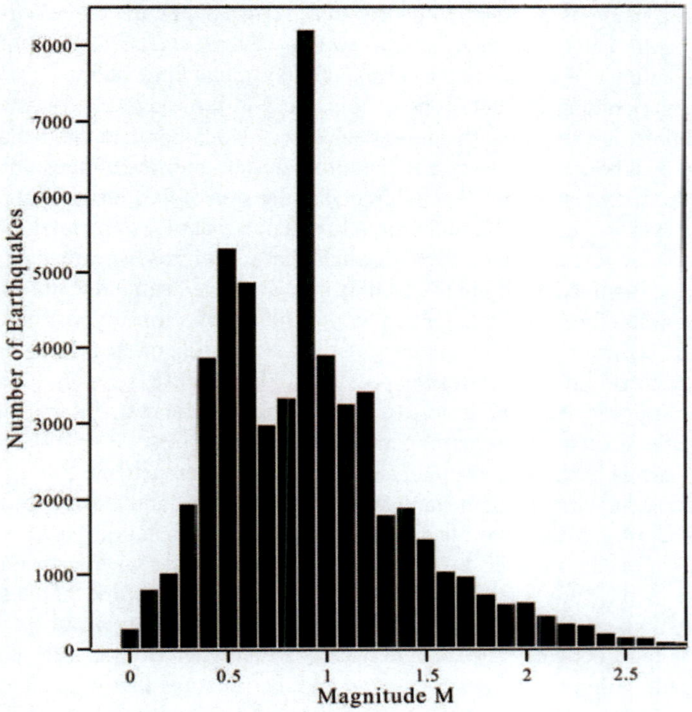

Fig. 6.2 Distribution of earthquakes' magnitudes in the Geysers geothermal field

Based on the seismic activity of the area and the location of the injection wells, three clusters that consisted of the spatial locations of the hypocenters of the microseismic events were selected and analyzed. One of the clusters was then divided into four subclusters, and each of them was also analyzed in order to delineate the possible finite-size effects.

A correlation function $C(r)$, introduced by Hirata et al. (1987) (see also Hirata 1989), was defined by

$$C(r) = \frac{2}{N(N-1)} N_r(R < r) , \tag{6.7}$$

where $N_r(R < r)$ is the number of pairs of events that have a spacing R less than r, and N is the total number of events within the region of interest. If the spatial distribution of the microseismic events has a fractal structure, $C(r)$ would follow a power law:

$$C(r) \sim r^{D_f} . \tag{6.8}$$

The fractal dimension D_f defined by Eq. (6.8) is also called the *correlation dimension*. The entire cluster of the events, as well as several smaller ones, as described earlier, were then analyzed, and the corresponding fractal dimensions D_f were computed. All the computed fractal dimensions were in the range $2.51 \le D_f \le 2.60$. These estimates are significantly larger than 2, and also larger than the fractal dimension of the backbone of the largest percolation clusters in three dimensions.

Recall from the last section that the fractal dimension of the cluster of hypocenters of tectonic events is close to 2, and probably equal to that of the fractal dimension of the backbone of percolation clusters. The fact that for microseismic events, $D_f > 2$ confirms that seismic activity in the GGF is likely to have been induced by the injection of cold water into the formation, rather than being of tectonic type. Therefore, the estimates of D_f provide significant insight into the structure of the fracture networks, as well as their origin. It is, therefore, possible to directly use the spatial distribution of microseismic events' hypocenters to map out the fracture network of a large-scale porous medium, such as geothermal or oil and gas reservoirs.

The estimates of the fractal dimension D_f for the microseismic events are completely consistent with that of the largest percolation cluster at the percolation threshold, $D_f \approx 2.53$, but they are not close to that of the percolation backbone. Thus, microseismic events induced by injection of a fluid, such as water or a chemical compound, into a porous formation happen on a fracture network that is similar to the largest 3D percolation clusters, whereas the tectonic events occur on the backbone of the fault network. A plausible reason is that when fluid is injected into the GGF, its path within the porous formation and the fractures that it generates within the rock are, due to the heterogeneity of the formation, a more or less random. Even if the path is not random, but contains extended correlations, the structure of the cluster at the largest length scale should still resemble that of a percolation cluster. The high-pressure fluid generates some fractures that are dead end, because the growth of such fractures stops when the pressure of the water cannot overcome the resistance offered by the rock. As a result, the network generated by the injection contains both dead-end, as well as multiply-connected fractures, i.e., it is the largest or sample-spanning percolation cluster.

In the next section more evidence for this picture of the tectonic versus induced seismicity is presented.

6.5 The Gutenberg–Richter Law and its Relation with Percolation

The distribution of energy released during earthquakes has been shown to follow the well-known Gutenberg–Richter (GR) law (Gutenberg and Richter 1956). The GR law was originally suggested as an empirical relation for the number of earthquakes of size greater than M and is given by

$$\log_{10} N_{\mathrm{GR}} = a - bM , \tag{6.9}$$

where N_{GR} is the cumulative number of earthquakes with a magnitude greater than M that occur in a specified area and time, and a and b are constants. The GR law is valid for earthquakes both regionally and globally. The constant b, which is usually referred to as the "b value" in the earthquake literature, varies from region to region but is typically very close to 1.0 (Frohlich and Davis 1993). The constant a is a measure of the regional level of seismicity.

Although the GR law was originally developed as an empirical relation, it is now recognized to be part of a broad range of natural phenomena that exhibit fractal scaling (Turcotte 1997). Thus, fractal scaling implies that

$$N_{\mathrm{GR}} = mS^{-b} , \tag{6.10}$$

where N_{GR} is the cumulative number of earthquakes with rupture areas greater than S that occur in a specified area and time, and m is a constant. Aki (1981) proposed an important relation between the fractal dimension D_f of a fault network and the b value of the GR law. If during an earthquake slip scales with the area of the active fault plane, then the Aki relation is given by, $D_f = 3b/c$, where c is a scaling constant that has a worldwide average of about 1.5 (Kanamori and Anderson 1975). Thus, the Aki relation simplies to $D_f \approx 2b$. Since, for earthquakes with a tectonic origin D_f is very close, or equal, to the fractal dimension of the backbone of percolation clusters, $D_f \approx 1.9$, the implication is that the b value should be about 0.95, in agreement with a vast amount of data that indicate (Frohlich and Davis 1993) that b is always nearly 1.0. This observation may connect the GR to the percolation statistics.

An equation similar to (6.9) also holds for microseismic events. Figure 6.3 presents the results for the data analyzed by Tafti et al. (2013). Analyzing eight sets of data, Tafti et al. (2013) estimated the b values to be 1.31 ± 0.04. Given that, as described earlier, the fractal dimension of the spatial distribution of the microseismic events is, $D_f \approx 2.55 \pm 0.05$, the b values for microseismic events also follow the Aki relation, and supports the proposal that the fractal dimension of the spatial distribution of induced microseismic events should be close or equal to that of the sample-spanning percolation cluster at the percolation threshold.

The energy E released during earthquakes (of tectonic origin) is believed to be related to M by

$$\log_{10} E = c - dM , \tag{6.11}$$

so that the GR law is essentially a power law connecting the frequency distribution of earthquakes with the energy release E (and, presumably, other physical quantities). If we combine Eqs. (6.9) and (6.11), we obtain

$$\log_{10} N = p - q \log_{10} E , \tag{6.12}$$

which is a power law expressing the fact that the larger the earthquakes, the fewer their occurrences, where p and q are two constants. Equation (6.12) is reminiscent of the

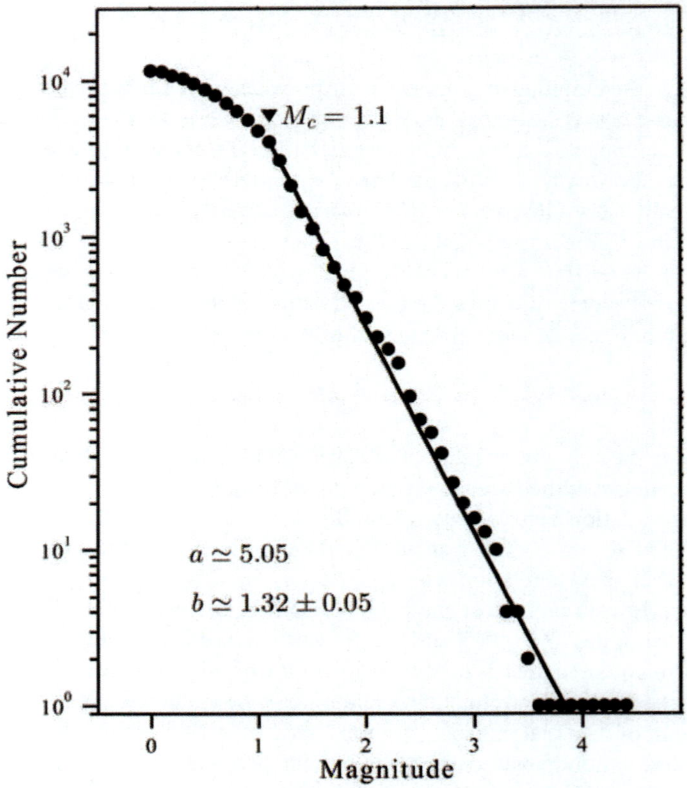

Fig. 6.3 The frequency–magnitude plot for extracting the b value for the entire seismicity catalogue of the Geysers geothermal field (after Tafti et al. 2013)

relationship between n_s, the number of percolation clusters of size s, in percolation theory described in Chap. 2, and is used again in the next section.

6.6 Percolation Model of Regional Seismicity

Otsuka (1971, 1972) and Vere-Jones (1976) appear to be the first to attempt relating percolation properties with seismicity. Scholz (1968) had already presented similar ideas, but with no explicit reference to percolation. Otsuka argued that the surface matter of the Earth's crust is always somewhat distorted, and that earthquakes occur when this distortion exceeds a certain threshold value. Moreover, he argued that each unit volume of rock contains an energy distortion (preceding an earthquake) of $E_d \approx 2 \times 10^4$ erg/cm^3 and, thus, he proposed that numerically $E \simeq V$, where V is the volume of the crust about to produce an earthquake. In this way, he drew a closer analogy between the occurrence of earthquakes and percolation, since now Eq. (6.12) can be rewritten in terms of the number and "volume" associated with the clusters of the severely distorted Earth's surface matter.

Using extensive simulations, Otsuka studied a growth model in which lattice sites were filled stochastically in sequence, with the clusters of filled sites being taken to represent rupture areas on the fault plane. The role of the probability p of percolation is played by the probability that a particular piece of the crust is severely distorted, which must depend on the specific sample region and the geological conditions during the period of observation. In particular, the power law of earthquake clusters will *not* be observed unless p takes on the values of the percolation threshold p_c. By comparing the earthquake frequency data from various regions of the world, Otsuka demonstrated that this was consistent with the data, i.e., power law (6.12) is an oversimplification. His results do seem to indicate that certain regions, e.g., northern Japan, correspond to $p < p_c$, whereas other regions, e.g., South America, perhaps correspond to $p > p_c$. Every earthquake is, of course, restricted to a finite region and, thus, no earthquake cluster can be considered as being infinitely large, even if $p > p_c$. Otsuka's quantitative results are not, however, directly comparable with percolation characteristics described in Chap. 2, as he did not use proper normalization; did not use suitable quantities to plot, and used a Bethe lattice of coordination number $Z = 3$.

Otsuka's simulation model became known in Japan as the *go-game model*. Saito et al. (1973) solved exactly a simple case of the "go-game model," which produced a b value of ≈ 0.75. Vere-Jones (1976) proposed independently a branching process model for crack formation, which also produced, $b \approx 0.75$. As pointed out by Vere-Jones (1977) and Murty (1983), both models correspond precisely to percolation on a Bethe lattice.

Chelidze (1979, 1982), Chelidze and Kolesnikov (1983), Lomnitz-Adler (1985), Trifu and Radulian (1989), and Stark (1991) also attempted to establish a link between the earthquakes statistics and percolation. The work by Chelidze (1979, 1982) was concerned with the structure of the fracture network that was described in Chap. 5. Chelidze (1979,1982) and Chelidze and Kolesnikov (1983) also compared qualitatively the relation between the predictions of percolation theory and various observational data for rock fracture.

Lomnitz-Adler (1985) proposed a model for earthquake occurrence in which percolation, motivated by an asperity model, was included to explain the frequency–magnitude law for earthquakes of sizes between a hypothetical "elementary faultlet" and an earthquake large enough to rupture the entire fault plane; see also the discussions by Bebbington et al. (1990).

Trifu and Radulian (1989) proposed a model for the seismic cycle based on the existence of an asperity space–strength distribution along the fault plane, and used percolation theory to model the active zone—the fault region that can generate a major earthquake—by percolation and is characterized by a specific background seismicity. Their model could explain the nonlinearity of the frequency–magnitude relationship—the GR law—and provide estimates of magnitude of major earthquake domain in each zone. In addition, the model, in agreement with historical data, could determine a complex form of the recurrence period of earthquakes in the major magnitude domain that ranged from about 40 to 100 years, regardless of the seismic zone.

Stark and Stark (1991) attempted to develop a quantitative assessment of seismic fluid flux based on percolation. They assumed that, during an earthquake, slipped regions experience transient high-permeability, high-porosity, and high-pressure gradients, which redistribute fluids rapidly. Percolation theory was used to model the size distribution of the slip planes. By making a number of assumptions regarding fluid flow, they derived an equation that linked fluid flux, slip rate, and fracture porosity, which was then used to compute the total seismic fluid throughput for a region. Good agreement with data was obtained.

Wu (1998) assumed that earthquakes' "nucleation" begins at the percolation threshold, and developed a model that explained the order of magnitude of the seismic moment release during the nucleation, the dependence of the seismic moment of the main shock on the duration of the nucleation process, and the observation that the fraction of the moment release during the nucleation has no systematic variation with the size of the main shock. His model also indicated that the source time function of the nucleation phase may be complex, and that not all earthquakes are accompanied by a nucleation process, consistent with some observations. By assuming that there exists scale invariance associated with the criticality, Wu (1998) derived a scaling model to describe the electromagnetic emission during earthquake rupture.

Perhaps the most promising percolation model for regional seismicity was suggested by Sammis and Sornette (2002). The finite-time singularity in Eqs. (6.2) and (6.3) can develop in a heterogeneous formation, driven by increasing stress, which happens when the stress shadow from a previous large earthquake decays due to tectonic loading. We represent the heterogeneity of the crust in the form of stochastic rupture thresholds $\sigma_c(\mathbf{r})$, distributed spatially according to some statistical distribution, with $\mathbf{r}$ being the spatial position. This is similar to models of rupture and fracture, first introduced by Sahimi and Goddard (1986). We assume that the thresholds are distributed either completely randomly, or that the extent of the spatial correlations between them is finite and short. Suppose that the last largest earthquake has cast a stress shadow $-\sigma_s$ over a very large area S, and assume that it is uniform in space (the last assumption can be relaxed). Then, the stress increases uniformly throughout the region by the tectonic loading with a rate $d\sigma_t/dt$. The stress-to-rupture at an arbitrary point $\mathbf{r}$ at time t is given by

$$\Delta\sigma(\mathbf{r}) = \sigma_t - \sigma_s + \left(\frac{d\sigma_t}{dt}\right) t \ . \tag{6.13}$$

We assume that only that part of the region for which $\Delta\sigma(\mathbf{r}) > 0$ is seismically active, and can eventually rupture and generate an earthquake. The field $\Delta\sigma(\mathbf{r})$ may be viewed as a highly rough stochastic landscape in which the line $\Delta\sigma(\mathbf{r}) = 0$ defines the boundary between the active and inactive ($\Delta\sigma(\mathbf{r}) < 0$) regions. Right after the last large earthquakes, only a few small "islands" or clusters on the landscape have $\Delta\sigma(\mathbf{r}) > 0$; the rest of the landscape is inactive with $\Delta\sigma(\mathbf{r}) < 0$ because, aside from the aftershocks, the seismic activity is weak, still localized, and incoherent. As the seismic activity increases, however, more islands with $\Delta\sigma(\mathbf{r}) > 0$ emerge on the landscape. At some point when the seismic activity is high enough, the islands begin to connect, and eventually form a "continent," or a sample-spanning area on the landscape.

It should be clear that the process just described is the complete analog of formation of percolation clusters in which the role of p, the probability of having an active bond or site in a percolation network, is played by the fraction of the areas for which $\Delta\sigma(\mathbf{r}) > 0$. Thus, the moment the "continent" has formed is when $p = p_c$, where p_c is the percolation threshold. In Chap. 2, we described the percolation cluster-size distribution in which n_s is the number of clusters of size s, which can be taken as representing the distribution of earthquake sizes in the percolation model just described. Recall from Chap. 2 that at or very near percolation threshold p_c, one has

$$n_s \propto s^{-\tau} f[(p - p_c)s^{\sigma}] , \tag{6.14}$$

where τ and σ are the percolation exponents defined in Chap. 2. Recall also from Chap. 2 that, the mean size of earthquakes $\langle s \rangle$ (clusters or islands), denoted in Chap. 2 by $S(p)$, is defined by

$$\langle s \rangle = \frac{\sum_s s^2 n_s}{\sum_s s n_s} \propto |p - p_c|^{-\gamma} , \tag{6.15}$$

with $\gamma = (3 - \tau)/\sigma$, so that average earthquake size diverges at the critical point, or the percolation threshold. The corresponding seismic energy E_r that is released per each earthquake is usually taken to be, $E_r \propto$ (rupture area)$^{3/2}$. If rupture area $\propto \langle s \rangle$, then

$$E_r \propto |p - p_c|^{-3\gamma/2} , \quad p \to p_c . \tag{6.16}$$

Equation (6.16) can then be written as

$$\frac{d\langle s \rangle}{dt} = \langle s \rangle^{(\gamma+1)/\gamma} . \tag{6.17}$$

The model, as described, is not free of problems, however, not only is the average seismic energy in a given time, dE/dt, divergent, but also is its cumulative (integral) value, which is the aforementioned cumulative Benioff strain. This is caused by the fact that, if we use the random percolation model described in Chap. 2, the exponent γ is always larger than one. It should, however, be clear that random percolation model is not completely suitable for highly heterogeneous media, such as rocks that contain long-range correlations that can often be described by self-affine stochastic functions, such as the fractional Brownian motion. Thus, if one uses the percolation model with such long-range correlations that was originally developed by Sahimi and Mukhopadhyay (1996; see Chap. 3), the percolation exponents vary with the strength and type of correlations, and the problem with the divergent cumulative Benioff strain is removed

Rundle et al. (2020) proposed a new model based on the invasion percolation algorithm (see Chaps. 3 and 16) in which cluster dynamics were characterized by bursts in space and time. Their model could reproduce quantitatively the observed frequency–magnitude scaling of earthquakes in the limit that the occupation probability approaches the bond percolation threshold in two dimensions, and provided estimates for the b-value.

Let us note that Nandan et al. (2021) disputed the notion that seismicity and faulting within the Earth's crust should be interpreted as indicating the existence of underlying physical mechanisms associated with a critical point in the sense of phase transitions, or percolation. The problem is, however, still wide open.

We end this chapter by noting two shortcomings of standard percolation in the context of fractures and earthquakes. One is that the model fails to incorporate the dynamics which should be an essential part of any realistic fracturing process. Perhaps, the IP model of Rundle et al. (2020) is a step in the right direction. The second shortcoming of the standard percolation model is the effect of direction associated with stress. Perhaps, one might be able to utilize directed percolation (see Chaps. 3 and 21) to address this aspect. The first step toward such a model was taken by Wanliss et al. (2017) who studied bursty multiscale energy dissipation from earthquakes in the Chilean central zone. They provided the evidence that earthquake-radiated energy and directed percolation belong to the same universality class.

Chapter 7
Flow and Transport Properties of Porous Materials

7.1 Introduction

As Chaps. 4–6 should have made it clear, one of the most important applications of percolation theory is to solve problems in porous materials and media, from smallest to the largest length scales. We already described in Chaps. 4 and 5 some of such applications to characterization of the morphology of porous media and of connectivity of fracture networks. In the present chapter, we describe application of percolation concepts to modeling of flow and transport in porous media and materials, including fluid flow and diffusion. In a separate chapter, we will take up the problem of mass transfer, solute transport, and mixing in porous media and how percolation has been used fruitfully to model these important phenomena that are relevant to groundwater flow, enhanced oil recovery, transport of contaminants in aquifers, and several other problems.

7.2 Percolation, Poor Man's Percolation, or Critical-Path Analysis?

At the outset a crucial question must be addressed: when and under what conditions should one use the effective-medium approximation (EMA)–the poor man's percolation described briefly in Chap. 2–the critical-path analysis (CPA), briefly mentioned in Chap. 5, or the power laws of percolation theory near the percolation threshold p_c that were discussed in Chap. 2? The answer is that the choice depends strongly on the strength of the heterogeneity of porous media as well as on the range of porosity (in single-phase flow) or fluid saturation (in multiphase flows) that is being investigated.

Let us first describe the basic idea behind the CPA, which was first proposed by Ambegaokar et al. (1971) and, independently, Pollak (1972) in order to estimate hopping conductivity of extremely disordered semiconductors, and was proven

M. Sahimi, *Applications of Percolation Theory*, Applied Mathematical Sciences 213,
https://doi.org/10.1007/978-3-031-20386-2_7

rigorously later on by Tyč and Halperin (1989); see also Černý (2004) who prove the validity of the CPA for a continuum percolation model. A good review of the properties of the CAP was given by Hunt et al. (2017), and is based on the following concept.

Suppose that the porous medium is represented by a pore network in which the pore (hydraulic, electric, or magnetic) conductances follow a probability distribution function (PDF) $h(g)$. We remove all the pore conductances from the network and, then, begin to fill up the network again by replacing the conductances, in their original locations, in the order of decreasing values by starting from the largest conductance. Clearly, at the beginning there is no sample-spanning cluster (SSC) of pore conductances, but as percolation theory has taught us, after we reinstate a sufficiently large fraction of the pores' conductances, a SSC is formed, and the macroscopic conductivity of the network rises from zero. The first pore conductance that completes the formation of the SSC is referred to as the *critical conductance* g_c, while the just formed cluster is called the *critical cluster*. Therefore, p_c, the bond percolation threshold of the pore network, is related to g_c by

$$p_c = \int_{g_c}^{\infty} h(g)dg \ . \tag{7.1}$$

Recall from Chap. 2 that for a $d-$dimensional pore network of average connectivity Z, the relation $p_c \approx d/[Z(d-1)]$ provides very accurate estimates of p_c, so that for a given $h(g)$ one can estimate the critical conductance g_c.

If $h(g)$ is very broad, varying over orders of magnitude, then, all the pore conductances that were reinstated before g_c are much larger than g_c, and are effectively in series with it, since it is g_c that controls the flow of hydraulic or electric current, as all the carriers passing through the pores with conductances much larger g_c must eventually pass through g_c and, therefore, the resistance of the larger conductors can be neglected. On the other hand, all the pore conductances that are reinstated with values smaller than g_c, i.e., after formation of the SSC, will be much smaller than g_c. After the remaining pore conductances are reinstated, we recognize that since they all are much smaller than g_c, they are essentially in parallel with g_c and play no significant role in the flow process and, hence, they can also be neglected. Therefore, the macroscopic conductivity is essentially g_c.

Numerical simulation of Berman et al. (1986) confirmed the accuracy of the CPA. Katz and Thompson (1986, 1987) applied the CPA to estimate the permeability and electrical conductivity of porous media (see below), followed by others (Le Doussal 1989; Friedman and Seaton 1998; Skaggs 2011; Ghanbarian et al. 2016; Ghanbarian 2020a, b; see below). Sahimi (1993a) used the CPA to estimate the effective permeability of porous media during the flow of power-law fluids (see below).

Since what the CPA does is reducing a highly heterogeneous medium in which percolation may seem to play no role to one in which it is dominated by percolation effect, one may argue that the two are in fact the two sides of the same coin. This is true, but with one caveat: the CPA usually gives rise to power laws for the properties of interest near the percolation threshold, but the exponents that characterize such power laws may be *non-universal*, whereas a strict application of percolation relies on its *universal* power laws near the percolation threshold.

On the other hand, when the porosity of porous materials, or the saturation of fluids that they contain is high so that the porous media are in a state far from their critical porosity, saturation, or percolation threshold, or their pore-size distribution is narrow—the medium is weakly disordered—use of the EMA is recommended.

Such statements can be quantified further. Consider the choice between the EMA and the percolation power laws near p_c. The critical region is the range of the percolation probability p in which the percolation power laws (Chap. 2) are valid and accurate. In fact, the power laws are, theoretically speaking, supposed to be valid *only* close to p_c. So, the question has always been, how close is close? As the distance from p_c increases, the critical region crosses over to one in which the EMA is applicable. This was first argued by Kirkpatrick (1973) and demonstrated numerically by Sahimi et al. (1983c).

Consider two-phase flow of air–water in soil or other types of porous media. A critical or threshold moisture content S_c (or air-filled porosity) may be estimated using a relation for bond percolation given in Chap. 2 and mentioned above, namely, $p_c = d/[Z(d-1)]$, which yields $S_c = \phi d/[(d-1)Z]$. Thus, for 3D porous media, $S_c = 3Z\phi/2$. While Z can, in principle, be as high as 15, it is often around 6, leading to values of S_c that range from 0.1ϕ, often seen in soil, to as high as 0.25ϕ for $Z = 6$. Much earlier, it had been estimated (Sahimi et al. 1983c) that the upper bound for the critical region is roughly $p \approx 2/Z$, leading to a crossover moisture content $S_x \approx 2\phi/Z$. In contrast, consider the air permeability K_a in soil. While the saturation dependence of K_a is not strongly influenced by the pore-size distribution, its value under dry conditions is best estimated by the CPA (Hunt 2005a).

Thus, as discussed by Hunt et al. (2017), the following conclusions emerge:

(i) Simple power-law relationships, predicted by percolation theory (Chap. 2), describe the data that are obtained for a number of properties under a wide range of conditions.

(ii) The connectivity of a fluid within a porous medium is much more directly relevant to the saturation dependence of flow and transport properties than the overall connectivity of the medium, unless the medium itself is poorly connected.

(iii) The pore-size distribution does not have a strong effect, except for the saturation dependence of the hydraulic conductivity (see below), and when it is very broad.

(iv) The region of relevance of percolation power laws to predicting the hydraulic conductivity is restricted to near the threshold, i.e., the critical region.

(v) Percolation is not relevant to predicting the properties of porous media at high porosities. Instead, poor man's percolation—the EMA—is relevant and often accurate.

(vi) In the case of thermal conductivity of porous materials, when both the solid and the fluid (such as brine) are conductive, many theories (Torquato 2002; Sahimi 2003a) that have been developed for materials that are in a state far from the percolation threshold are relevant.

7.3 Selecting the Correct Percolation Model of the Unsaturated Zone in Soil

If we are to develop a percolation model in order to compute flow and transport properties of soil, a most abundant type of porous media, we need to proceed prudently. Long before application of percolation concepts was developed and popularized, there were already empirical and semi-empirical relations in the literature that not only linked the flow and transport properties to saturation and porosity, but were also very suggestive of the possibility of applying the percolation concepts to modeling of flow and transport properties of soil. One such empirical correlation is what has been known for over 80 years as Archie's law (Archie 1942), already mentioned in Chap. 4, which expresses the effective electrical conductivity g_e of a porous medium by a power law in both the porosity and saturation, and is reminiscent of the percolation power laws, namely, $g_e \approx g_0 S_w^n \phi^m$, where m is usually called the Archie or cementation exponent. For laboratory-scale samples of porous media, the reported saturation exponent n is frequently near the universal percolation exponent, $\mu \approx 2$ for the effective conductivity, but m varies more widely. If resistivity logs of large-scale porous media are analyzed, however, both n and m vary (Dashtian et al. 2015). This particular difference between laboratory- and field-scale porous media, reflected in the values of the exponents n and m, is easily understood because it is related to the more highly constrained variation in water distribution within a pore space, compared to the potential variability in the structure of the pore space itself.

In response, as described in Chap. 3, theorists developed models (Kogut and Straley 1979; Sahimi et al. 1983b; Halperin et al. 1985; Feng et al. 1987; Hilfer 1991; Balberg 2009) that explained the non-universal values of conductivity exponent μ, known in percolation theory for both steady-state (Kogut and Straley 1979) and unsteady-state (Sahimi et al. 1983b) transport, as well as the permeability exponents, as describing continuum percolation.

The next question to address is: how does one apply percolation theory to the problem of correlating and predicting the effective permeability and electrical conductivity of porous media? Berkowitz and Balberg (1993) suggested the following algorithm:

(i) identify a variable analogous to p, the percolation probability;

(ii) determine the critical, or threshold, value of the variable, similar to the percolation threshold p_c; and

(iii) determine the appropriate power of the power laws similar to those in terms of $(p - p_c)$.

One point must be emphasized. Percolation requires that the permeability and electrical conductivity to vanish at the critical porosity or percolation threshold. Thus, if additional flow and transport mechanisms are at work, percolation cannot take them into account. Such additional mechanisms include, for example, flow of liquid films and vapor in the permeability problem, and the solid matrix of the pore space (or at least its internal pore surface) being conducting for the electrical conductivity prediction. If a porous medium is only mildly disordered, then the range of saturations in which percolation power laws predict the properties is narrow. If, on the other hand, the porous medium is highly heterogeneous, then the range of applicability of percolation often encompasses the entire saturation range.

For the application of percolation to unsaturated soils (with air being the second fluid phase; the case of two liquid phases will be described later in this book), the spatial distribution of the water content is also important, as it also has implications for the relevance of an important morphological characteristic of a porous medium, namely, its pore-size distribution. For example, the pore-size distribution has no relevance to the saturation dependence of the air permeability, and has been shown to follow the universal percolation power law (Hazlett and Furr 2000; Ghanbarian-Alavijeh and Hunt 2012a, b). On the other hand, the pore-size distribution strongly influences the hydraulic conductivity of unsaturated porous media. The difference is due to the fact that, for example, under desaturation, the largest pores are lost to the water flow paths, whereas during saturation, only the smallest pores are lost to the dominant air flow paths.

7.4 Diffusivity

Aside from direct numerical simulation, diffusivity of gases in porous media has been predicted and computed by a variety of theoretical approaches.

7.4.1 *Percolation Approach*

Depending on its porosity ϕ, the pore space of a porous medium can be below, at, or above its critical porosity ϕ_c or its percolation threshold. Thus, in general, diffusion of gases in porous media should also be considered in three distinct regimes:

(i) If $\phi < \phi_c$, then there is no SSC of connected pores. Thus, even though gas molecules can diffuse locally, there is no macroscopic diffusion across the pore space. The effective diffusion coefficient D_e at time t is given by, $D_e \propto \langle R^2(t)\rangle/t$, where $\langle R^2(t)\rangle$ is the mean-squared displacement of the gas molecules in the pore space. Below the critical porosity, the size of large and typical clusters of pores is the percolation correlation length ξ_p. But, only a fraction $P(\phi)$ of pores that are in the largest clusters actually contribute significantly to diffusion, where P is the percolation probability defined in Chap. 2. Near but close to the critical porosity we have $\xi_p \sim |\phi - \phi_c|^{-2\nu}$ and $P \sim |\phi - \phi_c|^{\beta}$ (see Chap. 2). Thus, in that region, $\langle R^2(t)\rangle \sim P(p)\xi_p^2 \propto |\phi - \phi_c|^{\beta-2\nu}$, and

$$D_e \propto t^{-1}|\phi - \phi_c|^{\beta-2\nu} , \tag{7.2}$$

which vanishes as $t \to \infty$ because, as discussed above, there is no macroscopic diffusion below the critical porosity.

(ii) If $\phi = \phi_c$, then the SSC is a statistically self-similar and fractal cluster. Diffusion in such clusters is *not* Fickian, and $\langle R^2(t)\rangle$ grows with the time t nonlinearly: $\langle R^2(t)\rangle \sim t^{2/D_w}$, where D_w is called the fractal dimension of the random walk. Then,

$$D_e \propto t^{(2-D_w)/D_w} . \tag{7.3}$$

In this case diffusion is *anomalous*. If diffusion is only in the SSC of percolation, then $D_w = 2 + (\mu - \beta)/\nu = 2 + \theta$ (Gefen et al., 1983), with μ, β, and ν being the percolation exponents defined in Chap. 2. If diffusion is in all the clusters, then $D_w = (2\nu + \mu - \beta)/(\nu - \beta/2)$. Thus, in both cases, $D_w > 2$ and although $\langle R^2(t)\rangle$ grows with t without bounds, D_e vanishes as $t \to \infty$.

(iii) If $\phi > \phi_c$, then there is a SSC and macroscopic diffusion occurs in the pore space. In this case diffusion in the SSC dominates, and $D_e \propto g_e$, where g_e is the electrical conductivity of the pore space saturated by an electrically conducting fluid (such as brine). Since near ϕ_c one has (see Chap. 2), $g_e \sim (\phi - \phi_c)^\mu$, we conclude that

$$D_e \propto (\phi - \phi_c)^\mu . \tag{7.4}$$

If the contribution of the finite clusters cannot be ignored one must multiply D_e by $X^A(\phi)$, the fraction of accessible pores. As (see Chap. 2), $X^A \sim (\phi - \phi_c)^\beta$ near ϕ_c, one obtains $D_e \propto (\phi - \phi_c)^{\mu-\beta}$. But, the contribution of such cluster above ϕ_c is usually negligible. If we renormalize D_e by its bulk value, i.e., in the limit $\phi \to 1$, then Eq. (7.4) is rewritten as

$$\frac{D_e}{D_0} = \left(\frac{\phi - \phi_c}{1 - \phi_c}\right)^\mu . \tag{7.5}$$

Recall from Chap. 2 that the universal value of μ in 3D is $\mu = 2$. The question then is whether gas diffusivity in porous media follows Eq. (7.4) or Eq. (7.5).

Ghanbarian and Hunt (2014) analyzed 71 sets of experimental data, containing 632 data points, for gas diffusivity in a variety of porous media, including repacked, undisturbed, and field measurements. They found that 66 sets of experimental data followed Eq. (7.5) with $\mu \approx 2$, in perfect agreement with percolation theory. Figure 7.1 presents one set of such data that they analyzed. Equally important is the fact that Eq. (7.5) appears to be accurate over the entire range of the porosity ϕ.

The percolation model also confirmed the existence of anomalous diffusion in a disordered porous material. Knacksted et al. (1995) prepared a porous material, a ternary microemulsion comprised of three components, didodecyl dimethyl ammonium bromide, water, and cyclohexane, which was a bicontinuous water–oil system, but by tuning the volume fraction of the three components could undergo, at high water contents, a structural transition to disconnected water droplets in oil, i.e., the water phase would undergo a percolation transition, thus allowing measurement of the various mechanical and transport properties, including the diffusivity.

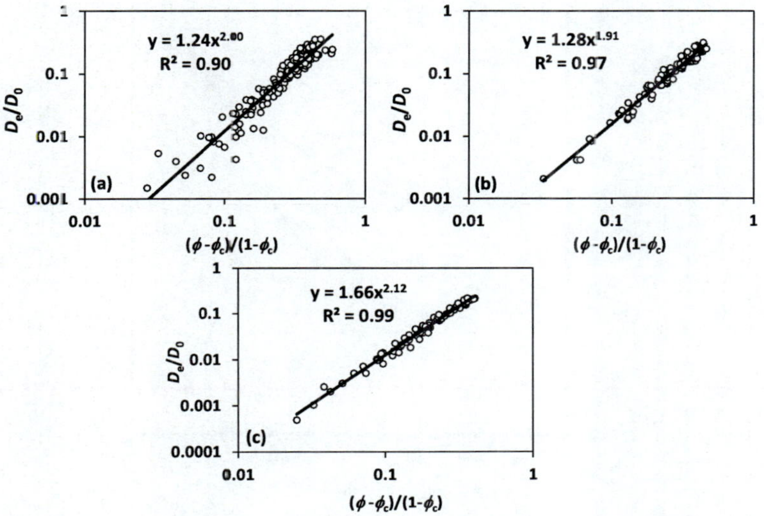

Fig. 7.1 Normalized gas diffusion coefficient D_e/D_0 versus relative air-filled porosity for **a** Moldrup et al. (2000a); **b** Moldrup et al. (2000b), and **c** Moldrup et al. (2004). The lines represent the best fit of the data (after Ghanbarian and Hunt 2014; courtesy of Dr. Behzad Ghanbarian)

Knackstedt et al. (1995) measured water self-diffusion D_e in the material by pulsed field gradient spin-echo technique at 25 ± 0.5°C. By varying the length of the gradient pulse $\mathcal{T}$ and maintaining a constant gradient pulse interval Δ, the diffusion coefficient was measured. The decay I of the echo density is given by

$$I = I_0 \exp[-D_e G^2 \mathcal{T}^2 \gamma^2 (\mathcal{T} - \Delta/3)] , \tag{7.6}$$

where G is the gradient strength, γ is the gyromagnetic ratio of the observed nucleus (^{1}H in this case), and I_0 is the signal intensity in the absence of a gradient pulse. The gradient strength was calibrated with a sample of H_2O for which D_e was known.

In their experiments, diffusion was anomalous for the intermediate values of the water volume fractions, 13.5%–14.1%, and diffusion corresponded to the case in which transport takes place on all the clusters. Therefore, the relevant equation is Eq. (7.3) with $D_w = (2\nu + \mu - \beta)/(\nu - \beta/2)$. Using the numerical values of the three percolation exponents given in Chap. 2, we obtain $D_w \approx 4.91$. A logarithmic plot of D_e versus t yielded a straight line with a slope consistent with Eq. (7.3).

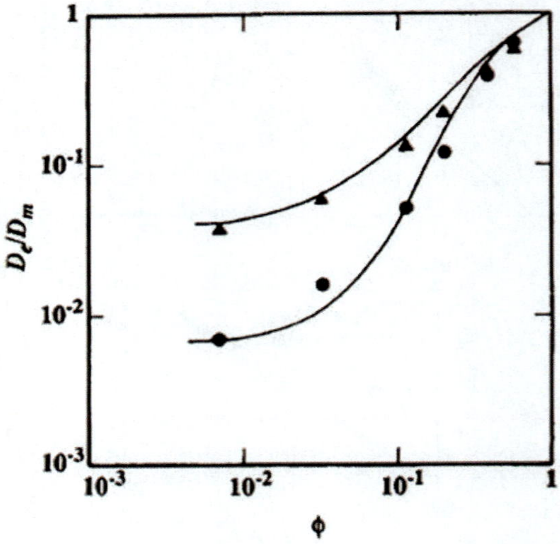

Fig. 7.2 Comparison of the predictions of the EMA for the porosity dependence of effective gas diffusivity in alumina catalyst particles (curves) and their comparison with the experimental data at 1 atm (tringles) and 10 atm (circles). The porous medium was modeled by a simple-cubic network of macro- and micropores. The upper curve is for the case in which the ratio R of the diffusivities in the macro- and micropores was $R = 0.025$, while the lower curve is for $R = 0.004$ (after Benzoni and Chang 1984)

7.4.2 Poor Man's Percolation: Effective-Medium Approximation

Benzoni and Chang (1984) used the poor man's percolation, i.e., the EMA of Chap. 2, to estimate the effective diffusivity of high-area alumina catalyst particles. Their predictions and the comparison with experimental data are shown in Fig. 7.2. The agreement between the two is excellent over the entire range of porosity.

7.4.3 Effective-Medium Approximation and Percolation

Equation (7.4) is not necessarily valid over the entire range of the porosity. In fact, as described in Chap. 2, the power laws of percolation are valid near the percolation threshold. Thus, Ghanbarian et al. (2015) combined the EMA and the power laws of percolation in order to predict the effective diffusivity of gases and solutes in partially saturated porous media. Consider the diffusion problem in a porous medium with porosity ϕ in the presence of two phases, water with saturation S_w and critical saturation (percolation threshold) S_c, and a gas at saturation $S_g = \phi - S_w$ with a critical saturation $S_g^c = 1 - S_c$, for example. We assume that for water saturations $S_w < S_x$ the power law of percolation, Eq. (7.5), describes the diffusivity of the solute, whereas for $S_w > S_x$ it is the EMA that accurately predicts the same, where S_x is the crossover saturation. Therefore, combining Eqs. (2.50) (Chap. 2) and Eq. (7.5), the solute diffusivity $D_s(S_w)$ at saturation S_w of water is given by

$$\frac{D_s(S_w)}{D_s(\phi)} = \begin{cases} \left(\frac{S_x - 2\phi/Z}{\phi - 2\phi/Z}\right)\left(\frac{S_w - S_c}{S_x - S_c}\right)^{\mu} & S_c \le S_w \le \S_x\,, \\ \frac{S_w - 2\phi/Z}{\phi - 2\phi/Z} & S_x \le S_w \le \phi\,, \end{cases} \tag{7.7}$$

where $D_s(\phi)$ is the solute diffusivity when the pore space is fully saturated by the water, i.e., when its gas content $\S_g = 0$ and, therefore, $S_w = \phi$. In a similar manner, one obtains the following expressions for the diffusivity $D_g(S_g)$ of the gas at saturation S_g:

$$\frac{D_g(S_g)}{D_g(\phi)} = \begin{cases} \left(\frac{S_g^x - 2\phi/Z}{\phi - 2\phi/Z}\right)\left(\frac{S_g - S_g^c}{S_g^x - S_g^c}\right)^{\mu} & S_g^c \le S_g \le S_g^x \\ \frac{S_g - 2\phi/Z}{\phi - 2\phi/Z} & S_g^x \le S_g \le \phi\,. \end{cases} \tag{7.8}$$

Here, S_g^x and S_g^c are, respectively, the gas saturations at the crossover point between percolation and EMA descriptions and the critical gas saturation. Clearly, $D_g(\phi)$ is the diffusivity when there is no water in the pore space. The rest of the notation is the same as before.

Ghanbarian et al. (2015) utilized Eqs. (7.7) and (7.8) to predict the diffusivities and compared the results with nine sets of data obtained by direct numerical simulations. Excellent agreement was found in all cases. Hunt et al. (2014b) used the same approach to predict diffusivity of a solute in porous media, and compared the results with experimental data for 106 experiments with 766 data points. Good agreement was found between the predictions and the data.

7.5 Permeability

There have been many approaches to estimating the effective permeability of porous media, including the CPA and the EMA, the poor man's percolation. Let us describe each.

7.5.1 Poor Man's Percolation: Effective-Medium Approximation

A porous medium consists of pore throats connected together at the pore bodies. The effective sizes of both the pore throats and pore bodies are distributed according to statistical distributions $f_t(r_t)$ and $f_b(r_b)$. It is, however, not straightforward to measure $f_b(r_b)$, which is why it is usually not available. Thus, since the macroscopic permeability is controlled by the pore throats, for convenience we refer to the pore throats as pores, and their distribution $f(r)$ as the pore-size distribution (PSD). As described in Chap. 2, in the EMA, a heterogeneous pore space is represented by a uniform medium with the size of all the pores being r_e. We assume that the pores are cylindrical. Then, for slow flow the flow conductance k_f is given by, $k_f \propto r^4$, while other pore shapes may also be considered. The EMA predicts that the macroscopic permeability K_e is given by (Doyen 1988; David et al. 1990)

$$K_e = \frac{\phi}{C_s\tau}\frac{r_e^4}{\langle r_b^2\rangle}\,, \tag{7.9}$$

with ϕ being the porosity, τ is the flow tortuosity for which various theories as well as empirical and semi-empirical relations have been developed (see Chap. 4), C_s is a geometrical factor with $C_s = 8$ for cylindrical pores in Hagen–Poiseuille (slow or laminar) flow, and r_b is the size of the pore bodies. Since the distribution $f_b(r_b)$ of the size of the pore bodies is typically not available, David et al. (1990) suggested that one should use $\langle r_b^2\rangle \simeq \langle r^2\rangle = \int_{r_m}^{r_M} r^2 f(r)dr$, with r_m and r_M being, respectively, the minimum and maximum pore radii. r_e^4 is computed by the EMA (see Chap. 2):

$$\int_{r_m}^{r_M} \frac{r_e^4 - r^4}{r^4 + (Z/2-1)r_e^4} f(r)dr = 0\,. \tag{7.10}$$

Here, Z is the coordination number or connectivity of the porous medium. Mukhopadhyay and Sahimi (2000) derived an EMA for predicting the direction-dependent macroscopic permeabilities of anisotropic porous media; see also Bernasconi (1974) for the analogous problem of conductivity of anisotropic materials.

Koplik et al. (1984) analyzed in detail a Massillon sandstone, mapped its pore space onto an equivalent random network, and employed the EMA to calculate its permeability and electrical conductivity (see below). They found that the predictions for K_e differ from the data by about one order of magnitude, while those for σ differ by a factor of about 2. They attributed the difference to the fact that most sedimentary rocks are highly heterogeneous and anisotropic, none of which is adequately taken into account by the EMA.

Doyen (1988) utilized the EMA to compute the permeability of Fontainebleau sandstone. He prepared a series of epoxy-impregnated thin sections of the sandstone from cores with porosity between 0.05 and 0.22 and, using an image analyzer, determined the distributions of the throat and pore sizes from the digitized and

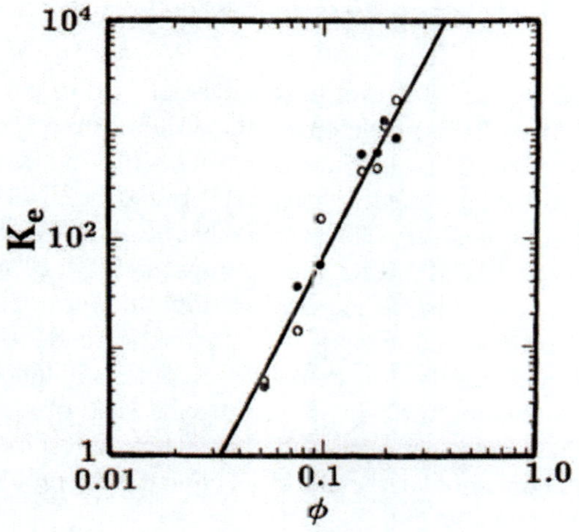

Fig. 7.3 Logarithmic plot of the permeability of Fontainebleau sandstone as a function of its porosity. Solid and open circles represent, respectively, the experimental measurements and the EMA predictions (after Doyen 1988)

segmented microsections. For each sandstone, the permeability was calculated based on the characteristic lengths, which were estimated directly from the experimental size histograms. Figure 7.3 presents his results for the permeability and compares them with the experimental data.

Richesson and Sahimi (2021a) used the EMA, Eqs. (7.9) and (7.10), to compute the effective permeability of 28 sandstone samples that were undergoing *deformation* under the effect of an external pressure. Despite the complexity of the problem due to the deformation of the pore space, they showed that when the EMA is combined with a model of deformation, it provides very accurate predictions for the experimental data.

Ghanbarian and Daigle (2016b) simulated fluid flow in binary mixtures of low- and high-permeability porous media, constructed with spheres and ellipsoids using the lattice-Boltzmann (LB) method, and computed their permeability. They then used the EMA to predict the permeability of the same model porous media. The EMA predictions agreed well with those obtained with the LB simulations in simple- and body-centered cubic packings of particles, if the permeability k_h of the high-permeability component is not substantially different from k_l, the low-permeability component. Overall, the EMA-predicted permeability was within a factor of two of the LB simulations.

Ghanbarian and Javadpour (2017) utilized the EMA to predict the pore pressure-dependent gas permeability of shales, using as the input the pore-throat size distribution, connectivity, tortuosity, porosity, and the gas characteristics. They compared their predictions with six datasets, three experiments, one computed by a pore-network model, and two obtained by lattice-Boltzmann simulations. The comparison indicated that the EMA predictions are accurate to within a factor of 3 of the measurements and simulations. A similar approach was used by Ghanbarian et al. (2019) to compute the gas permeability of tight sandstones.

7.5.2 Critical-Path Analysis

As discussed earlier in this chapter, Ambegaokar et al. (1971) and, independently, Pollak (1972) developed the concept of the CPA for estimating the electrical conductivity of highly heterogeneous semiconductors. The CPA may also be invoked to estimate the effective permeability and the electrical conductivity of a brine-saturated porous medium. The problem has been reviewed and discussed extensively by Ghanbarian (2021). Katz and Thompson (KT) (1986; 1987) were the first to extend the CPA to estimate the permeability and electrical conductivity of porous media. In a porous medium, the local hydraulic conductance is a function of the pore length l, and therefore g_c defines a characteristic length l_c. Since both flow and electrical conduction problems belong to the class of scalar percolation problems (Chap. 2), the length that signals the percolation threshold in the flow problem is the same as the threshold in the electrical conductivity problem. Thus, one has a trial solution for g_e given by

$$g_e = \phi g_c(l)\,[p(l) - p_c]^{\mu} \ , \tag{7.11}$$

where the porosity ϕ ensures a proper normalization of the fluid or the electric-charge density. The function $g_c(l)$ is equal to $c_f l^3$ for the flow problem and $c_c l$ for the conduction problem. For appropriate choices of the function $p(l)$, the conductance $g_e(l)$ achieves a maximum for some $l_{\max} \le l_c$. In general $l^f_{\max}$ for the flow problem is different from $l^c_{\max}$ for the conduction problem, because the transport paths have different weights for the two problems.

We describe the KT derivation of the CPA for the effective permeability here, and postpone their derivation of the electrical conductivity, which parallels precisely that of K_e, to the next section. If $p(l)$ allows for a maximum in the conductance, and if the maximum occurs for $l_{\max} \le l_c$, then we may write

$$l^f_{\max} = l_c - \Delta l_f = l_c\left[1 - \frac{\mu}{1+\mu+l_c\mu p''(l_c)/p'(l_c)}\right] . \tag{7.12}$$

If the pore-size distribution of the medium is very broad, then $l_c\mu p''(l_c)/p'(l_c) \ll 1$, and using $\mu \simeq 2$ for 3D percolation yields

$$l^f_{\max} = l_c\left(1 - \frac{\mu}{1+\mu}\right) \simeq \frac{1}{3} l_c \ . \tag{7.13}$$

Now, writing and

$$K_e = a_1\phi\left(l^f_{\max}\right)^2\left[p(l^f_{\max}) - p_c\right]^{\mu} \ , \tag{7.14}$$

we obtain to first order in Δl_c or in Δl_f,

$$p(l^f_{\max}) - p_c = -\Delta l_f\, p'(l_c) \ . \tag{7.15}$$

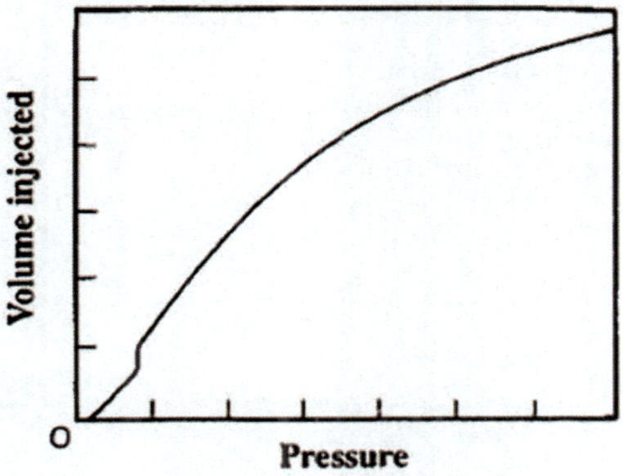

Fig. 7.4 Schematic volume–pressure curve during mercury porosimetry

To obtain the constant a_1, Katz and Thompson (1986) assumed that at a local level the rock conductivity is g_f, the conductivity of the fluid that saturates the pore space, and that the local pore geometry is cylindrical. These imply that $a_1 = g_f$. Therefore, one obtains

$$K_e = \frac{1}{226} l_c^2 g_e / g_f \tag{7.16}$$

and

$$\frac{g_e}{g_f} = \frac{l_{\max}^c}{l_c} \phi S(l_{\max}^c) , \tag{7.17}$$

where $S(l_{\max}^c)$ is the volume fraction (saturation) of the connected pore space involving pore widths of size $l_{\max}^c$ and larger.

Equation (7.16) [as well as Eq. (7.17); see below] involves no adjustable parameters, as they are all fixed and precisely defined. To obtain an estimate of the characteristic length l_c, KT proposed to use mercury porosimetry described in Chap. 4. Consider a typical mercury porosimetry curve in which the pore volume of the injected mercury is obtained as a function of the pressure; see Fig. 7.4. The initial portion of the curve is obtained *before* a SSC of pores, filled with mercury, is formed. There is also an inflection point beyond which the pore volume increases rapidly with the pressure. The inflection point signals the formation of the SSC. Therefore, from the Washburn equation (Chap. 4) we must have $l \geq -4\sigma \cos\theta / P_i$, for that portion of the curve that is beyond the inflection point, where P_i is the pressure at the inflection point, θ is the contact angle, and σ is the interfacial tension. Then, $l_c = -4\sigma \cos\theta / P_i$ defines the characteristic length l_c. Typically, l_c is estimated through mercury-injection curves with an error of at most 15 percent. Thus, the procedure to use the KT expression for the permeability is as follows. From a mercury porosimetry experiment, the parameter l_c is estimated, and then Eq. (7.16) is used to estimate K_e.

Figure 7.5 compares the permeability of a set of sandstones, predicted by Eq. (7.16), with the data. The dashed lines mark a factor of two. No adjustable parameter

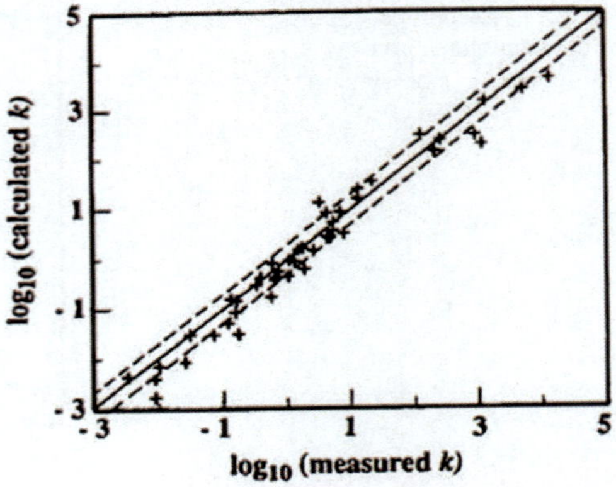

Fig. 7.5 Logarithmic comparison of the CPA-predicted effective permeability K_e (in millidarcy) with the experimental data (after Katz and Thompson 1986)

was used and the agreement between predictions and theory is very good. Similarly good agreement is obtained for the conductivity of the porous medium (Katz and Thompson 1987), which we will describe in the next section. Normally, note that the exponent μ used in Eq. (7.14) can take on its value for continuum percolation discussed in Chap. 3. This depends, of course, on the structure of the pore-size distribution of the pore space.

Ghanbarian (2020a) invoked the CPA to compute the effective permeability of packing of spherical particles. He compared the predictions with eight datasets, including 105 packings of particles, and demonstrated that the CPA provides accurate estimates for K_e. Note that Modaresi et al. (2019) improved the CPA predictions of the hydraulic conductivity by introducing a scale-dependent tortuosity into the model, although one reason for the popularity of the CPA is that it is free of adjustable parameters; see also Daigle (2016).

Some related relevant works should be mentioned here. Swanson (1981) had already recognized that during mercury injection (or flow of any non-wetting fluid in a porous medium) large pores dominate the flow paths, and that the inflection point in the pore volume–pressure curve signals the formation of a sample-spanning cluster. Thus, he postulated a relation between the effective permeability and the capillary pressure curve. He maximized the product of the length and saturation, both of which are estimated from the mercury-injection curve, and obtained

$$l \sim (S^2 l^2)_{\max} , \tag{7.18}$$

where $l_{\max}$ is a length scale very similar, both qualitatively and numerically, to the length scale l_c introduced by the KT. The agreement between the predictions of Eq. (7.18) and the data was very good, which is not entirely surprising, given the similarity between Swanson's $l_{\max}$ (not to be confused with KT's $l_{\max}$) and l_c. It should, however, be noted that the transport paths considered by Swanson were appropriate for electrical conduction, but not for the fluid flow problem.

Banavar and Johnson (1987) and Le Doussal (1989) calculated the constant 1/226 in Eq. (7.16) slightly differently than the way the KT did. For example, Banavar and Johnson (1987) estimated the constant to be $\approx 7.68 \times 10^{-3}$, which should be compared with the KT's estimate, $1/226 \simeq 4.42 \times 10^{-3}$. Banavar and Johnson's predictions are still within the error bars of Fig. 7.5. Nyame and Ilbston (1980) used an empirical parameter similar to l_c to describe the permeability of cement pastes. Hagiwara (1984) replaced S^2 in Eq. (7.16) with g_e to obtain $K_e \sim g_e l^2$, which appears again to agree with data.

Application of the CPA to a pore-network model of porous materials implies that most of the flow is restricted to the critical path, making the *effective coordination number* Z_e of the pore space relevant. In fact, Z_e describes how many sites (pore bodies in a pore space) should be connected to a given site in order to guarantee percolation across the sample. With $Z_e \approx 2.7$ (Seager and Pike 1974), and the aforementioned result by Sahimi et al. for the extent of the critical region (Sahimi et al. 1983c), one obtains $\phi_x \approx 2\phi/2.7 \approx 0.74\phi$ for the crossover porosity above which the percolation analysis would not be valid. Thus, a large coordination number or connectivity of a porous medium leads to a small threshold, but the relevance of a broad pore-size distribution can still make Z_e small enough for the dominant flow paths to generate a wide critical region. Analysis of diffusion in porous media suggests that an even larger crossover value, $\phi_x \approx 0.75$, is appropriate and in many cases (Freund and Sornette 2007; Ewing and Hunt 2006; Ghanbarian-Alavijeh and Hunt 2012a, b) the critical region effectively extends through the entire range of accessible conditions.

Friedman and Seaton (1998) carried out an extensive study of the effective permeability and electrical conductivity of 3D pore networks, using numerical simulations and the CPA. Hunt and Gee (2002a, b) used the CPA to study fractal porous media, and compared the results with the data from the Hanford Site (in Washington State). Hunt (2005a) utilized continuum percolation and the CPA to study saturation dependence of air permeability. The CPA may also be extended, in order to calculate the permeability of fractured rocks. Indeed, Charlaix et al. (1987a) used arguments very similar to those of KT to calculate the permeability of fracture networks with a broad distribution of fracture apertures.

Johnson et al. (1986) introduced a well-defined parameter Λ, defined by

$$\Lambda = 2\frac{\int |E(\mathbf{r})|^2 dV_p}{\int |E(\mathbf{r})|^2 dS_p} \neq \frac{V_p}{S_p}\,, \tag{7.19}$$

where $E(\mathbf{r})$ is the potential in the electrical conduction problem in a porous medium that is saturated with a conducting fluid (such as brine), and V_p and S_p are, respectively, the pore volume and the pore surface area. Note that V_p/S_p is a *geometrical* parameter that can be measured and is *independent* of any flow or transport process. On the other hand, Λ is a *dynamical property*, defined for the specific problem of electrical conduction and, therefore, cannot be measured by geometrical analysis alone. Since $E(\mathbf{r})$ vanishes in certain regions of the pore space (for example, in the isolated and dead-end region), Λ is roughly a measure of the *dynamically connected pores*, i.e., those that belong to the backbone of the flow or transport path. Johnson et al. (1986) proposed that for 3D porous media

$$K_e \simeq c\frac{\Lambda^2}{8F} = c\frac{\Lambda^2}{8}\frac{g_e}{g_f}\,, \tag{7.20}$$

where $c \sim \mathcal{O}(1)$, and $F = g_f/g_e$ is the formation factor. It should be pointed out that Eq. (7.20) is not *in general* exact, although due to the physical meaning of Λ, it represents an appealing relation between K_e and g_e. Note that Λ plays the same role that the critical length scale l_c does in the CPA (Martys and Garboczi 1992).

Note that Eqs. (7.16) and (7.20) both suggest a relationship between the effective permeability, and the electrical conductivity of fluid-saturated porous media. In general, one does not expect a direct link between the two, as g_e a dynamical property, whereas the permeability is a static characteristic of a pore space and, in fact, $\sqrt{K_e}$ represents a length scale. The possible relationship between the two quantities has been studied for decades; see, for example, Skaggs (2011) for a good discussion.

Various groups tested the accuracy of Eq. (7.20) (see, for example, Straley et al. 1987; Saeger et al. 1991; Kostek et al. 1992), using a variety of numerical and analytical methods, as well as experimental data for well-characterized porous media. It appears that (Kostek et al. 1992), unless a porous medium contains two widely different relevant length scales (such as micropores and macrospores), Eq. (7.20) should be very accurate. Arns et al. (2005) carried out numerical simulation of flow and mercury porosimetry in images of porous media, and studied the role of the critical length scale l_c in the CPA described above, and found good agreement between the CPA predictions and the direct numerical simulations.

The CPA may also be used for estimating the permeability of a porous medium saturated by a *non-Newtonian* fluid (Sahimi 1993a). For the flow of a power-law fluid in a cylindrical pore, the volume flow rate q is given by $q = g(\Delta P_p)^{1/n}$, where ΔP_p is the pressure drop along the pore, n is the power-law fluid index ($n = 1$ for Newtonian fluids), and g, the flow conductance, is given by

$$g = \left(\frac{n\pi}{3n+1}\right)\left(\frac{1}{2m}\right)^{1/n} l^3\,, \tag{7.21}$$

with l being the pore's length and m the usual parameter of power-law fluids ($m = \eta$ when $n = 1$, where η is the fluid's viscosity). Then, the CPA predicts that macroscopic volume flow rate Q is given by (Sahimi 1993a)

$$Q = (K_{\mathrm{KT}})^{(n+1)/2n}\left(\frac{C_n}{m}\right)^{1/n}\left(\frac{\Delta P}{L}\right)^{1/n}\,, \tag{7.22}$$

where K_{KT} is the permeability that was predicted by Katz and Thompson, Eq. (7.16), in the limit $n = 1$; L is the length of the medium; ΔP is the pressure drop across the pore space; and C_n is given by

$$C_n = \frac{(7.8)^{1-n}(23.6\mu_n)^{\mu_n}}{2^{2n+1}(3n+1)^n(\mu_n+1)^{\mu_n+2}} F^{2(1-n)/5}\phi^{(n-1)/10} . \tag{7.23}$$

The effective permeability is then defined by a generalized Darcy's law that relates Q and ΔP to each other. Here, μ_n is the generalized conductivity exponent for a percolation network of power-law conductors, first studied by Kenkel and Straley (1982), and is given by

$$\mu_n = (d-1)\nu + \frac{\zeta(n) - \nu}{n} , \tag{7.24}$$

where ν is the exponent for the percolation correlation length, and $\zeta(n)$ is the generalized resistivity exponent introduced in Chap. 2, Eq. (2.21), which depends on n. Note that in the limit $n = 1$, Eq. (7.24) reduces to Eq. (2.21). Meir et al. (1986) determined $\zeta(n)$ for many values of n.

7.5.3 Percolation-CPA Models of Hydraulic Conductivity and Air Permeability

We consider soil and other types of porous media that are either partially or completely saturated by a liquid, or are filled by air. The more general case of two-phase flow with two liquids (such as oil and water) will be considered later in this book. Thus, consider first the hydraulic conductivity of a fully saturated porous medium. This corresponds, in the case of the air permeability, to completely dry conditions. As the discussions in Chaps. 2 and 4 indicated, the percolation threshold of 3D porous media is low. Therefore, water (or air) can find an interconnected path through the medium that avoids all but the largest pores. For a given critical volume fraction v_c, the CPA predicts the largest possible value of the smallest pore radius r_c that resides on such a path,

$$v_c = S_c = \int_{r_c}^{r_{\max}} r^3 f(r)dr , \tag{7.25}$$

where for a liquid v_c is expressed as the moisture content S_c, whereas for the air permeability under completely dry conditions v_c is simply the critical value of the air-filled porosity ϕ_c^a. Note that the critical air fraction for percolation and the critical water content are not, in general, equal, although for simplicity of explanation no distinction may be made. In the case of the hydraulic conductivity, reducing the water content implies that the largest pores become filled with air. Thus, the upper limit of the integral in Eq. (7.25) can no longer include the largest pores, and in order for the integral to remain equal to v_c, the lower limit must be reduced and r_c must diminish with decreasing saturation. The reduction of r_c with diminishing saturation accounts for the influence of the pore-size distribution on the hydraulic conductance $K_e(S)$ and its dependence on saturation S.

On the other hand, reducing the air content does not affect the occupation of the largest pores by air (Hunt 2005a) and, therefore, the air permeability. The water—the wetting fluid—enters the smallest pores first, and the pore space defined by $r_c < r < r_{\max}$ remains air-filled. Since the air permeability at any saturation is always controlled by the same subset of pores that defines the property under completely dry conditions, Eq. (7.25) defining v_c provides the critical pore radius for air flow, *independent* of the air content, unless it falls below the threshold. Thus, the CPA reveals a fundamental difference in the saturation dependences of the hydraulic conductivity K_e and the air permeability K_a: for the former the bottleneck pore radius depends on the saturation, but not so for the latter.

Therefore, because the pore-size distribution has little, if any, influence on the saturation dependence of the air permeability K_a and its dependence on air-filled porosity are controlled by the topology of the air-filled pore space as described by percolation theory, and in particular the universal power law in terms of the air-filled porosity ϕ^a:

$$\frac{K_a(\phi^a)}{K_a(\phi)} = \left(\frac{\phi^a - \phi_c^a}{\phi - \phi_c^a}\right)^2 . \tag{7.26}$$

The power law (7.26) has been reported for many types of soil, as reviewed by Hunt et al. (2017). Note that the exponent of the power law (7.26) is equal to $\mu = e$, the exponents that characterize the power laws for the permeability and conductivity of percolation networks (Chap. 2). ϕ_c^a, the critical air fraction for percolation, is treated as a fitting parameter, but its value can be predicted using the water retention curves (Ghanbarian-Alavijeh and Hunt 2012a, b). Lack of comprehensive data in a few cases explains those for which the universal power law cannot be obtained. In addition to incomplete data, in all the cases where the universal power law could not be confirmed, the reduction in the water content was accompanied by compression (Tang et al. 2011), implying that power law (7.26) is not appropriate, since $K_a(\phi)$ is not a constant over the entire range of the air-filled porosities.

7.6 Electrical Conductivity

A good review of electrical conductivity of porous media is given by Cai et al. (2017). Similar to the diffusivity in porous media, as well as the permeability, percolation models have also been used to predict the electrical conductivity of brine-saturated porous media.

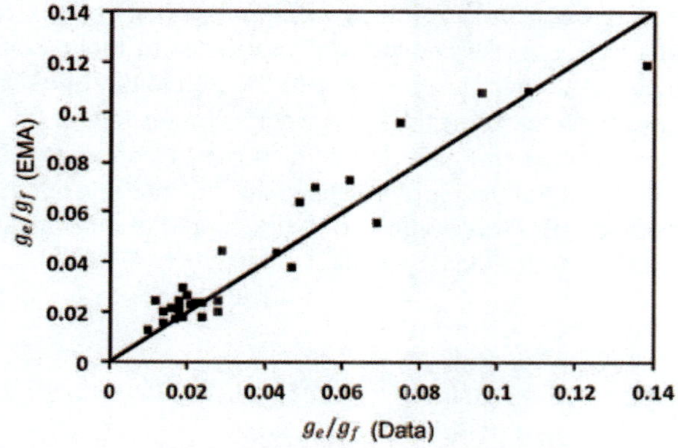

Fig. 7.6 Logarithmic plot of the formation factor of Fontainebleau sandstone as a function of porosity. Solid circles represent the experimental data, while open circles show the EMA predictions (after Doyen 1988)

7.6.1 *Effective-Medium Approximation*

The EMA for the electrical conductivity is very similar to the effective permeability, described in Sect. 7.4. Assuming that the pores are cylindrical of radius r, the electrical conductivity g of a pore filled by brine is given by $g \propto r^2$. One can, of course, consider other pore shapes. The EMA predicts that the macroscopic electrical conductivity g_e is given by (Doyen 1988; David et al. 1990)

$$g_e = \left(\frac{\phi}{\tau}\frac{r_e^2}{\langle r_b^2\rangle}\right) g_f \ , \tag{7.27}$$

where the notation is the same as before. As already explained in Sect. 7.4, since the distribution $f_b(r_b)$ of the size of the pore bodies is typically not available, David et al. (1990) suggested that one can use $\langle r_b^2\rangle \simeq \langle r^2\rangle = \int_{r_m}^{r_M} r^2 f(r)dr$, with r_m and r_M being, respectively, the minimum and maximum pore radii, and $f(r)$ is the pore-size distribution. Given Eq. (2.45), the pore-conductance distribution $w(g)$ can be converted to a pore-size distribution $f(r)$, since $w(g)dg = f(r)dr$ and $g \propto r^2$. Thus, the quantity r_e^2 is determined by the EMA:

$$\int_{r_m^2}^{r_M^2} \frac{r_e^2 - r^2}{r^2 + (Z/2-1)r_e^2}\frac{f(r^2)}{2r}dr = 0\ . \tag{7.28}$$

Doyen (1988) used Eq. (7.27) to compute the electrical conductivity of a series of Fontainebleau sandstones whose porosity ranged between 5 and 22 percent. Figure 7.6 compares Doyen's predictions of the formation factor, $F = g_f/g_e$ by the EMA with the experimental data for the Fontainebleau sandstones. The EMA predicts the conductivity to within a factor of 3 of the experimental data.

Richesson and Sahimi (2021b) used the EMA, Eqs. (7.27) and (7.28), to compute the effective electrical conductivity of 28 sandstone samples, saturated by brine, which were undergoing *deformation* under the effect of an external pressure. In this case, surface conduction contributes to the overall conductivity of the porous materials. Thus, in order to use the percolation power laws and/or the EMA, one must first determine the surface contribution and subtract it from the overall electrical conductivity, and then utilize the theory. Despite such complications, and the added complexity of the problem due to the deformation of the pore space, they showed that when the EMA is combined with a model of deformation, it provides very accurate predictions for the experimental data.

7.6.2 Combining Effective-Medium Approximation and Percolation

Similar to their approach for solute diffusion in porous media, Ghanbarian et al. (2014, 2015) combined the EMA and the power laws of percolation to develop a model for the electrical conductivity and, thus, the formation factor $F(\phi)$ as a function of the porosity ϕ:

$$F(\phi) = \begin{cases} \dfrac{(1-\phi_c)(\phi_x-\phi_c)}{(\phi-\phi_c)^{\mu}} & \phi_c \leq \phi \leq \phi_c \\ \dfrac{1-\phi_c}{\phi-\phi_c} & \phi_x \leq \phi \leq 1 \end{cases}, \tag{7.29}$$

where, as before, ϕ_c is the critical porosity (percolation threshold) and ϕ_x is the crossover porosity, i.e., the porosity at which the description of the electrical conductivity by the percolation power laws crosses over to what is predicted by the EMA. Note that in the limits, $\phi_c = 0$ and $\mu = 2$ one recovers the well-known Archie's laws (Archie 1942), namely, $F(\phi) \propto \phi^{-2}$, i.e., the Archie exponent m, defined by $F \propto \phi^{-m}$, is predicted to take on a universal value of 2. Analyzing extensive *field data*, Dashtian et al. (2015) pointed out, however, that in many cases the Archie exponent is non-universal.

Ghanbarian et al. (2014) compared the predictions of their model with several sets of experimental data, as well as the results of direct numerical simulations, and found satisfactory agreement between the two. The same type of approach was employed to study formation factor of clay-rich porous media by Daigle et al. (2015). In this case, surface conduction contributes to the overall conductivity of the porous materials. Thus, in order to use the percolation power laws and/or the EMA, one must first determine the surface contribution and subtract it from the overall electrical conductivity, and then utilize the theory (see also Richesson and Sahimi, 2021b). After doing so, Daigle et al. (2015) found good agreement between the data and the predictions.

Ghanbarian and Sahimi (2017) used the same theoretical approach to predict the electrical conductivity of two-phase flow in unconsolidated porous media, represented by packings of particles. They determined the pore-size distribution of the

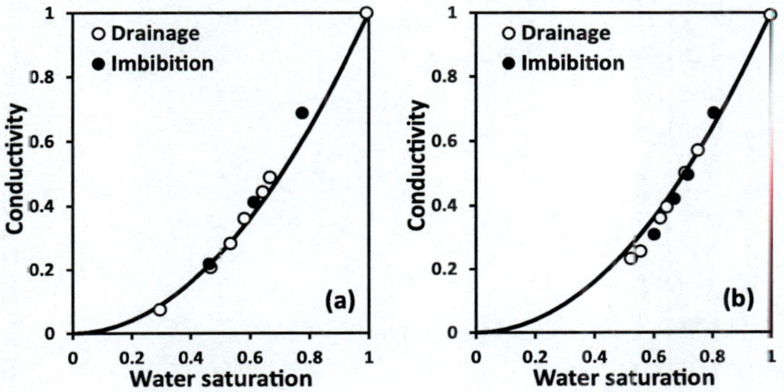

Fig. 7.7 Dependence of the relative electrical conductivity on the water saturation S_w with, **a** n-decane and **b** naphtha as the non-wetting fluid in a water-wet packing of uniform glass beads. The experimental data are from Sharma et al. (1991). The curves represent the predictions with a zero saturation threshold ($S_c = 0$) and the percolation conductivity exponent $\mu = 2$ (after Ghanbarian and Sahimi 2017)

porous media using the capillary pressure curves (see Chap. 4), and then used the above theory to predict the electrical conductivity of water phase in the porous media during both imbibition (invasion of the pore space by a wetting fluid) and drainage (invasion by a non-wetting fluid). Figure 7.7 presents their predictions and compares them with the experimental data of Sharma et al. (1991).

7.6.3 Critical-Path Analysis

The derivation of the CPA for the electrical conductivity of porous media by Katz and Thompson (1987) parallels that of the effective permeability described in Sect. 7.4. Thus, the analog of Eq. (7.12) for the conductivity is

$$l_{\max}^c = l_c - \Delta l_c = l_c\left[1 - \frac{\mu}{3 + \mu + l_c \mu p''(l_c)/p'(l_c)}\right], \tag{7.30}$$

while the analog of Eq. (7.13) reads

$$l_{\max}^c = l_c\left(1 - \frac{\mu}{3+\mu}\right) \simeq \frac{3}{5} l_c\,, \tag{7.31}$$

so that

$$p(l_{\max}^c) - p_c = -\Delta l_c p'(l_c)\,. \tag{7.32}$$

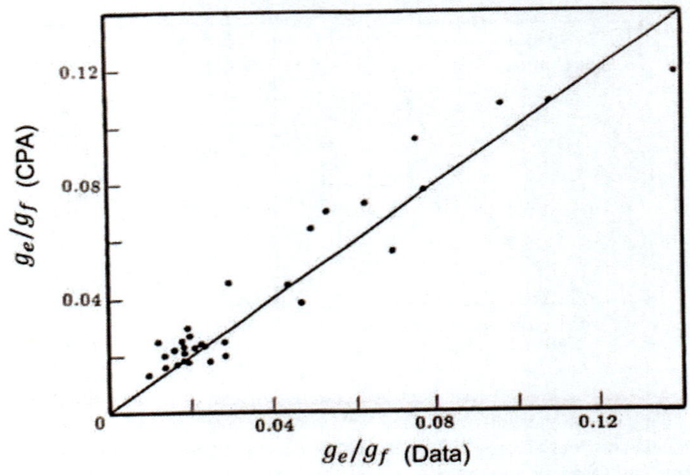

Fig. 7.8 Logarithmic comparison of the effective electrical conductivity of porous media, predicted by the CPA, with the experimental data (after Katz and Thompson (1987))

Then, one finally obtains Eq. (7.17) for the effective electrical conductivity. Figure 7.8 presents a comparison between the predictions of the CPA and the experimental data.

To improve the predictions of the CPA for the electrical conductivity, Tyč and Halperin (1989) considered a random conductance network in which the conductance of the ith bond g_i was distributed according to an exponential distribution, $g_i = g_0(\lambda x_i)$, with g_0 being a constant, x_i a random number distributed according to a distribution $X(x_i)$. According to the CPA and Eq. (7.1), the critical conductance g_c is defined such that the bond-percolation threshold of the network is given by

$$p_c = \int_{x_c}^{\infty} X(x)dx \ , \tag{7.33}$$

and all conductances $g \geq g_c$ form the SSC at p_c. In order to analyze the case in which the conductances are broadly distributed, Tyč and Halperin (1989) considered the limit of very large λ, and derive the following relation for the effective conductivity:

$$g_e = C\ell^{2-d} g_c \lambda^{-\alpha} \ , \tag{7.34}$$

which improves on the CPA prediction that takes g_c as the effective conductivity of the network. Here, C is a constant; ℓ is the length of the bonds; d is the space dimensionality of the network ($d = 2$ and 3); and α is an exponent such that for $d = 2$, $\alpha = 0$, and $0.2 \leq \alpha \leq 1.0$ for $d = 3$. Le Doussal (1989) suggested $\alpha = (d-2)\nu$, where ν is the critical exponent of percolation correlation length (Chap. 2) with

$\nu \approx 0.89$ for $d = 3$. This relation had already been suggested by Ambegaokar et al. (1971).

7.6.4 Renormalization Group Method

Deriving the EMA necessitates making a fundamental assumption: the fluctuations in the potential field are small, since we require the average of the fluctuations to be zero. If, however, the fluctuations are large, as in the case of, for example, a composite near its percolation threshold p_c, or a material with a broad distribution of the conductances, then the EMA loses its accuracy. In that case, a position-space renormalization group (PSRG) method is more appropriate because, in averaging the properties of the system, the method takes into account the properties of the *pre-averaged* material. It also predicts non-trivial power law (i.e., nonlinear) for the conductivity near p_c, a distinct advantage over the EMA that always predicts the critical exponents of the conductivity to be unity.

Assume, for example, that in a square or a cubic network each bond conducts with probability p. Since, in practice, the network is so large that we cannot calculate its properties exactly, we partition it into $b \times b$ or $b \times b \times b$ cells, where b is the number of bonds in any direction, and compute their effective properties, which are hopefully representative of the properties of the original network. The shape of the RG cell can be selected arbitrarily, but it should preserve, as much as possible, the symmetry properties of the original network. For example, since the square network is self-dual [the dual of the square network is obtained by connecting the centers of its neighboring square cells, which is again a square network] and because self-duality plays an important role in its percolation and transport properties, we use self-dual cells to represent this network (Bernasconi 1978). Figure 7.9 presents examples of the RG cells with $b = 2$ in the square and simple-cubic networks, where the 2D cell is self-dual.

Each RG cell is replaced by one bond in each principal direction; see Fig. 7.10. If each bond in the original network conducts with probability p, then the bonds that replace the cells will conduct with probability $p' = R(p)$, where $R(p)$ is called the *RG transformation* (which represents the spanning probability) and represents the *sum* of the probabilities of *all* of its conducting configurations, with some conducting and the rest insulating. To compute $R(p)$, we solve for the percolation and conduction problems in each cell by applying a fixed potential gradient across the cell in a given direction. For example, when considering conduction, the 2×2 RG cell of Fig. 7.9 is equivalent to the circuit shown there, which is called the Wheatstone bridge. Thus, for the $b = 2$ cell, we only need to deal with a circuit of five bonds, and for the $2 \times 2 \times 2$ cell we construct an equivalent 12-bonds circuit shown in Fig. 7.9. To compute $R(p)$, we determine all the possible conducting configurations of such circuits, with some bonds conducting and some insulating. Thus, for the 2×2 RG cell we obtain

$$p' = R(p) = p^5 + 5p^4q + 8p^3q^2 + 2p^2q^3 , \tag{7.35}$$

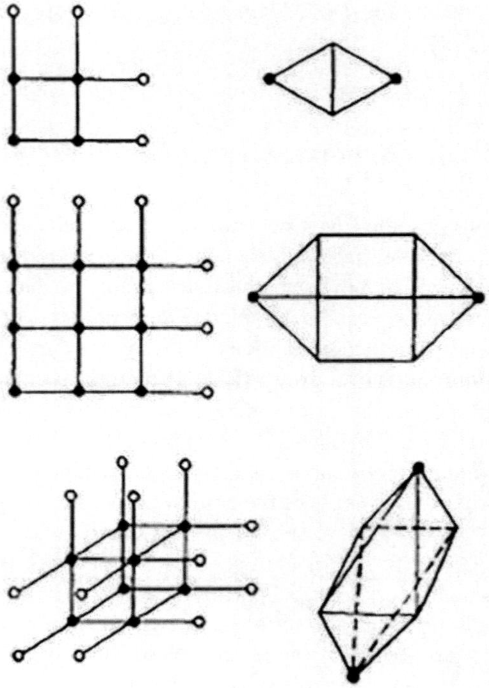

Fig. 7.9 Two- and three-dimensional RG cells with scale factor $b = 2$ and their equivalent circuit

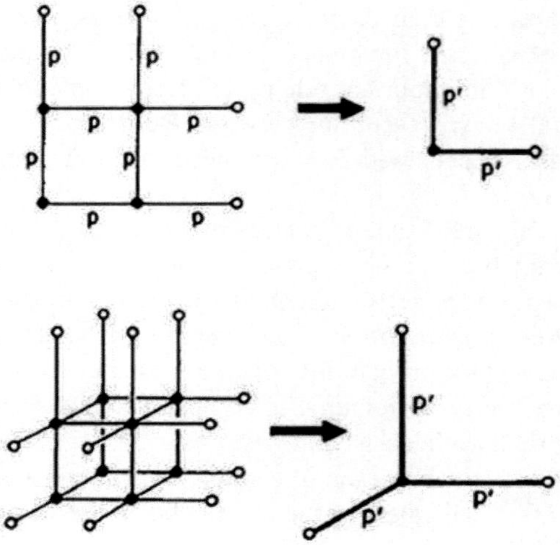

Fig. 7.10 Two- and three-dimensional RG cells with scale factor $b = 2$, in which each bond conducts with probability p, and their equivalent RG cell in which the probability that a bond conducts is p'

where $q = 1 - p$. It is easy to see how Eq. (7.35) was derived: there is only one conducting RG cell configuration with all the five bonds conducting (probability p^5), five conducting configurations with four bonds conducting and one bond insulating (probability $5p^4q$), and so on.

The SSC at p_c is self-similar, implying that the RG transformation should remain invariant at p_c. The invariance of $R(p)$ should also be true at $p = 1$ and $p = 0$, because under any reasonable RG transformation, full and completely empty (insulating) networks should be transformed to full and empty networks again. The points $p = 0$, 1, and p_c are thus called the *fixed points* of the transformation and are denoted by p^*. Since the RG transformation should not change anything at these points, the implication is that at the fixed points the probability p of having a conducting bond in the RG cell and that of a bond in the renormalized cell, $p' = R(p)$, should be the same. Thus, the fixed points are the solution of the polynomial equation

$$p^* = R(p^*) , \tag{7.36}$$

and indeed this equation usually has three roots that are $p^* = 0$, $p^* = 1$, and $p_c = p^*$, where p^* is the RG transformation prediction for p_c. For the RG cells of Fig. 7.9, we obtain $p^* = 1/2$ for both 2×2 and 3×3 cells, the exact result. In fact, it can be shown (Bernasconi 1978) that the RG transformation for such 2D self-dual cells for *any b* always predicts $p_c = 1/2$. With the $2 \times 2 \times 2$ RG cell we obtain $p^* \simeq 0.208$, which should be compared with the numerical estimate, $p_c \simeq 0.249$.

Given the simplicity of such "back of the envelope" calculations, we guess that we should calculate $p_c = p^*$ for several cell sizes b and then use finite-size scaling (Chap. 2) according to which

$$p_c - p_c(b) \sim b^{-1/\nu} , \tag{7.37}$$

to extrapolate the results to $b \to \infty$ and estimate p_c, where p_c is the true percolation threshold of the network, and ν is the critical exponent of percolation correlation length ξ_p. Indeed, Reynolds et al. (1980) used this method and obtained very accurate estimates of site and bond percolation thresholds of the square network. As always, however, life is more complex than we like it to be! Ziff (1992) showed that as $b \to \infty$, the probability $R(p_c)$ approaches a *universal* value of 1/2 for *all* 2D networks. This implies that, *asymptotically*, Eq. (7.36) is *wrong*, because p_c is *not* universal and depends on the structure of the network, whereas $R(p_c)$ *is* universal (as $b \to \infty$). Ziff's discovery does not violate the universality of $R(p_c)$; it only fixes its value. Stauffer et al. (1994) showed that $R(p_c)$ is also universal in 3D. Sahimi and Rassamdana (1995) showed, however, that although $R(p_c)$ is universal, the following equation

$$R(p_c) = \alpha \tag{7.38}$$

always provides an estimate of p_c in (0,1) for any $0 < \alpha < 1$. Moreover, it appears that, among all the possible values, $\alpha = p_c$ still provides the fastest convergence to the true value of p_c.

The critical exponent ν of the percolation correlation length can be estimated based on $R(p_c)$. Suppose that the percolation correlation lengths in the original and renormalized lattices (RG cells) are ξ_p and ξ'_p, respectively. Since each bond of the RG cells is replaced by another bond b times its length, we must have

$$\xi'_p = \frac{1}{b}\xi_p \, . \tag{7.39}$$

On the other hand, $\xi_p \sim (p - p_c)^{-\nu}$ and, because of the universality of the percolation exponents, $\xi'_p \sim [R(p) - R(p_c)]^{-\nu}$. If we linearize $R(p)$ and $R(p_c)$ around $p_c = p^*$, and use Eq. (7.39), we obtain

$$\nu = \frac{\ln b}{\ln \lambda} \, , \tag{7.40}$$

where $\lambda = dR(p)/dp$, evaluated at $p = p^*$. For example, the 2×2 and 3×3 self-dual RG cells yield, $\nu \simeq 1.43$ and 1.38, respectively, which should be compared with the exact value $\nu = 4/3$ (see Table 2.3). For the $2 \times 2 \times 2$ cell of Fig. 7.9, one obtains $\nu \simeq 1.03$, which should be compared with the numerical estimate, $\nu \simeq 0.88$, listed in Table 2.3.

The PSRG approach may also be used to compute the conductivity of a random resistor network (and the permeability and diffusivity) (Young and Stinchcombe 1975; Stichcombe and Watson, 1976, Payandeh 1980). One begins with the original probability density function $f_0(g)$ of the bond conductance g of the RG cell and replaces it with a new distribution $f_1(g)$, the probability distribution for the conductance of a bond in the renormalized cell, which is calculated by determining the equivalent conductance of the RG cell. Thus, for an $n-$bond RG cell, one obtains a recursion relation relating $f_1(g)$ to $f_0(g)$:

$$f_1(g) = \int f_0(g_1)dg_1 f_0(g_2)dg_2 \cdots f_0(g_n)dg_n \delta(g_p - g') \, , \tag{7.41}$$

where $g_1, \ldots, g_n$ are the conductances of the n bonds of the RG cell and g' is the equivalent conductance of the RG cell. For example, for the five-bond cell of Fig. 7.9, one has

$$g' = \frac{g_1(g_2g_3 + g_2g_4 + g_3g_4) + g_5(g_1 + g_2)(g_3 + g_4)}{(g_1 + g_4)(g_2 + g_3) + g_5(g_1 + g_2 + g_3 + g_4)} \, , \tag{7.42}$$

and if, $f_0(g) = (1 - p)\delta(g) + p\delta(g - g_0)$, then, using Eq. (7.41),

$$f_1(g) = [1 - R(p)]\delta(g) + 2p^3q^2\delta\left(g - \frac{1}{3}g_0\right) + 2p^2(1 + 2p)q^2\delta\left(g - \frac{1}{2}g_0\right) + 4p^3q\delta\left(g - \frac{3}{5}g_0\right) + p^4\delta(g - g_0) \, . \tag{7.43}$$

One now iterates Eq. (7.41) to obtain a new distribution $f_2(g)$ by substituting $f_1(g)$ in the right side of (7.39). The iteration process should continue until a distribution $f_\infty(g)$ is reached, the shape of which does not change under further iterations. This is called the *fixed-point distribution* and the conductance of the original network is simply an average of $f_\infty(g)$.

It is, however, difficult to analytically iterate Eq. (7.41) many times. The common practice is to replace the distribution after the ith iteration by an *optimized* distribution $f_i^o(g)$ which closely mimics the properties of $f_i(g)$. The optimized $f_i^o(g)$ is usually taken to have the following form:

$$f_i^o(g) = [1 - R(p)]\delta(g) + R(p)\delta[g - g^o(p)] , \tag{7.44}$$

where $g^o(p)$ is an *optimal* conductance. In the past, various schemes have been proposed for determining $g^o(p)$, one of the most accurate of which was proposed by Bernasconi (1978), according to which, if after i iterations of Eq. (7.35), $f_i(g)$ is given by

$$f_i(g) = [1 - R(p)]\delta(g) + \sum_i a_i(p)\delta(g - g_i) , \tag{7.45}$$

then $g^o(p)$ is approximated by

$$g^o(p) \simeq \exp\left[\frac{1}{R(p)}\sum_i a_i(p)\ln g_i\right] . \tag{7.46}$$

Once $g^o(p)$ is calculated, Eq. (7.41) is iterated again, the new distribution $f_{i+1}(g)$ and its optimal form $f_{i+1}^o(g)$ are determined, and so on. In practice, after a few iterations, even an initially broad $f_0(g)$ converges rapidly to a stable and narrow distribution with a shape that does not change under further rescaling. The conductivity of the resistor network is simply a suitably defined average of this distribution.

Renormalization methods are usually very accurate for 2D networks, and are flexible enough to be used for anisotropic systems as well. They have, however, two drawbacks for 3D systems. The first is that the predictions of the PSRG methods for percolation networks with *any* type of $b = 2$ RG cell are not accurate, and even after the first iteration of Eq. (7.41), the renormalized conductance distribution $f_1(g)$ is very complex. For example, if we begin with a binary distribution, $f_1(g)$ will have *seventy-three* components of the form $\delta(g - g_i)$, with $i = 1, \cdots, 73$. Hence, analytical calculation of $f_2(g)$ is very difficult, if not impossible. The second drawback is that, even for a $b = 3$ RG cell, analytical determination of the RG transformation becomes very difficult, because the total number of possible configurations of the RG cell is of the order of 10^{11}. Thus, one must resort to a Monte Carlo renormalization group method (Reynolds et al., 1980), which is, however, not any simpler than the simple Monte Carlo method.

7.6.5 Renormalized Effective-Medium Approximation

To address the difficulties that the PSRG methods encounter for 3D resistor networks, Sahimi et al. (1983c) and Sahimi (1984) proposed a new method that combined the EMA and PSRG methods, and is called the renormalized EMA (REMA). Their method took advantages of two facts.

(i) Each time a resistor network is renormalized, it becomes less critical in the sense that, its associated percolation correlation length ξ_p' is smaller than the original correlation length ξ_p by a factor of the cell size b; see Eq. (7.39).

(ii) As discussed and demonstrated above, the EMA is usually accurate away from the percolation threshold, if the morphology of the material does not contain any correlations.

Therefore, if one employs the EMA with the first iteration $f_1(g)$ of the original bond conductance distribution $f_0(g)$, instead of $f_0(g)$ itself, the performance of the EMA should improve. That is, in the REMA the conductance distribution that one uses is $f_1(g)$, instead of $f_0(g)$. Because the bonds of the renormalized resistor network are b times longer than the original ones, this necessitates a rescaling of conductivities of the renormalized network to replicate the old one: the REMA conductivity is taken to be the same as that for the original resistor network at $p = 1$.

The REMA is sharply more accurate than *both* the EMA and PSRG. For bond percolation, the REMA predicts the percolation threshold p_c to be the root of the following equation:

$$R(p_c) = \frac{2}{Z} \,, \quad \text{bond percolation,} \tag{7.47}$$

which should be compared with the prediction of the EMA. For example, with the 3D RG cell of Fig. 7.9, the RG transformation is given by

$$\begin{aligned} R(p) = {} & p^{12} + 12p^{11}q + 66p^{10}q^2 + 220p^9q^3 + 493p^8q^4 + 776p^7q^5 + 856p^6q^6 \\ & + 616p^5q^7 + 238p^4q^8 + 48p^3q^9 + 4p^2q^{10} \,, \end{aligned} \tag{7.48}$$

so that, using Eqs. (7.47) in (7.48) with $Z = 6$, one obtains $p_c \simeq 0.267$, only 7% larger than the numerical estimate (see Table 2.2), $p_c \simeq 0.249$ for the simple-cubic network. Moreover, so long as one uses the type of 2D self-dual RG cells that are shown in Fig. 7.9, the REMA predicts the exact bond-percolation threshold of the square lattice for *any* cell size b.

As for site percolation, the EMA is given by

$$\int \frac{g - g_e}{g + (\Gamma - 1)g_e} f(g)dg = 0 \,, \tag{7.49}$$

where $\Gamma = 1/(\gamma_2 - 2/Z)$, with

$$\gamma_2 = \frac{1}{2}\left\{\int_0^\infty \left[I_0^d(t) - 2I_0^{d-1}(t)I_1(t) - I_0^{d-2}I_2^2(t)\right]\exp(-dt)dt\right\} \,, \tag{7.50}$$

for a $d-$dimensional cubic lattice, with $I_m(t)$ being the modified Bessel function of order m. It is then not difficult to show that Eqs. (7.49) and (7.48) predict that the site-percolation threshold of a lattice is the root of the following equation:

$$R(p_c) = \Gamma^{-1/2} \quad \text{site percolation} \; . \tag{7.51}$$

Here, $R(p)$ is the RG transformation for site percolation. For example, if we use a $2 \times 2 \times 2$ RG cell for site percolation in the simple-cubic lattice, we obtain

$$R(p) = p^8 + 8p^7q + 28p^6q^2 + 56p^5q^3 + 54p^4q^4 + 24p^3q^5 + 4p^2q^6 \; , \tag{7.52}$$

which should be compared with Eq. (7.48), the RG transformation for a $2 \times 2 \times 2$ cell for bond percolation. Using Eq. (7.52) in (7.51), we obtain $p_c \simeq 0.367$ for site-percolation threshold of a simple-cubic lattice, which is 17% larger than the numerical estimate $p_c \simeq 0.3116$ (see Table 2.2), but represents a sharp improvement over $p_c \simeq 0.46$ that the EMA predicts. Finally, the REMA prediction for the correlated bond-percolation threshold of a simple-cubic lattice is $p_c \simeq 0.105$, in excellent agreement with the numerical estimate, $p_c \simeq 0.103$. Therefore, the REMA predictions for the percolation thresholds of various percolation models in both 2D and 3D lattices are mostly accurate and close to the numerical estimates currently available.

Using the REMA, Sahimi et al. (1983c); Sahimi (1984) and Sahimi (1988) also obtained very accurate predictions of the effective conductivity of various 2D and 3D networks. Figure 7.11 compares the predictions of the REMA for the effective conductivity of a square network in bond percolation with those of the EMA and the simulation results, while Fig. 7.12 shows the same for the simple-cubic network.

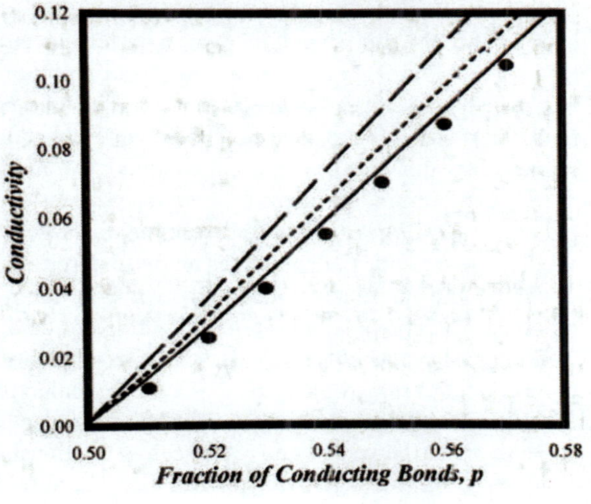

Fig. 7.11 Comparison of the Monte Carlo data for the effective conductivity of a square network (symbols) with the predictions of the EMA (the left-most curve) and the REMA with RG cell of size $b = 2$ (middle curve) and $b = 3$ (solid curve) (after Sahimi et al. 1983c)

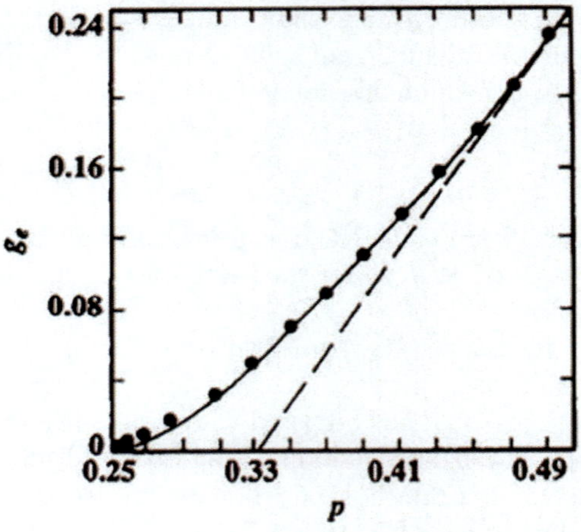

Fig. 7.12 Comparison of the Monte Carlo data for the effective conductivity of a square network (symbols) with the predictions of the EMA (dashed line) and the REMA with RG cell of size $b = 2$ (solid curve) (after Sahimi et al. 1983c)

7.6.6 *Percolation and CPA models of Electrical Conductivity and Electrokinetic Coupling*

The next property of soils that we consider are the electrical and hydraulic conductivity, as well as the electrokinetic coupling, since their models illustrate the subtle differences in the roles played by the pore-size distribution and fluid connectivity in their respective dependence on the saturation. The fact that the flux of water q through a given pore, $q \propto r^4/\ell$, depends much more strongly on the pore dimensions than does the electrical current, $i \propto r^2/\ell$, is important to the comparison between the two. As discussed by Hunt et al. (2017), the contrast between the two properties is illustrated best in a geometry that is simple, but general enough. The electromechanical coupling, on the other hand, involves a surface layer along the fluid–solid interface that, for a cylindrical pore, has a cross section proportional to r, and a conductance r/l. It is the difference between the r^4/ℓ, r^2/ℓ, and r/ℓ-dependence that causes the pore-size distribution to have a much greater impact on the hydraulic conductivity or permeability than on the electrical conductivity, and for the electrokinetic coupling to be essentially independent of the pore size. As argued by Friedman and Seaton (1998), if the CPA is applied to determine the largest possible (critical) value of the smallest pore radius r_c that defines an optimal (critical) path for flow, then the same r_c will be relevant to any conduction or flow property of the wetting fluid. In many models ℓ, the pore's length is often assumed to be a constant, but in a self-similar medium ℓ is proportional to r, in which case the critical hydraulic conductance is proportional to r^3 and the critical electrical conductivity is proportional to r, while the electrokinetic coupling coefficient is independent of pore size. But, regardless of the relationship between ℓ and r, the scaling relationships between the various critical conductances provide a first approximation to the scaling of the conductivities:

$$g_c^h = r_c^2 g_c^e = r_c^3 g_c^{\mathrm{ek}} . \tag{7.53}$$

The actual dependence on pore radius is affected by additional factors. For example, if the pores have rough internal surface, the fractal dimensionality of the surface will be a contributing factor.

Although in the traditional network models of porous media pore bodies and pore throats constitute two distinct entities, the dominant resistance to flow is offered by the pore throats, whereas pore bodies are where most of the water is stored. Calculation of the hydraulic conductance or permeability $K_e(S)$ based on the CPA ignores the distinction between the pore bodies and pore throats. An estimate of the pore size can be made if one can develop an expression for the dependence of r_c on the moisture content S. To simplify the calculations we assume strong wetting condition, which is equivalent in soil physics to the premise that film flow permits (Blunt and Scher 1995; Tokunaga and Wan 1997; Hashemi et al. 1998, 1999a, b) the porous medium to adjust to removal of water by evacuating all pores with radii larger than some equilibrium value r_h.

Following Hunt and Gee (2002b), the percolation condition relating the smallest (or critical) pore size to the critical volume fraction v_c, when the largest pore filled with water has $r = r_{\max}$, is obtained from Eq. (7.25) by substituting for the pore-size distribution $f(r)$ given earlier,

$$v_c = \left(\frac{3 - D_f}{r_{\max}^{3-D_f}}\right) \int_{r_c}^{r_{\max}} dr\, r^{2-D_f} = 1 - \left(\frac{r_c}{r_{\max}}\right)^{3-D_f} . \tag{7.54}$$

The percolation condition that relates the smallest (critical) pore size for the critical path to the critical volume fraction v_c when the largest pore filled with water has $r = r_h$ is expressed by

$$v_c = \left(\frac{3 - D_f}{r_{\max}^{3-D_f}}\right) \int_{r_c}^{r_h} dr\, r^{2-D_f} = \left(\frac{r_h}{r_{\max}}\right)^{3-D_f} - \left[\frac{r_c(S)}{r_{\max}}\right]^{3-D_f} . \tag{7.55}$$

The two equations have the same form, but the upper limit in Eq. (7.55) has been reduced from $r_{\max}$ to r_h, which produces a corresponding reduction in $r_c(S)$, hence requiring its representation as a function of S to indicate partial saturation. In order to calculate the hydraulic conductivity as a function of S, we should use a threshold moisture content S_c, in place of the critical volume fraction v_c. Combining Eqs. (7.25), (7.54), and (7.55) and replacing v_c with S_c yield as expression for r_c for the unsaturated conditions in terms of the same for the saturated conditions:

$$r_c(S) = r_c(\phi) \left[\frac{1 - \phi + (S - S_c)}{1 - S_c}\right]^{1/(3-D_f)} . \tag{7.56}$$

To the lowest order approximation the scaling of the critical hydraulic conductance also represents that of the hydraulic conductivity, and is given in a self-similar porous medium:

$$g_c^h(S) = g_c^h(\phi)\left[\frac{1-\phi+(S-S_c)}{1-S_c}\right]^{3/(3-D_f)}, \tag{7.57}$$

$$K_e(S) = K_0\left[\frac{1-\phi+(S-S_c)}{1-S_c}\right]^{3/(3-D_f)}. \tag{7.58}$$

Note that only when the smallest pore size $r_0 \to 0$, implying that $\phi \to 1$, can one express $K_e(S)$ as a power of $(S - S_c)$. Averaging over all the conductances on the critical path replaces (Hunt 2004a, b; Balberg 2009) $3/(3 - D_f)$ with $D_f/(3 - D_f)$, a result that holds all the way to the percolation threshold. In this regard, the CPA is more accurate than other methods (Ghanbarian and Hunt, 2014) for predicting the saturation dependence of $K_e(S)$ except, of course, at low saturations where flow of thin films is important. The scaling of the critical value of the electrical conductance is

$$g_c^e(S) = g_c^e(\phi)\left[\frac{1-\phi+(S-S_c)}{1-S_c}\right]^{1/(3-D_f)}, \tag{7.59}$$

$$g_e(S) = g_0\left[\frac{1-\phi+(S-S_c)}{1-S_c}\right]^{1/(3-D_f)}. \tag{7.60}$$

As the saturation decreases, so also does the effect of the pore-size distribution on $K_e(S)$, and connectivity and tortuosity become the dominant factors that are represented by the fundamental power laws of percolation theory. Eventually, $K_e(S)$ vanish in the limit of a finite moisture content at which the capillary flow network becomes disconnected. The hydraulic conductivity does not actually vanish at the critical moisture content since film flow can take place even when only water films exist without water-filled pores. But, if this feature is neglected, the approach underestimates K_e at low moisture content. Excluding film flow from the analysis allows, however, use of the percolation power law for low S,

$$\frac{K_e(S)}{K_0} = \frac{g_e(S)}{g_0} \propto (S - S_c)^{\mu}. \tag{7.61}$$

In Eq. (7.61), we should use the universal value, $\mu \approx 2$, because the non-universal values of the exponents, including μ, are predicted by the CPA, if appropriate. The electrokinetic coupling is expected to follow the same power law throughout the range of accessible saturation. Note that Eqs. (7.61) predicts a very rapidly varying K_e or g_e with the moisture contents near the threshold saturation. At moisture contents approaching the porosity, however, Eq. (7.61) leads to a relatively slowly varying conductivity. The importance of determining the prefactor is perhaps greater than the ability to determine the appropriate value of the exponent (Berkowitz and Balberg 1993).

Thus, we may summarize the results of this section as follows:

(i) The crossover moisture content in typical soils is near the threshold for K_e, preventing general use of a universal power law of percolation.

(ii) The crossover moisture content in typical soils is near full saturation for g_e, justifying use of Archie's law for its saturation dependence with the exponent $n = 2$ (and, in many cases, also for m, the Archie or cementation exponent).

(iii) Both g_e and K_e reduce to known results for non-universal behavior in the limit $\phi \to 1$ for which the conductance distribution reduces to a power law. Use of a model (Ghanbarian and Hunt, 2012) that produces the same soil–water retention curve as the pore–solid fractal (the RS) model generates the non-universal exponents of percolation theory in a more physically feasible model for which the porosity ϕ need not approach 1.

(iv) The universal expression for the electrical conductivity is sufficiently accurate to make it possible to diagnose experimental difficulties. Many porous media, which do not have such complicating features as a contact resistance (Kechavarzi and Soga 2002) or dissolution of ions at low brine salinity (Abu-Hassanein et al. 1996; Roberts and Lin 1997), do follow the percolation power law. In the former case, percolation power law is recovered at smaller S where the contact resistance is negligible, whereas in the latter case at large S, where the influence of dissolution of ions on brine salinity can be neglected, the power law is recovered.

7.7 Thermal Conductivity

If a porous medium is filled by air or any gas whose thermal conductivity is much smaller than that of the solid matrix of the medium, then one has a percolation problem because one needs a SSC of conducting grains or solid phase in order for the macroscopic thermal transport to occur. But if the porous medium is saturated completely or partially by any fluid that has a significant thermal conductivity, then given that the solid matrix of the medium is also conducting, one always has a SCC of conducting regions throughout the porous medium, and the problem is no longer a percolation one. In addition, one must also consider the resistance at the fluid–solid interface and, if the porous medium is unconsolidated, at the contact area between the grains.

But the SSC can only be made of the fluid phase, or form only through the solid phase, or a mixture of both. Therefore, one should observe a type of a transition as the SSC switches from one in which the fluid phase is dominating to one in which the solid matrix controls conduction. The transition should occur at or very close to the percolation threshold of the fluid or the solid phase although, strictly speaking, it is not a percolation transition since there is no diverging correlation length in the system. This type of transition has indeed been observed in experiments.

Many models, some empirical and some theoretical have been proposed for computing the thermal conductivity. These models were reviewed by Ghanbarian and

Daigle (2016a), and will not be repeated here. Here, we describe two approaches that are consistent with the theme of this chapter and book.

7.7.1 Percolation Model

There have been some efforts to invoke percolation theory to study thermal conductivity of various disordered materials. Such materials were, however, composite solids. We discuss this briefly at the end of this chapter, but postpone the main discussions of such efforts to Chap. 12.

7.7.2 Effective-Medium Approximation

Let us first point out that relatively sharp upper and lower bounds have been derived for the thermal conductivity of various types of porous media. These bounds were reviewed comprehensively by Torquato (2002) and Sahimi (2003a). Davis and Artz (1995) utilized the EMA to compute the thermal conductivity of metal-matrix composites. Such materials are used in electronic packaging. They compared the EMA predictions with direct numerical simulation of the conduction equation based on the finite-element techniques. An important conclusion of their work was that the thermal boundary resistance, which occurs at the interface between the metal and the included phase, typically ceramic particles, has a large effect on the thermal conductivity, if the particles are small.

7.7.3 Percolation-Modified Effective-Medium Approximation

Consider a porous medium that contains low- and high-conductivity portions with volume fraction, respectively, of ϕ and $(1-\phi)$. Then, the EMA, Eq. (2.45), is simplified to

$$\phi\frac{K_l - K_e}{K_l + (Z/2-1)K_e} + (1-\phi)\frac{K_h - K_e}{K_h + (Z/2-1)K_e} = 0\,, \tag{7.62}$$

where K_l and K_h are the low and high conductivities. As described in Chap. 2 and earlier in this chapter, the EMA does not reproduce the correct power law of percolation near the percolation threshold. To remedy this, McLachlan (1987) modified Eq. (7.64) and rewrote it as

$$\phi \frac{K_l^{1/\mu} - K_e^{1/\mu}}{K_l^{1/\mu} + \left(\frac{1-\phi_c}{\phi_c}\right) K_e^{1/\mu}} + (1-\phi) \frac{K_h^{1/\mu} - K_e^{1/\mu}}{K_h^{1/\mu} + \left(\frac{1-\phi_c}{\phi_c}\right) K_e^{1/\mu}} = 0 \,, \qquad (7.63)$$

where μ is the usual conductivity exponent near the critical threshold ϕ_c. Note also that since the EMA predicts a critical threshold, $\phi_c = 2/Z$, the term $(Z/2 - 1)$ has been rewritten in terms of ϕ_c, which allows one to treat it as an adjustable parameter.

Deprez et al. (1988) used Eq. (7.63) to fit the thermal conductivity data for sintered nickel. Their fit yielded, $\mu = 1.8 \pm 0.3$, supporting the accuracy of Eq. (7.63), although we should keep in mind that the modification of Eq. (7.64) to (7.63) is purely empirical.

If a porous medium is fully saturated by a fluid, then Eq. (7.63) may be used to estimate K_e, provided that K_l and K_h are taken, respectively, as the thermal conductivity of the fluid and the solid matrix. If, however, the pore space is partially saturated by, say water at saturation S_w, and the porosity of the pore space is ϕ, then Eq. (7.63) is rewritten as

$$(\phi - S_w) \frac{K_d^{1/\mu}(\phi) - K_e^{1/\mu}(S_w)}{K_l^{1/\mu}(\phi) + \left(\frac{\phi - S_c}{S_c}\right) K_e^{1/\mu}(S_w)} + S_w \frac{K_s^{1/\mu}(\phi) - K_e^{1/\mu}(S_w)}{K_s^{1/\mu}(\phi) + \left(\frac{\phi - S_c}{S_c}\right) K_e^{1/\mu}(S_w)} = 0 \,, \qquad (7.64)$$

where S_c is the critical water saturation (percolation threshold), K_d represents the thermal conductivity of the dry phase (air and the solid matrices), while K_s is the conductivity of the saturating fluid and the solid matrix.

Ghanbarian and Daigle (2016a) utilized Eq. (7.64) to fit the data for a variety of porous media. While they obtained good fits for all the cases that they examined, the fitted estimate of the conductivity exponent μ was always less than 1.0, whereas $\mu \approx 2$ for 3D percolation. Therefore, percolation-modified EMA does not explain and fit the data completely and consistent with percolation.

7.7.4 *Universality Across Flow and Transport Properties*

We now address the dependence of the thermal conductivity of porous media on saturation. Both the solid matrix and water (in the pores) conduct thermally and, therefore, in a saturated medium every portion conducts. Furthermore, the thermal conductivity of the most conductive portion—usually the solid—tends to exceed that of the liquid by a factor less than 10, meaning that the large conductivity contrasts required to associate the percolation probability p with just one constituent of the medium are not present. Consequently, if one is to use percolation at all, p must account for both the water and the solid phases. But, this means that even for relatively small moisture content (for the typical soil porosity, $\phi = 0.4$; the porosity of consolidated rock is clearly smaller) the relatively generous condition, $p < 0.75$ for

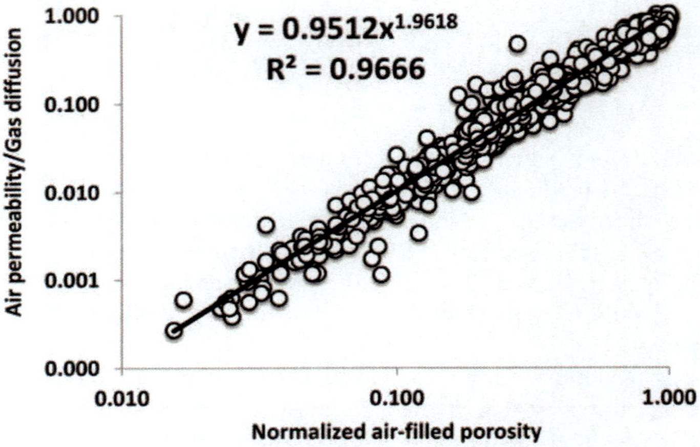

Fig. 7.13 Universality of air permeability and gas diffusivity. Both follow a power law in the porosity with an exponent as close to 2

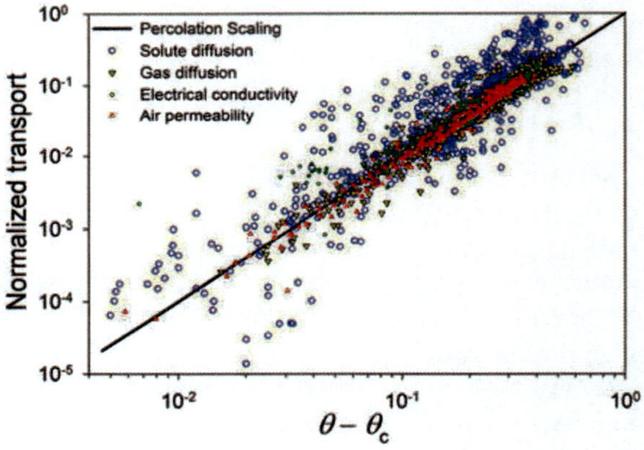

Fig. 7.14 Universality of multiple flow and transport properties. All follow a power law in the porosity difference $(\phi - \phi_c)$ [equal to $(\theta - theta_c)$ in the figure] with an exponent as close to 2

the validity of percolation power laws can be violated. In particular, for the moisture content $S = 0.15$, one has $p = 0.75$. Thus, for less than half the range of the moisture contents, could one expect a quadratic dependence predicted by percolation theory, and for more than half, the linear dependence that the EMA provides is applicable (Ghanbarian and Sahimi 2017). Moreover, angular contacts between grains lead to large contact resistances, but just a small amount of moisture can increase the contact area greatly and produce a much smaller resistance, reducing the range of p-values for which percolation power laws are appropriate even further.

The simultaneous relevance of percolation scaling to many processes allows their representation in the same graph. Figure 7.13 (Hunt et al. 2014b) demonstrates this

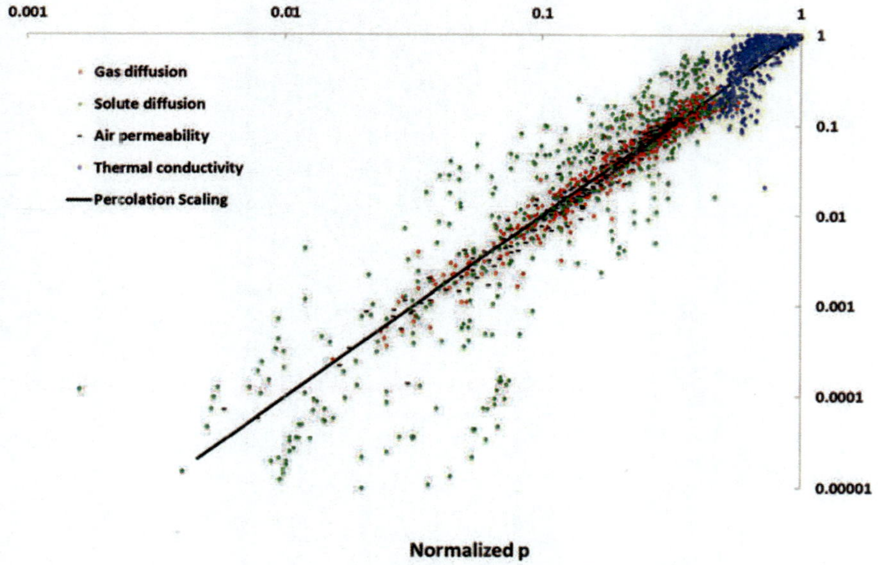

Fig. 7.15 Universality of multiple flow and transport properties, including thermal conductivity. All follow a power law in the difference $(1 - p)$ with an exponent ≈ 2.02

first for the air permeability and the gas diffusion coefficient. The same procedure can be extended to include solute diffusion and electrical conductivity as in, for example, Fig. 7.14. One can also add in the thermal conductivity, as shown in Fig. 7.15. Careful inspection indicates that the thermal conductivity tends to cross over from a linear dependence at high p to a quadratic dependence, but then fall rapidly when the contact resistance grows rapidly at reduced saturations. In spite of the complication of the contact resistance, a plot of 168 thermal conductivity values (Côté and Konrad 2005) under dry conditions generated (Hunt et al. 2014b) a regression proportional to $(1 - \phi)^{2.02}$ with an R^2 value of 0.79, compatible with the universal scaling of percolation. This result is analogous to the Archie exponent m.

Chapter 8
Mixing and Dispersion in Flow Through Porous Media

8.1 Introduction

In Chap. 7, we described and discussed flow and transport processes in porous media that involve only one fluid and one fluid phase. In the present chapter, we take up the flow and transport phenomena that are at the next level of complexity, namely, those that involve at least two fluids, but still one fluid phase. A prominent example is hydrodynamic dispersion that involves two miscible fluids. In the present chapter, we describe how percolation theory provides deep insight into this phenomenon.

8.1.1 *The Phenomenon of Dispersion*

When two miscible fluids are brought into contact, with an initially sharp front separating them, a transition zone develops across the initial front, the two fluids slowly diffuse into one another, and after some time develop a diffused mixed zone. If we assume that the volumes of the two fluids do not change upon the mixing, then the net transport of the solute across any arbitrary plane is represented by Fick's[1] second law of diffusion

$$\frac{\partial C}{\partial t} = D_m \nabla^2 C \ . \tag{8.1}$$

Here, C is the solute concentration at time t and D_m is the molecular diffusivity of the solute in the solvent. The mixing of the two fluids is independent of whether

[1] **Adolf Eugen Fick** (1829–1901) was a German physician and physiologist who in 1855 introduced Fick's law of diffusion of a gas across a membrane. He was also the first to measure cardiac output, which is now known as the *Fick principle*. Fick's work led to the development of the direct Fick method for measuring cardiac output, setting the foundation for dialysis and extracorporeal membrane oxygenation.

M. Sahimi, *Applications of Percolation Theory*, Applied Mathematical Sciences 213,
https://doi.org/10.1007/978-3-031-20386-2_8

or not there is flow through the medium. If, however, the solvent is also flowing, then there will be some additional mixing, caused by flow, which in turn is due to a nonuniform velocity field. The spatial distribution of the fluid velocities is controlled by the morphology of the medium, the Reynold number (fast or slow flow), and the chemical or physical interactions with the solid surface of the medium, i.e., reaction and adsorption. The resulting mixing is called *hydrodynamic dispersion* or dispersion for brevity.

Dispersion is important to a wide variety of processes, such as miscible displacements in enhanced oil recovery, salt water intrusion in coastal aquifers where fresh and salt waters mix by dispersion, and in situ study of the characteristics of aquifers, where a classical method of determining the characteristics is injecting fluid tracers into the aquifers and measuring their travel times. In addition, dispersion plays a fundamental role in pollution of the surface waters due to industrial and nuclear wastes. In particular, burying nuclear wastes deep in the rock has been suggested as a way of sequestering the pollution that they may cause, but leakage of the wastes into groundwater aquifers has been a major concern. Dispersion also occurs in flow and reaction in packed-bed chemical reactors and has been studied extensively by chemical engineers for a long time; see, for example, Bernard and Wilhelm[2] (1950) as one of the earliest papers on the subject.

8.2 Mechanisms of Dispersion

A most important characteristic of dispersion of a solute in flow through a disordered porous medium is the *first-passage time* (FPT) of a fluid particle or tracer between the entrance and exit planes. The FPT depends on the path, or streamline, that a tracer follows through the pore space. A population of tracers passing the entrance plane at the same instant will arrive at the exit plane by a set of streamlines with a distribution of transit times. Thus, a solute concentration front will spread in the mean-flow direction as it passes through the medium, giving rise to the FPT distribution (FPTD), which is a measure of longitudinal dispersion in pore space.

Likewise, a population of tracers, passing simultaneously through a restricted area of the entrance plane, will not follow entirely the mean flow to the exit plane, but will spread in the transverse directions, the direction perpendicular to the mean flow. The distribution of the FPTs for crossing the porous medium at a given transverse plane is a measure of transverse dispersion in the porous medium.

Dispersion is driven by two fundamental mechanisms in macroscopically homogeneous, microscopically disordered porous media. The two mechanisms arise in the pore-level velocity field caused by the irregularity of the pore space.

---

[2] **Richard Herman Wilhelm** (1909–1968), Chairman of Chemical Engineering Department at Princeton University, made significant contributions to fixed-bed catalytic reactors, fluidized transport, and separation processes. The R.H. Wilhelm Award of the American Institute of Chemical Engineers for contributions to reaction engineering is named in his honor.

(i) Streamtubes branch out and rejoin repeatedly at the junctions of flow passages in the interconnected pore space. The tangling and divergence of streamlines is strengthened by the widely varying orientations of flow passages and the local connectivity of the pore space, giving rise to wide variations in the lengths of the streamlines and their downstream transverse separations. This is a *kinematic* mechanism.

(ii) The speed for traveling along a given flow passage depends on the flow resistance or hydraulic conductance of the passage, its orientation, and the local pressure field. This is a *dynamic* mechanism of mixing.

The two mechanisms suggest two geometrical aspects of dispersion, defined with respect to direction of the mean velocity: a longitudinal effect due to the difference between the velocity components in the direction of the mean flow, and a transverse effect due to the differences between *local* velocities orthogonal to the direction of the mean flow. The net result is a broad FPTD between the entrance and exit planes of a porous medium.

Note that the two mechanisms do *not* depend on molecular diffusion, which, however, affects mixing by moving the solute from one streamline to another, as well moving it along the streamlines relative to the average velocity, which is typically weak. In addition, the solid surface of the pores acts locally as a separator of the streamlines and, thus, as a barrier to diffusion.

8.3 The Convective–Diffusion Equation

Dispersion in microscopically disordered and macroscopically isotropic and homogeneous porous media is usually described by the convective–diffusion (CD) equation:

$$\frac{\partial C}{\partial t} + \mathbf{v} \cdot \nabla C = D_L \frac{\partial^2 C}{\partial x^2} + D_T \nabla_T^2 C \ , \tag{8.2}$$

where $\mathbf{v}$ is the macroscopic mean-flow velocity, C is the solute's mean concentration, and ∇_T^2 is the Laplacian in the transverse directions. Thus, the process is represented by *anisotropic* diffusional spreading of the solute concentration, with the effective diffusivities being the longitudinal and transverse dispersion coefficients, D_L D_T. One important goal of any study of dispersion in porous media is to investigate the conditions under which it *cannot* be represented by the CD equation.

Dispersion is said to be *Gaussian* if it is described by the CD equation. If a population of the solute particles is injected into the medium at $\mathbf{r}_0 = (x_0, y_0, z_0)$ at $t = 0$ [i.e., $C(x_0, y_0, z_0, t) = C_0\delta(t)$], for diffusive dispersion the probability density $P(\mathbf{r}, t)$ follows the Gaussian distribution

$$P(\mathbf{r}, t) = (8\pi^3 D_L D_T^2 t)^{-3/2} \exp\left[-\frac{(x - x_0 - \mathrm{v}t)^2}{4D_L t} - \frac{(y - y_0)^2}{4D_T t} - \frac{(z - z_0)^2}{4D_T t}\right] , \tag{8.3}$$

where $P(\mathbf{r}, t)d\mathbf{r}$ is the probability that a solute particle is in a plane between $\mathbf{r}$ and $\mathbf{r} + d\mathbf{r}$ at time t, with $\mathbf{r} = (x, y, z)$. $P(\mathbf{r}, t)$ is proportional to C/C_0 and, therefore, Eq. (8.3) represents a solution of Eq. (8.2). If $\mathcal{Q}(\zeta - \zeta_0, t)dt$ is the probability that the solute, beginning in the plane at ζ_0, crosses *for the first time* a plane at ζ between t and $t + dt$, then, since $\mathcal{Q}$ and P are related:

$$P(\zeta - \zeta_0, t) = \int_0^t P(\zeta - \zeta_1, t - \tau)\mathcal{Q}(\zeta_1 - \zeta_0, \tau)d\tau \ , \tag{8.4}$$

we have

$$\mathcal{Q}(\zeta - \zeta_0, t) = |\zeta - \zeta_0|(4\pi D_\zeta t^3)^{-1/2} \exp\left[-\frac{(\zeta - \zeta_0 - \mathrm{v}_\zeta t)^2}{4D_\zeta t}\right] , \tag{8.5}$$

with D_ζ and v_ζ being, respectively, the dispersion coefficient and the mean-flow velocity in the ζ-direction.

The moments of $\mathcal{Q}$ yield information about the flow field and the dispersion. In particular, the first two moments of the FPTD in the longitudinal direction are given by

$$\langle t \rangle = \frac{L}{\mathrm{v}}, \tag{8.6}$$

and

$$\langle t^2 \rangle = \langle t \rangle^2 \left(1 + \frac{2D_L}{L\mathrm{v}}\right) , \tag{8.7}$$

where $L = \zeta - \zeta_0$. In general, one can easily show that for large L and to the leading order, one has

$$\langle t^n \rangle \sim \langle t \rangle^n \ , \tag{8.8}$$

where $n > 1$ is an integer. Equation (8.8) holds true, if dispersion is described by the CD equation. Therefore, one way of showing that the CD equation does not describe dispersion in a porous medium is by showing that $\langle t^n \rangle / \langle t \rangle^n$ $(n > 1)$ is *not* a constant, i.e., the ratio depends on t, and one requires more information than just $\langle t \rangle$ to describe the moments of the FPTD.

Experimental measurement of the longitudinal dispersion coefficient D_L is routine and straightforward, whereas the same is not true about D_T. Nuclear magnetic resonance (NMR) offers an efficient and accurate method for measuring D_T. It is a tracer method in that, every single molecule is tagged noninvasively by its local precession frequency in a nonuniform magnetic field (Callaghan 1991; Seymour and Callaghan 1997). Magnetic resonance imaging makes it possible to measure the distribution of the local velocities $\mathbf{v}(\mathbf{r}, t)$ (see, for example, Guilfoyle et al. (1992); Tessier and Packer (1998)). Thus, the two methods together provide an accurate method for measuring D_L and D_T and their relations with the mean-flow velocity. These matters were discussed comprehensively by Sahimi (2011).

8.4 Taylor–Aris Dispersion in a Capillary Tube

Dispersion in a capillary tube, as a model of a pore, is relevant to the same phenomenon in a porous medium. Although Griffiths (1911) was the first to report some experimental data that demonstrated the essence of the dispersion in a tube with the effect of molecular diffusion being present, it was Taylor (1953)[3] who first carried out a rigorous analysis of dispersion in a capillary tube. His analysis was approximate, but his final result for the dispersion coefficient D_L turned out later to be correct.

Aris (1956)[4] studied the same problem without making any of the approximations that Taylor had made. They both considered dispersion in a cylindrical capillary tube of radius R. Starting from a CD equation for a tube

$$\frac{\partial C}{\partial t} + 2\mathrm{v}_m \left[1 - \left(\frac{r}{R}\right)^2\right] \frac{\partial C}{\partial x} = D_m \left(\frac{\partial^2 C}{\partial r^2} + \frac{1}{r}\frac{\partial C}{\partial r} + \frac{\partial^2 C}{\partial x^2}\right) , \tag{8.9}$$

where v_m is the mean-flow velocity in the tube, and defining a *mean* concentration C_m by

$$C_m = \frac{\displaystyle\int_0^{2\pi}\int_0^R C(r, x) r dr d_\theta}{\displaystyle\int_0^{2\pi}\int_0^R r dr d\theta} = \frac{2}{R^2}\int_0^R C r dr . \tag{8.10}$$

Taylor and Aris showed that in the limit of long times,

$$\frac{\partial C_m}{\partial t} = D_L \frac{\partial^2 C_m}{\partial x_1^2} , \tag{8.11}$$

where $x_1 = x - \mathrm{v}_m t$ is the moving coordinate with respect to the mean-flow velocity, and

$$D_L = D_a + \frac{R^2 \mathrm{v}_m^2}{48 D_r} , \tag{8.12}$$

---

[3] Sir **Geoffrey Ingram Taylor** (1886–1975) was a British physicist and mathematician who made seminal contributions to the theory of turbulence and fluid dynamics, wave theory, solid mechanics, and dispersion among other phenomena. He has been described as "one of the greatest physical scientists of the 20th century."

[4] **Rutherford "Gus" Aris** (1929–2005) was Regents Professor of Chemical Engineering at the University of Minnesota. A member of the National Academy of Engineering, he finished a mathematics degree from the University of London at the age of 16, and received a doctorate by correspondence, writing his thesis in 6 weeks. He made seminal contributions to control theory, reaction engineering, and catalysis. One of his best known quotes, the basis of dynamic programming is, "If you don't do the best you can with what you happen to have, you'll never do the best you might have done with what you should have had." The author had the pleasure and honor of taking courses with him, and having him serving on his Ph.D. dissertation committee in 1984.

with the subscripts a and r signifying the fact that D_a and D_r are the contributions of the axial and radial molecular diffusion, respectively. That is, if we delete $\partial^2 C/\partial x^2$ from Eq. (8.8), i.e., neglect axial diffusion, D_a will also be deleted from Eq. (8.12). Numerically, of course, $D_a = D_r = D_m$. Note that in Taylor–Aris dispersion theory D_L depends *quadratically* on v_m.

8.5 Classification of Dispersion Regimes in Porous Media

Due to the significance of dispersion, ample experimental data have been accumulated over the past several decades, particularly for dispersion in beadpacks, unconsolidated sandpacks, and sandstones. Taken together, the data indicate that dispersion in consolidated porous media is similar to that in unconsolidated ones. Figure 8.1 collects experimental data for D_L/D_m for sandpacks, indicating that there are five discernable regimes of dispersion.

For unconsolidated porous media, the Péclet[5] number is defined by $\mathrm{Pe} = d_g \mathrm{v}/D_m$, where d_g is frequently taken to be the average diameter of a grain or bead. One may use a similar definition of Pe, with d_g replaced by the typical pore size. Then, the five dispersion regimes, indicated by Fig. 8.1, are as follows.

(i) $\mathrm{Pe} < 0.3$ defines the *diffusion* regime in which convection is so slow that molecular diffusion controls the mixing almost completely. In this regime, we have *isotropic* dispersion, i.e., one in which $D_L = D_T$, such that

$$\frac{D_L}{D_m} = \frac{D_T}{D_m} = \frac{1}{F\phi}, \tag{8.13}$$

where F is the formation factor (see Chap. 4) and ϕ is the porosity of porous media. The quantity $1/(F\phi)$, related to the tortuosity of the pore space (see Chap. 4), varies commonly between 0.15 and 0.7, depending on the type of porous media.

(ii) For $0.3 < \mathrm{Pe} < 5$, dispersion is in the *transition* zone in which convection contributes to mixing, but the effect of diffusion is still quite strong. D_L/D_m appears to increase with Pe, although it is difficult to say how.

(iii) For $5 < \mathrm{Pe} < 300$, one has the *power-law* dispersion regime. Convection dominates dispersion, but the effect of molecular diffusion cannot be neglected, and one writes

[5] **Jean Claude Eugene Péclet** (1793–1857) was a French physicist who was among the first students to study at the Ecole Normale Supérieure, and was also one of the founders in 1829 of the Ecole Centrale des Arts et Manufactures, a private engineering school in Paris whose aim was to train new types of engineers in the context of the ongoing industrial revolution. Péclet is notably famous for his Traite de la chaleur consideree dans ses applications in which he introduced a dimensionless number that now bears his name (the Péclet number) to quantify the ratio of convective and diffusive heat transfer, which has been generalized to mass transfer as well.

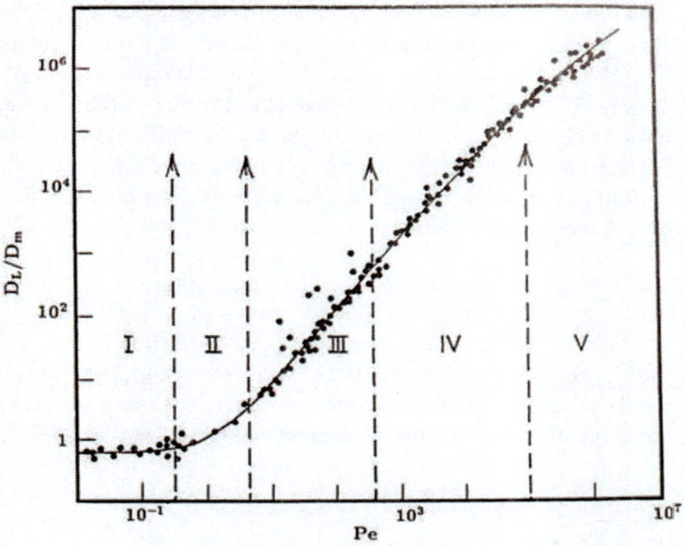

Fig. 8.1 Dependence of the longitudinal dispersion coefficient D_L on the Péclet number Pe, and the various dispersion regimes. $D_m = D$ is the molecular diffusivity

$$\frac{D_L}{D_m} = \frac{1}{F\phi} + a_L \text{Pe}^{\beta_L}, \tag{8.14}$$

$$\frac{D_T}{D_m} = \frac{1}{F\phi} + a_T \text{Pe}^{\beta_T}. \tag{8.15}$$

The *average* values of β_L and β_T from all the available experimental data are $\beta_L \simeq 1.2$ and $\beta_T \simeq 0.9$. This regime may also be referred to as the *boundary-layer* dispersion Koch and Brady (1985), because it is consistent with the existence of a diffusive boundary layer near the pores' solid surface, first identified by Saffman (1980), where molecular diffusion transfers the solute from the very slow regions near the solid walls to faster streamlines, where the flow fluid takes it away. In practice, β_L and β_T are often taken to be unity.

(iv) The *convective regime*, also called *mechanical dispersion*, is defined by $300 < \text{Pe} < 10^5$. In this regime, dispersion is simply the result of a stochastic fluid velocity field induced by the randomly distributed pore boundaries. Since diffusion no longer plays any role, simple dimensional analysis indicates that [ignoring the $1/(F\phi)$ term, which is small]

$$\frac{D_L}{D_m} \sim \text{Pe}, \tag{8.16}$$

$$\frac{D_T}{D_m} \sim \text{Pe}. \tag{8.17}$$

(v) *Turbulent dispersion* is defined by $\mathrm{Pe} > 10^5$, although the Péclet number is no longer the only correlating parameter, as the Reynolds number should also be considered. For flow through porous media, however, this regime is not of interest.

A sixth dispersion regime, which is not evident in Fig. 8.1, is the so-called *holdup dispersion* (Koch and Brady 1985), first studied by Aris (1959) and others. In this regime, the solute may be trapped for a long time in the dead-end pores or inside the solid grains, from which it can escape only by very slow molecular diffusion. One has [ignoring again the $1/(F\phi)$ term]

$$\frac{D_L}{D_m} \sim \mathrm{Pe}^2 \,. \tag{8.18}$$

In low-porosity porous media that are barely connected, with their porosity being close to the critical porosity or the percolation threshold, there are many dead-end pores and, therefore, this is a relevant dispersion regime.

8.6 Percolation and Pore-Network Models

The great advantage of pore-network models (PNMs) is that it allows to explicitly remove the mean-field nature of the flow field and the absence of the heterogeneity that are assumed in many other fluid-mechanical models of dispersion. Sahimi et al. (1982) were the first to develop the PNMs of dispersion in porous media. In their model, one first determines the flow field in the network. Then, solute particles are injected into the network at random at the upstream plane at $x = 0$. Each particle selects a streamline at random. The convective travel time for a pore is given by $t = l/\mathrm{v}_p$, where l is the pore's length and v_p is the flow velocity in the pore. Complete mixing at the nodes was assumed and, therefore, the probability that a pore was selected, once a particle arrived at a node, is proportional to the flow rate in that pore. The FPTD $\mathcal{Q}$ for the particles are computed by fixing the longitudinal or lateral positions and calculating the time at which the solute arrives at the fixed positions for the first time. D_ζ, the dispersion coefficient in the ζ-direction, is given by

$$D_\zeta = \int_0^\infty \mathcal{Q}(\zeta - \zeta_0, t)\frac{s_\zeta^2}{2t}dt \,, \tag{8.19}$$

where $\zeta_0(x_0)$ is the starting position of the particles, $s_x^2 = (x - x_0 - \mathrm{v}t)^2$, and $s_\zeta^2 = (\zeta - \zeta_0)^2$ for $\zeta = y$ and z.

The model is, however, appropriate for purely mechanical dispersion, since it ignores the pore-scale molecular diffusion. To include the effect of molecular diffusion and simulate the boundary-layer dispersion, the following method may be adopted (Sahimi and Imdakm 1988). The convective time t_c for traveling along a streamline in a pore is calculated. If $t_c \gg t_r$, where t_r is the time scale for significant radial diffusion, then one sets $t = t_c + t_r$, since the tracer has enough time to diffuse to a faster streamline. To simulate the holdup dispersion, the solute particles are allowed to diffuse into the dead-end pores. Transport in such pores is only by molecular diffusion. In a series of papers, Sahimi et al. (1982, 1983a, 1986a, b) and Sahimi and Imdakm (1988) demonstrated that such a PNM reproduces and simulates all the regimes of dispersion described above, see also Hasan et al. (2019).

A more refined PNM of dispersion using random-walk simulations was developed by Bijeljic et al. (2004). They mapped a Berea sandstone onto an equivalent pore network with square cross sections, and utilized the random-walk simulation, similar to that developed by Sahimi and co-workers. To include the effect of molecular diffusion, the particles were moved randomly, as defined by the coordinates of a spherical coordinate system. In a time step Δt, the diffusion distance is $r_d = \sqrt{2D_m \Delta t}$. Thus, the coordinates of the solute particle are given by the usual relation, $x = r_d \cos\theta \sin\varphi$, $y = r_d \sin\theta \sin\varphi$, and $z = r_d \cos\varphi$. The transition probabilities for moving the particles from one pore to another depended on the last step of the particles. If they arrived at a node of a network as a result of convection, then the next pore was selected with a probability proportional to the volume flow rate in that pore. If the solute particles arrived at a node as a result of a diffusion step, then the probability of selecting the next pore was taken to be proportional to the cross-sectional area of that pore. The longitudinal dispersion coefficient D_L was then computed by the method that Sahimi and co-workers used. PNM proposed by Bijeljic et al. correctly reproduced the dependence of D_L on the Péclet number Pe, as shown in Fig. 8.1. A somewhat similar model was also used by Bruderer and Bernabé (2001) in their studies of the transition from Taylor–Aris dispersion to purely mechanical dispersion in very heterogeneous porous media; see also Acharya et al. (2007). A similar model was used (Bijeljic and Blunt, 2007) for computing the transverse dispersion coefficient D_T. As in the case of D_L, the computed dependence of D_T on Pe was found to be in agreement with experimental data.

8.7 Effect of Percolation and Connectivity

Let us consider dispersion in porous media with percolation disorder, modeled by PNMs in which a fraction of the pores (bonds) are cut at random. According to the critical-path analysis described in Chap. 7, flow in a porous medium with a broad pore size or permeability distribution may be mapped onto an equivalent percolation problem. The same is true about dispersion in such porous media, as a broad pore-size distribution gives rise to a broad distribution of pore flow velocities that, in turn, affects dispersion. There are in fact two features of percolating pore networks that influence dispersion. One is the fact that such pore networks have a large number of dead-end pores (bonds) near the percolation threshold p_c and, thus, holdup dispersion described above may be important. The second feature is that for length scales shorter than the percolation correlation length ξ_p, the sample-spanning cluster (SSC) and its backbone are fractal objects (see Chap. 2) and, thus, dispersion is not expected to be described by a CD equation. We call this regime anomalous dispersion.

To study dispersion in percolation pore networks, two important characteristic quantities are essential: the *dispersivities*, $\alpha_L = D_L/\mathrm{v}$ and $\alpha_T = D_T/\mathrm{v}$, which are proportional to each other, but α_L is usually larger than α_T. Physically, the dispersivities represent the length scale over which a CD equation can describe dispersion and, thus, in some sense they are similar to the percolation correlation length ξ_p, because dispersion in a percolation network can be described by a CD equation only if the dominant length scale of the system is larger than ξ_p.

8.7.1 Scaling Theory

In their simulations of dispersion in percolation networks, Sahimi et al. (1982, 1983a, 1986a, b) found that as p_c is approached, the dispersion coefficients also increase. The increase may be attributed to the fact that near p_c the transport paths are very tortuous, resulting in broad first-passage time distributions (FPTDs) and, hence, extensive mixing zone of the two fluids. Figure 8.2 shows their results for dispersion in a percolating square network.

de Gennes (1983b) studied dispersion near p_c, and showed that in calculating D_L, the average flow velocity that one must use must be based on the *total* travel time of the solute in the sample-spanning cluster, rather than the travel time along the backbone *alone*. Indeed, in experimental measurements of the concentration profiles and D_L, it would be very difficult to measure the travel times along the backbone alone. Instead, what is routinely measured is the total travel or transit time.

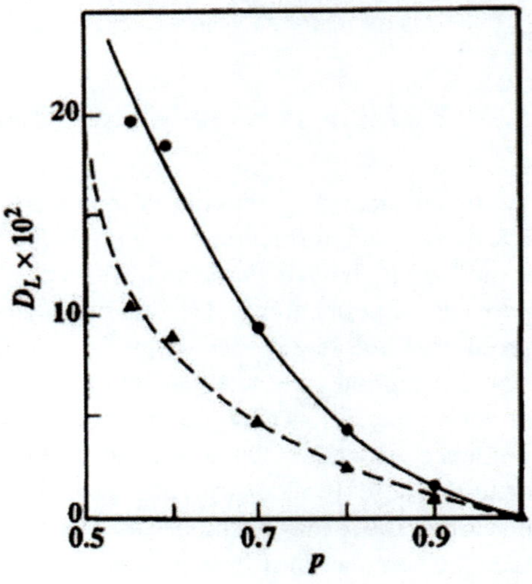

Fig. 8.2 Dependence of the longitudinal dispersion coefficient D_L on the fraction p of flow-carrying bonds in a square network. Circles denote the results for the case in which diffusion into the dead-end pores was accounted for, whereas triangles show the results for dispersion in the backbone of the network (after Sahimi and Imdakm 1988)

Thus, the discussions so far must have made it clear that D_L and D_T are sensitive to the structure of a porous medium. Similar to anomalous diffusion described in Chap. 7, we may define a crossover time τ_{co} such that for times $t \gg \tau_{co}$ dispersion is Gaussian or diffusive and follows a CD equation, whereas for $t \ll \tau_{co}$ it is anomalous, with the crossover between the two regimes taking place at about $t \simeq \tau_{co}$. For dispersion near p_c, this time scale is estimated from

$$\tau_{co} \sim \frac{\xi_p^2}{D_L} , \tag{8.20}$$

since the dominant length scale in the system is ξ_p. To derive the power laws for D_L and D_T, we must consider separately the various dispersion regimes described in Sect. 8.7. Let us first introduce two random-walk fractal dimensions by

$$\langle \Delta x^2 \rangle \sim t^{2/D_w^l} , \tag{8.21}$$

$$\langle y^2 \rangle \sim \langle z^2 \rangle \sim t^{2/D_w^t} , \tag{8.22}$$

where $\langle \Delta x^2 \rangle = \langle (x - \langle x \rangle)^2 \rangle = \langle x^2 \rangle - \langle x \rangle^2$. Equations (8.21) and (8.22) are defined for length scales $L \ll \xi_p$. Two average flow velocities may also be defined. One is an average velocity v_c, defined in terms of the travel time in the SSC. Then, $v_c \sim K_e / X^A$, where K_e is the effective permeability of the pore network (porous medium), and X^A is the percolation accessible fraction described in Chap. 2. Thus, near p_c

$$v_c \sim (p - p_c)^{T-\beta} \sim \xi_p^{-\theta_c} , \tag{8.23}$$

where $\theta_c = (T - \beta)/\nu$, and T, ν, and β are, respectively, the critical exponents of the permeability, the correlation length, and the accessible fraction, defined in Chap. 2. On the other hand, if an average solute velocity v_B is defined in terms of the travel times along the backbone, then $v_B \sim K_e / X^B$, where X^B is the percolation backbone fraction defined in Chap. 2. Thus, near p_c

$$v_B \sim (p - p_c)^{T-\beta_{bb}} \sim \xi_p^{-\theta_{bb}} , \tag{8.24}$$

where $\theta_B = (T - \beta_{bb})/\nu$, and β_{bb} is the critical exponent of X^B defined in Chap. 2. For length scales $L \ll \xi_p$, one should replace ξ_p in Eqs. (8.23) and (8.24) by L and, therefore, $v_c \sim L^{-\theta_c}$ and $v_B \sim L^{-\theta_{bb}}$, respectively. We also define a *macroscopic* Péclet number by

$$\mathrm{Pe}_M = \frac{v \xi_p}{D_e} , \tag{8.25}$$

where D_e is the effective diffusivity of the system and v is either v_c or v_B. For $L \ll \xi_p$, we replace ξ_p in Eq. (8.25) by L. Having defined the essential quantities, we now investigate the power laws that the dispersion coefficients and τ_{co} follow near p_c.

(i) Consider, first, the small Péclet number regime. In this case, convection has no effect and, $D_L \sim D_T \sim D \sim (p - p_c)^{\mu-\beta}$, as derived in Chap. 7, where μ is the critical exponent of the electrical conductivity of the (fluid-saturated) system. For $L \ll \xi_p$, we have anomalous diffusion, and based on the discussions in Chap. 7, we can immediately write

$$D_w^l = D_w^t = D_w = 2 + \theta , \tag{8.26}$$

where $\theta = (\mu - \beta)/\nu$, and D_w is the fractal dimension of the random walk defined in Chap. 7. Moreover, according to Eq. (8.20)

$$\tau_{\rm co} \sim (p - p_c)^{-\mu-2\nu+\beta} \sim \xi_p^{2+\theta} , \tag{8.27}$$

so that, $\tau_{\rm co} \sim L^{2+\theta}$ for $L \ll \xi_p$. Equation (8.26) describes dispersion coefficients in the entire SSC in the limit $\mathrm{Pe}_M \to 0$. For dispersion in the backbone θ in Eqs. (8.26) and (8.27) should be replaced with $\theta_{\rm bb} = (\mu - \beta_{\rm bb})/\nu$. Moreover, for the nth moment of the FPTD, we have $\langle t^n \rangle \sim \langle t \rangle^n$, where $n > 1$ is an integer and, therefore, for $L \ll \xi_p$, we have $\langle t^n \rangle \sim (L^2/D_e)^n$, so that

$$\langle t^n \rangle \sim L^{n(2+\theta)} . \tag{8.28}$$

(ii) Although any porous medium has a large number of dead-end pores near its critical porosity, as the experiments of Charlaix et al. (1988a) indicated, the porous medium must be extremely close to p_c if the dead-end pores are to have significant effect. Thus, it is worthwhile to consider dispersion only in the backbone, for which we have $D_L/D_e \sim \mathrm{Pe}_M$, and $D_T/D_e \sim \mathrm{Pe}_m$ (i.e., mechanical dispersion), which means that $D_L \sim D_T \sim \xi_p \mathrm{v}_B \sim \xi_p^{1-\theta_{\rm bb}} \sim (p - p_c)^{T-\beta_{\rm bb}-\nu}$.

Using the numerical values of T, $\beta_{\rm bb}$, and ν given in Table 2.3, we obtain $D_L \sim D_T \sim (p - p_c)^{-0.56}$ in 2D, and $D_L \sim D_T \sim (p - p_c)^{0.04}$, so that D_L and D_T *diverge* in 2D, but *vanish* very weakly in 3D. For $L \ll \xi_p$, we replace ξ_p with L and, thus,

$$D_L \sim L^{1-\theta_{\rm bb}} , \tag{8.29}$$

with a similar power law for D_T. As all the length scales of the system must be proportional to each other (and to L), Eq. (8.29) may be rewritten as $D_L \sim \langle \Delta x^2 \rangle^{(1-\theta_{\rm bb})/2}$. Using a fundamental property of random-walk processes, $D_L \sim d\langle \Delta x^2 \rangle/dt$, we obtain $d\langle \Delta x^2 \rangle/dt \sim \langle \Delta x^2 \rangle^{(1-\theta_{\rm bb})/2}$, which after integration yields

$$\langle \Delta x^2 \rangle \sim t^{2/(1+\theta_{\rm bb})} , \tag{8.30}$$

with similar equations for $\langle y^2 \rangle$ and $\langle z^2 \rangle$. Equation (8.30) implies that

$$D_L \sim t^{(1-\theta_{\rm bb})/(1+\theta_{\rm bb})} , \tag{8.31}$$

and that

$$D_w^l = D_w^t = 1 + \theta_{\rm bb} \; . \tag{8.32}$$

Equation (8.30) implies that in 2D, $\langle \Delta x^2 \rangle \sim \langle y^2 \rangle \sim t^{1.26}$, and in 3D, $\langle \Delta x^2 \rangle \sim \langle y^2 \rangle \sim \langle z^2 \rangle \sim t^{0.97}$. That is, dispersion is *superdiffusive* in 2D, so that the mean-square displacements of the solute grow with time faster than linearly, whereas it is *subdiffusive* in 3D, so that the mean-square displacements grow slower than linearly with time.

On the other hand, according to Eq. (8.24) for $L \ll \xi_p$ we have $v_B \sim L^{-\theta_{\rm bb}} \sim \langle x \rangle^{-\theta_{\rm bb}}$, and since $v_B \sim d\langle x \rangle/dt$, we obtain, after integration

$$v_B \sim t^{-\theta_{\rm bb}/(1+\theta_{\rm bb})} \; , \tag{8.33}$$

which is in sharp contrast with diffusive dispersion for which v_B is *constant*. Finally, since $\alpha_L = D_L/v_B$, we obtain

$$\alpha_L \sim t^{1/(1+\theta_{\rm bb})} \; , \tag{8.34}$$

which means that for nondiffusive dispersion, the dispersivity depends on t. The time scale $\tau_{\rm co}$ is given by

$$\tau_{\rm co} \sim (p - p_c)^{-T+\beta_{\rm bb}-\nu} \sim \xi_p^{1+\theta_{\rm bb}} \; , \tag{8.35}$$

and for $L \ll \xi_p$ we have $\tau_{\rm co} \sim L^{1+\theta_{\rm bb}}$. Equation (8.35) should be compared with Eq. (8.27). It is not too difficult to show that $\langle t^n \rangle \sim (p - p_c)^{-\nu-n(T-\beta_{\rm bb})} \sim \xi_p^{1+n\theta_{\rm bb}}$, and $\langle t \rangle \sim \xi^{1+\theta_{\rm bb}}$, so that, $\langle t^n \rangle/\langle t \rangle^n \sim \xi_p^{1-n}$. That is, from the scaling of $\langle t \rangle$ alone one *cannot* obtain the scaling of $\langle t^n \rangle$ for $n > 1$. For $\xi_p \gg L$, we have

$$\langle t^n \rangle \sim L^{1+n\theta_{\rm bb}} \; , \tag{8.36}$$

which should be compared with Eq. (8.28) if we replace θ by $\theta_{\rm bb}$.

(iii) For holdup dispersion, one has $D_L \sim (v_c\xi_p)^2/D_e$, which is the same as Eq. (8.18) in which the length scale is ξ_p and molecular diffusivity D_m has been replaced by the effective diffusivity D_e in the porous medium, as suggested by de Gennes (1983b). We therefore write, $D_L \sim \xi_p^{2-2\theta_c+\theta} \sim (p - p_c)^{-2\nu-\mu-\beta+2T}$, with a similar result for D_T, and thus $D_L \sim (p - p_c)^{-1.5}$ and $D_L \sim (p - p_c)^{-0.17}$ in 2D and 3D, respectively. That is, as p_c is approached the dispersion coefficients *diverge*, which undoubtedly is due to the contribution of the dead-end pores and the long times that the solute particles spend there. For $L \ll \xi_p$, we have

$$D_L \sim L^{2-2\theta_c+\theta} \; . \tag{8.37}$$

Using the same type of reasoning as before, we find that

$$\langle \Delta x^2 \rangle \sim t^{2/(2\theta_c-\theta)} \; , \tag{8.38}$$

with similar scalings with the time t for $\langle y^2 \rangle$ and $\langle z^2 \rangle$. Thus,

$$D_L \sim t^{(2-2\theta_c+\theta)/(2\theta_c-\theta)} \tag{8.39}$$

and

$$D_w^l = D_w^t = 2\theta_c - \theta \;, \tag{8.40}$$

and, therefore, $\langle \Delta x^2 \rangle \sim t^{2.3}$ and $\langle \Delta x^2 \rangle \sim t^{1.1}$ in 2D and 3D, respectively. That is, dispersion *is always superdiffusive* when the holdup dispersion, i.e., trapping of the solute particles in the stagnant regions of the pore space, is dominant. It is then straightforward to show that

$$\mathrm{v}_c \sim t^{-\theta_c/(1+\theta_c)} \;, \tag{8.41}$$

and therefore

$$\alpha_L \sim t^{(2+\theta)/[(2\theta_c-\theta)(1+\theta_c)]} \;. \tag{8.42}$$

The time scale τ_{co} is given by

$$\tau_{co} \sim (p - p_c)^{-2T+\mu+\beta} \sim \xi_p^{2\theta_c-\theta} \;, \tag{8.43}$$

and $\tau_{co} \sim L^{2\theta_c-\theta}$ for $L \ll \xi_p$. Therefore, we obtain $\langle t^n \rangle \sim (p-p_c)^{-n\nu(\theta+2)+\nu} \sim \xi_p^{n(\theta_p+2)-1}$, so that $\langle t^n \rangle / \langle t \rangle^n \sim \xi_p^{n-1}$ and, therefore, the scaling of $\langle t \rangle$ alone is *not* enough for obtaining the scaling of the higher moments $\langle t^n \rangle$ of the FPTD for any $n > 1$. In the $L \ll \xi_p$ regime, we have

$$\langle t^n \rangle \sim L^{n(\theta+2)-1} \;. \tag{8.44}$$

The above scaling relations were derived by Sahimi (1987) and refined later on by Sahimi (2012), and were confirmed by the PN simulations of Sahimi and Imdakm (1988) and Koplik et al. (1988). The fact that $\langle t^n \rangle / \langle t \rangle^n$ depends on n means that *there is no unique time scale for characterizing non-Gaussian dispersion* in porous media with percolation disorder. Koplik et al. (1988) also proposed a general scaling equation for the FPTD, given by

$$\mathcal{Q}(t) = \frac{1}{t_d} F_s\left(\frac{t}{t_d}, \frac{t_c}{t_d}\right) \;, \tag{8.45}$$

where t_d and t_c are the diffusion and convective time scales, respectively, and $t_d \sim \xi_p^{2+\theta}$, as Eq. (8.27) indicates. t_d is also the largest time that the solute particles spend in the dead-end pores, since the length of the longest dead-end branches is on the order ξ_p. The scaling function F_s has the following limiting behavior:

$$F_s(x, y) \to \begin{cases} F_1(x) & \text{as } y \to \infty \;, \\ yF_2(x) & \text{as } \; y \to 0 \;. \end{cases} \tag{8.46}$$

The limit of pure diffusion corresponds to $y = t_c/t_d \to \infty$, whereas the convective limit corresponds to $y \to 0$.

Related pore-network simulations were carried out by Andrade and co-workers (Lee et al. 1999; Makse et al. 2000; Andrade et al. 2000). They utilized pore networks at the percolation threshold and computed several important properties, including the minimum traveling time $t_{\min}$ and the most probable time t^* between two points separated by a distance r, the length $\ell_{\min}$ of the path corresponding to the time $t_{\min}$, and ℓ^* corresponding to t^*. For 2D pore networks at the percolation threshold, they found that $t_{\min} \sim r^{1.33}$, $t^* \sim r^{D_{\rm bb}} \sim r^{1.64}$, $\ell_{\min} \sim r^{D_{\min}} \sim r^{1.13}$, and $\ell^* \sim r^{D_{\rm op}} \sim r^{1.22}$. Here, $D_{\rm bb}$ is the fractal dimension of the backbone (the flow-carrying part of the network; see Chap. 2), $D_{\min}$ is the fractal dimension of the *shortest path* between the two points, and $D_{\rm op}$ is the fractal dimension of the *optimal path* between the two points. Although seemingly different from the above scaling relations, these results are virtually identical with those derived above for 2D networks, when the numerical values of the various fractal dimensions are used. In 3D, however, the two sets of scaling relations differ, the reason for which is not clear.

8.8 Critical-Path Analysis

In Chap. 7, we described the critical-path analysis (CPA) and its applications to estimating flow and transport properties of porous media and materials. The CPA was invoked by Hunt et al. (2011) to study dispersion in heterogeneous porous media and to derive expressions for the properties of interest. Recall that in the CPA transport in a highly heterogeneous porous medium is mapped onto a percolation system at or very near the percolation threshold p_c. Thus, Hunt and Skinner (2008) defined a tortuosity τ for the SSC of percolation, and in particular that of the backbone—the flow-carrying part of the cluster. Near p_c the length of the shortest path $\xi_{\min}$ on the backbone follows a power law,

$$\xi_{\min} \sim |p - p_c|^{-\nu_m} , \tag{8.47}$$

where ν_m is a universal exponent. The (relative) tortuosity was then defined by Hunt and Skinner as $\tau = \xi_{\min}/\xi_p$, where ξ_p is the correlation length of percolation. Since near p_c, $\xi_p \sim |p - p_c|^{-\nu}$, where ν is the associated universal exponent, one has

$$\tau \sim |p - p_c|^{\nu - \nu_m} . \tag{8.48}$$

On the other hand, as mentioned earlier, Lee et al. (1999) studied the statistics of the travel times of solute particles in the percolation clusters near p_c, and showed that the time t^* for traveling along the most probable path between two points separated by a distance x scales as

$$t^* \sim x^{D_{\rm bb}} , \tag{8.49}$$

where D_{bb} is the fractal dimension of the backbone (see Chap. 2).

These concepts were utilized by Hunt et al. (2011) in order to study dispersion and derive analytical expressions for the longitudinal dispersion coefficient D_L. The general procedure that they developed is as follows:

(i) Assume that the number of bonds is b. One adapts the cluster statistics, given in terms of b and the usual percolation variable $(p - p_c)$ to the linear extent of the porous medium, and to specific values of the smallest conductance g on such a cluster, as well as its critical conductance g_c that the CPA identified (see Chap. 7). To do so, one equates the cumulative distribution that ranges from $g(g_c)$ to ∞ with the corresponding probability $p(p_c)$. In practice, however, the volume fraction V and its critical value V_c are utilized, instead of p and p_c (see Chap. 7).

(ii) One then integrates the cluster statistics over all clusters larger than or equal to the system size in order to determine the probability density function $W_p(g; x)$, the probability that a system of length x is spanned by an interconnected cluster of conductances with minimum conductance g.

(iii) One determines the time $t_g(g)$ required for the solute to traverse the clusters.

(iv) One determines $W_p(t; x) = [gW_p(g; x)][dg/dt_g(g)]$. Note that the extra factor g is due to the fact that the flux of a uniform concentration of the solute must be proportional to the solvent flux, and the probability that a solute particle arriving at a given distance differs from the probability that such a cluster actually exists.

To apply the procedure, one must specify a model of the pore-size distribution of the porous media. Hunt et al. (2011) used a fractal model for the pore-size distribution proposed by Rieu and Sposito (1991). In such fractal porous media, the length of a pore is proportional to it radius. Thus, the probability density function representing the fraction of pore volume in pores of radius r is $W(r) = (3 - D_f)r^{2-D_f}/r_M^{3-D_f}$, where D_f is the fractal dimension of the pore space and r_M is the maximum pore size. Therefore, the porosity ϕ is given by

$$\phi = \int_{r_m}^{r_M} W(r)dr = 1 - \left(\frac{r_m}{r_M}\right)^{3-D_f} , \tag{8.50}$$

where r_m is the minimum pore size. The two radii r_m and r_M correspond to hydraulic conductances K_m and K_M and, therefore, $r_m/r_M = K_m/K_M$.

Next, step (i) above is carried out. One needs a specific value V_c for percolation. Given the discussion in Chaps. 4 and 7, let us take V_c to be the moisture content S_c of a porous medium at which the effective diffusivity vanishes. For slow flow, the flow conductance of a pore is $g \propto r^4/l$, and since $l \propto r$, we obtain $g \propto r^3$. Thus, the size distribution $W(r)$ is converted to a conductance distribution $W(g)$, and

$$V_c = \int_{g_c}^{g_M} W(g)dg , \tag{8.51}$$

so that if V_c is known, so will also be g_c.

Step (ii), i.e., determining $W_p(g;x)$, is carried out next. First, note that $V - V_c$ can be transformed to $g^{1-D_f/3} - g_c^{1-D_f/3}$, as shown by Hunt and Skinner (2008), who also showed that

$$W(g;x) \propto \frac{1}{c} Ei\left[a\left(\frac{x}{L}\right)^c\right], \tag{8.52}$$

where $Ei(y) = \int_y^\infty [\exp(-t)/t]dt$ is the exponential integral, $a = |1-(g/g_c)^{(3-D_f)/3}|^2$, and $c = 2/\nu$, with ν being the critical exponent of percolation correlation length ξ_p, defined in Chap. 2. Then, the time $t(g)$ that a solute particle takes to enter the pore space on one face and exit it on the opposite first, i.e., its first-passage time [step (iii) above] is given by (Hunt and Skinner 2008).

$$t(g) = t_0\left(\frac{x}{L}\right)^{D_{bb}}\left(\frac{1}{3-D_f}\right)\left[\frac{1}{(1-S_c)^{\nu(D_{bb}-1)}}\right]\left[\left(1+\frac{S_c}{1-S_c}\right)\left(\frac{g}{g_c}\right)^{1-D_f/3} - 1\right]$$
$$\times\left[\frac{1}{\left|\left(\frac{g}{g_c}\right)^{1-D_f/3} - 1\right|}\right]^{\nu(D_{bb}-1)} \equiv t_g\left(\frac{x}{L}\right)^{D_{bb}}. \tag{8.53}$$

Here, t_0 is the typical time for the solute to travel along a single pore, and D_{bb} is the fractal dimension of the backbone of the SSC of percolation (see Chap. 2).

Having determined $t_g(g)$, step (iv) of the above procedure can be taken by determining $W_p(t;x) = [gW_p(g;x)][dg/dt_g(g)]$ and, therefore, its variance, $\sigma_x^2 = \langle x^2\rangle - \langle x\rangle^2$. The longitudinal dispersion coefficient D_L is then given by

$$D_L = \frac{1}{2}\frac{d\sigma_x^2}{dt}. \tag{8.54}$$

As described earlier, for Gaussian dispersion D_L is a constant in the limit of long times, whereas for anomalous dispersion D_L is always a function of time. One may also determine a dispersivity α_L by

$$\alpha_L = \frac{\sigma_x^2}{\langle x\rangle}, \tag{8.55}$$

which in the limit $t \to \infty$ is constant, if dispersion is Gaussian and, therefore, the properties of α_L may be studied as a way of studying anomalous dispersion.

In order to see whether the predictions by the new expressions agree with the experimental data, one must select its various parameters carefully. One such parameter is the critical volume fraction V_c that, as described in Chap. 3, takes on a value, $V_c \simeq 0.15$ for 3D continuum percolation. The numerical value of ν, the critical exponent for 3D percolation, was given in Chap. 2, $\nu \simeq 0.89$. The fractal dimension D_{bb}

of the backbone depends, however, on the percolation model considered (Sheppard et al. 1999). Invasion percolation (IP) (see Chaps. 3 and 16) in which the defending fluid can remain trapped in clusters (or, in practice, in pockets, such as air bubbles) is called the IP with trapping, and models slow displacement of an incompressible fluid by another immiscible fluid, which is the most relevant to the present problem. The relevance of the IP with trapping to the present case is because its backbone tends to be less tortuous than in random percolation, hence producing a smaller dispersivity, but a larger dispersion coefficient D_L, because with a minimally tortuous path the fluid velocity hardly diminishes with distance. Finally, one must choose the fractal dimension D_f, which is *not* the same as that of the SSC of percolation, but rather is related to the pore-size distribution $W(r)$ mentioned earlier.

In the next section, we will compare the predictions with the experimental data and the results of precise computer simulations.

8.9 Comparison with Experiments

The increase in the dispersion coefficients near p_c that is shown in Fig. 8.2 was confirmed by the experiments of Charlaix et al. (1987b, 1988a) and Hulin et al. (1988a), who studied dispersion in model porous media and measured D_L. Charlaix et al. (1988a) constructed 2D hexagonal networks of pores with diameters that were of the order of millimeters. They reported that as the fraction of the open pores decreased, D_L increased sharply, and that Eq. (8.14) seems to agree with their data. But, even when the dispersion coefficients were measured quite close to p_c, the quadratic dependence of D_L/D_m on Pe, Eq. (8.18), was *not* satisfied (although the fraction of the dead-end pores is quite large near p_c), presumably because the exchange time between the flowing fluids and the dead-end regions was so long that could not be detected during the experiment.

Hulin et al. (1988a) measured D_L in bidispersed sintered glass materials, prepared from mixtures of two sizes of beads. They reported that when the porosity was decreased from 0.3 to 0.12, D_L increased by a factor of 30. The results of the two studies also indicated that dispersion is more sensitive to *large-scale* inhomogeneities of a porous medium than to its detailed *local* structure. Somewhat similar results were obtained by Charlaix et al. (1987b).

Gist et al. (1990) studied dispersion in a variety of sandstones and carbonate rocks and used the CPA model of Katz and Thompson (1986, 1987), described in Chap. 7, to quantify their results. Following Sahimi et al. (1982, 1983a, 1986a, b) they argued that the fundamental quantity to be considered is the ratio ξ_p/d_g, where d_g is the mean grain size. Since $\xi_p/d \sim (X^A)^{-\nu/\beta}$, where X^A is the percolation accessible fraction (see above and also Chap. 2), because $\nu/\beta \simeq 2.0$ (see Chap. 2), one writes $\xi_p/d_g \sim (X^A)^{-2}$. Because X^A is roughly proportional to the fluid saturation S, we obtain

$$\frac{\xi_p}{d_g} \sim S^{-2} . \tag{8.56}$$

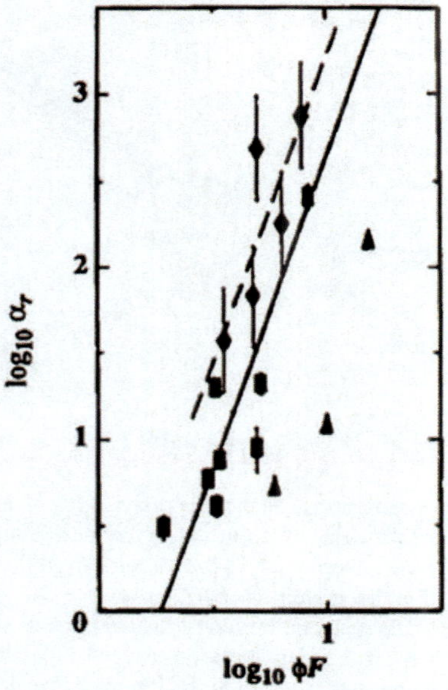

Fig. 8.3 Dependence of the reduced dispersivity, $\alpha_r = \alpha_L/d_g$ on ϕF, where ϕ is the porosity and F is the formation factor, in sandstone (squares), epoxies (triangles), and carbonates (diamonds). The solid line is the fit of the data for sandstones, while the dashed line represents the fit for the carbonates (after Gist et al. (1990))

Gist et al. (1990) derived a relation between $\alpha_r = \alpha_L/d_g$ and ξ_p/d_g using the percolation model of Sahimi et al. Their final result is

$$\alpha_r \sim \left(\frac{\xi_p}{d_g}\right)^{2.2} \sim S^{-4.4} . \tag{8.57}$$

Experimental data of Gist et al. for sandstones, epoxies, and carbonates supported the validity of Eq. (8.57); see Fig. 8.3, hence confirming the relevance of percolation to dispersion in a heterogeneous porous medium.

The CPA analysis of Hunt et al. (2011) is also supported by a vast amount of data. Figure 8.4 compares the length scale dependence of the CPA-predicted dispersivity $\alpha_L(x)$ with the experimental data over orders of magnitude variations in the length scale x.

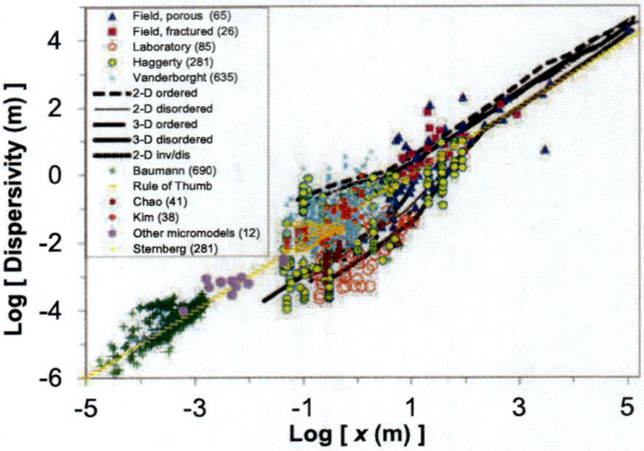

Fig. 8.4 Comparison of CPA predictions for the dispersivity α_L as a function of length scale x, showing clearly that the trend of the data is the same over the entire range of length scales and does not change from a slope of 1.5 to a slope of 1 in the vicinity of 100 m. A slope of 1.5 for length scales below 100 m would lead to dispersivities of 10^{-8} m at $x = 10^{-4}$ m, instead of the measured values more nearly 10^{-5} m. The data at smaller length scales are not only spread over many orders of magnitude of the spatial scale, but are also numerous (over 1300 data points). The inset refers to the references for the experimental data, for which the interested reader should consult Hunt et al. (2011) (After Hunt et al. (2011); courtesy of Professor Allen Hunt)

Therefore, once again percolation theory provides a power tool for analyzing a complex phenomenon in porous media and makes accurate predictions for the important quantities of interest.

Chapter 9
Multiphase Fluid Flow in Porous Media

9.1 Introduction

Two- and three-phase flows in porous media and materials constitute a set of important phenomena that occur in a wide variety contexts, from enhanced recovery of oil, extraction of thermal energy from geothermal reservoirs, and storage and sequestration of CO_2 in geological formations, to transmission of water by building materials, flow and evaporation of brine in soil and the resulting salt precipitation in the pore space, packed-bed chemical reactors, and many more. A large number of factors affect this class of phenomena, including capillary, viscous, and gravitational forces, the viscosities of the fluids and the interfacial tension separating them, the morphology of the pore space, the wetting properties of the fluids and pores' surface, and others.

But, as described in Chaps. 2 and 4, in a disordered multiphase system the connectivity of the phases plays a critical role in determining its macroscopic properties, and multiphase flows in porous media are no exception. Although the classical continuum models have provided a wealth of information and insights into multiphase flows in porous media, they cannot predict the relative permeabilities (RPs) to the flowing phases. Assuming that fluid flow is slow enough so that Darcy's law for the fluid velocity, $\mathbf{v} = -(K_e/\eta)\nabla P$, holds, where K_e, η, and P, are, respectively, the effective permeability of the pore space, the fluid's viscosity, and the pressure, one generalizes Darcy's law to phase i of a two-phase (or three-phase) flow system by writing, $\mathbf{v}_i = -(K_i/\eta_i)\nabla P_i = -k_{ri}(K_e/\eta_i)\nabla P_i$, where $k_{ri} = k_i/K_e$ is the RP to phase i. Estimating K_e by various theoretical approaches was described in Chap. 7.

The reason for the inability of the continuum model for predicting the RPs, particularly when one or more fluid phases are barely connected, is that they cannot account for the effect of the connectivity of the pore space of the fluid phases. In fact, the RPs and capillary pressure represent the input to such models. Therefore, one must develop an independent way of computing the RPs and the capillary pressure. Pore-network (PN) models, of the type described in the previous chapters, together with

M. Sahimi, *Applications of Percolation Theory*, Applied Mathematical Sciences 213,
https://doi.org/10.1007/978-3-031-20386-2_9

the concepts of percolation theory, currently represent the most promising approach to the task of computing the RPs and the capillary pressure. Computing the latter was described in Chap. 4. In the present chapter, we describe the PN models of multiphase flows and the computations of the RPs, and the prominent role that percolation theory plays in it.

But, before embarking on describing the various percolation models of two-phase flows in porous media, it should be emphasized that a fundamental assumption in *all* the percolation models of two-phase flow in porous media is that the occupation probability p of pore throats (bonds) or pore bodies (sites or nodes), i.e., the probability that a bond or site is filled by a fluid, is proportional to the capillary pressure needed for entering that bond or site. Without making the assumption, it would be difficult to make a one-to-one correspondence between a percolation model and the two-phase flow problem in porous media. Although in some percolation models, such as invasion percolation (IP) (see below and also Chap. 3), the occupation probability is not explicitly defined, its analog can be readily calculated.

Over the past four decades, several PN models of two-phase flow in porous media have been developed that were based explicitly on the concepts of the classical percolation and its variants (see Chap. 3). Strictly speaking, such models are applicable only when the capillary number Ca, Eq. (4.18), is very small. Other models, though more general and applicable when Ca is not too small, are still based on the percolation concepts, as they invoke the concept of macroscopic connectivity of the pore space.

9.2 Random Percolation Model

Let us first mention that detailed pore-network simulations of two- and three-phase flows in porous media have been carried out by a large number of groups and, in particular, by Valvatne and Blunt (2004), and Piri and Blunt (2005a, b). Such simulations, without invoking explicitly the concepts of percolation—even though these are percolation problems—have provided considerable insights into multiphase flows in porous media, although the associated computations are typically highly intensive.

Larson (1977) was first to develop a random percolation model of two-phase flow in porous media. The details of his model were explained in a series of papers by Larson (1977), Larson et al. (1981a, b). Larson et al. (1981a) described a random percolation model for drainage, i.e., displacement of a wetting fluid by a non-wetting one, in the limit of low capillary number Ca. As usual, they represented a porous medium by a simple-cubic network of bonds and sites whose sizes were distributed according to a pore-size distribution. A bond (throat) next to the interface between the two fluid phases could be penetrated by the displacing fluid, if the capillary pressure at that point exceeded a critical value, implying that the bond's radius must exceed a critical radius $r_{\min}$, the same radius that is defined by Eq. (4.36), implying that during drainage the *largest* pore throats are filled by the non-wetting fluid.

All the bonds that are connected to the (non-wetting) displacing fluid by a path of pores or bonds with effective radii larger than $r_{\min}$ are considered as *accessible*, with the accessibility defined in the percolation sense described in Chap. 2. Larson et al. (1981a) also assumed that all the accessible bonds with radii that are at least as large as $r_{\min}$ are filled with the non-wetting fluid. Strictly speaking, the assumption is not, of course, correct, because an interface that starts at one external face of a porous medium must travel along a certain path of eligible connected pores and throats before it reaches an accessible bond that can be penetrated. Larson et al. (1981a) also assumed that the displaced fluid is compressible, so that even if a cluster of pore throats filled by the fluid is surrounded by the displacing fluid, it can still be invaded. As we discuss below, the assumption of compressibility does not, however, result in a serious error.

In another paper, Larson et al. (1981b) proposed a percolation model of imbibition, in order to calculate the residual non-wetting phase saturation $S_{\rm rnw}$ and its dependence on Ca. To do so, they modeled the creation of isolated blobs of the non-wetting fluid by a random-site percolation (see Chap. 2), and calculated the fraction $\hat{g}(s)$ of the active sites at the site percolation threshold that are in clusters of length s in the direction of flow. They then argued that the quantity represents the desired blob-size distribution. To compute $S_{\rm rnw}$, Larson et al. assumed that once a blob is mobilized, it is permanently displaced. But in reality this is not always the case, because a blob can get trapped again, can join another blob to create a larger one, and so on.

Since the goal of such models is to simulate slow displacement of one fluid by another immiscible fluid, the fundamental assumption in all percolation models is that pore-level events are controlled by the capillary forces. Simple scaling analysis enables us to estimate the range of the capillary number for which the assumption is satisfied. The capillary pressure across the interface is proportional to

$$P_c \sim \frac{\sigma \cos\theta}{d_g}\,, \tag{9.1}$$

where d_g is a typical grain size, σ is the interfacial tension, and θ is the contact angle. On the other hand, the viscous pressure drop is proportional to

$$P_\mu \sim \frac{\eta_w v d_g}{K_e}\,, \tag{9.2}$$

with η_w being the viscosity of the wetting phase. Therefore,

$$\frac{P_\mu}{P_c} \sim \frac{\rm Ca}{K_d}\,, \tag{9.3}$$

where $K_d = K_e/d_g^2$ is a dimensionless permeability, which is small (on the order of 10^{-3} or smaller) because K_e, the effective permeability of the porous medium, is controlled by the narrowest throats in the medium. For capillary-controlled displacements, one must have ${\rm Ca} \ll 1$. In practice, one has ${\rm Ca} \sim 10^{-6} - 10^{-8}$. Experimental data (Le Febvre du Prey 1973; Amaefule and Handy 1982; Chatzis and Morrow 1984) seem to support the estimate, since they indicate that S_{rnw} is constant for $Ca < Ca_c$, where Ca_c is the critical value of Ca for capillary-controlled displacement, whereas S_{rnw} decreases only when $Ca > Ca_c$. Larson et al. (1981b) compiled a wide variety of experimental data and compared them with their predictions; they found qualitative agreement with their predictions.

Heiba et al. (1982, 1983, 1984, 1992)[1] further developed the random percolation model and used it to compute the RPs for the various regimes of the wettability. They distinguished between pore throats that are *allowed* to a fluid—i.e., can potentially be filled by that fluid—and those that are actually occupied by it. Then, given the pore-size distribution of the pore space, they derived the same distribution for the allowed and occupied pores.

Assume, for example, that one fluid is strongly wetting, while the other one is completely non-wetting. Then, according to the percolation model of Heiba et al. (1982, 1992), during the primary drainage the pore-size distribution of the pores occupied by the non-wetting fluid is given by Eq. (4.32), since the *largest* throats are occupied by the non-wetting fluid, whereas during imbibition, displacement of a non-wetting fluid by a wetting one, the pore-size distribution of the pores occupied by wetting fluid is given by Eq. (4.35), because the *smallest* pore bodies are occupied by the wetting fluid. In a similar fashion, one can derive expressions for the pore-size distribution of the pores occupied by the displacing and displaced fluids during the secondary imbibition and drainage, i.e., those that begin at the end of primary imbibition and drainage. Once the pre-size distributions are determined, calculating the permeability of each fluid phase and, therefore, its RP reduces to a problems described in Chap. 7, namely, computing the effective permeability. In particular, when the permeability of a given fluid phase is computed, the conductance of the bonds occupied by the second fluid is set to zero, as the two phases are immiscible. This assumption neglects, of course, the presence of a thin film of the wetting fluid on the pores' surface that is occupied by the second fluid. Therefore, calculating the RPs of the two fluid phases is a percolation problem, explained in Chap. 7, and any of the methods described there may be utilized for calculating the RPs to the fluid phases. Note that, as described in Chap. 4, drainage is better described by bond percolation, whereas imbibition is more complex and represents a site-percolation problem.

Heiba et al. (1983) extended the random percolation model to the case in which the porous medium is intermediately wet, or has mixed wettability characteristics. Consider, for example, an intermediately wetted porous medium for which both the

---

[1] As mentioned in Chap. 4, the original random percolation model of Heiba et al. was developed in early 1982, with the results presented as a preprint at the 1982 Annual Conference of the Society of Petroleum Engineers (SPE), and submitted as a paper to the SPE. The manuscript was accepted for publication in the same year, subject to some minor clarifications. It took, however, 10 years to make the clarifications, since the first two authors had moved on with their careers, and the last two were preoccupied with other problems! In the meantime, an entire "industry" was created based on the SPE preprint! The paper was eventually published in 1992, setting a world record for procrastination!

primary and secondary displacements are considered as a drainage process. Therefore, the formulae developed by Heiba et al. (1982, 1992) for the drainage can be modified for such porous media.

Ramakrishnan and Wasan (1984) used similar ideas and developed expressions for the RPs. They also considered the effect of the capillary number Ca on the RPs. Just as the residual saturations S_r depend on the Ca (in fact, $S_r \to 0$ as Ca $\to \infty$), the RPs also depend on the Ca. Normally, if Ca is small, the RPs do not exhibit great sensitivity to it. Evidence for this assertion was provided by the experimental data of Amaefule and Handy (1982). As Ca increases, however, the RP curves lose their curvature and in the limit Ca $\to \infty$ they become straight lines. Ramakrishnan and Wasan (1984) developed formulae that took such effects into account.

9.3 Computing the Relative Permeabilities with Random Percolation Model

The most stringent test of any theory is its ability to predict accurately the quantities of interest, such that they would be in agreement with experimental data. A most important property of two-phase flow in porous media is the RPs of the two fluid phases. The random percolation model that was described above has been used to compute the RPs by various approaches, which are described next.

9.3.1 *Use of the Bethe Lattices*

The above random percolation model of two-phase flow was utilized by Heiba et al. (1992) and Sahimi et al. (1986a). The latter group simulated the model in a simple-cubic network, in order to compute the RPs, while Heiba et al. (1982) used a Bethe lattice (see Chap. 2). In a Bethe lattice, the microscopic conductivity g_m follows the power law, $g_m \sim (p - p_c)^2$ near the percolation threshold, which is in agreement with the power law for the macroscopic conductivity g_e of three-dimensional (3D) percolation networks. Both g_m and g_e also vary linearly with p, the fraction of conducting bonds, away from p_c. Moreover, the percolation threshold of a Bethe lattice of coordination number Z is given by (see Chap. 2), $p_c = 1/(Z - 1)$, so that a Bethe lattice of coordination number 5 has the same bond percolation threshold of a simple-cubic network with $Z = 6$. Thus, by calculating g_m and adjusting the coordination number of a Bethe lattice such that its percolation threshold is close to those of 3D porous media, one obtains very accurate estimates of the conductivity and permeability and, therefore, the RPs of porous media.

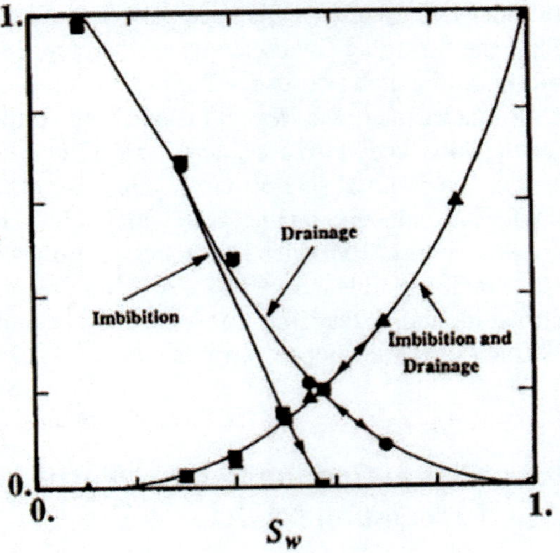

Fig. 9.1 Relative permeability k_r and its dependence of the wetting-phase saturation S_w for strongly wetting, completely non-wetting porous media, computed by the pore-network simulation of Sahimi et al. (1986a)

Figure 9.1 presents the results of pore-network simulations of Sahimi et al. (1986a). A comparison between Fig. 9.1 and typical experimental data shown in Fig. 9.2 indicates that all the qualitative aspects of the experimental data are reproduced by the model. In fact, numerically, the results differ from the experimental data by at most 10 percent. Heiba et al. (1983) also showed that their model can predict all the relevant experimental features of the RPs and capillary pressure for intermediately wetted porous media, as well as to the case in which there are three fluids in the porous medium, such as, for example, oil, water, and gas (Heiba et al. 1984).

9.3.2 *Poor Man's Percolation: Effective-Medium Approximation*

Ghanbarian et al. (2016) used a combination of percolation power laws near p_c and the effective-medium approximation (EMA) to compute the relative permeability of water in soil. The approach was already described in Chap. 7 and, therefore, only the specific features of their work are mentioned here. The distribution of the throat sizes that they used was of the fractal type, already mentioned and utilized in Chap. 8:

$$f(r) = \left(\frac{D_f}{r_m^{-D_f} - r_M^{-D_f}}\right) r^{-1-D_f} , \tag{9.4}$$

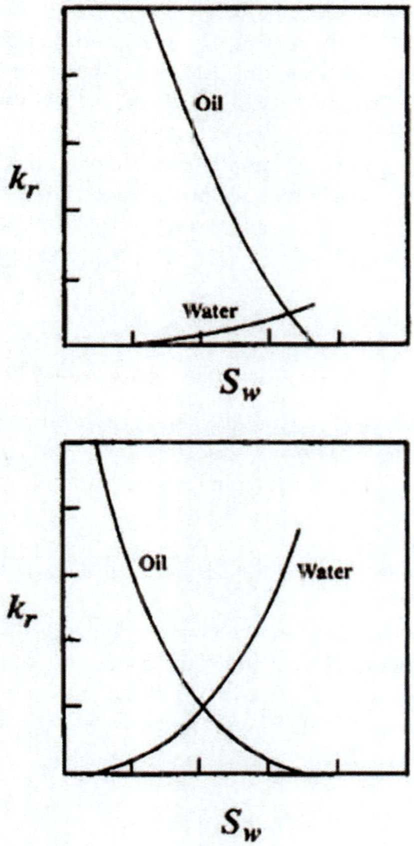

Fig. 9.2 Typical relative permeabilities for a water-wet (top) and oil-wet (bottom) porous medium

with r_m and r_M being the minimum and maximum pore-throat sizes, and D_f the fractal dimension of the pore space. The porosity ϕ of the pore space is then given by

$$\phi = \int_{r_m}^{r_M} sr^3 f(r)dr = \left(\frac{s}{3 - D_f}\right)\left(\frac{D_f}{r_m^{-D_f} - r_M^{-D_f}}\right)\left(r_M^{3-D_f} - r_m^{3-D_f}\right) , \quad (9.5)$$

in which s is a shape factor. The water volumetric content S_w of soil in all pore with a size $r \leq r_m$ is given by Eq. (9.5) in which r_M is replaced by r. Since the capillary pressure P_c is given by (see Chap. 4), $P_c = A/r$, with A being a constant, the capillary pressure–saturation relation is given by

$$S_w = \frac{\theta}{\phi} = 1 - \left(\frac{\beta}{\phi}\right)\left(1 - \left|\frac{P_c}{P_a}\right|^{D_f - 3}\right) , \quad |P_a| \leq |P_c| \leq |P_M| , \quad (9.6)$$

where $P_a = A/r_M$ is the air pressure, $P_M = A/r_M$, and $\beta = \phi r_M^{3-D_f}/(r_M^{3-D_f} - r_m^{3-D_f})$. Note that r_M and r_m are the largest and smallest *accessible* pore throats, with the accessibility defined in the percolation sense.

To use the EMA, one must first derive the pore-conductance distribution. The flow conductance of a pore of radius r is given by, $g \propto r^4/l$, where l is the pore's length. For example, as mentioned in Chap. 8, in porous media whose pore-throat size distribution follows a fractal law, $l \propto r$, so that $g \propto r^3$. Let us write this relation in a more general form, $g \propto r^\gamma$. If $h(g)$ is the pore-conductance distribution, then, given Eq. (9.4) and the fact that, $h(g)dg = f(r)dr$, we obtain

$$h(g) = \left(\frac{1-\alpha}{g_M^{1-\alpha} - g_m^{1-\alpha}}\right) g^{1-\alpha} , \tag{9.7}$$

where $\alpha = (\gamma + D_f)/\gamma$. As described in Chaps. 2 and 7, according to the EMA the effective flow conductance $g_e(S_w)$ of the pore space at any water saturation S_w is the solution of the following equation:

$$\int_{g_c}^{g} \frac{g_e(S_w) - g}{g + [(1 - S_{wc})/S_{wc}]g_e(S_w)} h(g)dg = 0 , \tag{9.8}$$

while the effective flow conductance $g_e(S_w = 1)$ at fully saturated condition is given by Eq. (9.8) in which the upper limit of the integral is replaced by g_M, corresponding to r_M. In Eq. (9.8), S_{wc} is the critical or residual saturation of water, corresponding to a critical radius r_c and a critical conductance g_c. The RP of water is then given by

$$k_{rw}(S_w) = \frac{g_e(S_w)}{g_e(S_w = 1)} , \tag{9.9}$$

Equation (9.8) and those associated with it are accurate if S_w is not too close to S_{wc}, the critical or residual saturation that represents the percolation threshold of the water phase. Close to S_{wc} one has the percolation power law of Chap. 2:

$$k_{rw} = k_0(S_w - S_{wc})^2 , \quad S_{wc} \le S_w \le S_{wx} , \tag{9.10}$$

where S_{wx} is a crossover saturation at which the power law (9.10) merges with the EMA description. Power law (9.10) has been confirmed by experiments; see, for example, Sahni et al. (1998) and Dicarlo et al. (2000). Therefore, two parameters, k_0 and S_{wx}, must be determined. The former is estimated by setting the EMA equation and Eq. (9.9) equal. S_{wx} is estimated by setting the derivatives of the two equations equal.

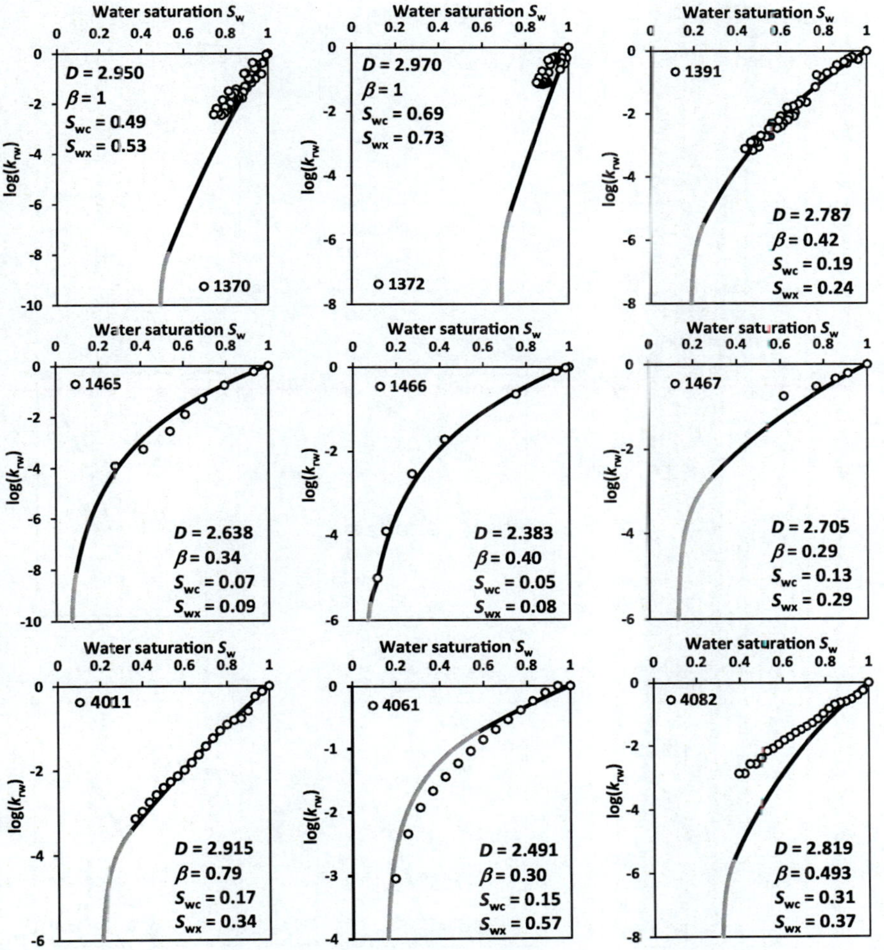

Fig. 9.3 Comparison of the predicted saturation dependence of water relative permeability, computed based on the desaturating capillary pressure curve, and the experimental data (symbols) for 21 soil samples from the UNSODA database. Light gray and black curves represent, respectively, the predictions of the EMA and percolation power laws (after Ghanbarian et al. 2015)

Figure 9.3 compares the predictions of the EMA percolation with the experimental data for 21 soil samples obtained from Unsaturated Soil Hydraulic Database (UNSODA). The agreement is excellent. It should, however, be pointed out that the EMA-percolation approach could not predict some other experimental data, the reason for which is not yet well understood; see Ghanbarian et al. (2016). It must also be mentioned that Levine and Cuthiell (1986) used the EMA and the random percolation model of Heiba et al. to calculate the RPs to two-phase flows in porous media,

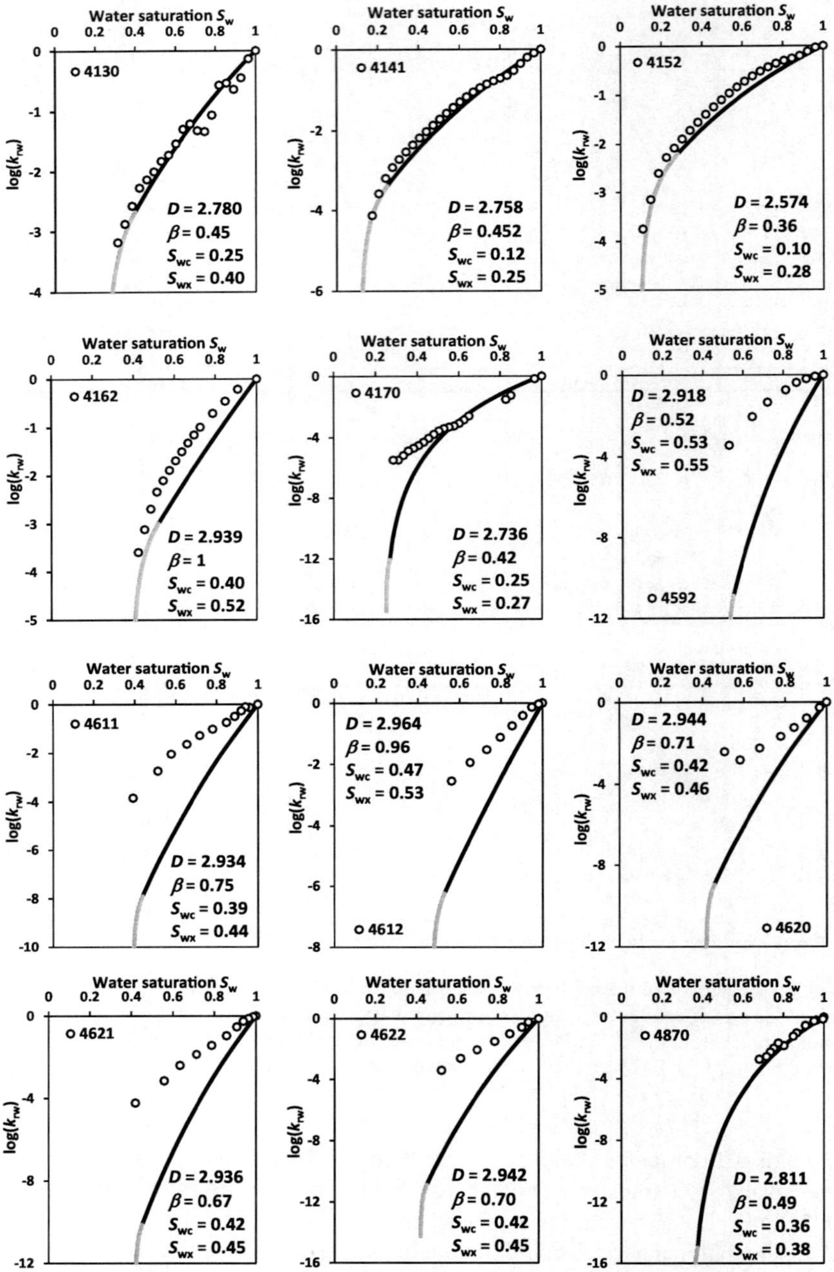

Fig. 9.3 (continued)

although their EMA differs from what is described here, and the authors made no comparison with experimental data.

Salimi et al. (2020) used the EMA to compute the relative permeabilities of two-phase flow in water-wet porous media. The main difference between their work and those of others mentioned earlier was that they used non-circular pores with various shapes and combinations. A finite-element approach was utilized to compute the hydraulic conductivity of arbitrarily shaped prisms, which were partly filled with oil and water, and the results were utilized in the EMA formulation. The predictions were also compared with direct simulations with PN models.

9.4 Random Site-Correlated Bond Percolation Models

Chatzis and Dullien (1982) developed a PN model in which the sites represented the pore bodies with random radii. The pore throats were represented by the bonds with effective radii that were correlated with those of the pore bodies. Chatzis and Dullien (1982, 1985), Diaz et al. (1987), and Kantzas and Chatzis (1988) computed the RPs and capillary pressure curves for sandstones. On the other hand, Wardlaw et al. (1987) determined experimentally the correlations between the sizes of the pore bodies and pore throats, and found that there are little, if any, such correlations in Berea sandstones, but that there may be some correlations between the two in the Indiana limestone.

Li et al. (1986), Constantinides and Payatakes (1989) and Maier and Laidlaw (1990, 1991b) also proposed PNMs in which the sizes of the pore bodies and throats were correlated. In spite of the fact that the correlated model is much more detailed than the random-bond model, its predictions for the RPs are not fundamentally different from those of random percolation model.

9.5 Invasion Percolation

Invasion percolation (IP) and its efficient simulation were already described in Chap. 3. From a conceptual point of view, the IP is perhaps a more appropriate model of capillary-controlled displacements than the random percolation models, with the most obvious reason being the fact that there is a well-defined interface between the immiscible fluid that enters a porous medium from one side and displaces the defender, the second immiscible fluid, in a systematic way. Thus, the concepts of history and the sequence of pore invasion according to a physical rule are naturally built into the model, which is also supported by ample experimental evidence.

9.5.1 Comparison with Experimental Data

Lenormand and Zarcone (1985a) displaced oil (the wetting fluid) by air (the non-wetting fluid) in a large and transparent 2D etched network. Their data led to, $D_f \simeq 1.82$ for the fractal dimension of the invasion cluster at the breakthrough point, which is consistent with what 2D computer simulations of the IP with trapping yield (see Chap. 3).

Jacquin (1985) and Shaw (1987) performed experiments that provided strong support to the validity of the IP. Shaw (1987), for example, showed that if a porous medium, filled with water, is dried by hot air, the dried pores, i.e., those filled with air, form a percolation cluster with the same D_f as that of the IP.

Stokes et al. (1986) used a cell packed with glass beads, an essentially 3D pore space. The wetting fluid was either water or a water–glycerol mixture, while the non-wetting fluid was oil. When the oil displaced water (drainage), the resulting patterns were consistent with an IP process.

Chen and Wada (1986) used a technique in which one uses index matching of the fluids to the porous matrix to "look" inside the porous medium. Their data were consistent with the IP model. Chen and Koplik (1985) used small 2D etched networks with oil and air as the wetting and non-wetting fluids, respectively, and found that their drainage patterns were consistent with the assumptions and results of the IP. Lenormand and Zarcone (1985b) used 2D etched networks and a variety of wetting and non-wetting fluids, and showed that their drainage experiments were completely consistent with an IP description of the phenomenon.

9.6 Flow of Thin Wetting Films in Pores

If pore network simulation of two- and three-phase flow is to mimic experiments, simulating imbibition must account for the effect of thin wetting fluids on the pores' surface, filled by the non-wetting fluid. The films preserve the wetting fluid's continuity in the pore space. The important point to remember is that cylindrical pore throat cannot support flow of thin wetting fluids on their internal surface. Hashemi et al. (1998, 1999a, b) described in detail how the flow of thin wetting fluid in pore networks can be simulated, utilizing pore throats with square cross sections. In addition, in the presence of thin films, one must carefully define what constitutes a cluster of the wetting fluid. Hashemi et al. defined such a cluster as a set of nearest neighbor pore bodies— the network's nodes—that are filled by that fluid and are connected to each other either by thin films that flow in the crevices of the throats, i.e., the corner areas on the throats' walls with square cross sections that connect the nearest neighbor pores, or by the throats themselves if they are filled by the wetting fluid.

Simulation of imbibition entails consideration of several types of displacement mechanisms. One is a pistonlike displacement—one with a flat velocity profile—in the throats, which is typically followed by several types of pore filling by the invading

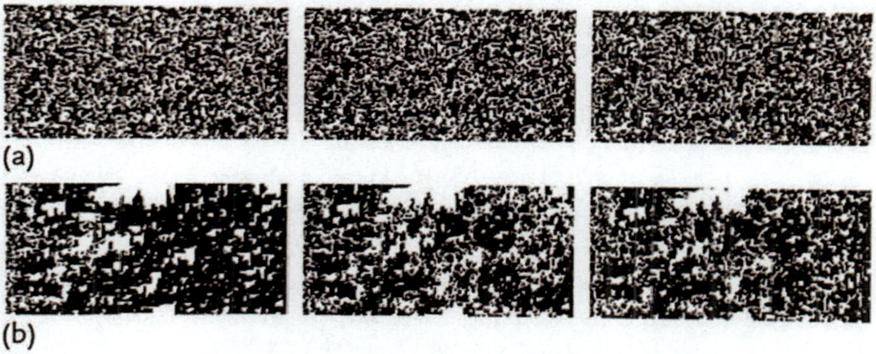

Fig. 9.4 Snapshots of the 2D pore space during invasion by two fluids, preceded by imbibition for wetting-phase fractional flow, $f_w = 0.5$, and **a** Ca $= 10^{-5}$ and **b** Ca $= 10.^{-1}$ Black and white areas represent the wetting and non-wetting fluids, respectively. Time increases from left to right (after Hashemi et al. 1999a)

fluid, the mechanisms of which depend on the pore's number of nearest neighbor pores that are already filled with the fluid. For a network of coordination number Z, there are Z pore-filling displacement mechanisms, denoted by $\mathcal{D}_0$ to $\mathcal{D}_{Z-1}$ (Blunt and Scher 1995), representing filling of a pore with 0 to $Z - 1$ connecting throats that contain the non-wetting fluid; see Fig. 9.4. To determine how the displacement proceeds, one needs to calculate the capillary pressure required for each mechanism. In general, $P_c(\mathcal{D}_0) > P_c(\mathcal{D}_1) > \cdots > P_c(\mathcal{D}_{Z-1})$. Moreover, the process is limited by the largest radius of curvature necessary to fill the pore, which depends on the number of surrounding throats filled with the invading fluid. Displacement of type $\mathcal{D}_0$ can occur only if the non-wetting fluid is compressible, since in this case it is this fluid that is trapped in a pore surrounded by throats that are filled by the wetting fluid; see Fig. 9.4. Calculating $P_c(\mathcal{D}_i)$ and taking into account the effect of the shapes and sizes of the pores and throats, as well as the contact angle is, however, very difficult, and even if it were not, the computations would be prohibitive.

To circumvent the difficulty, a parameterization of $P_c(\mathcal{D}_i)$ is used for describing the advancement of the fluid, which is as follows. The mean radius of curvature R_i for filling by the $\mathcal{D}_i$ mechanism is (Blunt and Scher 1995)

$$R_i = R_0 + \sum_{j=1}^{i} A_j x_j \ , \tag{9.11}$$

where x_j is a random number distributed uniformly in (0,1), A_j an input parameter, and R_0 the maximum size of the adjacent throats. A_j emulates the effect of the pore space variables and determines the relative magnitude of $P_c(\mathcal{D}_i)$. For example, if for $j \geq 2$ we set $A_j = 0$, then the pore-filling events become independent of the number of the filled throats and, hence, the fluid advance is similar to the IP. If, on the other hand, A_j are relatively large for large values of j, then the model emulates the case of small or medium values of the ratio of pore and throat radii. R_i, computed by Eq. (9.11), is taken to be the pore radius for the $\mathcal{D}_i$ mechanism of pore filling. Thus, if one fluid is strongly wetting, the critical capillary pressure P_c for pore filling, when the pore has i adjacent unfilled throats, is taken to be

$$P_c = \frac{2\sigma}{R_i}\,, \tag{9.12}$$

where σ is the interfacial tension between the two fluids.

During imbibition one must also consider the snap-off mechanism of throat filling: as P_c decreases, the radius of curvature of the interface increases, and the wetting films occupying the crevices of the walls swell. At some point, further filling of the crevices causes the radius of curvature of the interface to decrease, leading to instability and spontaneous filling of the center of the throat with the wetting fluid. The critical capillary pressure for snap-off is given by

$$P_c = \frac{\sigma}{r_t}\,, \tag{9.13}$$

which is always smaller than $P_c = 2\sigma/r_t$ for pistonlike displacement, implying that snap-off can occur only if pistonlike advance is topologically impossible, because there is no neighboring pore filled with the invading fluid.

Therefore, to simulate imbibition and the flow of thin wetting fluid films, consider a pore or a throat filled with the non-wetting fluid, and assume that the wetting fluid is supplied by thin film flow along a crevice of length ℓ. We ignore any pressure drop in the non-wetting fluid and in the portions of the network completely saturated by the wetting fluid—a valid assumption if the capillary number is small—but the pressure drop along the thin wetting layers in the crevices of the network is significant and cannot be neglected. The relation between the wetting fluid flow rate q and the pressure gradient for the thin films in the corners is

$$q = -\frac{r^4}{\beta_s \eta_w}\frac{dP_w}{dx}\,, \tag{9.14}$$

where η_w and P_w are the viscosity and local pressure of the wetting fluid, r is the local radius of curvature in the corner, and β_s is a dimensionless factor that depends on the shape of the cross section of the throat and the boundary conditions at the phase boundary. For example (Blunt and Scher 1995), for square crevices $\beta_s = 109$, if there is no slip at the interface between the two fluids. More generally, β_s varies anywhere between 15 and 290 (Ransohoff and Radke 1988). We assume that, locally, the interface is in capillary equilibrium and, hence, $P_w = -P_c = -\sigma/r$, and that flow of the thin films (in the crevices), as well as β_s, is independent of the time. Then, for the thin films, $q = \sigma^4/(\eta_w \beta_s P_c^4) dP_c/dx$, i.e., flow of the thin wetting films is driven by a capillary pressure gradient. Integrating the equation in $(0, \ell)$, we obtain

$$P_{c0} = P_{cl}\left(1 + 3\frac{\beta_s P_{cl}^3 q \eta_w \ell}{\sigma^4}\right)^{-1/3}, \tag{9.15}$$

where P_{c0} is the capillary pressure of the element (pore or throat) at the inlet of the crevice and P_{cl} is the local capillary pressure where the element is being filled, either by pistonlike throat filling, or throat filling by snap-off, or by the $\mathcal{D}_i$ mechanism of pore filling.

9.7 Invasion Percolation with Two Invaders and Two Defenders

The models of two-phase flow and displacement that have been described so far involve only one displacing fluid—the invader—and one displaced fluid—the defender—whereas in many problems, both at laboratory and field scales, one has a situation in which there are at least *two* invaders and *two* defenders. For example, at the end of the displacement processes, the displaced fluid resides only in isolated blobs or clusters of finite sizes that can no longer be displaced by any of the displacement processes described so far. In order to mobilize and displace such blobs, the capillary number Ca must be significantly increased, which would then give rise to three other displacement processes that are quasi-static and dynamic displacement of blobs, both of which are time-dependent phenomena, and steady-state dynamic displacement. The last process can be carried out if the displacing and displaced fluids are *simultaneously* injected into the porous medium. Thus, simulating displacements with two invading and two defending fluids is of practical importance. But, such a model is also motivated by other practical considerations.

For example, in a typical experiment that measures the RPs of two-phase flows, the porous medium is initially saturated with the non-wetting (or the wetting) fluid. Then a *mixture* (not a solution) of both fluids of a given composition is injected into the sample at a constant flow rate. The two fluids are uniformly distributed at the entrance to the medium, and the flow is maintained until steady state is reached at which point the pressure drop along the sample is recorded that, together with the flow rates of the two fluids and Darcy's law, yields estimates of the RPs. Thus, measurement of the RPs involves simultaneous invasion of a porous medium by two immiscible fluids.

As another example, consider two-phase flow in fractures. As described in Chap. 5, a natural fracture usually has a rough self-affine surface. Experimental observations in horizontal fractures indicate (Glass and Norton 1992; Glass and Nicholl 1995; Glass et al. 1995) that, in two-phase flow through a fracture, when an invasion front encounters a zone characterized on average by much smaller apertures (the zone of the wetting fluid) or much larger ones (the zone of the non-wetting fluid), the invading front advances into that region at the expense of the *already invaded region*. In other words, some apertures in the already invaded region are spontaneously re-invaded by the defending fluid to provide invading fluid for the newly encountered zone. In addition, it has been observed that at low flow rates, gravity-driven fingers drain a distance behind the invading finger tip. Thus, two-phase fluid invasion in horizontal fractures involves *simultaneous* imbibition and drainage of the apertures within the fractures.

Hashemi et al. (1998, 1999a, b) developed an IP model with two invaders and two defenders. In simulating such a model, one must recognize that when the fluid clusters are pushed one after another, ℓ in Eq. (9.15) represents the minimum distance between the element to be filled *by the thin wetting films* and the point at the interface between the invading and defending clusters, where the path length through the elements completely filled with the wetting fluid is zero. Therefore, when a cluster of the non-wetting fluid pushes a wetting fluid cluster, one has $\ell = 0$, as there is no path of pores and throats that contain thin films of non-wetting fluid in the defender cluster. However, $\ell \neq 0$, if a cluster of the wetting fluid pushes a cluster of the non-wetting fluid, since in this case a thin film of the wetting fluid can participate in the displacement. If each throat is a channel of radius r and characteristic length d (e.g., the distance between two neighboring nodes), we rewrite Eq. (9.15) in dimensionless form using $\ell^* = \ell/d$, $u = q/d^2$, and $P_c^* = P_c r/\sigma$, to obtain

$$P_{c0}^* = P_{c\ell}^* \left(1 + 3C P_{c\ell}^{*3} \ell^*\right)^{-1/3} , \tag{9.16}$$

where $C = \beta_s (d/r)^3 (u\mu_w/\sigma) = \beta_s (d/r)^3 \mathrm{Ca}$, with Ca being the capillary number for the flow of thin wetting films. Assuming that the flow rate in each crevice is constant, we consider the filling of an element with the wetting fluid and use Eq. (9.16), replacing ℓ with the minimum distance between the filling element and the inlet of the filled one, where the path length through the throats completely filled with the wetting fluid is zero (i.e., only the paths that consist of the thin wetting films are counted in calculating ℓ).

At time $t = 0$ the network is filled with the non-wetting fluid, assumed to be incompressible so that its trapping by the wetting fluid is possible. A trapped non-wetting fluid cluster can be broken up into pieces only by flow of the wetting films (see below). A site is selected at random on one face of the network for injection, and another site on the opposite face for production. Thus, the boundary condition in the direction of macroscopic displacement is a constant injection rate. The motion of the injected fluid is represented by a series of discrete jumps in which, at each time step, the invader displaces the defender from the available pores through the least resistance path. In the presence of film flow through the crevices and strongly wetting condition, all the pores are accessible to the wetting fluid, while in the case of no film flow, only the interface pores are accessible. At each stage, the pressure needed for flow of the invader from the injection site to the production site through *all the accessible pores* is calculated by summing up the capillary pressure differences of the pores through *all the possible paths* (see below). The path with the least pressure is then selected. This is completely different from the usual IP models in which only the interface pores are considered. If the wetting fluid is displacing the non-wetting one, then after the element is filled, for every possible new path the corresponding ℓ is recomputed (in the case of the non-wetting fluid displacing the wetting fluid, one has $\ell = 0$). If the number of filled throats adjacent to any of the pores has increased, the pore-filling capillary pressures are updated to represent the proper $\mathcal{D}_i$ mechanism.

Consider, for example, a situation in which the non-wetting fluid is in a cluster that is connected to the inlet of the network, and is in contact with three wetting clusters, referred to as WF1, WF2, and WF3. Suppose that the non-wetting fluid tries to displace the wetting fluid. Since there are three wetting fluid clusters that are also in contact with one or several other non-wetting fluid clusters in "front" of them, which in turn touch other non-wetting fluid clusters, including one that is connected to the outlet of the network, one must consider all the possible paths of the clusters that are pushing one another, and identify the one that requires the minimum pressure for the displacement from the point of the contact between the inlet non-wetting fluid cluster and the network's outlet. To do so, one must identify all the non-wetting clusters that are in contact with WF1, WF2, and WF3, referred to as the secondary clusters. Then, the tertiary clusters that are in contact with the secondary clusters must also be identified, and so on, until all the possible paths from the inlet non-wetting cluster to the outlet are listed. As such, this is a problem in combinatorial mathematics.

One then calculates the minimum pressure ΔP for displacing a wetting fluid cluster by a non-wetting cluster, and vice versa. Consider, first, the case in which the non-wetting fluid, connected to the inlet of the network, tries to displace a wetting cluster. Thus, only the elements (pores and throats) of the wetting cluster that are at the interface between the two types of clusters are accessible for displacement, since there is no flow of the non-wetting films in any element. Let 1 denote an interfacial element between the invader (the non-wetting fluid) and the defender cluster (the wetting fluid), and 2 represent the outlet point of the wetting cluster to another non-wetting cluster. Recall that the pressure drop between the inlet and outlet elements is, $\Delta P_{12} = P_{nw1} - P_{w2}$, and that, $P_{nw1} - P_{w1} = P_{c1}$, so that, $\Delta P_{12} = P_{w1} - P_{w2} + P_{c1}$. If we assume that there is no pressure drop in the wetting cluster, which is true at low-to-moderate Ca, then $\Delta P_{12} \simeq P_{c1}$. Thus, we calculate ΔP_{12} for all the interfacial elements between any two non-wetting and wetting clusters that are in contact. Then, for a path that starts from the inlet non-wetting cluster to the outlet of the network, we write the total required pressure drop $(\Delta P)_{\rm path}$ as a sum over all such ΔP_{12}s for the non-wetting and wetting clusters that are in contact, being pushed by one cluster on one side (the inlet point 1) and pushing another cluster at another side (the outlet point 2), and belonging to the path:

$$(\Delta P)_{\rm path} = (\Delta P_{12})_{\rm cluster\ 1} + (\Delta P_{12})_{\rm cluster\ 2} + (\Delta P_{12})_{\rm cluster\ 3} - \cdots = \sum_i (P_{c1})_{\rm cluster\ i}\ , \tag{9.17}$$

where cluster 1 is in contact with cluster 2, which is in contact with cluster 3, and so on. In dimensionless form, each term of Eq. (9.17) is calculated using Eq. (9.16). One then selects the path for which $(\Delta P)_{\rm path}$ is minimum, and keeps in mind that when a non-wetting fluid cluster pushes a wetting fluid one, one must calculate the capillary pressure for *all* the interfacial elements (i.e., all the P_{c1}) between the two clusters, in order to determine the minimum ΔP_{12}.

Consider, next, the case in which the wetting fluid, connected to the network inlet, attempts to push a non-wetting cluster. The general method of selecting the displacement path is the same as before, but two distinct cases must be considered.

(i) Suppose that there is no flow of thin wetting films in the crevices. Using the same arguments and notation as above, it is straightforward to show that for any non-wetting cluster, which is pushed by the wetting fluid and is also pushing another wetting cluster, $\Delta P_{12} = -P_{c1}$, assuming again that the pressure in the non-wetting cluster is the same everywhere. The minimum pressure is again determined by considering $(\Delta P)_{\rm path}$ for all the possible paths.

(ii) If there is flow of thin wetting films, then the wetting fluid can reach any part of any cluster of the defending non-wetting fluid. One must now calculate the minimum P_c for filling *every* element within the non-wetting fluid cluster, as well as those that are at the interface with the wetting fluid. Let 1 denote an interfacial element between the invading wetting cluster and the defending non-wetting cluster, 2 be a point inside (a pore of) the non-wetting cluster, and 3 be the outlet element of the non-wetting cluster next to another wetting cluster. Since the minimum of such P_cs corresponds to the minimum ℓ [see Eq. (9.16)], then point 2 is the location of the element within the non-wetting cluster and 1 is the point at the interface between the wetting and non-wetting fluids that has the smallest distance to that element. Thus, one has, $\Delta P_{13} = P_{w1} - P_{nw3}$ and $P_{c1} = P_{nw1} - P_{w1}$, and therefore $\Delta P_{13} = P_{nw1} - P_{nw3} - P_{c1}$. Recall that P_{c1} is related to the capillary pressure of the element located at 2 through Eq. (9.16). Since $P_{nw1} = P_{nw2} = P_{nw3}$, we obtain $\Delta P_{12} = -P_{c1}$. Therefore, in all the cases the individual ΔP_{12}s depend only on the inlet condition of the clusters, and are independent of the outlet conditions. Note that in both cases a distinct ℓ is associated with each possible path.

At the end of imbibition, the network is invaded with *both* the wetting and non-wetting fluids, which initiates the fractional flow displacement (FFD). The fraction f_w of the wetting fluid in the injected mixture is fixed. At the early stages of the FFD, there is a continuous path of the wetting fluid through the network, while the non-wetting fluid remains entrapped in its isolated clusters, so that the injected wetting fluid can exit from the network's opposite face, whereas the non-wetting fluid accumulates in the network. After successive injections, the injected non-wetting fluid joins its entrapped clusters and, thus, larger clusters of the non-wetting fluid are generated progressively. The invader fluids push many clusters of the non-wetting and wetting fluids, and the wetting fluid clusters that are already in the network.

To determine the path of the mobilized clusters from the inlet to the outlet, we first identify all the clusters and their neighbors. Then, all the possible paths of the clusters from the injection site to the production site are identified and stored in a list. Using Eq. (9.16), the pressure required to mobilize each of the clusters in the list is calculated for each path. The path that requires the least pressure is selected as the flow path between the inlet and the outlet of the network. This represents an important difference between the model for the FFD and the usual IP models in which only the throat with the smallest resistance at the interface is considered. The difference is due to the fact that mobilization of an entrapped fluid cluster is different from simple displacement of one fluid by another in a pore or throat.

9.8 Random Percolation with Trapping

Random percolation with trapping was developed first by Sahimi (1985) and Sahimi and Tsotsis (1985) to model catalytic pore plugging of porous media. In the problem that they studied, the pores of a porous medium are plugged as the result of a chemical reaction and deposition of the solid products on the surface of the pores. Large pores take a long time to be plugged, and if they are surrounded by small pores that are quickly plugged, they become trapped and cannot be reached by the reactants penetrating the porous medium from outside.

Accurate computer simulations of Dias and Wilkinson (1986), who proposed the same model for two-phase flow problems in porous media, indicated that most properties of random percolation with trapping in both 2D and 3D are the same as those of random percolation without trapping. The pore-size distribution (or threshold capillary pressures for pore invasion) that was considered by Dias and Wilkinson was, however, narrow [a uniform distribution in (0,1)]. If the pore-size distribution is broad (as in the problem studied by Sahimi and Tsotsis), percolation with and without trapping may not necessarily have the same scaling properties.

9.9 Roughening and Pinning of a Fluid Interface

The structure of the fluid interface during imbibition is interesting and different from that during drainage. The invading fluid cluster during imbibition is compact, but capillary forces lead to random local pinning of the interface that results in an interface with a rough self-affine structure, which was demonstrated by several sets of experiments. But before embarking on an analysis of the experiments, let us review briefly the dynamics of rough surfaces and interfaces.

Let us first define the width $w(L)$ of an interface, which is given by

$$w(L) = \langle [h(x) - \langle h \rangle_L]^2 \rangle^{1/2} , \tag{9.18}$$

where h is the height of the interface at position x (relative to a flat surface of length L) and $\langle h \rangle_L$ its average over a horizontal segment of length L. According to the scaling theory of Family and Vicsek (1985) for growing rough surfaces, one has the following scaling form at time t:

$$h(x) - \langle h \rangle_L \sim t^{\beta} S_1(x/t^{\beta/\alpha}) , \tag{9.19}$$

where α and β are two exponents that satisfy

$$\alpha + \frac{\alpha}{\beta} = 2 , \tag{9.20}$$

and the scaling function $S_1(u)$ has the properties that $|S_s(u)| < c$ for $u \gg 1$, and $S_1(u) \sim L^{\alpha} S_1(Lu)$ for $u \ll 1$, where c is a constant. It is then straightforward to see that

$$w(L, t) \sim t^{\beta} S_2 \left(t/L^{\alpha/\beta}\right) , \tag{9.21}$$

where $S_2(u)$ is another scaling function, and therefore

$$w(L, \infty) \sim L^{\alpha} . \tag{9.22}$$

Note that $w(L, t)$ is a measure of the correlation length along the direction of interface growth. For 2D systems (a film grown on a line), $\alpha = 1/2$ and $\beta = 1/3$.

9.10 Comparison with Experimental Data and Relation with Directed Percolation

Experiments of Rubio et al. (1989) and Horváth et al. (1991a, b) seemed to confirm the above scaling picture. Rubio et al. (1989) carried out their experiments in a thin (2D) porous medium made of tightly packed clean glass beads of various diameters. Water was injected into the porous medium to displace the air in the system. The motion of the interface was recorded and digitized with high resolution. The experiments of Horváth et al. (1991a, b) were very similar (see below).

Analyzing their experiments, Rubio et al. (1989) determined that $\alpha \simeq 0.73$, significantly different from $\alpha = 1/2$ mentioned above. Horváth et al. (1990) reanalyzed data of Rubio *et al.* and reported that $\alpha \simeq 0.91$. They also carried out their own experiments in a Hele-Shaw cell—the narrow gap between two flat and parallel surfaces—packed randomly and homogeneously with glass beads, displaced the air in the pore space with glycerol–water mixture, and reported $\alpha \simeq 0.81$ and $\beta \simeq 0.65$.

A possible connection between the roughening and pinning of the interface during imbibition and directed percolation (see Chaps. 3 and 21) was proposed by Buldyrev et al. (1992a, b) and Tang and Leschhorn (1992). Recall from Chap. 3 that in directed percolation the bonds of a network are directed and diode-like. Flow and transport along such bonds is allowed in only one direction. If the direction of the external potential is reversed, no transport in the diode-like bonds takes place. This induces a macroscopic anisotropy, such that one needs *two* correlation lengths for characterizing the system. One, ξ_L, is for the longitudinal (external potential) direction, while the other, ξ_T, is for the transverse (perpendicular to the external potential) direction. The percolation thresholds p_c of directed networks are much *larger* than those of random percolation described in Chap. 2. Near p_c one has

$$\xi_L \sim (p - p_c)^{-\nu_L} , \tag{9.23}$$

$$\xi_T \sim (p - p_c)^{-\nu_T} . \tag{9.24}$$

Buldyrev et al. (1992a, b) carried out an interesting experiment in which they formed an interface by dipping paper (a 2D porous medium) into a fluid, and allowing it to invade the paper. It was argued by them and Tang and Leschhorn (1992) that the width scales as $w(L) \sim \xi_T$, while $L \sim \xi_L$, i.e., ξ_T is the characteristic length scale for the width, while ξ_L sets the characteristic scales for *both* the distance parallel to the interface and the time. Therefore,

$$\alpha = \frac{\nu_T}{\nu_L} , \tag{9.25}$$

which for $d = 2$ with $\nu_L \simeq 1.73$ and $\nu_T \simeq 1.1$ yields, $\alpha \simeq 0.63$, in perfect agreement with their measurement. But their estimate of α is lower than that of the experiments by Rubio et al. (1989) and Horváth et al. (1991a, b). Note that Herrmann and Sahimi (1993) also suggested a link between invasion percolation in radial geometry and directed percolation.

9.11 Finite-Size Effects and Devil's Staircase

Strictly speaking, all the results of percolation are applicable to systems that are essentially of infinite extent. If they have a finite size, the dependence of their macroscopic properties on their linear size L can be investigated using the finite-size scaling described in Chap. 2. So, let us describe the effect of the linear size of a finite porous sample on its capillary pressure and the RPs.

Thompson et al. (1987b) measured the electrical resistance of a porous medium during mercury injection (drainage), and showed that the resistance decreases (the permeability increases) during mercury injection in steps on the so-called Devil's staircase.[2] Their data are shown in Fig. 9.5. The steps were irreversible in that, small hysteresis loops did not retrace the steps, and were not reproduced on successive injections. When the number $N_{\Delta R}$ of the resistance steps larger than ΔR was plotted versus ΔR, a power-law relation was found

$$N_{\Delta R} \sim (\Delta R)^{\lambda_R} , \tag{9.26}$$

[2] Quoting from Wikipedia, the Devil's staircase is "a function that is continuous, but not absolutely continuous. It is a notorious counterexample in analysis, because it challenges naive intuitions about continuity, derivative, and measure. Though it is continuous everywhere and has zero derivative almost everywhere, its value still goes from 0 to 1 as its argument reaches from 0 to 1. Thus, in one sense the function seems very much like a constant one which cannot grow, and in another, it does indeed monotonically grow, by construction." https://en.wikipedia.org/wiki/Cantor_function.

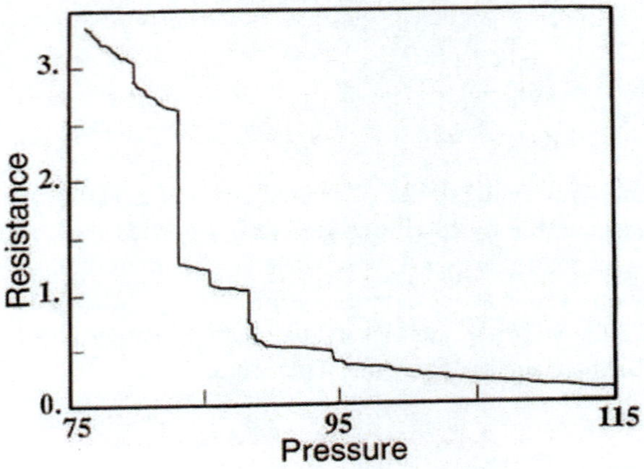

Fig. 9.5 Resistance of a sandstone versus the injection pressure during mercury porosimetry (after Thompson et al. 1987b)

with $0.57 \leq \lambda_R \leq 0.81$. The magnitude of λ_R presumably depends on the strength of the competition between the capillary and gravitational forces, such that $\lambda_R \simeq 0.57$ signifies the limit of no gravitational forces, whereas $\lambda_R \simeq 0.81$ represents the limit in which gravitational forces are prominent. Based on the stepwise decrease of the resistance and the apparent first-order (discontinuous) phase transition, Thompson et al. (1987b) concluded that mercury injection should not be modeled by percolation that usually represents a *second-order phase transition*, which is one that is characterized by a continuous vanishing or divergence of a physical quantity, such as the permeability or conductivity, as the percolation threshold is approached.

Simulation of the same process by Katz et al. (1988), Roux and Wilkinson (1988), and Sahimi and Imdakm (1988), and related simulation of Batrouni et al. (1988) indicated, however, that such a stepwise decrease in the resistance can be predicted by a (random or invasion) percolation model. The reason for the stepwise decrease in the sample resistance is that, in a *finite sample* penetration of any pore by mercury causes a *finite* change in the resistance. As the sample size increases, however, the size of the step changes decreases, such that for a very large sample the steps would vanish and the resistance curves become continuous and smooth. In fact, using a percolation model, Roux and Wilkinson (1988) showed that for a 3D porous medium of linear size L

$$N_{\Delta R} \sim L^{3(\mu-\nu)/(\mu+3\nu)}(\Delta R)^{-3\nu/(\mu+3\nu)}\,, \tag{9.27}$$

where μ and ν are, respectively, the critical exponents of the conductivity and the percolation correlation length near p_c (see Chap. 2). Note that, $\lambda_R = 3\nu/(\mu+3\nu) \simeq 0.57$ agrees well with the experimental data in the absence of gravity.

9.12 Displacement Under the Influence of Gravity: Gradient Percolation

In all the discussions so far, the effect of gravity on immiscible displacements was ignored. In 3D porous media, however, the effect of gravity cannot be neglected. The hydrostatic component of the pressure adds to the applied one, which then creates a vertical gradient in the effective injection pressure. Due to the gradient, the fraction of pores that are accessible to the displacing fluid decreases with the height of the system. A modification of the IP model proposed by Wilkinson (1984), and developed further by Sapoval et al. (1985) and Gouyet et al. (1988), succeeded in taking into account the effect of gravity.

The competition between gravity and capillary forces is usually quantified by the Bond number Bo,[3] also called the Eötvös number,[4] defined by

$$\mathrm{Bo} = \frac{\Delta\rho g d_g^2}{\sigma} , \tag{9.28}$$

where $\Delta\rho$ is the density difference between the two fluids, g is the gravity, and d_g is the typical size of the grain. Wilkinson (1984) showed that, unlike in random percolation and in the IP that are characterized by a diverging correlation length ξ_p (see Chap. 2), the correlation length ξ_g in his percolation model for immiscible fluid displacement under the influence of gravity does *not* diverge, but that it reaches a maximum given by

$$\xi_g \sim \mathrm{Bo}^{-\nu/(1+\nu)} , \tag{9.29}$$

where, as before, ν is the exponent that characterizes the divergence of the percolation correlation length ξ_p, so that $\xi_g \sim \mathrm{Bo}^{-0.47}$, and $\xi_g \sim \mathrm{Bo}^{-4/7}$ in 3D and 2D, respectively. In 3D there is a transition region where both fluids (displacing and displaced) may percolate in a porous medium, the width w of which is given by

$$w \sim \mathrm{Bo}^{-1} . \tag{9.30}$$

Similar results were derived by Sapoval et al. (1985) and Gouyet et al. (1988) in the context of *gradient percolation*, which is a model in which a gradient G is imposed on the occupation probability p in one direction of the network. The model

[3] **Wilfrid Noel Bond** (1897–1937) was an English physicist and engineer known for his contributions to fluid mechanics. After receiving his Doctor of Science from the University of London, he was a Lecturer at the University of Reading from 1921 until he passed away. The Bond number, used to characterize the shape of bubbles and drops, is named after him.

[4] **Baron Loránd Eötvös de Vásárosnamény** (1848–1919), also called **Baron Roland von Eötvös** in English literature, was a Hungarian physicist, who made fundamental contributions to gravitation and surface tension, and the invention of the torsion pendulum. A university, Eötvös Loránd University, and a research institute, Eötvös Loránd Institute of Geophysics, both in Hungary, as well as the Eötvös crater on the Moon, the asteroid 12301 Eötvös, the mineral lorándite, and peak (Cime Eotvos) in the Dolomites are named after him.

had, in fact, been considered earlier by Trugman (1983) who referred to it as the *graded percolation*. Sapoval et al. and Gouyet et al. carried out a scaling analysis similar to Wilkinson's to show that

$$\xi_g \sim G^{-\nu/(1+\nu)} , \tag{9.31}$$

which is completely similar to Eq. (9.29) in which the Bond number Bo has been replaced with G.

Wilkinson (1984) also derived an important result regarding the effect of gravity on the residual oil saturation (ROS). He showed by a scaling analysis that the difference $S_{ro} - S^0_{ro}$, where S_{ro} is the ROS for Bo $\neq 0$ and S^0_{ro} is the corresponding quantity when Bo = 0 is given by

$$S_{ro} - S^0_{ro} \sim \mathrm{Bo}^{\lambda_B} , \tag{9.32}$$

where $\lambda_B = (1+\beta)/(1+\nu)$, with β being the standard percolation exponent for the fraction of accessible pores near the percolation threshold, or the ROS (see Chap. 2). Thus, scaling law (9.32) predicts that in a 3D porous medium, $S_{ro} - S^0_{ro} \sim \mathrm{Bo}.^{0.74}$. In addition, Wilkinson (1984) proposed a simple model for simulating the IP under the influence of gravity.

9.12.1 Comparison with Experiments

Clément et al. (1987) and Hulin et al. (1988b) injected Wood's metal, a low-melting point liquid alloy, into the bottom of a vertical and evacuated crushed-glass column. The experiments were carried out at low capillary number Ca by controlling the flow velocity v. After the front reached a given height, the injection was stopped and the liquid was allowed to solidify. The horizontal sections of the front corresponding to various heights were then analyzed, and the correlation function $C(r)$ (see Chap. 4) of the metal distribution in the horizontal planes was determined, in order to see whether a fractal structure had been formed.

Another series of experiments were carried out by Birovljev et al. (1991) in a 2D porous medium. They used transparent models consisting of a monolayer of 1 mm glass beads placed at random and sandwiched between two plates. The system was filled with a glycerol–water mixture, which was displaced by air invading the system at one end.

The 3D experiments of Hulin et al. (1988b) and the 2D experiments of Birovljev et al. (1991) were completely consistent with the predictions of percolation. For example, Birovljev et al. (1991) reported that, $\xi_g \sim \mathrm{Bo}^{-0.57}$, so that the exponent 0.57 agrees perfectly with the prediction, $\nu/(1+\nu) = 4/7 \simeq 0.57$.

9.13 Immiscible Displacements at Finite Capillary Numbers: Site-Bond Invasion Percolation

So far, we have described capillary-controlled displacements, i.e., those in which viscous forces do not play any important role, so that the capillary number Ca is very small, on the order of 10^{-4} or smaller. In practice, however, particularly in enhanced oil recovery processes, it is often true that for, for example, waterflooding Ca is relatively large, so that viscous forces are important. Early simulation work on this problem included those of Singhal and Somerton (1977), Mohanty et al. (1980) and Payatakes et al. (1980). Their works were followed by detailed models of that were developed by Koplik and Lasseter (1985), Dias and Payatakes (1986a, b); Leclerc and Neale (1988), and Lenormand et al. (1988).

In Koplik and Lasseter's work, the pore space was modeled by a 2D, but non-planar network of cylindrical pore throats and spherical pore bodies with distributed effective sizes. The local coordination number of the network was distributed stochastically. The equations to solve in the pore network were those for the pressure field throughout the network, and for the saturations of the fluid phases. In a pore throat, the pressure drop ΔP is given by

$$\Delta P = -\frac{1}{g_{p1}}Q + P_c - \frac{1}{g_{p2}}Q , \tag{9.33}$$

where Q is the volume flow rate and g_{pi} is the single-phase flow conductance of fluid region i. In order to justify use of Eq. (9.33), it was assumed that the flow field in each fluid region away from the interface between the two immiscible fluids is unaffected by the other fluid. Equation (9.33) implies that one has a nonlinear fluid flow problem, if the radius of the meniscus between the two fluids and, thus, the capillary pressure P_c change as the meniscus passes from a pore body or pore throat into the contiguous pore throat or pore body, which is normally the case. The nonlinear problem was, however, converted into a constrained linear one, by assuming that the meniscus stops at the interface during the passing period, meaning that there is no flow in the pore throat until the constraints are violated and the meniscus either moves forward into the pore body or returns to the pore throat.

If a pore body or pore throat is filled by only one of the two fluids, then P_c is dropped from Eq. (9.33). Hence, Eq. (9.33) together with mass conservation for every pore body, $\sum_i Q_i = 0$, yields a set of equations for the pressure at the *center* of the pore bodies that is solved numerically, then, in a time step Δt, a meniscus m with velocity v_m moves a distance $v_m \Delta t$, generating a new fluid distribution and repeating the procedure.

Dias and Payatakes (1986a, b) developed a somewhat more advanced model, which at the same time was simple enough that one could carry out computations with large networks. The pores had converging–diverging segments with a sinusoidal profile. The volume flow rate for single-phase flow through such a pore is given by

$$Q = \frac{\pi c_0 d_p^3}{4\mu(-\Delta P_1)} \Delta P_{cd} \; , \tag{9.34}$$

where c_0 is a constant, d_p is the smallest diameter of the pore (at the minimum of the sinusoidal profile), ΔP_1 is a dimensionless pressure drop along the pore (which is a function of d_p) when the flow is creeping and the Reynolds number is unity, and ΔP_{cd} is the pressure drop along the converging–diverging pore. The solution of two-phase flow in the same pore was given by Sheffield and Metzner (1976). The Washburn approximation (see Chap. 4) was used for the capillary pressure.

Lenormand et al. (1988) developed a more realistic model (see also Leclerc and Neale 1988) in which they solved the actual nonlinear problem, whereas Koplik and Lasseter, and Dias and Payatakes had replaced the nonlinear flow problem by a sequence of linear problems. Consider a pore (throat) between nodes i and j with radius r_{ij} for which, assuming slow fluid flow, the volume flow rate Q_{ij} is given by

$$Q_{ij} = \frac{\pi r_{ij}^4}{8\ell\eta_{ij}} (P_i - P_j - P_{cij})^+ \; , \tag{9.35}$$

where ℓ is the pore's length, P_i and P_j are the pressures at i and j, and P_{cij} is the capillary pressure in the pore. The mixture viscosity η_{ij} was assumed to be given by $\eta_{ij} = \frac{1}{2}[\eta_2(\alpha_i + \alpha_j) + \eta_1(2 - \alpha_i - \alpha_j)]$, where α_i is the fraction of the pore occupied by fluid i. The plus sign in Eq. (9.35) implies that $Q_{ij} = 0$, if $P_i - P_j < P_{cij}$, due to which Eq. (9.35) is a nonlinear relation between Q and the nodal and capillary pressures.

Figure 9.6 presents the displacement patterns computed for several values of the Ca and M, the viscosity ratio. Only drainage was studied and, therefore, very low Ca corresponds to the IP, whereas very large Ca represents miscible displacements. The results are also in excellent agreement with the experiments of Lenormand et al. (1983) in 2D etched networks.

Lenormand et al. (1988) ignored the effect of pore bodies in their model and simulations. Sahimi et al. (1998) generalized model of Lenormand et al. in order to take into account the effect of the pore bodies, since they represent most of the fluid capacity of porous media. In model of Sahimi et al., the size of the throats was distributed according to a statistical distribution. Consider a pore body i and all the throats that are connected to it. Since the pore body's size must be larger than r_{tm}, the size of the largest throat that is connected to it, one takes the size of the pore body i to be mr_{tm}, where $m > 1$ is any suitable numerical factor. The network is initially saturated by the wetting fluid. The non-wetting fluid invades the network from one of its faces, entering the largest pore throat connected to the injection face, i.e., the throat that offers the least resistance to the invading fluid with the required capillary pressure for its invasion being $P_c = 2\sigma/r_t$. After the non-wetting fluid fills the throat, the pore body that is connected to its end is also filled. One then checks all the throats that are available at the new interface between the two fluids and select again the largest throat for displacement. If a throat, or a cluster of throats, which is

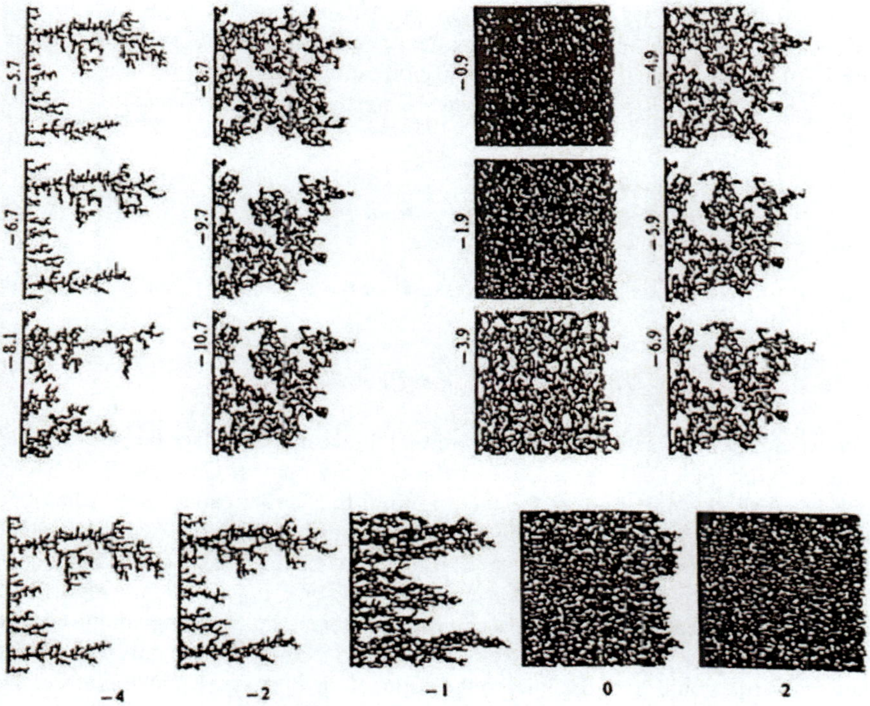

Fig. 9.6 Displacement patterns produced by simulating model of two-phase flow by Lenormand et al. Numbers in the horizontal and vertical directions refer, respectively, to log M and logCa, with M and Ca being the viscosity ratio and the capillary number (after Lenormand et al. 1988)

filled by the wetting fluid is surrounded by the non-wetting fluid, it remains trapped for the rest of the simulation. The procedure is repeated until the invaded throats form a sample-spanning cluster.

Note that because the pressure distribution is calculated explicitly, and that the pore bodies have finite volumes, one must not only consider the possibility of displacing the wetting fluid from a throat, but that it may also have to expel the non-wetting fluid from a pore body. To see this, consider a throat filled by the wetting fluid, and assume that its two end pore bodies are both filled by the non-wetting fluid. Sahimi et al. (1998) showed that such configurations prevent formation of closed loops. If the non-wetting fluid attempts to enter the throat from one end to expel the wetting fluid, then the wetting fluid will also attempt to expel the non-wetting fluid that is in the pore body at the other end of the throat. Due to this constraint, flow calculation for the model of Sahimi et al. (1998) is different from that of Lenormand et al. (1988) and, in contrast with theirs, is capable of handling trapping of the clusters of the wetting fluid, as well as their mobilization.

Consider a throat of radius r_{ij} between pore bodies i and j. If the non-wetting fluid is displacing the wetting fluid, then the volume flow rate in the throat is given by Eq. (9.35) in which P_{cij} is the capillary pressure in the throat between the two fluids, i.e., $P_{cij} = P_{nw} - P_w$. If, however, the wetting fluid attempts to displace the non-wetting fluid, then

$$Q_{ij} = \frac{\pi r_{ij}^4}{8\ell\mu}(P_i - P_j + P_{cpj})^+ , \tag{9.36}$$

where $P_{cpj} = 2\sigma/r_{pj}$ is the capillary pressure for the pore body j, with r_{pj} being its radius, $+$ means that

$$Q_{ij} = 0 , \quad \text{if} \quad P_{cpj} < P_j - P_i < P_{cij} . \tag{9.37}$$

If constraint (9.37) is satisfied, then the wetting fluid cannot displace the non-wetting fluid from the pore body. In practice, if the capillary number Ca is small, (9.37) is always satisfied, implying that the wetting fluid in a throat cannot expel the non-wetting fluid from one of its end pore bodies and, therefore, it is trapped in that throat. The rest of the procedure is similar to that of Lenormand et al. (1988).

Sahimi et al., in their model, represent a *site-bond invasion percolation* algorithm. Its most striking feature is that, at low capillary numbers, the sample-spanning cluster of the non-wetting fluid contains no closed loops. Moreover, there is only one path of the non-wetting fluid from the inlet to the outlet of the pore space and, therefore, the backbone (the flow-carrying part) of the cluster is simply a long strand. The fractal dimension $D_{\rm bb}$ of the backbone is $D_{\rm bb} \simeq 1.14$, completely different from $D_{\rm bb} \simeq 1.64$ for 2D percolation (see Chap. 2), which is indicative of the strand-like structure of the backbone. Moreover, there is a unique path of the non-wetting fluid from the top to the bottom of the pore network.

9.14 Evaporation and Drying in Porous Media

Another two-phase flow phenomenon in porous media for which percolation theory has provided rational explanations and predictions is drying. The phenomenon is important to numerous problems in science and technology (see, for example, Or et al., 2013), ranging from agriculture and soil physics (Aydin et al. 2008; Ben-Noah and Friedman 2018), to remediation and recovery of soil contaminated with hydrocarbons (Nadim et al. 2000; Shokri et al. 2010; Jambhekar et al. 2015; Hosseini and Alfi 2016; Soltanian et al. 2016; Lu et al. 2017), recovery of volatile hydrocarbons from oil reservoirs, cosmetics, building restoration, and such material processing as the production of food, wood, paper, textiles, pharmaceuticals, and washing powders. In addition, water evaporation disturbs the availability, transport, and partitioning of nutrients, as it controls the transfer of heat, air, and humidity between the atmosphere and subsurface. Thus, understanding evaporation and drying of porous media and factors that affect them has been a long-standing problem and studied for decades.

Similar to other flow and transport processes in heterogeneous porous materials and media described so far in this book, the spatiotemporal dynamics of a fluid—liquid or vapor—distribution and the drying front in porous are controlled by two sets of distinct factors:

(i) One set includes the morphological properties of the porous medium and, if they are present, the fractures (Prat 2002; Fantinel et al. 2017; Borgman et al. 2017). Such properties include the pore-size distribution, pore connectivity, pore surface structure, and correlations between the sizes of the pores. Large-scale porous media in which evaporation occurs are also normally anisotropic, which is caused by stratification, a key feature of sedimentary rock on a scale from micrometer up to hundreds of meters. Stratification is caused by a variety of processes, such as weathering, deposition, cementation, and compaction, that together form sedimentary rock. Small-scale stratification is due to alternately operating depositional processes.

Zhang et al. (2013) showed, through experimental and numerical studies, that such small-scale stratification has a dominant effect on the capillary pressure field, CO_2/brine saturation patterns, the flow regime, and the apparent relative permeability of sandstone. Shokri et al. (2010) developed a model to describe the drying of stratified porous media, and verified the model with experimental data. They showed that presence of interfaces in stratified porous media affects the fluid phase distribution, the drying rates, and other aspects of evaporation in porous formations. Vorhauer et al. (2017) showed that the impact of nonuniform distribution of evaporation rate becomes increasingly less important with increasing the number of porous layers. The transition from thin porous media, in which drying is highly dependent on the local structure of the external mass transfer, to thick porous media in which it is much less sensitive to the details of the external mass transfer at the surface, is progressive under isothermal conditions.

(ii) The second set includes dynamic factors, such as capillarity-driven flow (Scherer 1990; Shokri et al. 2008b; Le et al. 2009), pore-scale evaporation and vapor diffusion (Miri et al. 2015; Lehmann and Or 2009; Hajirezaie et al. 2017), and formation of liquid films on the pores' surface (Chauvet et al. 2009; Lehmann and Or 2009; Yiotis et al. 2001). In addition, wettability also plays a prominent role (Bergstad and Shokri 2016). The combination of the effect of all the main factors mentioned in (i) and (ii) gives rise to ramified preferential flow paths and the spatial distribution of the fluids.

Despite being a very complex phenomenon, evaporation consists generally of two stages with a transition between the two. Stage-1 evaporation is controlled mostly by the external conditions and occurs over a short time. It represents the period when the saturated zone at the bottom of the porous medium is connected to the surface via capillary-induced liquid flow, and is limited by the evaporative demand—the external condition. The duration of the first stage is influenced by the pore-size distribution, the correlations between the pore sizes, the pore connectivity, and transport properties of porous media. Stage-1 evaporation ends and the transition period begins when connection of the liquid with the surface begins to break down. The transition period lasts much longer than stage-1 evaporation (Shokri et al. 2008a, b; Lehmann and Or 2009; Chauvet et al. 2009). It ends when all the connections of the liquid are disrupted from the surface, marking the onset of stage-2 evaporation, which is limited by vapor diffusion through the pore space.

Evaporation from a pore space is a displacement process in which the liquid phase in the pore space recedes as a result of vaporization into a vapor phase. Thus, drying is essentially a drainage process (with one caveat; see below) in which the non-wetting vapor displaces the evaporating wetting liquid. Then, in the absence of viscous effects, i.e., when the interface between the liquid and vapor moves slowly, the receding interface is described by an IP process (Shaw 1994). Taking advantage of the link between drying and the IP, pore-network (PN) models have been developed for studying evaporation and drying in porous media (Prat 1995, 2002, 2011; Bray and Prat 1999; Yiotis et al. 2001; Surasani et al. 2008). The models range from simple to relatively advanced PNs, although most of them have been developed for 2D porous media (Yiotis et al. 2001; Rahimi et al. 2016; Le et al. 2017; Dashtian et al. 2018), although three-dimensional models have also been developed (Le et al. 2009; Yiotis et al. 2015; Attarimoghaddam et al. 2017; Vorhauer et al. 2017; Dashtian et al. 2019).

9.14.1 Pore-Network Simulation

Let us describe how PN simulation of drying of porous media is carried out, and how the percolation concepts help to understand the phenomenon better. Yiotis et al. (2001) showed that both advection and capillarity affect the drying patterns and rates. Thus, we assume slow (laminar) flow of the liquid in the PN. Mass transfer in the vapor phase is by diffusion, for which the driving force is the concentration gradient alongside the drying front. The PN's top surface is assumed to be a layer whose throats are initially empty and through which air (or another gas) flows into the pore space, while the other sides of the PN are sealed. The liquid flows in the pore space, while the vapor diffuses to the boundary layer and induces further evaporation.

The advancement of the drying front in the pore space is simulated by the IP processes in which a gas (the non-wetting phase as the agent for drying) invades the liquid-saturated porous medium throat-by-throat according to the lowest capillary pressure needed for entering a throat at the interface between the vapor and liquid. The capillary pressure for entering a throat of radius r_{ij} between pore bodies i and j at the liquid–vapor interface is given by (see above and Chap. 4).

$$P_c = \frac{2\sigma}{r_{ij}}, \tag{9.38}$$

where σ is the surface tension. At every step of the simulation, we identify the state of the pores, which is determined by the filling state of the pore throats connected to them: if at least one neighboring throat does not contain the liquid, it is a vapor-filled pore; otherwise, it is considered as a liquid-filled pore. A throat can be filled with the liquid, fully or partially, or contain its vapor that is leaving it through a path of pores and throats that are also filled by the vapor. A vapor-filled pore can be either at equilibrium or at an unknown vapor pressure, to be determined. In reality, equilibrium vapor pressure prevails only at the menisci between liquid and the vapor, but we assume that a pore is at equilibrium vapor pressure if at least one of its neighboring throats still contains the liquid, even if it is only partially so. Therefore, diffusion of the vapor between menisci of partially filled throats and the adjoining vapor-filled pore happens without any resistance. Numerical simulation indicates that the assumption leads to slightly—about 5 percent—overestimates of the drying rate, but simplifies the computations greatly.

Once the states of the pores and throats are identified, the second stage of the simulation, namely, modeling of flow of the liquid throughout the PN begins. After one or a few steps of the IP that advance the drying front, we solve the governing flow equations. The liquid flow rate $Q_{ij}^{(l)}$ in a throat ij is given by the Hagen–Poiseuille equation,

$$Q_{ij}^{(l)} = \frac{\pi r_{ij}^4}{8\eta^{(l)}\ell_{ij}}\left[P_i^{(l)} - P_j^{(l)}\right], \tag{9.39}$$

where the liquid viscosity $\eta^{(l)}$ is assumed to be constant, as we simulate an isothermal process. Here, $P_i^{(l)}$ is the liquid pressure in pore i and ℓ_{ij} is the length of the throat ij. Since there is no liquid accumulation in the pores, we must have

$$\sum_{\{ij\}} \rho^{(l)} Q_{ij}^{(l)} = 0, \tag{9.40}$$

with $\rho^{(l)}$ being the liquid's mass density. The sum is over the set of the throats $\{ij\}$ that are connected to pore i. Substituting for $Q_{ij}^{(l)}$ and writing the mass balance for every interior node of the PN in the liquid phase result in a set of linear equations, $\mathbf{GP} = \mathbf{b}$, where $\mathbf{G}$ is the conductance matrix that depends only on the morphology of the network, $\mathbf{P}$ is the nodal pressure vector to be calculated, and $\mathbf{b}$ is a vector related to the external boundary conditions. The set of the equations can be solved by the standard conjugate-gradient method.

As for the vapor, we use the equation for evaporation in the so-called Stefan tube, as described by Bird et al. (2007)[5] in order to compute its mass flow rate in the

[5] **Robert Byron Bird** (1924–2020) was Professor of Chemical Engineering at the University of Wisconsin-Madison. He made fundamental contributions to the theory of transport phenomena, flow of non-Newtonian fluids, kinetic theory of polymers, and rheology. Together with Warren E. Stewart and Edwin N. Lightfoot, Bird published the extremely influential classic textbook ***Transport Phenomena***. He was a member of the National Academy of Engineering, National Academy of Sciences, Royal Netherlands Academy of Arts and Sciences, and Royal Belgian Academy of Sciences, and was also a Fellow of the American Academy of Arts and Sciences, the American Physical Society, the American Academy of Mechanics, and the American Association for the Advancement of Science. He was also granted the Dutch title Knight of the Order of Orange-Nassau for his "exceptional contributions to the promotion of Dutch language and culture in the United States."

throats. Ignoring the pressure in the vapor-filled throats, vapor mass conservation at the pores must hold, implying that,

$$\sum_{\{ij\}} \rho^{(v)} Q_{ij}^{(v)} = \sum_{\{ij\}} S_{ij} \frac{D^{(v)}}{\ell_{ij}} \frac{PM^{(v)}}{RT} \ln \left[\frac{P - P_i^{(v)}}{P - P_j^{(v)}} \right] = 0 , \tag{9.41}$$

where $D^{(v)}$, $M^{(v)}$, and $\rho^{(v)}$ are, respectively, the diffusion coefficient, molecular weight, and mass density of the vapor, P is the total pressure in the vapor phase, and S_{ij} is the cross-sectional area of throat ij. The vapor concentration at the vapor–liquid interface is equal to the equilibrium concentration, which is zero outside the boundary layer of the PN. Writing down Eq. (9.41) for all the vapor-filled pores and throats results in another set of equations that govern the pressures in such pores, which can be solved by the standard conjugate-gradient method. The pressure at the liquid–vapor interface is assumed to be equal to the equilibrium vapor pressure. At a meniscus between the liquid and the vapor, the Kelvin equation describes the dependence of the parameters on the vapor pressure:

$$\ln \left[\frac{P^{(v)}}{P^*} \right] = -\frac{2M^{(v)}\sigma}{RTr_{ij}\rho^{(l)}} = -\frac{P_c M^{(v)}}{RT\rho^{(l)}} , \tag{9.42}$$

where R is the gas constant. Once the set of the equations is solved, the vapor flux at any particular stage of the drying is computed. Thus, all the properties of interest may be computed.

9.14.2 *Scaling Theory of Drying*

Since drying is a drainage process—with the caveat that it is not driven by external injection, but by internal diffusion in the gas phase—one in which a non-wetting fluid (air or any other gas) expels a wetting one (water, brine, or other liquids) from a pore space, percolation theory provides a scaling analysis of the phenomenon. First, consider the case in which viscous forces are not important, and recall the Bond number Bo, defined by Eq. (9.28). Then,

(i) if Bo > 0 (downward displacement of a heavier fluid by a lighter one), then width w of the interface is given by Eq. (9.29). In this case, as Tsimpanogiannis et al. (1999) discussed, the front width is finite and self-affine, whereas within the front the pattern possesses the fractal characteristics of percolation clusters. At the upstream of the front, however, the structure is compact. If we define a mean front position, it would represent the transition of the displacement from an IP to a pistonlike pattern;

He received the National Medal of Science, the Bingham Medal for his outstanding contributions to the field of rheology, and the Eringen Medal, among other awards.

see the aforementioned experiments of Hulin et al. (1988b). We refer to this as the IP in a stabilizing gradient (IPSG), which we already described.

(ii) If, however, $\mathrm{Bo} < 0$ (downward displacement of a lighter fluid by a heavier one), then the displacement is IP in a destabilizing gradient (IPDG), and the pattern has features of capillary fingering in which the displacement occurs by invading fingers of a mean width still given by Eq. (9.28), whose local characteristics are still controlled by percolation. This regime has been discussed in detail in Frette et al. (1992) and Meakin et al. (1992).

As Tsimpanogiannis et al. (1999) discussed, the effect of viscous forces is more complex. At larger scales viscous effects predominate, and two limiting patterns are expected, namely, pistonlike displacement and viscous fingering, depending on whether the viscosity ratio $M = \eta_w/\eta_{nw}$ between the viscosities of the two fluids, is smaller or greater than 1, respectively. At smaller scales capillary forces are important, and the problem is similar to IPSG (the limit of small M) or IPDG (with M being large); see Yortsos et al. (1997) and Xu et al. (1998). In particular, in the former case, fully developed drainage generates an advancing front of a finite width w_V, followed by a more compact pattern, such that the width can be shown to scale with the front capillary number $\mathrm{Ca_F}$, based on the viscosity η_{nw} of the non-wetting phase (Wilkinson 1986),

$$w_V \sim \left(\frac{\mathrm{Ca_F}}{\sigma^2_{\mathrm{PSD}}}\right)^{-\nu/[1+\mu+\nu(D_f-1)]}. \tag{9.43}$$

Here, σ^2_{PSD} is the dimensionless variance of the pore-size distribution of the pore space, and μ and D_f are the usual conductivity exponent and fractal dimension of the largest percolation cluster at the threshold. For a comprehensive discussion, numerical simulations, and comparison with experimental data see Tsimpanogiannis et al. (1999). The effect of film flow on the drying patterns in porous media was studied by Yiotis et al. (2004) and by Wu et al. (2020). It was shown that film flow is, in fact, a major mechanism for transport during drying, with its effect being dominant when capillarity controls the process, which is the case in typical applications. By contrast, viscous flow in the bulk contributes negligibly.

Chapter 10
Percolation in Evolving Media

10.1 Introduction

In the previous chapters, we described the applications of percolation theory to flow and transport in porous media. The phenomena that we described were passive in the sense that they did not cause any changes in the morphology of the porous media. In the present chapter, we address the problem of application of percolation to several phenomena in composite materials and porous media that cause dynamical changes in the morphology of their pore space. The phenomena that we will discuss have wide applications in various industries and, therefore, the application of percolation theory to their modeling takes on added practical significance.

10.2 Noncatalytic Gas–Solid Reactions with Fragmentation

We describe the application of percolation to the phenomenon through a concrete example. Gasification of a single char particle in CO_2, which is described by the so-called Boudouard reaction[1]

$$C(s) + CO_2(g) \rightarrow 2CO(g) \tag{10.1}$$

is an important phenomenon that has been studied for decades. The char particles are porous and contain a large number of small pores. Two important regimes to be considered are the *kinetic*, or reaction-limited regime, in which the concentration of CO_2 is the same everywhere, with diffusion playing no significant role, and the *diffusion-limited regime* in which the reactants must diffuse in the pore space before they can react.

[1] Named after **Octave Leopold Boudouard** (1872–1923), a French chemist who discovered it.

M. Sahimi, *Applications of Percolation Theory*, Applied Mathematical Sciences 213,
https://doi.org/10.1007/978-3-031-20386-2_10

10.2.1 The Reaction-Limited Regime

Many studies have indicated that a Langmuir[2]–Hinshelwood[3] kinetic expression can correlate the experimental data on reaction (10.1), and that the expression reduces to a first-order reaction in CO_2, if the gas' partial pressure is low. Therefore, the reaction rate $\hat{R}$ per unit accessible area is written as

$$\hat{R} = k_0 \exp(-E/R_g T), \tag{10.2}$$

where k_0 is the pre-exponential factor, E is the activation energy, R_g is the gas constant, and T is the temperature of the system. In the kinetic regime, the concentration of CO_2 is the same everywhere. Therefore, knowledge of the rate of consumption of carbon is enough for describing the gasification process. A mass balance on a volume element of carbon C yields

$$\frac{d\phi}{dt} = \left(\frac{M}{\rho}\right) \hat{R} S^A, \tag{10.3}$$

where ϕ is the porosity of the carbon particle, M and ρ are, respectively, its molecular weight and density, and S^A is the particle's accessible surface area. We assume that the carbon particles are pure. It should be clear that CO_2 can only reach that part of the carbon particle that is accessible from the outside and, therefore, the accessibility that we invoke here is exactly in the percolation sense defined in Chap. 2. This makes it clear that coal gasification in the kinetic regime is a percolation phenomenon. Moreover, as Eq. (10.3) indicates, to study the problem we only need to keep track of the evolution of the perimeter of the external surface of the particle, i.e., a purely geometrical problem similar to percolation.

An important consequence of gasification is the phenomenon of fragmentation, which can occur both in the kinetic and diffusion-limited regimes, if diffusion is not very fast. If the reaction–consumption of the carbon particle continues, its porosity increases, and, at some well-defined value of ϕ, the particle breaks into several fragments. In some cases, such as fragmentation during devolatilization, one might have a pressure-induced fracture and fragmentation of the particle. If the consumption rate of the particle is low enough, however, as is the case in many practical applications, fracture-induced fragmentation can be safely neglected. Fragmentation in the diffusion-limited regime is usually called *perimeter fragmentation*, since it occurs mainly on the external surface or perimeter of the particle.

---

[2] **Irving Langmuir** (1881–1957) was an American chemist, physicist, and engineer. He made fundamental contributions to the understanding of surface chemistry, sorption, and thin films, for which he was awarded the Nobel Prize in Chemistry in 1932. *Langmuir*, the American Chemical Society journal for surface science, is named in his honor.

[3] Sir **Cyril Norman Hinshelwood** (1897–1967) was a British physical chemist who made fundamental contributions to chemical kinetics and reaction mechanisms, for which he received the Nobel Prize in Chemistry in 1956.

Fragmentation of coal particles is believed to be responsible for decreased burnout times, enhanced production of fly ash, weight loss in coal combustors, and increased emission of submicron unburnt carbon and NO_x from pulverized coal combustion systems. This phenomenon was studied experimentally by Sundback et al. (1984); Kerstein and Niksa (1984), and Sadakata et al. (1984). Particle fragmentation can be inferred from image analysis of char particles, following coal devolatilization (Sundback et al. 1984), or from size measurements of char particles retrieved from fluidized beds and char particle breakup (Kerstein and Niksa, 1984). In the early studies of gasification, fragmentation was either ignored completely or was dealt with empirically by treating the porosity ϕ as an adjustable parameter of the model without any regard for the particle's morphology. For example, based on experimental observations, Gavalas (1981) assumed that, $\phi_f \approx 0.8$, which, interestingly enough, implies that the critical volume fraction of the solid matrix for fragmentation is about 0.2, close to 0.17, the critical volume fraction for percolation in continuous media discussed in Chap. 3.

10.2.2 Diffusion-Limited Regime

In the diffusion-limited regime, transport of the reactants into the pore space, both by ordinary and pressure diffusion, controls the overall mass transfer of the reactants and the gasification process. We introduce two effective transport coefficients, L_{ij}^D and L_{ij}^P, which describe the contributions of diffusion and permeability of the pore structure of the particle, where ij refers to transport of component i in its mixture with j. The two coefficients are given by (Jackson 1977)

$$L_{ij}^D = \frac{1}{1/D_{Ki} + (1 + \delta_{ij} X_i)/D_{ij}(P)}, \tag{10.4}$$

$$L_{ij}^P = \frac{1 + D_{Ki}/D_{ij}(P)}{1/D_{Ki} + (1 + \delta_{ij} X_i)/D_{ij}(P)} + \frac{k_p P}{\eta_m}, \tag{10.5}$$

which are derived based on the Maxwell[4]–Stefan[5] equations of transport for a multicomnponent mixture (Krishna and Wesselingh 1997). Here, the effective

[4] **James Clerk Maxwell** (1831–1879) was a Scottish mathematician who developed the classical theory of electromagnetic radiation, unifying electricity, magnetism, and light as different manifestations of the same phenomenon, which is called the "second great unification in physics," with the first one realized by Isaac Newton.

[5] **Josef Stefan** (1835–1893) was an ethnic Carinthian Slovene physicist and mathematician, as well as a poet in the Austrian Empire. He made numerous fundamental contributions to a variety of scientific fields. Stefan–Boltzmann law for black-body radiation, Stefan–Boltzmann constant, the Stefan problem between two phases with the interface between them moving (such as ice melting), Stefan's equation for the dependence of ice-cover thickness on the temperature history, Stefan's formula for the specific surface energy at a given interface, Stefan flow that describes the movement

Knudsen and pressure diffusivities at pressure P, X_i is the mole fraction of component i, $\delta_{ij} = (M_i/M_j)^{1/2} - 1$, which essentially represents the ratio of the molecular weights of i and j, k_p is the permeability of a single pore, and η_m is the effective viscosity of the mixture.

The evolution of the system is governed by a diffusion–reaction equation that describes how the reactant is transported within the pore space, and how the particle is consumed. Moreover, one also needs a mass balance at the external surface of the particle to express the fact that the diffusive flux there is equal to the rate of mass transfer, Eq. (10.9) below. Thus, for a spherical particle, the concentration C_i of component i at time t evolves according to

$$\frac{\partial C_i}{\partial t} = \left(\frac{1}{R_g T}\right)\frac{1}{r^2}\frac{\partial}{\partial r}\left(r^2 L_{ij}^D P\frac{\partial X_i}{\partial r} + r^2 L_{ij}^P X_i \frac{\partial P}{\partial r}\right) + s_i \hat{R} S^A \tag{10.6}$$

with the initial condition

$$C_i(r, 0) = 0, \tag{10.7}$$

and the boundary condition at $r = 0$,

$$\frac{\partial C_i}{\partial r} = 0, \tag{10.8}$$

and a mass-transfer coefficient K defined by

$$K(C_{ib} - C_i^f) = \left(\frac{1}{R_g T}\right)\left(L_{ij}^D P\frac{\partial X_i}{\partial r} + L_{ij}^P X_i \frac{\partial P}{\partial r}\right). \tag{10.9}$$

Here, s_i is the stoichiometric coefficient of component i in the reaction, and C_{ib} and C_i^f are, respectively, the concentration of i in the bulk, and at a radial position r_f where fragmentation occurs. If the process is highly nonisothermal, then the transport coefficients depend on temperature, and we must add the energy equation to the above model. Because during gasification the pore space evolves, with its porosity ϕ increasing, the transport coefficients also depend on ϕ. Thus, we must also estimate them as a function of ϕ, a problem described in Chap. 7 using percolation theory.

Next, one writes down mass balance for a particle of radius R:

$$\frac{\partial \phi}{\partial t} = \left(\frac{M}{\rho}\right)\hat{R} S_a + \left(\frac{\partial \phi}{\partial r}\right)\left(\frac{\partial R}{\partial t}\right), \tag{10.10}$$

with the initial condition, $\phi(r, 0) = \phi_0$, with ϕ_0 being the initial porosity of the system. Note that the second term of the right side of Eq. (10.10) is due to the shrinkage of the particle with time as a result of the reaction on its external surface

of a chemical species by a flowing fluid, induced to flow by the production or removal of the species at an interface, and many others are all named after him.

and perimeter fragmentation. The contribution of the reaction can easily be accounted for by writing a mass balance on the external surface of the particle, which yields

$$\frac{dr_f}{dt} = -\left(\frac{M}{\rho}\right)\hat{R}, \tag{10.11}$$

with $r_f(0) = r_0$. If perimeter fragmentation starts at a time t_f, then

$$\frac{dr_f}{dt} = -\left(\frac{M\hat{R}}{\rho S^A}\right)\frac{\partial\phi}{\partial r}, \tag{10.12}$$

with the continuity condition, $r_f(t_f^-) = r_f(t_f^+)$. Clearly, for $t < t_f$, Eq. (10.12) should not be used, because fragmentation has not begun yet.

10.3 Percolation Models of Coal Gasification with Fragmentation

How does one model coal gasification with fragmentation? Theoretical investigation of the phenomenon goes back to Petersen (1957) who modeled the pore structure as an idealized network of randomly intercepting cylindrical pores. Many other models have been developed since Petersen's work, which were comprehensively described by Sahimi et al. (1990). Here, we are mainly interested in the percolation models. There have been three types of such models for reactive porous media and, in particular, gasification.

10.3.1 The Reaction-Limited Regime

In a series of papers, Kerstein and co-workers (Grant et al. 1989; Fletcher et al. 1990, 1992) developed a percolation model for devolatilization of coal. As described above, the main assumption is that labile bonds break during pyrolysis and yield smaller molecular fragments that constitute the light gas and tar fractions. Char formation, on the other hand, is associated with the formation of stable, nonlabile bonds during coal devolatilization. The relative amounts of gas, tar, and char are governed by the reaction rates (see above), identified with the labile bond breaking and stable bond-forming processes. The former concept is actually quite old, going back to Pitt (1962) who represented coal as a collection of a large number of species that decompose by parallel first-order reactions.

As discussed by Grant et al. (1989), in a series of papers Solomon and co-workers (Solomon and King 1984; Squire et al. 1986; Serio et al. 1987) developed a detailed chemical model for the release of both tar and light gases during devolatilization.

They listed a total of nineteen first-order, distributed-energy rate expressions for the release of various light gases, and derived a set of differential equations to solve for the various properties of interest. Solving such a large number of differential equations with 19 chemical reactions is, however, not practical (or, at least, it was not in the 1980s), particularly if one is to extend the approach to gas release during the production of tar with its very large number of distinct molecular species. Therefore, Solomon and co-workers characterized tar production with a single first-order, distributed-energy rate expression with chain statistics and, using Monte Carlo simulations, lattice statistics, which opened the door to the percolation models.

The work of Solomon and collaborators on pyrolysis demonstrated that many of the mechanistic features of the time-dependent conversion of coal macromolecule to molecular fragments, such as tar, depend upon some form of lattice statistics. The statistics can be obtained by Monte Carlo simulations, or by using percolation theory that provides analytical expressions for the statistics of bridge dissociations involved in devolatilization.

In order to use a percolation model, one must determine the compositions of various chemical moieties within coals, and use them so as to provide some of the chemical information needed in the model of devolatilization. For example, the ratio of aromatic bridgehead carbons to total aromatic carbons provides a functional measure of the number of aromatic carbon atoms in a typical cluster of fused aromatic rings (Grant et al. 1989). In addition, the number of peripheral carbon atoms per cluster that are involved in side chains or bridges provide rough estimates of the average molecular weight of a cluster and the number of branching sites for the cluster, which are important measures for characterizing lattice statistics and percolation models.

As described in Chap. 2, and also discussed by Grant et al. (1989), percolation describes the size distribution of finite clusters of sites, or bonds, joined by intact bridges, but isolated from all remaining sites by broken bridges. Moreover, percolation predicts the existence of a percolation threshold—the critical bridge population in the present problem—that depends only on the coordination number, and above which an infinite cluster (or array) coexists with clusters of finite size.

To adapt the fundamental features of percolation theory for modeling coal combustion, as well as char and tar pairs obtained during pyrolysis, we identify an infinite lattice used in percolation theory as macroscopic unreacted coal and/or char, whereas relatively small tar molecules are identified with the finite clusters. Given such identifications, the path to the development of a percolation model opens up.

10.3.1.1 Statistics of Percolation

Grant et al. (1989) used a Bethe lattice as the model of unreacted coal. The motivation for the assumption was that all the percolation properties of a Bethe lattice have been derived analytically and, therefore, there is no need for Monte Carlo calculations. Moreover, as mentioned in Chap. 9, many of percolation properties of the Bethe lattices are qualitatively similar to those of three-dimensional (3D) media. We also adopt the language of Grant et al. in bonds—bridges—and assume that the coordination number Z of the Bethe lattice is $Z = \sigma + 1$. During the gasification, p, the fraction of unbroken (unreacted) bridges (bonds) varies dynamically with time. P_n, the probability that a given site belongs to a cluster of n sites and s bridges (bonds), is given by

$$P_n(p) = nb_n p^s (1-p)^\tau , \tag{10.13}$$

where $s = n - 1$ and $\tau = n(\sigma - 1) + 2 = n(Z - 2) + 2$. Note that τ represents the number of broken bridges on the perimeter of an s-bridge cluster. In Eq. (10.13), nb_n is the number of distinct possible configurations for a cluster of size n that contains a given site, and b_n is the same quantity, but expressed on a per site basis, which is given by

$$nb_n = \left(\frac{\sigma+1}{n\sigma+1}\right)\binom{n\sigma+1}{n-1}, \tag{10.14}$$

where $(\cdot)$ indicates the binomial coefficient. Given $P_n(p)$, the total fraction $P(p)$ of the sites that are in all the clusters is given by

$$P(p) = \sum_{n=0}^{\infty} P_n(p) = \left(\frac{1-p}{1-p^*}\right)^{\sigma+1}, \tag{10.15}$$

with p^* being the root of the equation, $p^*(1-p^*)^{\sigma-1} = p(1-p)^{\sigma-1}$. Note that, as mentioned in Chap. 2, the percolation threshold of the Bethe lattice is at $p_c = 1/(Z-1) = 1/\sigma$. For $p > p_c$, $P(p)$ is not unity, but is given by $P(p) = 1 - R(p)$, where, by definition, $R(p)$ is the fraction of the sites in the lattice. In addition, we define $Q_n(p) = P_n(p)/n = b_n p^{n-1}(1-p)^{n(\sigma-1)+2}$, which represents the number density of $n-$site clusters per site. Moreover, $G(p)$, the configuration generating function, is defined by

$$G(p) = \sum_{n=0}^{\infty} Q_n(p) = \left[1 - \left(\frac{\sigma+1}{2}\right)p^*\right]\left(\frac{p^*}{p}\right)^{(\sigma+1)/(\sigma-1)}, \tag{10.16}$$

10.3.1.2 Chemical Reactions and Kinetics

Grant et al. proposed a simple sequence of reactions that started with cutting a chemical bond in a labile bridge to form a highly reactive bridge intermediate, such as two free-radical side chains that are temporarily trapped in the reaction cage, which is consumed rapidly by one of two competitive processes. The reactive bridge material may either be released as a light gas with the concurrent relinking of the two associated sites within the reaction cage to produce a stable or charred bridge, or it may be stabilized to produce side chains from the reactive bridge fragments. The stabilized side chains may be converted eventually into light gas fragments through a subsequent, slower reaction. Therefore, the following scheme was proposed to represent the devolatilization process of coal:

(i) A reactive bridge intermediate is formed by a labile bridge through a relatively slow step and a reaction rate coefficient k_b:

$$\mathcal{L} \rightarrow \mathcal{L}^* \tag{10.17}$$

(ii) A char bridge and gas g_2 are formed from the reactive intermediate with a reaction rate coefficient k_c:

$$\mathcal{L}^* \rightarrow C + 2g_2. \tag{10.18}$$

(iii) A side chain is also formed by the reactive intermediate with a reaction rate coefficient k_δ:

$$\mathcal{L}^* \rightarrow 2\delta. \tag{10.19}$$

(iv) The side chains are converted into light gases with a reaction rate coefficient k_g:

$$\delta \rightarrow g_1. \tag{10.20}$$

In this scheme, gas is produced by the charring process to produce g_2, and by detachment of the side chains to produce g_1. Thus, a bridge yields two side chains, and ultimately two gas molecules, which is why a factor of 2 was used in reactions (10.18) and (10.19). Assuming that the symbols for the various species also represent their fractional abundance, expressed as normalized bridge parameters, the governing equations for the reactions are given by

$$\frac{d\mathcal{L}}{dt} = -k_b\mathcal{L}, \tag{10.21}$$

$$\frac{d\mathcal{L}^*}{dt} = k_b\mathcal{L} - (k_c + k_\delta)\mathcal{L}^*. \tag{10.22}$$

Therefore, assuming steady-state approximation for $\mathcal{L}^*$, we obtain the following set of equations:

$$\frac{dC}{dt} = \frac{k_b\mathcal{L}}{\rho_r + 1}, \tag{10.23}$$

$$\frac{d\delta}{dt} = \left(\frac{2\rho_r k_b\mathcal{L}}{\rho_r + 1}\right) - k_g\delta, \tag{10.24}$$

$$\frac{dg_1}{dt} = k_g\delta, \tag{10.25}$$

$$\frac{dg_2}{dt} = 2\frac{dC}{dt}, \tag{10.26}$$

with $\rho_r = k_\delta/k_c$. Note that the fraction p of the unbroken bridges is given by, $p = \mathcal{L} + C$, linking directly a percolation variable to measurable quantities. The initial conditions at $t = 0$ are, $C = C_0$, $\mathcal{L} = \mathcal{L}_0$, $\delta = 2(1 - p_0) = 2(1 - C_0 - \mathcal{L}_0)$, and

$g = g_1 = g_2 = 0$. Note also that mass conservation implies that, $g = g_1 + g_2$, $g_1 = 2(1-p) - \delta$, and $g_2 = 2(C - C_0)$.

Next, we relate these quantities to measurable ones. Thus, (Grant et al. 1989), the total mass m_t per lattice site is

$$m_t(t) = m_a + \frac{1}{2} m_b (1 - C_0)(\sigma + 1), \tag{10.27}$$

where m_a is the average mass of a site. The second term on the right side of Eq. (10.27) includes mass m_b of the bridges, corrected by the fraction $(1 - C_0)$ of bridges that might have already stabilized at $t = 0$. Note that $(\sigma + l)/2$ is the ratio of the bridges and the sites, which converts a bridge parameter, such as $(1 - C_0)$ to a site quantity. The mass $m_g(t)$ per lattice site of the gas released up to time t is given by

$$m_g(t) = \frac{1}{4} m_b g(\sigma + 1). \tag{10.28}$$

Note that $(\sigma + 1)/4 = (1/2)(\sigma + 1)/2$, and that the factor 1/2 is for converting m_b into the half-bridge mass assigned to the average mass of side chains and of light gases released. The mass $m_{\rm tar}(t; n)$ per lattice site of the tar clusters of size n is given by

$$m_{\rm tar}(t; n) = \left[n m_a + \frac{m_b \mathcal{L}(n-1)}{p} + \frac{m_b \tau \delta}{4(1-p)} \right] Q_n(p), \tag{10.29}$$

hence linking directly the percolation statistics $Q_n(p)$ to the total mass of tar. Therefore, the total tar mass $m_{\rm tar}(t)$ of all the tar clusters is given by

$$m_{\rm tar}(t) = \sum_{n=1}^{\infty} m_{\rm tar}(t; n) = m_a \Phi \sum_{n=1}^{\infty} n Q_n(p) + m_b \Omega \sum_{n=1}^{\infty} Q_n(p), \tag{10.30}$$

or

$$m_{\rm tar}(t) = m_a \Phi P(p) + m_b \Omega G(P), \tag{10.31}$$

with

$$\Phi = 1 + r \left[\frac{\mathcal{L}}{p} + \frac{\delta(\sigma - 1)}{4(1-p)} \right], \tag{10.32}$$

$$\Omega = \left[\frac{\delta}{2(1-p)} - \frac{\mathcal{L}}{p} \right], \tag{10.33}$$

and $r = m_b/m_a$. Finally, the mass fractions of gas, tar, and char are calculated by, $f_g(t) = m_g(t)/m_t(t)$, $f_{\rm tar} = m_{\rm tar}(t)/m_t(t)$, and, $f_{\rm char}(t) = 1 - f_g(t) - f_{\rm tar}(t)$. The formulation of the problem as one of the percolations is complete.

10.3.1.3 Comparison with Experimental Data

Figure 10.1 compares the predictions of the percolation model of Grant et al. (1989) with experimental data for volatilization yields of char, tar, and light gases versus time for Illinois No. 6 high-volatile bituminous coal, reported by Serio et al. (1987), while Fig. 10.2 makes the same comparison for Montana Rosebud subbituminous coal. The agreement between the predictions and the data is excellent, indicating the power of percolation theory for describing the dynamic evolution of a complex material.

Fletcher et al. (1990) used the percolation model to study the effect of temperature and heating rate on the product yields. To do so, consider the quantity $\rho_r = k_\delta / k_c$. Assuming that

$$k_b = A_b \exp[-(E_b \pm V_b)/(R_g T)], \tag{10.34}$$

$$k_\delta = A_\delta \exp[-(E_\delta \pm V_\delta)/(R_g T)], \tag{10.35}$$

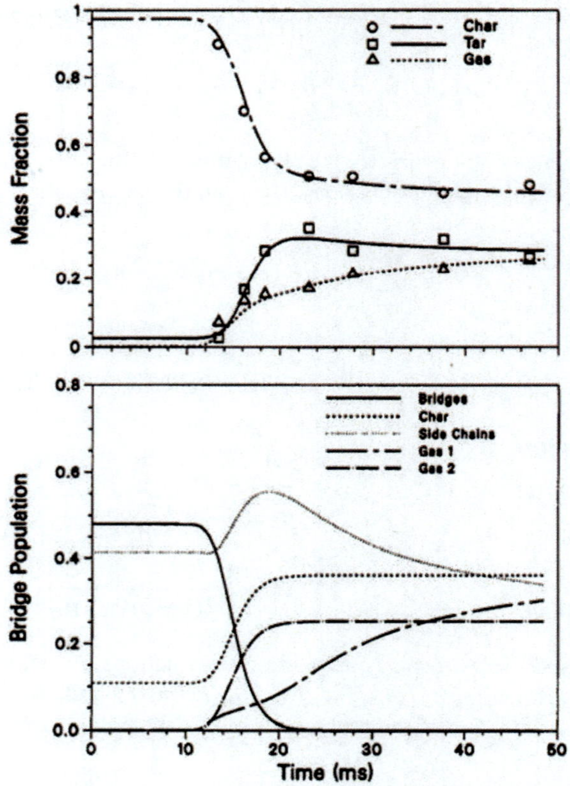

Fig. 10.1 Top: The results of numerical simulations of devolatilization for dynamic evolution of yields of char, tar, and light gases for Illinois No. 6 high-volatile bituminous coal, and their comparison with experimental data. Bottom: bridge dynamical variables, δ, g_1 and g_2 (divided by 2) on a per site basis (after Grant et al. 1989)

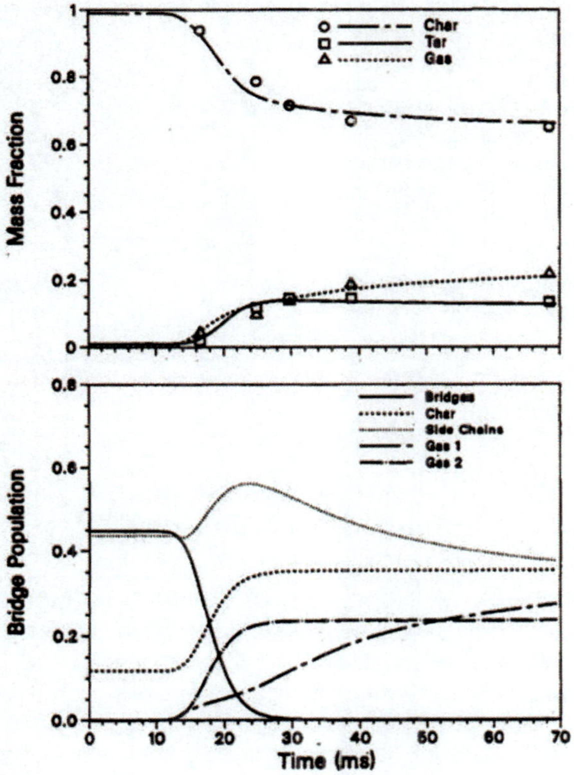

Fig. 10.2 Top: The results of numerical simulations of devolatilization for dynamic evolution of yields of char, tar, and light gases for Montana Rosebud subbituminous coal. Bottom: bridge dynamical variables δ, g_1 and g_2 (divided by 2) on a per site basis (after Grant et al. 1989)

where A_i, E, and V_i are, respectively, the pre-exponential frequency factor, the activation energy, and the distributed variation in the activation energy for the ith process, one obtains

$$\rho_r = \frac{k_\delta}{k_c} = A_{\rho_r} \exp[-(E_{\rho_r} \pm V_{\rho_r})/(R_g T)], \tag{10.36}$$

with $A_{\rho_r} = A_\delta / A_b$, $E_{\rho_r} = E_\delta - E_c$, and V_{ρ_r} is the corresponding distributed activation term. Figure 10.3 compares the predictions of the percolation model with the experimental data for the total volatiles yield for APCSB Pittsburgh No. 8 coal. Once again, the agreement is excellent. In another paper, Fletcher et al. (1992) used the same percolation model to describe the generation of tar precursors of finite size, based on the number of cleaved labile bonds in the infinite coal lattice, including treatment of vapor–liquid equilibrium and a cross-linking mechanism.

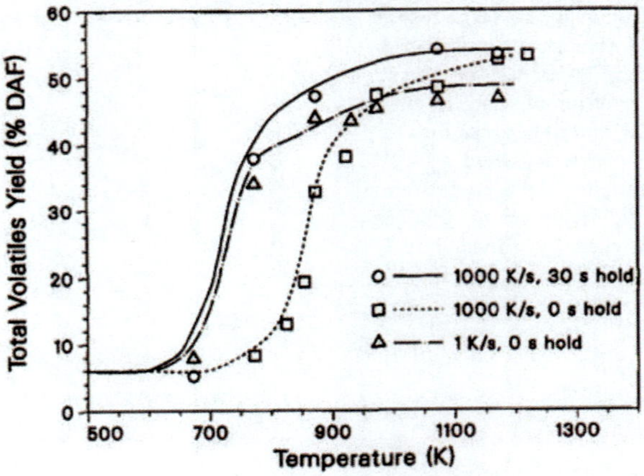

Fig. 10.3 Comparison of the predictions of the percolation model (continuous curves) for the total yield of volatiles with the experimental data for APCSB Pittsburgh No. 8 coal, at various maximum temperatures and hold times (after Fletcher et al. 1990)

10.3.2 Hybrid Percolation-Continuum Models of Diffusion-Limited Regime

Percolation models for coal gasification, when diffusion of the reactants must be taken into account, are based on the numerical solution of Eqs. (10.6)–(10.9) and calculating the consumption rate of the coal particles. Since the porosity of coal increases with time, the effective morphological properties, such as the accessible surface area, and diffusion coefficients also vary with porosity and time. Their functional dependence on the porosity is derived by using percolation theory, as described in Chap. 7. To include the percolation effect in the model, one assumes, similar to the percolation model for the kinetic regime, a certain model of the pore space and calculates all the percolation quantities that are needed for use in Eqs. (10.6)–(10.9). As such, such models are a hybrid of the classical continuum equations of transport and reaction, and percolation approach for estimating the effective transport properties of a disordered medium, described in Chap. 7.

Mohanty et al. (1982) were the first to develop a percolation model of reactive porous media. They represented the porous medium by a cubic tessellation in which a randomly selected fraction p of the cubic polyhedra represented the pores. The increase in the porosity as a result of gasification was modeled by simply increasing p. That is, the gasification was modeled as a random percolation process. Similar to two-phase flow and displacement of one fluid by another immiscible invading fluid, described in Chap. 9, in which there was a well-defined interface between the displaced and displacing fluids, there is also a reaction front between the reactants and the solid matrix in the present problem. As a result, the increase in the porosity should, in principle, be modeled by consuming the solid polyhedra adjacent to the percolating pore cluster or the reaction front.

Reyes and Jensen (1986) improved the model of Mohanty et al. (1982) by representing the pore space as a Bethe lattice of cylindrical pores with distributed radii, in

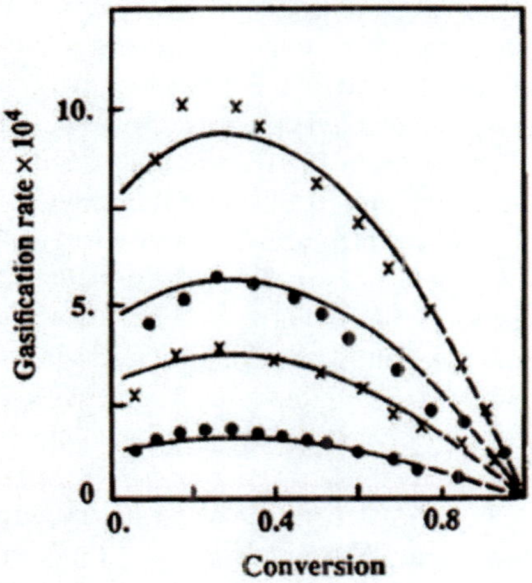

Fig. 10.4 Comparison of the predictions (curves) of the percolation model of char-air gasification rate (in s^{-1}) with the experiments (symbols). The pore space was represented by a Bethe lattice of connectivity $Z = 7$. The curves, from top, are for $T = 753$ K, 728 K, 703 K, and 673 K (after Reyes and Jensen 1986)

order to take into account the effect of the pore-size distribution of the coal particles, but, similar to the percolation model in the kinetic regime described above, they also modeled the porosity increase by random percolation. Reyes and Jensen (1986) obtained a good agreement for the kinetic regime between the predicted reaction rates at various values of conversion—the consumed fraction of the particle—and the experimental data. This is shown in Fig. 10.4.

Shah and Ottino (1987a) also modeled gasification by solving Eqs. (10.6) and (10.9), but representing the pore space by a cubic network. Unlike the earlier model of Mohanty et al. (1982) and Reyes and Jensen (1986), Shah and Ottino (1987a) modeled the increase in the porosity in a physical manner, namely, by monitoring the evolution of the coal's morphology and the reaction front that was constantly moved to the edge of the remaining matrix.

Let us mention that Schiulaz et al. (2018) studied the problem of combustion in disordered explosive media, a long-standing problem that has been known to occur after local regions of the material, called hot spots, are ignited. Using a continuum model of the sort described above, they showed that a system of randomly distributed hot spots exhibits a dynamic phase transition that, depending on various physical factors of the system, can be either first or second order, with the latter case resembling a percolation transition. The two regimes are separated by a tricritical point.

10.3.3 Pore-Network Models

Pore-network models (PNMs) were originally developed by Kerstein and Bug (1986); Kerstein and Edwards (1987), and Sahimi and Tsotsis (1987, 1988), and Sahimi (1991). The deatils of their models are not the same, but the main ideas are essentially identical. One begins with a network in which a fraction p of the bonds represents the pores, with the remaining bonds representing the solid matrix. One then identifies the solid bonds on the external perimeter of the pore space. They are consumed and redesignated as pores, the process time is increased by one unit, the perimeter solid bonds in the new configuration of the network are identified, and so on. If the consumption of the solid bonds is continued, then at a porosity ϕ_f the solid particle disintegrates into finite fragments (clusters) with a wide variety of shapes, sizes, and masses. The precise value of ϕ_f depends on the microscopic details of the solid matrix and its chemical composition, just as the percolation threshold p_c depends on the coordination number Z of the network. For example, if a portion of the matrix is nonreactive, fragmentation would occur at a lower ϕ_f. The presence of such impurities can easily be accounted for by designating a fraction of the bonds as nonreactive.

Figure 10.5 presents the dependence of the reaction rate, i.e., the fraction of consumed solid bonds per unit time, on the total conversion in a 3D network with $Z = 14$. There is a maximum in the reaction rate for small values of p, which is observed frequently in experiments. The maximum is caused by two competing geometrical factors. At low values of conversion, the consumption of the matrix increases the length of the perimeter of the solid matrix—the cluster of solid bonds—which causes an increase in the reaction rate, which is proportional to the perimeter length. At high conversions, however, the length of the perimeter decreases with consumption, causing the reaction rate to decrease monotonically. Kerstein and Bug (1986) showed that in such a model the solid matrix is consumed in a time t_c, with

$$t_c \sim \ln(\epsilon^{-1}), \tag{10.37}$$

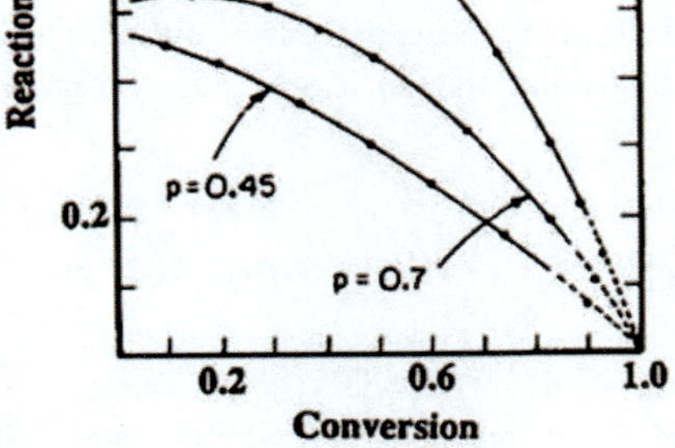

Fig. 10.5 Dependence of the reaction rate on the initial fraction p of solids in a pore-network percolation model of gasification (after Sahimi 1991)

with $\epsilon = \phi_f - \phi$, so that one has a simple way of estimating the burning time as a function of porosity.

How do we account for the effect of diffusion? Kerstein and Edwards (1987) assigned to each solid bond a random burning time selected from an exponential distribution, $f(t) = \lambda \exp(-\lambda t)$, and related λ to the mass m of the solid matrix via, $\lambda = am - b$, where b is an adjustable parameter. The idea was that different portions of the particle have different burning times, because the diffusing reactants reach them at different times. After fragmentation, a distinct value of λ is used for each fragment. Although the model was somewhat ad hoc, it did provide a reasonable fit to some experimental data.

In the model of Sahimi and Tsotsis (1987), diffusion was represented by random walk of the reactants that are injected into the system at the external surface of the network. The random walk was executed on the bonds that represent the pores, and the probability of selecting any pore was proportional to its conductance. When the reactant hit a solid bond, it reacted with it with probability, $p_r = \hat{R}/k_0$, i.e., the normalized reactivity of the matrix. Thus, the effect of pore-size distribution, temperature, and other influencing factors were taken into account. If the reaction did occur, the reactive molecule would disappear and a fraction of the solid bond would be consumed. Since one uses finite networks to represent char particles of micron size, each bond represents a macroscopic mass unit. Thus, each solid bond must be hit by the diffusing molecule, in a reactive collision, at least n_c times, with n_c being a parameter of the model, before it is totally consumed. If $p_r = 1$, then only the most exposed part of the matrix is consumed at the initial stages of the process. The net effect is the removal of all the irregularities of the external surface of the matrix, making its shape smooth. Thus, at most, perimeter fragmentation can occur.

On the other hand, if p_r is small, the reactive molecules penetrate deep into the pore space and consume the weak points of the matrix. As a result, matrix fragmentation occurs and many fragments with a wide variety of shapes and sizes appear. Thus, the limits $pr_{=}1$ and 0 represent, respectively, the diffusion-limited and the kinetic (reaction-limited) regimes, and varying p_r between 0 and unity enables one to study a variety of situations in which the relative importance of the kinetic and transport effects can be varied.

10.3.3.1 Fragmentation During Gasification

To describe the evolution of fragment size distribution, a dynamic scaling of the following form was proposed: (Sahimi and Tsotsis 1987; Sahimi 1991)

$$n_s \sim t^w s^{-\tau} f(s/t^z), \tag{10.38}$$

where n_s is the number of fragments (per network bond) of s bonds, w, z, and τ are dynamical exponents, and $f(x)$ is a scaling function. The exponents are presumably independent of the pore-size distribution of the particle, but may depend on other factors. The exponent w is positive before a characteristic time t_m is reached at which

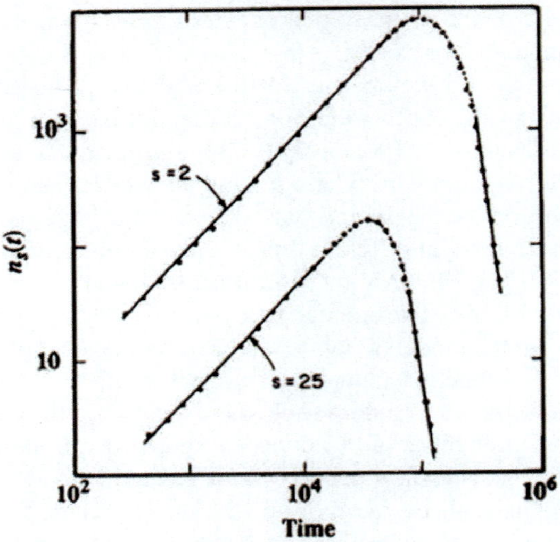

Fig. 10.6 Dynamic evolution of number of fragmets $n_s(t)$ of size $s = 2$. The reactive porous medium was represented by a simple cubic network at its site percolation threshold of 0.311 (after Sahimi 1991)

the number of fragments reaches a maximum, and is negative for $t > t_m$, with t^w representing the dynamic evolution of the number of fragments that distinguishes Eq. (10.38) from that of percolation, Eq. (2.14). A similar equation was used in polymer-chain degradation, and even porous rock fragmentation (Englman et al., 1984). Thus, Eq. (10.38) may be a general property of many fragmentation processes. Figure 10.6 presents the evolution of $n_s(t)$, calculated for a cubic network, which indicates that $w = 1$. Cai et al. (1991) developed a linear rate equation for describing fragmentation with mass loss and derived Eq. (10.38).

We note that Dunn-Rankin and Kerstein (1987) reported the results of extensive numerical simulations in order to understand the sensitivity of measured coal particle-size evolution to fragmentation. To do so, they used two distinct approaches. One was based on the percolation model described above, which built the evolution of the distribution based on an ensemble of individual particle size time histories. The second simulation was a population balance approach in which the entire distribution was treated as a unit.

10.4 Noncatalytic Gas–Solid Reactions with Pore Blocking

This class of phenomena is encountered in several important chemical processes. For example, limestone is used to control SO_2 emissions from fluidized-bed combustors, while metal oxides are used in coal gas desulphurization. In both cases, solid products are formed by noncatalytic gas–solid reactions. When the stoichiometric volume ratio, i.e., the ratio of the molar volume of the solid product and stoichiometrically equivalent volume of the solid reactant, exceeds one, the solid product occupies more space than the original solid, the porosity of the porous material decreases as the reaction proceeds, and pore plugging occurs, implying that this is a percolation phenomenon. Similar to gasification with fragmentation, this problem has also been investigated by many authors, using a variety of ideas and methods, which were reviewed by Sahimi et al. (1990). As usual, however, we are only interested in the percolation models of the phenomena, which we now describe.

During any noncatalytic gas–solid reaction with solid products in a porous material, as well as in catalyst deactivation described in the next section, all pores of the porous material can be divided into three distinct groups (Sahimi and Tsotsis, 1985).

(i) Pores that are completely plugged,

(ii) pores that are partially plugged (with reduced radii), but are still accessible to the reactant, and

(ii) pores that are not completely plugged yet, but are surrounded by plugged ones and cannot be reached by the reactants. Thus, the first difference between this phenomenon and random percolation emerges: in the present problem, some of the open pores may become trapped by the plugged pores and, therefore, this constitutes a percolation phenomenon with trapping described in Chap. 9.

We consider a specific example and describe a percolation model for it. Consider the sulphation of calcined limestone particles, which occurs according to the following reaction:

$$CaO(s)+SO_2(g)+\tfrac{1}{2}O_2(g)\rightarrow CaSO_4(s). \tag{10.39}$$

Experimental data indicate that, during the reaction the solid volume increases by a factor of about 3.1, and that in excess O_2, the reaction is first order with respect to the partial pressure of SO_2. Similar to gasification with fragmentation, we may consider both the kinetic and diffusion-limited regimes. But, in practice, the kinetic regime is hardly ever realized and, therefore, only the diffusion-limited problem is of practical importance; see Yortsos and Sharma (1986); Yu and Sotirchos (1987), and Sotirchos and Zarkanitis (1993) for the percolation models in the kinetic regime.

In the diffusion-limited regime, the governing equations for a spherical porous particle are given by (Reyes and Jensen 1987)

$$\frac{\partial C_1}{\partial t}=\frac{1}{r^2}\frac{\partial}{\partial r}\left(r^2 D_1\frac{\partial C_1}{\partial r}\right)-\hat{R}C_2S^A. \tag{10.40}$$

with the boundary conditions that

$$\frac{\partial C_1}{\partial r}=0,\quad r=0 \tag{10.41}$$

$$D_1\frac{\partial C_1}{\partial r}=K(C_{1b}-C_{R_0}),\quad r=R, \tag{10.42}$$

where D_1 is the effective diffusivity of the pair SO_2-O_2, C_1, and C_2 are, respectively, the SO_2 concentrations at the gas–solid and solid–solid interfaces, and the rest of the notation is as before. Equations (10.40)–(10.42) are similar to (10.6)–(10.9), except that in the present problem the contribution of pressure diffusion is negligible. One also needs an equation for C_2 to express its variations within the solid layer that builds up in every pore. Since transport within the layer is by diffusion, one has

$$\frac{\partial}{\partial \zeta}\left(\zeta^{\delta-1} D_\ell \frac{\partial C_2}{\partial \zeta}\right) = 0, \tag{10.43}$$

with the moving boundary conditions that,

$$D_\ell \frac{\partial C_2}{\partial \zeta} = \hat{R} C_2, \tag{10.44}$$

$$C_2 = C_1, \tag{10.45}$$

at $\zeta = r_{ss}$, with r_{ss} being the position of solid–solid interface. Here, D_ℓ is the diffusivity within the layer, ζ is measured in the direction perpendicular to the reacting surface, and δ is a geometrical factor such that $\delta = 1, 2$, and 3 correspond, respectively, to slabs, cylinders, and spheres. Experimental data for D_ℓ are used in Eqs. (10.43) and (10.44), because its theoretical estimation is difficult.

One also needs an equation that describes the consumption rate of CaO. Suppose that ϕ_s is the volume fraction of CaO in the particle. Then, a simple mass balance yields

$$\frac{d\phi_s}{dt} = -\left(\frac{M_s}{\rho_s}\right) \hat{R} C_2(\zeta, t) S_a, \quad \zeta = r_{ss}, \tag{10.46}$$

where M_s and ρ_s are the molecular weight and density of the solid, respectively, and $\phi_s(t = 0) = \phi_0$. Due to the formation of solid products, the pore radius $R(r, t)$ continuously shrinks, and because the total volume is fixed, $R(r, t)$ is related to the initial pore radius R_0 and r_{ss} through the following equation:

$$R(r, t) = e_r R_0 + (1 - e_r) r_{ss}(r, t), \tag{10.47}$$

where e_r is the expansion coefficient whose numerical value, as mentioned above, is about 3.1. The last step of the modeling is writing down an evolution equation for the pore-size distribution $f(r, t)$ of the porous material. Only accessible pores are of interest to us and, therefore,

$$\frac{\partial}{\partial t}[\phi_a f(r, t)] + \frac{\partial}{\partial r}\left[\phi_a f(r, t)\frac{dR}{dt}\right] = \left(\frac{d\phi_a}{dt}\right) f(r, t), \tag{10.48}$$

where ϕ_a is the accessible porosity of the material, $f(r, 0) = f_0(r)$ (the initial pore-size distribution) and $f(r, \infty) = 0$. The second boundary condition arises because, after a long time, all the pores will be plugged.

By solving the governing equations, one computes the evolution with time or position of all the quantities of interest. Percolation enters the model in two distinct ways. One is through D_1, the effective diffusivity of the pair SO_2–O_2. Since the pores

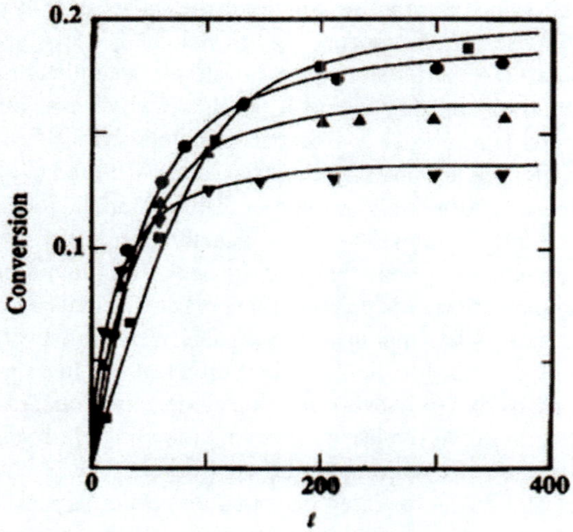

Fig. 10.7 Comparison of the predictions of the percolation model for conversion of CaO (curves) and their predictions with experimental data (symbols). The results, from top, are for $T = 1023$ K, 1113 K, 1163 K, and 1213 K (after Reyes and Jensen 1987)

plug, D_1 must be calculated for a percolation network whose porosity is shrinking, a problem described in Chap. 7. The second percolation aspect of the problem is the change in the accessible porosity ϕ_a (accessible fraction of the pores) that also decreases as the pore space of the material evolves. Thus, it should be clear how to implement the model: assume a network model of the pore space and estimate its ϕ_a and D_1 as the pore space evolves (as the porosity decreases), utilize the results in the governing equations, and solve for the properties of interest.

The predictions of the model are reasonably accurate. For example, Fig. 10.7 compares the predicted dynamic conversion of CaO with the experimental data. The pore-space model was assumed to be a Bethe lattice with a coordination number of $Z = 6$. For short times, the agreement is excellent, but at longer times the predictions and the experimental data exhibit some differences. Other variations of this basic model, as well as other approaches to this phenomenon, were reviewed by Sahimi et al. (1990).

10.5 Percolation Models of Catalyst Deactivation

An important phenomenon for which the concepts of percolation theory have been invoked in order to obtain a better understanding of deactivation of porous catalysts. Good reviews of the subject were given by Hughes (1984) and Froment (2001). The phenomenon is typically caused by a chemical species that is adsorbed on and poisons the catalyst's surface and its active sites, where catalytic reactions occur, and frequently blocks its pores due to the formation of undesirable products, such as coke. One often finds that the reactants, products, and reaction intermediates, as well as various reactant stream impurities, also serve as poisons and/or poison precursors. As a result, the morphology of the catalyst evolves with time, and its pore volume decreases. After sometime, no sample-spanning cluster of open pores exists, and the catalyst deactivates and loses its effectiveness, implying that the deactivation is a percolation process. Because of its industrial significance, numerous theoretical and experimental investigations have been devoted to catalyst deactivation. As usual, we are interested only in the percolation models of catalyst deactivation.

Early studies of catalyst deactivation either used a bundle of parallel capillary tubes to represent the pore space, or employed continuum equations of transport and reaction in which the transport coefficients and other important parameters were inputs to the model. Such models were, however, inadequate because they could not account for the important effect of the interconnectivity of the catalyst's pores and its dynamic evolution during deactivation. As an improvement to such models, beginning in the late 1970s, an interesting class of catalyst deactivation models was developed by Froment and co-workers (see Nam and Froment (1987); Beeckman et al. (1987), for references to their earlier papers) in which probabilistic arguments were used to derive expressions for the activity or reaction rate of the catalyst. To represent the pore space, they used a variety of models, such as a single pore, the bundle of parallel pores, and a Bethe lattice with coordination number $Z = 3$. Using the Bethe lattice, Froment and co-workers were the first to recognize that, in agreement with the experimental observations, the catalytic activity of a catalyst undergoing deactivation vanishes much earlier than what is predicted by the bundle of parallel tube model. This is, of course, caused by the interconnectivity of the pores, the main idea in percolation, but was not recognized as such by them.

Catalyst deactivation was first identified as a percolation phenomenon by Sahimi and Tsotsis (1985). To describe percolation models of catalyst deactivation, we again distinguish between the kinetic and diffusion-limited regimes.

10.5.1 Reaction-Limited Deactivation

For concreteness, we describe the model that was developed by Sahimi and Tsotsis (1985). A porous catalyst, in an isothermal reactor, with a catalytically active material, uniformly distributed in its pores at an initial concentration C_0, reacts, while simultaneously undergoes slow deactivation. The overall reaction rate $\hat{R}$ in a single pore of radius R and length l_p of the catalyst is given by

$$\hat{R} = 2\pi R l_p \hat{R}_0 (C_0 - C_s)^m, \tag{10.49}$$

where $C_0 - C_s$ is the concentration of the catalyst's active sites at time t, $\hat{R}_0$ is the reaction rate per unit area, and m is the order of the reaction. A parallel deactivation–reaction process results in the deposition of a contaminant, a deposit with an average volume b per unit weight of the sites, which poisons the active sites, while simultaneously blocks part of the pore volume. The rate of change in the concentration of the poisoned sites is given by

$$\frac{dC_s}{dt} = \hat{R}_d(C_0 - C_s)^n, \tag{10.50}$$

where $\hat{R}_d$ is the deactivation–reaction rate per unit area, and n is the reaction order. The chemistry of the reactions is embedded in $\hat{R}_0$ and $\hat{R}_d$. A partially plugged pore with an initial radius R_0 has an effective radius R_e, which is the radius of a cylindrical pore with the same open volume and length, defined by

$$R_e = R_0\left[1 - \left(\frac{\alpha}{R_0}\right)g(\theta)\right], \tag{10.51}$$

with

$$g(\theta) = \begin{cases} 1 - [1 + (n-1)\theta]^x & n < 1,\ \theta < x, \\ 1 - \exp(-\theta) & n = 1, \\ 1 - [1 + (n-1)\theta]^x & n > 1, \end{cases} \tag{10.52}$$

where $x = 1/(1-n)$, and

$$\alpha = 2C_0 b, \tag{10.53}$$

$$\theta = \hat{R}_d C_0^{n-1}. \tag{10.54}$$

Here, θ is the dimensionless time, and α is the effective size of the deposit. Note that if $\alpha < R_0$, then $R_e > 0$ at all times, whereas for $\alpha \geq R_0$ we have $R_e = 0$ after a finite time θ_p, the time for plugging the pore. This establishes a one-to-one correspondence between R_0 and its plugging time θ_p. A pore of initial radius R_0 is plugged at time θ_p as long as $R_0 < \alpha g(\theta)$.

As the deactivation proceeds, larger and larger isolated islands of partially plugged pores, which are surrounded by the completely plugged pores, emerge. Such pores can no longer contribute to the catalyst's catalytic activity. Intuitively, one expects that at or below a critical volume fraction of the plugged pores no sample-spanning cluster of open pores should exist. In practice, the assumption of kinetic control may not be valid near this point, because after some pores are plugged, the reactants must diffuse in the pore space to reach the active sites of the catalyst. Thus, the model is oversimplified, but it is still useful because it clearly demonstrates the role of percolation during catalyst deactivation.

It should be emphasized that, similar to non-catalytic reactions with solid products, catalyst deactivation is a percolation phenomenon with trapping. In both phenomena, however, the effect of trapping is significant only if the pore-size distribution of the pore space is broad enough to allow the pore space to have pore sizes that vary over orders of magnitude. Otherwise, the effect of trapping is small and the deactivation process is essentially the same as the random bond percolation described in Chap. 2.

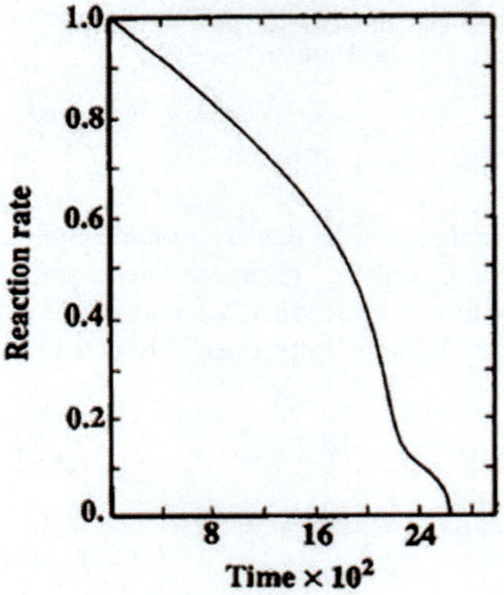

Fig. 10.8 Reaction rate (normalized by its value when no pore has been plugged) versus (dimensionless) time during a reaction-limited catalyst deactivation, represented by a simple cubic network with a log-normal pore-size distribution (after Sahimi and Tsotsis 1985)

Simulation of the percolation model proceeds as follows. One begins with a pore network with distributed pore sizes, estimates the effective radii of the pores at any given time (which decrease with time) using Eq. (10.52), and identify the completely plugged pores, the partially plugged but accessible pores, and the trapped ones. In the subsequent steps of simulations, the trapped pores are ignored, since they cannot be reached by the reactant. The simulation continues until a sample-spanning cluster of open pores no longer exists. Normally, the reaction rate of the catalyst is proportional to its accessible surface area.

Figure 10.8 presents the dynamic evolution of the reaction rate of a catalyst, calculated with a simple-cubic network as the model of the catalyst and a log-normal pore-size distribution. It is clear that the reaction rate vanishes at a finite plugging time θ_p. The knee of the curve is typical of deactivation processes in the kinetic regime, and has been observed in experiments.

10.5.2 Diffusion-Limited Deactivation

Catalyst deactivation is usually accompanied by diffusional limitations. Similar to gasification, percolation models of catalyst deactivation in this regime are also divided into two groups: the hybrid continuum models and the pore-network models.

10.5.2.1 Hybrid Continuum Models

In this approach, one uses differential equations of diffusion and reaction, similar to those for gasification and noncatalytic gas–solid reactions with solid products. For concreteness, we discuss and develop a model of catalyst deactivation for hydrodemetallation (HDM). During catalytic hydroprocessing for sulfur and nitrogen removal, nickel and vanadium porphyrins, which are the principal metals in petroleum and other metal-bearing molecules, undergo the HDM reactions, leading to metal deposition and catalyst poisoning. It has been suggested that the mechanism of HDM involves an intermediate B, such that one has

$$A \rightleftharpoons B \rightarrow C, \tag{10.55}$$

with the kinetic coefficient for the forward reaction $A \rightarrow B$ neing k_1, for the backward reaction $B \rightarrow A$ being k_2, while the reaction $B \rightarrow C$ is governed by a kinetic coefficient k_3. Here, A represents the metalloporphyrin compound in oil, and C is the metal deposited on the pore surface. The reactions are generally believed to be first order and, therefore, the concentrations C_A and C_B of A and B satisfy the following diffusion–reaction equations:

$$\frac{\partial C_A}{\partial t} = \nabla \cdot (D_A \nabla C_A) - k_1 C_A + k_2 C_B, \tag{10.56}$$

$$\frac{\partial C_B}{\partial t} = \nabla \cdot (D_B \nabla C_B) + k_1 C_A - (k_2 + k_3) C_B. \tag{10.57}$$

These equations, in spherical coordinates, are similar to Eqs. (10.6) and (10.40). The main difference between catalyst deactivation and noncatalytic gas–solid reactions with solid products is that the former occurs only at the active sites of the catalyst and, therefore, the deposition of the solid products on the pore surfaces is not as simple as that in the noncatalytic case.

During deactivation, the deposition mode in any pore can be either uniform or discrete. Strictly speaking, uniform deposition occurs only in the kinetic (reaction-limited) regime, but to simplify computations in the diffusion-limited regime one may assume that the thickness of the solid layer formed on the pore surface is the same everywhere. This implies that the fraction of the active sites covered by the deposits increases rapidly from zero to one. Physically, this is plausible if the deposit is porous enough to allow the reactants to reach the active sites, or if it contains catalytically active materials itself.

In discrete deposition, on the other hand, one assumes that a fraction of the catalyst's active sites become nucleated by the product, after which deposition occurs only on such sites. Therefore, one no longer has a uniform layer of deposits on the pores' surface, but a series of discrete lumps. Once the type of pore deposition is spacified, one calculates various properties of the catalyst as the deposition proceeds. For example, in uniform pore deposition, the pore radius R changes with time according to

$$\frac{\partial R}{\partial t} = -\frac{M_d \hat{R}_d \gamma \theta_c}{\rho_d}, \tag{10.58}$$

with the initial condition that, $R(t = 0) = R_0$, where M_d and ρ_d are the molecular weight and density of the deposits, respectively, $\hat{R}_d$ is the rate of deposit production per unit area, γ is the degree of catalytic activity relative to the original surface, and θ_c is the fraction of the active sites covered by the deposits. The assumption is that the pore is cylindrical, and that the reaction is first order.

Therefore, the simulations proceed as follows. One assumes a model of the pore space to determine its effective diffusivity during various stages of deactivation. One then solves Eqs. (10.56) and (10.57) numerically, and uses Eq. (10.48) to keep track of the evolution of the pore sizes and, thus, update the structure of the pore network. Numerical simulation of Eqs. (10.56) and (10.57) stops when the pore network reaches its percolation threshold. This approach, a hybrid continuum model, was developed by Shah and Ottino (1987b); Melkote and Jensen (1989), Beyne and Froment (1990a,1993), and Chigada et al. (2010), and was relatively successful.

10.5.2.2 Pore-Network Models

In this approach, one represents the pore space as a network of interconnected pores with distributed pore sizes. If the pores are cylindrical, and diffusion in the radial direction is ignored, and is assumed to be mainly in the axial direction, then the governing equations for diffusion and reaction in a pore are simplified to

$$\frac{\partial C_A}{\partial t} = D_{Ap} \frac{\partial^2 C_A}{\partial x^2} - k_1 C_A + k_2 C_B, \tag{10.59}$$

$$\frac{\partial C_B}{\partial t} = D_{Bp} \frac{\partial^2 C_B}{\partial x^2} + k_1 C_A - (k_2 + k_3) C_B, \tag{10.60}$$

which are valid in every pore, provided that for every pore,$l_p/R \gg 1$. Normally, a steady- or quasi-steady-state condition is assumed because the rate of change in the catalyst pore space is much smaller than the rate of diffusion and reaction. Note that, whereas D_A and D_B are macroscopic diffusivities, D_{Ap} and D_{Bp} represent microscopic or pore-scale diffusivities. Equations (10.59) and (10.60) are then solved with the boundary conditions that, $C_A(x = 0) = C_{A0}$, $C_A(x = l_p) = C_{Al}$, $C_B(x = 0) = C_{B0}$, and $C_B(x = l_p) = C_{Bl}$, but we should keep in mind that the boundary concentrations at both ends of any pore are still unknown and are computed through numerical simulations. To obtain the concentration profile of species C in the entire network, we write down a mass balance for every node of the network

$$\sum_{j\in\{i\}} S_{ij}(J_A)_{ij} = 0. \tag{10.61}$$

$$\sum_{j\in\{i\}} S_{ij}(J_B)_{ij} = 0, \tag{10.62}$$

where S_{ij} is the cross-sectional area of pore ij, and J_A is the diffusive flux of A. The sums are over pores j that are directly connected to i. Using the analytical solutions of Eqs. (10.50) and (10.60), the fluxes J_A and J_B are determined and substituted into Eqs. (10.61) and (10.62). These fluxes contain the nodal concentrations C_{A0}, C_{Al}, C_{B0}, and C_{Bl}. Writing Eqs. (10.61) and (10.62) for every node of the network results in a set of $2n$ simultaneous equations for the nodal concentrations in a network of n internal nodes, from the numerical solution of which the concentration profiles of A and B and, thus, $C_c = \tau_m k_3 C_B$ in the entire network are obtained, where τ_m is the measurement time. Once the concentration profile of C in every pore is computed, a pore deposition mode is assumed and the deposit profile in every pore is computed.

The reactive molecules are usually relatively large with an effective size R_m comparable to that of the pores. If so, the pore diffusivities D_{Ap} and D_{Bp} are a function of $\lambda = R_m/R$, and since metals or coke continuously deposit on the pore surface and reduce R, the pore diffusivities also decrease continuously. In this case, one may use an equation that relates the pore diffusivity to λ, a well-known example of which is given by and Brenner[6] and Gaydos (1977),

$$\frac{D_{Ap}}{D_\infty} = 1 + 1.125\lambda \ln\lambda + 0.416\lambda + 2.25\lambda^2 \ln\lambda, \tag{10.63}$$

which is accurate for $\lambda \le 0.4$, where D_∞ is the bulk diffusivity in an unbounded system.

The network model is valid both near and far from the percolation threshold of the catalyst, i.e., the deactivation point, and is free of the limitations of the hybrid continuum models. It was developed by Arbabi and Sahimi (1991a) and can be used for any pore structure or pore-size distribution, and any diffusion–reaction mechanism. Figure 10.9 compares the weight profile of C in a catalyst particle, represented by a cubic network of linear size L, with the experimental data for a HDM process. Given the complexities of such phenemona, the agreement is excellent. Arbabi and Sahimi (1991b) also showed that their model can be used for studying deactivation with more complex kinetic schemes than (10.55).

[6] **Howard Brenner** (1929–2014) was a professor of chemical engineering who taught in several universities, the last of which from 1981 was Massachusetts Institute of Technology. He made profound contributions to fluid dynamics, microfluidics, complex liquids, interfacial transport process, emulsion rheology, and multiphase flows. A member of the National Academy of Engineering and National Academy of Sciences, Brenner was also a fellow of the American Academy of Arts and Sciences, the American Association for the Advancement of Science, and the American Institute of Chemical Engineers, and received the Fluid Dynamics Prize from the Division of Fluid Dynamics of the American Physical Society, the Bingham Medal of the Society of Rheology, and the Warren K. Lewis Award from the American Institute of Chemical Engineers.

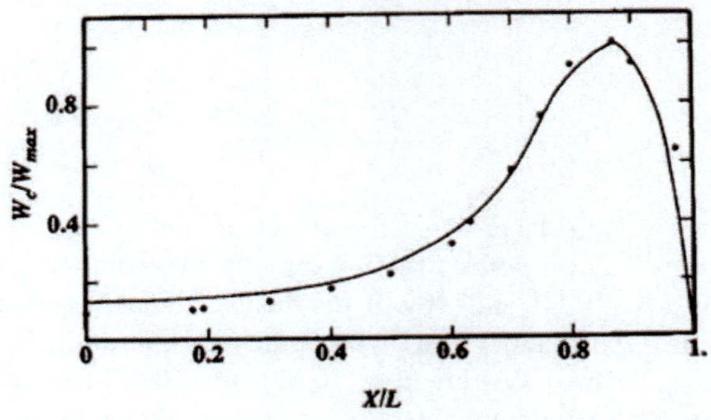

Fig. 10.9 Comparison of predictions f pore-network percolation model for the weight W of deposit C (normalized by its maximum value $W_{\max}$) during an HDM process with the experimental data (circles). X is the distance from the center of the catalyst, represented by a $L \times L \times L$ cubic network (after Arbabi and Sahimi 1991a)

The significance of pore blocking, a purely percolation effect, to catalyst deactivation was further demonstrated in a study by Ye et al. (2016). They used a pore-network model, validated by comparing the simulation results with experimental data, as well as those obtained by a continuum model (see above) for hydrogenation of benzene to cyclohexane in Pd/alumina catalysts. Their simulations demonstrated that pore blocking has a significant effect on the effectiveness factor of the catalyst, contributing to up to 50 percent of the hysteresis loop area for multiphase reactions in porous catalysts. In addition, the simulations indicated that pore blocking is significantly enhanced when the pore network is poorly connected—it is near its percolation threshold—and the pore-size distribution is wide, whereas it is insensitive to the volume-averaged pore size. These results imply that multiphase catalyst material characterization and design must account for the pore blocking effect.

The same type of pore-network model, but more detailed and complex than the original model of Arbabi and Sahimi (1991a, b), and Dadvar et al. (2001); Dadvar and Sahimi (2002, 2003), was developed by Ye et al. (2019), who studied deactivation by coking in a single catalyst particle. As discussed above, such detailed models provide detailed insights into the progression of deactivation from active site, to pore and particle, which can be used for catalyst design. When the model was applied to deactivation by coking during propane dehydrogenation in a Pt-Sn/Al_2O_3 catalyst particle, it was found that the deactivation process can be separated into two stages when there exist severe diffusion limitation and pore blockage, and the toxicity of coke formed in the later stage is much stronger than of coke formed in the early stage.

10.5.3 Catalyst Deactivation at the Reactor Scale

A widely used method for converting glucose to fructose is by enzymatic isomerization, which uses immobilized glucose isomerase and takes place in a packed-bed reactor that consists of microporous particles with a range of pore sizes, characterized by a pore-size distribution. The micropores are also interconnected, giving rise to a

3D network of pores with distributed sizes and connectivities. The particles themselves generate a 3D pore network at the reactor level with distributed pore sizes, but with a fixed connectivity. The efficiency of the process is significantly affected by deactivation of the microporous particles. While transport through the micropores is by diffusion, one has coupled flow and diffusion in the reactor's pores between the particles. Thus, the effects of both diffusion and convection must be taken into account.

In a series of papers Dadvar et al. (2001); Dadvar and Sahimi (2002, 2003) developed a multiscale modeling approach to the problem, beginning with the relevant phenomena at the scale of the micropores, and integrating them into the particle and reactor length scales, taking into account the effect of deactivation at all the relevant length scales. Both the micropores and the reactors were represented by pore networks. The reactor was represented by a real random packing of particles, originally constructed by Finney (1970), whose pore space was mapped onto an equivalent 3D Voronoi network in which the pores were represented by the edges of the Voronoi polyhedra. Good agreement was obtained between the prediction of the model for conversion of glucose to fructose and the experimental data.

Izadbakhsh and Khatami (2011) developed a reactor model to predict the deactivation behavior of a methanol-to-olefin catalyst in a fixed-bed reactor. The effect of coking on molecular transport in the pore structure of the catalysts, $(SiO_2)_x(Al_2O_3)_y(P_2O_5)_z$, known as SAPO-34, was taken into account by using the percolation theory. They used a Bethe lattice to compute the percolation properties of interest.

10.6 Reaction Kinetics and Diffusion-Controlled Annihilation

All the reaction processes that we have considered so far in this chapter change the morphology of the materials, either by clogging the pores, or by enlarging them. We now consider the effect of the morphology of the materials on the kinetics of the reactions, without any changes in the structure of the pore space. Consider the reaction

$$A + A \rightarrow 0, \tag{10.64}$$

where molecules A, initially distributed randomly in the system, diffuse until they collide with each other and disappear. In the classical kinetics, reaction (10.64) is considered as being second order, implying that the rate of reaction R is given by

$$R = KC_A^2, \tag{10.65}$$

where K is a constant. On the other hand, by definition, the rate of reaction is the rate of change of concentration C_A with respect to time, i.e., $R = -dC_A/dt$, which, when substituted into Eq. (10.65) and integrated, yields

$$\frac{1}{C_A} - \frac{1}{C_{A0}} = Kt, \tag{10.66}$$

where C_{A0} is the initial concentration of A. According to Eq. (10.66), the dimensionality of the system plays no role in the concentration distribution. This represents a mean-field approximation and, indeed, implicit in the derivation of Eq. (10.66) is that the system is represented by a *mean* concentration, which would be true if it is *well mixed*, i.e., if C_A is uniform everywhere. If the uniformity cannot be achieved, however, then, Eq. (10.66) is not expected to hold, at least for low-dimensional systems, because the fluctuations of C_A are large and cannot be ignored. In this case, Kang and Redner (1985) showed that

$$C_A = C_{A0}t^{-d/2}, \tag{10.67}$$

for $d \le 2$. For $d \ge 2$, we have $C_A \sim t^{-1}$, which is essentially the same as Eq. (10.66).

What happens if reaction (10.64) occurs in a material whose connectivity is described by percolatiuon statistics? It can be shown that at the percolation threshold p_c, or above p_c over length scales smaller than the percolation correlation length ξ_p, the concentration C_A is given by

$$C_A \sim t^{-d_s/2} \sim \frac{1}{S(t)}. \tag{10.68}$$

Here, $d_s = 2D_f/D_w$, with D_f being the fractal dimension of the sample-spanning (SSC) cluster (see Chap. 2), and D_w is the fractal dimension of the random walk (diffusion) of the reactants on the cluster; see Sect. 7.4.1. d_s is called the *fracton* (Alexander and Orbach 1982) or *spectral dimension*, whose physical significance will be described in detail in Chap. 14. $S(t)$ represents the mean number of distinct sites visited by the diffusing reactants at time t (Montroll and Weiss 1965; Hughes 1995), which for the SSC at p_c, or above p_c over length scales $L \ll \xi_p$, is given by (Rammal and Toulouse 1983)

$$S(t) \sim t^{d_s/2}. \tag{10.69}$$

Equation (10.68) is valid if $d_s < 2$. Alexander and Orbach (1982) showed that for percolation, $d_s \simeq 4/3$, independent of the dimensionality of the system. Therefore, Eq. (10.68) takes on a simple form, $C_A \simeq t^{-2/3}$, which was confirmed by computer simulations (Meakin and Stanley 1984). Equation (10.68) also implies that the reaction rate coefficient K is no longer a constant. Instead, one writes (de Gennes 1983a)

$$K \sim \frac{dS(t)}{dt}, \tag{10.70}$$

which, together with Eq. (10.69), yields

$$K \sim t^{d_s/2-1}. \tag{10.71}$$

Note that K is a constant only if $d_s = 2$. If we now use $d_s \simeq 4/3$, we obtain

$$K \sim t^{-1/3} \sim C_A^{1/2}. \tag{10.72}$$

Equations (10.68) and (10.71) were confirmed by experiments. Here, we describe the experiments briefly, and refer the reader to Kopelman (1988) for the complete details. In one experiment, an exciton–fusion reaction was studied in which two triplet excitations fused and produced a singlet excitation. The reaction occurred inside a mixed crystal of naphthalene alloy, made of $C_{10}H_8$ and $C_{10}D_8$. The naphthalene molecules were distributed randomly among the alloy's lattice sites, and the excitons were restricted to the $C_{10}H_8$ clusters. If the mole fraction x of $C_{10}H_8$ is not large enough, the excitons would be restricted to small clusters and cannot diffuse very far. If, however, x is larger than the percolation threshold x_c, which in this case was, $x_c \simeq 0.08$, then the excitons explore a large cluster. At or very close to x_c, we expect Eq. (10.72) to hold, whereas, for $x \gg x_c$, one recovers the classical result that K is independent of time. Figure 10.10 shows the data obtained by the experiments at $x_c \simeq 0.08$, the percolation threshold of the naphthalene clusters. The slope of the fitted curve is 0.32 ± 0.03, in complete agreement with Eq. (10.71).

Another interesting experiment that confirmed Eq. (10.71) is naphthalene photodimerisation in a porous membrane. In this case, one has the reaction, $A^* + A^* \rightarrow A^{**} \rightarrow A + A$ + a *photon*, where A^* is a naphthalene molecule excited to its first triplet state, and A^{**} is the intermediate dimer in its first excited singlet state. The experiments were carried out in a solution embedded in many types of porous membranes. The slope of the ln K versus ln t plot was again about 1/3, in good agreement with Eq. (10.71).

What happens if reaction (10.64) is carried out under *steady state* condition, when C_A does not depend on time? In this case, the classical kinetics does not have much to offer, but it still tells us that the reaction is second order. Suppose, however, that reaction (10.64) is carried out at steady state in a material with a fractal morphology. If we write

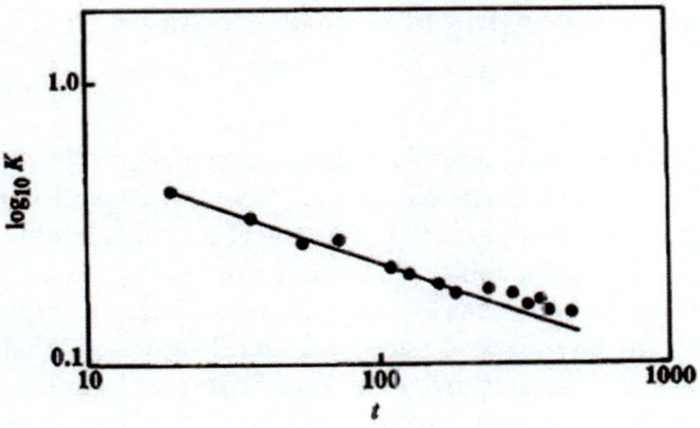

Fig. 10.10 Logarithmic plot of the reaction rate coefficient K versus time t (in ms) for the exciton fusion reaction in isotropically mixed naphthalene crystals at the percolation threshold. The slope of the line is, $d_s/2 - 1 = -0.32 \pm 0.03$ (after Kopelman (1988))

$$R = -\frac{dC_A}{dt} = KC_A^2 = K_0 C_A^2 t^{d_s/2-1}, \tag{10.73}$$

where K_0 is a constant, and integrate the equation, we obtain

$$\frac{1}{C_A} - \frac{1}{C_{A0}} = \frac{2K_0}{d_s} t^{d_s/2}. \tag{10.74}$$

Thus, as $C_A \to 0$, or as $t \to \infty$, we have

$$t \sim (\frac{d_s}{2K_0})^{2/d_s} C_A^{2/d_s}, \tag{10.75}$$

which, when substituted into Eq. (10.73), yields

$$R = K_0 C_A^n, \tag{10.76}$$

with

$$n = 1 + \frac{2}{d_s}. \tag{10.77}$$

With $d_s \simeq 4/3$, one obtains $n = 5/2$, distinctly different from the classical value, $n = 2$. Note that Eq. (10.77) implies that if the reaction is carried out in a one-dimensional system for which $d_s = 1$, such as a molecularly thin wire, its order would be *three*, although the reaction is still bimolecular.

There are also fractal systems whose fractal dimension is *less* than one. They were dubbed *dust fractals* by Mandelbrot (1983). A practical example may be catalytic islands on a noncatalytic support, which are in fact quite common. For dust fractals, we have $0 < d_s < 1$, and hence, $3 < n < \infty$, i.e., one may have a reaction with a very large order. Equation (10.77) was also confirmed experimentally; see Kopelman (1988).

10.7 Clogging of Porous Media by Precipitation of Particles Suspended in a Flowing Fluid

So far, we have described phenomena that change the structure of a porous medium by chemical reactions. There is, however, another important class of phenomena that do the same, not by a chemical reaction, but by physical interactions between a fluid and its content flowing in the pore space, and the surface of the pores. This class of phenomena includes deep-bed filtration, migration of fines (small, solid, and electrically charged particles), and flow of stable emulsions in a porous medium, which occur in many processes of industrial significance, such as gel permeation and enhanced recovery of oil. Since the three phenomena are in many ways similar, we only describe fines migration, and only note its differences from the other two.

Fines migration is caused by the contact of the solid matrix of a porous medium with an incompatible brine solution. If, for example, a sandstone core, saturated with a brine solution, is flushed with brine, followed by the injection of distilled water, the permeability of the core decreases. The reason is that fines, in the form of clays or other small particles, deposit on the surface of the pores and reduce the porosity and, thus, the permeability. This causes severe problems for enhanced recovery of oil as the reduction in the permeability decreases its amount that can be produced.

Stable emulsions, on the other hand, are generated in situ in a porous medium, or are sometimes injected into it. In miscible displacement of oil by an agent, if the viscosity of the displacing fluid is less than that of oil, the process would be unstable and very inefficient (see Sahimi 2011, for a comprehensive discussion). To make the process more stable and efficient, stable emulsion droplets are used to increase the viscosity of the displacing fluid. The droplets can, however, deposit on the surface of the pores and reduce the permeability of the pore space. As such, the phenomenon is very similar to fines migration. The main difference between the two is that emulsions only form monolayers on the pore surface, whereas fines can deposit and form a multilayer structure. In both problems, however, the reduction in the permeability of the pore space is a percolation process, because it is the loss of the pore connectivity that leads to the reduction in the effective permeability.

The percolation problem that one encounters in the flow of fines and their precipitation in the pore space is, however, different from the type of percolation processes described so far. This is because fines can form a porous "cake" on the external surface of a porous medium, and prevent other incoming fines from penetrating the pore space. As such, it is perhaps wise to call the phenomenon a "screened percolation" problem, since a significant portion of the pore space may be blocked—or screened—by the cake formation.

How do we model this class of phenomena? As with the other three phenomena discussed so far in this chapter, there have been many studies of these processes by a variety of methods and techniques. Most of them used a continuum approach in which the phenomena are modeled by differential equations of mass and momentum conservation with no regard for the evolving pore space structure. These studies were reviewed by Tien and Payatakes (1979), and are not of interest to us. As in the case of most flow-related problems in porous media, a few models have been developed in which the pore space is represented by a network of interconnected pores, in order to take into account the effect of the pore space topology and percolation. These models have been reviewed and discussed by Sahimi et al. (1990), and are mentioned here briefly.

Todd et al. (1984) were probably the first to use a pore-network model to simulate what the petroleum industry calls "formation damage," which is caused by particle precipitation during flow in porous media. They used a cubic network and modeled capture of the particles in a pore throat by a simple statistical function. Their model was unable, however, to predict the reduction in the permeability of porous medium, when compared with experimental core floods.

Here, we describe the most comprehensive of such models, namely, the pore-network percolation model that was first developed by Imdakm and Sahimi (1991), a summary of which was given by Sahimi and Imdakm (1991). A somewhat similar model for granular media was developed by Burganos et al. (1992), and much later by Yang and Balhoff (2017). Feng et al. (2015) also presented a model very similar to that of Imdakm and Sahimi (1991). In the model of Imdakm and Sahimi (1991), the porous medium was represented by a pore network of interconnected cylindrical pores with distributed radii. Unlike all the previous simulations of flow in pore networks (see Chaps. 4 and 7–9), Imdakm and Sahimi (1991) assumed that the surface of the pores is not smooth. It is well known that (see, for example, Thompson et al. 1987a) the surface of many porous media, such as sandstone, is rough and, at least over certain length scales, is characterized by a surface fractal dimension $D_s > 2$. For a realistic simulation of flow of suspended particles in porous media, one must incorporate some measure of surface roughness into the model.

To do so, it was assumed that the surface of each pore is covered by overhangs of various heights h, where typically, $h/R \leq 0.05$. The overhangs provide large surface area for attraction, adsorption, and retention of the suspended particles. The calculations are carried out in several steps:

(i) First, the initial flow field within the pore network is computed.

(ii) Next, the first batch of M particles, assumed to be spherical with distributed radii r_p and selected from a particle-size distribution, are injected simultaneously into the pore network at time $t = 0$, another M particles at time Δt, and so on. If the particles and/or the surface of the pores are electrically charged, their electrical charges q_p and q_s are also specified.

(iii) The exact trajectory of each particle within a pore is then determined. To do so, one writes a force balance for each particle a force balance, $\sum_i F_i = 0$, and a torque balance, $\sum_i I_i = 0$, with F_i and I_i being, respectively, the various forces and their corresponding torques acting on the particles. Such forces include gravitational, molecular dispersion, double-layer interaction, and drag forces, and the corresponding torques. Because fluid flow within the pores is slow enough, and the particles are not typically too large, the flow field near the particles can be represented by a linear shear flow for which exact expressions for drag force F_D and torque were derived by Goldman et al. (1967), which were utilized in the model.

Combining the force and torque balances yields a differential equation of the form, $H(y, dy/dz, d^2y/dz^2$, with z being the axial position of the center of the particle, and $y = R - r$, where r is the radial position of the particle's center. This equation is solved numerically for every particle within every pore, in order to determine their exact path within the pore.

(iv) If a particle's path does take it to the surface of a pore, then, depending on the signs of the charges q_p and q_s, one of the two events may take place. Suppose that q_p and q_s are of opposite signs, and consider a particle on the rough surface of the pore in the vicinity of an overhang. It can be shown that, if

$$F_s(2r_p h - h^2)^{1/2} > 10.205\pi r_p^2 \tau_s (r_p - h) + 3.776\pi r_p^3 \tau_s, \tag{10.78}$$

is satisfied, then the particle will attach itself to the surface. Here, F_s is the attractive surface interaction force that, for a particle sufficiently close to the surface, is dominated by the London force F_L, and τ_s is the local shear stress at the pore's surface. The first term on the right side of the inequality is the contribution of net drag (hydrodynamic) force, acting through the center of the particle, whereas the second term is the contribution of the moment M_D around the center of the particle. Without any surface roughness and associated overhangs, inequality (10.78) is always violated, i.e., no permanent attachment or adsorption occurs, and a particle will simply roll on the surface and leave the pore. Thus, for every particle that arrives on the surface of a pore, inequality (10.78) is checked. If it is satisfied, then the particle is attached to the surface and its position is recorded.

(v) Fluid flow exerts a drag force on the deposited particles that increases the resistance to the flow in that pore, which is equivalent to decreasing the effective radius of the pore. Thus, after each deposition event, the pore radius is updated. The new effective radius of the pore is estimated based on the change in the pressure drop along the pore, which was determined by Goldman et al. (1967), and is given by

$$\pi R_a^2 \Delta P_a = \pi R_b^2 \Delta P_b + 10.205 r_p^2 \tau_s, \tag{10.79}$$

where ΔP_a (R_a) is the pressure drop (pore radius) after deposition of the particle, while ΔP_b (R_b) is its corresponding value before deposition. Thus, using Eq. (10.79), we compute the new effective radius R_a of the pore.

(vi) If, however, q_s and q_p have the same sign, then, there will be repulsive forces between the particles and the surface at a certain separation distance δ_c between the pore surface and that of the particle. δ_c is determined as the distance from the pore surface at which the net force acting on the particle changes sign (becomes repulsive). If $\delta_c > h$, where h is the height of the overhang in the vicinity of the particle, the particle can have no contact with the pore surface (overhang) and will pass through the pore. If, however, $\delta_c < h$, then, the inequality (10.77), in which h is replaced by $h - \delta_c$, is checked to see whether the conditions for particle deposition are favorable. The question of detachment of particles from rough surfaces was addressed experimentally by Das et al. (1994), which confirmed the role of surface roughness in the phenomenon.

(vii) If a particle does not deposit on a pore's surface, it would travel through the pore and arrives at a node (or pore body). It then selects its next pore with a probability proportional to the volume flow rate in that pore. Thus, the particles' motion is biased by the flow field in favor of the pores that carry a larger flux of fluid. If the effective radius r_p of the particles is larger than that of the pore, R, the entrance to that pore is completely plugged by the particle, and the pore will no longer contribute to fluid flow. The radius of the pore is set to zero for the rest of the simulations. This is the so-called *size-exclusion* (SE) phenomenon (Imdakm and Sahimi, 1987), which is routinely observed in many processes involving the flow of suspended particles in porous media. If, however, $r_p < R$, the trajectory of the particle within the new pore is calculated and the process is continued.

(viii) The SE phenomenon and particle precipitation on the pore surface are also responsible for the decrease in the permeability of the medium, and the subsequent reduction in the efficiency of such processes. The reduction in the permeability is due to the loss of pore connectivity and, therefore, is a percolation process. Although at the beginning of the process most particles are typically too small to cause any pore blocking by the SE mechanism, the gradual deposition and the resulting reduction in the effective radii of the pores increase the likelihood of pore blocking by the SE at later stages of the process.

The SE, as well as deposition of the particles and decreasing the pores' radii, also alter the flow field in the pore network. Therefore, each time a few particles are deposited on the pore surfaces, the flow field in the entire network is recalculated, based on which and using Darcy's law, the permeability of the pore space at that time was computed.

(ix) The change in the morphology of the network by particle deposition and pore blocking also results in a change in the hydrodynamic forces acting on the precipitated particles. Therefore, at every stage of the simulation, inequality (10.77) is checked for all the deposited particles. Its violation for such particles implies that the hydrodynamic forces have become strong enough to resuspend the particles in the flowing flow. If deposition of emulsions is modeled, only monolayer formation is allowed. The prediction for the resuspension, made by Imdakm and Sahimi (1991) using the pore-network model, was confirmed 30 years later by the elegant experiments of Bizmark et al. (2020).

Figure 10.11 compares the predicted permeabilities for both fines migration and flow of stable emulsions with the experimental data, and the agreement is excellent. Also shown are the predicted permeabilities, if the pore surface were smooth.

The model is quite general and applicable to a wide class of problems that involve flow of suspended particles through a porous medium. Not only is the effect of pore space morphology taken into account but also the chemistry of the flowing fluid and

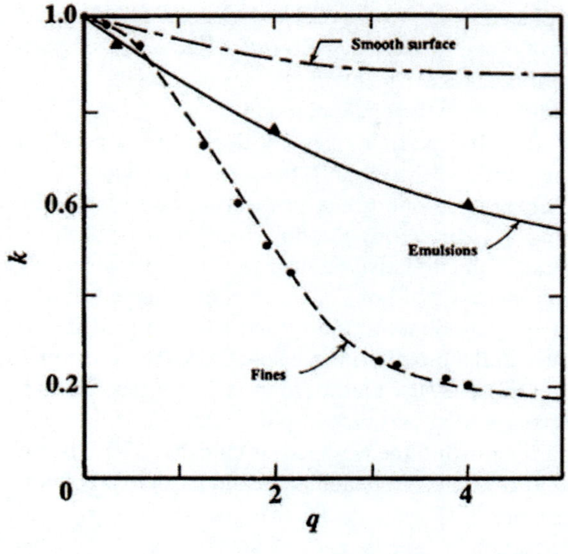

Fig. 10.11 Comparison of the predictions of the pore-network percolation model for reduction of the permeability k and its dependence on the injected pore volume q with the experimental data (symbols). k has been normalized by its initial value at $q = 0$ (after Sahimi and Imdakm 1991)

all the important forces between the particles and the surface are also accounted for. For example, in order to account for the pH and ionic strength of the solution, the double-layer interaction force $F_{\rm DL}$, which is given by

$$F_{\rm DL} = r_p \gamma \kappa (q_s^2 + q_p^2) \left[\frac{2q_s q_p}{q_s^2 + q_p^2} - \exp(-\kappa\delta) \right] \left\{ \frac{\exp(-\kappa\delta)}{2\left[1 - \exp(-2\kappa\delta)\right]} \right\}, \quad (10.80)$$

is adjusted, where γ is the fluid's dielectric constant, and κ is the Debye–Huckle reciprocal length, which is given by

$$\kappa = \left[\left(\frac{4\pi e^2}{\phi k_B T} \right) \sum_i C_{-} z_i \right]^{1/2}. \quad (10.81)$$

Here, k_B is the Boltzmann's constant, T is the absolute temperature, z_i is the valence f the ith ionic species, C_i is its concentration, and e is the electron's charge. Thus, one may adjust both the ionic strength and the pH of the solution.

Chapter 11
Vector Percolation and Rigidity of Materials

11.1 Introduction

In this chapter, elastic percolation models of disordered materials, which represent a significant extension of the resistor and pore-network models that are used for computing electrical conductivity, permeability, and diffusivity of disordered materials and porous media, are described and discussed. We pay particular attention to elastic percolation networks (EPNs), which, as mentioned in Chap. 2, are networks in which each bond is an elastic element. Such models are examples of *vector percolation*, as opposed to *scalar percolation* models for the effective conductivity and other transport properties.

The EPNs are discrete models of disordered two-phase materials in which the contrast between the elastic properties of the two phases is large. As shown in this chapter, in at least one type of vector percolation, namely, the *rigidity percolation* model, the deformation of the material represents a linear, vector, but *non-local* transport process that possesses complex properties that are absent in many other discrete (or continuum) models of scalar transport in disordered materials. As discussed in this chapter, in rigidity percolation the mere fact that there is a sample-spanning cluster (SSC) is not sufficient for supporting the transmission of stress and strain across the material, because there are long-range correlations in the system, giving rise to stress transport and the aforementioned non-locality in the EPNs.

11.2 Derivation of Elasstic Networks From Continuum Elasticity

Elastic percolation networks can be derived based on the continuum equations of linear elasticity. The equilibrium state of a linearly elastic material is described by the standard equation,

$$\nabla \cdot \boldsymbol{\sigma} = \mathbf{0}, \tag{11.1}$$

M. Sahimi, *Applications of Percolation Theory*, Applied Mathematical Sciences 213,
https://doi.org/10.1007/978-3-031-20386-2_11

where $\boldsymbol{\sigma}$ is the stress tensor defined by the standard relation,

$$\boldsymbol{\sigma} = \lambda_e(\nabla \cdot \mathbf{u})\mathbf{U} + \mu_e\left[\nabla\mathbf{u} + (\nabla\mathbf{u})^{\mathrm{T}}\right], \tag{11.2}$$

with μ_e and λ_e being the Lamé constants of the material, $\mathbf{U}$ the identity tensor, and $\mathbf{u}$ the displacement field (presumed infinitesimal) of a point from its original position. The Galerkin finite-element approximation transforms the continuum medium into a discrete mesh and Eq. (11.1) into a discrete system of linear equations governing the displacements of the grid points in the mesh. If φ is a trial function, then one forms the following equation

$$\int_\Omega \varphi(\nabla \cdot \boldsymbol{\sigma})\, d\Omega = \mathbf{0}, \tag{11.3}$$

where Ω is the volume of the space over which the integration is carried out. Using the divergence theorem, Eq. (11.3) is converted to

$$\int_A (\mathbf{n} \cdot \varphi\boldsymbol{\sigma})\, dS - \int_\Omega (\nabla\varphi \cdot \boldsymbol{\sigma})\, d\Omega = \mathbf{0}, \tag{11.4}$$

where A is the external surface of the system, and $\mathbf{n}$ is the unit normal vector pointed outward. As the boundary condition, one can either take the stress to be zero along the surface, or fix the displacement thereby restricting the class of functions used to represent $\mathbf{u}$. Hence, the surface integral in Eq. (11.4) is eliminated and will not appear anymore. In Galerkin's method $\mathbf{u}$ is represented by, $\mathbf{u} = \sum_i \mathbf{u}_i\varphi_i$, where $\{\varphi_i\}$ is a set of basis functions, and the sum is over all the basis functions in the set. Thus,

$$\boldsymbol{\sigma} = \lambda_e\left(\sum_i \mathbf{u}_i\nabla\phi_i\right)\mathbf{U} + \mu_e\left(\sum_i \mathbf{u}_i\nabla\phi_i + \sum_i \nabla\phi_i\mathbf{u}_i\right). \tag{11.5}$$

If Eq. (11.5) is substituted into Eq. (11.4), and the basis functions are normalized, so that, $\sum_i \varphi_i = 1$, then, after some considerable algebra, the governing equation for the displacement $\mathbf{u}_i$ is obtained (see Jerauld 1985; Sahimi 2003a for details):

$$\sum_{j\in\{i\}} \mathbf{W}_{ij} \cdot (\mathbf{u}_i - \mathbf{u}_j) = \mathbf{0}, \tag{11.6}$$

where the sum is over all the nearest neighbors j of i, $\{i\}$, and

$$\mathbf{W}_{ij} = \int_\Omega \left[(\lambda_e + \mu_e)\nabla\varphi_i\nabla\varphi_j + \mu_e(\nabla\varphi_i \cdot \nabla\varphi_j)\mathbf{U}\right] d\Omega.8 \tag{11.7}$$

It is clear that $\mathbf{W}_{ij}$=$\mathbf{W}_{ij}^{\mathrm{T}}$=$\mathbf{W}_{ji}$ and, therefore, Eq. (11.6) is equivalent to having a total elastic energy $\mathcal{H}$ of the material, in discrete form, given by

$$\mathcal{H} = \frac{1}{2}\sum_{ij}(\mathbf{u}_i - \mathbf{u}_j)\cdot \mathbf{W}_{ij}\cdot(\mathbf{u}_i - \mathbf{u}_j). \tag{11.8}$$

Since the sum is over all nearest-neighbor pairs ij, $\mathbf{W}_{ij} \neq \mathbf{0}$ can be thought of as forming a bond between i and j. Therefore, Eq. (11.8) describes an elastic network in which the elastic properties of the bonds are distributed according to a statistical distribution function.

11.3 The Born Model

Consider the elastic model expressed by Eq. (11.8) on a triangular lattice. The simplest basis functions are

$$\varphi_1 = \frac{1}{2} + x - \frac{y}{\sqrt{3}}, \quad \varphi_2 = \frac{1}{2} - x - \frac{y}{\sqrt{3}}, \quad \varphi_3 = \frac{2y}{\sqrt{3}},$$

and are defined on the equilateral triangle $\{(1/2, 0), (-1/2, 0), (0, \sqrt{3}/2)\}$. With the basis functions, Eq. (11.7) becomes

$$\mathbf{W}_{ij} = \frac{\mu_{ij}[4(1+\nu_p)\mathbf{R}_{ij}\mathbf{R}_{ji} + (1-3\nu_p)\mathbf{U}]}{4(1-\nu_p)}, \tag{11.9}$$

and the associated elastic energy is given by

$$\mathcal{H} = \frac{1+\nu_p}{1-\nu_p}\sum_{ij}\mu_{ij}[(\mathbf{u}_i - \mathbf{u}_j)\cdot\mathbf{R}_{ij}]^2 + \frac{1-3\nu_p}{4(1-\nu_p)}\sum_{ij}(\mathbf{u}_i - \mathbf{u}_j)^2, \tag{11.10}$$

where ν_p is the Poisson's ratio, μ_{ij} is the shear modulus of bond ij, and $\mathbf{R}_{ij}$ is the unit vector along the line from i to j. The first term of Eq. (11.10) is the energy of a network of *central-force* springs (Hooke's law), i.e., springs that transmit force only in the $\mathbf{R}_{ij}$ direction, but do not transmit shear forces, whereas the second term is a contribution analogous to scalar transport (for example, the power dissipated in conduction), since $(\mathbf{u}_i - \mathbf{u}_j)^2$ represents the *magnitude* of the displacement difference $\mathbf{u}_i - \mathbf{u}_j$.

The model expressed by Eq. (11.10) is referred to as the Born[1] model, and may be viewed as an analog of a three-dimensional (3D) solid in plane-stress with holes normal to the (x, y) plane, or as a 2D solid with the Poisson's ratio defined as the negative ratio of the strain in the y−direction and that in the x−direction, when a stress is applied in the x−direction, but none is applied in the y−direction. Any

[1] **Max Born** (1882–1970) was a great German physicist and mathematician who played a key role in the development of quantum mechanics, and made important contributions to solid-state physics and optics. He was awarded the 1954 Nobel Prize in Physics for his "fundamental research in quantum mechanics, especially in the statistical interpretation of the wave function."

predictions for a 3D solid in plane-strain can be generated from those of the Born model using the transformation $\nu_p' = \nu_p/(1+\nu_p)$, where ν_p' is the Poisson's ratio for the plain strain.

The Born model suffers from some peculiarities. It is not, for example, difficult to show (although at first glance it may not be obvious) that the elastic energy $\mathcal{H}$ defined by Eq. (11.10) is *not* invariant with respect to arbitrary rigid body rotations, a fundamental requirement for any reasonable model of elastic properties of materials, except when, $\nu_p = 1/3$, which reduces the model to a network of central-force springs (see Sahimi 2003a, for a comprehensive discussion). In his original formulation of the model (11.10), Born inserted the scalar-like part of the elastic energy (11.10) as a substitute for the many-body, angular, and bending terms (see below) that normally arise in describing the elastic properties of materials, because the expansion of such scalar two-body terms is much simpler and more convenient than expanding the many-body terms that they replace. When viewed in this way, the coefficients of the model should be treated as fitting parameters. Hence, we rewrite Eq. (11.10) as

$$\mathcal{H} = \frac{1}{2}\alpha_1 \sum_{ij}[(\mathbf{u}_i - \mathbf{u}_j)\cdot \mathbf{R}_{ij}]^2 + \frac{1}{2}\alpha_2 \sum_{ij}(\mathbf{u}_i - \mathbf{u}_j)^2, \tag{11.11}$$

so that α_1 and α_2 represent two adjustable parameters (without any regard to their theoretical relation to the Poisson's ratio ν_p). Viewed in this way, one may use the Born model for modeling and fitting elastic properties of certain materials, a subject that will be discussed later in this book. Note that, so long as $\alpha_2 > 0$, the scalar-like term of Eqs. (11.10) or (11.11) is the dominant contributing factor to the elastic energy $\mathcal{H}$. This implies immediately that, although the Born model is a vector model, it behaves effectively as a scalar model.

If percolation disorder is present in the model, i.e., if μ_{ij} in Eq. (11.10) is nonzero with probability p, and zero with probability $(1-p)$, then, one must also determine the percolation threshold p_{ce} of the model. At this point, we have no reason to believe, *a priori*, that, in general, the percolation threshold p_{ce} is the same as p_c, the threshold for the scalar percolation, i.e., it is not yet clear whether the percolation threshold p_c at which the effective conductivity, permeability, and diffusivity of a material vanish is exactly the same as p_{ce}, the threshold for the elastic moduli of an elastic percolation network.

But, in the particular case of the Born model, the dominance of the scalar term of the Hamiltonian (11.11) implies that the percolation threshold of the model at which its elastic moduli vanish is the same as that of scalar percolation models, $p_c = p_{ce}$, and that near the percolation threshold the elastic moduli of the model follow Eqs. (2.17) or (2.18). We will return to this point shortly.

11.4 The Central-Force Networks

Consider the limit $\nu_p = 1/3$ of Eq. (11.10), i.e., a network of central-force (CF), or Hookean springs. We write, $\mu_{ij} = e_{ij}\alpha$, where e_{ij} is distributed according to a probability density function, and α is the CF constant. Thus, the elastic energy of the model is given by

$$\mathcal{H} = \frac{1}{2}\alpha \sum_{ij} [(\mathbf{u}_i - \mathbf{u}_j) \cdot \mathbf{R}_{ij}]^2 e_{ij}. \tag{11.12}$$

One can determine straightforwardly the elastic moduli of CF networks, if no percolation effect is present, i.e., no bond is broken and $e_{ij} \neq 0$ everywhere. We collect some of such results here. For the triangular network,

$$\begin{cases} K_e = (\sqrt{3}/2)\alpha \\ \mu_e = (\sqrt{3}/4)\alpha \end{cases} \quad \text{triangular lattice} \tag{11.13}$$

Note that according to Eq. (11.13), Poisson's ration ν_p of a uniform triangular network is, $\nu_p = (K_e - \mu_e)/(K_e + \mu_e) = 1/3$, as expected.

In a square network, the change in the elastic energy density ΔV as a result of deformation of a square cell is given by

$$\Delta V = \frac{1}{2}K(u_{xx} - u_{yy})^2 + \frac{1}{2}\mu_p(u_{xx} - u_{yy})^2 + 2\mu_s u_{xy}^2, \tag{11.14}$$

where μ_p and μ_s are the shear moduli in pure and simple shear, respectively. If, however, there is no restoring force between adjacent network elements, i.e., if the energy does not depend even implicitly upon the angle between neighboring bonds, then, an undeformed square cell and its deformed configurations in simple shear have the same energy, implying that their deformation costs no energy and, therefore, the effective shear moduli $\mu_s = 0$, even when all the bonds of the network are connected. We refer to such a state as *floppy*. A similar reasoning would then yield the other two elastic moduli. Thus,

$$\begin{cases} K_e = \mu_p = \frac{1}{2}\alpha \\ \mu_s = 0 \end{cases} \quad \text{square network} \tag{11.15}$$

Note that, because the square network is elastically anisotropic, its Poisson's ratio is not unique and depends on how it is defined. Similar calculations yield,

$$K_e = \begin{cases} \dfrac{1}{3\ell_0}\alpha, & \text{simple-cubic lattice,} \\ \dfrac{1}{\sqrt{3}\ell_0}\alpha, & \text{BCC lattice,} \\ \dfrac{2\sqrt{2}}{\sqrt{3}\ell_0}\alpha, & \text{FCC lattice,} \end{cases} \tag{11.16}$$

where ℓ_0 is the initial length of the bonds. A simple-cubic lattice does not possess a shear modulus in simple shear. It should be emphasized that Eqs. (11.13), (11.15), and (11.16) are valid at *zero temperature* when the external stress is zero (or infinitesimally small).

11.4.1 Elastic Networks in Biological Materials

Two-dimensional (2D) elastic networks of the type just described arise in many biological cells, where they are attached to the cells' plasma or nuclear membrane. Alternatively, they may be wrapped around the cells, acting as their walls. The human red blood cells contain neither a nucleus nor other cytoskeletal components (for example, microtubules). Instead, they possess only a membrane-associated cytoskeleton composed of tetramers of the protein spectrin. The erythrocyte cytoskeleton is highly convoluted *in vivo*, but when stretched by about a factor of about seven in area, it reveals its structure in the form of a relatively uniform 2D network with a coordination number of between four and six. The spectrin tetramers are attached, about midway along their contour length, to the plasma membrane by the protein ankyrin using, as an intermediary, a protein called band 3. Typically, 120,000 tetramers cover the 140 μm^2 membrane area of a typical erythrocyte, which corresponds to a tetramer density of 800 μm^{-2}. The tetramers are attached to one another at junction complexes that contain actin segments of about 35 nm long. The erythrocyte is covered by about 35,000 junction complexes (that is, nodes in the language of networks) with an average separation (bond length) of about 75 nm.

Lateral cortex of the auditory outer hair cells provides an example of a 2D biological network with a low (less than six) coordination number. In such networks, the nodes are connected to only four of their neighbors. In guinea pigs, for example, outer hair cells are roughly cylindrical with a diameter of about 10 μm, with the lateral cortex attached to the plasma membrane. The cortex consists of parallel filaments with a thickness of about 6 nm, that are spaced about 60 nm apart, and wound around the cylinder's axis. The filaments are crosslinked at intervals of about 30 nm by thinner filaments. Experiments indicate that the network is elastically anisotropic, as the circumferential filaments are actin while the crosslinks are spectrin.

Although such 2D materials can also be represented as solid sheets, it is better to view them as 2D networks, which are what they really are. Some key questions regarding these networks are as follows.

(i) What happens when filaments, both soft and stiff, are linked together to form an elastic network?

(ii) To what extent do the network's elastic (and other) properties represent those of the individual filaments?

(iii) What is the role of interactions among the filaments?

(iv) How do the elastic properties of biological networks depend on stress and temperature?

11.5 Static and Dynamic Rigidity and Floppiness of Networks

We already showed that a square lattice of CF springs has no *shear rigidity* in the plane, i.e., in simple shear, and referred to it as *floppy*. Moreover, one can rotate the horizontal edges of all the squares belonging to the same column by one angle and the vertical edges in each row by another angle, resulting in a network configuration that can also be obtained from the original lattice by a continuous deformation of its boundaries, which is not a simple shear. On the other hand, if one fixes the external boundary of the network in any one of the sets of allowed shapes, then, the entire network will be completely rigid. Thus, the position of *all* the interior sites are determined by fixing the boundary. In this sense, the floppiness of the square network in CF percolation is *macroscopic*, because its boundaries are floppy. It is also important, from a practical view point, to distinguish between macroscopic and *microscopic* (or local) floppiness, as the latter refers to the floppiness of a small portion of the network. Microscopic floppiness can emerge in a trivial way when the positions of some of the sites are not completely determined by the bonds that are attached to them.

The opposite of microscopic floppiness is the concept of *local rigidity*, the requirement that the position of every site be fixed completely by the positions of the other sites to which it is directly bonded. This is a necessary, but not sufficient, condition for microscopic rigidity. Thus, although a material may be macroscopically rigid, some of its local regions can still be floppy. In this case, some of the microscopic degrees of freedom remain free and are not determined by the bonding structure, even if the external boundaries are fixed.

Such considerations lead to the concept of a *geometrically rigid network*, one in which the configuration of the sites cannot be deformed without changing the lengths of at least some of its bonds. According to this definition, the network rigidity depends only on its bonding structure, i.e., its topology. The question of whether a network is geometrically rigid is equivalent to asking whether the configuration of a network of N sites can determine all the $\frac{1}{2}N(N-1)$ distances r_{ij} between the sites, a conceptually simple question but, nevertheless, a quite complex problem. Networks that do not satisfy the geometrical rigidity requirements are called *geometrically floppy*.

Exact conditions exists for 2D networks that relate rigidity to connectivity (Laman 1970; Lovász and Yemini 1982; Recski 1992; Hendrickson 1992). According to Laman's theorem,

A random lattice consisting of N Sites and B bonds so that $2N - B = 3$ is rigid, if and only if there is no subset of the lattice, consisting of n sites connected by b bonds for which the relation $2n-b = 3$ is violated.

This constraint can be used for constructing 2D CF percolation networks and determining whether they are rigid *without solving for the stress or strain distribution in the network*. It is, however, restricted to 2D systems, and its 3D version,

$3N - B = 6$, is known (Hendrickson 1992) not to be a generally necessary and sufficient condition for rigidity.

The main problem with the geometrical criteria, at least as far as using them in practice is concerned, is that they are very difficult, if not impossible, to utilize for checking the rigidity of a material, because such criteria are not directly accessible to experiments. What can be verified experimentally, such as vanishing of the shear modulus or of the velocity of sound in some directions, is related to *dynamical stability* of the material that, in turn, gives rise to the concept of *dynamical floppiness*. In turn, dynamical floppiness is related to the *stability* of the expansion of the energy $\mathcal{H}(\{\mathbf{r}\})$ of the material around a reference configuration $\{\mathbf{R}\}$.

We consider a network of N sites and energy $\mathcal{H}(\{\mathbf{r}\})$, with the reference positions of the sites being the set $\{\mathbf{R}\} = \{\mathbf{R}_1, \mathbf{R}_2, \cdots, \mathbf{R}_N\}$. We expand $\mathcal{H}$ in the Nd dimensional space of positions $\{\mathbf{r}\} = \{\mathbf{r}_1, \mathbf{r}_2, \cdots, \mathbf{r}_N\}$ around the reference state. The question of rigidity of the network or material reduces to the question of stability of the expansion, with the simplest way of investigating it being through a linear stability analysis. To do so, consider a set of displacement deviations $\{\mathbf{d}\} = \{\mathbf{d}_1, \mathbf{d}_2, \cdots, \mathbf{d}_N\}$, where $\mathbf{d}_i$ represents the deviation of the site i from its equilibrium position $\mathbf{R}_i$. The harmonic expansion of $\mathcal{H}$ is then written as

$$\mathcal{H}_h = [\mathbf{d}]\mathbf{D}[\mathbf{d}], \tag{11.17}$$

where $\mathbf{D}$ is the $Nd \times Nd$ *dynamical matrix* of the expansion, defined by

$$\mathbf{D}_{ii} = [\nabla_i \cdot \nabla_i]\mathcal{H}(\{\mathbf{r}\})|_{\{\mathbf{R}\}}, \tag{11.18}$$

$$\mathbf{D}_{ij} = [\nabla_j \cdot \nabla_i]\mathcal{H}(\{\mathbf{r}\})|_{\{\mathbf{R}\}}, \tag{11.19}$$

with $\mathbf{D}_{ii}$ and $\mathbf{D}_{ij}$ being $d-$dimensional matrices.

The dynamical matrix $\mathbf{D}$ has $Nd = N_1 + N_2$ eigenvalues, of which $N_1 = \frac{1}{2}d(d+1)$ correspond to the rigid-body degrees of freedom of the system and, therefore, must vanish because of the macroscopic translational and rotational invariance of the system. The remaining N_2 eigenvalues λ_α are the force constants for the internal harmonic eigenmodes $\{\mathbf{d}^\alpha\}$ that satisfy, $\mathbf{D}[\mathbf{d}^\alpha] = \lambda_\alpha[\mathbf{d}^\alpha]$, with $[\mathbf{d}^\alpha] = \{\mathbf{d}_1^\alpha, \mathbf{d}_2^\alpha, \cdots, \mathbf{d}_N^\alpha\}$, where $\mathbf{d}_i^\alpha$ is a $d-$dimensional vector that describes the deviations of site i in mode α.

If the equilibrium state $\{\mathbf{R}\}$ is dynamically stable, all the force constants must be positive, $\lambda_\alpha > 0$, for all values of α. There are restrictions for geometrically floppy networks on the dynamical matrix $\mathbf{D}$ that may reduce its rank and cause some of its eigenvalues to vanish. Since such networks have free degrees of freedom, it is natural to assume that there are also free eigenmodes $\{\mathbf{d}^\beta\}$ corresponding to the free degrees of freedom. There are no restoring forces for such free modes and, therefore, $\lambda_\beta = 0$ for all free modes β. This implies immediately that the equilibrium reference configuration $\{\mathbf{R}\}$ of floppy networks is not stable with respect to such free eigenmodes.

11.6 Rigidity Percolation

Now that the concepts of rigidity and floppiness have been introduced, we turn our attention to an important problem. If in Eq. (11.12) e_{ij} is nonzero with probability p, and is equal to zero with probability $(1 - p)$, then one obtains an elastic percolation network. Percolation on such CF networks is called *rigidity percolation*, although percolation in more general elastic networks (see below) has also been referred to as such.

Such networks are of both theoretical and practical interest. In many engineering problems, structures composed of bars or beams connected at nodes, which are called trusses, acquire their rigidity mainly from the tensile and compressive stiffness of the beams, and these are CF type of contributions. In contrast, those in which angular forces, e.g., covalent bonds at molecular scale, are the most important are usually referred to as frames. For example, in the absence of friction between the particles of a granular packing, which is a reasonable model of unconsolidated porous materials (such as powders), the mechanical behavior of the packing is similar to those of rigidity percolation. It is not difficult to see that rigid systems in which angular forces dominate their behavior support an applied stress, as long as they are simply connected. In contrast, the CF systems require higher degrees of connectivity, and because of their importance to granular media, glasses, and gels, they must be studied carefully.

It was established in mid 1980s (Feng and Sen 1984; Jerauld 1985) that the bond percolation threshold p_{ce}^B of rigidity percolation is much larger than p_c^B, the corresponding threshold of scalar percolation, which also represents the threshold in random resistor networks. The reason for the difference is clear: A CF percolation network may contain a cluster of intact springs that, although sample-spanning, is not rigid, i.e., its deformation does not cost any energy and, hence, the deformation does not change the elastic energy of the system (see the discussion in Sect. 11.3). Due to this phenomenon, it is necessary to precisely define the rigidity of lattices. This will be addressed shortly.

11.7 Elastic and Superelastic Percolation Models

Having introduced rigidity percolation, let us define more precisely an elastic percolation network (EPN) by recalling our discussions in Chap. 2. Consider disordered lattices or networks in which each bond represents an elastic element, such as a spring (or beam) with an elastic constant e, selected from a probability density function $\psi(e)$. In the most general model, one may consider a two-phase network in which

$$\psi(e) = ph_1(e) + (1 - p)h_2(e), \tag{11.20}$$

implying that the elastic constants of phases 1 and 2 are selected from probability density functions $h_1(e)$ and $h_2(e)$. The simplest case when $h_(e) = \delta(e - a)$ and $h_2(e) = \delta(e - b)$, i.e., when the network is a mixture of two types of bonds with elastic constants a and b, is the discrete analogue of two-phase composites described in Chap. 2. If $h_2(e) = \delta_+(0)$,

$$\psi(e) = ph_1(e) + (1 - p)\delta_+(0), \tag{11.21}$$

i.e., if a fraction p of the bonds have an elastic constant e, selected from a normalized probability density function $h_1(e)$, while the elastic constant of the rest of the bonds is zero, representing, for example, voids or pores in the material, then, the model is what we refer to as the EPN.

As the percolation threshold p_{ce} of an EPN is approached, all the elastic moduli E_e of the network decrease and eventually vanish at $p = p_{ce}$. Near p_{ce}, all the elastic moduli of an EPN follow the power law,

$$E_e \sim (p - p_{ce})^f, \tag{11.22}$$

where f is a percolation critical exponent. If, on the other hand, $h_1(e) = \delta(e - \infty)$,

$$\psi(e) = p\delta(e - \infty) + (1 - p)h_2(e), \tag{11.23}$$

implying that a fraction p of the bonds are superrigid, while the rest are comparatively "soft" with a finite elastic constant selected from a normalized probability density function $h_2(e)$, then, the model is usually called a *superelastic percolation network* (SEPN). As p_{ce} is approached from below—as the fraction of superrigid bonds increases and approaches p_{ce}—all the elastic moduli E_e of the SEPN increase and eventually diverge at $p = p_{ce}$, where an SSC of the superrigid bonds is formed for the first time. Near p_{ce}, one has

$$E_e \sim (p_{ce} - p)^{-\chi}, \tag{11.24}$$

where χ is another percolation critical exponent. It should be clear that the exponents f and χ are the elastic analogs of μ and s, the exponents that characterize the conductivity of conductor–insulator, and conductor–superconductor percolation composites. Later in this chapter, we will describe and discuss the conditions under which f and χ may be expected to be universal, i.e., to be independent of the morphology of the network *and* the constitutive equation that describes the elastic behavior of the network's bonds.

Before embarking on describing the properties of the EPNs, let us recall the number of independent elastic moduli of a 2D or 3D network, since most of the networks that have been used in the numerical simulations or theoretical analyses of elastic properties of composite materials are *not*, in the sense of symmetry group of crystals, isotropic. As is well known, the number of independent elastic moduli of isotropic materials is two. One may measure, for example, the effective Young's modulus Y_e, the shear modulus μ_e, and the bulk modulus B_e, and any two of the moduli suffice for characterizing the elastic properties and rigidity of an isotropic network or material.

The triangular network, which has been used extensively in the numerical simulations of elastic properties of materials, is isotropic elastically. All 3D topologically ordered lattices are elastically anisotropic. For the cubic family of lattices, namely, the square, simple-cubic, body-centered (BCC), and face-centered (FCC) lattices, which have been extensively used in the numerical simulations, three independent elastic moduli are necessary in order to characterize the elastic behavior. In terms of the components of the elastic stiffness tensor **C**, the three main elastic moduli for a $d-$dimensional material are given by

$$\mu_e = \begin{cases} C_{44}, & \text{simple shear,} \\ \frac{1}{2}(C_{11} - C_{12}), & \text{pure shear,} \end{cases} \tag{11.25}$$

$$Y_e = \frac{(C_{11} - C_{12})(C_{11} + 2C_{12})}{C_{11} + C_{12}}, \tag{11.26}$$

$$K_\epsilon = \frac{1}{d}(C_{11} + 2C_{12}). \tag{11.27}$$

The equations for the shear modulus are applicable when the lattice is elastically anisotropic. An example is the square lattice that can have two different shear modes, the simple shear (used in most theoretical discussions) and pure shear. This means that the Young's and shear moduli of the cubic family of lattices are *not* isotropic in the sense that, they depend on the direction. Thus, unless otherwise specified, whenever in this chapter we refer to Y_e and μ_e, we mean the Young's and shear moduli in the principal directions of the network. On the other hand, the bulk modulus of the same lattices is isotropic.

11.7.1 Numerical Simulation and Finite-Size Scaling

To calculate the elastic moduli of a lattice, one minimizes its total elastic energy $\mathcal{H}$ with respect to $\mathbf{u}_i$, the displacement vector of node i of the network. Thus, one has, $\partial\mathcal{H}/\partial\mathbf{u}_i = 0$. Writing down this equation for every interior node of the lattice results in a dN simultaneous linear equations for the nodal displacements $\mathbf{u}_i$ of a $d-$dimensional network of N internal nodes, if no rotational motion is involved. The boundary conditions depend on the elastic modulus to be calculated. For example, to calculate the elastic constant C_{11}, i.e., a tensile mode, one stretches two opposite faces of the network by a given strain ϵ and imposes periodic boundary conditions in the other directions. The resulting set of linear equations is solved by either Gaussian elimination (for small networks), or by an efficient iterative method, such as the conjugate-gradient method. The convergence criterion for the iterative method is that for *all* sites i, $|\mathbf{u}_i^{(k)} - \mathbf{u}_i^{(k-1)}|/|\mathbf{u}_i^{(k-1)}| < \delta$, where $\mathbf{u}_i^{(k)}$ is the displacement of site i after the kth iteration, and δ is a small number on the order of $10^{-3} - 10^{-5}$. From the solution of the set of equations for the nodal displacements, one computes the elastic energy $\mathcal{H}$ and, hence, the elastic modulus $C_{11} = 2\mathcal{H}/\epsilon^2$.

11.7.2 The Correlation Length of Rigidity Percolation

A correlation length ξ_e is defined for rigidity percolation that, similar to the correlation length ξ_p of scalar percolation, may diverge as p_{ce} is approached:

$$\xi_e \sim (p - p_{ce})^{-\nu_e}. \tag{11.28}$$

Early computer simulations of 2D rigidity percolation (Sahimi and Goddard 1985; Lemieux et al. 1985; Arbabi and Sahimi 1988a) yielded, $\nu_e \simeq 1.1$ for bond percolation. Moreover, these simulations, based on the finite-size scaling analysis, indicated that for rigidity percolation in the triangular network one has, $f/\nu_e \simeq 1.45$, an interesting result due to the fact that it differs from the critical exponent μ of the effective conductivity of percolation networks for which, $\mu/\nu \simeq 0.973$ (see Table 2.3). These studies did not hint at the possibility of the existence of significant corrections-to-scaling.

Roux and Hansen (1988) and Hansen and Roux (1988, 1989) used a transfer-matrix method to study bond percolation in the CF triangular network, demonstrating that the location of the percolation threshold p_{ce} and the scaling exponents of the CF model are very sensitive to whether one accounts for the effect of the correction-to-scaling (see the next section). They estimated that for the triangular network, $p_{ce}^B \simeq 0.642$ and $f/\nu_e \simeq 3.0$. Although their estimate of p_{ce}^B is only 1.2% less than the earlier estimates, it apparently causes a dramatic shift in the value of f/ν_e from 1.45 to 3.0. Moreover, it appears that while the corrections-to-scaling are large at $p_{ce}^B \simeq 0.642$, they are practically non-existent at $p \simeq 0.65$, which is in agreement with the findings of the early simulations and explains why a much lower value of f/ν_e had been obtained by such simulations. It is, therefore, clear that accurate estimation of the critical exponents of rigidity percolation requires very accurate estimates of the percolation thresholds, and also taking into account the effect of corrections-to-scaling. We now discuss such aspects of the problem.

11.7.3 Finite-Size Scaling Analysis

Finite-size scaling analysis of elastic moduli of the EPNs and SEPNs are similar to that for the effective conductivity and permeability of a material, described in Chap. 2. We recall from Chap. 2 that, according to the finite-size scaling theory, for any effective elastic modulus E_e of an EPN of linear size L that follows a power law in $|p - p_{ce}|$ at the percolation threshold (of the same network when $L \to \infty$), one has

$$E_e(L, p_{ce}) \sim L^{-\tilde{f}}\left[a_1 + a_2 h_1(L) + a_3 h_2(L)\right], \tag{11.29}$$

where $h_1(L)$ and $h_2(L)$ represent the non-analytical and analytical correction-to-scaling terms, respectively, which are particularly important for small to moderate values of L, and $\tilde{f} = f/\nu_e$. A similar equation may be used for finite-size scaling analysis of the SEPNs with the exponent $-\tilde{f}$ replaced by $\tilde{\chi}$. For example, an equation similar to (11.29) has been used for estimating the exponents f and χ by studying the elastic moduli of finite-size networks at p_{ce}.

If one is interested in studying the scaling behavior of the topological properties P of the EPNs, such as the accessible fraction of the elastic bonds, and the backbone of the rigid SSC, then the functions $h_1(L)$ and $h_2(L)$ are usually combined into a single correction-to-scaling function $h(L) = a_2 L^{-\omega}$, so that Eq. (11.29), written for P, is given by

$$P(L, p_{ce}) \sim L^{-\tilde{\zeta}}\left(a_1 + a_2 L^{-\omega}\right), \tag{11.30}$$

where ω is the leading correction-to-scaling exponent, and is presumably universal, with $\tilde{\zeta} = \zeta/\nu_e$, where ζ is the exponent associated with $P(p)$. On the other hand, if one is interested in studying the scaling properties of the elastic moduli and estimating the associated exponents f and χ, then the most accurate results are obtained (Sahimi and Arbabi 1991) if one uses

$$h_1(L) = (\ln L)^{-1}, \tag{11.31}$$

$$h_2(L) = L^{-1}. \tag{11.32}$$

11.8 The Force Distribution

A fruitful way of investigating an EPN is (Sahimi and Arbabi, 1989) by studying its force distribution (FD)—the distribution of the forces that the intact (uncut) bonds of an elastic network suffer. In order to calculate the FD, one imposes a given boundary condition on the network and determines the nodal displacements $\mathbf{u}_i$ from which the total force F_i, exerted on bond i and, thus, its distribution, are calculated. Of particular interest are the moments M_q of the FD, defined by

$$M_q = \sum_i n_{F_i} F_i^q, \tag{11.33}$$

where n_{F_i} is the number of bonds that suffer a force with magnitude F_i. It should be clear that the FD for EPNs is the analog of the distribution of currents in random resistor networks. In the EPNs near p_{ce}, the moments M_q follow the power law:

$$M_q \sim (p - p_{ce})^{\tau(q)} \sim \xi_e^{-\tau(q)/\nu_e}, \tag{11.34}$$

where all the $\tau(q)$ are distinct (Sahimi and Arbabi 1989). For length scales $L \ll \xi_e$, one should replace ξ_e by L in Eq. (11.34) and, therefore,

$$M_q \sim Ł^{-\tilde{\tau}(q)}, \tag{11.35}$$

where $\tilde{\tau} = \tau/\nu_e$. Only non-zero values of F_i are included in Eq. (11.33) and, therefore, M_0 is simply the total number of bonds that carry a non-zero force, i.e., the total number of bonds in the backbone of the EPN. Therefore, $-\tilde{\tau}(0)$ is simply the fractal dimension $D_{\rm bb}^e$ of the backbone of the EPN, which is not necessarily equal to $D_{\rm bb}$, the fractal dimension of the backbone of scalar percolation or conductance and pore networks described in Chaps. 2 and 7. On the other hand, M_1 is the average force that a bond carries, while M_2 is proportional to and, in some cases, is exactly equal to, the elastic moduli of the network; therefore, $\tau(2) = f$.

11.9 Determination of the Elastic Percolation Threshold

The macroscopic floppiness of the square network in the CF models may be extended to any $d-$dimensional cubic network. The floppiness also implies that, for a $d-$dimensional simple-cubic network one has, $p_{ce}^B = p_{ce}^S = 1$, where p_{ce}^S is the site percolation threshold of the system, whereas, for scalar percolation in the square network, $p_c^S \simeq 0.5928$ and $p_c^B = 1/2$ and, for the simple-cubic network, $p_c^S \simeq 0.3116$ and $p_c^B \simeq 0.2488$ (see Tables 2.1 and 2.2). Therefore, a meaningful study of rigidity percolation is restricted to certain lattices, e.g., the triangular and BCC networks, as well as random lattices with no regular topology.

Because accurate estimation of the critical exponents that characterize various properties of rigidity percolation near p_{ce} depends sensitively on precise estimates of p_{ce}, the important question is how to estimate p_{ce} with enough precision.

11.9.1 Estimating the Percolation Threshold by Moments of the Force Distribution

One method of obtaining precise estimates of p_{ce} is based on the moments of the FD. Arbabi and Sahimi (1993) proposed that if, in a network of linear size L with a fraction p of intact springs, one calculates the ratio $r = M_2/M_1$ for various values of L and p, then, at the true value of p_{ce} a plot of $\ln r$ versus $\ln L$ should be a straight line since, if one writes down power laws similar to (11.29) for the moments M_q, then, at the true value of p_{ce} the contributions of the corrections-to-scaling to both moments should be of the same order of magnitude and would, therefore, cancel one

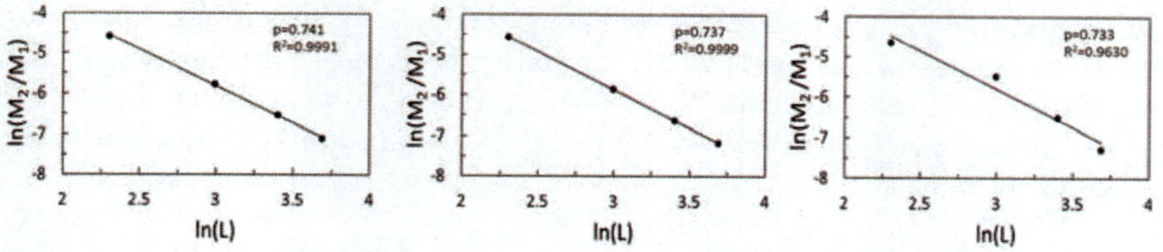

Fig. 11.1 Logarithmic plot of M_2/M_1, the ratio of the first two moments of the force distribution, versus the linear size L of the BCC lattice, for three values of p, together with the associated values of the goodness of the fit R^2. Note the deviation from linearity for $p = 0.733$ (after Arbabi and Sahimi 2021)

another. In principle, this should be true for any ratio, $r = Mq/M_{q-1}$. For $q > 2$, however, the moments of the FD are subject to large uncertainties. This idea was used (Arbabi and Sahimi 1993) on the triangular network with the modest network sizes, $L = 25, 35, 45$, and 55, and a few values of $0.636 \le p \le 0.65$. A plot of $\ln r$ versus $\ln L$ turned out to be a straight line only if $0.640 < p < 0.642$. Thus, p_{ce}^B was estimated to be

$$p_{ce}^B \simeq 0.641 \pm 0.001, \quad \text{triangular lattice,} \tag{11.36}$$

The same method also yields (Arbabi and Sahimi, 1993, 2021)

$$p_{ce}^S \simeq 0.713 \pm 0.002, \quad \text{triangular lattice.} \tag{11.37}$$

With pure elongational shear along the cubic axis, the BCC lattice with CFs would have no restoring elastic constant. The BCC is, however, rigid with respect to a transverse shear, as well as a volume change and, therefore, it should possess a well-defined $p_{ce} < 1$. The moment method was also used to study bond percolation on the BCC lattice with CFs (Arbabi and Sahimi, 1993, 2021), by computing M_1 and M_2 were calculated for various network sizes L in the range, $0.72 \le p \le 0.75$. A plot of $\ln(M_2/M_1)$ versus $\ln L$ for various values of p would produce a straight line if $p_{ce}^B \simeq 0.737$, but, not for $p < 0.733$ or $p > 0.741$; see Fig. 11.1. Therefore,

$$p_{ce}^B \simeq 0.737 \pm 0.004, \quad \text{BCC lattice,} \tag{11.38}$$

much larger than $p_c^B \simeq 0.1795$ for scalar bond percolation on the BCC network (see Chap. 2).

11.9.2 Constraint-Counting Method

The percolation threshold of the CF networks can also be estimated by a constraint-counting argument (Feng et al. 1985). When p, the fraction of uncut springs, is small, the CF network consists of disconnected floppy regions and, hence, has many zero-frequency modes, the number of which is given by the degrees of freedom $N_f = Nd$ minus the number of constraints $N_c = ZNp/2$, where N is the number of nodes of the lattice, and Z is its coordination number. Thus, the fraction f_0 of zero-frequency modes is given by

$$f_0 = \frac{Nd - ZNp/2}{Nd} = 1 - \frac{Zp}{2d}, \tag{11.39}$$

which vanishes at $p = p_{ce}^B$ and, therefore

$$p_{ce}^B = \frac{2d}{Z}. \tag{11.40}$$

As we show shortly, Eq. (11.40) is identical with the bond percolation threshold that the effective-medium approximation predicts for the CF networks. The constraint-counting method has also been used for determining the percolation threshold of a variety of other EPNs (Schwartz et al. 1985). Note, however, that if the same constraint-counting arguments are used for estimating the site percolation threshold of the CF networks (Thorpe and Garboczi 1987), it results again in Eq. (11.40) and, thus, in this sense, the arguments fail.

11.9.3 The Pebble Game

Estimating p_{ce} by the moments of the FD is a general method, applicable to any lattice of any dimensionality. For 2D lattices, however, Laman's theorem has been used fruitfully by Moukarzel and Duxbury (1995) and Jacobs and Thorpe (1995, 1996) to estimate p_{ce} of the CF networks. One starts with an empty lattice and adds the bonds one by one at a time. Moukarzel and Duxbury (1995) utilized the theorem as follows. For a given p one identifies the configuration of the cluster of occupied sites and, from it, the configuration of the cluster of occupied (uncut) bonds. Each rigid cluster is a body with three degrees of freedom, so that one must generalize the constraint $2n - b = 3$ of the original lattice to $3n_b - b = 3$, where n_b is the number of bodies or rigid clusters in a configuration. The algorithm was made very efficient by the realization (Hendrickson 1992) that one can determine whether a bond (or bar) is *redundant* with respect to the bonds that are already in the lattice. If the bonds are CF (Hookean) springs, a redundant bond violates $3n_b - b = 3$, leading to internal stresses in the lattice. Thus, if a redundant spring is added to a cluster of bonds that is already *rigid* (i.e., it satisfies the constraint $3n_b - b = 3$), the algorithm identifies the bonds that become internally stressed when the extra spring is added, in which case the bonds are given the same label as that of the cluster. In this way, one determines whether a given CF percolation network is rigid.

Jacobs and Thorpe (1995) used a somewhat different algorithm to implement Laman's theorem, calling it a *pebble game*. In general, one can "play" an (d_1, d_2)

pebbel game in which d_1 denotes the number of local degrees of freedom. while d_2 represents the number of trivial *global* degrees of freedom, which depends on the boundary conditions. A free pebble is one that is on a site, while an anchored one covers a bond. A free pebble represents a single motion that a site can execute. If two additional free pebbles are found at a different site, then the distance between these sites is not fixed. If we place a bond between the two sites, it will constrain their distance of separation. This independent constraint is recorded by anchoring one of the four free pebbles to the bond, which must always remain covered. In this way, one constructs efficient algoritms for determining whether a cluster of CF springs is rigid.

Thus, in the pebble game for rigidity percolation, one begins with a network of n sites, selects at random nearest-neighbor bonds, and examines them. The pebbles are then shuffled around the network in order to free two of them at each site at the ends of the test bonds, and also release an anchored pebble from covering a bond by anchoring a neighboring free pebble to that bond. It is always possible to free three pebbles, because they correspond to the rigid body motion of that bond. A redundant bond is identified when the search for the fourth pebble results in a closed loop back to the sites at the ends of the test bond, indicating that the distance between the incident sites is already fixed, and the redundant bond is not covered. When a redundant bond is found, the set of sites searched in the failed trial to free the fourth pebble defines an overconstrained region, and the sites are mutually rigid. A systematic search is then undertaken for mapping out all rigid clusters after building the network. An important aspect of the method is that it should be used with random lattices that are called (Jacobs and Thorpe 1995) *generic lattices* , which are lattices of a given topology but with no special symmetry. For example, by taking a regular lattice and randomly displacing each site's location by a small amount, thereby eliminating the connected collinear and parallel bonds, such as those in the triangular lattice. For a generic network built on the triangular lattice, p_{ce}^B was estimated to be (Jacobs and Thorpe 1995), $p_{ce}^B \simeq 0.6602 \pm 0.0003$, somewhat larger than that of the regular triangular network, Eq. (11.36).

11.10 Elastic Percolation Networks with Bond-Bending Forces

It is well known that microscopic *many-body interactions* (as opposed to two-body interactions in the CF models), and in particular three-body bending and four-body twisting, have important consequences for the elastic moduli of solids, especially for weakly bonded tenuous materials, such as certain types of aggregates and gels. If such many-body interactions are incorporated in lattice models, they reveal an important advantage of the discrete models. In a continuum model, the distinction between two-body and multi-body interactions can become obscure, whereas, in a lattice model, the sites that interact with each other directly leave a clear signature and distinguish themselves from any other types of interactions.

The simplest EPN with many-body interactions is one in which the CFs, as well as bond-bending or angle-changing forces that represent three-body interactions are present. The percolation threshold of the model can be equal to that of scalar percolation, if any deformation of the lattice is done at some costs to its elastic energy. The elastic energy of such models is given by (Kantor and Webman 1984)

$$\mathcal{H} = \frac{1}{2}\alpha\sum_{\langle ij\rangle}[(\mathbf{u}_i - \mathbf{u}_j)\cdot\mathbf{R}_{ij}]^2 e_{ij} + \frac{1}{2}\gamma\sum_{\langle jik\rangle}(\boldsymbol{\delta\theta}_{jik})^2 e_{ij}e_{ik}, \tag{11.41}$$

where α and γ are, respectively, the central and bond-bending force constants, and $\langle jik\rangle$ indicates that the sum is over all triplets in which the bonds $j - i$ and $i - k$ form an angle with its vertex at i. The first term on the right-hand side of Eq. (11.41) represents the usual contribution of the CFs, while the second term is due to bond-bending, or angle-changing, forces. The precise form of $\boldsymbol{\delta\theta}_{jik}$ depends on the microscopic details of the model. One may consider at least two classes of such models which we now describe.

If bending of the bonds that make an angle of 180° with one another (i.e., the collinear bonds) is not allowed, then

$$\delta\theta_{jik} = (\mathbf{u}_i - \mathbf{u}_j)\cdot\mathbf{R}_{ik} + (\mathbf{u}_i - \mathbf{u}_k)\cdot\mathbf{R}_{ij}, \tag{11.42}$$

which is referred to as the Kirkwood–Keating (KK) model. Kirkwood (1939) used the model to study vibrational properties of rod-like molecules, while Keating (1966) studied the elastic properties of covalent crystals. In the latter case, bond stretching represented a repulsive interaction to perturbations from the equilibrium length of covalent bonds, whereas bond bending was a repulsive interaction to perturbations from the equilibrium tetrahedral angle.

If bending of the collinear bonds is allowed, then (Wang 1989; Arbabi and Sahimi 1990a)

$$\boldsymbol{\delta\theta}_{jik} = \begin{cases} (\mathbf{u}_{ij}\times\mathbf{R}_{ij} - \mathbf{u}_{ik}\times\mathbf{R}_{ik})\cdot(\mathbf{R}_{ij}\times\mathbf{R}_{ik})/|\mathbf{R}_{ij}\times\mathbf{R}_{ik}|, & \mathbf{R}_{ij}\text{ not parallel to }\mathbf{R}_{ik},\\ |(\mathbf{u}_{ij}+\mathbf{u}_{ik})\times\mathbf{R}_{ij}|, & \mathbf{R}_{ij}\text{ parallel to }\mathbf{R}_{ik}, \end{cases} \tag{11.43}$$

where $\mathbf{u}_{ij} = \mathbf{u}_i - \mathbf{u}_j$. For *all* 2D systems, Eq. (11.43) is simplified to

$$\boldsymbol{\delta\theta}_{jik} = (\mathbf{u}_i - \mathbf{u}_j)\times\mathbf{R}_{ij} - (\mathbf{u}_i - \mathbf{u}_k)\times\mathbf{R}_{ik}, \tag{11.44}$$

and is referred to as the bond-bending (BB) model. For most materials, one has $\gamma/\alpha \le 0.3$.

11.10.1 The Percolation Threshold

Phillips and Thorpe (1985) used a constraint-counting analysis to predict the bond percolation threshold $(p_c^B)_{kk}$ of the d-dimensional KK model in a lattice with coordination number Z. The number of constraints N_c associated with the central and BB forces is, $N_c = N_\alpha + N_\gamma$, where

$$N_\alpha(Z) = \frac{1}{2}Z, \quad N_\gamma(d, Z) = \frac{1}{2}(d-1)(2Z-d), \tag{11.45}$$

where all the sites with $Z \leq d-2$ must be eliminated from the counting. Since in a percolation network each bond is present with probability p, we must write, $N_c(p) = N_\alpha(pZ) + N_\gamma(d, pZ)$ and, therefore, $(p_c^B)_{kk}Z + (d-1)[2(p_c^B)_{kk}Z - d] = 2d$, which, when solved for the percolation threshold, yields

$$(p_c^B)_{kk} \simeq \frac{1}{Z}\frac{d(d+1)}{2d-1}, \tag{11.46}$$

which should be compared to Eq. (11.40) for the CF model. Recall from Chap. 2 that, for scalar percolation, one has (Shante and Kirkpatrick 1971)

$$p_c^B \simeq \frac{1}{Z}\frac{d}{d-1}. \tag{11.47}$$

Thus, for 2D lattices, $(p_c^B)_{kk} = p_c^B \simeq 2/Z$, and simulations have confirmed that the percolation threshold of the KK model in 2D is the same as that of scalar percolation. For 3D lattices, however, Eq. (11.46) predicts that

$$(p_c^B)_{kk} \simeq \frac{2.4}{Z}, \tag{11.48}$$

so that the average coordination number of the network at the percolation threshold, i.e., $\langle Z\rangle = (p_c^B)_{kk}Z$, should be about 2.4, whereas Eq. (11.47) predicts, $p_c^B \simeq 1.5/Z$, so that in scalar percolation the average coordination number of the SSC at p_c is about 1.5. Note a striking feature of Eq. (11.46), namely, the percolation threshold does *not* depend on α and γ.

The BB model in $d-$dimensions has the same percolation threshold as the scalar percolation, if each site of the network interacts with at least $\frac{1}{2}d(d-1)$ of its nearest neighbors. In practice, this can be easily achieved and, therefore, by using suitable three-body interactions between the sites of the lattice, one obtains EPNs with percolation thresholds that are equal to those of scalar percolation, a highly desirable property as far as many practical applications are concerned. Hence, in what follows we restrict our discussions to the BB model. Figure 11.2 presents the typical dependence of the elastic moduli of the BB model in a simple-cubic network on the fraction p of the intact bonds.

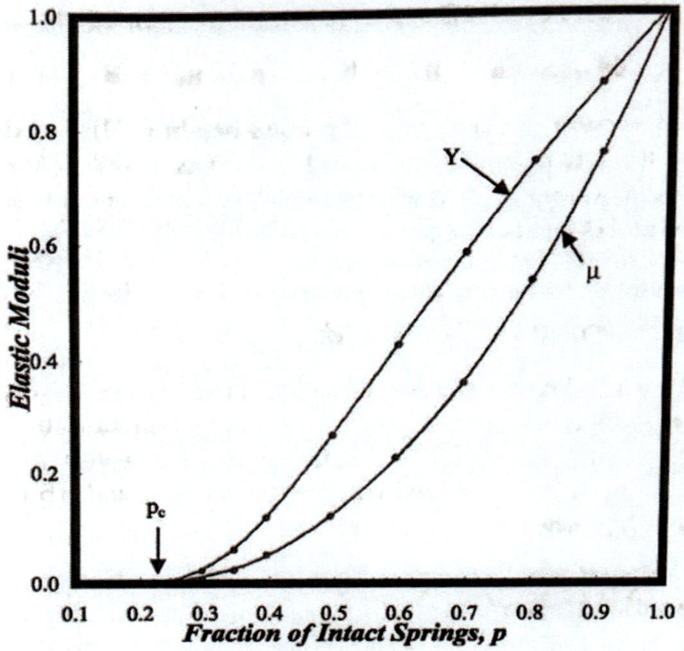

Fig. 11.2 Dependence of the Young's modulus Y and shear modulus μ of the simple-cubic lattice, in the bond-bending model, on the fraction of p of the intact bonds in bond percolation, with $\gamma/\alpha = 1/4$

11.10.2 The Force Distribution

The force distribution (FD) in the BB models has also been computed and studied (Sahimi and Arbabi, 1989, 1993). Unlike the CF model, there are some subtleties in FD of the BB model. Far from p_c, the contributions of the CFs totally dominate the elastic energy $\mathcal{H}$, in which case we may expect a unimodal FD, similar to that of rigidity percolation far from p_{ce}. If, however, we decrease γ, holding α fixed, bending of two bonds with respect to each other becomes easier, implying that the contribution to $\mathcal{H}$ of the BB forces increases. If γ/α is lowered to a small value, say of the order of 0.01, the contributions of the central and BB force become comparable, and, therefore, the FD takes on a distinct bimodal shape in which the smaller maximum of the distribution is due to the BB forces. Further decrease in γ/α means that the BB contributions become large enough that one ignores the CFs and, therefore, we may expect the distribution to take on a unimodal shape again.

At p_c and for fixed γ/α, the BB contributions are *always* larger than those of the CFs and depend only weakly on γ/α. As a result, although the FD is bimodal, the magnitude of the maximum due to the BB forces is much larger than that of the CFs.

11.11 Nature of Phase Transition in Elastic Percolation Netwoks

Although rigidity percolation in 2D lattices has been studied extensively, the nature of the phase transition is still to some extent controversial. In a d-dimensional CF lattice, no element of the structure can be moved in any direction with respect to the rest of the lattice, nor can it be rotated along $\frac{1}{2}d(d-1)$ independent axes (Obukhov 1995), implying that there are $\frac{1}{2}d(d+1)$ constraints on the possible motion of each element of the rigid lattice. Therefore, there are long-range, non-decaying orientational correlations in rigidity percolation.

The existence of such long-range correlations provides an important hint about the nature of phase transition in the sample-spanning rigid cluster at p_{ce}, because such correlations give rise, in the thermodynamic limit, to clusters that, after averaging, are translationally, but not scale, invariant (Sahimi and Mukhopadhyay 1996; Knackstedt et al. 2000). Since scale-invariant structures at a transition point represent the signature of second-order phase transitions, the conclusion is that the rigidity transition may be first order and discontinuous, rather than representing a second-order continuous transition.

Moukarzel et al. (1997a) provided further numerical evidence in support of the conclusion. Moreover, by solving the problem on the Bethe lattices, which corresponds to the mean-field limit of percolation at its upper critical dimension, they found (Moukarzel et al. 1997b) that, at least in some cases, the percolation transition in the sample-spanning rigid cluster at p_{ce} is of first-order type. But, if the topological properties of sample-spanning rigid cluster with CFs undergo a first-order phase transition at p_{ce}, then interpretation of the power laws that govern the elastic moduli of the CF percolation networks that are characteristic of second-order phase transitions becomes problematic.

A possible resolution of the contradiction may be as follows: Although the sample-spanning CF cluster may be compact (and, hence, the phase transition in its properties may be first order), its *backbone* is not (Moukarzel et al. 1997a), and may in fact be a fractal object with a well-defined fractal dimension, $D_{\rm bb}^e \approx 1.78$, for 2D CF model. If so, the fractality of the rigid backbone implies that the rigidity transition, *defined as the point at which the elastic moduli vanish*, is also second order.

Jacobs and Thorpe (1995, 1996) computed the critical exponents ν_e and β_e that characterize, respectively, the divergence of the elastic correlation length ξ_e, defined earlier, near p_{ce}, and vanishing of the percolation fraction $P(p)$ (the fraction of bonds in the sample-spanning cluster) near p_{ce}, $P(p) \sim (p - p_{ce})^{\beta_e}$, as well as the fractal dimensions D_f^e and $D_{\rm bb}^e$, with D_f being the fractal dimension of the rigid sample-spanning cluster at p_{ce}. Their estimates for 2D networks were, $\nu_e \simeq 1.21, \beta_e \simeq 0.175$, and $D_f^e \simeq 1.86$, consistent with a geometrical second-order phase transition at p_{ce}. Their estimate, $D_{\rm bb}^e \simeq 1.8$, was consistent with that of Moukarzel et al. (1997a). But there are still 2D cases for which the rigidity percolation transition at p_{ce} is believed to be first order, such as the random-bond model (Duxbury et al. 1999) and networks with chemical order (Chubynsky and Thorpe 2002).

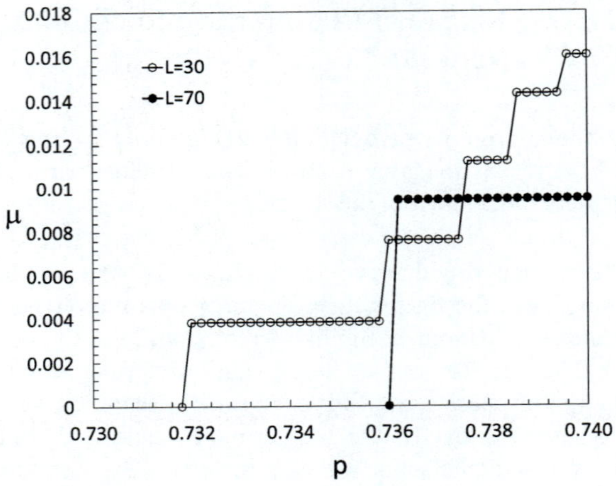

Fig. 11.3 Comparison of the stepwise decrease in the shear modulus μ very near the rigidity threshold p_{ce} for two lattice sizes L. Note the single step in the lattice of size $L = 70$, after which μ sharply drops to zero, and compare it with the multiple steps for $L = 30$ (after Arbabi and Sahimi 2021)

Two- and three-dimensional EPNs with central and BB forces have been studied extensively, and all the evidence suggest that the percolation transition is continuous and second order. In contrast to the 2D EPNs with the CFs that have also been studied extensively, 3D rigidity percolation with only CFs has not been extensively (Arbabi and Sahimi, 1993, 2021; Chubynsky and Thorpe 2007). Early simulations with relatively small BCC lattice indicated (Arbabi and Sahimi 1993) that the transition is continuous and second order, but the most recent simulations (Arbabi and Sahimi 2021) with large BCC lattices provided convincing evidence that the transition is, in fact, first order. Figure 11.3 demonstrates clearly the first-order phase transition of the shear modulus.

Liu et al. (2019) (see also Henkes et al. 2016) introduced two new complementary concepts, namely, *frictional* rigidity percolation and *minimal rigidity proliferation*, in order to identify the nature of the frictional jamming transition in packings of paricles. They constructed rigid clusters using a (3, 3) pebble game (see above) for sliding and frictional contacts and simulated it on a honeycomb lattice with next-nearest neighbors, as well as on a hierarchical lattice. They found, for both lattices, a *continuous* rigidity transition, but the associated critical exponents that they computed numerically for frictional rigidity percolation on the honeycomb lattice were distinct from those of CF rigidity percolation. To explain this, they proposed that localized motifs, such as hinges connecting rigid clusters that are allowed only with friction, may give rise to the new frictional universality class that they identified. At the same time, the difference between the exponents characterizing the spanning rigid cluster for frictional that they computed and those for the CF rigidity percolation was small, posing the question of whether they could in fact be identical in certain limits.

Thus, they developed an algorithm for constructing a minimally rigid cluster (i.e., a rigid cluster that if one removes any edge or bond from it, it becomes non-rigid or flexible) by invoking generalized Henneberg moves,[2] dubbed the aforementioned minimal rigidity proliferation. Their simulations indicated that such clusters for both frictional and the CF rigidity percolation may be in the same universality class as the connectivity percolation, hence suggesting the existence of *superuniversality* between all three transitions for the minimally rigid clusters.

We should also point out that packings of particles also undergo a rigidity transition as the particles' packing fraction reaches a critical point (Durian 1995; O'Hern et al. 2003; Majmudar et al. 2007) and, unlike random EPNs described so far, they also feature contact network rearrangements. Such a rigidity transition is called the *jamming transition*, which describes the state of the system as it moves from a zero to a nonzero bulk and shear modulus with increasing packing fraction, and the emergence of sample-spanning rigid cluster. To demonstrate this, the contact network of a 2D frictionless—one with the CFs only—packing of particle was extracted at the jamming point. When the rigid clusters were constructed based on this network using the (2,3) pebble game, it was found at the onset of rigidity or jamming every particle participating in the contact network was part of one rigid cluster, implying that the spanning rigid cluster was bulky at the transition (Ellenbroek et al. 2015).

11.12 Scaling Properties of the Elastic Moduli

Given the preceding discussions, we summarize the exponents f and χ that characterize the power-law behavior of the elastic moduli of the EPNs and SEPNs in 2D and 3D lattices. Note that, as discussed in the last section, the rigity transition in 3D is first order and, therefore, the exponent f and χ for that model are not defined.

Rigorous upper and lower bounds were derived for the exponent f by Kantor and Webman (1984) and Roux and Guyon (1986) that link f with the topological exponents of scalar percolation defined and described in Chap. 2. These bounds are given by

$$1 + \nu d < f < \nu(D_{\min} + d), \tag{11.49}$$

where $D_{\min}$ is the fractal dimension of the shortest path of a percolation cluster defined in Chap. 2. These bounds yield $11/3 \simeq 3.66 < f < 4.17$ and $3.64 < f < 3.85$ for 2D and 3D materials, respectively, and are relatively sharp.

Sahimi (1986b) proposed that

$$f = \mu + 2\nu, \tag{11.50}$$

[2] A bar-and-joint framework in the plane with degree of freedom 1 is called a *mechanism*. The operations of 0-extension and 1-extension are called the Henneberg moves, named after German mathematician **Ernst Lebrecht Henneberg** (1850–1933), which can always be performed on a framework such that its degree of freedom is preserved.

Table 11.1 Estimates of the critical exponents f and χ of the elastic moduli of the EPNs and SEPNs

Exponent	CF (2D)	CF (3D)	BB (2D)	BB (3D)
f	3.92	–	3.96	3.76
χ	1.23	–	1.24	0.72

where μ is the critical exponent of the effective conductivity of percolation networks. The predictions of Eq. (11.50) are in excellent agreement with the numerical results listed in Table 11.1 and, thus, Eq. (11.50) is likely to be an exact scaling law. The physical origin of Eq. (11.50) is clear: elasticity introduces two extra factors of ξ_p, the correlation length of percolation and, therefore, one must have

$$\frac{\text{elastic moduli}}{\text{conductivity}} \sim \xi_p^2,$$

which immediately results in Eq. (11.50). Note that Bergman (2002) argued that for 2D SEPNs, $\chi = s$, where s is the critical exponent of superconducting percolation networks defined in Chap. 2. But, it has also been proposed (Arbabi and Sahimi 1990b; Sahimi and Arbabi 1993) that

$$\chi = \nu - \frac{1}{2}\beta. \tag{11.51}$$

The predictions of Eq. (11.51) are in better agreement with the numerical estimates of χ listed in Table 11.1 than with Bergman's suggestion, although in any case, the difference is not large.

More generally, let us rewrite Eq. (11.41) in a slightly different form,

$$\mathcal{H} = \frac{1}{2}\sum_{\langle ij\rangle}[(\mathbf{u}_i - \mathbf{u}_j)\cdot \mathbf{R}_{ij}]^2 k_{ij} + \frac{1}{2}\sum_{\langle jik\rangle}(\delta\boldsymbol{\theta})^2 m_{jik}, \tag{11.52}$$

where the meanings of k_{ij} and m_{jik} are clear. Consider a two-component EPN of linear size L with the two types of bonds, having (k_1, m_1) and (k_2, m_2) as their CF and BB force constants, where $k_1 < k_2$ and $m_1 < m_2$. The fractions of the two types of bonds are $(1-p)$ and p, respectively. If

$$\tilde{m} = \frac{m_1/m_2}{L^{-(f+\chi)/\nu}}, \qquad \tilde{k} = \frac{k_1/k_2}{L^{-(f'+\chi)/\nu}}.$$

then Duering and Bergman (1988) showed that, depending on the values of the elastic constants, one may have four distinct scaling regimes that are as follows:

$$\mu_e \simeq \begin{cases} \mu_1 L^{\chi/\nu} h_1(k_1/m_1), & \tilde{k} \ll 1 \text{ and } \tilde{m} \ll 1, \\ \mu_1 h_2(k_1/m_1)(k_1/k_2)^{-\chi/(f'+\chi)}, & \tilde{k} \gg 1 \text{ and } \tilde{m} \ll 1, \\ \mu_1 h_3(k_1/m_1)(m_1/m_2)^{-\chi/(f+\chi)}, & \tilde{k} \ll 1 \text{ and } \tilde{m} \gg 1, \\ \mu_1 h_4\left[(k_1/k_2)^{f'+\chi}(m_2/m_1)^{f+\chi}, k_1/m_1\right](k_1/k_2)^{-\chi/(f'+\chi)}, & \tilde{k} \gg 1 \text{ and } \tilde{m} \gg 1. \end{cases} \tag{11.53}$$

Here, h_1 h_2 h_3; and h_4 are universal scaling functions, and μ_1 is the shear modulus of a homogeneous material with microscopic force constants k_1 and m_1. Observe that in the last three cases the shear modulus does not depend on the linear size L of the network. Moreover, in the fourth case, the scaling function h_4 depends on two variables, whereas, for the first three cases, the scaling functions depend only on a single scaling variable. Similar scaling laws can also be written down for the other elastic moduli of the network. The scaling representations (11.54) have been developed in analogy with those for the effective conductivity and the dielectric constant that were described in Chap. 2. Although, intuitively, one expects to have $f = f'$, Duering and Bergman (1988) found that, at least in 2D, $f' = \mu$, where μ is the critical exponent of the effective conductivity of the network near p_c.

11.13 Poor Man's Percolation: Effective-Medium Approximation

The problem of computing the effective elastic moduli of elastic networks, and in particular those of the EPNs, may be addressed by an approach that parallels that of conduction in the same networks. In particular, the derivation of such formulation for the effective elastic moduli of the Born model, which, in the limit $\nu_p \to 1/3$, reduces to the rigidity percolation problem, in terms of an appropriate Green function, is straightforward, albeit tedious. Once a perturbation expansion is developed for the problem, which parallels completely that of conduction. In the same network (Sahimi et al., 1983a), an effective-medium approximation (EMA) is derived.

The effective medium is a lattice in which the $\mathbf{W}_{ij}$ in Eq. (11.6) is all replaced with $\mathbf{W}_{ij}^e$, its value in the effective medium. Thus, Eq. (11.6) becomes

$$\sum_j \mathbf{W}_{ij}^e \cdot (\mathbf{u}_j^e - \mathbf{u}_i^e) = \mathbf{0}, \tag{11.54}$$

where $\mathbf{u}_i^e$ is the displacement of site i in the effective medium. If we subtract Eq. (11.54) from (11.6), after some rearrangements, we obtain

$$\sum_j \mathbf{W}_{ij}^e \cdot [(\mathbf{u}_j - \mathbf{u}_j^e) - (\mathbf{u}_i - \mathbf{u}_i^e)] = -\sum_j \mathbf{\Delta}_{ij} \cdot (\mathbf{u}_j - \mathbf{u}_i), \tag{11.55}$$

where $\mathbf{\Delta}_{ij} = \mathbf{W}_{ij} - \mathbf{W}^e_{ij}$. A *vector* Green function is now introduced by

$$\sum \mathbf{W}^e_{ij} \cdot (\mathbf{G}_{jm} - \mathbf{G}_{im}) = -\delta_{ij}\mathbf{u}, \tag{11.56}$$

with the aid of which the *exact* but the implicit solution of Eq. (11.6) is obtained

$$\mathbf{u}_i = \mathbf{u}^e_i + \sum_k \sum_j \mathbf{G}_{ij} \cdot \mathbf{\Delta}_{jk} \cdot (\mathbf{u}_j - \mathbf{u}_k). \tag{11.57}$$

Because the elastic moduli appear as the coefficients of $\mathbf{u}_{ij} = \mathbf{u}_i - \mathbf{u}_j$, it is more convenient to work with $\mathbf{u}_{ij}$. We thus obtain

$$\mathbf{u}_{ij} = \mathbf{u}^e_{ij} + \sum_l \sum_k (\mathbf{G}_{ik} - \mathbf{G}_{jk}) \cdot \mathbf{\Delta}_{lk} \cdot \mathbf{u}_{lk}. \tag{11.58}$$

Since $\mathbf{u}_{ij} = -\mathbf{u}_{ji}$, and $\mathbf{\Delta}_{lk} = \mathbf{\Delta}_{kl}$, we rewrite Eq. (11.58) in a more compact form

$$\mathbf{u}_{ij} = \mathbf{u}^e_{ij} + \sum_l \sum_k \gamma_{ijkl} \cdot \mathbf{\Delta}_{lk} \cdot \mathbf{u}_{lk}, \tag{11.59}$$

where $\gamma_{ijkl} = (\mathbf{G}_{il} - \mathbf{G}_{jl}) - (\mathbf{G}_{ik} - \mathbf{G}_{jk})$.

To derive the EMA, the poor man's version of percolation, we set all the $\mathbf{\Delta}_{lk} = \mathbf{0}$, except for only one bond mk, for which $\mathbf{\Delta}_{mk} \neq \mathbf{0}$, then Eq. (11.59) becomes

$$\mathbf{u}_{mk} = (\mathbf{U} - \gamma_{mkmk} \cdot \mathbf{\Delta}_{mk})^{-1} \cdot \mathbf{u}^e_{mk}. \tag{11.60}$$

In the EMA approach, one demands that $\langle \mathbf{u}_{mk} \rangle = \mathbf{u}^e_{mk}$, where the averaging is taken with respect to the statistical distribution of the heterogeneities, which then yields,

$$\langle (\mathbf{U} - \gamma_{mkmk} \cdot \mathbf{\Delta}_{mk})^{-1} \rangle = \mathbf{U}. \tag{11.61}$$

All we need now is the Green functions $\mathbf{G}_{ij}$, which depend on the topology of the lattice, which is done most conveniently by discrete Fourier transformation. We do note that, because the Green functions depend only on distance and orientation, we must have $\mathbf{G_{jm}} = \mathbf{G}(\mathbf{j} - \mathbf{m})$, and $\mathbf{W}^e_{\mathbf{jm}} = \mathbf{W}^e(\mathbf{j} - \mathbf{m})$. Moreover, because of translational invariance, $\mathbf{m}$ may be set to zero without loss of generality.

11.13.1 The Born Model

We first consider the Born model on a $d-$dimensional simple-cubic lattice. For simplicity, we rewrite Eq. (11.9) as

$$\mathbf{W}_{ij} = a_1 \mathbf{R}_{ij} \mathbf{R}_{ij} + a_2 \mathbf{U}, \tag{11.62}$$

so that we must only determine $\gamma = \gamma_{\mathbf{0k0k}} \cdot \mathbf{W}_{\mathbf{0k}}$. In general,

$$\gamma_{\mathbf{0k0k}} = -\frac{2}{(2\pi)^d} \int_{-\pi}^{\pi} \int_{-\pi}^{\pi} \int_{-\pi}^{\pi} [1 - \cos(\mathbf{k} \cdot \boldsymbol{\theta})] \, \hat{\mathbf{G}}(\boldsymbol{\theta}) d\boldsymbol{\theta}, \tag{11.63}$$

where $\hat{\mathbf{G}}(\boldsymbol{\theta})$ is the Fourier transform of the Green function. With $\boldsymbol{\theta} = (\theta_1, \cdots, \theta_d)$, for a $d-$dimensional cubic lattice, one has

$$2\hat{\mathbf{G}}(\boldsymbol{\theta}) = \begin{pmatrix} [da_2 + a_1(1 - \cos\theta_1)]^{-1} & 0 & \cdots & 0 \\ 0 & [da_2 + a_1(1 - \cos\theta_2)]^{-1} & \cdots & 0 \\ \cdot & \cdot & \cdot & \cdot \\ 0 & 0 & \cdots & [da_2 + a_1(1 - \cos\theta_d)]^{-1} \end{pmatrix}. \tag{11.64}$$

Without loss of generality, $\mathbf{k}$ may be taken to be $(1, 0, \cdots, 0)^{\mathrm{T}}$, as all the bonds are equivalent, where T denotes the transpose operation. Thus, we only need γ_{0k0k} which requires only two quantities,

$$\gamma_{11} = -\frac{2}{(2\pi)^d} \int \cdots \int \frac{(a_1 + a_2)(1 - \cos\theta_1)}{da_2 + a_1(1 - \cos\theta_1)} d\theta_1 = -\left(1 + \frac{a_2}{a_1}\right)\left[1 - \frac{d}{\sqrt{d(d + 2a_1/a_2)}}\right], \tag{11.65}$$

and

$$\gamma_{kk} = -\frac{2}{(2\pi)^d} \int \cdots \int \frac{a_2(1 - \cos\theta_1)}{da_2 + a_1(1 - \cos\theta_k)} d\theta_1 d\theta_k = -\frac{1}{\sqrt{d(d + 2a_1/a_2)}}, \tag{11.66}$$

where both integrals are over the first Brillouin zone, $(-\pi, \pi)$. Therefore, if the effective values of a_1 and a_2 are, respectively, a_{1e} and a_{2e}, then, with a simple percolation distribution, $\psi(x) = (1 - p)\delta_+(0) + p\delta(x - a_i)$ ($i = 1$ and 2), and $c_e = a_{e1}/a_{e2}$, one obtains

$$\frac{a_{2e}(1 + 1/c_e)}{a_2(1 + 1/c)} = \frac{p + \gamma_{11}}{1 + \gamma_{11}}, \tag{11.67}$$

$$\frac{a_{2e}}{a_2} = \frac{p + \gamma_{22}}{1 + \gamma_{22}}, \tag{11.68}$$

where $c = a_1/a_2$.

It is then straightforward to show that the EMA, Eqs. (11.67) and (11.68), predict that the elastic moduli vanish at $p_{ce}^B = 1/d = 2/Z$, the same as in scalar transport derived in Chap. 2. We already explained that this is not surprising because the presence of the scalar-like contribution to the elastic energy (11.10) means that the percolation threshold of the system is p_c not p_{ce}. Moreover, if we set $a_2 = 0$ (i.e., the Poisson's ratio $\nu_p = 1/3$), we find that the EMA predicts the exact, albeit trivial, result that removal of any springs at all will cause a $d-$dimensional cubic lattice of CF springs to have a zero effective spring constant, which was also already pointed out.

Next, we consider the Born model on the triangular lattice. In this case, there are three distinct $\mathbf{W}_{ij}$ that correspond to three directions along three vectors with components $(1,0)$, $(1/2, \sqrt{3}/2)$, and $(-1/2, \sqrt{3}/2)$, which can, however, be transformed into one another upon coordinate rotation. Moreover,

$$\hat{\mathbf{G}}(\boldsymbol{\theta}) = \begin{pmatrix} S_{11} & -S_{12} \\ -S_{12} & S_{22} \end{pmatrix}, \tag{11.69}$$

where

$$S_{11} = (5 - \nu_p) - (1 - 3\nu_p)\cos^2\theta_1 - 4\cos\theta_1\cos\theta_2, \tag{11.70}$$

$$S_{22} = (7 - \nu_p) - (5 + \nu_p)\cos^2\theta_1 - 2(1 - \nu_p)\cos\theta_1\cos\theta_2, \tag{11.71}$$

$$S_{12} = \sqrt{3}(1 + \nu_p)\sin\theta_1\sin\theta_2. \tag{11.72}$$

The Green function γ is then given by

$$\gamma = -\frac{2}{(2\pi)^2}\int_{-\pi}^{\pi}\int_{-\pi}^{\pi} 2(1 - \cos^2\theta_1)\hat{\mathbf{G}}(\boldsymbol{\theta})d\boldsymbol{\theta}. \tag{11.73}$$

It is straightforward to show that for the triangular lattice, $\gamma_{12} = \gamma_{21} = 0$ and, therefore, one needs only to compute γ_{11} and γ_{22}. In the limit $\nu_p = -1$ one recovers the random resistor network model with a percolation threshold at $p_c = 2/Z$, which is also the percolation threshold of the Born model (within the EMA), while $\nu_p = 1/3$ reduces Eq. (11.61) to the EMA for the elastic moduli of the triangular network with the CF springs.

11.13.2 Central-Force Percolation

If only the CFs are present, then $a_2 = 0$ ($\nu_p = 1/3$), and $\mathbf{W}_{ij} = a_1\mathbf{R}_{ij}\mathbf{R}_{ij}$. Therefore, in this limit,

$$\gamma_{kk} = -\frac{2}{(2\pi)^d}\int\cdots\int\left\{\sum_{\mathbf{l}}[1 - \cos(\mathbf{l}\cdot\boldsymbol{\theta})]\,\mathbf{ll}\right\}^{-1}\cdot[1 - \cos(\mathbf{k}\cdot\boldsymbol{\theta})\mathbf{kk}]\,d\boldsymbol{\theta}. \tag{11.74}$$

Equation (11.74) can be integrated case by case for a variety of Bravais lattices in order to obtain the Green functions for constructing the EMA. Since only γ_{11} is needed for deriving the EMA, it is not difficult to show that $\gamma_{11} = -2d/Z$ and, therefore, for a $d-$dimensional network of coordination number Z, the effective force constant α_e is given by

$$\int_0^\infty \frac{\alpha_e - \alpha}{\alpha + \alpha_e(Z/2d - 1)}\psi(\alpha)d\alpha = 0, \tag{11.75}$$

which is very similar to Eq. (2.45) for the effective conductivity of random resistor networks. Here, $\psi(\alpha)$ is the probability density function of the force constant α. As pointed out earlier in this chapter, αe_{ij} may be interpreted as the force constant of the bond ij, so that any statistical distribution of e_{ij} may be converted to one for α. Given the effective force constant α_e, one calculates the elastic moduli of the network. For example, the effective shear modulus of the triangular network is given by Eq. (11.13) in which α is replaced by the effective α_e.

To test the accuracy of Eq. (11.75), its predictions in certain well-known limits are examined. It is not difficult to see that with a percolation-type distribution, $\psi(\alpha) = (1-p)\delta_+(0) + p\delta(\alpha - \alpha_1)$, Eq. (11.75) predicts a bond percolation p_{ce}^B given by Eq. (11.40). According To Eq. (11.40), the average coordination number $\langle Z\rangle$ of a $d-$dimensional CF network at the rigidity percolation threshold p_{ce}^B is, $\langle Z\rangle = p_{ce}^B Z = 2d$, independent (approximately, of course) of the coordination number of the network itself. This implies that, for a network of the CF springs to be rigid, one must have

$$Z > 2d, \quad \text{macroscopic rigidity}, \tag{11.76}$$

hence explaining why CF percolation in simple-cubic lattices, as well as in the hexagonal lattice (in 2D) and the diamond lattice (in 3D), is trivial. The inequality (11.76) expresses the condition for *macroscopic* rigidity of the network. It is straightforward to show that the condition for *microscopic* rigidity of the spring networks is

$$Z > d, \quad \text{microscopic rigidity}. \tag{11.77}$$

The CF networks with $Z = 2d$, which include $d-$dimensional cubic lattices, are usually called *marginally bonded* networks.

As usual, Eq. (11.75), also predicts that the critical exponents f and χ of the elastic moduli are unity, in disagreement with the numerical estimates listed in Table 11.1. Because the predicted percolation threshold differs very little from its actual value, however, the predictions of Eq. (11.75) are in very good agreement with numerical simulation results, except for a very small region near p_{ce}^B, (Jerauld 1985; Feng et al. 1985; Garboczi and Thorpe 1985, 1986a, b; Thorpe and Garboczi 1987; Pla et al. 1990). A typical example of the comparison between the numerical results and the EMA predictions is shown in Fig. 11.4.

11.13.3 The Bond-Bending Models

Derivation of an analytical EMA for the BB model is far more complex than for the Born and the CF models. Schwartz et al. (1985) and Mall and Russel (1987) derived various EMAs for several types of the BB model, including the following models:

(i) The beam model: This is closely related to the KK and BB models. In this model, each bond of the lattice is a beam, whose elastic behavior of is governed by three material-dependent constants, $a_1 = l/(YA)$, $a_2 = l/(\mu A)$, and $a_3 = l^3/(YI)$,

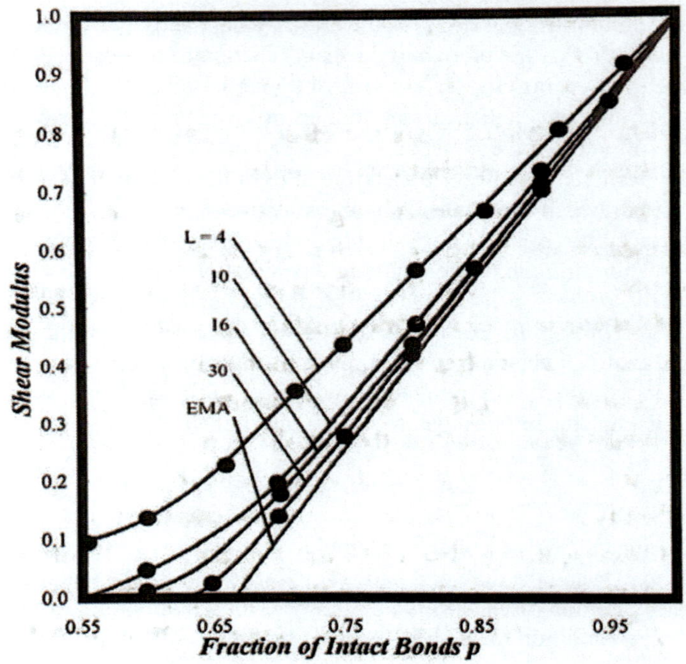

Fig. 11.4 Dependence of the shear modulus of a triangular network, in rigidity percolation, on the fraction of p of the intact bonds. Shown are the results for several network linear sizes L, as well as the predictions of the EMA [after Sahimi (2003a), originally reproduced from Jerauld (1985)]

where Y and μ are the Young's and shear moduli, A the cross section of the beam, and I the moment of inertia for flexion. Each site of the network is characterized by a displacement vector $\mathbf{u}_i$ and a rotational angle φ_i. If a site is rotated, the beams bend accordingly and, thus, the local momenta are taken into account.

Consider, for example, a square network. For a horizontal beam that connects sites i and j, one has a longitudinal force F_{lj} acting at j given by

$$F_{lj} = \alpha_1(\mathrm{u}_{ix} - \mathrm{u}_{jx}), \tag{11.78}$$

a shear force

$$F_{sj} = \alpha_2(\mathrm{u}_{iy} - \mathrm{u}_{jy}) + \frac{1}{2}\alpha_2 l(\varphi_i + \varphi_j), \tag{11.79}$$

and a flexural torque

$$\tau_j = \frac{1}{2}\alpha_2 l(\mathrm{u}_{iy} - \mathrm{u}_{jy} + l\varphi_j) + \alpha_3 l^2(\varphi_i - \varphi_j), \tag{11.80}$$

where $\alpha_1 = 1/a_1$, $\alpha_2 = 1/(a_2 + a_3/12)$, and $\alpha_3 = \alpha_2(a_2/a_3 + 1/3)$. In a similar way, analogous equations for the vertical direction can be written down. In mechanical equilibrium, the sum of all the forces and torques acting on a site must be zero.

Elastic properties of lattices of beams were studied extensively several decades ago (see, for example, Kaliski 1963; Lewinski 1985), but such studies did not include disorder in the model. In the context of disordered lattices, and in particular elastic percolation networks, Roux and Guyon (1985) studied the model. The percolation threshold of the model is the same as that of scalar percolation. Numerical simulations of Roux and Guyon (1985) also indicated that the scaling properties of the beam model are identical to those of the BB model.

(ii) The granular (disk) model: The model was first introduced by Schwartz et al. (1984). The elastic energy of the system is given by

$$\mathcal{H} = \frac{1}{2}\sum_{\langle ij\rangle} e_{ij}\left\{\alpha[(\mathbf{u}_i - \mathbf{u}_j)\cdot\mathbf{R}_{ij}]^2 + \gamma\{\mathbf{R}_{ij}\times[\mathbf{R}_{ij}\times(\mathbf{u}_i - \mathbf{u}_j)]\right.$$
$$\left. + a(\varphi_i + \varphi_j)\mathbf{z}\times\mathbf{R}_{ij}\}^2 + \delta(\varphi_i - \varphi_j)^2\right\};, \qquad (11.81)$$

where a is the radius of the disks, δ is the elastic constant associated with the restoring torque when two disks counterrotate, and $\mathbf{z}$ is a unit vector normal to the plane of the disks. The rest of the notations is the same as before. With the couplings, any relative displacement or rotation of two neighboring disks is done at a cost to the energy $\mathcal{H}$ of the system. Therefore, the percolation threshold of this model is the same as that of scalar percolation. The model was studied by Schwartz et al. (1985) and Limat (1988a, b, c, 1989) who reported that the scaling properties of the elastic moduli of the model are similar to those of the BB model.

11.13.4 Filamentous Networks

Mao et al. (2013a, b) derived an EMA for the elastic moduli of filamentous materials, modeled by either the triangular or the kagome lattice, in which the bonds are CF springs, with the BB forces being present between neighboring bonds on the same filament. Such models are important to the study of biopolymer networks and gels that are at the heart of understanding the mechanical properties of the cytoplasm, because the rigidity of the intracellular material is governed by the cytoskeleton, which is a complex network of filamentous proteins, crosslinks, and other associated proteins (see, for example, Alberts et al., 1994).

A realistic model of 2D filamentous networks is one in which (Head et al. 2003; Wilhelm and Frey 2003) semiflexible filaments of length L are placed on a 2D surface, with the position of their center-of-mass and the orientation of the rods being selected at random. The points at which two filaments cross are joined in frictionless crosslinks. This is similar to a random network of beams or rods, which Mao et al.

approximated it with the trinagular or kagome lattice. The relative rotation of two rods about a crosslink can be done at no energy cost, whereas bending the rods at crosslinks does change the total energy of the system. If the average network or mesh size is L_c, i.e., if the average crosslink separation $L_c > L_s$ along a filament, with L_s being the shortest distance between crosslinks, then, the model is characterized by the ratio $\mathcal{R} = L/L_c$. If $\mathcal{R} \to \infty$, all filaments traverse the sample, and the system has finite, γ−independent shear and bulk moduli, where γ is the BB force constant, Eq. (11.41). That is, there is effectively a CF rigidity transition at an average coordination number of $Z = 4$ as $\mathcal{R}$ is decreased from infinity. There is also a transition at $\mathcal{R} = \mathcal{R}_c \approx 5.9$ from a floppy to a rigid state, as well as a wide crossover region between $\mathcal{R}_c$ and $\mathcal{R} = \infty$ in which the shear modulus changes from being bend dominated, nonaffine, and nearly independent of the force constant α at small $\mathcal{R}$ to being stretch-dominated, nearly affine, and nearly independent of γ at large $\mathcal{R}$.

We do not present the details of Mao et al.'s EMA, as it is long and complex, but parallels very much what was described above for deriving the EMA for the born model, as well as the work of Mall and Russel (1987), which Mao et al. did not cite. Mao et al. (2013a) showed that their EMA for a filamentous triangular network in which each bond is present with a probability p predicts a rigidity threshold that has the same value for all positive bending rigidity and a crossover characterizing bending, stretching, and bend-stretch coupled elastic regimes, controlled by the CF rigidity percolation point at $p_{ce} \approx 2/3$ (see above) of the lattice, when fiber bending rigidity vanishes.

Mao et al. (2013b) then extended their analysis to the kagome lattice that has a maximum coordination number of $Z = 4$. Their analysis indicated that at $p = 1$, the elastic response of the network is purely affine (the strain field is homogeneous), and the macroscopic elastic modulus E_e is independent of γ. If $\gamma = 0$, one has a first-order rigidity–percolation transition at $p = 1$. If, however, $\gamma > 0$, E_e decreases continuously as p decreases, eventually vanishing at a continuous rigidity–percolation transition at $p_c \approx 0.605$, which is the same for all nonzero values of γ. The EMA of Mao et al. (2013b) predicted scaling forms for E_e, exhibiting crossover from a bending-dominated response regime for small γ/α to stretching-dominated response for large γ/α near both $p = 1$ and p_c.

11.14 Critical-Path Analysis

As described in Chaps. 7 and 8, a powerful method for analyzing transport processes in heterogeneous materials is the critical-path analysis (CPA). When the CPA is applied to CF networks Garboczi (1988), it would predict that the effective force constant α_e is approximately equal to the critical force constant α_c, defined similarly to g_c, the critical conductance described in Chap. 7. We should keep in mind, however, that when the CPA is applied to the CF networks, the critical force constant α_c is defined as that value of α at which a sample-spanning *rigid* cluster is formed for the first time. In other words, α_c should not be taken as the value of the force constant when a cluster has formed at p_c and is merely connected, rather it should also be rigid, which happens for the first time at p_{ce}, since for $p_c < p < p_{ce}$ such a cluster is still floppy.

Therefore, suppose that the force constants are distributed uniformly between α_1 and α_2. Beginning with α_2 and working down, a fraction p of the bonds will be in place with the value of their force constants being between α_2 and α_1. Therefore,

$$\alpha_c = \alpha_2 - p(\alpha_2 - \alpha_1) = \alpha_1 + (1 - p)(\alpha_2 - \alpha_1). \tag{11.82}$$

Garboczi (1988) showed that the predictions of Eq. (eq11.83) are in good agreement with the results of computer simulations with the triangular lattice. as well as the predictions of the EMA for the CF model.

11.15 Fixed Points of Elastic Percolation and University of the Poisson's Ratio

Bergman and Kantor (1984) showed that the ratio K/μ of the bulk and shear modulus of EPNs appears to approach a constant value as the percolation threshold is approached. Based on an EMA, they conjectured that $K/\mu = 4/d$ represents an exact and universal value. A universal value of K/μ at p_c would represent a type of a fixed point, much like those in the renormalization group transformations described in Chap. 7. Further studies and simulations of Schwartz et al. (1985) and Arbabi and Sahimi (1988b) indicated that, in the lattice models the ratio of the two moduli does appear to approach an apparently universal value, and that this type of fixed point behavior seems to hold for the ratios of other elastic moduli as well.

Let us point out that, a universal value of the ratio of the elastic moduli appears to also exist in brittle fracture of materials, near the point at which a sample-spanning crack forms (Sahimi and Arbabi 1992). The existence of such a fixed point, which has been demonstrated both by computer simulations and experiments, has important practical implications for brittle fracture of disordered materials.

Chapter 12
Transport Properties of Composite Materials

12.1 Introduction

Composite materials constitute a large class of natural or man-made disordered media, which includes granular materials, composite solids, metal–insulator films, porous media, polymers and aggregates, colloids, and many others. Currently, composite materials are the subject of intensive research, not only for the fundamental scientific questions that have been raised by their interesting structural and transport properties but also for their practical applications. In this chapter, we describe the application of percolation concepts to predicting the effective properties of composite media. As predicting and calculating the effective properties of porous media was already described in Chap. 7, and those of polymers and gels will be described in Chap. 13, we do not consider them in the present chapter.

As far as the electrical conductivity of composite materials is concerned, Ziman (1968) was probably the first who invoked percolation to explain the conduction process. He suggested that the mobility of an electron in a semiconducting material should be given by a solution of the percolation problem. Eggarter and Cohen (1970) suggested that the mobility should be proportional to the percolation probability $P(p)$, described in Chap. 2. But, in an elegant and simple experiment, Last and Thouless (1971) [1] demonstrated that although the phenomenon is of percolation type, the electrical conductivity is not proportional to $P(p)$. They measured the macroscopic conductivity of a sheet of colloidal graphite paper with holes randomly punched in it. As the surface fraction of the holes increased, the electrical conductivity of the sheet decreased, and finally vanished at a well-defined critical fraction of the holes, but the conductivity curve was completely dissimilar to $P(p)$.

[1] **David James Thouless** (1934–2019) was a British physicist who was a professor of mathematical physics at the University of Birmingham in the United Kingdon, and later at the University of Washington in Seattle. He made many fundamental contributions to science, including topological phase transitions and topological phases of matter, for which he shared the 2016 Nobel Prize in Physics with **F. Duncan M. Haldane** and **J. Michael Kosterlitz** (his postdoctoral student).

M. Sahimi, *Applications of Percolation Theory*, Applied Mathematical Sciences 213,
https://doi.org/10.1007/978-3-031-20386-2_12

A more recent set of experimental data for the same type of 2D sheets was reported by McLachlan (1988). He measured the effective conductivity of a 2D square tessellation of a sheet in which a randomly selected fraction p of the squares were made of graphite—the low-conductivity phase—while the rest were made of the high-conductivity phase, copper. Figure 12.1 presents his data. Even though the sheet is always conducting, the effect of percolation, i.e., connectivity of the high-conductivity phase is clearly seen. Moreover, when μ, the conductivity critical exponent for the power-law behavior of the effective conductivity near the percolation threshold p_c was treated as an adjustable parameter, it yielded, $1.19 \leq \mu \leq 1.3$, which is consistent with what is expected for 2D percolation, $\mu \simeq 1.3$ (see Chap. 2).

Far from the percolation threshold p_c, the effective properties of composite materials are accurately predicted by the effective-medium approximation (EMA), described in the previous chapters. Thus, we restrict our discussion to mostly the scaling behavior of the composites in the critical region near p_c, where percolation effects are dominant and universal properties, independent of the microscopic details of the materials, may prevail. As we shall see later in this chapter, however, it is often true that the critical region is quite broad and, therefore, such universal scaling laws can be very useful for predicting the effective properties of a composite material over a broad range of the volume fraction of the components constituting the composite.

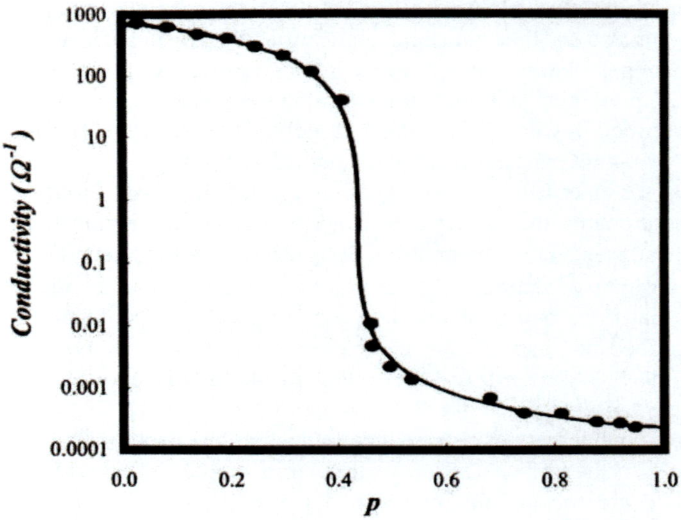

Fig. 12.1 Effective conductivity of a two-dimensional square tessellation of a sheet in which a randomly selected fraction p of the squares were made of graphite, while the rest were made of copper (the high-conductivity phase) (after McLachlan 1988)

12.2 Electrical Conductivity of Powders and Polymer Composites

There have been numerous experimental measurements of the percolation properties, and electrical conductivity of polymer composites and packing of particles over the past 50 years, demonstrating the relevance of percolation theory to explaining the data and predicting the properties. In the discussion that follows, we describe several sets of such experiments, but remind the reader that a complete list of such experiments is simply too long to be given here.

One of the simplest three-dimensional (3D) composites is a random close packing of hard spheres. The spheres can be considered as one phase of the material, while the host medium between the spheres (the matrix) constitutes another phase. Although such a two-phase medium may seem too simple, its effective properties are directly relevant to those of a wide variety of systems of practical importance. For example, if the spheres conduct electricity, but the matrix does not, the packing is a model for studying the electrical conductivity of disordered composites, such as powders. If the space between the particles represents the pore space of a porous medium, the packing is a good model for unconsolidated porous rock. Rheological properties of a system in which hard spheres are suspended in a fluid have been of interest for a long time. It has been found that the viscosity η of the suspension follows the following power law:

$$\eta \sim (\phi_c - \phi)^{-m} , \tag{12.1}$$

where ϕ is the volume fraction of the solid spheres, and ϕ_c is its critical value at which the viscosity diverges. Thus, percolation is certainly relevant to the modeling of the viscosity of a suspension.

12.2.1 Powders

Historically, the work of Malliaris and Turner (1971) was probably the first in which the electrical conductivity of a compacted powder of spherical particles was measured, and a clear reference was made to percolation. They prepared powder samples of high-density polyethylene particles (with radius $R_p = 150\,\mu\text{m}$) and nickel particles (with radius $R_n = 4 - 7\,\mu\text{m}$), measured their electrical conductivity, and reported that the powder was non-conducting unless "the composition of the metal reached a critical value." The critical value, which is clearly the percolation threshold of the metallic phase, was found to depend on R_p/R_n. Moreover, the electrical resistivity of the powder dropped *twenty* orders of magnitude at the percolation threshold $p_c \simeq 0.06$ for samples with nickel segregated on the outside of the polyethylene spheres, and at $p_c \simeq 0.35$ for samples prepared such that the nickel penetrated the spheres. Malliaris and Turner (1971) remarked that, "The critical composition for a sudden increase in the electrical conductivity of the system was assumed to correspond to the first nonzero probability for infinitely long chains of contiguously occupied lattice sites," a clear reference to the formation of a sample-spanning cluster (SSC) of the metallic particles. Although they did not refer to percolation explicitly, they used bond percolation thresholds of various lattices computed by Domb and Sykes (1961) to interpret their experimental data.

As part of a freshman physics course at Harvard University, Fitzpatrick et al. (1974) were the first to explicitly study percolation and electrical conduction in a random close packing of equal-size spheres. The conducting spheres were made of aluminum, the insulating ones of acrylic plastic. The system was pressed between two parallel electrodes made of crumpled aluminum, and the measurements were made with an ohmmeter at high resistances and with a simple Wheatstone bridge (see Chap. 7) at low resistances. The overall conductivity of the packing was measured as a function of the volume fraction of the conducting spheres. It was observed that the conductivity vanishes at a finite value of the volume fraction of the conducting spheres. Independently, Clerc et al. (1975) studied mixtures of conducting and insulating confectioners' dragées, which were neither spherical nor closely calibrated.

The first systematic and extensive study of percolation and conduction properties of packings of particles appears to have been carried out by Ota et al. (1978). They carried out experiments with molded plastic spheres, all having the same diameter, in which a fraction of the sphere were electroplated with a copper coating to make them conducting. The spheres were poured into a cylinder equipped with a pair of electrodes, one on the rigid bottom of the cylinder, and the other one at the top. Their data, shown in Fig. 12.2, indicate that the effective conductivity g_e vanishes at $p_c \simeq 0.29 \pm 0.02$. Since the filling factor f_l (see Chap. 2) of such packings is about 0.6, one obtains the critical volume fraction to be, $\phi_c = f_l p_c \simeq 0.17$, in agreement with the theoretical prediction of Scher and Zallen (1970) for the percolation threshold of random 3D percolating continua described in Chap. 3. The data of Ottavi et al. (1978) also indicated that near p_c

$$g_e(p) \sim (p - p_c)^{\mu} , \tag{12.2}$$

with $\mu \simeq 1.7 \pm 0.2$, relatively close to $\mu \simeq 2.0$ for 3D percolation (Chap. 2).

Troadec et al. (1981) measured thermal and electrical conductivity of powders of conducting WTe_2 and semiconducting WSe_2, characterized by both a geometrical anisotropy (grain shapes and sizes) and by anisotropy in a transport property. Such mixtures belong to the family of dichalcogenides of transition metals, TX_2, with a layered structure and hexagonal arrangement within each plane, making the powder anisotropic. The main difference between the crystallographic structures of WSe_2 and WTe_2 is in the coordination number of the W atoms. In the former, W is at the center of a trigonal prism of Se atoms, whereas in WTe_2 W has an octahedral environment, but is not precisely at the center of the octahedron. Moreover, WSe_2 is semiconducting, in contrast with the semimetallic nature of WTe_2.

Troadec et al. used powder grains that were single crystal platelets having a thickness of about 1–2 μm and a horizontal hexagonal shape with the largest dimension

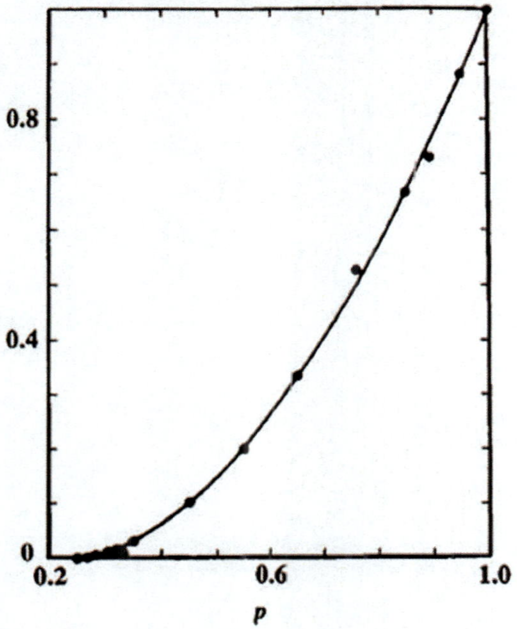

Fig. 12.2 Dependence of the effective electrical conductivity g_e of a mixture of conducting and non-conducting spheres on the fraction p of the conducting particles (after Ottavi et al. 1978)

being about 20 μm. The mixed powder was outgassed under secondary vacuum for 2 hours before being sintered in the same pressure and temperature conditions. The packing fraction was about 92%. Figure 12.3 shows their measured anisotropy in the electrical conductivity, expressed as $g_{e\ell}/g_{et} - 1$, where $g_{e\ell}$ (g_{et}) is the conductivity in the direction parallel (perpendicular) to the applied potential, and p is the fraction of WTe_2 in the mixture. The anisotropy exhibits a minimum at the percolation threshold of the WTe_2 phase, because as the percolation threshold of the phase is approached, its conducting paths become so tortuous that the current cannot distinguish between distinct directions. Indeed, Shklovskii (1978) suggested that, near p_c, one must have

$$\frac{g_{e\ell}}{g_{et}} - 1 \sim (p - p_c)^{\lambda_p} , \tag{12.3}$$

where λ_p is a universal critical exponent. If the system always remains conducting, then the anisotropy should be minimum at p_c, and this is consistent with the data shown in Fig. 12.3. The conductivity of anisotropic systems can also be predicted by the EMA described in Chaps. 2 and 7 (Bernasconi 1974; Mukhopadhyay and Sahimi 2000) and, indeed, the EMA predicts that the conductivity anisotropy vanishes at p_c.

Using extensive numerical simulations and finite-size scaling, Mukhopadhyay and Sahimi (1994) estimated the exponent λ_p in both 2D and 3D. The results were $\lambda_p \simeq 0.82$ and 0.52 for, respectively, 2D and 3D materials. They also suggested

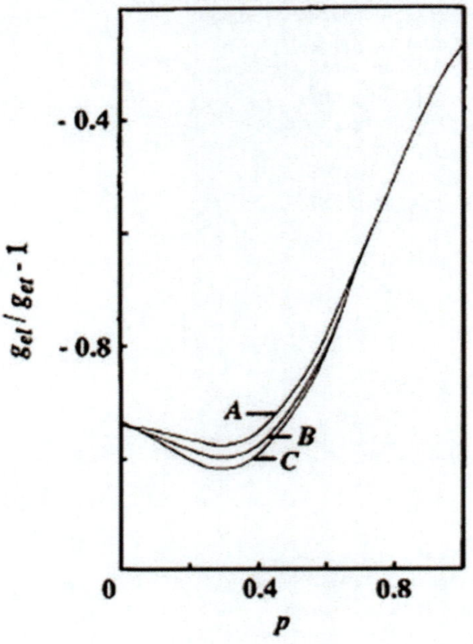

Fig. 12.3 Measured conductivity anisotropy of a mixture of WTe_2 and WSe_2 powders as a function of the fraction p of WTe_2 at temperatures (A) 300 K; (B) 200 K, and (C) 100 K (after Troadec et al. 1981)

that, $\lambda_p = \mu - \beta_{\mathrm{bb}}$, where β_{bb} is the critical exponent of the backbone of percolation clusters, defined in Chap. 2.

The effect of anisotropy is not restricted to powders. For example, in an early experiment Smith and Lobb (1979) measured the conductivities of 2D conductor–insulator networks, generated photolithographically from laser speckle patterns. Their measured conductivity of the isotropic samples vanished at $p_c \simeq 0.59$, in agreement with site percolation threshold of a square network, with a critical exponent $\mu \simeq 1.3$, in perfect agreement with 2D percolation conductivity (Chap. 2). But the measured conductivity anisotropy of the samples, measured in terms of the difference between the conductivities of the material in the two directions, decreased as p_c was approached.

A different set of experiments was carried out by Deptuck et al. (1985). Sintered, submicron silver powder with a volume fraction $\phi \sim 0.4$ is commonly used for millikelvin and submillikelvin cryostats. The sinter remains elastic and percolating for ϕ as low as 0.1. Submicron copper oxide–silver powder is routinely used in heat exchangers for optimizing heat transfer to dilute liquid ^{3}He-^{4}He mixtures. In these composites, silver component behaves as a percolating phase with volume fraction $\phi_c < 0.1$. Deptuck et al. carried out a systematic study of the electrical conductivity and Young's modulus of such powders and measured them in the *same* range of ϕ, thus making it possible to compare their critical behavior in the same range of ϕ. Their data, shown in Fig. 12.4, indicated that the conductivity vanishes with a critical

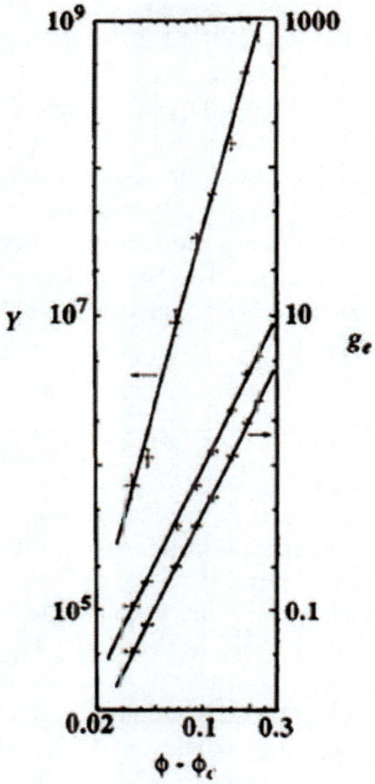

Fig. 12.4 Conductivity and Young's modulus of sintered, submicron, silver-powder beams as a function of $(\phi - \phi_c)$, with $\phi_c = 0.062$. The two conductivity curves are for temperatures 78 K (upper curve) and 300 K (after Deptuck et al. 1985)

exponent $\mu \simeq 2.15 \pm 0.25$, completely consistent with $\mu \simeq 2.0$ for 3D percolation. while Young's modulus Y vanishes according to the scaling law

$$Y \sim (\phi - \phi_c)^f \ , \tag{12.4}$$

with $f \simeq 3.8$, in excellent agreement with the critical exponent $f \simeq 3.75$ of elastic percolation networks with stretching and bond-bending forces described in Chap. 11. As Fig. 12.4 indicates, the critical region is quite broad. We will come back to their experiments later in this chapter.

Similar data were reported by Lee et al. (1986), who used silver-coated glass–Teflon composites. The Teflon powder was composed of particles with diameters of about 1 μm. The powder was then mixed with glass spheres coated with a 600 °A silver layer which provided high conductivity. The conductivity of the mixture was then measured as a function of the volume fraction ϕ of the conducting particles. Lee et al. found that $\phi_c \simeq 0.17$, in good agreement with the percolation threshold of 3D continuum percolation predicted by Scher and Zallen (1970). Moreover, when they fitted their conductivity data to Eq. (12.2), they found, $\mu \simeq 2.0 \pm 0.2$, in excellent agreement with that of 3D percolation conductivity. Lee et al. also found that the critical region in which Eq. (12.2) was applicable was quite broad. Deprez et al.

(1989), who measured the electrical conductivity of sintered nickel samples, also found that Eq. (12.2) (with $\mu \simeq 2.0$) was applicable over the *entire range* of the volume fraction.

Most recently, Pokhrel et al. (2021) reported on electrical conductivity and percolation threshold of a binary packing of disordered spherocylinders. They measured the percolation threshold p_c and the conductivity critical exponent μ for composites of conducting (CrO_2) and insulating (Cr_2O_3) rodlike nanoparticles; the two types of particles were geometrically identical. They reported that, $p_c \simeq 0.305 \pm 0.026$, and $\mu \simeq 2.52 \pm 0.03$. The conductivity exponent is consistent with what is expected for continuum percolation in 3D (Chap. 3).

12.2.2 Polymer Composites

Let us first point out that conductive polymer composites are ubiquitous in technological applications (see, for example, Huang and Zhi 2016), which is why the development of strategies for improving their electrical conductivity at a given filler concentration has been a topic of ongoing research for decades.

Fabrication of electrically-conductive polymers, in which percolation—formation of a SSC for flow of charge across the composite—plays the leading role, has been accomplished by adding conducting fillers to a polymer matrix. Such fillers include graphite (Nagata et al. 1999), carbon black (for a review see Balberg 2002), carbon fibers (Gordeyev et al. 2000), metal powders (Mamunya et al. 2002), and, more recently, carbon nanotubes (CNTs; see below), graphene (see below), and metal nanowires (Gelves et al. 2006) in an insulating polymer matrix. The percolative conductivity and dielectric constant of such composites were reviewed by Nan et al. (2010).

Examples of such polymer composites are abundant. An early work that reported on the fabrication of electrically conductive polymer composite with carbon black is that of Balberg et al. (1983), who measured the resistivity of a polymer composite composed of elongated carbon black aggregates, embedded in an insulating plastic, polyvinylchloride. Due to the elongation of the aggregates, the composite was anisotropic. Measurements of the anisotropy of the material produced results very similar to those shown in Fig. 12.2, except that the curves ended at the percolation threshold, since the plastic matrix was insulating, in agreement with Eq. (12.3).

Another early study of polymer composites was undertaken by Hsu and Berzins (1985) who studied percolation effects in a polymer composite, namely, perfluorinated ionomers that have the general formula $[(CF_2)_n\ CF]_m$–O–R–SO_3X, where n is typically 6–13, R is a perfluoro alkylene group that may contain ether oxygene, and X is any monovalent cation. They are comprised of carbon–fluorine backbone chains with perfluoro side chains containing sulfonate, carboxylate, or sulfonamide groups, possess exceptional transport, chemical, and mechanical properties, and have been used as membrane separators, acid catalysts, and polymer electrodes. Percolation effects are important in such composites due to a spontaneous phase separation occurring in the wet polymer, which is the conductive aqueous phase that is distributed randomly in the insulative fluorocarbon phase. Thus, the effective properties of such polymers are dominated by the distribution and connectivity of the clusters of the conductive phase in the midst of an insulating phase, which is clearly a percolation phenomenon. Hsu and Berzins (1985) measured the electrical conductivity and elastic moduli of the polymer composites, and reported their measurements to be consistent with the predictions of percolation theory. For example, they found that for $10^{-2} \leq p - p_c \leq 0.8$, a very broad region, the conductivity of the polymers followed Eq. (12.2), which again indicates the broad applicability and usefulness of the percolation power universal laws.

Mamunya et al. (2002), for example, added Cu and Ni microparticles into epoxy resin and poly(vinyl chloride) polymer matrices. Li et al. (2006) fabricated a stainless steel fiber/polyvinylidene fluoride composite by simple blending and hot pressing. The percolation threshold of the composite, 0.094 volume fraction, was much lower than that of the common two-phase metal particle–polymer composite.

While there are many examples of such polymer composites, what should be clear by now is that, for a filler volume fraction ϕ larger than the volume fraction ϕ_c, which represents the percolation threshold of the conducting filler, an SSC of connected filler particles emerges throughout the polymer matrix, and the effective electrical conductivity of the composite suddenly increases by orders of magnitude. Moreover, percolation theory tells us that this phenomenon occurs within a narrow range of the filler volume fraction

Even when the morphology of the material is such that it gives rise to non-universal power laws for the material's conductivity near the percolation threshold, experimental data and the predictions of percolation theory that were discussed in Chap. 3 have been found to be consistent with each other. An example is the work of Lee et al. (1986) who measured the effective conductivity of both sliver-coated glass–Teflon composites and indium–glass composites. In one experiment with the conducting sphere–insulator composites, the silver-coated glass spheres were randomly distributed inside the Teflon host, with the conduction occurring through the contact points between the spheres. The conductivity exponent was measured to be, $\mu \simeq 2.0 \pm 0.2$, in excellent agreement with that of 3D percolation. In the second set of experiments with the insulating sphere–conductor composites, the hard glass spheres were randomly distributed inside the indium and, therefore, conduction took place in the channels between the spheres. Scanning-electron microscope images showed that the narrow necks of the conducting paths do exist. In this case, the estimated conductivity exponent based on fitting the data was, $\mu \simeq 3.1 \pm 0.3$, which agrees with the prediction that (see Chap. 3), with a conductance distribution $f(g) = (1 - p)\delta(g) + p(1 - \alpha)g^{-\alpha}$, one has, $\mu_\alpha \simeq \mu + \alpha/(1 - \alpha)$. Since for this composite, $\alpha \simeq 1/2$, one obtains $\mu_\alpha \simeq 3.0$, in agreement with the measurements.

Michels et al. (1989) reported extensive data for the electrical conductivity of a well-known class of composites, namely, carbon black (CB)/polymer composites. From a master batch of well-dispersed CB, casting of crosslinked samples were prepared and, for each CB concentration, the effective conductivity was measured. At a critical volume fraction well below 2 weight percent of the CB, the composite's conductivity exhibited a sharp rise to values 11–14 orders of magnitude larger than that of the matrix, which was a thermosetting polymer, hence indicating the formation of a SSC of the CB. Near the critical volume fraction of the CB, the effective conductivity of the sample was found to follow Eq. (12.2) with, $\mu \simeq 2.1 \pm 0.2$, in excellent agreement with the numerical estimate of μ for 3D percolation (see Chap. 2).

Andrade et al. (1998) reported experimental data on the conductivity of pure vapor-grown carbon fibers (VGCF) under compression as a conducting component of a polymer composite. They measured the dependence of the effective conductivity g_e on the apparent density ϕ of unmixed VGCF, when submitted to a uniaxial and low-speed compaction process. Near the percolation threshold, the contact resistance in the conductivity measurement device plays a significant role. Figure 12.5 presents their data. It is seen that the change in the conductivity of VGCF during compression indicates the existence of two regimes, where a characteristic non-linear dependence of the effective conductivity g_e on the density ϕ at low consolidation levels is followed by a roughly linear dependence of g_e on ϕ at higher pressures. The linear regime, far from $\phi_c \simeq 0.019$, is accurately predicted by the EMA. If the data of Fig. 12.5

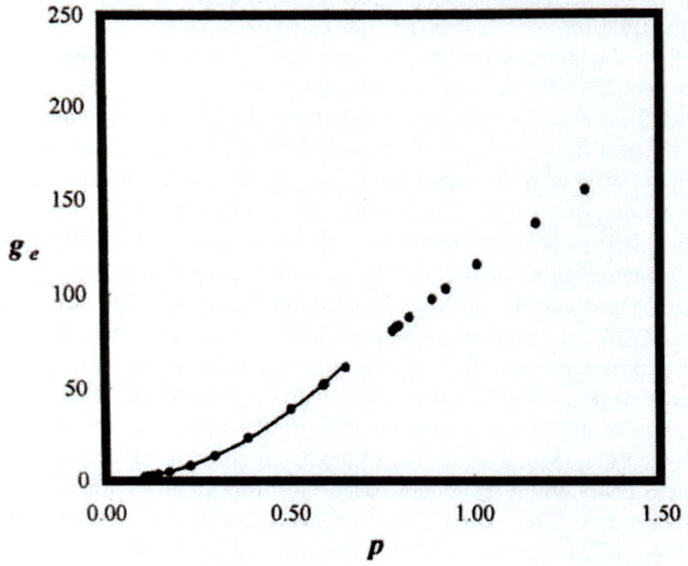

Fig. 12.5 Effective conductivity of compressed vapor-grown carbon fibers (VGCF) (in $\Omega^{-1}\text{cm}^{-1}$) as a function of the density p (in gr/cm^3) of the unmixed VGCF. The solid curve shows the percolation prediction, $g_e \sim (p - p_c)^{\mu}$ with $p_c \simeq 1.91 \pm 0.06$ (after Andrade et al. 1998)

near ϕ_c are fitted to Eq. (12.2), with p replaced by ϕ, one obtains, $\mu \simeq 1.91 \pm 0.06$, consistent with the conductivity exponent of 3D percolation systems, $\mu \simeq 2.0$. The same behavior was observed when VGCF was used as a conducting filler for a polymer composite, poly-propylene.

One strategy to improve the electrical conductivity of polymer composites relies on formation of *double-percolated networks*, which is induced by immiscible polymer blends, and mixtures of fillers in a single polymer matrix. The strategy enhances interparticle connectivity and, thus, the conductivity. Thongruang et al. (2002) combined the two strategies and studied quaternary composites of high-density polyethylene (HDPE), ultrahigh molecular weight polyethylene (UHMWPE), graphite (G), and carbon fiber (CF), and examined the electrical conductivity, morphology, thermal signature, and mechanical properties of HDPE/UHMWPE/G composites that exhibit evidence of double percolation. When the CF was added, orders of magnitude increase in the conductivity were obtained, due to its role in spanning non-conductive regions and enhancing the connectivity of conduction pathways. When the CF concentrations were above the percolation threshold, adding G promoted increases in the conductivity.

12.2.3 Polymer–Graphene Composites

Another class of electrically conductive polymer composites that have attracted wide attentions consists of a polymer and graphene. The latter is a 2D sheet of carbon atoms that are sp^2-bonded into a hexagonal arrangement (Suarez-Martinez et al. 2011) with excellent electrical conductivity if it is in a suitable isolated environment. In addition, there is also a family of graphene-related materials that differ from pure graphene by having, for example, multiple stacked layers—graphene nanoplatelets—or even different chemical structures, such as graphene oxide (Gao et al. 2014). It is due to such properties that use of graphene and related materials as fillers in a polymeric matrix can make the composite conducting, and has attracted wide attention. Such composites are very versatile with a wide range of potential applications. Similar to other polymer composites, the conductivity of polymer–graphene composites depends on the polymeric matrix and the type of graphene, as well as the processing methodology, and post-production treatments.

In general, there are three regimes in the electrical conductivity of such composites. One is the usual regime, namely, when there is no SSC for conduction through the filler and, thus, no charge can flow and the composite is insulating. The second regime is when the volume fraction of the filler is at or above the percolation threshold and forms an SSC, hence allowing electrons to flow through the cluster and, therefore, the composite becomes conductive. The third regime is between the first two: the fillers are not in direct contact, but are connected via electrons tunneling through an interface formed between the filler and the polymeric matrix. In this case, the conductivity of the composite is lower than when an SSC of the filler has formed. The tunneling is, of course, a quantum phenomenon, and has been directly observed in a carbon-nanofiber polymer composite (Balberg et al. 2004; Ambrosetti et al. 2010).

Percolation conductivity in such composites is also affected by the fillers forming aggregated structures that become interconnected by individual filler particles. The resulting agglomeration promotes percolation. Similarly, the fillers may form a phase-separated, co-continuous morphology that consists of graphene-rich and poor phases within the composite, leading to what is referred to as *selective localization* (see, for example, Kim et al. 2011; Gao et al. 2014), and often producing a conductive composite at lower loadings. It is, therefore, clear that percolation plays a key role in the electrical conductivity of such polymer composites.

Two-dimensional polymer composites of this type were fabricated by Pang et al. (2010), who reported on the fabrication of a graphene nanosheet/ultra-high molecular weight polyethylene composite with a segregated structure. They used water/ethanol solvent-assisted dispersion and hot compression at 200 °C. At a percolation threshold as low as 0.070 vol. percent the electrical conductivity of the composite increased by orders of magnitude, which was due to the formation of a 2D SSC of the conducting graphene.

A 3D example of such composites, made of polyimide and graphene as the conductive nanoparticle, was fabricated by Yoonessi et al. (2017), who reported on the preparation of a class of lightweight polyimide-graphene nanocomposites with low, 0.015 volume percent and tunable electrical conductivity, which exhibited an ultra-low graphene percolation threshold of 0.03 vol% and maximum DC conductivity of 0.94 S/cm. They attributed the high conductivity to excellent dispersion, extraordinary electron transport in the well-dispersed graphene, high number density of graphene nanosheets, and the $\pi - \pi$ interactions between the aromatic moieties of the polymide and the carbon rings in graphene. The DC conductivity was shown to follow the power-law dependence on the graphene volume fraction near the percolation threshold. The exponent μ of the conductivity was estimated to be, $\mu \simeq 3.8$, which is in the range of non-universal values of μ in continuum percolation (see Chap. 3).

The problem of designing highly conductive polymer–graphene composites is an active area of research, and has been comprehensively reviewed by Marsden et al. (2018).

12.2.4 Polymer–Carbon Nanotube Composites

Another type of polymer composites that have attracted tremendous attentions consist of a polymer with the CNTs. The reason for the interest is the structural properties of CNTs. Such tubes are hexagonal networks of carbon atoms rolled up into seamless, hollow cylinders, with each capped with half of a fullerene molecule (Dresselhaus et al. 1998). They are also isotropic materials, distinguishing nanotubes from other carbon structures. The CNTs may be single-walled nanotubes (SWCNTs), or multi-walled nanotubes (MWCNTs) with a collection of several concentric graphene cylinders held together by weak van der Waals forces. Aside from such desirable properties, the CNTs have high flexibility and mechanical strength with large elastic moduli, both of which are absent in graphite fibers. A CNT is very stiff for small loads, but it turns soft for larger loads, implying that it can be deformed without breaking. Their deformation is also reversible, which is why they can be flattened, twisted, and buckled.

In addition, surface properties of CNTs affect positively their composites. When many CNTs are mixed in the polymeric matrix, there will be strong van der Waals interactions between them, forcing them to form bundles of nanotubes with diameters in the range of 10–100 nm. The bundles can break if sheared along the axis, causing failure between nanotubes but at stresses much below the intrinsic capabilities of a single CNT

Such polymer–CNT composites are of great interest due to the efficiency of the tubes as a conductive additive. They have excellent electrical properties, large aspect ratios (ratio of length and diameter), and the tendency to become entangled into a 3D, interconnecting network in molten plastic. It is now possible to use 3D printing to produce such composites (Mora et al. 2020).

Coleman et al. (1998) were probably the first to report observation of percolation conductivity in a composite made of a polymer with CNTs. They mixed poly(*p*phenlyenevinylene-co-2,5-dioctoxy-*m*-phenylenevinylene), a variation on the more widely studied poly(*para*phenylenevinylene, with a CNT powder, and measured its electrical conductivity. At a mass fraction of the nanotube of about 8.4%, the electrical conductivity increased sharply by *ten* orders of magnitude, signaling a percolation transition.

In addition to the relevance of the percolation threshold ϕ_c, experiments have indicated that the conductivity of such composites near ϕ_c also follows the power law predicted by percolation theory. For example, Gao et al. (2008) reported on the low percolation threshold in conductive polymer composites with a segregated structure in which the conductive particles, including carbon black and the CNTs, were only on the interface between the polymer matrix and the particles, rather than being randomly distributed in the entire composite. They fabricated a polymer composite with multiwalled CNTs using alcohol-assisted dispersion under ultrasonication and intense mechanical mixing. The polymer, which had an ultrahigh molecular weight, was polyethylene. The experimental data were fitted to the power law for the conductivity near the percolation threshold, which was only 0.072 vol. percent, to estimate the critical exponent μ, which turned out to be 1.3, in perfect agreement with 2D percolation, as the composite had formed a 2D conductive network.

Using continuum percolation (Chap. 3), Kyrylyuk and van der Schoot (2008) studied the effect of bending flexibility, length polydispersity, and attractive interactions between CNTs on the percolation threshold in polymeric and colloidal background, demonstrating that continuum percolation provides a predictive tool for designing nanotube-based composites. Their study indicated that the host matrix controls the percolation threshold through the the interactions that it induces between the CNTs, not only during processing but also through the degree of connectedness that must be set by the tunneling distance of electrons during percolating conduction.

Kyrylyuk et al. (2017) reported on the fabrication of polymer–CNT composites and measuring its electrical conductivity, and showed that the percolation threshold can be significantly lowered by adding small amounts of a conductive polymer latex. They showed that, when colloidal particles of various sizes and shapes, which in their study were spherical latex particles and rod-like nanotubes, are mixed, it introduces competing length scales that strongly influence the formation of the SSC and significant electrically conductivity. See also Kazemi et al. (2021), who prepared a polymer composite with CNT and nanoclays.

Since the early works, a very large number of papers have been published, reporting on the fabrication of such polymer–CNT composites, as well as their theoretical modeling, long list of which are given by Bauhofer and Kovacs (2009) and Kyrylyuk et al. (2017), who reviewed the research field.

12.2.5 *Effect of Thickness and Particle-Size Distribution*

The thickness of a composite material affects its percolation properties. It is not difficult to show that

$$p_c(\ell) - p_c(\infty) \sim c\left(\frac{R_p}{\ell}\right)^{1/\nu_3}, \tag{12.5}$$

where $p_c(\ell)$ is the effective percolation threshold of a sample of thickness ℓ, $p_c(\infty)$ is the corresponding threshold for $\ell \to \infty$, c is a constant, and ν_3 is the correlation length exponent for 3D percolation. Equation (12.5) is completely similar to finite-size scaling analysis described in Chap. 2. Clerc et al. (1980) studied this effect by varying the thickness ℓ of a material made of layers of particles with a diameter R_p.

Figure 12.6 shows their results for various thicknesses l from which one obtains $\nu_3 \simeq 0.85$, in very good agreement with $\nu(d = 3) \simeq 0.89$ for 3D percolation

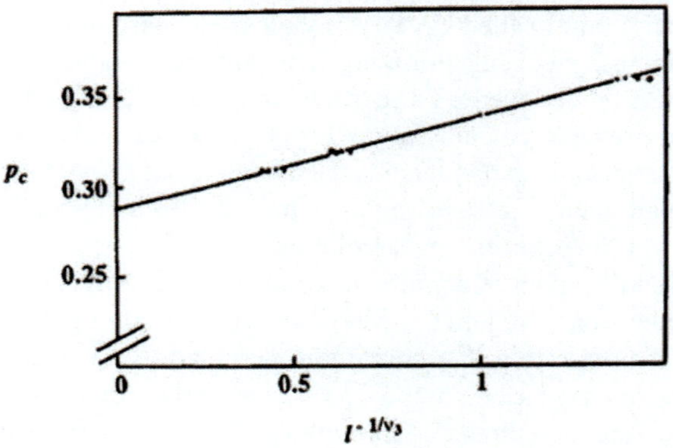

Fig. 12.6 Dependence of the percolation threshold p_c of a mixture of conducting and insulating spheres on the thickness l of the sample, with ν_3 being the critical exponent of 3D percolation correlation length (after Clerc et al. 1980)

(Chap. 2). The finite thickness of the sample also affects the conductivity of the material. It is not difficult to show that

$$\frac{g_{e1}}{g_{e2}} \sim \left(\frac{\ell_2}{\ell_1}\right)^{(\mu_3-\mu_2)/\nu_3}, \tag{12.6}$$

where g_{e1} (g_{e2}) is the conductivity of a sample of thickness ℓ_1 (ℓ_2), and μ_2 (μ_3) is the critical exponent of conductivity for 2D (3D) percolation. The experimental data of Clerc et al. (1980) for the conductivity of their samples were also in agreement with Eq. (12.6).

The size distribution, as well as the shapes of the particles or filler, may also affect the transport properties. For example, instead of having particles or fillers of the same size, one may have particles of a variety of diameters, in which case the resulting composite will be even more disordered than one with monosize particles. Since formation of the SSC of the fillers in a composite matrix is critical to its transport properties, the issue has been studied extensively. As already discussed, some of the fillers, such as the CNTs and graphene nanosheets are flexible enough that they can become wavy or crumpled, which in turn affects the formation of the SSC.

Kwon et al. (2016) studied the problem by considering the flexibility of different kinds of fillers, and carrying out extensive Langevin dynamics simulations. The three types of fillers had different shapes, namely, nanospheres, nanorods, and nanoplates. Their study indicated that when the sizes or radii of gyration of the fillers were comparable, the nanorods form an SSC at a lower volume fraction than nanoplates, whereas nanospheres required the highest volume fraction to form the percolating network. The percolation threshold ϕ_c of the nanospheres increased with an increase in their radius of gyration R_g, whereas ϕ_c for nanorods and nanoplates decreased with R_g. The effect of flexibility on ϕ_c was much more significant for nanoplates than for nanorods. Their study also indicated that the electric conductivity of the composites follows the power law of percolation, albeit with a critical exponents μ that may depend on the shape and flexibility of the filler.

12.3 Metal–Insulators Composites

Thin metal films are two-phase mixtures of metal and nonmetal components that have interesting structural and transport properties. We already described in chapter two how the percolation properties of their morphology can be and were measured by Voss et al. (1982) and Kapitulnik and Deutscher (1982), whose results were published in two back-to-back papers in *Physical Review Letters*.

Thick films of metallic and nonmetallic components, i.e., those with a thickness greater than ξ_p, also have interesting properties, and have been studied for a long time. They can also be produced by co-sputtering or co-evaporation of two components that are insoluble in each other, one being the nonmetallic component and the other one the metallic phase. They have been studied for their novel superconducting and magnetic properties. However, the discovery by Abeles et al. (1975) that the electrical conductivity of such films near p_c follows Eq. (12.2) made them the subject of many studies. They are mainly composed of a metal with a bulk lattice structure. The non-metallic phase is amorphous and often in the form of isolated islands. They are often called *granular* metal films because inspection of their structure indicates that the metal grains are often surrounded by very thin amorphous layers of the nonmetallic component, so that the metallic grains remain separated. Granular composites always have high percolation thresholds, which can be close to the random close-packing fractions, if the grain size is constant and the insulator is relatively thick. Examples of such composites are Ni–SiO_2, Al–Si, and Al–Ge. Unlike thin metal films, one cannot easily find clusters of metallic grains in granular metals and, therefore, it is more difficult to interpret their structure in terms of percolation properties. There are also strong correlations in the structure of thick metal films, so that the random percolation model described so far may not be totally suitable. Deutscher et al. (1978) proposed that thick metal films have a granular structure if their nonmetallic phase is amorphous, but should be considered as random (continuous) composites if the insulating phase is crystalline. By random, we mean a composite in which the two phases cannot be identified from their structures, so that they both play a symmetric role. Since metal–insulator composites have been studied for decades, we begin our discussion by describing several properties of metal films, relate them to percolation quantities, and describe some key papers published over the years.

The first application of percolation to the interpretation of various properties of very thin (2D) metal films appears to have been made by Liang et al. (1976). They prepared ultrathin films that were 1.5 mm wide with pre-evaporated indium electrodes spaced 25 mm apart. The substrate used was SiO films, deposited on a microscope slide glass. Several kinds of metallic compounds were evaporated and studied. Liang et al. reported that the sudden drop in the resistivity of semimetal bismuth ultrathin films was very steep, signaling a percolation transition. When they plotted the conductivity of the composite versus the area fraction ϕ of the conducting phase, they found that near the percolation threshold the conductivity follows Eq. (12.2) with $\mu \simeq 1.15$, reasonably close to $\mu \simeq 1.3$ for 2D percolation (Chap. 2). They found, however, that $\phi_c \simeq 0.67$, quite different from $\phi_c = 0.5$, the prediction of Scher and Zallen (1970) for 2D continuum percolation (Chap. 3). The difference may be attributed to the existence of negative correlations that are usually present in such films, but were absent in the Scher–Zallen theory for random and continuum percolation.

Papandreou and Ne'dellec (1992) fabricated Pd films with a typical thickness of about 100 $^\circ A$ by depositing them by electron gun evaporation on quartz substrate, in order to measure their electrical conductivity, and on NaCl substrate for TEM observations. The substrates were coated with a SiO layer of 150 $^\circ A$, which ensured similar growth conditions, regardless of the substrate below SiO. The samples were then irradiated under normal incidence with 100 keV Xe ion beam, after which their conductivities were measured. They reported that as the percolation threshold was approached, the sample's conductivity vanished according to Eq. (12.2) with $\mu \simeq 1.3$, in complete agreement with the prediction of 2D percolation.

Let us now describe some of the experimental results for thick films. In a seminal paper, Abeles et al. (1975) studied the growth of the grains of the metals W or Mo in the insulators Al_2O_3 or SiO_2 using co-sputtering. They changed the volume fraction of the metal in the mixture in the range $0.1 < \phi \leq 1$ and obtained very finely dispersed grain structures. With annealing, one can change the size of the metal grains over a wide range. The advantage of using the W or Mo is that annealing does not cause the metal to precipitate. The metal remains uniformly dispersed within the insulator. The conductivity of the films was determined for various values of the metallic phase volume fraction. Due to correlations, the metal–insulator transition occurred at $\phi_c \simeq 0.47$, considerably higher than the Scher–Zallen prediction for 3D continuum percolation, $\phi_c \simeq 0.15 - 0.17$, which may be attributed to some sort of bias towards the formation of isolated metallic regions. Figure 12.7 shows the conductivity of the samples measured by Abeles et al. as a function of the volume fraction of the metallic phase. The straight line has a slope $\mu \simeq 1.9 \pm 0.2$, in very good agreement with $\mu \simeq 2.0$ for the critical exponent of 3D percolation conductivity (Chap. 2).

Using percolation networks of resistors, Cohen et al. (1978) showed that one may simulate systems that have high percolation thresholds, such as granular composites. In their model, one starts with an initial network already containing a certain fraction p_0 of metallic bonds, arranged in a configuration of noncontacting metallic regions. One then replaces at random the nonconducting bonds with metallic ones and estimates the conductivity and percolation threshold of the system. In this manner, correlations are introduced into the model, and there is a systematic bias toward the formation of isolated metallic regions, which is what was observed in the films studied by Abeles et al. Clearly, the percolation threshold of the network depends on p_0, and Cohen et al. showed that the conductivity data of Abeles et al. can be quan-

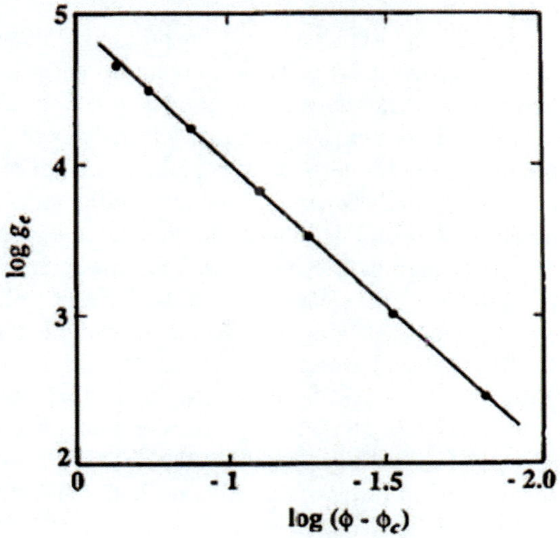

Fig. 12.7 Logarithmic plot of the conductivity of $W–Al_2O_3$ films versus departure from the percolation threshold, $\phi_c - \phi$ (after Abeles et al. (1975))

titatively predicted by their model. Webman et al. (1975) used percolation networks of resistors to predict the electrical conductivity of several types of disordered materials, such as metal–ammonia solutions, and alkali–tungsten–bronzes that undergo metal–insulator transitions, similar to what was found in the study of Abeles et al. (1975), and showed that such percolation networks provide quantitative predictions of the electrical conductivity in all cases that they studied.

Experimental data similar to those of Abeles et al. were also reported by Kapitulnik and Deutscher (1982). They prepared Al–Ge films by co-evaporating them onto glass substrates from two electron beam guns through a mask with slits, and measuring their electrical conductivity. The conductivity of the thick samples near the metal–insulator transition followed Eq. (12.2) with $\mu \simeq 2.1 \pm 0.5$, completely consistent with the prediction of 3D percolation. For thin samples, the corresponding exponent was found to be, $\mu \simeq 0.9 \pm 0.25$, smaller than that of 2D percolation, $\mu \simeq 1.3$, but almost consistent with it. The crossover between 2D and 3D films were also studied by Kapitulnik and Deutscher (1983). They found that the percolation threshold of the composite depended on the thickness of the sample and followed Eq. (12.3), as expected. Experimental data for Al–Ge films were reported by Kapitulnik et al. (1990), and for In–Ge thin films by Tessler and Deutscher (1989).

The number of papers that have been published over the last five decades in which link between percolation and metal–insulator composites has been studied and quantified is simply too large to be described here. We describe briefly some of them here. Das Sarma et al. (2005) analyzed the low density dependence of 2D low-temperature conductivity in undoped high-mobility n-GaAs heterostructures, and concluded that the 2D metal–insulator transition in the 2D composite was a density inhomogeneity-driven percolation transition. They argued that this was due to the breakdown of screening in the randomly charged impurity disorder background. The measured conductivity, when fitted to the power law (12.2) yielded a conductivity exponent, $\mu \approx 1.4$, only 8% larger than the 2D percolation exponent, $\mu \simeq 1.3$ at low temperatures.

Adam et al. (2008) studied transport in graphene nanoribbons with an energy gap in the spectrum in the presence of random charged impurity centers. They established that at low carrier density the composite exhibited a density inhomogeneity-driven 2D metal–insulator transition that was in the percolation universality class. They also predicted that, for very narrow graphene nanoribbons, there should be a dimensional crossover to the 1D percolation universality class with observable signatures in the transport gap, and that there should also be a crossover to the Boltzmann transport regime at high carrier densities. Their measured conductivity exponent μ and the critical density ϕ_c were consistent with the percolation transition.

Lee et al. (2019) reported on the measurement of the temperature dependence of the local resistance and the nanoscale domain distribution of $NdNiO_3$ areas between Au contacts gapped by 40–260 nm. A sharp resistance drop emerged below the bulk metal–insulator transition temperature at about 105 K, whose amplitude scaled inversely with the nanogap width. Lee et al. used X-ray photoemission electron microscopy in order to directly correlate the resistance drop to the emergence and distribution of individual metallic domains at the nanogap, hence providing deeper insight into percolation at the metal–insulator transition in such rare-earth nickelates.

Finally, we should mention that optical properties of percolating metal films have also been measured and explained by using percolation theory. Some of such properties are related to the dielectric constant of materials, and will be described shortly. For a comprehensive discussion, the interested reader is referred to Sahimi (2003a).

12.4 Tunneling Versus Percolation

According to percolation theory, the conductivity of a metal–insulator composite must vanish, if the volume fraction ϕ of the metallic phase is less than the percolation threshold. Over the years many sets of experimental data have been reported that have indicated that some metal–insulator composites have a finite conductivity for $\phi < \phi_c$, even though the material is in the dielectric regime. Such a non-zero conductivity is attributed to interparticle tunneling (see, for example, Chui 1991; Pollak and Adkins 1992), and various theories have been proposed to explain the electrical conductance in dielectric regime as being the result of the tunneling, but the global electrical connectedness of the tunneling conduction network remained unexplained for quite some time. In addition, even with a full understanding of macroscopic connectivity of tunneling conduction, one must still explain why tunneling conduction seems to follow percolation power law, whereas the conducting particles are not geometrically connected and do not form an SSC.

A resolution of the issue was offered by Toker et al. (2003). They prepared Ni–SiO_2 films with a thickness of 2.5–5.5 μm and width of 5 mm, which were fabricated by depositing on Pyrex substrates by co-sputtering of Ni and SiO_2, in which the volume fraction of Ni varied between 0.24 and 0.92. Various images of the composite were analyzed in which "conducting islands" were identified. The lowest current that they considered, referred to as the "cut-off current" I_{co}, was varied. Therefore, the total area A_i covered by the conducting islands depended on I_{co}. Note that the larger I_{co}, the smaller is A_i, because only lower resistance paths of the conduction network are included. Then, Toker et al. (2003) discriminated between various current maps by determining their fractal dimensions.

The fractal dimension D_f was estimated by measuring A_i as a function of the "window length" L. Near ϕ_c the L values considered were much larger than the particle's size, but much smaller than ξ_p, the correlation length of the percolation network. One expects to have, $D_f(d = 3) = D_i + 1$, where D_i is the fractal dimension of the conducting islands corresponding to A_i (i.e., in a 2D cut of the 3D structure). In that regime, Toker et al. estimated that, $D_f \simeq 2.5$ ($D_i \simeq 1.5$), in agreement with 3D percolation (Chap. 2). But, as the overall system departed from the vicinity of the percolation threshold, it became homogeneous, lost its fractal nature and, therefore, D_f approached 3. Hence, as the cutoff current I_{co} was increased, D_f decreased.

Therefore, the type of the current network in the composite corresponding to each I_{co} is different, such that for large I_{co} only the inter-grain tunneling resistors that are smaller than a given value are included in the conducting network. As a result, the images corresponding to large I_{co} represent the contribution of the lowest resistors of the network, i.e., to currents that flow through the sample by a route that consists mainly of the particles that are separated by the typical tunneling distance (here, "nearest neighbors"). But the low I_{co} values represent current paths that are added to the above inter-adjacent grain paths, and correspond to the larger (tunneling) resistors that are either removed "nearest neighbors," or originate from higher order neighbors. Hence, the network defined by the highest values of I_{co} is similar to the one encountered in fused-particles composites, i.e., a bona fide percolation network.

12.5 AC Conductivity and Dielectric Properties of Heterogeneous Composites

Consider a regular or random network in which each bond has either a conductance g_1 with probability p, or g_2 with probability $q = 1 - p$. As described in Chap. 2, the effective conductivity of such a network has interesting properties described by Eqs. (2.6) and (2.7). We also discussed in Chap. 2 two limiting cases of the system, namely, when $g_2 = 0$ and g_1 is finite—a conductor–insulator mixture—and one in which $g_1 = \infty$ and g_2 is finite—a superconductor–conductor mixture. Both cases correspond to $h = g_2/g_1 = 0$ and, therefore, the point $h = 0$ at $p = p_c$ is particularly important. In the critical region near $h = 0$ and $p = p_c$, where both $|p - p_c|$ and h are small, the effective conductivity g_e follows the scaling law (2.22), where, in addition to the critical exponent μ, s describes the effective conductivity of conductor–superconductor networks near p_c, and Φ_+ and Φ_- are two homogeneous functions corresponding, respectively, to the regions above and below p_c. Equation (2.22) was first proposed by Efros and Shklovskii (1976). Similar to μ and s, the two scaling functions are universal and do not depend on the network type, once h and $p - p_c$ are fixed.

For any fixed and nonzero value of h, the effective conductivity g_e depends smoothly on $p - p_c$, since Eq. (2.22) may be rewritten in a different form expressed by Eq. (2.23), where $\Psi(x) = x^t\Phi_+(x^{-\mu-s}) = (-x)^\mu\Phi_-[(-x)^{-\mu-s}]$. If we expand the scaling function $\Psi(x)$ as a Taylor series around $x = 0$, $\Psi(x) = \Psi(0) + \Psi_1 x + \Psi_2 x^2 + \cdots$, at $p = p_c$ and for $|h| \ll 1$ we must have

$$g_e \sim \Psi(0)(g_1^s g_2^\mu)^{1/(\mu+s)} \equiv \Psi(0) g_1 h^u , \tag{12.7}$$

where

$$u = \frac{\mu}{\mu + s} . \tag{12.8}$$

Equation (12.8) implies that $\Phi_-(x) \sim \Phi_+(x) \sim \Psi(0)x^u$, hence demonstrating clearly the homogeneous nature of such functions. Many other properties of these scaling functions are discussed by Clerc et al. (1990) and Sahimi (2003a), to whom the interested reader is referred.

12.5.1 AC Conductivity

The AC conductivity and dielectric properties of a disordered composite near p_c has been studied (Efros and Shklovskii 1976; Bergman and Imry 1977), and all such efforts have been reviewed comprehensively by Clerc et al. (1990) and Sahimi (2003a). We summarize the main theoretical results, and discuss their experimental verification.

If g_1 and g_2 are viewed as complex conductances, then, a percolation model may be developed for the AC conductivity and dielectric properties. Consider a percolation network in which a fraction p of the bonds are purely resistive, while the remaining fraction $q = 1 - p$ behaves as perfect capacitors. Thus, we set, $g_1 = 1/R$ and $g_2 = iC\omega$, where R is the resistance of the bond, C is the capacitance, ω is the frequency, and $i = \sqrt{-1}$. The conductance ratio h is then given by, $h = i\omega/\omega_0$, where $\omega_0 = 1/(RC)$. In the static limit ($\omega = 0$) the capacitors become insulators, the model reduces to the usual percolating conductor–insulator mixture, and is usually referred to as the $R - C$ model. The key result is that if $p \approx p_c$, the $R - C$ model possesses scaling properties, and the effective conductivity $g_e(p, \omega)$ of the mixture follows the following scaling law:

$$g_e(p, \omega) \sim \frac{1}{R}|p - p_c|^{\mu}\Phi_{\pm}\left(\frac{i\omega}{\omega_0}|p - p_c|^{-\mu-s}\right), \tag{12.9}$$

which follows directly from Eq. (2.22). An immediate consequence of Eq. (12.9) is the existence of a time scale t_s that diverges as p is approached from either side of p_c:

$$t_s \sim \omega_0^{-1}|p - p_c|^{-(\mu+s)}. \tag{12.10}$$

The significance of t_s is discussed below.

12.5.2 Dielectric Constant

The frequency-dependent complex dielectric constant $\epsilon(p, \omega)$ of the mixture is now defined by the following equation:

$$\epsilon(p, \omega) = \frac{g_e(p, \omega)}{i\omega}, \tag{12.11}$$

which is a generalization of the usual static dielectric constant ϵ_0. It is straightforward to show that for an insulating dielectric medium $g_e \simeq i\omega\epsilon_0$ as $\omega \to 0$. Using general analytic properties of the effective complex dielectric constant of a random mixture, Bergman and Imry (1977) derived the following power laws that are obtained from Eqs. (12.9) and (12.11):

$$g_e(p_c, \omega) \sim \omega^x , \tag{12.12}$$

$$\epsilon(p_c, \omega) \sim \omega^{-y} , \tag{12.13}$$

where the exponents x and y satisfy the following relation:

$$x + y = 1 . \tag{12.14}$$

Equation (12.14) is a direct consequence of the complex conductivity being an analytic function of $i\omega$. Bergman and Imry (1977) argued that the main contribution to the AC properties is due to polarization effects between various percolation clusters in the disordered mixture, and proposed that

$$x = \frac{\mu}{\mu + s} , \tag{12.15}$$

$$y = \frac{s}{\mu + s} , \tag{12.16}$$

so that in 2D where $\mu = s$ (see Chap. 2), one has $x = y = 1/2$. Gefen et al. (1983), who studied anomalous diffusion on percolation clusters (see Chap. 7), argued that the fractal nature of diffusion and that of percolation clusters at length scales up to ξ_p, the correlation length of percolation, dominates the contributions to the AC properties and proposed instead that

$$x = \frac{\mu}{\nu(2 + \theta)} , \tag{12.17}$$

$$y = \frac{2\nu - \beta}{\nu(2 + \theta)} , \tag{12.18}$$

which also satisfy Eq. (12.14), where $\theta = 2 + (\mu - \beta)/\nu$. We shall discuss the experimental verification of these results shortly.

An important consequence of Eq. (12.9) is the power law for ϵ_0 in the critical region near p_c. Using Taylor's expansion of $\Phi_\pm$ described above, it is not difficult to show that

$$\epsilon_0 \sim A_\pm C|p - p_c|^{-s} . \tag{12.19}$$

That is, the static dielectric constant *diverges* as p_c is approached from *both sides*, and the critical exponent that characterizes this divergence is s, the critical exponent of a percolation network of superconductors–conductors. This spectacular result was first derived by Efros and Shklovskii (1976). In Eq. (12.19), A_+ and A_- represent, respectively, the amplitudes of ϵ_0 above and below p_c. Although Eq. (12.19) is supposedly valid only in the limit, $\omega = 0$, its validity extends to higher frequencies as long as $\omega \ll 1/t_s$, where t_s is the time scale given by Eq. (12.10). An important property of the amplitudes $A_\pm$ is that their ratio A_+/A_- is a universal quantity, independent of the microscopic details of the medium.

We now write, $g_e(p,\omega) = g' + ig'' = i\omega\epsilon(p,\omega) = i\omega(\epsilon' - i\epsilon'')$. Then, a loss angle δ is defined by

$$\tan\delta = \frac{g'}{g''} = \frac{\epsilon''}{\epsilon'}\ , \tag{12.20}$$

and it is clear that $0 \le \delta \le \pi$. Note that this loss angle is defined in a manner completely analogous to that for gelling solutions, described in Chap. 13.

Next, consider the effective conductivity of the $R - C$ model at p_c. According to Eq. (12.7),

$$g_e(p,\omega) \sim \frac{\Psi(0)}{R}\left(\frac{i\omega}{\omega_0}\right)^u\ , \tag{12.21}$$

so that at p_c the loss angle δ_c is *universal* and is given by

$$\delta_c = \frac{\pi}{2}(1-u) = \frac{\pi}{2}\frac{s}{\mu+s}\ , \tag{12.22}$$

Although Eq. (12.22) is supposedly valid exactly at $p = p_c$, it is important to keep in mind that the universal loss angle δ_c can also be measured in a broad frequency range if $|p - p_c|$ is small enough, implying that $1/t_s \ll \omega \ll \omega_0$. Note that, since, for *any* 2D percolating medium, we have $s = \mu$, we must have $\delta_c = \pi/4$, another remarkable result.

12.5.3 Experimental Verification

Let us now discuss the experimental verification of the predictions of the percolation models of the AC conductivity and dielectric constant. One of the earliest experimental studies of dielectric properties was reported by Castner et al. (1975), who had measured the static dielectric constant of n-type silicon. About 1200 °A of Au was evaporated on thin-disk samples that consisted of two imperfect Schottky[2] barriers[3] with thin (about 5–10 °A thick) oxide barriers, bounding from 0.2 to 1.0 mm

[2] **Walter Hans Schottky** (1886–1976) was a German physicist who played a major early role in developing the theory of electron and ion emission phenomena. He invented the screen-grid vacuum tube, co-invented the ribbon microphone and ribbon loudspeaker, and made many significant contributions in the areas of semiconductor devices. Schottky diode, Schottky effect, Schottky barrier, Schottky defect, Schottky–Mott rule, and Mott–Schottky equation are all named after him.

[3] A Schottky barrier is a potential energy barrier for electrons formed at a metal–semiconductor junction, and has rectifying characteristics, suitable for use as a diode.

of bulk semiconductor. They varied the concentration c_d of the donor, and showed that ϵ_0 diverges as c_d approaches a critical concentration from the insulating side. Although percolation was not mentioned in that work, the divergence of ϵ_0 was a clear indication of the percolation transition indicated by Eq. (12.19).

To explain the data, Dubrov et al. (1976) developed a model in which each bond of a percolation network represented a 300 ohm resistor and a 0.5 μF capacitor in parallel. Starting with a square network with only capacitors, they added resistors to the network at randomly selected bonds, and made measurements of the conductivity of the network at very low frequencies. As the fraction of the resistors approached the percolation threshold of the network, the dielectric constant of the network appeared to diverge. Dubrov et al. did propose a scaling law for the divergence that was similar to Eq. (12.19), although they did not attempt to estimate the associated critical exponent, since the network they used was too small.

A definitive experimental study of dielectric constant of composite materials near p_c was undertaken by Grannan et al. (1981). The composite consisted of small spherical Ag particles, randomly distributed in a nonconducting KCl host. The metal particles were prepared by evaporating Ag in the presence of argon gas and a small amount of oxygen. They were polydisperse, with their sizes varying between 60 and 600 $^\circ A$, with the overall size distribution being log-normal. The composite was prepared by mixing a given amount of Ag particles and KCl powder, then compressed into a solid pellet under high pressure. The material's dielectric constant was measured by a capacitance bridge operated at 1 kHz. Figure 12.8 presents the data as a function of the volume fraction of Ag. The dielectric constant appears to diverge at $p_c \simeq 0.2$, somewhat larger than $\phi_c \simeq 0.15$–0.17 for 3D percolating continua predicted by Scher and Zallen (1970).

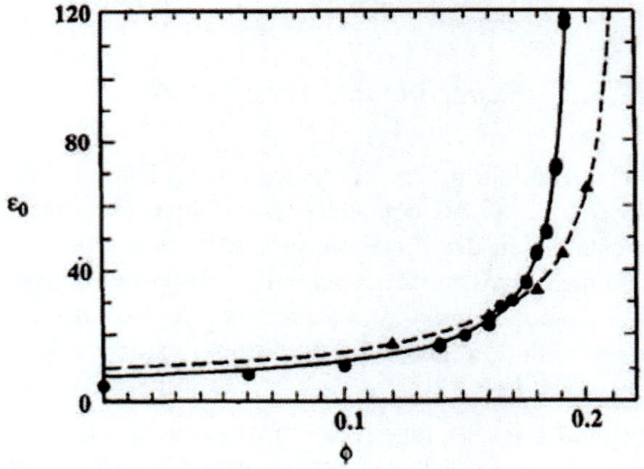

Fig. 12.8 Static dielectric constant ϵ_0 of two series of Ag–KCl composites versus the volume fraction ϕ of Ag, with the solid curves representing the best fit of the data to Eq. (12.20) (after Grannan et al. 1981)

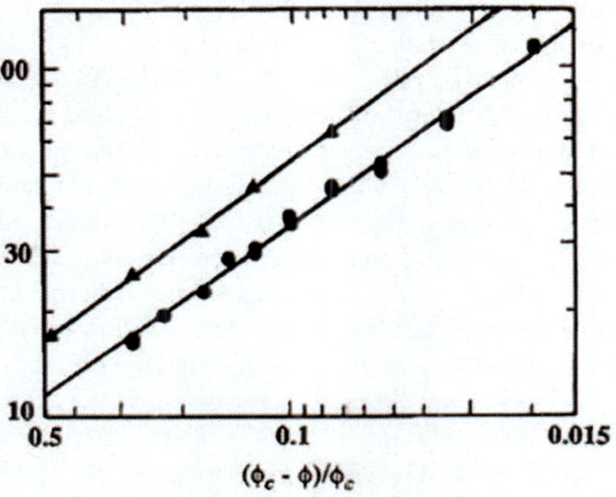

Fig. 12.9 Logarithmic plot of the data presented in Fig. 12.12 (after Grannan et al. 1981)

Figure 12.9 shows the logarithmic plot of the same data, all of which appear to lie on a straight line and indicating that ϵ_0 diverges as p_c is approached with an exponent $s \simeq 0.73 \pm 0.01$, in perfect agreement with the percolation prediction $s \simeq 0.735$ (Chap. 2). Similar data were reported by Niklasson and Granqvist (1984).

Laibowitz and Gefen (1984) prepared a series of samples of Au films on Si_3N_4 with varying thicknesses, which were selected to span the entire metal–insulator transition. Insulating samples below p_c were easily achievable, indicating that the contribution of tunneling and hopping to the conductivity could be ignored in the more metallic samples. The AC conductivity and capacitance of the sample, which is proportional to the dielectric constant, samples were then measured. When the data were fitted to Eqs. (12.12) and (12.13), they obtained $x \simeq 0.95 \pm 0.05$ and $y \simeq 0.13 \pm 0.05$, in rough agreement with Eq. (12.14). Equations (12.15) and (12.16) predict, however, that for 2D media for which $\mu = s$, $x = y = 1/2$, whereas Eqs. (12.17) and (12.18) predict that, $x \simeq 0.34$ and $y \simeq 0.66$, neither of which agree with the experimental data.

Song et al. (1986) measured the AC electrical properties of a powder mixture of amorphous carbon and Teflon in the frequency range 10 Hz–13 MHz. Due to its stability, Teflon powder was used as the insulating component. Moreover, the low conductivity of the amorphous carbon powder made it possible to easily observe the change of the conductivity as a function of p. The samples were prepared by mixing the carbon and Teflon powder to the desired volume fraction, which were then compressed. The electrical conductivity was then measured near p_c and was found to follow power law (12.2) with $\mu \simeq 1.85 \pm 0.25$, in good agreement with that of 3D percolation conductivity. The dielectric constant, in the static limit, was found to diverge according to Eq. (12.19) with $s \simeq 0.68 \pm 0.05$, in reasonable agreement with the theoretical expectation. The AC conductivity and the dielectric constant were also measured, from which it was estimated that $x \simeq 0.86 \pm 0.06$ and $y \simeq 0.12 \pm 0.04$, which do not agree with Eqs. (12.15) and (12.16) that predict, $x \simeq 0.6$ and $y \simeq 0.4$, nor with the predictions of Eqs. (12.17) and (12.18), $x \simeq 0.73$ and $y \simeq 0.27$.

The resolution of this apparent disagreement between the theory and the experimental data was provided by Hundley and Zettl (1988). They measured the AC conductivity and dielectric constant of thin Au films, similar to those of Laibowitz and Gefen (1984), but extended the frequency range to between 100 Hz and 1 GHz. Their measurements, shown in Fig. 12.10, indicated that in the intermediate frequency regime, corresponding to the experiments of Laibowitz and Gefen, as well as Song et al., $x \simeq 1.0$ and $y \simeq 0$, in agreement with the data of Laibowitz and Gefen (1984), whereas, at higher frequencies $x \simeq 0.32$, in excellent agreement with the prediction of Eq. (12.17), and $y \simeq 0.8$, in rough agreement with the prediction of Eq. (12.18). Thus, the fractal nature of percolation clusters seems to play an important role in the AC conductivity and dielectric properties of disordered materials at high frequencies.

Laugier et al. (1986b) measured AC conductivity of random mixtures of glass microbeads, a varying function of them having their surface coated with silver that, however, did not change appreciably the density of the powder. The average diameter of the beads was about 30 μm, and the frequencies used were as large as 50 MHz. The loss angle δ was also measured. Equation (12.22) predicts that at p_c one must have $\tan\delta_c \simeq 0.45$, while the measured value was $\tan\delta_c \simeq 0.5$, in good agreement with the prediction.

Another class of disordered systems whose AC conductivity and dielectric properties have been measured is microemulsions, which are thermodynamically stable, isotropic, and transparent dispersions of two immiscible fluids, such as water

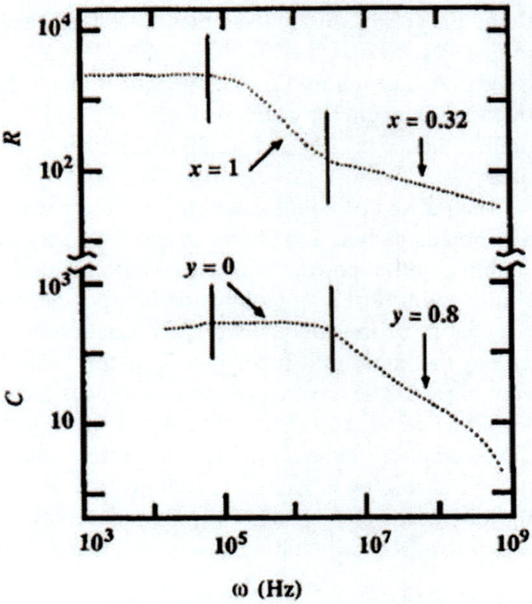

Fig. 12.10 Frequency dependence of the resistance R and capacitance C of an Au fractal film at 300 K (after Hundley and Zettl 1988)

and oil, with one or more surfactants being surface active. A water in oil (W/O) microemulsion usually consists of small spherical water droplets surrounded by a monomolecular layer of surfactant and dispersed in a continuous oil phase. The W/O microemulsions usually have a small macroscopic conductivity, because the water droplets are separated by the surfactant layers and the oil phase. Ionic surfactants can donate an ion to the water phase and increase its conductivity. If the volume fraction ϕ_w of the water phase exceeds a critical value $\phi_{\rm wc}$, the conductivity increases sharply, usually by several orders of magnitudes, which is due to the fact that charge carriers are able to move along connected paths in the microemulsion. Thus, the conductivity transition in microemulsions is a percolation phenomenon, which can also be induced by increasing the temperature, holding ϕ_w constant.

van Dijk (1985) was probably the first to measure the dielectric constant of a microemulsion at $\phi_{\rm wc}$. His fluid was a microemulsion of AOT, sodium di-2-ethylhexylsulfosuccinate, an anionic surfactant with a SO_3^- head group and two hydrocarbon tails, water and iso-octane. The volume fraction of water was changed by varying the amount of oil, keeping the molar ratio water/AOT constant. Figure 12.11 presents the data, indicating a sharp peak for the dielectric constant and the dramatic increase of the electrical conductivity, both at $\phi_{\rm wc}$, in agreement with the predictions of the percolation model. Moreover, according to Eq. (12.22), at $\phi_{\rm wc}$ the loss angle δ_c is independent of frequency, and Fig. 12.12 indicates that this is indeed the case over more than one order of magnitude variations of the frequency. From Fig. 12.11, one also obtains, $u \simeq 0.62 \pm 0.02$, reasonably close to the percolation prediction $u = \mu/(\mu + s) \simeq 0.73$. More extensive measurements for the same microemulsions were reported by van Dijk et al. (1986).

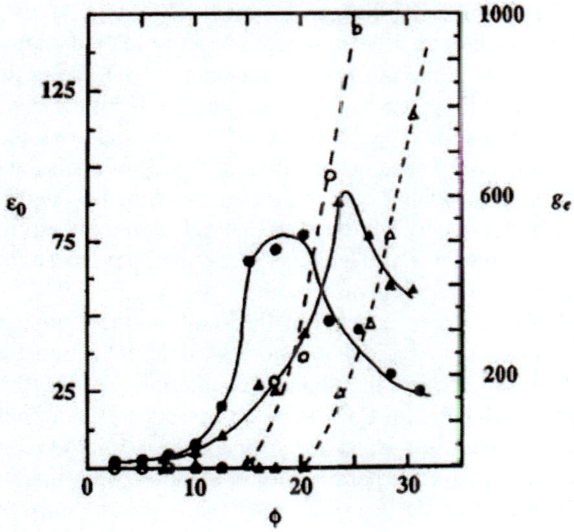

Fig. 12.11 Static dielectric constant ϵ_0 and DC conductivity g_e of two different microemulsions as functions of the water volume fraction ϕ. The data are for [H_2O/AOT] = 25 (circles) at 318 K, and for [H_2O/AOT] = 35 at 313 K (after van Dijk 1985)

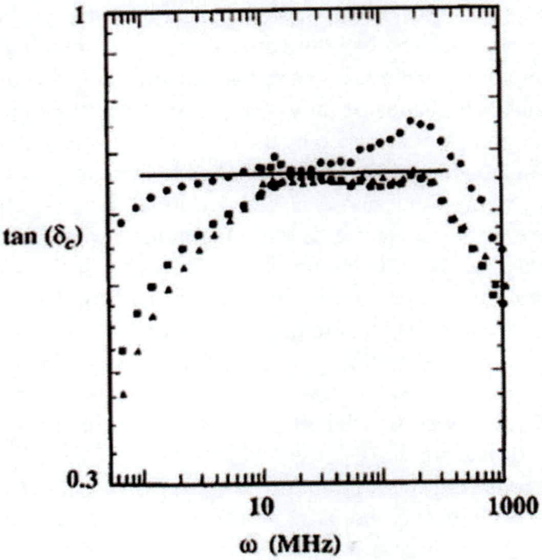

Fig. 12.12 Logarithmic plot of the loss angle δ versus frequency ω for three microemulsions. The horizontal line indicates, $\tan(\delta_c) = 0.67$ (after van Dijk 1985)

Moha-Ouchane et al. (1987), Clarkson and Smedley (1988), and Peyrelasse et al. (1988) all measured the AC conductivity and dielectric constant of the same microemulsion fluids as that of van Dijk et al. They found, however, that the static dielectric constant diverges at ϕ_{wc} with an exponent $s' \simeq 1.6$, significantly larger than $s \simeq 0.73$, predicted by Eq. (12.19). Grest et al. (1986) argued that, in microemulsions, one must take into account the effect of cluster diffusion that rearranges the mixture and dynamically changes its structure (for a good discussion of this important point see Bordi et al. 1996). Thus, they proposed a dynamic percolation model in which the percolation clusters diffuse randomly in the network. Grest et al. (1986) then suggested that the exponent s, characterizing the divergence of the static dielectric constant, should be replaced by $s' = 2\nu - \beta$. Note that, Eq. (12.19) also implies that s should be replaced with s'. Even $s' \simeq 1.35$ is in disagreement with the measurements of Moha-Ouchane et al., Clarkson and Smedley, and Peyrelasse et al.. However, the data reported by Ponton et al. (1991) for a ternary microemulsion composed of water, iso-octane, and AOT yielded, $s' \simeq 1.35$, in perfect agreement with the predictions of the dynamic percolation model.

The difference between the static and dynamic percolation models of dielectric constant of microemulsions was nicely demonstrated by Capuzzi et al. (1999). They exchanged the sodium counterion Na^+ of the sodium-based AOT surfactant with calcium Ca^{2+} in order investigate the effect of the counterion charge on the structure of water in normal decane microemulsions. The Ohmic conductivity and dielectric permittivity of samples were measured at a constant mole ratio, $[\text{water}]/[\text{Ca(AOT)}_2] = 26.6$. When the volume fraction ϕ of the dispersed phase was increased, a percolation transition was observed at a constant temperature of 25 °C at $\phi_c \approx 0.15$, obtained by studying the Ohmic conductivity and the static dielectric permittivity, as we all as by the frequency dependence of the complex permittivity. Capuzzi et al. also estimated the critical exponents μ and s, or s'. A comparison of the data for the two counterions, Ca^{2+} and Na^+, in AOT surfactant water in n-decane microemulsions revealed an important difference between the two. The percolation phenomenon below ϕ_c was dynamic for the Na^+-based microemulsions, giving rise to the formation of clusters of droplets, whereas the Ca^{2+}-based microemulsions exhibited signature of the static percolation.

12.6 Hall Conductivity

The Hall effect, first discovered by the American physicist Edwin Hall[4], is the result of a voltage difference across an electrical conductor, referred to as the Hall voltage, which is transverse to an electric current in the conductor and to an applied magnetic field perpendicular to the current. The resulting Hall conductivity provides an effective tool for way understanding electrical transport in a composite material. In the presence of the magnetic field, the conductivity of the material is described by a tensor $\mathbf{g}_e$ that has non-zero off-diagonal components, even when the material is isotropic. Some of the off-diagonal terms are symmetric and even in the magnetic field H, while others are antisymmetric and odd in H. If we choose H to be, for example, parallel to the $z-$direction, then $\mathbf{g}_e$ will have the following form:

$$\mathbf{g}_e = \begin{pmatrix} g & 0 & 0 \\ 0 & g & 0 \\ 0 & 0 & g \end{pmatrix} + \begin{pmatrix} 0 & g_h & 0 \\ -g_h & 0 & 0 \\ 0 & 0 & 0 \end{pmatrix} + \begin{pmatrix} \delta g_\perp & 0 & 0 \\ 0 & \delta g_\perp & 0 \\ 0 & 0 & \delta g_\parallel \end{pmatrix} \tag{12.23}$$

where g is the Ohmic conductivity (at $H = 0$), g_h is the Hall conductivity, and $\delta g_\perp$ and $\delta g_\parallel$ are the transverse and longitudinal magnetoconductivities, respectively. Strictly speaking, Eq. (12.23) is applicable only to an isotropic material, so that even crystals with cubic symmetry are excluded, for which the form of the antisymmetric part of the equation is also correct at low magnetic fields.

Suppose now that an electric field $\mathbf{E}$ is imposed on the material. Then, the material has an Ohmic conductivity g_e, and a Hall conductivity $\mathbf{g}_{eh}$. For an isotropic material and small H, $\mathbf{g}_{eh}$ is proportional to H and is defined by Kirchhoff's law relating the current density $\mathbf{I}$ and the electric field $\mathbf{E}$:

$$\mathbf{I} = g_e\mathbf{E} + \mathbf{E} \times \mathbf{g}_{eh} \ . \tag{12.24}$$

If Eq. (12.24) is inverted for small H, the Hall coefficient $R_H = g_{eh}/(Hg_e^2)$ appears along the Ohmic resistivity $R_e = 1/g_e$: $\mathbf{E} = R_e\mathbf{I} + R_H(\mathbf{H} \times \mathbf{I})$. Thus, the main goal

[4] **Edwin Herbert Hall** (1855–1938) was an American physicist and a Professor at Harvard University.

is to calculate the Hall conductivity and Hall coefficient in disordered composites. Volger (1950) and Juretchke et al. (1956) pioneered such calculations by treating various types of disorder in 3D composites. In particular, Juretchke et al. (1956) derived the exact general solution for the Hall effect at low H in isotropic 2D composite films. Their results were, however, forgotten for a long time, until they were rediscovered in the 1970s.

To make further progress, we restrict ourselves to low H, so that the three terms of the sum in Eq. (12.23) are proportional to successively higher powers of H, namely, H^0, H^1, and H^2. This allows one to solve for the electric field by a perturbation method, leading to the following expressions for g_{eh} and $\delta \mathbf{g}_e$:

$$g_{eh} = \frac{1}{\Omega}\int g_h(\mathbf{r})\frac{[\mathbf{E}^{(x)}\times\mathbf{E}^{(y)}]_z}{E_0^2}\,d\Omega\,, \tag{12.25}$$

$$(\delta g_e)_{ab} = \frac{\langle E^{(a)}\cdot\delta\mathbf{g}\cdot E^{(b)}\rangle}{E_0^2} + \frac{1}{\Omega E_0^2}\int d\Omega\int[\mathbf{g}_h(\mathbf{r})\times\mathbf{E}^{(a)}(\mathbf{r})]\cdot\nabla\nabla' G(\mathbf{r},\mathbf{r}'|g)\cdot[\mathbf{g}(\mathbf{r}')\times\mathbf{E}^{(b)}(\mathbf{r}')]\,d\Omega'\,. \tag{12.26}$$

Here, Ω is the volume of the material, $\mathbf{E}^{(a)}(\mathbf{r})$ is the local electric field in the Ohmic problem (i.e., when $H=0$) in the composite when the average of applied field $\mathbf{E}_0$ lies along the $a-$axis, and G is the Green function for the same problem, which is the solution of the following boundary value problem:

$$\nabla\cdot[g(\mathbf{r})\nabla G(\mathbf{r},\mathbf{r}'|g)] = -\delta(\mathbf{r}-\mathbf{r}')]\,,\quad G=0 \text{ on the boundaries}\,. \tag{12.27}$$

Although one is able to write down Eqs. (12.25)–(12.27) formally, in practice neither the Green function G nor the local fields $\mathbf{E}^{(a)}(\mathbf{r})$ are straightforward to derive for any but the most trivial systems. Nevertheless, these expressions are useful for a variety of reasons.

12.6.1 Poor Man's Percolation: Effective-Medium Approximation

Stachowiak (1970) carried out a mean-field-type computation of the effective conductivity of a composite material in the presence of a magnetic field. Cohen and Jortner (1973) developed an EMA for the problem of Hall conductivity of disordered materials, an extension of the EMA described earlier in this book. The goal is to compute the effective Hall conductivity $\mathbf{g}_{eh}$ by an EMA, because we already know how to estimate the Ohmic conductivity g_e by the EMA. Thus, Cohen and Jortner considered a composite material in which any portion of it had either a magnetoconductivity $\mathbf{g}_{1h}(\mathbf{H})$ or $\mathbf{g}_{2h}(\mathbf{H})$ with probabilities p and and $1-p$, respectively.

Similarly, the Ohmic conductivity ($H = 0$) of any portion of the material took on values g_1 and g_2 with probabilities p and $1 - p$, respectively. The EMA equation that Cohen and Jortner derived is given by

$$\left\langle [g_d\mathbf{1} - \frac{1}{3}(\mathbf{g} - \mathbf{g}_i)]^{-1} \cdot (\mathbf{g}_i - \mathbf{g}) \right\rangle = 0 , \tag{12.28}$$

where the averaging is over the distribution of the local Ohmic and hall conductivities. If Eq. (12.28) is solved to the first order in the field H for $\mathbf{g}_h$, one obtains

$$\mathbf{g}_h = \frac{\langle \mathbf{g}_{ih}/(g_i + 2g)^2 \rangle}{\langle (g_i + 2g)^{-2} \rangle} , \tag{12.29}$$

Stroud (1980) extended the EMA of Cohen and Jortner to composite materials that were composed of normal and superconducting components. He ignored quantum effects and, in order to take into account the possibility of the material being anisotropic, considered ellipsoidal inclusions, rather than the spherical ones considered by Cohen and Jortner. In his model, the (volume) fraction of the normal conductor was $1 - p$ with a conductivity tensor $\mathbf{g}_m(B)$, where B is the magnitude of the local magnetic induction. The components of the tensor are given by

$$g_m^{xx} = g_m^{yy} = \frac{g_0}{1 + \tilde{H}^2} , \quad g_m^{zz} = 0 , \tag{12.30}$$

$$g_m^{xy} = -g_m^{yx} = \frac{g_0 \tilde{H}}{1 + \tilde{H}^2} , \tag{12.31}$$

where $\tilde{H}$ is the dimensionless field, and g_0 is a constant conductance. The remaining components of the tensor are zero. The (volume) fraction of the superconducting component of the material is p. The composite is characterized by an effective conductivity tensor $\mathbf{g}_e$ whose components are predicted by the EMA to be given by

$$\begin{aligned} g_e^{xx} + p\Gamma_{xx}^{-1} &= g_m^{xx} , \\ g_e^{zz} + p\Gamma_{zz}^{-1} &= g_m^{zz} , \\ g_e^{xy} = -g_e^{yx} &= g_m^{xy} , \end{aligned} \tag{12.32}$$

with $g_e^{yy} = g_e^{xx}$, where

$$\Gamma_{xx} = \frac{1}{2g_e^{zz}\mathcal{R}} \left[1 - \frac{1}{\sqrt{\mathcal{R}(1 - \mathcal{R})}} \sin^{-1} \sqrt{\mathcal{R}} \right] , \tag{12.33}$$

$$\Gamma_{zz} = -\frac{1}{g_e^{zz}\mathcal{R}} \left[1 - \sqrt{\frac{1 - \mathcal{R}}{\mathcal{R}}} \sin^{-1} \sqrt{\mathcal{R}} \right] , \tag{12.34}$$

where $\mathcal{R} = 1 - g_e^{xx}/g_e^{zz}$ is the anisotropy. Equations (12.32) can be derived in a straightforward manner from the EMA for anisotropic materials, developed by Stroud (1975). It can be shown that the result $g_e^{xy} = g_m^{xy}$ is in fact *exact*.

Interestingly enough, Eqs. (12.32) reveal that, in the presence of a strong magnetic field, addition of a superconducting component to a normal conductor *increases* the composite's resistivity. This effect arises from current distortions near the superconducting inclusions that cause extra dissipation in the normal conductor, hence increasing its resistivity, and is in qualitative agreement with the experimental data. Moreover, according to the EMA, at a fixed value of p, the (volume) fraction of the superconductors, the resistance of the material linearly rises at first with the field, with the slope being proportional to p, and then saturates at large fields. If we define a Hall coefficient $R_H(p, \tilde{H}) = 1/(g_e^{xx}\tilde{H})$ (which is slightly different from what was introduced earlier), then according to the EMA,

$$R_H(p,0) = \lim_{\tilde{H}\to 0}\{\tilde{H}^{-1}g_e^{xy}(p,\tilde{H})/[g_e^{xx}(p,\tilde{H})]^2\} = \frac{R_H(0,0)}{p_c^2}(p_c - p)^2 , \quad (12.35)$$

where p_c is the EMA's prediction for the percolation threshold of the system. That R_H vanishes at p_c (of the superconducting component) is expected, since for $p > p_c$ the composite cannot support a potential difference across itself. In strong fields,

$$R_H \sim \frac{g_e^{xy}}{(g_e^{xx})^2}\tilde{H}^{-1} \sim \tilde{H}^{-2} , \quad (12.36)$$

since g_e^{xx} saturates and g_e^{xy} decreases as $\tilde{H}^{-1}$. Moreover, $R_H(p, 0)$ satisfies the *exact* relation

$$R_H(p,0)g_e(p)^2 = \lim_{\tilde{H}\to 0}\frac{g_m^{xy}}{\tilde{H}} = \text{constant} , \quad (12.37)$$

where $g_e(p)$ is the effective conductivity of the composite in zero field. Since, $g_e(p) \sim (p_c - p)^{-s}$, where s is the superconductivity exponent defined and discussed earlier in this chapter and, in Chap. 2, we obtain

$$R_H \sim (p_c - p)^{2s} , \quad (12.38)$$

which is not only valid in any dimension but also consistent with the EMA prediction, Eq. (12.32), since the EMA predicts that $s = 1$.

Unlike Ohmic conductivity g_e, numerical simulation of the Hall conductivity based on a network model and calculating g_{eh} and R_H are quite complex, since the general circuit element of a bond of the network cannot be a simple resistor, but must be a conductance *matrix*. Bergman et al. (1990) proposed a network model for calculating g_{eh} and R_H that satisfies all the necessary requirements for being an appropriate model.

12.6.2 Scaling Properties

A complete scaling theory of Hall conductivity was developed by Bergman and Stroud (1985). Consider again the two-component network in which each bond has good Ohmic and Hall conductivities g_1 and g_{1h} with probability p, and poor Ohmic and Hall conductivities g_2 and g_{2h} with probability $1 - p$. Bergman and Stroud (1985) proposed that for $g_{2h}/g_{1h} \ll 1$ and for p close to p_c

$$\frac{g_{eh} - g_{2h}}{g_{1h} - g_{2h}} = |p - p_c|^{\tau} F\left(\frac{g_{2h}/g_{1h}}{|p - p_c|^{\mu+s}}\right) , \tag{12.39}$$

where μ and s are the usual conductivity exponents, and τ is a new critical exponent characterizing the power-law behavior of g_{eh} near p_c when $p > p_c$, $g_2 = 0$, and $g_{2h} = 0$, i.e.,

$$g_{eh} \sim (p - p_c)^{\tau} . \tag{12.40}$$

Equation (12.40) is analogous to Eq. (2.23). The scaling function $F(x)$ has the properties that,

$$F(x) \sim \begin{cases} \text{constant if } x \ll 1 \text{ and } p > p_c, & \text{regime I} , \\ x^2 \quad \text{if } x \ll 1 \text{ and } p < p_c, & \text{regime II} , \\ x^{\tau/(\mu-s)} \text{ if } x \gg 1 \text{ and } p \simeq p_c, & \text{regime III} . \end{cases} \tag{12.41}$$

Another way of understanding Eq. (12.40) is in terms of the Hall coefficient R_H. To each bond of a network, we assign a Hall coefficient $R_{1h} = g_{1h}/g_1^2$ with probability p or $R_{2h} = g_{2h}/g_2^2$ with probability $1 - p$. Then, according to the scaling theory of Bergman and Stroud (1985), one has

$$R_H \sim \begin{cases} a_1 R_{1h}|p - p_c|^{-\psi} + b_1 R_{2h}(g_2/g_1)^2|p - p_c|^{-2\mu} & \text{regime I} , \\ a_2 R_{1h}|p - p_c|^{-\psi} + b_2 R_{2h}|p - p_c|^{2s} & \text{regime II} , \\ a_3 R_{1h}(g_2/g_1)^{-\psi/(\mu+s)} + b_3 R_{2h}(g_2/g_1)^{2s/(\mu+s)} & \text{regime III} , \end{cases} \tag{12.42}$$

where a_i and b_i are the constants, and ψ is a critical exponent that characterizes the power-law behavior of R_H as p_c is approached from below, and is related to μ and τ by

$$\psi = 2\mu - \tau . \tag{12.43}$$

As pointed out by Bergman and Stroud (1985), the important point to remember is that in regime I the ratio of the second and the first terms is on the order of $(g_{2h}/g_{1h})|p - p_c|^{-\tau}$ and, therefore, depending on the morphology of the material, either term may dominate, so that the experimental verification of these scaling laws is not straightforward.

12.6.3 Comparison with Experimental Data

Early experimental studies of Hall conductivity and Hall coefficient in a material with percolation disorder were carried out by Levinshtein et al. (1975). Their data confirmed the divergence of R_H as p_c is approached. Palevski et al. (1984) showed that the Hall coefficient does indeed remain finite in a 2D (thin) metal film. Dai et al. (1987) studied the same problem in a 3D Al–Ge metal–insulator composite. In regime I, the poor conductor was Al-doped-Ge that was found to dominate the Hall effect. To make the first term of Eq. (12.42) dominant in regime I, Dai et al. dissolved the metallic aluminum in KOH, which left the doped Ge as the good conductor, the poor conductor role now being played by the vacuum. Then, the first term of Eqs. (12.42) in regime I is dominant and, therefore, the critical exponents ψ and τ can be directly measured.

Figure 12.13 presents the measured Hall coefficient R_H versus $p - p_c$, where p is the volume fraction of Al, and $p_c \simeq 0.15$ is the percolation threshold. The data yielded, $\tau \simeq 3.8 \pm 0.2$ and $\psi \simeq 0.38 \pm 0.05$, which should be compared with accurate simulation results based on percolation networks, $\tau \simeq 3.56 \pm 0.06$ and $\psi \simeq 0.49 \pm 0.06$. Thus, theory and experiment are in general agreement with each other. Since the EMA of Cohen and Jortner (1973) and of Stroud (1980) can be used for accurately estimating g_{eh} and R_H away from p_c, we have a fairly complete theory of Hall effect in composite materials.

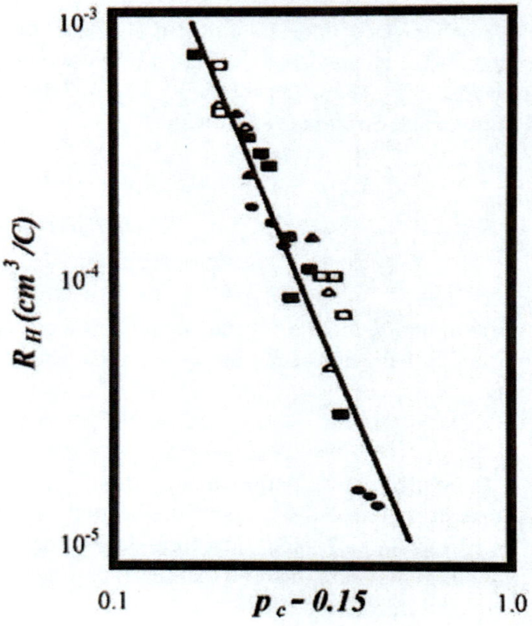

Fig. 12.13 Logarithmic plot of the effective Hall coefficient R_H versus $p - 0.15$, where p is the volume fraction of Al. Different symbols refer to various substrates (after Dai et al. 1987)

Rohde and Micklitz (1987) reported on the measurement of the Hall conductivity of the metallic side of the metal–insulator transition in a percolating system as a function of the metal volume fraction p. Their data indicated that the Hall coefficient R_H of granular Sn:Ar mixtures diverges according to power law (12.42) with the critical volume fraction, $p \approx 0.26 \pm 0.01$ and the critical exponent, $\psi \approx 0.49 \pm 0.07$, which is in agreement with the prediction of percolation theory.

12.7 Mechanical Properties of Heterogeneous Materials

So far, we have described the application of percolation theory to scalar transport properties of disordered materials, namely, electrical conductivity and the dielectric constant. But, as Chap. 11 made it clear, vector transport, i.e., transport of stress, strain, or force in disordered materials and media gives rise to a new model of percolation whose properties, especially near the percolation threshold, are distinct from those of the conductivity. In this section, we describe applications of percolation theory to modeling of elastic properties of heterogeneous media, including porous materials. Chapter 13 will describe the applications to polymers, gels, and glasses and, in particular, their elastic and rheological properties.

12.7.1 Foams

Foams are a special type of material that consist of a random distribution of gas bubbles in a much smaller volume of liquid (see, for example, Weaire and Hutzler 1999). If the mixture contains enough stabilizing surfactants, then the morphology of the material will be essentially stable and constant over time scales that can be as long as hours. Under these conditions, and due to their practical importance, understanding the rheology and mechanical properties of foams are important issues to be addressed. Consider, for example, the response of foams to an applied force. Although foams are mostly made of gas bubbles, they can support deformation and shear forces, much like a solid material, which is due to the increase in the gas–liquid interface surface area, a result of deforming the tightly packed gas bubbles and the corresponding energy cost by the surface tension. If the applied force is small, then the mechanical response of foams is linear, and their shear modulus is simply the stress per unit strain, regardless of whether the experiment is carried out under controlled stress or controlled strain conditions. With increasing the applied stress the response of the system crosses over to a non-linear regime and becomes irreversible as topological changes are induced under which a few bubbles in a finite region suddenly change their neighbors. At still higher applied stresses, the rearrangement of the bubbles becomes more frequent, but the resulting strain remains finite.

Eventually, a critical yield stress is reached at which the foam begins to flow indefinitely at a non-zero strain rate by a never-ending series of neighbor-switching rearrangements. If the strain rate is low, the rearrangements are discrete avalanche-like events, but if the strain rate is high, the deformation is continuous and resembles a viscous liquid. Clearly, the transition from a non-flowing system to a flowing one is a non-linear phenomenon, characterized by the yield stress threshold (Sahimi 1993a). In this section, we are mainly interested in the mechanical response of foams before they begin to flow, and the relation between this phenomenon and vector percolation.

Durian (1997) developed a realistic model of foams in order to study their deformation and mechanical response. To understand his model, consider the first two gas bubbles in a sea of background liquid that are brought into contact with each other. Since the ranges of van der Waals force (which is attractive and originates from the dielectric mismatch between the gas and the liquid) and of the electric double-layer force (which is repulsive and is due to the adsorbed surfactants) are typically small—less than 100 nm—compared with a typical size of gas bubbles which is about 20 μm, the two bubbles experience no force until they actually touch each other. When pushed into each other, the shape of the two bubbles is distorted, and the region of their contact flattens out into a soap film with a thickness that depends on the combination of the applied, van der Waals, and double-layer forces. The formation of the soap film and the increase in the total surface area give rise to a mutually repulsive force that is proportional to the liquid–gas surface tension, and is nearly harmonic, i.e., it is of central-force type. Thus, Durian assumed this force to be harmonic. Then, the effective central force for bubble i is proportional to σ/R_i, where σ is the surface tension at the liquid–gas interface, and R_i is the bubble's radius. For two mutually repulsing bubbles with their centers at $\mathbf{r}_i$ and $\mathbf{r}_j$, one has $|\mathbf{r}_i - \mathbf{r}_j| < R_i + R_j$, and the individual springs are added in series, so that the effective spring constant is $F_0/(R_i + R_j)$, where F_0 is the force constant that is of the order of, $F_0 \simeq \sigma\langle R\rangle$, with $\langle R\rangle$ being the average bubble size. Therefore, the repulsive force $\mathbf{F}^r_{ij}$ acting on the center of bubble i due to bubble j is given by

$$\mathbf{F}^r_{ij} = F_0\left[\frac{1}{|\mathbf{r}_i - \mathbf{r}_j|} - \frac{1}{R_i + R_j}\right](\mathbf{r}_i - \mathbf{r}_j) \, . \tag{12.44}$$

One must also take into account the dissipation effect, since when a foam is strained at a non-zero rate, energy is dissipated due to shear flow of the viscous liquid within the soap film by flow within the adsorbed surfactant films, as well as other mechanisms. The drag force generated by the dissipation is then assumed to be given by

$$\mathbf{F}^d_{ij} = -b(\mathbf{v}_i - \mathbf{v}_j) \, , \tag{12.45}$$

where b is a proportionality constant, assumed to be the same for all the bubbles. Thus, writing a force balance for (the center of) each bubble and setting the result to zero, one obtains

$$\mathbf{v}_i = \langle \mathbf{v}_j \rangle + \frac{F_0}{b} \sum_j \left[\frac{1}{|\mathbf{r}_i - \mathbf{r}_j|} - \frac{1}{R_i + R_j} \right] (\mathbf{r}_i - \mathbf{r}_j) + \frac{1}{b} \mathbf{F}_i^a , \tag{12.46}$$

where $\mathbf{F}_i^a$ is the applied force. The only relevant time scale in the system is $t_d = b\langle R \rangle / F_0$, which is set by the average bubble size and the competition between dissipation and storage of energy; the numerical values of b and F_0 are irrelevant.

Bolton and Weaire (1990) were in fact the first to point out the relevance of vector percolation to modeling mechanical properties of foams. To simulate the response of foams, one places the bubbles on a grid, e.g., a square lattice, and solves the equation of motion—the coupled force-balance equations for all the bubbles. Durian (1997) carried out such simulations, and provided their technical details. His simulations indicated that He has a broad bubble size distribution that depends on the gas volume faction. Figure 12.14 presents the shear modulus of the system as a function of the gas volume fraction. The general trends in the figure are, on one hand, entirely similar to central-force percolation and, on the other hand, to the experimental data of Mason et al. (1995). The percolation threshold, i.e., the critical gas volume fraction at which the shear modulus vanishes, depends weakly on the broadness of the bubble size distribution, but is around a value of 0.84, which is identical with the dense random packing of hard disks.

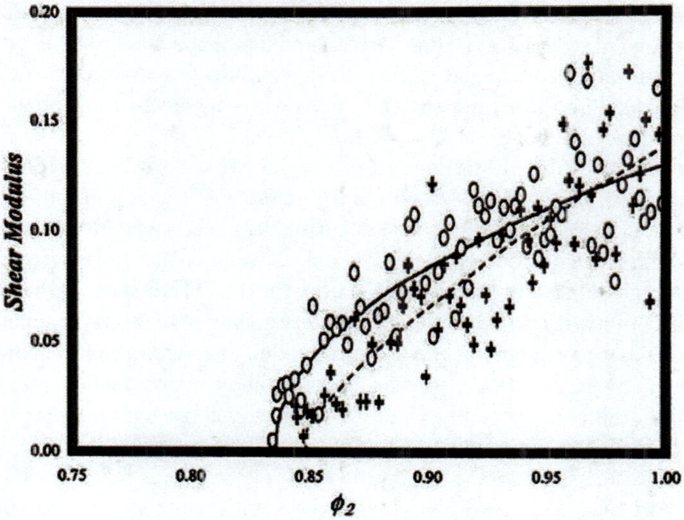

Fig. 12.14 Shear modulus of foam versus the gas fraction ϕ_2, which vanishes at the rigidity percolation threshold. Circles and + denote, respectively, the data for uniform bubble-size distributions with widths of 0.75 and 0.1 (after Durian 1997)

12.7.2 Porous Materials

As usual, we focus mostly on the behavior of materials in the vicinity of their percolation threshold, and in particular the power laws that govern the elastic moduli, since the critical exponents that characterize such power laws are largely independent of many details of the materials' morphology, whereas the numerical value of the elastic properties and also the percolation thresholds of materials at which an elastic modulus vanishes (or diverges) are non-universal and vary greatly from one material to another. Indeed, the agreement between the measured critical exponents and the predictions of vector percolation models is the most stringent test of the applicability of such models.

As already mentioned, Deptuck et al. (1985) measured Young's modulus of a sintered, submicrometer silver powder, commonly used for millikelvin and submillikelvin cryostats. The sinter remains elastic and percolating even if the volume fraction ϕ of the silver particles is as low as 0.1. They also used submicrometer copper oxide–silver powder, routinely used in heat exchangers for optimizing heat transfer to dilute ^{3}He–^{4}He mixtures. In these composites, the silver component acts as the percolating phase with a percolation threshold ϕ_c less than 0.1. Measurements of Deptuck et al. (1985) indicated that, over a broad range of the volume fraction ϕ, Young's modulus of the powder follows the power law, $Y \sim (\phi - \phi_c)^f$ with $f \simeq 3.8$, in excellent agreement with the critical exponent f of the 3D bond-bending model (see Table 11.1). Moreover, their measurements indicated that the critical region in which this power law is valid is quite broad, and hence the power law is quite useful even for quantitative estimation of the elastic moduli. Other experimental data on sintered powder samples of copper and silver were presented by Maliepaard et al. (1985), who reported that $f \simeq 3.6$.

Measurements of the elastic moduli of 2D porous composites were also carried out by Benguigui (1984, 1986), Lobb and Forrster (1987), Sofo et al. (1987), and Craciun et al. (1998). Benguigui's first experiments (1984) were done with sheets of metal in which holes had been punched; his measured exponent was, $f \simeq 3.5 \pm 0.5$, which, although consistent with the value of f for the 2D BB models, was criticized because his composite represented a *continuum* equivalent to the Swiss-cheese model described in Chap. 3, as opposed to a lattice model, and, according to Halperin et al. (1985) and Feng et al. (1987), the critical exponent f for such continua should be larger than the corresponding value for lattice systems by about 3/2 (see Chap. 3). More careful measurements by Benguigui (1986) then yielded $f \simeq 5.0 \pm 0.5$, consistent with the prediction of Halperin et al. (1985), $f \simeq 5.5$. Measurements of Lobb and Forrester (1987) were also carried out with metal sheets into which holes had been punched. They reported $f \simeq 4.95 \pm 1.1$, again consistent with Halperin et al.'s prediction. Sofo et al. (1987) used randomly holed metalized Mylar and reported, $f \simeq 5.3 \pm 0.7$, in even better agreement with the prediction of Halperin et al. (1985).

On the other hand, Craciun et al. (1998) measured the elastic moduli of porous ceramics, an important class of materials that have wide applications. Their estimated exponent of the elastic moduli was, $f \simeq 2.0$, which they attributed it to the predominance of central forces in their composites.

Another interesting set of experiments was carried out by Allen et al. (1988) who measured dynamical (frequency-dependent) shear modulus of a model 2D percolation lattice, which can be thought of as the lattice analogue of a porous material, with the cut bonds representing the pore space of the material. Their measurements were made by exciting small torsional linear oscillations on a cylindrical screen as links were cut to approach p_c. Their measurements yielded $f \simeq 3.6 \pm 0.2$, relatively close to the corresponding exponent of the 2D BB model, $f \simeq 3.96$.

An interesting study of mechanical properties of two types of compressed expanded graphite was reported by Celzard et al. (2001). Such materials are prepared from natural graphite platelets that are inserted into sulfuric acids, which results in an intercalated compound. The resulting material is then submitted to a thermal shock by passing it through a flame which induces swelling of the platelets. The samples are then compressed by uniaxial compaction. Celzard et al. (2001) measured the electrical conductivity and elastic moduli of two samples of these materials. Figure 12.15 presents their data in logarithmic scales, as functions of the apparent density of the

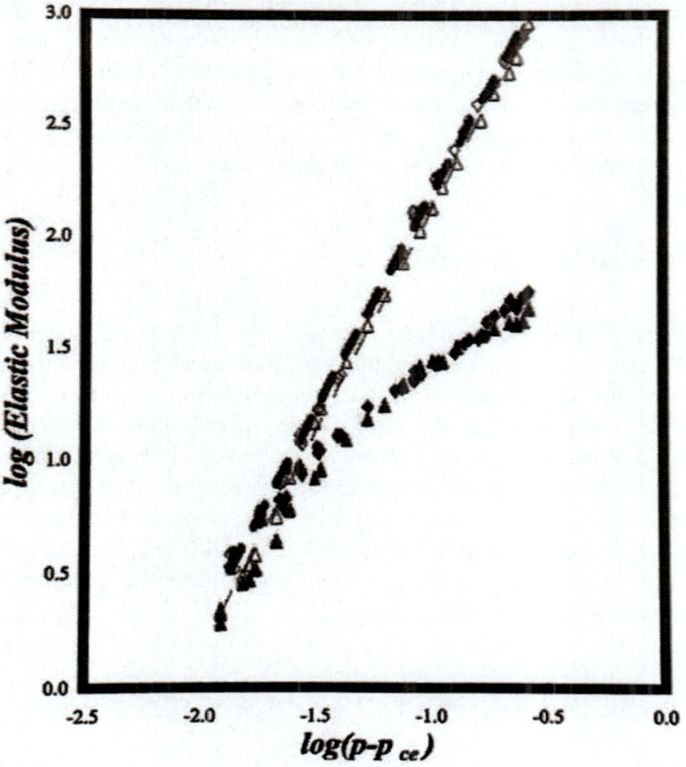

Fig. 12.15 Elastic moduli of four cubic samples of expanded graphite. The lines indicate the fit of the data to the percolation power law (after Celzard et al. 2001; courtesy of Dr. Alain Celzard)

conducting material. The electrical conductivity and the elastic moduli of the samples vanish at different percolation thresholds. From Eqs. (11.46) and (11.47) one finds that

$$\frac{(p_c^B)_{kk}}{p_c^B} = \frac{d^2 - 1}{2d - 1} ,$$

where $(p_c^B)_{kk}$ is the percolation threshold of the BB Kirkwood-Keating model at which the elastic moduli vanish, p_c^B is the bond threshold of the scalar percolation at which the electrical conductivity vanishes, and d is the dimensionality of the sample material. Thus, for $d = 3$, one obtains $(p_c^B)_{kk}/p_c^B = 8/5$, which was indeed confirmed by the experiments of Celzard et al. (2001), hence confirming the validity of vector percolation for predicting the elastic properties of composite materials, even as far as their percolation threshold for various transport properties is concerned.

On the other hand, Amaral-Labat et al. (2013) prepared rigid foam using a thermoset resin, which was gradually converted into reticulated vitreous carbon foam by pyrolysis at increasingly higher heat-treatment temperatures (HHT). The corresponding changes in the Young's modulus Y and the compressive strength σ of the materials were measured over a wide range of porosities. The materials exhibited a percolation transition with a *zero* percolation threshold. At very low densities, Young's modulus and the compressive strength appear to follow the power laws predicted by percolation theory near the percolation threshold. But, whereas the exponent f associated with the power-law behavior of Y significantly varied with the material's density and the HHT, the exponent associated with σ does not change much.

12.7.3 Superrigid Materials

Benguigui and Ron (1993) constructed a composite of a very soft material—in their case a gel—and a very rigid material that was alumina and zirconia, and measured Young's modulus of the material, as the volume fraction of the rigid component approached ϕ_c, the percolation threshold, or the critical volume fraction, for formation of a SSC of the rigid material. Their measurements indicated that the Young's modulus Y diverges at ϕ_c, and that, near the percolation threshold, it follows the power law, $Y \sim (\phi_c - \phi)^{-\chi}$, with $\chi \simeq 0.66 \pm 0.04$, in excellent agreement with the value of superelasticity exponent for the 3D BB model (see Table 11.1). Thus, there remains little, if any, doubt that the linear mechanical properties of disordered solids are described by vector percolation models.

12.8 Percolation Aspects of Classical Superconductivity

Let us now briefly describe what may be called the classical properties of superconducting composites, since in discussing such properties the quantum mechanical effects are ignored. Thus, we do not discuss the quantum properties that depend on the continuity of the superconducting wave function and its behavior in a magnetic field. If the quantum mechanical effects are ignored, then, percolation theory explains several aspects of the observed properties of superconductors. Of course, once we speak of percolation, we mean some type of disorder, the interplay of which with superconductivity is an intriguing problem. This is because disorder enhances the electrical resistance of a material, whereas superconductivity implies a zero-resistance state.

In general, we may divide superconductors into two groups, namely, granular and homogeneous. Experiments have shown that homogeneous samples exhibit sharp superconducting transitions and the crossover between insulating and superconducting behavior occurs at a resistance that seemingly takes on a universal value of 6.5 kΩ (Goldman and Markovic 1998). But, for granular superconductors—those that are composed of grains separated by thin insulating particles that are called the *Josephson junctions*[5] and are characterized by very broad tails in the temperature-dependent resistance $R(T)$—the transition between the insulating and superconducting phases appear to be not so universal (see, for example, Dynes et al. 1978; Jaeger et al. 1989; Frydman et al. 2002; Khan et al. 2000; Bose et al. 2005). Although, as Anderson (1959)[6] predicted, superconductivity can persist even in the presence of disorder (what he referred to as *dirty superconductors*), experiments on thin films (Goldman and Markovic 1998) demonstrated a transition from a superconducting to an insulating state with an increasing disorder or a magnetic field. Using extensive numerical simulations, Dubi et al. (2007) demonstrated that, for weak disorder, or high electron density, increasing the magnetic field results in the eventual vanishing of the amplitude of the superconducting order parameter, leading to an insulating state. At lower electron densities or higher disorder, however, increasing the magnetic field suppresses the correlations between the phases of the superconducting order parameter in various islands or clusters, leading to a distinct type of superconductor–insulator transition. In addition, they showed that in the high disorder regime, there are still superconducting clusters in the sample, even on the insulating side of the transition, which is consistent with experiments (Kowal and Ovadyahu 1994; Crane et al. 2007).

For the role that the percolation process plays in the emergence of superconductivity in cuprates[7] see Pelc et al. (2018). In this section, we consider percolation properties of granular superconductors, which are characterized by two parameters,

---

[5] Named after **Brian David Josephson** (1940–) who won the Nobel Prize in Physics in 1973, a Josephson junction consists of two or more superconductors coupled by a weak link that can be a thin insulating barrier, a short section of non-superconducting metal, or a physical constriction that weakens the superconductivity at the point of contact, and has many important applications in quantum–mechanical circuits.

[6] **Philip Warren Anderson** (1923–2020) was a theoretical physicist who made fundamental contributions to the theories of localization, antiferromagnetism, symmetry breaking, and high-temperature superconductivity. He also proposed the name for a field of physics now known as condensed matter physics. He received numerous awards, including the Nobel Prize in Physics in 1977.

[7] Cuprate superconductors are a family of high-temperature materials that are made of layers of copper oxides, CuO_2, alternating with layers of other metal oxides that act as charge reservoirs. They are, at ambient pressure, one of the highest temperature superconductors known, first discovered by Bednorz and Müller (1986); see Chap. 1.

the size of their grains, and the energy barrier between the grains. The grain-size distribution is measured by electron microscopy, while the properties of the barriers are deduced from measurements of normal-state (non-superconducting) resistivity. These two parameters determine the Josephson energy coupling E_j. Garland (1989) reviwed many granular properties of superconducting materials, while Zeimetz et al. (2002) reviewed applications of percolation theory to granular superconductors.

Granular superconductors have been modeled by 2D random conductance networks. For example, Strelniker et al. (2007) utilized such 2D networks in which the resistance between two grains was governed either by Josephson junction coupling, or by quasiparticle tunneling. They showed that their model can explain the experimental results over a wide range of temperatures and resistances. In addition, they also developed an EMA for the problem.

12.8.1 Magnetoconductivity

As usual, let us consider a disordered binary composite in d dimensions ($d = 2$ or 3), described by a conductivity field $g(\mathbf{r})$ that can take on two values, $g_1 = \infty$ with probability p and a finite value g_2 with probability $1 - p$. The effective conductivity $g_e(p)$ must therefore diverge when $p > p_c$, the percolation threshold for the superconducting component. For $p < p_c$, g_e is finite, with an expected asymptotic behavior given by Eq. (2.37) and characterized by the exponent s that depends on both the dimensionality and the microstructure of the composite; its values for 2D and 3D materials are listed in Table 2.3. In the Swiss-cheese model (see Chap. 3) that consists of a random distribution of overlapping $d-$dimensional spheres of normal conductors in a superconducting matrix, or in the inverted Swiss-cheese model (Chap. 3) in which the roles of the matrix and the spheres are reversed, the numerical value of s may be different from those associated with the lattice models, listed in Table 2.3.

Apart from an experimental study by Deutscher and Rappaport (1979), however, there appears to be little experimental evidence confirming this picture. In both 2D and 3D, an experimental test of the prediction would be complicated by the proximity effect, which causes the superconducting region to grow a distance ξ_s into the surrounding normal region, where ξ_s, is the temperature-dependent normal-conductor coherence length.

A better realization of such classical effects may be obtained upon the application of a magnetic field. How does a magnetic field affect the behavior of a granular superconductor? There are two types of homogeneous superconductors that behave quite differently in an external magnetic field.

(i) *Type*-I superconductors remain superconducting and expel the magnetic field up to a critical field H_c, since below H_c the flux is excluded from the superconductor (except for a thin layer). They then abruptly cross over to the normal state.

(ii) *Type*-II superconductors allow the magnetic field to penetrate into the system if H is greater than a lower critical field H_{c1}, but lose their superconductivity at an *upper critical field* H_{c2}. Between H_{c1} and H_{c2}, the flux penetrates partially and inhomogeneously, typically forming a lattice of quantized flux lines. Such a field fully penetrates the superconducting composite, and, to a first-order approximation, the magnetic induction B can be assumed to be uniform and equal to the applied H in the normal component. In a magnetic field, the conductivity g_N of the normal component becomes a tensor, as does g_e. Among the fields satisfying this condition, there are two ranges to consider.

(i) At sufficiently low fields, the off-diagonal elements of both g_e and g_N are linear in H, while the diagonal elements are unchanged to first order in H. Under such conditions, it can be shown that $g_{e,ij} = g_{N,ij}$ for $i \neq j$, leading to the interesting prediction that, near p_c, the Hall coefficient of the composite below the percolation threshold for superconductivity follows the relation:

$$R_H(p) \equiv \frac{g_{e,12}}{H g_{e,11}^2} \sim R_H(1)(p_c - p)^{2s} . \tag{12.47}$$

This prediction also apparently remains to be tested in 2D or 3D.

(ii) At higher fields, one expects both a transverse and a longitudinal magnetoresistance in the composite when a superconducting material is added, even if the pure normal conductor has no magnetoresistance because, as discussed above, the superconducting inclusions distort the nearby current lines, leading to increased dissipation. This conclusion was confirmed by measurements of Resnick et al. (1979) on a composite of Pb in an Al matrix and, as discussed above, was also confirmed by the EMA calculations of Stroud (1980).

12.8.2 Magnetic Properties

The starting point for studying the magnetic properties of superconductors is the London penetration depth of a composite that contains a volume fraction p of a superconductor and $1 - p$ of a normal conductor. In the London[8] theory of superconductivity, the basic electrodynamic properties are described by two macroscopic equations:

$$\mathbf{E} = \frac{\partial}{\partial t}\left(\frac{m}{n_s q^2}\mathbf{I}_s\right) , \tag{12.48}$$

$$\mathbf{B} = -c\nabla \times \left(\frac{m}{n_s q^2}\mathbf{I}_s\right) . \tag{12.49}$$

[8] The brothers **Fritz Wolfgang London** (1900–1954) and **Heinz London** (1907–1970) made fundamental contributions to understanding electromagnetic properties of superconductors. Fritz, who was a professor at Duke University, also made seminal contributions to the theories of chemical bonding and of intermolecular forces, including London dispersion forces. Heinz taught at the University of Bristol and Oxford University.

Equation (12.48) simply describes the undamped response of n_s particles per unit volume, each of mass m and charge q, to the force generated by an electric field $\mathbf{E}$. If we combine Eq. (12.48) with the Maxwell equation

$$\nabla \times \mathbf{B} = \frac{4\pi}{c}\,\mathbf{I}, \tag{12.50}$$

we obtain

$$\left(\nabla^2 - \frac{1}{\Lambda^2}\right)\mathbf{B} = 0\,, \tag{12.51}$$

where

$$\Lambda = \sqrt{\frac{mc^2}{4\pi n_s q^2}} \tag{12.52}$$

is the *London penetration depth* of the superconductor. Physically, Eq. (12.51) implies that the magnetic field decays exponentially within the superconductor with a decay length Λ.

To obtain an analogous quantity in a normal conductor–superconductor composite, we write Eq. (12.48) in the frequency domain as

$$\mathbf{I}_S = g_S\mathbf{E}\,, \tag{12.53}$$

$$g_S = \frac{iA}{\omega}\,, \tag{12.54}$$

where $A = c^2/(4\pi\Lambda^2)$, and g_S is clearly an imaginary, frequency-dependent (inductive) conductivity of the superconducting component.

12.8.3 Comparison with Experimental Data

Let us now consider some percolation properties of superconducting materials that have been tested either by experimental observations or by numerical simulations.

12.8.3.1 The London Penetration Depth

For $p > p_c$, where p_c is the percolation threshold of the superconducting component, the conductivity of the composite at sufficiently low frequencies is dominated by g_S. Therefore, we deduce that

$$g_e \equiv \frac{iA_e}{\omega} \sim g_S(p - p_c)^{\mu}\,, \tag{12.55}$$

which implies that $A_e \sim A(p - p_c)^{\mu}$, or

$$\Lambda_e \sim (p - p_c)^{-\mu/2} . \tag{12.56}$$

Experimental test of Eq. (12.56) would require making a series of samples with various values of p, and then measuring the penetration depth of each sample. Similar to the measurement of the exponent s below p_c, such a direct test would be complicated by the proximity effect. Measurements of the temperature-dependent penetration depth in a single sample of composite superconductor have also been interpreted in terms of a temperature-dependent volume fraction $p(T)$ of the superconductor. Assuming that $p(T)$ is smooth, the normal conductor–superconductor transition occurs at a critical temperature T_c such that, $p(T_c) = p_c$. If we expand $p(T)$ around T_c, we obtain $p(T) - p(T_c) = \alpha(T - T_c)$, where $\alpha < 0$. This argument then yields

$$\Lambda(T) \sim (T_c - T)^{-\mu/2} , \tag{12.57}$$

near T_c, in contrast to the usual Bardeen[9]–Cooper[10]–Schrieffer[11] temperature dependence, $\Lambda(T) \sim (T_c - T)^{-1/2}$. It also predicts that for the resistivity above T_c, $R(T) \propto (T - T_c)^s$, which differs from the predictions of superconducting fluctuation theory. Both predictions rely, of course, on a somewhat arbitrary (linear) transformation from volume fraction p to temperature T.

12.8.3.2 The Specific Heat

Deutscher et al. (1980) proposed a percolation model for the onset of superconductivity by assuming that the grains are coupled if $E_j > k_B T$, where k_B is the Boltzman's constant, and E_j is the Josephson energy coupling. Because the coupling energy depends on T, more and more grains become coupled as T is lowered. The cou-

---

[9] **John Bardeen** (1908–1991) was an American physicist and engineer, and the only scientist to be awarded the Nobel Prize in Physics twice; first in 1956 with **William Bradford Shockley** (1910–1989) and **Walter Houser Brattain** (1902–1987) for the invention of the transistor, and again in 1972 with Leon N. Cooper and John Robert Schrieffer for a fundamental theory of conventional superconductivity, which is now known as the BCS theory.

[10] **Leon N. Cooper** (1930–) is an American physicist who, in addition to winning the Nobel Prize in Physics for developing the BCS theory of superconductivity, is also known for Cooper pair—a pair of electrons bound together at low temperatures in a certain manner, and the BCM theory of learning in the visual cortex.

[11] **John Robert Schrieffer** (1931–2019) was an American physicist who, with John Bardeen and Leon Cooper, received the 1972 Nobel Prize in Physics for developing the BCS theory of superconductivity. He also developed the Schrieffer–Wolff transformation, which is a unitary transformation for perturbatively diagonalizing a system's Hamiltonian to first order in the interaction, which represents an operator version of second-order perturbation theory; the Su–Schrieffer–Heeger (SSH) model, a one-dimensional lattice model for explaining the properties of conducting polymers, and paramagnons, magnons in magnetic materials that are in their high temperature, disordered (paramagnetic) phase.

pling is obtained randomly with a probability that, similar to what was discussed above, depends on T. This assumption defines a percolation process because when the coupling probability is equal to the percolation threshold, an SSC of coupled superconducting grains is formed. If the temperature is still lowered, the SSC of the superconducting grains grows in size. The distributions of the grain sizes and the junction resistances give rise to randomness in the coupling energy.

In Deutscher et al.'s model, the coupling between the grains can be very strong, giving rise to a broad grain-size distribution, in which case the randomness in the coupling energy is due to the dependence on the grain size of the temperature at which a grain becomes superconducting. If, on the other hand, the grain size distribution is narrow, the coupling would be weak, which was also the limit that Deutscher et al. considered. They used a dense random packing of hard spherical grains with a size of about 30 $^\circ A$, small enough that only the SSC of the superconducting grains was the main contributor to the specific heat of the system, the main property that was calculated. The coupling energy was estimated by assuming that the normal-state resistances of the junctions follow a Gaussian distribution with a variance proportional to the mean. The mean of the distribution was treated as the only adjustable parameter of the model. Based on this model, the temperature dependence of the specific heat of the system was calculated, and was found to be in excellent agreement with the experimental data for granular Al-Al_2O_3 films.

12.8.3.3 The Critical Current

In Josephson tunneling through the junctions, there is no potential difference between the grains up to a *critical current* I_c. If the temperature of the system is lowered below the critical temperature T_c of the system, the number of superconducting links N_ℓ, per unit cross-sectional area, increases. Percolation theory provides estimates for N_ℓ, the critical current I_c, and the critical temperature T_c at which the material becomes superconducting. In the percolation model of Deutscher et al. (1980) T_c is close to the temperature at which the specific heat attains its maximum. Alternatively, T_c may be defined as the temperature at which 50% of the transition in the resistivity occurs. Near p_c, one has the power law

$$I_c \sim (p - p_c)^v , \tag{12.58}$$

where v is a new critical exponent, and the granular superconductor can be viewed as a network of superconducting blobs connected by links (see Chap. 2) that are quasi-linear chains of Josephson junctions. The current that flows through parallel links is inversely proportional to N_ℓ, and the distance between the links is roughly proportional to the percolation correlation length ξ_p. Thus, for a d-dimensional system, N_ℓ is on the order $1/\xi_p^{d-1}$. Since $I_c \sim N_\ell$, we obtain

$$I_c \sim (p - p_c)^v \sim (p - p_c)^{\nu(d-1)} , \tag{12.59}$$

so that υ is a purely topological exponent. Equation (12.59) was confirmed by Monte Carlo simulations in 2D (Kirkpatrick 1979; Lobb and Frank 1979) and 3D percolation networks (Kim and Stroud 1992), and by careful experiments of Deutscher and Rappaport (1979) who prepared thin (2D) superconductor–semiconductor films of Pb–Ge and measured their critical currents. They reported $\upsilon \simeq 1.3 \pm 0.1$, in excellent agreement with the prediction for 2D systems, $\upsilon = \nu = 4/3$.

We also note that if the grains make a structure similar to the Swiss-cheese or the inverted Swiss-cheese model (see Chap. 3), then, one can show that (Octavio et al. 1988), $\upsilon = (m + \nu)(d - 1)$, where m is the power that related the conductance of the necks to its width (see Chap. 3). Numeical simulation of Octavio et al. (1988) was in excellent agreement with this prediction.

12.8.3.4 The Critical Fields

The dependence of the magnetic induction B on the magnetic field H in typical type-I and type-II superconductors is well known. This behavior can be interpreted in terms of a field-dependent magnetic permeability $k_e = B/H$ that, for $H < H_c$ (or $H < H_{c1}$) is zero. Above H_c in type-I superconductors, k_e rises discontinuously to unity. In type-II superconductors, on the other hand, k_e rises continuously above H_{c1}, reaching unity at H_{c2}. Thus, according to this analysis, H_c or H_{c1} may be viewed as a type of breakdown field above which the superconducting component begins to become a normal conductor. Thus, this problem is seemingly similar to the phenomenon of dielectric breakdown in a network of insulating and conducting bonds below the conductivity threshold. The field $\mathbf{H}$ is curl free, i.e., $\nabla \times \mathbf{H} = \mathbf{0}$—because the screening currents are viewed as magnetization currents and, therefore, do not constitute a source for $\mathbf{H}$. In a similar manner, the analogue of $\mathbf{I}$ is the divergenceless field $\mathbf{B}$.

If this analogy is valid, one may deduce $H_{c1}(p)$ from the known results of the dielectric breakdown problem. Thus, at fixed composite volume, we expect to have

$$H_{c1}(p) \sim H_{c1}(0)\left[p_c^{(n)} - p\right]^y , \tag{12.60}$$

where $y = \nu$. The derivation of this equation is described in detail by Sahimi (2003b). Here, $p_c^{(n)}$ is the volume fraction of the superconducting material at which the normal conductor first forms a SSC, and ν is the usual percolation correlation length exponent. In 3D, as well as in some 2D systems, $p_c^{(n)} > p_c$ and, therefore, there exists a finite range of volume fractions in which the composite is an electrically perfect conductor, but has zero critical field, since at such volume fractions the composite is bicontinuous and, hence, can transport, both supercurrent through the superconducting component, and flux through the normal component. In a real composite, the validity of this analysis presumably depends on the manner in which the magnetic field is introduced. If a bicontinuous composite is cooled in a field, the flux, expelled from the superconducting component, might still continue to thread the composite even below T_c, just as in a single superconducting loop. But if the field is turned on below T_c, it will be screened out by the induced supercurrents. Moreover, if the flux through a given link of normal conductor is sufficiently small, and the superconducting grains are all large compared to the penetration depth Λ_S of the superconducting material (as assumed in the above discussion), this model must be modified by the requirement of flux quantization through each link.

A characteristic length scale of granular superconductors is their superconducting coherence length ξ_s, which can be defined as the diffusion length over a characteristic time t_c for the relaxation of the order parameter at a given T. Therefore, one has $\xi_s \sim \sqrt{Dt_c}$, where D is a diffusion coefficient, and $t_c \sim (T_c - T)^{-1}$. If $\xi_s > \xi_p$, one is in the homogeneous regime in which $H_{c2} \sim (T_c - T)$ near T_c. The upper critical field of a type-II superconductor is proportional to ξ_s^{-2}. In the inhomogeneous regime, where $\xi_s \ll \xi_p$, one has fractal or anomalous diffusion (see Chap. 7), instead of Fickian diffusion. Since ξ_s is defined as a diffusion length scale, ξ_s^2 should be proportional to $\langle R^2(t)\rangle$, the mean square displacement of a diffusing particle at time t which, for anomalous diffusion, is given by, $\langle R^2(t)\rangle \sim t^{2/(2+\theta)}$, where $\theta = (\mu - \beta)/\nu$, with μ, β and ν being the usual percolation exponents. Therefore, one obtains, $H_{c2} \sim \xi_s^{-2} \sim t_c^{-2/(2+\theta)}$, implying that

$$H_{c2} \sim (T_c - T)^{2/(2+\theta)} . \tag{12.61}$$

Equation (12.61) was confirmed experimentally by Gerber and Deutscher (1989), who measured the upper critical field of thin semicontinuous lead films on Ge substrates, and found that $H_{c2} \sim (T_c - T)^{\gamma}$ with $\gamma \simeq 0.66 \pm 0.06$. Since in 2D, $\theta \simeq 2.87$, the percolation prediction, $\gamma = 2/(2+\theta) \simeq 0.7$, is in very good agreement with the measured value.

12.8.3.5 Differential Diamagnetic Susceptibility

The last property we consider is the differential diamagnetic susceptibility, $\chi = \partial\mathbf{M}\partial H$ (where $\mathbf{M}$ is the susceptibility), of a superconducting composite near the percolation threshold p_c. It is known that in bulk superconductors near T_c, χ increases sharply in magnitude due to diamagnetic fluctuations, i.e., momentary fluctuations of the normal conductor into the superconducting state. A loosely analogous phenomenon occurs in composites, as was first discussed by de Gennes (1981) who noted that for $p < p_c$, the superconducting component is present only in the form of finite clusters, some of which contain closed loops. In the presence of an applied DC magnetic field, diamagnetic screening currents can flow in these loops, giving rise to a finite diamagnetic susceptibility χ. As $p \to p_c$, the loops become larger and larger, as does χ. One, therefore, expects χ to diverge at p_c according to a power law:

$$\chi_d \sim (p_c - p)^{-b} , \tag{12.62}$$

where b depends on the dimensionality of the system, and is a measure of the number and size of loops of the network. Equation (12.62) was first proposed by de Gennes (1981), who also suggested that $b = 2\nu - \mu$ in all dimensions. Rammal et al. (1983) proposed an alternative relation, $b = 2\nu - \mu - \beta$, where μ, β, and ν are the usual percolation exponents. The de Gennes formula agrees slightly better with the numerical simulations in 2D, while that of Rammal et al. agrees better in 3D. Lagar'kov et al. (1987) suggested yet another relation, $b = \nu$ in 2D and $b = 0$ (i.e., a logarithmic divergence) in 3D, and also argued that the divergence is removed if one considers the inductive interaction between screening currents induced in different clusters, an effect omitted in all the other estimates that considered such clusters as independent.

12.9 Thermal Percolation

As discussed earlier in this chapter, since most polymers have very low electrical and thermal conductivities, which limit their usefulness for a variety of applications, polymer composites with excellent electrical conductivity have been fabricated by incorporating electrically conductive fillers, such as graphite, carbon black, carbon fibers, metal powders, carbon nanotubes (CNTs), graphene, and metal nanowires in an insulating polymer matrix. By now we know that when the critical volume fraction, or the percolation threshold of the fillers is reached, an SSC of fillers is formed and the macroscopic electrical conductivity of the composite increases by orders of magnitude within a narrow range of the filler's volume fraction. The research field is still highly active, and new data and results appear at a rapid pace (see, for example, Kinloch et al. 2018; Huang et al. 2020).

The same has not turned out to be true, on at least a consistent basis, for the thermal conductivity of polymer composites. When various thermally conductive fillers, such as ceramic, metal, or carbon-based particles were added to a polymer matrix, the thermal conductivity of the resulting composite was not necessarily high (see, for example, Burger et al. 2016). This led to the suggestion that thermal conductivity of polymer composites may not have percolation characteristics, characterized by sudden sharp changes at the percolation threshold. The issue was analyzed and discussed by Forero-Sandoval et al. (2022).

Lack of a percolation transition in the thermal conductivity of a composite material is, to some extent, surprising, since both electrical and thermal conduction in a material are described by mathematically equivalent equations, and since the SSC of the fillers endows the composites with significant electrical conductivity, one would expect the same for thermal conductivity. Mamunya et al. (2002) added Cu and Ni microparticles to epoxy resin and poly(vinyl chloride) polymer matrices. Although electrical percolation was reached rather simply when the fillers' volume fraction was in the range 0.04–0.08%, the same was not observed even when the volume fraction of the fillers was high as 0.40.

Then, the realization that the CNTs, both of the single- and multi-walled types, have high intrinsic electrical and thermal conductivities motivated many researchers to attempt fabricating CNT–polymer nanocomposites. The electrical conductivity of such composites was already described and discussed above. But, the thermal conductivity of the same composites did not exhibit a percolation transition as the volume fraction of the filler exceeded the percolation threshold of disordered continua (see Chap. 3). Biercuk et al. (2000) reported five orders of magnitude increase in the electrical conductivity of SWCNT–epoxy nanocomposites at a weight percent 0.15 of the SWCNT, but the corresponding thermal conductivity did not exhibit a percolation transition, with the thermal conductivity enhancement being only about a factor of two.

Chalopin et al. (2009) and Prasher et al. (2009) offered one possible resolution of the problem. They proposed that CNT assemblies may perhaps be thermal *insulators*, rather than conductors, because the CNT–CNT junctions are point contacts and, therefore, are very inefficient in thermal transport as they possess very low thermal conductance.

On the other hand, using special fabrication techniques for constructing an effective network of fillers within a polymer matrix, several studies have reported the occurrence of thermal percolation. Bonnet et al. (2007), for example, suggested the possibility of thermal percolation in SWCNT–polymethylmethacrylate (PMMA) nanocomposite films, and proposed that the electrical and thermal percolation thresholds should coincide with each other. Although thermal percolation transition in SWCNT–epoxy (Yu et al. 2006) and epoxy–graphite nanoplatelet (GNP) nanocomposites (Yu et al. 2007) was not observed, evidence for the percolation transition in a SWCNT–GNP–epoxy composite was reported (Yu et al. 2008). A sharp increase in the thermal conductivity of the ternary composite was reported and attributed to more efficient formation of network that combined 1D and 2D nanofillers. Shtein et al. (2015a, b) also reported observation of thermal percolation transition in graphene–epoxy nanocomposites. Importantly enough, Shtein et al. reported the thermal percolation threshold at volume fraction of 0.17, much higher than the corresponding electrical percolation threshold, 0.05.

However, when the data for thermal conductivity near the transition point were fitted to the universal power law of percolation for the conductivity near the percolation threshold, Eq. (2.15), Shtein et al. obtained, $\mu \approx 0.84$, which is neither close to 1.3 for 2D networks, nor to 2.0 for 3D Ones (see Chap. 2). Kim et al. (2013, 2014) reported data for thermal percolation in nanocomposites of functionalized MWCNTs in reactive ethylene terpolymer, But the critical exponent was again too low, $\mu \approx 0.1$, which is indicative of a logarithmic increase, rather than a power-law one. Kargar et al. (2018) reported observation of thermal percolation transition in epoxy nanocomposites with graphene and hexagonal boron nitride (h-BN) additives. They reported that for the graphene–epoxy nanocomposite, the electrical percolation threshold was at about 0.1 volume fraction, significantly lower than their reported thermal percolation threshold at 0.3.

What is the reason for the discrepancy between polymer composites that do manifest thermal percolation and those that do not? A variety of reasons come to mind:

(i) Finite-size effects that smear the transition.

(ii) Relatively modest difference between the thermal conductivities of the fillers and the polymer matrix.

(iii) High resistance at the contact points that reduces the overall thermal conductivity.

(iv) Lack of high-resolution data, as well as lack of enough data points for fitting the data in the vicinity of the possible transition point to the power law for the conductivity, Eq. (2.15).

Chapter 13
Rheology and Elastic Properties of Network Glasses, Branched Polymers, and Gels

13.1 Introduction

One of the most important applications of percolation theory has been to describing rheology and elastic properties of a wide variety of polymers. Polymers, whether natural or synthetic, are everywhere and it is hard to imagine everyday life without them. They also have many interesting, and in many cases unusual, properties that justify their study.

Three seminal works in the early to mid 1970s demonstrated clearly how polymers can be studied, both theoretically and experimentally through statistical mechanics and critical phenomena, and eventually percolation theory. The first was by de Gennes (1972) who showed that there is a close connection between linear polymers—those whose monomers have functionality $Z = 2$—and a statistical mechanical model, namely, the $n-$vector[1] model. If no two polymer parts occupy the same point in space, the linear polymer corresponds to the limit $n \to 0$. If this restriction is ignored, the polymer corresponds to a random walk, or the $n = -2$ limit of the model. de Gennes' discovery made it possible to apply modern methods of statistical mechanics, such as the renormalization group theory, to the study of linear polymers.

The second work, on the experimental side, was by Cotton (1974) who used small-angle neutron scattering and a labeling technique in which the hydrogen atoms along the polymer chain were replaced by deuterium, to demonstrate that it is possible to detect one polymer chain among many others in a solution. Thus, one can analyze a single polymer chain and compare the experimental results with the theoretical predictions.

The works of de Gennes and Cotton were restricted to linear polymers, i.e., those in which each monomer is connected to two neighboring monomers. If, however, the

[1] The $n-$vector or O(n) model is a system of interacting spins on a crystalline lattice, developed by Harry Eugene Stanley, a Professor of physics at Boston University, as a generalization of the Ising model, in which $n-$component unit-length classical spins $\mathbf{s}_i$ are placed on the vertices of a $d-$dimensional lattice.

M. Sahimi, *Applications of Percolation Theory*, Applied Mathematical Sciences 213,
https://doi.org/10.1007/978-3-031-20386-2_13

monomers have a functionality $Z > 2$, so that each monomer is connected to up to Z neighboring monomers, then at least two other classes of polymers can be obtained.

(i) If the reaction time t is relatively short and below, but close to, a characteristic time t_g, then one obtains branched polymers in the solution, usually called a *sol*, which form a viscous solution. Such branched polymers are large but finite clusters of monomers.

(ii) On the other hand, if the reaction time is larger than t_g, a very large solid network of connected monomers appears that is usually called *chemical gel*, or simply gel. The gel network has interesting structural, mechanical, and rheological properties that will be described and discussed in this chapter. The characteristic time t_g is called the *gelation time* , and the point at which the gel network appears for the first time is called the *gel point* (GP). Most of us are already familiar with such sol–gel transformations in our daily lives, since we all know about milk-to-cheese transition, pudding, gelatine, etc.

As mentioned in Chap. 2, in the 1940s Paul Flory and Walter Stockmayer had derived an exact solution for the problem of polymerization. But their theory was developed based on a special lattice, namely, the Bethe lattice, an endlessly branching structure without any closed loops (see Fig. 2.1). Stauffer published the third important piece of work on polymers in the 1970s. He demonstrated the importance of the deviations from the Bethe lattice solution of Flory and Stockmayer for polymerization, and proposed to replace it by percolation on 3D lattices.

The discussion so far should have made it clear that there is clearly a phase transition as the liquid sol passes to the solid gel. The transition, which is characterized by a critical point—the GP—has been modeled by percolation models. In this chapter we describe how such models have been used to predict rheology of the sol phase and the elastic properties of the gel phase. But we emphasize that the applications of percolation theory to rigidity and elastic properties of heterogeneous materials that are described in this chapter are by no means exhaustive, because the number and variety of materials to which such models may be applicable is very large. In fact, concepts that are directly or indirectly related to vector percolation and the rigidity properties of materials (see Chap. 11) have been utilized to explain seemingly unrelated phenomena.

One striking example is the work of Hammonds et al. (1997) that showed that the existence of adsorption sites in zeolites, an important class of nanoporous catalytic materials that are used heavily in the chemical industry is related to the existence of floppy modes that were described in Chap. 11. They demonstrated that zeolite frameworks can support large bands of rigid unit modes in the **k** (wave vector) space. Localized rigid unit modes are formed within such frameworks that do not distort the constituent tehrahedra to any significant degree. Such localized modes enable cations at certain sites to pull the framework, with essentially zero cost in elastic energy, in such a way that oxygen-cation bonding distances become exactly optimal for the cation of interest.

Another striking example of the relevance of rigidity percolation is the experiments of Petridou et al. (2021). Phase transitions are abundant in non-living systems, but the surprising feature of the experiments of Petridou et al. was that the percolation rigidity transition occurred in a *living* system, which was an embryo whose morphogenesis is influenced by dynamic changes in the properties of the tissue, which have been attributed to occurrence of processes that are akin to a phase transition. Petridou et al. (2021) showed that rigidity percolation provides a robust theoretical framework for predicting phase transitions in the material and the associated structural changes of embryonic tissues based on the local cell connectivity. They monitored dynamic changes in the tissue's rheology and cell contact mechanics and demonstrated, based on rigidity percolation, that the zebrafish blastoderm undergoes a genuine rigidity transition, caused by a small reduction in adhesion-dependent cell connectivity below a critical value.

A third example, which is described in some detail in the present chapter, is network glasses. In 1932 Zachariasen[2] introduced the notion of continuous-random networks for studying network glasses. He proposed that the structure of a glass consists of "an extended three-dimensional network lacking periodicity with an energy content comparable with that of the corresponding crystal," and argued that for the energy of the glass to be comparable with that of the crystal, the coordination number of the polyhedra and the manner by which they are connected must be the same in a glass and in the corresponding crystal. Zachariasen (1932) stated, however, that, unlike crystals, the relative orientations of adjacent polyhedra "vary within rather wide limits," leading to a lack of long-range order in glasses. He envisaged such networks maintaining local chemical order but, by incorporating small structural disorder, having a non-crystalline topology. Warren (1934) coined the term *random networks* for such disordered materials, while Gupta and Cooper (1990) proposed the term *topologically disordered* for describing the structure of glasses as it conveys the essence of Zachariasen's view of a glass structure. Careful diffraction experiments, from which the radial distribution function can be determined, have confirmed the correctness of the model and, therefore, it is now widely accepted.

Examination of the amorphous structure of covalent glasses leads to classification of their building blocks into two groups. In one group are the bonding structures that consist of covalent bonds with densities that are on the order of $10^{22} - 10^{23}$ cm^{-3}, and can be specified by their chemical and topological structures. Defects, such as impurities, dangling bonds, and "wrong" bonds—for example, homopolar bonds in stoichiometric alloys—are in the second group, with their density being typically two orders of magnitude less than that of the covalent bonds. Therefore, the covalent bonds are primarily responsible for such electronic properties as the band-gap energy,[3] in contrast to gap states that are caused by the defects. Hence, topological properties play the most important role in the glassy characters and, therefore, the application

[2] **William Houlder Zachariasen** (1906–1979) was a Norwegian-American physicist, specializing in X-ray crystallography, who made fundamental contributions to understanding of the structure of glass. As the Professor and Chair of physics department at the University of Chicago, Zachariasen was instrumental in its development as a world-class center of research.

[3] A band-gap energy is an energy range in a solid where no electronic states can exist. In solids' graphs of the electronic band structure, the gap represents the energy difference between the top of the valence band and the bottom of the conduction band in insulators and semiconductors.

of percolation theory to characterization of many important properties of network glasses is natural.

In addition to the fact that percolation theory was first developed 25 years after Zachariasen's work, its link with the percolation transition remained unclear for decades, because percolation typically represents structures in equilibrium, whereas the liquid-glass phase transition is the result of the system going out of a complete metastable equilibrium. The connection between the two was first suggested by Cohen and Grest (1979) who invoked the concept of *free volume*. Fox and Flory (1951), among others, had already proposed that the liquid-glass transition resulted from disappearance of the free volume of the amorphous phase at some temperature. The basic assumptions of the Cohen-Grest model were as follows.

(i) With each molecule one can associate a local volume v of molecular scale.

(ii) When the local volume reaches a critical value v_c, the excess value is the free volume.

(ii) Molecular transport takes place only when the total free volume, having a volume greater than some critical value, is approximately equal to the molecular volume formed by the redistribution of the free volume. The redistribution of the free volume requires no local free energy.

The assumptions are valid if each molecule is restricted to move mainly within a cell or cage defined by its nearest neighbors. There is considerable experimental evidence in dense liquids that indicates that this is indeed the case. If the temperature T of the system is near or above the glass transition temperature T_g, then the local free energy E_i of a cell depends only on its volume v_i. Near T_g, the function $E_i(v)$ is approximated by distinct expressions Cohen and Grest (1979), which depend on v_c, a critical volume whose existence enables one to classify the cells into two groups. Those with $v > v_c$ are liquid-like and have a free volume $v_f = v - v_c$, whereas cells with $v < v_c$ are solid-like. In effect, the free volume is the difference between the glass and crystalline volumes. If $f(v)$ is the probability distribution of having a cell with volume v, then the fraction p of the liquid-like cells is clearly

$$p = \int_{v_c}^{\infty} f(v)dv \, , \tag{13.1}$$

Free exchange of free volume can, however, occur only between liquid-like cells that are nearest neighbors and have a sufficiently large number of nearest-neighbor liquid-like cells and, therefore, this is a correlated site percolation Stanley (1979) described in Chap. 3. Moreover, there is a critical value $p_c(Z)$ of the probability p, with Z being the coordination number, such that, for $p > p_c$ one has a sample-spanning cluster (SSC) of liquid-like cells, whereas for $p < p_c$ the liquid-like cells form only isolated clusters and, therefore, the system as a whole is in a solid glassy state. The connection between percolation and liquid-glass transition made it possible to calculate the thermodynamic properties of the system. This aspect of the problem was reviewed in detail by Grest and Cohen (1983). Since their work, it has become clear that their percolation model—a scalar model—is applicable to metallic glasses.

What we are mainly interested in this chapter is the mechanical properties of network glasses.

As discussed in Chap. 11, the rigidity and linear elastic properties of two-phase materials that are far from their percolation threshold, or those in which the contrast between the properties of the two phases is not large, are well-described and predicted by poor man's percolation model, namely, the effective-medium approximation (EMAs). But the effective properties of two-phase materials in which the contrast between the properties of the two phases is large, particularly those in which one or both phases form large clusters and are, therefore, near the percolation threshold, deviate greatly from the predictions of the EMA and other analytical approximations. It is the description of this type of two-phase materials that is best done by vector percolation, and in particular models that are based on elastic and superelastic percolation networks (EPNs and SEPNs, respectively), the focus of this chapter.

13.2 Connectivity of Network Glasses and Rigidity Transition

We first pose an interesting question: why is a glass rigid and has non-zero mechanical properties? At a fundamental level, the question is also important because we know that while in a rigid material a local disturbance results in a bulk response, the same is not true of liquids. For example, if we apply a small force to a crystal, it will lead to its collective motion, because it is energetically favorable for the crystal's atoms to maintain their *relative* positions. But if the same force is applied to a liquid, it will generate only local rearrangements of the atoms, and after a short time the liquid "forgets" the structure that it had before the force was applied.

Unlike a liquid though, the low-temperature state of a glass depends on its history, implying that its response to a shear deformation is not simple. If the glass contains N atoms, then the number of its possible initial low-temperature configurations grows exponentially with N. Moreover, on short time scales, a glass "remembers" its initial configuration when it is exposed to an applied force, whereas at much longer times, the glass "flows" and, hence, it forgets its past, which is usually referred to as *ageing*. Thus, a glass has certain properties that distinguish it from a liquid.

What does distinguish a glass from a crystal? One difference is that if both have N atoms, their number of states is, respectively, $\exp(N)$ and $O(1)$. The second difference is that a crystal breaks translational symmetry and possesses long-range spatial ordering, whereas there exists no such simple long length scales associated with a glass. The third difference between the two is that the response of a crystal to a shearing force is invariant under time translation, whereas for long measurement times a glass loses its rigidity and, therefore, does not exhibit the time invariance.

In an attempt to explain the strong glass-forming propensity of some chalcogenide materials, Phillips (1979,1981) proposed that covalent bonding in glasses can be optimized when the average coordination number $\langle Z\rangle$ of the three-dimensional

(3D) covalent network in mechanical equilibrium is about 2.4. In order to provide a firm theoretical foundation for his proposal, however, Phillips invoked the scalar percolation model. But as discussed in Chap. 2, the average coordination number in random scalar percolation at the percolation threshold p_c is about 1.5 and not 2.4. Thus, Phillips himself rejected the idea!

Phillips' idea was significantly refined by Thorpe (1983) who established the connection between this phenomenon and a vector percolation model. According to the Phillips–Thorpe (PT) idea, in covalent networks with stretching and Keating-type, or bond-bending (BB) forces described in Chap. 11, for average coordination numbers $\langle Z\rangle < 2.4$, the number N_c of interatomic forces per atom that act as constraints is less than the network dimensionality $d = 3$, which is the same as the number of degrees of freedom N_d per atom. Such networks deform easily under a shearing force, and possess a finite number N_0 of zero-frequency modes, $N_0 = N_d - N_c$. When $\langle Z\rangle > 2.4$, N_c is larger than N_d and, thus, such networks are macroscopically rigid and possess non-vanishing elastic moduli. Hence, when $\langle Z\rangle = 2.4$, one has $N_c = N_d$ and, therefore, $N_0 = 0$ represents the vector percolation threshold of the system.

We point out, however, the analysis leading to $\langle Z\rangle \simeq 2.4$ is of mean-field type. But, as mentioned in Chap. 11, computer simulations of He and Thorpe (1985) confirmed the accuracy of this mean-field analysis. Moreover, $\langle Z\rangle \simeq 2.4$ agrees with the predictions of Eqs. (11.46) and (11.48) and, thus, represents the average coordination number of 3D Kirkwood–Keating models at their percolation threshold.

13.2.1 Experimental Confirmation of the Phillips–Thorpe Theory

Experimental confirmation of the PT proposal has been provided by a wide variety of experiments, measuring several distinct properties of network glasses. The list of the papers and the data that they reported that we describe here is by no means exhaustive, but does present what we view as the most illuminating tests of the accuracy of the PT proposal.

Perhaps the first of evidence was provided by Ota et al. (1978), although no mention was made of vector percolation because it had not even been developed yet. They reported measurements of the bulk, shear and Young's moduli of Ge_xSe_{1-x} glasses, which are perhaps the simplest of such materials, over a wide range of the composition x. Over the composition range $0 < x < 1/3$, the coordination number of Ge and Se are 4 and 2, respectively and, thus, the average coordination number of the network glass is, $\langle Z\rangle = 2(x + 1)$. Therefore, an average coordination number $\langle Z\rangle \simeq 2.4$ corresponds to a critical composition $x_c \simeq 0.2$. Measurements of Ota et al. (1978) indicated that at around $x \simeq 0.2 - 0.25$, the three elastic moduli increase significantly, thus signaling a rigidity transition.

Bresser et al. (1986) carried out experiments with the same type of glass in the composition range $0 < x < 1/3$ and ^{129}I Mössbauer emission spectroscopy, and presented strong evidence that at $x_c \simeq 0.23 \pm 0.02$ the Mössbauer-site intensity ratio I_B/I_A exhibits dramatic threshold behavior. At $x = 0$ (i.e., g–Se) only B sites were observed and, thus, I_B/I_A is infinite. For $0 < x < 0.15$ the ratio I_B/I_A decreased with increasing x, reaching a minimum around $x \simeq 0.15$, but as $x \to x_c \simeq 0.23$, it increased and reached a maximum. Beyond x_c, I_B/I_A decreased monotonically with x again. The molar volume V_m of Ge_xSe_{1-x} also exhibited a threshold behavior, such that for $0 < x < 1/3$ it decreased monotonically with x, reaching its minimum at $x_c \simeq 0.23 \pm 0.02$, beyond which it increased again monotonically as x does. Note that $x_c \simeq 0.23 \pm 0.02$ agrees reasonably well with the PT estimate, $x_c = 0.2$, which corresponds to an average coordination number of about 2.4. Similar threshold behavior for the molar volume was also reported by Sreeram et al. (1991) who measured V_m for several chalcogenide glasses incorporating Ge, Sb, Se, As, and Te, and observed the threshold behavior at $\langle Z\rangle \simeq 2.4$.

Halfpap and Lindsay (1986) reported further experimental evidence for the relevance of vector percolation to network glasses by measuring the elastic moduli C_{11} and C_{44} of the Ge-As-Se glasses over a wide range of composition. The coordination number of As is 3 and, therefore, if the glass has the composition $Ge_xAs_ySe_{1-x-y}$, its average coordination number is $\langle Z\rangle = 4x + 3y + 2(1 - x - y) = 2x + y + 2$. Thus, by varying x and y one obtains glasses over a wide range of compositions. Despite different local structures of the glasses that are obtained when x and y are varied, measurements of Halfpap and Lindsay (1986) indicated that at $\langle Z\rangle \simeq 2.4$ there is a dramatic increase in the elastic moduli C_{11} and C_{44}. An example is shown in Fig. 13.1. Further evidence was provided by the experimental data of Mahadevan and Giridhar (1989), who also reported data for the same elastic moduli of various glasses formed with As, Ge, S, Sb, Se, and Te, and found essentially the same behavior that had been reported by Halfpap and Lindsay (1986).

Experiments of Tatsumisago et al. (1990) shed further light on the problem. They showed that the link between vector percolation and glass formation is better understood by studying the *liquid-state* behavior of glass-forming systems. They measured the elastic moduli, the viscosity, the thermal-expansivity, heat capacity, and the glass transition temperature T_g of various glasses formed with As, Ge, and Se, and showed that T_g is a universal function of $\langle Z\rangle$, i.e., independent of the composition of the glass, up to $\langle Z\rangle \simeq 2.4 - 2.5$. It is only above this point that the effect of the composition emerges. This is shown in Fig. 13.2. There is, however, no sudden upturn in the $\langle Z\rangle$–dependence of T_g, but the difference between the shapes of the heat capacity curves below and above the critical value $\langle Z\rangle \simeq 2.4$ is quite pronounced, which is also shown in Fig. 13.2. Other interesting features included a minimum in the activation energy at $\langle Z\rangle \simeq 2.4$ for the viscosity data, and a minimum in the heat capacity jump at $\langle Z\rangle \simeq 2.4$ and T_g. All such data indicate a structural transition at $\langle Z\rangle \simeq 2.4$. Similar data were also presented by Böhmer and Angell (1992) and by Wagner and Kasap (1996).

Another evidence for the PT transition is provided by the fact that the very small mass of H causes vibrational modes of H or H_2O to split off from the main vibrational

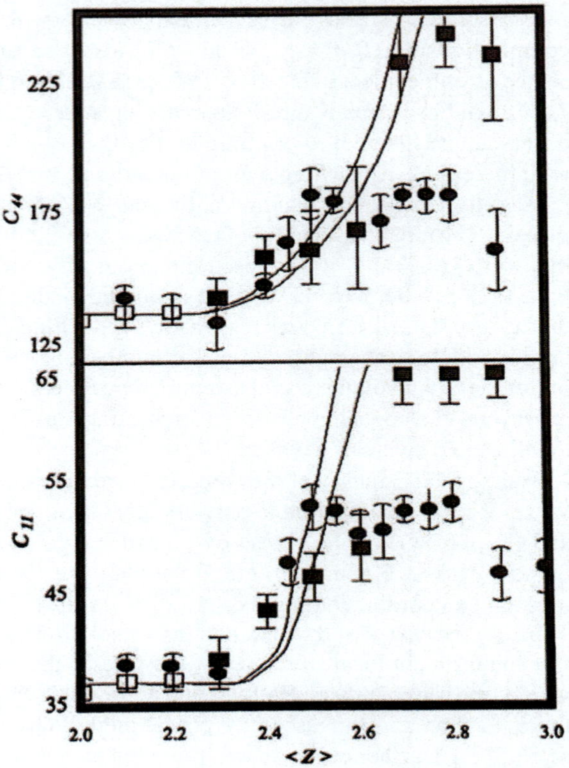

Fig. 13.1 Threshold behavior of the elastic moduli (in kbar) of glasses versus their average coordination number $\langle Z \rangle$. Circles and squares correspond, respectively, to high- and low-As series of glasses. The theoretical curves on the left are for amorphous Ge, while those on the right are for crystalline Ge (after Halfpap and Lindsay, 1986)

bands of Ge-As-Se chalcogenide glasses, and induces such excitations to have long lifetimes, which can be measured with the infrared hole-burning method. This is a technique that unravels the features of the homogeneous line shape of the transition involved, which is obscured by the presence of strong inhomogeneous broadening. At the frequency that saturation is performed, a dip–the hole–appears in the spectrum (see, for example, Uebbing and Sievers 1996; Wagie and Geissinger 2012). Such experiments indicate linear dependence of the relaxation rates on the average coordination number $\langle Z \rangle$, with a clear change in the slope at $\langle Z \rangle \simeq 2.4 \pm 0.1$, accompanied by a small but unambiguous cusp.

Wang et al. (2000) studied Raman spectra in chalcogenide glasses, GeS_{1-x} and Ge_xSe_{1-x} with $0 \leq x \leq 0.42$. Their data resulted in an average coordination number $2 \leq \langle Z \rangle \leq 2.84$, and indicated that the relative degree of fragility, representing the structural relaxation of the network, is larger for $\langle Z \rangle < 2.4$ than for $\langle Z \rangle > 2.4$, thus signaling a rigidity transition at $\langle Z \rangle \simeq 2.4$.

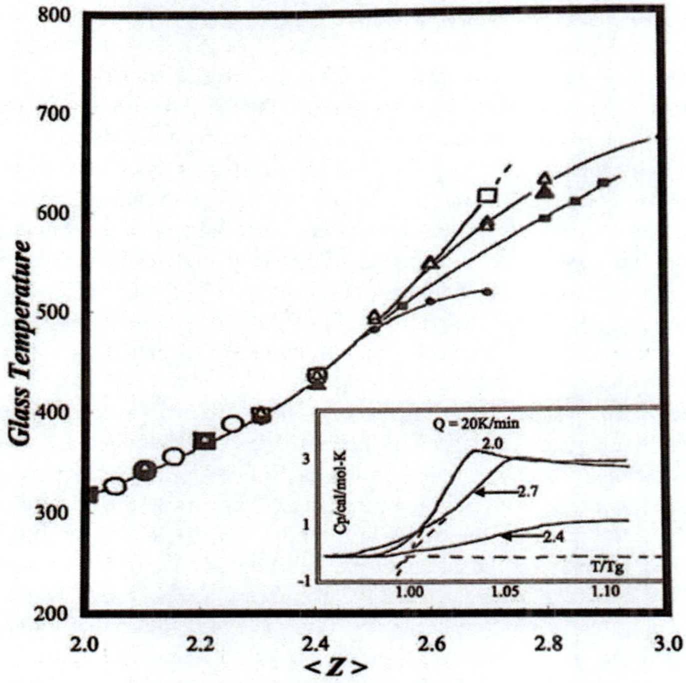

Fig. 13.2 Glass temperature T_g versus the average coordination number $\langle Z\rangle$ of $Se_{1-x}(Ge_yAs_{1-y})_x$. Symbols are the data for various values of y. The inset shows the specific heat capacity of the glasses, with the numbers on the curves giving the value of $\langle Z\rangle$ (after Tatsumisago et al. 1990)

13.2.2 Network Glasses with High Coordination Numbers

Despite the success of the PT theory, it is not free of ambiguities. Yun et al. (1989) reported the absence of any threshold behavior at $\langle Z\rangle \simeq 2.4$ in the elastic properties of Ge_xSe_{1-x}, contradicting the earlier data of Bresser et al. (1986) described above. Some other glasses (see below) exhibit an onset of vector percolation at values *higher* than $\langle Z\rangle \simeq 2.4$. Some of such differences may be attributed to the oxide contamination effects that adversely affect the chemical structure of the glasses. It could also be that in real materials the van der Waals forces between unbounded neighbors can smear out the percolation transition.

Feng et al. (1997) reported Raman scattering data for Ge_xX_{1-x} glasses ($X = S$ or Se), indicating that there is a stiffness threshold at a mean coordination number $\langle Z\rangle \simeq 2.46 \pm 0.01$ for the shear (transverse) optic mode, only slightly larger than the predicted value of 2.4. But Phillips (1999) suggested that although the critical connectivity for the hydrostatic is 2.4, it is about 2.45 for the shear modulus, in agreement with the data of Feng et al. (1997).

Srinivasan et al. (1992) reported experimental data for the thermal diffusivity of two sets, $Ge_xSb_ySe_{1-x-y}$, $Ge_xAs_yTe_{1-x-y}$, and $Si_xAs_yTe_{1-x-y}$ glasses, using a photoacoustic (PA) technique. In these experiments, a thin disk of the sample with the appropriate thickness (about 150 μm) is glued to the PA cell with silver paste, which acts as a good thermal backing medium. The sample is then irradiated with intensity modulated light, and the amplitude $\mathcal{A}$ of the PA signal is recorded as a function of the modulation frequency ω. The thermal diffusion length in the sample varies with ω, and from a plot of $\log \omega$ versus $\log \mathcal{A}$ a characteristic frequency ω_c is determined as one at which the slope of the plot changes. The thermal diffusivity D_T is then given by, $D_T = \omega_c \ell^2$, where ℓ is the thickness of the sample. Measurements of Srinivasan et al. (1992) indicated that the thermal diffusivity of the three glasses increases monotonically with the average coordination number $\langle Z \rangle$, reaching a maximum at $\langle Z \rangle \simeq 2.6$, beyond which it decreases again.

The data of Srinivasan et al. and similar ones may be explained based on a modified picture of network glasses proposed by Tanaka (1989), who studied network glasses $Ge(Si)_xAs_yS(Se)_{1-x-y}$ and proposed, following Zallen (1983), that the dimensionality d of the network plays an important role in glasses' properties. Zallen suggested that $d = 1$, 2 and 3, respectively, for amorphous Se, $As_2S(Se)_3$, and Si(Ge). For example, $d = 1$ for Se implies a chain-like morphology in which entangled chain molecules are held together by van der Waals intermolecular forces, whereas $d = 3$ implies 3D continuous-random networks. Tanaka (1989) found that the molar volumes for various glasses exhibit two extrema: They reach their minimum at an average coordination number $\langle Z \rangle \simeq 2.4$, which is in agreement with the data of Bresser et al. (1986) and Sreeram et al. (1991), whereas their maximum is reached at $\langle Z \rangle \simeq 2.67$. The optical band-gap energies of the glasses also exhibited a maximum at $\langle Z \rangle \simeq 2.67$, while their elastic moduli indicate two upturns, one at $\langle Z \rangle \simeq 2.4$ and another one at $\langle Z \rangle \simeq 2.67$. Similar data had earlier been reported by Tanaka et al. (1986). To explain such data, the following scenario was proposed by Tanaka (1989):

(i) In glassy Se(S) with coordination number $Z = 2$, the network is roughly one dimensional.

(ii) When As and/or Fe atoms are introduced into such chalcogen glasses, the 1D molecules are crosslinked, so that the structure gradually transforms for $d = 1$ to $d = 2$. At $\langle Z \rangle \simeq 2.67$, the layer structures, which are segmental, are fully evolved and, hence, $d = 2$.

(iii) With further increase in the local connectivity, the structures undergo a transition to 3D networks.

The prediction, $\langle Z \rangle \simeq 2.67$, may be explained based on a constraint-counting argument developed by Tanaka (1989). Let us first recall the PT constraint-counting method again. The number of constraints N_c in 3D covalent glasses is $N_c = \frac{1}{2}Z + (2Z - 3)$, where $\frac{1}{2}Z$ represents the stretching (radial) constraints, and $(2Z - 3)$ is the number of the BB valence-force constraints, with 2 corresponding to the two degrees of freedom in a spherical representation (r, θ, φ) of an atom bonded to another atom located at the origin, and the 3 being related to the system's rotation around the three main axes. Since the total number N_d of degrees of freedom in 3D is 3, and because at the rigidity transition we must have $N_c = N_d$, we obtain the PT estimate $Z = \langle Z \rangle = 2.4$.

Using a simple modification of the PT analysis, the value of the average coordination number, $\langle Z \rangle \simeq 2.67$, for the glasses that were studied by Tanaka (1989) and Srinivasan et al. (1992) can be predicted. In this case, the number of constraints is, $N_c = \frac{1}{2}Z + (Z - 1)$, so that the number of the angular constraints is reduced from $(2Z - 3)$ for fully 3D glasses to $(Z - 1)$, due to the assumption of having network glass configurations that are planar. As the number of degrees of freedom is still $N_d = 3$, at the transition one must have, $N_c = N_d$, and, thus, $Z = \langle Z \rangle = 8/3 \simeq 2.67$, in agreement with Tanaka's and Srinivasan et al.'s data. Therefore, it appears that in these experiments a glass is fixed stably in a 3D space, if $\langle Z \rangle \simeq 2.67$.

Liu and Löhneysen (1993) measured the low-temperature specific heat C_p of g-As_xSe_{1-x} and fitted their data to $C_p = a(x)T + b(x, T)T^3$. They found that both the PT rigidity transition at $\langle Z \rangle \simeq 2.4$ and the Tanaka transition at $\langle Z \rangle \simeq 2.7$ contribute to the fitted functional form, with a double maximum in $a(x)$. They also found that if they subtract the contribution of the second term of the fitted equation, then $a(x)$, which is a measure of the two-level point defects, vanishes at $\langle Z \rangle \simeq 2.4$, hence proving that, in the absence of interlayer cross-linking, one would have a crystal. One may also explain these data by the fact that the defects are all associated with chemical disorder (for example, homopolar bonds), and that the concentration of these is minimized at $x = 0.4$.

13.2.3 *Network Glasses with Dangling Atoms*

Boolchand and Thorpe (1994) refined the idea that network glasses must become rigid when their average coordination number is about 2.4, in order to analyze networks that have onefold-coordinated (OFC) atoms, i.e., dangling ends. One might believe that they can be removed as in, for example, H in a−Si, a−C, and a−Ge-Si networks (see, for example, Mousseau and Thorpe (1993) and references therein), but it is known that the presence of H in a−Si network degrades the network's elastic moduli. Therefore, it is instructive to explicitly consider the effect of OFC atoms. Boolchand and Thorpe (1994) used a constraint-counting method of the type described above.

Consider a 3D covalent network with N atoms, N_Z of which having a coordination number of Z. Following exactly the same argument that was utilized for deriving Eq. (11.45) for $Z \geq 2$, one has $N_\alpha = Z/2$ constraints (corresponding to the stretching force of the Kirkwood–Keating model) and $N_\gamma = 2Z - 3$ constraints (corresponding to the BB term of the model). For an atom with $Z = 1$, one has $N_\alpha = 1/2$ and $N_\gamma = 0$. Therefore, the total number of constraints is given by

$$N_c = \frac{1}{N}\left[N_1 \times \frac{1}{2} + \sum_{Z\geq 2} N_Z \left(\frac{1}{2}Z + 2Z - 3\right)\right], \tag{13.2}$$

and the number N_0 of the zero-frequency modes per atom is

$$\begin{aligned}N_0 = N_d - N_c &= 3 - \left(\frac{1}{2}\frac{N_1}{N} + \sum_{Z\geq 2}\frac{5}{2}Z\frac{N_Z}{N} - \sum_{Z\geq 2}3\frac{N_Z}{N}\right) \\ &= 3 - \left(\frac{1}{2}\frac{N_1}{N} + \sum_{Z\geq 1}\frac{5}{2}\frac{ZN_Z}{N} - \frac{5}{2}\frac{N_1}{N} - \sum_{Z\geq 1}3\frac{N_1}{N} + 3\frac{N_Z}{N}\right) \\ &= 6 - \frac{5}{2}\langle Z\rangle - \frac{N_1}{N}\,,\end{aligned} \tag{13.3}$$

where $\langle Z\rangle = \sum_{Z\geq 1} ZN_Z/N$. As the general condition for the onset of rigidity is $N_0 = 0$, we obtain

$$\langle Z\rangle = 2.4 - 0.4\frac{N_1}{N}\,. \tag{13.4}$$

Equation (13.4) implies that the critical average coordination number at the onset of rigidity depends on the fraction of the OFC atoms, and that their presence reduces the critical value. An example of a glass to which this model may be applicable is the ternary glass $Ge_xS_{1-x-y}I_y$, formed by a fast quench over a wide range of compositions. The Ge, Se, and I atoms possess a coordination number of 4, 2, and 1, respectively, and bond in conformity to the $8 - \mathcal{N}$ rule. For a given x, one can determine the critical value y_c of iodine concentration y at which Eq. (13.4) is satisfied. Since for this glass, $\langle Z\rangle = 4x + 2(1 - x - y_c) + y_c = 2(x - y_c + 1)$, one obtains,

$$y_c = \frac{1}{3}(10x - 2)\,. \tag{13.5}$$

Experimental data of Dembovskii et al. (1971) indicated that the maximum iodine concentration at which a glass can form is $y_c \simeq 0.55$ at $x = 0.386$, whereas, given this value of x, Eq. (13.5) predicts that $y_c \simeq 0.62$, slightly higher than the experimental value. On the other hand, had one ignored the effect of OFC atoms, the predicted critical concentration (i.e., at $\langle Z\rangle = 2.4$) would have been completely wrong.

13.2.4 Stress Transition in Network Glasses

Georgiev et al. (2000) reported experimental data for bulk As_xSe_{1-x} that indicated that a stress-free rigidity transition occurs at $\langle Z\rangle = 2.29 \pm 0.01$, and another transition occurs in a stressed phase at $\langle Z\rangle = 2.37 \pm 0.01$, which are below the $\langle Z\rangle \simeq 2.4$. To explain such data, Thorpe et al. (2000) considered two types of models. In one, they constructed random bond networks by randomly positioning points in a plane. When such networks are very large, they contain no closed loops and, therefore, are similar to the Bethe lattices (Chap. 2). Thorpe et al. found that the rigidity transition in such networks is *first order*, which is similar to the central-force percolation transition on the Bethe lattices and in BCC lattice (Arbabi and Sahimi 2021). The rigidity transition in this network occurred at $\langle Z\rangle = 2.3893$.

In the second model, Thorpe et al. began with a smaller coordinated floppy network and added bonds that led to rigid regions. So long as such bonds are unstressed, i.e., they can have their natural length (angle) without being forced to change by their surroundings, they are retained in the network. If addition of a bond results in its being redundant and creates a stressed region, it is removed. If only central forces operate in the system, one can continue building up this network until it is entirely unstressed, in which case the percolation transition occurs at $\langle Z\rangle = 2.375 \pm 0.001$. The transition at this point is second order. Due to the presence of the BB forces, however, it may happen that while one part of the set of constraints associated with a given bond removes floppy modes, the other part introduces strain and, therefore, there is a *stress transition* at which a single additional bond causes a large stressed region to appear, so that the associated stress percolation transition is expected to be first order. In Thorpe et al.'s simulations, this type of transition occurred at $\langle Z\rangle \simeq 2.392$. Although these values are not in numerical agreement with the data of Georgiev et al. (2000), they do point out the possibility of developing an appropriate vector percolation model for explaining their data.

13.3 Branched Polymers and Gels

An old, but still quite useful review of the link between polymers and gels, on the one hands, and percolation and critical phenomena, on the other hand, is that of Stauffer et al. (1982). Materials that contain gels, or use their specific properties, are numerous. An important example is eye humor, a transparent water-like fluid similar to plasma that contains low protein concentrations that fills both the anterior and the posterior chambers of the eye. In addition, gels play an important role in laboratory technology, such as gel chromatography, in the fabrication of a wide variety of products, such as glues, cosmetics, contact lenses, etc., and in food technology. Most importantly, the sol–gel transition is a general phenomenon that has been utilized for producing a variety of ceramic materials (Brinker and Scherer 1990).

Chemical reactions are responsible for the interconnectivity of the monomers in chemical gels, of which there are three types:

(i) *Polycondensation,* in which one begins with either bifunctional units A-A or trifunctional ones B_3, or more generally $Z-$functional units B_Z. The reaction links the A units with the B units, with each elementary reaction being accompanied by the elimination of a molecule between units of A and B. Thus, a polymer network is formed in which the chains are terminated by either A or B. No two units of the same class participate in a reaction with each other and, therefore, there is always exactly one bifunctional unit between polyfunctional units in the polymer network.

(ii) *Vulcanization,* which starts with long linear polymer chains in a solution that are crosslinked by small units. A good example is rubber, the elasticity of which is due to the introduction of S-S bonds between polyisoprene chains. Only a small number of bonds is needed to crosslink the chains and form an interconnected polymer network.

(iii) *Additive polymerization,* whose initial solution contains two types of units, the A = A units that are bifunctional when the double bond opens, and B = D = B units that are quadrifunctional when the two double bonds open independently. If the reaction polymerizes the A = A units, one obtains A-A-A-A-··· chains, whereas reaction between the A units and the B = D = B units reticulates the network. The length of the chains between two reticulation points is not fixed, but depends crucially on the initiation process, as well as the relative concentrations of the bi-and quadrifunctional units.

In *physical gels*, on the other hand, the monomers or particles are attached to each other by relatively weak and reversible association, or such physical processes as entanglement. A well-known example is silica aerogel (see below). Another example is a solution of gelatin in water below a certain critical temperature, where a coil-to-helix transition takes place, with bonds appearing to form by winding of helixes of two adjacent chains. Physical gels can be made and also destroyed by thermal treatment.

As mentioned in Chap. 2, as well as earlier in the present chapter, although Flory (1941) and Stockmayer (1943) developed the theory of polymerization in the 1940s, their model is equivalent to percolation on Bethe lattices. Stauffer (1976), as well as de Gennes (1976a, b), pointed out the importance of the deficiencies of the Bethe lattice solution, and suggested instead that one must to use 3D lattices. This aspect of the problem has been described by the scalar percolation or a variant of it, and is now well-understood.

de Gennes (1976a) also proposed that the elastic and viscoelastic properties of the gel and sol phases can be described by appropriate random resistor network models. His suggestion was widely accepted for a long time and was utilized for interpreting experimental data. It was recognized in the 1980s that, while de Gennes' suggestion may be applicable to certain classes of materials, more general models are needed for at least several other important classes of polymeric materials, leading to the development of vector percolation models that were described in Chap. 11 and are invoked in the present chapter.

13.3.1 Percolation Model of Polymerization and Gelation

To understand the connection between gelation and percolation, consider a solution of molecules or monomers with functionality $Z \geq 3$. We assume that the monomers occupy the sites of a periodic lattice, and react with probability p with their nearest-neighbor monomers and form a chemical bond. If p is small, only small polymers—clusters of reacted monomers—are formed. As p increases, more monomers react and larger and larger polymers—clusters of reacted monomers—are formed with a broad size distribution. The mixture of clusters of reacted monomers and the isolated unreacted monomers represents the sol phase.

For $p > p_c$, where p_c is a characteristic value that depends on Z, an "infinite" cluster of reacted monomers is formed that represents the solid gel network. Near the gel point (GP) the gel usually coexists with a sol solution such that the finite polymers are trapped in its interior. As $p \to 1$, almost all monomers react, and the sol phase disappears. Thus, p_c signals a *connectivity* transition: For $p > p_c$, an infinite cluster, together (possibly) with a few finite-size clusters of reacted monomers, exist and, thus, the system is dominated by the solid gel. The fraction of chemical bonds formed at the GP is the analog of the bond percolation threshold p_{cb} (Chap. 2). Thus, formation of branched polymers and gels is a percolation process. In practice, the monomers do not react randomly; there are usually some correlations in the reaction of the monomers, but they do not change the essence of the process.

13.3.2 *Morphology of Branched Polymers and Gels*

Experimental studies of sol-gel transition during chemical reactions usually proceed by measuring the time evolution of the rheological properties, such as the viscosity of the sol phase, or the elastic moduli of the gel phase, assuming that the experimental parameter—time or frequency—and the theoretical one—the number of crosslinks—are linearly related in the vicinity of the GP. Rheological measurements are usually performed by using a cone and plate rheometer, or by the more accurate magnetic sphere rheometer. The ranges of shear rates, deformations, and times of measurements of such devices allow the determination of the steady-state zero-shear viscosity and the steady-state linear elastic moduli up to the vicinity of the phase transition at the GP, but it is almost impossible to do such measurements *at* the GP (see below).

The correlation or connectivity length ξ of branched polymers diverges as p_c is approached according to the power law

$$\xi \sim |p - p_c|^{-\nu_p} , \tag{13.6}$$

which is the analog of power law (2.11). Above the GP the correlation length of polymers is interpreted as the mesh size of the gel network. For any length scale greater than ξ the gel network is essentially homogeneous. Below the GP the correlation length is the typical radius of the finite polymers in the sol phase. In this case, however, the polymers with radii much larger than ξ have completely different characteristics. Therefore, we discuss such polymers separately, and we refer to them as the branched polymers.

13.3.2.1 Gel Polymers

Let us consider several important morphological properties of branched polymers and gel networks. The *gel fraction* $P_g(p)$ is the fraction of the monomers that belong to the gel network, which is measured by simply weighing the solid gel at different times during polymerization. It should be clear that $P_g(p) > 0$ only if $p > p_c$, and that P_g is the analog of percolation fraction or percolation probability $P(p)$ defined in Chap. 2. Of particular interest to us is the behavior of $P_g(p)$ near p_c, which has been shown to follow a power law:

$$P_g(p) \sim (p - p_c)^{\beta_p} , \tag{13.7}$$

which is completely similar to power law (2.2).

At the GP the gel network is *not* homogeneous, but is a self-similar fractal network with a fractal dimension D_p that in d−dimensions is given by

$$D_p = d - \beta_p / \nu_p , \tag{13.8}$$

which is the same as Eq. (2.27). The number distribution of the polymers, i.e., the probability $Q(s, \epsilon)$ that a polymer of the sol phase contains s monomers at a distance $\epsilon = |p - p_c|$ from the GP, is clearly the analog of the quantity n_s defined in Chap. 2 and, therefore, in analogy with Eq. (2.14) we write

$$Q(s, \epsilon) \sim s^{-\tau_p} h_1(\epsilon s^{\sigma_p}) , \tag{13.9}$$

where h_1 is a universal scaling function. Based on Eq. (13.9), we compute two distinct mass averages. One, the *weight-average molecular weight*, is defined by

$$M_w = \frac{\int s^2 Q(s, \epsilon) ds}{\int s Q(s, \epsilon) ds} \sim \epsilon^{-\gamma_p} \sim |p - p_c|^{-\gamma_p} , \tag{13.10}$$

where $\gamma_p + 2\beta_p = \nu_p d$, with d being the dimensionality of the material. In the polymer literature M_w is called the *degree of polymerization*. The second mass average is defined by

$$M_z = \frac{\int s^3 Q(s, \epsilon) ds}{\int s^2 Q(s, \epsilon) ds} \sim \epsilon^{-1/\sigma_p} \sim |p - p_c|^{-1/\sigma_p} , \tag{13.11}$$

where $\sigma_p = (\tau_p - 2)/\beta_p$, as in percolation (see Chap. 2). Note, however, that the average $\langle M \rangle = \int s Q(s, \epsilon) ds / \int Q(s, \epsilon) ds$ does *not* diverge at the GP.

13.3.2.2 Experimental Confirmation of the Percolation Model

A rational way of confirming the percolation model is through the universal critical exponents and fractal dimensions introduced above, since they do not depend on the details of the morphology of polymers or the location of the GP. Thus, we focus on a comparison between the measured exponents for polymers, such as β_p, γ_p and σ_p, and the values of the corresponding exponents for the power laws of percolation theory.

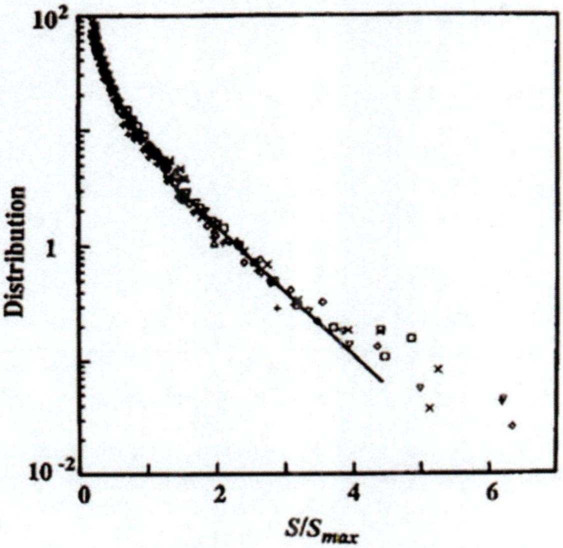

Fig. 13.3 Normalized polymer-size distribution as a function of polymer-size s. Percolation theory predicts the slope to be $1-\tau \simeq -1.3 \pm 0.1$ (after Leibler and Schosseler 1985)

In their experiments with irradiated polystyrene solution in cyclopentane, Leibler and Schosseler (1985) coupled gel permeation chromatography and light scattering to deduce the polymer-size distribution that provides a direct means of measuring the exponent τ_p. Figure 13.3 presents their data from which one obtains, $\tau_p \simeq 2.3 \pm 0.1$, consistent with the percolation prediction of 2.18 (see Table 2.3). Lapp et al. (1989) further checked the result by carrying out similar experiments in a system made by chemical end-linking of polydimethylsiloxane, and Patton et al. (1989) carried out experiments in a system in which polyester was made by bulk condensation polymerization. The measurements of both groups were consistent with the percolation prediction for the exponent τ_p. Note that since the polymer fractal dimension D_p is related to τ_p by

$$D_p = \frac{d}{\tau_p - 1}, \tag{13.12}$$

the aforementioned data imply that $D_p \simeq 2.5$, in agreement with the percolation value, $D_f \simeq 2.53$, listed in Table 2.3.

Candau et al. (1985) and Adam et al. (1987) tested the accuracy of Eq. (13.10). The latter carried out static light scattering measurements on a polyurethane sol, whereas the former performed experiments on polystyrene systems crosslinked with divinylbenzene. Figure 13.4 depicts the results of Adam et al. (1987) from which one obtains, $\gamma_p \simeq 1.71 \pm 0.06$, only 5% smaller than the percolation value of 1.82. A similar estimate was reported by Candau et al. (1985).

On the other hand, if we combine Eqs. (13.7) and (13.10), we obtain, $M_w \sim P_g^{-\gamma_p/\beta_p}$ and, thus, a plot of $\log(M_w)$ versus $\log(P_g)$ yields an estimate of γ_p/β_p.

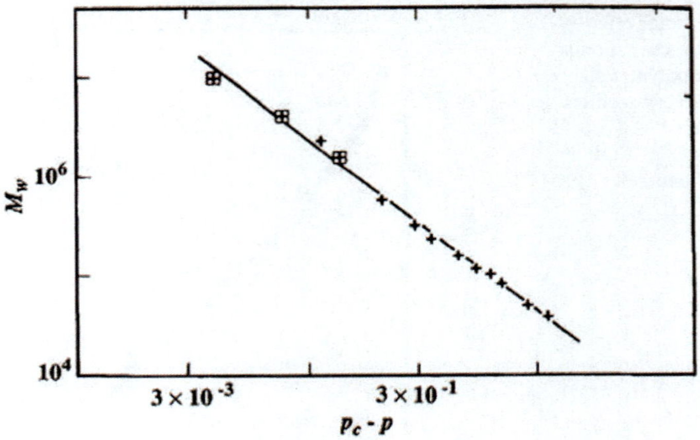

Fig. 13.4 Dependence of degree of polymerization M_w on $p_c - p$ for a polyurethane sol. The slope of the curve is $-\gamma_p$ (after Adam et al. 1987)

Schmidt and Burchard (1981) (see also Trappe et al. 1992) carried out anionic copolymerization of divinylbenzene with styrene and obtained both branched polymers and gels. Light scattering was used to measure various properties of interest. When $\log(M_w)$ was plotted versus $\log(P_g)$, as shown in Fig. 13.5, a straight line with the slope $\gamma_p/\beta \simeq 4.5$ was obtained, in good agreement with the percolation value of 4.44.

Therefore, 3D percolation provides a very accurate description of the universal morphological properties of gel polymer networks, whereas the predictions of the Flory–Stockmayer theory of polymerization, $D_p = 4$, $\gamma_p = \beta_p = 1$, do not agree with the experimental data described above.

13.3.2.3 Lattice Animal Model of Branched Polymers

After a polymer is formed by a chemical reaction, experimentalists usually analyze its structure by diluting it in a good solvent. The branched polymers in a dilute solution of a good solvent may swell and have a radius *larger* than their extent at the end of the chemical reaction. Thus, it is important to consider both the typical polymers and the swollen ones. A percolation model for the structural properties of branched polymers in a dilute solution of a good solvent has also been developed and has received experimental confirmation.

Consider a swollen branched polymer in a good solvent with a radius larger than the polymer correlation length ξ. The structural properties of such branched polymers are described by *lattice animals*,[4] which are percolation clusters below

[4] A lattice animal is a finite set of connected sites of a regular lattice, which is known as polyominoes on the square lattice and polycubes on the cubic lattice.

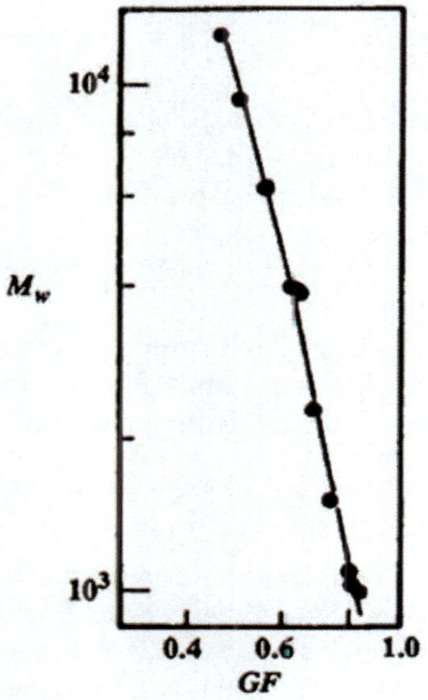

Fig. 13.5 Dependence of degree of polymerization M_w on the gel fraction P_g during anionic copolymerization of divinylbenzene with styrene. The slope of the curve is $-\gamma_p/\beta$ (after Schmidt and Burchard 1981)

p_c with radii larger than the percolation correlation length ξ_p. Their statistics are, however, completely different from those of percolation clusters. To see this, let us first define a few key properties of lattice animals.

Suppose that $A_s(p)$ is the average number, per lattice site, of the clusters, and a_{sm} is the total number of geometrically different configurations for a cluster of s sites and perimeter m. Thus, $A_s(p) = \sum_m a_{sm} p^s (1-p)^m$. For large s, the asymptotic behavior of $A_s(p)$ is described by the following power law

$$A_s(p) \sim \lambda_g^s s^{-\theta} , \tag{13.13}$$

where θ is a universal exponent, independent of the coordination number of the lattice, whereas the *growth parameter* λ_g is not universal. Moreover, for large s a fractal dimension D_a, defined by

$$s \sim R^{D_a} , \tag{13.14}$$

describes the structure of the animals or the branched polymers, with R being their radius. Lubensky and Isaacson (1978) and Family and Coniglio (1980) showed that the exponents θ and D_a are not related to any of the percolation exponents defined in Chap. 2. Moreover, Parisi and Sourlas (1981) showed that

$$\theta = \frac{d-2}{D_a} + 1 , \tag{13.15}$$

and that

$$D_a = 2\ , \quad d = 3\ . \tag{13.16}$$

A pair correlation function $C(r)$, i.e., the probability that two monomers or sites, separated by a distance r, belong to the same polymer or cluster, may also be defined. For a $d-$dimensional branched polymer and for large r, the correlation function decays as

$$C(r) \sim r^{D_a - d}\ . \tag{13.17}$$

The Fourier transform of $C(r)$ is proportional to the scattered intensity $I(q)$ in an X-ray or a neutron scattering experiment, where q is the magnitude of the scattering vector. If we Fourier transform Eq. (13.17), we obtains

$$I(q) \sim q^{-D_a}\ . \tag{13.18}$$

Two other key differences distinguish lattice animals from percolation clusters. One is that the exponents θ and D_a defined above are applicable for *any* $p < p_c$ (recall that the percolation exponents are defined for $p \simeq p_c$), so long as $R \gg \xi$. The other difference is that the upper critical dimension for lattice animals or branched polymers, i.e., the dimension at which the mean-field approximation—the solution obtained with the Bethe lattice—to the critical exponents becomes exact, is 8, two more than that of percolation.

In practice, however, polymer solutions are almost always polydispersed and contain polymers of all sizes with radii that are smaller or larger than the correlation length ξ. Thus, one must define *average* properties, with the average taken over the polymer-size distribution. An average polymer radius is defined by,

$$\langle R\rangle = \frac{\sum_s s^2 R(s) Q(s,\epsilon)}{\sum_s s^2 Q(s,\epsilon)}\ , \tag{13.19}$$

which, when combined with Eqs. (13.9) and (13.14), yields a relation between s and $\langle R\rangle$ (Daoud et al. 1984):

$$s \sim \langle R\rangle^{D_a(3-\tau_p)}\ , \tag{13.20}$$

so that, in analogy with Eq. (13.14), an *effective fractal dimension*, $D_a^e = D_a(3-\tau_p)$, may be defined. In effect, Eq. (13.20) mixes the branched polymer fractal dimension D_a with the gel network exponent τ_p. Thus, provided that the percolation description of polymerization is valid, which the experiments described above confirm it to be the case, we should have, $\tau_p(d=3) \simeq 2.18$, implying that

$$D_a^e \simeq 1.64\ , \quad d = 3\ , \tag{13.21}$$

indicating that the effective fractal dimension is smaller than that of a single branched polymer. Since we have defined an effective fractal dimension for a dilute polydisperse polymer solution, the scattering intensity for a solution of such polymer should also be modified to

$$I(q) \sim q^{-D_a(3-\tau_p)} . \quad (13.22)$$

In practice, Eq. (13.22) is used in a scattering experiment to estimate D_e.

13.3.2.4 Experimental Confirmation of Lattice Animal Model of Branched Polymers

Experimental evidence for Eq. (13.16) is actually provided through Eq. (13.18). Bouchaud et al. (1986) carried out small-angle neutron scattering experiments on a monodisperse polyurethane sample and measured the scattered intensity as a function of q. Figure 13.6 presents the data for the polymer from which one obtains, $D_a = 1.98 \pm 0.03$, in excellent agreement with Eq. (13.16). Bouchaud et al. (1986) also synthesized a natural polydisperse polyurethane sample and carried out small-angle neutron scattering on a dilute solution of it. Figure 13.6 also presents the data from which one obtains, $D_a^e \simeq 1.6 \pm 0.05$, in good agreement with the theoretical prediction given by Eq. (13.21).

Adam et al. (1987) carried out static light scattering experiments in dilute polydisperse polyurethane solutions and reported that, $D_a^e \simeq 1.62 \pm 0.08$, again in good agreement with (13.21). Leibler and Schosseler (1985) measured the average radius of polystyrene, crosslinked by irradiation by elastic light scattering and found that, $D_a^e \simeq 1.72 \pm 0.09$, consistent with Eq. (13.21). Patton et al. (1989) performed both quasi-elastic and elastic light scattering experiments on branched polyesters and reported that, $D_a^e \simeq 1.52 \pm 0.1$, somewhat lower than the prediction, but still consistent with it.

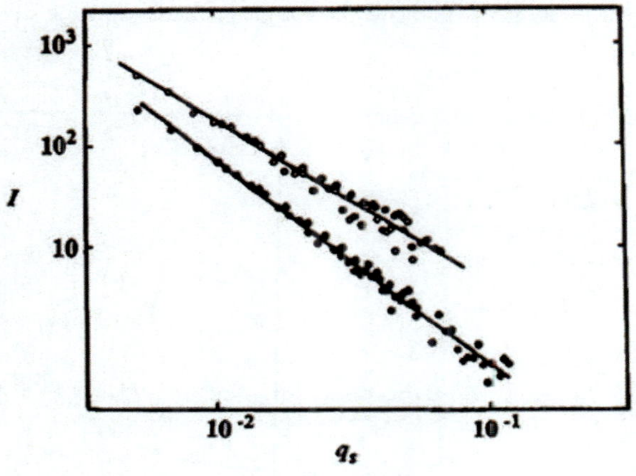

Fig. 13.6 Small-angle neutron scattering data for branched polymers. The upper curve is for a polydisperse polymer solution with a slope −1.6. The lower curve is for a single polymer in a good dilute solvent with a slope −1.98 (after Bouchaud et al. 1986)

A different experiment was carried out by Dubois and Cabane (1989) on a silica gel, a physical gel that has a morphology more complex than the chemical gels used in the experiments described above (see the discussion below). Despite significant differences, Dubois and Cabane reported that, $D_a^e \simeq 1.58$, quite close to the theoretical prediction, Eq. (13.21).

13.4 Elastic Moduli of Critical Gels

Studying the viscosity of the sol phase and the elastic moduli of the gel phase as the GP is approached from either side enables one to understand the evolution of a polymer's molecular structure during the gelation. This is shown schematically in Fig. 13.7. Initially, when no monomer has reacted yet (corresponding to $p = 0$, with p being the fraction of reacted monomers), the liquid solution has a steady shear viscosity η that increases with the extent of the reaction, which also increases the average molecular weight M_w of the evolving polymer. At the GP the viscosity and the longest relaxation time $t_{\max}$ both diverge. For $p > p_c$ the equilibrium elastic moduli increase until they attain their highest values, which is when the reaction is brought to completion and $p \to 1$.

All the experimental data for the elastic moduli of gel networks above, but close to the GP indicate that they follow a power law given by

$$\text{elastic moduli} \sim (p - p_c)^z\ , \quad p > p_c\ . \tag{13.23}$$

On the other hand, the viscosity of the sol phase diverges as

$$\eta \sim (p_c - p)^{-k}\ , \quad p < p_c\ , \tag{13.24}$$

while, as shown below, the longest relaxation time diverges as

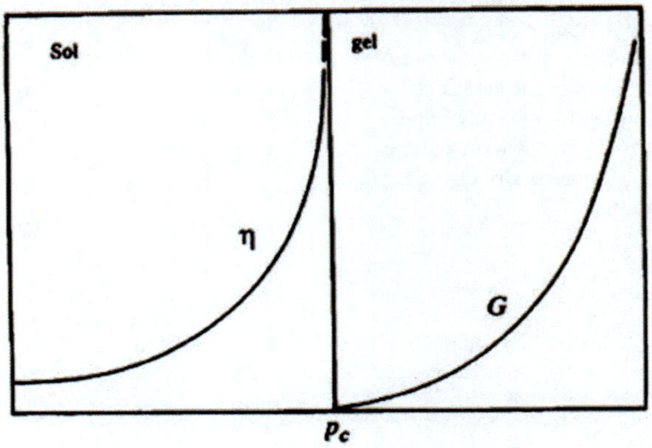

Fig. 13.7 Typical variations of viscosity η below the gel point, and of the elastic moduli above the gel point, with the fraction of the reacted monomers

$$t_{\max} \sim |p - p_c|^{-k-z} \; , \quad |p - p_c| \ll 1 \; . \tag{13.25}$$

In practice, it is precisely the divergence of η that signals the formation of a gel network. Divergence of $t_{\max}$ implies that measurements of the viscosity and elastic moduli at p_c is essentially impossible, because the steady-state condition cannot be reached in a finite time. In addition, precise measurement of the GP is often very difficult.

To overcome such problems, *dynamic-mechanical experiments* in which the sample is exposed to a periodically varying stress field, are carried out. For example, a tensile stress is used that, at time t, is given by

$$\sigma_{zz}(t) = \sigma_{zz}^0 \exp(i\omega t) \; . \tag{13.26}$$

The tensile stress results in a time-dependent longitudinal strain $\epsilon_{zz}(t)$ that varies with the frequency of the stress, but exhibits, in general, a phase-lag φ such that

$$\epsilon_{zz}(t) = \epsilon_{zz}^0 \exp[-i(\omega t - \varphi)] \; . \tag{13.27}$$

We may then employ a *dynamic tensile modulus* $E^*(\omega)$, defined as

$$E^*(\omega) = \frac{\sigma_{zz}(t)}{\epsilon_{zz}(t)} = E' + iE'' \; . \tag{13.28}$$

Analogous experiments can, of course, be carried out for other types of mechanical loading. Of particular interest are measurements for simple shear that determine the relation between the shear strain ϵ_{zx}, yielding the displacement along the x direction, per unit distance normal to the shear plane $z =$ constant, and the shear stress σ_{zx} that acts on the shear plane along x.

Such dynamic-mechanical experiments measure the small amplitude oscillatory shear properties of continuously evolving gels with *no* singularity), although even such experiments cannot entirely overcome the difficulties in the determination of the viscosity η in the critical region and the associated exponent k, because the measurements cannot be carried out *at* the GP and in the limit of *zero frequency*. To estimate k, one usually measures the frequency-dependent complex modulus, $G^*(\omega) = G'(\omega) + iG''(\omega)$. At the GP and for low frequencies, one has

$$G' \sim G'' \sim \omega^u \; , \quad p = p_c \; , \tag{13.29}$$

with

$$u = \frac{z}{z+k} \; , \tag{13.30}$$

where G' (storage modulus) and G'' (loss modulus) describe storage and dissipation in an oscillating strain field of constant amplitude. Typical variations of G' and G'' with ω are shown in Fig. 13.8 for a polycondensed gel very close to its GP.

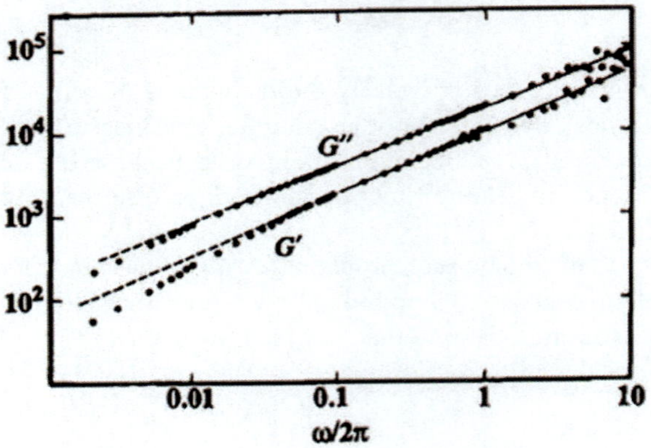

Fig. 13.8 Frequency-dependence of the storage modulus G' and loss modulus G'' for a polycondensed gel close to the gel point (after Durand et al. 1987)

The complex modulus $G^*(\omega)$ is sometimes written as $G^* = G + i\omega\eta$, for which Durand et al. (1987) proposed the following scaling form:

$$G^*(\omega, \epsilon) \sim \epsilon^z h_2(i\omega\epsilon^{z+k}) , \tag{13.31}$$

where $\epsilon = |p - p_c|$, and $h_2(x)$ is a universal scaling function. The significance of scaling equation (13.31) is that it enables one to collapse the data for all values of ϵ and ω onto a single universal curve—called the *master curve* by polymer researchers. In the low-frequency regime, we do not expect G^* to depend on ϵ, but only on ω, in which case one finds that $G^* \sim (i\omega)^u$, which is equivalent to Eq. (13.29). Moreover, a loss angle δ is defined by, $\tan\delta = G'/G''$. The remarkable property of δ is that it takes on a value δ_c at the GP, given by

$$\delta_c = \frac{\pi}{2}(1-u) = \frac{\pi}{2}\frac{k}{z+k} . \tag{13.32}$$

so that, if the critical exponents z and k are universal, so will also be the loss angle δ_c.

Accurate determination of the GP—the percolation threshold of the polymer network—is important, either for avoiding it in order to prevent gelation, so that a branched polymer with specific properties can be produced, or for making materials very close to GP, since they have unusual properties. Similar to the percolation threshold, the GP depends on the functionality of the polymer and decreases with increasing functionality. This implies that polymers with crosslinks of high functionality gel very early.

Holly et al. (1988) argued that since as the GP is reached, the loss angle $\tan\delta$ becomes independent of the frequency, if one plots $\tan\delta$ versus time at various frequencies, the intersection of the various curves should be at the GP. Figure 13.9, taken from Lin et al. (1991), demonstrates the implementation of the proposed method.

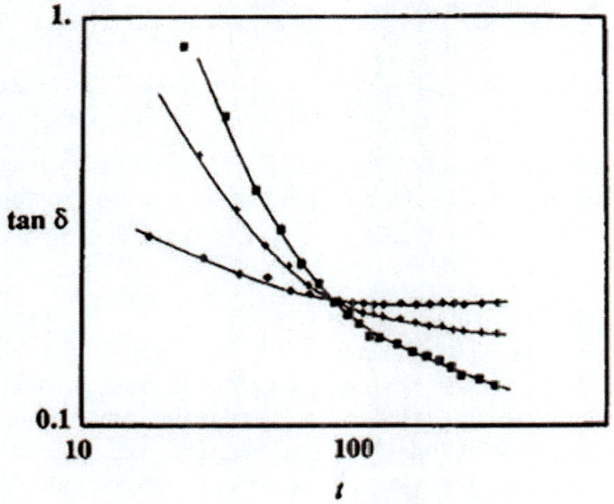

Fig. 13.9 Determination of gel point from data for loss angle δ. Time is in minutes. The data are for frequencies 31.6 rad s^{-1} (diamonds), 1.0 rad s^{-1} (+) and 0.0316 rad s^{-1} (squares) (after Lin et al. 1991)

13.4.1 The Spectrum of the Relaxation Times

Time-dependence of any macroscopic relaxation process is always indicative of the underlying microscopic dynamics and gels polymers are no exception. Thus, one may develop suitable equations that correctly describe dynamic evolution of the observed or measured responses of the material. In the simplest case, there is only a single characteristic time τ, whose origin goes back to Debye[5] who proposed it in his seminal work on the dielectric response of polar liquids. If we define a shear compliance $J(t)$ by

$$J(t) = \frac{\epsilon_{zx}}{\sigma_{zx}^0} ,$$

then, applying an oscillatory shear stress

$$\sigma_{zx}(t) = \sigma_{zx}^0 \exp(i\omega t) ,$$

to a polymer means imposing on the sample an oscillatory strain $\epsilon_{zx}(t) = \Delta J \sigma_{zx}(t)$, where ΔJ is the *relaxation strength* of the material. The governing equation for

[5] **Peter Joseph William Debye** (1884–1966) was a Dutch physicist and physical chemist. He developed the application of dipole moments to the charge distribution in asymmetric molecules, and equations that linked the moments to temperature and dielectric constant. The units of molecular dipole moments are called debyes in his honor. He also extended what is now called the Debye model, which is an extension to lower temperatures of Albert Einstein's theory of specific heat, and also extended Niels Bohr's theory of atomic structure by introducing elliptical orbits. Debye also calculated the effect of temperature on X-ray diffraction patterns of crystalline solids, the so-called Debyre–Waller factor. Together with his assistant Erich Hückel, Debye developed an improvement of Svante Arrhenius' theory of electrical conductivity in electrolyte solutions. He received numerous awards, including the Nobel Prize in Chemistry in 1936.

$\epsilon_{zx}(t)$ is then given by

$$\frac{d\epsilon_{zx}(t)}{dt} = -\frac{1}{\tau}\left[\epsilon_{zx}(t) - \Delta J \sigma_{zx}^0 \exp(i\omega t)\right] . \tag{13.33}$$

Assuming a solution, $\epsilon_{zx}(t) = \sigma_{zx}^0 J^*(\omega)\exp(i\omega t)$, where $J^*(\omega)$ is the complex shear compliance, substituting it into Eq. (13.33) and solving it, yield

$$J^*(\omega) = \frac{\Delta J}{1 + i\omega\tau} , \tag{13.34}$$

which is usually referred to as a *Debye process*.

Polymers are, however, complex materials and their dynamics in a reaction bath cannot be described by a single relaxation time, but rather by a distribution of such characteristics times. For the shear properties, for example, we write the dynamic compliance $J^*(\omega)$ as a sum of Debye processes with relaxation times τ_i and relaxation strengths ΔJ_i, so that

$$J^*(\omega) = J_u + \sum_i \frac{\Delta J_i}{1 + i\omega\tau_i} , \tag{13.35}$$

which is usually represented by an integral,

$$J^*(\omega) = J_u + \int \frac{\mathcal{R}(\tau)}{1 + i\omega\tau} d\tau , \tag{13.36}$$

where $\mathcal{R}(\tau)$ is called the *retardation time spectrum* of the shear compliance J^*. These results may also be written in terms of the complex modulus $G^*(\omega) = 1/J^*(\omega)$, which then yields

$$G^*(\omega) = G_u - \int \frac{H(\tau)}{1 + i\omega\tau} d\tau , \tag{13.37}$$

where $H(t)$ is called the *relaxation time spectrum* of the complex modulus $G^*(\omega)$ [or $G(t)$] (Ferry, 1980).

The observed power-law behavior of G' and G'', Eq. (13.29), implies that the relaxation time spectrum is self-similar in time, i.e., it follows a power law:

$$H(t) = \frac{G_0}{\Gamma(u)}\left(\frac{t}{t_0}\right)^{-u} , \tag{13.38}$$

where G_0 is a characteristic modulus, t_0 is the shortest relaxation time, and $\Gamma(x)$ is the gamma function. The modulus of a fully crosslinked polymer network is typically $10^6 - 10^7$Pa, while the relaxation time of a network strand is about $10^{-7} - 10^{-4}$ sec. The relaxation time spectrum extends from the shortest time at which strands are beginning to be probed, to the infinite relaxation time of the diverging (largest) polymer cluster. The parameters G_0 and t_0 are material characteristics of the gel.

Most gels seem to possess the same value of u, for which the universality of vector percolation provides an explanation (see below). There are, however, gels that exhibit no universality in the value of u.

The moments of $H(t)$ are directly related to the viscosity and the elastic moduli. Using Eq. (13.31), we can back-calculate $H(t)$ (Daoud 1988):

$$H(t) \sim t^{-u} h_3(t\epsilon^{z+k}) , \tag{13.39}$$

where h_3 is another universal scaling function. Equation (13.39), which indicates that, in the scaling regime near the GP, the relaxation time distribution is a slowly decaying power law, generalizes Eq. (13.38) to any value of p, the extent of the polymerization. Daoud (1988) pointed out that two distinct average or characteristic times may be defined. One is

$$t_1 = \frac{\int H(t)dt}{\int [H(t)/t]dt} \sim \epsilon^{-k} \sim \eta , \tag{13.40}$$

while the second one is given by

$$t_2 = \frac{\int tH(t)dt}{\int H(t)dt} \sim \epsilon^{-z-k} . \tag{13.41}$$

Note that t_2 is in fact identical with $t_{\max}$, the longest relaxation time of the gel system; see Eq. (13.25). Because of the existence of $H(t)$ and its scaling form, *any* relaxation property in the intermediate time or frequency range is *not* exponential, but follows a *power law*.

13.4.2 Experimental Confirmation

To validate the vector percolation models for the rheology of gels and their elastic properties, we compare the experimental data for the scaling properties of the elasticity of gels by dividing the gels into two groups. What follows is a brief description of each.

13.4.3 Chemical Gels

Numerous papers have reported experimental data for the elastic moduli of chemical gel networks and the associated exponent z. Examples include the data for hydrolyzed polyacrylamide (Allain and Salomé, 1987b,1990), Tetraethyl orthosilicate reactions (Hodgson and Amis 1990; Takahashi et al. 1994), gelatin solutions

(Djabourov et al. 1988), polycondensation of polyoxypropylated trimethylolpropane with hexamethylenediisocyanate (Durand et al. 1987), and several other measurements (Gauthier-Manuel and Guyon 1980; Adam et al. 1981; Gordon and Torkington 1981; Tokita et al. 1984; Fadda et al. 2001). Such data yielded estimates of z in the range 1.9–2.4, which do not agree with the critical exponent f for the 3D BB model, $f \simeq 3.76$. In fact, if the size of a chemical gel network is large enough, the BB forces may not play an important role in determining the elastic properties of the near critical gels, and therefore the only important forces between the monomers are the central (stretching) forces (CFs). But, as described in Chap. 11, the CF percolation model in 3D appears to undergo a first-order transition and, therefore, the exponent f is not well-defined for it.

The experimental estimate of z does appear to agree with that of μ, the conductivity critical exponent, which had been suggested by de Gennes (1976a). We will return to this point shortly. Alexander (1984) argued that there are terms in the elastic energy of some gels and rubbers, which are under internal or external stresses, which are similar to the Born model described in Chap. 11. As discussed there, the critical exponent of the elastic moduli of the Born model is equal to the conductivity exponent μ, and in particular in 3D, $f = \mu \simeq 2.0$, because near the percolation threshold p_c, the contribution of the purely scalar term of the total elastic energy of the Born model dominates that of the CFs.

There are also a few relatively precise sets of experimental data that seem to support de Gennes' conjecture. For example, Axelos and Kolb (1990) measured the rheological properties of pectin biopolymers that consist of randomly connected $\alpha(1-4)$ $D-$galacturonic acid units and their methyl esters. If the methyl ester content is low, pectin forms thermoreversible gels upon the addition of cations, such as calcium. Axelos and Kolb measured the frequency-dependence of the storage modulus $G'(\omega)$ and loss modulus $G''(\omega)$ and reported that, $z \simeq 1.93$, $k \simeq 0.82$, and $u \simeq 0.71$. Their elasticity exponent is close to the conductivity exponent $\mu \simeq 2.0$ for 3D scalar percolation (resistor networks). Less precise data, but still supportive of de Gennes' proposal, were reported by Adam et al. (1997) for complex modulus of end-linked poly(dimethylsiloxane) pregel polymer clusters, quenched at various distances from the gelation threshold. They reported that $z \simeq 1.9 \pm 0.15$, consistent with the 3D value of the conductivity exponent μ.

At first glance, de Gennes' suggestion that $z = \mu$ may seem incorrect, since z characterizes the power-law behavior of a vector property, whereas μ is associated with a scalar property. To justify his proposal, de Gennes introduced the notion of an elastic chain between neighboring nodes or monomers that are the analog of quasi-1D strands that percolation clusters possess near the percolation threshold. He then argued that if such chains are elongated, then their nodes carry an extra amount of energy. If we assume that the blobs (i.e., the multiply-connected parts of the SSC; see Chap. 2) in the large cluster of monomers do not contribute significantly to the elastic moduli, then one must only consider the energetics of the links or the chains. If the extra energy of such chains is larger than $k_B T$, then, as Daoud (2000) argued, one obtains de Gennes' proposal, $\mu = z$.

As for Alexander's proposal, rubbers and gels differ from the Born model in several important ways, such as the presence of non-linear terms in their elastic energy, and the possibility of negative as well as positive Born coefficients α_1 and α_2 in Eq. (11.11). Therefore, while one may use the Born model to *fit* the experimental data, it is not clear, at a fundamental level, that the Born model can actually describe the elastic properties of such gels, since its elastic energy is not rotationally invariant, even though the numerical value of the critical exponent of its elastic moduli might to be close to measured values for some chemical gels.

13.4.4 Enthalpic Versus Entropic Elasticity

There is yet another way of rationalizing the experimental data for the elastic properties of chemical gels near the GP. We first note that several sets of data for the elastic moduli of chemical gels and the associated exponent z deviate significantly from all the data described above. Examples include the measurements of Adam et al. (1985) for polycondensation, $z \simeq 3.3 \pm 0.5$; those of Martin et al. (1988) and Adolf et al. (1990) for gels made from 89% (by weight) of the diglycidyl ether of bisphenol A, cured with 11 percent (by weight) of diethanolamine that yielded $z \simeq 3.3 \pm 0.3$, and the data reported by Colby et al. (1993) for polyester gels that have been argued to lie in the middle of the static crossover between the Flory–Stockmayer theory and the percolation model. Colby et al. reported the estimate, $z \simeq 3.0 \pm 0.7$, which is inconsistent with the BB model, although one might argue that their reported estimate of z is agreement with $\mu = 3$, the mean-field value of the effective conductivity exponent (see Table 2.3). Measurements of the shear modulus of an end-linking polymer gel network by Takahashi et al. (1994) yielded $z \simeq 2.7$, which is again in the range of the above data.

One possible explanation for such data is that the elasticity of such gels is entropic rather than enthalpic (or energetic). Plischke and Joós (1998), Plischke et al. (1999), and Farago and Kantor (2000) argued that the CF and the BB models are applicable to gels at temperature $T = 0$, and that for $T \neq 0$ there is a contribution to the shear modulus that is entropic in nature. In analogy with the physics of rubber elasticity, Plischke et al. (1999) argued that near p_c or the GP, the polymer network consists essentially of long chains of singly connected particles (monomers or sites), linked to each other at various junction point, and are similar to the polymer chains that are crosslinked in rubber in order to produce a rigid amorphous material. Deformation of the material changes the distance between junctions points or crosslinks, as a result of which the entropy is generically decreased, resulting in an increase of free energy and a restoring force. When a SSC, or a large polymer network, is formed, there is a net shear restoring force, implying that the connecting chain of particles acts as a stretched spring. Molecular dynamics simulations of Plischke and co-workers and computer simulations of Farago and Kantor (2000), who used a model consisting of hard spheres in which a fraction p of the neighbors were tethered by inextensible bonds, both yielded $z \simeq \mu$.

On the other hand, del Gado et al. (1999) proposed a model, purported to be appropriate for entropic gels, in which one begins with a random collection of monomers with concentration p, each pair of which link up with probability p_b and form permanent bonds. Varying p_b produces a distribution of clusters of linked monomers and, hence, eventually a SSC. The monomers and the clusters then diffuse according to the bond fluctuation algorithm of Carmesin and Kremer (1988) in which they diffuse randomly in the solution, satisfying the excluded-volume interaction—no two monomers occupy the same point in space. Due to the random motion, the bonds may have to change their length in a set of allowed values and, thus, they may have to bend and take on various values of the angles between the bonds, which generates a wide variety of polymer conformations. The mean-squared displacement $\langle R^2(t)\rangle$ of the system's center of mass is then calculated that, due to the elastic potential that reduces the fluctuations proportionally to the effective elastic constant α_e, is given by

$$\langle R^2(t)\rangle \propto \alpha_e^{-1} \ , \tag{13.42}$$

and, therefore, the elastic constant and its power-law behavior near p_c and, hence, the exponent z can be computed. Simulations of del Gado et al. (1999) in 2D yielded a value $z \simeq 2.7 \pm 0.1$, disagreeing with the results of Plischke and co-workers, and Farago and Kantor.

If the elasticity of such gels is due to entropic effects, then, as Daoud and Coniglio (1981) argued (see also Martin et al., 1988), the elastic free energy per unit volume $\mathcal{H}$ must be given by

$$\mathcal{H} \sim \xi_p^{-d} \alpha_\xi \xi_p^2 \ , \tag{13.43}$$

where ξ_p is the correlation length of percolation, and α_ξ is the effective elastic constant of a long chain of length ξ_p connecting two nodes. Since $\alpha_\xi \sim \xi_p^{-2}$, we obtain

$$z = \nu d \ . \tag{13.44}$$

Equation (13.44) predicts that for 2D systems, $z \simeq 2.66$, quite different from the exponents f of vector percolation, as well as the conductivity exponent $\mu \simeq 1.3$ in 2D. It is, however, in agreement with the numerical simulations of del Gado et al. (1999). Daoud (2000) argued that Eq. (13.44) is valid when the energy of the chains is of the same order of magnitude as the thermal energy k_BT.

Equation (13.44) also predicts that in 3D, $z \simeq 2.64$, which is consistent with the experimental data of Adam et al. (1985); Martin et al. (1988); Adolf et al. (1990); Colby et al. (1993) and Takahashi et al. (1994) mentioned above, all of whom reported z values in the range $2.7 - 3.3 \pm 0.5$. Therefore, while such experimental data may be explained by the entropic effects and Eq. (13.44), the numerical results of Plischke and co-workers, and those of Farago and Kantor (2000) do not agree with the prediction of Eq. (2.44). Indeed, while the main argument of Plischke et al. (1999) was that the entropic effects are important at temperatures $T \neq 0$ where one should see a crossover to the conductivity exponent, all of the above experiments were carried out at finite temperatures, yet they did not produce $z \simeq \mu$.

13.4.5 Physical Gels

Two examples of physical gels are gelation of silica particles in NaCl solutions and in pure water (Gauthier-Manuel et al. 1987), and silica aerogels (Woignier et al. 1988). As discussed earlier, the attachment of the particles in such gels is by relatively weak association. The BB forces are important in such gels since when touching particles that form long chains are deformed, they roll on top of one another. Such motion and the displacement of the centers of any three mutually-touching particles generate forces that are equivalent to the BB forces.

Experimental measurements (Gauthier-Manuel et al. 1987; Woignier et al. 1988) of the elastic moduli of such gels support this: The exponent z was found to be about 3.8, in excellent agreement with the critical exponent f of the 3D BB model. But measurements by Devreux et al. (1993) indicated a crossover between the prediction of the BB model and another regime with a much smaller value of z. They measured the complex modulus G^* of silica gels formed by hydrolysis-condensation of a silicon alkoxide. For a restricted region near the GP, they reported $z \simeq 2.0 \pm 0.1$, close to the exponent μ of percolation conductivity, whereas for a region beyond the first one they reported $z \simeq 3.6 \pm 0.1$, which is close to the elasticity exponent of 3D BB model.

13.5 Viscosity of Near Critical Gelling Solutions

Extensive experimental data have been reported for the scaling behavior of the viscosity η of the sol phase near the GP, whose interpretation was debated for a long time. It was proposed by Sahimi and Goddard (1985) (see also Arbabi and Sahimi 1990b; Sahimi and Arbabi 1993) that the power-law behavior of η near the GP is analogous to that of the shear modulus of a superelastic percolation network (SEPN) near p_c. To further examine the proposal, we establish a rigorous relationship between the linear elasticity and the theory of viscous fluids.

Consider a general time-dependent system. The equation of motion for a macroscopically homogeneous material in terms of the displacement field $\mathbf{u}$ is given by

$$(\lambda + \mu)\nabla(\nabla \cdot \mathbf{u}) + \mu\nabla^2\mathbf{u} + \mathbf{F} = \rho\frac{\partial^2\mathbf{u}}{\partial t^2} , \tag{13.45}$$

where ρ is the mass density, and t is the time. The rest of the notation is the same as in Chap. 11. For an *incompressible* material, i.e., one whose bulk modulus K and the Lamé constant λ are divergent, one has the solenoidal condition,

$$\nabla \cdot \mathbf{u} = 0 . \tag{13.46}$$

Due to the incompressibility condition, the first term of Eq. (13.45) is indeterminate, and can be written in terms of the reactive hydrostatic pressure P, yielding

$$-\nabla P + \mu\nabla^2\mathbf{u} + \mathbf{F} = \rho\frac{\partial^2\mathbf{u}}{\partial t^2} . \tag{13.47}$$

On the other hand, the Navier–Stokes equations of motion for an incompressible and Newtonian viscous fluid is given by,

$$-\nabla P + \eta\nabla^2\mathbf{u} + \mathbf{F} = \rho\left(\frac{\partial\mathbf{v}}{\partial t} + \mathbf{v}\cdot\nabla\mathbf{v}\right) , \tag{13.48}$$

where $\mathbf{v}$ is the fluid velocity field, $\mathbf{F}$ is an external (body) force, and η is the fluid's dynamic viscosity. For an incompressible fluid, the the continuity equation is given by

$$\nabla\cdot\mathbf{v} = 0 . \tag{13.49}$$

For slow fluid flow, i.e., when the Reynolds number Re $\ll 1$, the non-linear inertial term, $\mathbf{v}\cdot\nabla\mathbf{v}$, is very small and can be neglected, which means that the Navier–Stokes equations reduce to

$$-\nabla P + \eta\nabla^2\mathbf{v} + \mathbf{F} = \rho\frac{\partial\mathbf{v}}{\partial t} . \tag{13.50}$$

Thus, we see that, under steady-state conditions, and when the flow of the fluid is slow, the governing equations for the displacement field $\mathbf{u}$ and the velocity field $\mathbf{v}$ are exactly identical, provided that there is a one-to-one correspondence between the shear modulus μ and the dynamic viscosity η.

In addition, under certain conditions, the effective viscosity η_e of a suspension of completely rigid spheres in an incompressible fluid of viscosity η_1 under creeping (very slow) flow conditions is related to the *steady-state* effective shear modulus μ_e of a two-phase material composed of the same completely rigid spheres in an incompressible matrix with shear modulus μ_1. In this case, the working equation is given by

$$\frac{\eta_e}{\eta_1} = \frac{\mu_e}{\mu_1} . \tag{13.51}$$

Equation (13.51) is exact if, regardless of the configuration of the particles, hydrodynamic interactions between the particles can be neglected, which is the case when the system is infinitely dilute, so that the volume fraction ϕ_2 of the particles approaches zero. If the system is non-dilute, Eq. (13.51) would still be exact, if the configurations of the particles in the flow and elasticity problems are identical.

Given the theoretical connection between the viscosity η of a solution and the shear modulus of an appropriate two-phase material, the one-to-one correspondence between η and the shear modulus of a SEPN should be clear because, (a) η and μ both diverge at p_c (the GP), and (b) percolation models predict accurately the morphology and elastic moduli of gel polymer networks. On the other hand, de Gennes (1979) [see also Allain and Salomé (1990)] suggested an analogy between η and the effective conductivity of a conductor-superconductor percolation networks whose effective conductivity diverges at p_c according to the power law, $g_e \sim (p_c - p)^{-s}$ (see Chap. 2), so that $k = s$.

A one-to-one correspondence between η and the shear modulus of a SEPN is not, however, enough to explain the scaling of η of a gelling solution near the GP, because most of the experimental data indicate that the value of k is either in the range 0.6–0.9 (see, for example, Adam et al. (1979, 1985); Allain and Salomé (1987a); Durand et al. (1987)), or in the range 1.3–1.5 (see, for example, Djabourov et al. 1988; Martin et al. 1988; Martin and Wilcoxon 1988), whereas the power-law behavior of the shear modulus of a 3D SEPN near p_c is characterized by a *unique* value of the exponent χ defined by Eq. (11.24).

The reason for the two distinct values of k is (Arbabi and Sahimi 1990b; Sahimi and Arbabi 1993) that the dynamics of the sol solutions that yield the two distinct values of k may be different. In one case, the solution may be such that there is little or very slow polymer diffusion in the reaction bath. Hence, a SEPN, a *static* system with fixed rigid clusters, may be better suited for describing the system. If so, the estimate $\chi = \nu - \frac{1}{2}\beta \simeq 0.65$ for 3D SEPNs, where ν and β are the usual percolation exponents), is consistent with the experimental data that produce a viscosity exponent in the range 0.6–0.9.

On the other hand, the gelling solution may be such that the polymers (clusters of monomers) can diffuse significantly in the reaction bath. To simulate this regime the following model was proposed (Arbabi and Sahimi 1990b; Sahimi and Arbabi 1993). We consider a SEPN in which each cluster of the totally rigid bonds represents a finite polymer. Due to the randomness of percolation networks, there is of course a wide distribution of such polymers or clusters in the network. The soft bonds (those with a finite elastic constant) represent the "liquid" solution in which the rigid clusters move randomly, with equal probability, in any direction of the network. This simulates the diffusion of the polymers in the reaction bath. Two rigid clusters cannot overlap, but they can temporarily join and form a larger cluster, which can also be broken up again at a later time. It was shown (Arbabi and Sahimi 1990b; Sahimi and Arbabi 1993) that the shear modulus of this dynamic SEPN diverges with an exponent χ' given by

$$\chi' \simeq 2\nu - \beta \,. \tag{13.52}$$

Equation (13.52) predicts that in 3D, $\chi' \simeq 1.35$, consistent with the set of experimental data that produce a viscosity exponent k in the range 1.3–1.5.

Daoud (2000) argued that, similar to the case of the elastic moduli described above, one must have two regimes. If the elastic chains carry an energy of the same order of magnitude as the thermal energy k_BT, then the exponent χ is given by Eq. (11.51). On the other hand, if the elastic chains are stretched and have an extra energy larger than k_BT, then one recovers Eq. (13.52), which had also been conjectured by de Gennes (1979), based on the analogy between the viscosity and the effective conductivity of conductor-superconductor percolation networks.

Similar to the elastic moduli of gel polymers and their associated exponent z, there are also some experimental data that indicate some deviations of the viscosity exponent k from χ or χ'. As pointed out earlier, however, experimental determination of k, as well as the elasticity exponent z, entails measuring the complex modulus $G^*(\omega)$ at a series of frequencies ω. Moreover, strictly speaking, the scaling laws for the elastic moduli of EPNs and the SEPNs are valid only in the limit, $\omega \to 0$, whereas in practice it is very difficult to reach such a limit and, therefore, the measured values of k may exhibit some deviations from χ or χ', which are probably transient effects that should diminish as very low frequencies are accessed.

Chapter 14
Vibrational Density of States of Heterogeneous Materials

14.1 Introduction

An important property of solids is their vibrational density of states. Consider a $L \times L \times L$ network (or a $L \times L$ in 2D) in which each site represents a particle of mass m, with the nearest-neighbor particles connected to each other by springs. Suppose that $\mathbf{u}_i = (u_{ix}, u_{iy}, u_{iz})$ is the vector displacement of the particle at i, which satisfies the equation of motion, Newton's second laws of motion,

$$m\frac{\partial^2 \mathbf{u}_i}{\partial t^2} = \sum \mathbf{F} , \tag{14.1}$$

where $\mathbf{F}$ is any force that acts on the particle i. If the springs connecting the particles are harmonic and can support only stretching forces, then $\sum \mathbf{F}$ is given by Hooke's law (force = spring constant $\times$ displacement), and the equation of motion becomes

$$m\frac{\partial^2 \mathbf{u}_i}{\partial t^2} = \sum_{<ij>} k_{ij} \left[(\mathbf{u}_j - \mathbf{u}_i) \cdot \mathbf{R}_{ij}\right] , \tag{14.2}$$

with $\mathbf{R}_{ij}$ being a unit vector from i to j, k_{ij} the spring (elastic) constant of the bond ij, and the sum is over all the bonds ij that are connected to i. For simplicity we take $m = 1$.

The standard method of analyzing Eq. (14.1) or (14.2) for the vibrational properties is by assuming that, $\mathbf{u}_i = \mathbf{A}_i \exp(-i\omega t)$, where ω is the frequency of the vibrations, and $\mathbf{A}_i$ is an unknown vector to be determined. Substituting the expression into Eq. (14.1) yields a set of $n = L^d$ simultaneous linear equations for the $\mathbf{A}_i$'s that has n positive eigenvalues $\omega_1^2, \omega_2^2, \cdots$, and n eigenvectors $\mathbf{A}_{e1}, \mathbf{A}_{e2}, \cdots$ Then, the solution to $\mathbf{u}_i$ is given by

$$\mathbf{u}_i = \mathrm{Re}[\sum_j c_j \mathbf{A}_{ej} \exp(-i\omega_j t)] , \tag{14.3}$$

M. Sahimi, *Applications of Percolation Theory*, Applied Mathematical Sciences 213,
https://doi.org/10.1007/978-3-031-20386-2_14

where c_j are complex numbers that must be determined from the initial conditions. Given the solution for $\mathbf{u}_i$'s, we can compute $\mathcal{N}(\omega)$, the *vibrational density of states* (DOS), where $\mathcal{N}(\omega)d\omega$ is the number of vibrational modes with a frequency between ω and $\omega + d\omega$. The DOS is important for determining the specific heat and thermal conductivity of a solid material, as well as the study of proteins and other biomaterials (see for example, Elber and Karplus 1986), and is computed from the distribution of the eigenmodes ω_i. Under certain conditions, however, there is a much simpler way of computing the DOS.

If we take the Fourier transform of Eq. (14.2), we obtain,

$$-\omega^2\hat{\mathbf{u}}_i = \sum_{<ij>} k_{ij}[(\hat{\mathbf{u}}_j - \hat{\mathbf{u}}_i)\cdot\mathbf{R}_{ij}] . \tag{14.4}$$

Writing Eq. (14.4) for a principal direction of the lattice, say x, yields,

$$-\omega^2\hat{u}_{ix} = \sum_{<ij>} k_{ij}(\hat{u}_{jx} - \hat{u}_{ix}) . \tag{14.5}$$

Consider the diffusion equation in discretized form,

$$\frac{\partial P_i}{\partial t} = \sum_j W_{ij}[P_j(t) - P_i(t)] , \tag{14.6}$$

which is usually called the *master equation*, where W_i is the transition rate between sites i and j, and $P_i(t)$ is the probability of finding a diffusing particle at site i at time t. If we take the Laplace transform of Eq. (14.6) and ignore the term that arises from the initial condition, we obtain,

$$\lambda\tilde{P}_i(\lambda) = \sum_{<ij>} W_{ij}(\tilde{P}_j - \tilde{P}_i) , \tag{14.7}$$

which is completely similar to Eq. (14.5), with the role of $\hat{u}_{ix}$ being played by $\tilde{P}_i$, and that of $-\omega^2$ by λ. We may then interpret λ as the frequency for diffusion, just as ω is the frequency for vibrations. This analogy was first exploited by Alexander et al. (1978) for computing the DOS. This is, of course, an approximation because $P_i(t)$ is a scalar quantity, whereas $\mathbf{u}_i$ is a vector, which is why computing the DOS based on the diffusion analogy is referred to as the *scalar approximation*.

To compute the DOS by the scalar approximation, consider $\langle P_0(t)\rangle$, the average probability that a diffusing particle, moving randomly in the lattice, is at the origin of its motion at time t, with the averaging taken over all the initial positions of the walker. It can be shown (Hughes 1995) that for classical diffusion in $d-$dimensional lattices,

$$\langle S(t)\rangle \sim t^{d/2} , \tag{14.8}$$

where $\langle S(t)\rangle$ is the mean number of distinct sites visited by the diffusing particle at time t. In the theory of random walks on lattices, it has also been shown that, $\langle P_0(t)\rangle = [\langle S(t)\rangle]^{-1}$, so that

$$\langle P_0(t)\rangle \sim t^{-d/2} . \tag{14.9}$$

It is then not difficult to show that in the scalar approximation $\mathcal{N}(\omega)$ and $\langle P_0(t)\rangle$ are related to each other through the following equation (Alexander et al. 1978)

$$\mathcal{N}(\omega) = -\frac{2\omega}{\pi}\mathrm{Im}\langle\tilde{P}_0(-\omega^2)\rangle . \tag{14.10}$$

If we take the Laplace transform of Eq. (14.9) and substitute the result into Eq. (14.10), we obtain

$$\mathcal{N}(\omega) \sim \omega^{d-1} , \tag{14.11}$$

which is the well-known result for the DOS in the Debye regime, i.e., at low frequencies, or long wavelengths/large length scales over which the material is homogeneous, such that $\omega < \omega_{co}$, where ω_{co} is a cutoff or crossover frequency to be described below.

It can be shown that if in the Debye regime we compute $\mathcal{N}(\omega)$ based on the solution of Eq. (14.1), it would still follow Eq. (14.11), provided that the material is macroscopically homogeneous. The general form of Eq. (14.11) is also consistent with this claim: There is nothing in (14.10) that indicates whether $\mathcal{N}(\omega)$ was computed from the solution of Eq. (14.1) or (14.5); only the dimensionality d of the material has entered (14.11). Vibrational states in homogeneous materials that are expressed by Eq. (14.11) are usually called *phonons*.

There are many methods for numerical computation of the DOS (see for example, Beniot et al. 1992; Royer et al. 1992). Our interest in this chapter is in the role that percolation plays in vibrational DOS of heterogeneous materials. Thus, let us describe and discuss this role.

14.2 Scalar Percolation Approximation for the Density of States

In this section we describe and discuss the application of the scalar percolation approximation for computing the DOS of heterogeneous materials, and compare the predictions with the results of large-scale numerical simulations, as well as experimental data.

14.2.1 Poor Man's Percolation: Effective-Medium Approximation

The DOS of a disordered network can be computed by an effective-medium approximation (EMA) (Derrida et al. 1984; Sahimi 1984). We do not give the complete derivation of the prediction of the EMa for the DOS, but present only the final result. Consider a $d-$dimensional percolation lattice with a percolation-type distribution of the transition rates W_{ij}, $f(W_{ij}) = (1-p)\delta(0) + p\delta(W_{ij}-1)$. Then, the EMA predicts that

$$\mathcal{N}(\omega) = \frac{2}{\pi\omega}\left[\frac{1}{2}I_w(d)(d-1)\right]^{1/2}(\omega^2-\omega_{co}^2)^{1/2}\ , \quad \omega > \omega_{co}\ , \quad 2 < d < 4\ , \tag{14.12}$$

where $\omega_{\rm cp}$ is a crossover frequency with, $\omega_{\rm co}^2 = -\lambda_{\rm co}$, and

$$\lambda_{co} = -\frac{d^2}{2(d-1)I_w(d)}(p-p_c)^2\ . \tag{14.13}$$

Here $I_w(d)$ is the Watson integral, which for the main three 3D cubic lattices is given by

$$\begin{aligned} I_{\rm SC} &= \frac{1}{\pi^3}\int\int_0^\pi\int \frac{dudvdw}{3-\cos u-\cos v-\cos w} \\ &= \frac{\sqrt{6}}{96\pi^3}\Gamma(\frac{1}{24})\Gamma(\frac{5}{24})\Gamma(\frac{7}{24})\Gamma(\frac{11}{24}) \simeq 0.50546\ , \end{aligned} \tag{14.14}$$

$$I_{\rm BCC} = \frac{1}{\pi^3}\int\int_0^\pi\int \frac{dudvdw}{1-\cos u\cos v\cos w} = \frac{1}{4\pi^3}\Gamma(\frac{1}{4})^4 \simeq 1.3932\ , \tag{14.15}$$

and

$$\begin{aligned} I_{\rm FCC} &= \frac{1}{\pi^3}\int\int_0^\pi\int \frac{dudvdw}{3-\cos u\cos v-\cos v\cos w-\cos w\cos u} \\ &= \frac{3}{2^{14/3}\pi^4}\Gamma(\frac{1}{3})^6 \simeq 0.4482\ , \end{aligned} \tag{14.16}$$

where $\Gamma(x)$ is the gamma function. According to Eq. (14.12), in the vicinity of (or at) p_c, the DOS approaches a constant for $\omega \gg \omega_{\rm co}$ (when $\omega_{\rm co}/\omega \ll 1$). As will be discussed in the next section, Alexander and Orbach (1982) showed that, *within the scalar approximation*, the vibrational DOS of a material at its percolation threshold p_c, or at $p > p_c$ but over length scales much smaller than ξ_p, the correlation length of percolation, is given by

$$\mathcal{N}(\omega) \sim \omega^{D_s-1}\ , \quad \omega \gg \omega_{co}\ , \tag{14.17}$$

where

$$D_s = 2\frac{D_f}{D_w} \tag{14.18}$$

is the spectral or fracton dimension of the network, with D_f and D_w being, respectively, the fractal dimension of the sample-spanning cluster (SSC) of percolation, and the fractal dimension of the diffusing particle on the SSC defined by its mean-squared displacements (see Chap. 7), $\langle R^2(t)\rangle \sim t^{2/D_w}$. For ordinary or Fickian diffusion, $D_w = 2$.

Equations (14.17) and (14.18) are actually supposed to be more general and valid for any material that has a fractal morphology with a fractal dimension D_f and a random walk fractal dimension D_w. Since according to the EMA, near or at p_c the DOS $\mathcal{N}(\omega)$ approaches a constant for $\omega \gg \omega_{co}$, the implication is that the EMA predicts the spectral dimension to be

$$D_s = 1\,, \quad 2 < d < 4\,. \tag{14.19}$$

It can be shown that (Sahimi et al. 1983b) the EMA also predicts that for the SSC at p_c, $D_w = 4$, and, therefore, we also obtain an EMA estimate for the fractal dimension D_f of the SSC, $D_f = 2$ for $2 < d < 4$, which should be compared with the numerical estimate of D_f in 3D, namely, $D_f \simeq 2.53$ (see Table 2.3).

For the times $t \gg t_{\rm co}$ transport in percolation networks is diffusive, while it is anomalous or fractal for $t \ll t_{\rm co}$, where $t_{\rm co}$ is the crossover time. It should be clear to the reader that the time scale $t_{\rm co}$ is nothing but $|\lambda_{\rm co}^{-1}|$ and, therefore,

$$t_{\rm co} = 2d^{-2}(d-1)I_w(d)(p-p_c)^{-2}\,. \tag{14.20}$$

Equation (14.20) should be compared with the correct scaling of $t_{\rm co}$, given by

$$t_{\rm co} \sim (p-p_c)^{-(2\nu+\mu-\beta)}\,, \tag{14.21}$$

where ν, μ and β are, respectively, the critical exponents for percolation correlation length, the conductivity, and percolation probability, as described in Chap. 2. Recall that in 3D, $\nu \approx 0.88$, $\mu \approx 2.0$, and $\beta \approx 0.41$. Thus, according to power law (14.21), we have, $t_{\rm max} \sim (p-p_c)^{-3.35}$, which should be compared with (14.20). For the DOS the role of $t_{\rm co}$ is played by the crossover frequency $\omega_{\rm co}$ defined above. Hence, for $\omega < \omega_{\rm co}$ and $p > p_c$ the mechanical vibrations of disordered materials, represented by a percolation network, are of ordinary, Debye type, and follow Eq. (14.11), whereas for $\omega \gg \omega_{\rm co}$ they are described by a new regime characterized by Eq. (14.17).

The EMA is also capable of predicting the DOS in the Debye or phonon regime, with the result being,

$$\mathcal{N}(\omega) = \frac{2}{\pi}J_d\omega^{d-1}\sin[(d/2-1)\pi]\left\{1-\frac{\omega^2}{2d}\left[1-\frac{1}{d}-\frac{\omega^2 I_w(d)}{2d}\right]^{-1}\right\}\,, \tag{14.22}$$

where

$$J_d = \frac{S_d}{(2\pi)^d} \int_0^\infty \frac{x^{d-1}}{x^2(x^2+1)} dx \ , \tag{14.23}$$

with $S_d = 2(\pi)^{d/2}/\Gamma(d/2)$ being the surface of the unit sphere in d dimensions. One can also derive an expression for the DOS for $1 < d < 2$. The result for $p = p_c$, or for $p > p_c$ and $\omega > \omega_{\rm co}$, is given by

$$\mathcal{N}(\omega) \sim \omega^{(d-2)/(d+2)} \ , \quad 1 < d < 2 \ , \tag{14.24}$$

which, when compared with Eq. (14.18), implies that

$$D_s = \frac{2d}{d+2} \ , \quad 1 < d < 2 \ . \tag{14.25}$$

The corresponding equation for the Debye or phonon regime is

$$\mathcal{N}(\omega) \sim \frac{\omega^{d-1}}{(p-p_c)^{d/2}} \ , \quad \omega \ll \omega_{co} \ , \tag{14.26}$$

which, in terms of frequency dependence of the DOS, is similar to Eq. (14.11), confirming once again that the EMA produces exact results, or provides very accurate approximations, for low-dimensional systems.

The predictions of the EMA are in qualitative agreement with those of the scaling theory that will be described in the next section, and also with direct numerical simulations (see for example, Li et al. 1990), which again indicates the usefulness of the EMA. The predictions are not, however, very accurate near p_c, which is the region where the EMA begins to be inaccurate. An obvious method of improving the performance of the EMA for predicting the DOS near p_c is to use higher order or cluster EMAs. The development of such cluster EMAs was described and implemented by Sahimi (1984).

14.2.2 *Scaling Theory of Phonons and Fractons*

In addition to the fractal dimension D_f of the morphology of heterogeneous material, the spectral, or fracton, dimension D_s is a key quantity for describing the dynamical properties of such materials, particularly those with a self-similar fractal morphology. Alexander and Orbach (1982) mapped the problem of anomalous diffusion onto the problem of vibrational DOS using the scalar approximation, and showed that the basic properties of vibrations of fractal materials, such as their DOS, the dispersion relation, and localization are all characterized by the spectral dimension D_s.

Rammal and Toulouse (1983) derived D_s via a scaling analysis, and showed that various random walk properties, such as the probability $P_0(t)$ of returning to the origin of the walk, and the mean number $\langle S(t)\rangle$ of the visited sites, are governed by the spectral dimension. The linear size of a region of sites visited by a diffusant or random walker is, $R = \langle R^2(t)\rangle^{1/2} \sim t^{1/D_w} \sim t^{1/(2+\theta)}$, where, $\theta = (\mu - \beta)/\nu$. The mean number of site visited $\langle S(t)\rangle$ is, therefore, proportional to the volume R^{D_f} of the region, $\langle S(t)\rangle \sim R^{D_f}$ and, thus (Rammal and Toulouse 1983)

$$\langle S(t)\rangle \sim t^{D_s/2}\,, \quad D_s \leq 2\,. \tag{14.27}$$

Therefore,

$$\langle P_0(t)\rangle \sim t^{-D_s/2}\,, \tag{14.28}$$

with D_s given by Eq. (14.18). In view of Eq. (14.10), we immediately obtain Eq. (14.17), the main prediction of Alexander and Orbach (1982). In analogy with the usual Debye DOS, Alexander and Orbach (1982) called the related excitations "fractons" and D_s the "fracton dimension." D_s was called the "spectral dimension" by Rammal and Toulouse (1983) because it represents the DOS for the vibrational excitation spectrum.

Rammal and Toulouse (1983) also derived the vibrational DOS of fractal materials using a finite-size scaling approach (see Chap. 2 for a detailed discussion). Consider a fractal material of linear size L with fractal dimension D_f. Its DOS per particle at the lowest frequency $\Delta\omega$ is defined by

$$\mathcal{N}(\Delta\omega, L) = (L^{D_f}\Delta\omega)^{-1}\,. \tag{14.29}$$

If we assume that the dispersion relation[1] for $\Delta\omega$ is given by

$$\Delta\omega \sim L^{-a}\,, \tag{14.30}$$

where a is an exponent to be determined, we can eliminate L from Eq. (14.29) to obtain

$$\mathcal{N}(\Delta\omega) \sim \Delta\omega^{D_f/a-1}\,. \tag{14.31}$$

The explicit expression for the exponent a is obtained by the mappings, $t \to 1/\Delta\omega^2$ (see above) and $\langle R^2(t)\rangle^{1/2} \to L$. We thus obtain

$$L \sim \Delta\omega^{-2/D_w}\,. \tag{14.32}$$

Therefore

$$a = \frac{2\nu + \mu - \beta}{2\nu} = \frac{D_w}{2} = \frac{D_f}{D_s}\,. \tag{14.33}$$

[1] A dispersion relation relates the wavelength or wavenumber of a wave to its frequency.

Since the material is fractal and self-similar, $\Delta\omega$ can be replaced by an arbitrary frequency ω and, therefore, we recover Eq. (14.17). Equation (14.32) also defines a dispersion relation,

$$\omega \sim [L(\omega)]^{-D_f/D_s} . \tag{14.34}$$

Alexander and Orbach (1982) observed that, if we use the most accurate estimates of the percolation exponents μ, β and ν (Table 2.3), then $D_s = 2D_f/D_w = 2(\nu d - \beta)/(2\nu + \mu - \beta)$ takes on a numerical value almost independent of the dimensionality d of the system, given by

$$D_s \simeq 4/3 , \quad \text{all } d .$$

It was also conjectured that, $D_s = 4/3$ is an exact result, which would have led to an exact scaling relation between the dynamical exponent μ and the static exponents ν and β. This is not, however, the case; $D_s = 4/3$ is only a very accurate approximate estimate.

14.2.3 Crossover from Phonons to Fractons

If the wavelengths λ of excited modes in a disordered material with percolation disorder at $p > p_c$ are larger than the percolation correlation length $\xi_p(p)$, the material is homogeneous on such a length scale and its vibrational excitations are weakly-localized phonons. This is because the scattering is determined by the square of the mass-density fluctuations, averaged over regions of volume λ^d. Hence, even if the short-range disorder is strong, its effective strength for phonons with $\lambda \gg \xi_p$ is very weak.

If, on the other hand, $\lambda \leq \xi_p$, the fractality of the material's morphology becomes relevant. Thus, there must be a crossover in the nature of the vibrational excitations when $\lambda(\omega_{\rm co}) \sim \xi_p$. From Eq. (14.34) the crossover frequency $\omega_{\rm co}$ is obtained

$$\omega_{co} \sim (p - p_c)^{\nu D_f/D_s} . \tag{14.35}$$

If we substitute $\omega_{\rm co} = v_s(p)k$ ($k \sim 1/\xi_p$) in Eq. (14.35), the $p-$dependence of the phonon velocity is obtained:

$$v_s(p) \sim (p - p_c)^{\nu D_f/D_s - \nu} \sim (p - p_c)^{(\mu-\beta)/2} . \tag{14.36}$$

For all space dimensionality d one has, $\mu - \beta > 0$. Thus, $v_s(p) \to 0$ as $p \to p_c$. The results for the DOS are then summarized as

$$\mathcal{N}(\omega) \sim \begin{cases} \omega^{d-1}/[v_s(p)]^d , & \omega \ll \omega_{\rm co} , \\ \omega^{D_s-1} , & \omega \gg \omega_{\rm co} , \end{cases} \tag{14.37}$$

whereas the dispersion relations become

$$\omega \sim \begin{cases} v_s(p)k \ , \ \omega \ll \omega_{\mathrm{co}} \ , \\ k^{D_f/D_s} \ \ , \ \omega \gg \omega_{\mathrm{co}} \ , \end{cases} \tag{14.38}$$

where, for $\omega \gg \omega_{\mathrm{co}}$, k does not represent the wave number due to the lack of the translational symmetry of the system, but rather it describes the inverse of a characteristic length Λ^{-1}. Note that, within the scale approximations, the phonons are always localized.

14.2.4 *Experimental and Numerical Confirmation of the Scalar Approximation*

The predictions of the scalar approximation of the vibrational DOS of heterogeneous materials have been supported by both large-scale numerical simulations and experimental data. In what follows we describe both.

14.2.4.1 Large-Scale Numerical Simulations

Initially, there were considerable difficulties in computing the DOS in large-scale latticess (Webman and Grest, 1984), but with the emergence of array—processing supercomputers, several numerical methods were developed to overcome such difficulties; for a comprehensive review see Nakayama et al. (1994).

Yakubo and Nakayama (1989) and Yakubo et al. (1990a, b) carried out numerical simulations of percolation lattices with $N = 10^6$ sites, using the forced oscillator method of Williams and Maris (1985). This algorithm is based on the principle that a complex mechanical system, when driven by a periodic external force of frequency ω, will respond with a large amplitude in those eigenmodes that are close to this frequency.

Figure 14.1 presents the DOS at the site percolation threshold $p_c \simeq 0.593$ of the square lattice (Yakubo and Nakayama 1989). The line through the solid circles has a slope of about 1/3, which holds even in the low-frequency region, because the lower cutoff frequency ω_L is determined by the finite size of the percolation clusters. The DOS of a square network with a fraction $p = 0.67$ of the open sites is shown in Fig. 14.2 (Yakubo and Nakayama 1989).

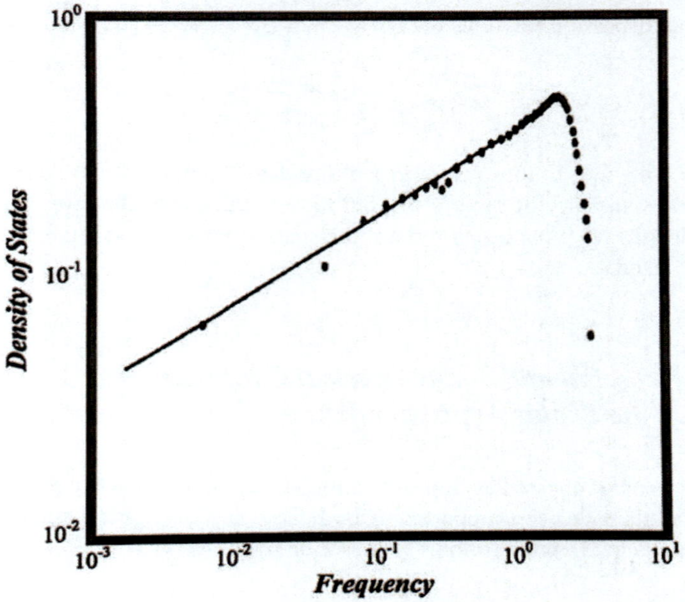

Fig. 14.1 Density of states at the site percolation threshold of the square lattice, $p_c \simeq 0.593$. The straight line represents $\mathcal{N}(\omega) \propto \omega^{1/3}$ (after Yakubo and Nakayama 1989)

The results presented in Figs. 14.1 and 14.2 indicate that the frequency dependence of the DOS is characterized by two regimes. For frequencies $\omega_{\rm co} \ll \omega \ll 1$, the DOS is closely proportional to $\omega^{1/3}$, in agreement with Eq. (14.17) with $D_s \simeq 4/3$. The crossover frequency $\omega_{\rm co}$ corresponds to the mode of wavelength λ, which is equal to the percolation correlation length ξ_p. Therefore, for $\omega \ll \omega_{\rm co}$ the DOS should be given by the conventional Debye law, Eq. (14.11), and by $\mathcal{N}(\omega) \sim \omega^{D_s-1}$ for $\omega \gg \omega_{\rm co}$. The numerical results shown in Figs. 14.1 and 14.2 are consistent with this picture, because the frequency dependence of the DOS for low frequencies, ($\omega \ll \omega_{\rm co}$), clearly follows $\mathcal{N}(\omega) \sim \omega$, corresponding to $d = 2$ and, thus, vibrational excitations behave as phonons. The region in the vicinity of $\omega_{\rm co}$ is the crossover region between phonons and fractons, in which the DOS varies smoothly, exhibiting neither a notable steepness, nor a hump in the vicinity of $\omega_{\rm co}$.

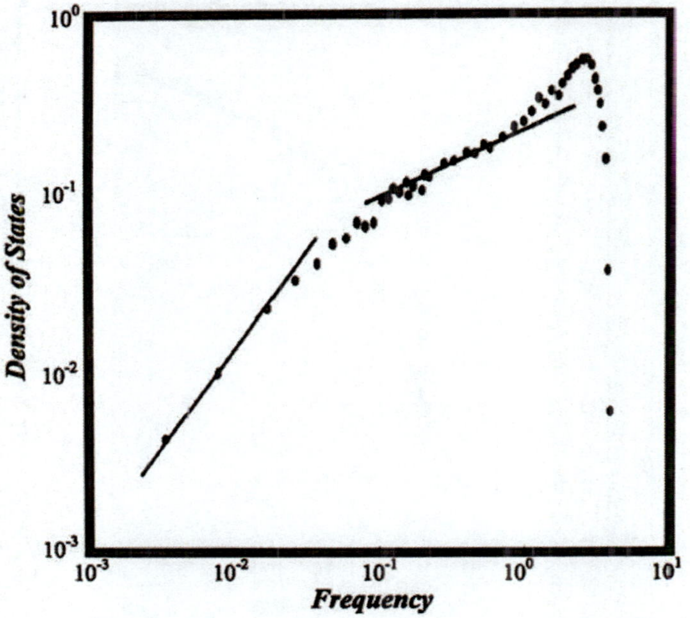

Fig. 14.2 Density of states of the square lattice at $p = 0.67$. The straight line on the left corresponds to $\mathcal{N}(\omega) \propto \omega$, while the one on the right represents $\mathcal{N}(\omega) \propto \omega^{1/3}$ (after Yakubo and Nakayama 1989)

The DOS of a simple-cubic network with percolation disorder ($p_c \simeq 0.3116$) at $p = 0.4$, computed for a 70^3 lattice, is shown in Fig. 14.3. At $p = 0.4$ the phonon-fracton crossover frequency $\omega_{\rm co}$ is about 0.1. The DOS in the range $0.1 < \omega < 1$ is again proportional to $\omega^{1/3}$ (since $D_s \simeq 4/3$ for both $d = 2$ and $d = 3$), whereas in the low-frequency regime, $\omega \ll 0.1$, it follows the Debye law, Eq. (14.11) with $d = 3$, $\mathcal{N}(\omega) \sim \omega^2$. The sharp peak at $\omega = 1$ is attributed to vibrational modes of a single site connected by a single bond to a relatively rigid part of the network. It is clear that there is no steepness or hump of the DOS in the crossover region in the vicinity of $\omega_{\rm co}$. The behavior of the DOS at the phonon-fracton crossover was determined based on a mean-field analysis (Loring and Mukamel 1986; Korzhenevskii and Luzhkov 1991). Loring and Mukamel (1986) suggested a smooth transition of the DOS at the phonon-fracton crossover, in contrast to the prediction of the EMA described above, or that of the scaling theory (Aharony et al. 1985).

The DOS calculated with bond percolation networks exhibits some interesting differences with site percolation networks, because the latter generates short-range correlations that are absent in the former. Recall that bond percolation a bond is open or active only if its end sites are also so.

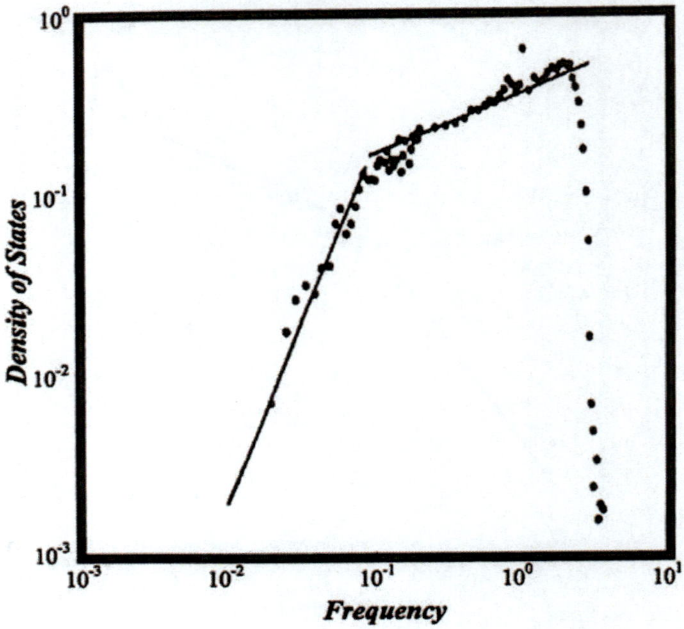

Fig. 14.3 Density of states for site percolation in the simple-cubic network at $p = 0.4$. The straight line on the left corresponds to $\mathcal{N}(\omega) \propto \omega^2$, while the one on the right represents $\mathcal{N}(\omega) \propto \omega^{1/3}$ (after Yakubo and Nakayama 1989)

14.2.4.2 Experimental Data

Most of the experimental studies of the fracton dynamics and the corresponding DOS involve aerogels and, therefore, we first briefly discuss their preparation and properties. Aerogels are highly porous solid materials with a very tenuous structure. Their porosity can be as high as 99%, which is why they often have very unusual and unique properties. They can, for example, be prepared in transparent form; have very small thermal conductivity and solid-like elasticity and, due to their large porosity, they possess a large internal surface area. They also have a wide range of applications, from catalyst supports to thermal insulators and radiators, and detectors of Cerenkov radiation (see for example, Fricke 1988; Courtens et al. 1989; Brinker and Scherer 1990).

Aerogels are prepared by a variety of methods using various materials. Silica aerogels, which have received the widest attention, are produced by hydrolysis of $Si(OR)_4$, where R represents either CH_3 or C_2H_5. A catalyst, which is either an acid or a base, is also used that strongly influences the reaction. The degree of hydrolysis is controlled by the concentration ratio $Si(OR)_4/H_2O$, whereas the final density of the aerogel is controlled by $Si(OR)_4/ROH$. Because of the acid or base catalyst, the pH of the solution also has a strong effect on the morphology of the gel. Hydrolysis produces -SiOH groups that polymerize into -Si-O-Si-, after which the particles begin to grow in the liquid solution and, after some time, form a gel network. The solvent is then removed to obtain the solid porous structure. Aerogels are obtained if the solvent is removed at a temperature *above* its critical point.

Silica aerogels appear to have the structure of percolation networks. Evidence for this comes from two different directions. One was provided by the elastic moduli of the material. Woignier et al. (1988) reported Young's modulus Y of silica aerogels as a function of their density and volume fraction. Figure 14.4 presents their data. Over nearly one order of magnitude variation, Y scales with the volume fraction ϕ of the solid as

$$Y \sim (\phi - \phi_c)^f \ , \tag{14.39}$$

where ϕ_c is the critical volume fraction or the percolation threshold of the network. As Fig. 14.4 indicates, $f \simeq 3.8$, which is in excellent agreement with that of 3D elastic percolation networks with the CF and BB forces, described in Chap. 11.

The second piece of evidence was provided by data on the morphology of silica aerogels. It has been reported (Schaefer and Keefer 1986; Courtens and Vacher 1987; Vacher et al. 1988; Woignier et al. 1990; Posselt et al. 1991) that aerogels may have a fractal morphology. Bourret (1988) and Duval et al. (1992) reported high-resolution electron microscopy observations that were compatible with a fractal morphology for these materials. Ferri et al. (1991) used light scattering techniques for characterizing the geometrical features of the materials.

Small-angle scattering techniques using neutrons and x-rays are very well-suited for systematically investigating the morphology of silica aerogels using such techniques. To characterize the structure of aerogels, the structure factor $S(q)$, which describes the correlation between the particles in a cluster, was measured, which is obtained from the Fourier transform of the particle density-density correlation function $C(\mathbf{r})$:

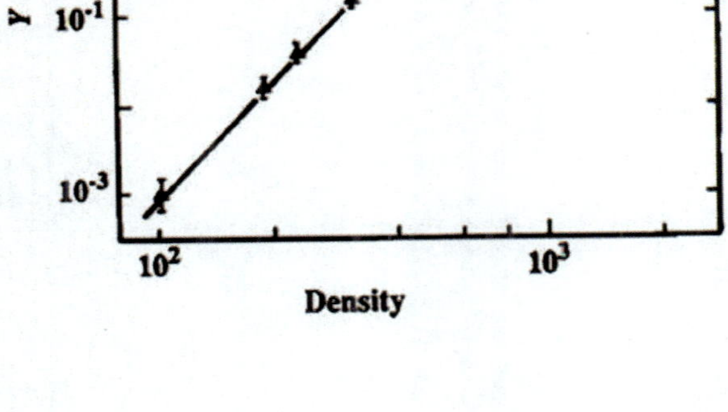

Fig. 14.4 Young's modulus Y of silica aerogels in (GPa) versus their apparent density (in kg.m^{-3}) (after Woignier et al. 1988)

$$S(\mathbf{q}) = 1 + \frac{N}{\Omega}\int_{\Omega} |C(\mathbf{r}) - 1| e^{i\mathbf{q}\cdot\mathbf{r}} d\mathbf{r} \ , \tag{14.40}$$

where $q = |\mathbf{q}| = (4\pi/\lambda)\sin(\theta/2)$, with λ being the wavelength of the radiation scattered by the material through an angle θ. If the self-similarity of a material extends up to a correlation length ξ_{ss}, then the correlation function $C(r)$ will be given by

$$C(r) \sim r^{D_f - 3} \exp(-r/\xi_{ss}) \ . \tag{14.41}$$

For small values of q with $q\xi_{ss} \ll l$, $S(q)$ is almost independent of q. When $q\xi_{ss} \gg 1$, one obtains, by substituting Eq. (14.41) into Eq. (14.40)

$$S(q) \sim q^{-D_f} \ . \tag{14.42}$$

According to Eq. (14.42), D_f can be estimated from the slope of the logarithmic plot of the measured corrected intensity versus q.

An example is presented in Fig. 14.5 for silica aerogels (Vacher et al. 1988), with the various curves labeled by the sample's macroscopic density ρ, and represent the best fits to the data. They have been extrapolated into the particle regime

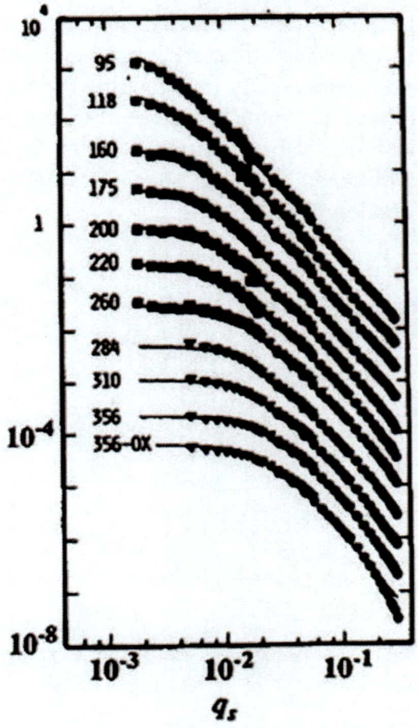

Fig. 14.5 Scattered intensities, for eleven silica aerogels, versus the magnitude q of the scattering vector in Å^{-1}. The top ten curves are for untreated and neutrally reacted samples of increasing density, while the bottom one is for an oxidized sample. The curves are labeled with the density in kg/m^3 (after Vacher et al. 1988)

($q > 0.15$ Å^{-1}) to emphasize that the fits do not apply in that region, particularly for the denser samples. Remarkably, the fractal dimension $D_f \simeq 2.4 \pm 0.03$ is independent of samples' density to within experimental accuracy. Furthermore, ξ_{ss} scales with ρ as $\xi_{ss} \sim \rho^{-1.67}$. The departure of $S(q)$ from the q^{-D_f}-dependence at large q indicates the presence of particles with radii of gyration of about a few angströms. Thus, silica aerogels exhibit three different length scale regions. (a) At short distances, elementary particles aggregate together to form clusters with a linear dimension ξ_{ss} at intermediate sizes, forming a gel by connecting the the clusters at larger distances. (b) at intermediate length scales, the clusters possess fractal morphology, and (c) at large length scales the gel is a homogeneous porous glass; see Courtens and Vacher (1987) and Kjems (1993) for more details.

Incoherent inelastic neutron scattering experiments measure the amplitude-weighted DOS. The scattered intensity is given by

$$I(q,\omega) \sim q^2 \frac{k_s}{k_i} \frac{n(\omega)}{\omega} \sum_i e^{-2W_i} \mathcal{N}_i(\omega), \tag{14.43}$$

where $n(\omega)$ is the Bose–Einstein distribution function,[2] the wave vectors k_i and k_s correspond to the incident and scattered neutrons, respectively, and $\mathbf{q} = \mathbf{k}_s - \mathbf{k}_i$. Here $\mathcal{N}_i(\omega)$ and W_i are, respectively, the DOS and the Debye–Waller factor[3] characteristic of the ith site. The sum is over the different sites (the atoms), each of which contributes to the incoherent-scattering with an intensity proportional to the amplitude of vibration at frequency ω. Incoherent neutron scattering from protons that are chemically bonded to the particle surfaces can be used to determine the DOS in porous materials (Richter and Passell 1980). The incoherent inelastic scattering by aerogels have been measured by several groups, a list of which is too long to be given here. Buchenau et al. (1992) and Kjems (1993) reviewed the experiments and their implications.

The phonon-fracton crossover in silica aerogels has been studied using such inelastic neutron scattering experiments on back scattering spectrometers (see for example, Pelous et al. 1989; Conrad et al. 1990), and also the spin-echo technique (Courtens et al. 1990; Schaefer et al. 1990). The former technique has the advantage that it enables one to observe the low-frequency Debye regime as a constant-intensity level, extending from the elastic line to the crossover frequency, and thus any excess modes at the phonon-fracton crossover should show up as a peak in the scattering intensity at that frequency, although no such peak has yet been observed in the backscattering experiments. Rather, a gradual decrease is observed as one passes through the crossover

---

[2] The Bose–Einstein distribution function $n(x)$ is given by

$$n(x) = [c\exp(x) - 1]^{-1},$$

where c is a constant.

[3] The Debyre–Waller factor is given by, $W_i = \langle \exp(i\mathbf{q} \cdot \mathbf{u}_i)\rangle^2$, where i is the basic imaginary number, $\mathbf{u}_i$ is the displacement of site i, and $\langle\cdot\rangle$ denotes either thermal or temporal averaging.

regime. The neutron scattering spin-echo technique has the advantage that its larger spectral range makes it a suitable tool for determining the spectral dimension D_s.

One way of studying the crossover is by plotting the crossover or cutoff frequency $\omega_{\rm co}$ as a function of q. Since q is inversely proportional to the length scale L, Eq. (14.34) tells us that

$$\omega_{\rm co} \sim q^{D_f/D_s} , \tag{14.44}$$

so that $\omega_{co} \sim q^{1.88}$, if silica aerogels are percolation fractals and $D_s \simeq 4/3$. Figure 14.6, taken from Courtens and Vacher (1989), shows such a plot for a series of aerogels with various densities. The data were obtained by Brillouin scattering of visible light and indicate that, $\omega_{\rm co} \sim q^{1.9}$, which is in excellent agreement with the theoretical prediction.

Some studies at higher frequencies (Reichenauer et al. 1989; Vacher et al. 1989) exhibited a change of slope in the logarithmic of the DOS versus ω at $\omega = 200$ GHz, giving a stronger increase with frequency at higher frequencies, which was interpreted as a crossover from fractons to vibrational modes within the particles. The effective slope of the DOS, i.e., $D_s - 1$, was about 1.5, which appears to have originated from the contributions by both the surface, which is proportional to ω, and the bulk that is proportional to ω^2, for particle modes. The energy resolution was, however, not high enough to observe the crossover to the long-wavelength phonon regime. Measurements of Vacher et al. (1989) at higher resolutions confirmed the existence of the extended fracton region. To further discuss this phenomenon, one must in fact take into account the effect of the vector nature of the vibrational modes, which will be described later in this chapter.

Neutron-scattering experiments on other disordered materials have also been analyzed in terms of the fracton theory. Freltoft et al. (1987) measured the low-frequency DOS for fractal silica aggregates by inelastic neutron scattering and estimated their spectral dimensions. Page et al. (1989) carried out inelastic neutron scattering experiments on fumed silica and compared the results with analogous results for amorphous quartz. They reported no evidence for a hump in the DOS near the phonon-fracton crossover, nor did they find that the temperature and wave-vector dependence of the

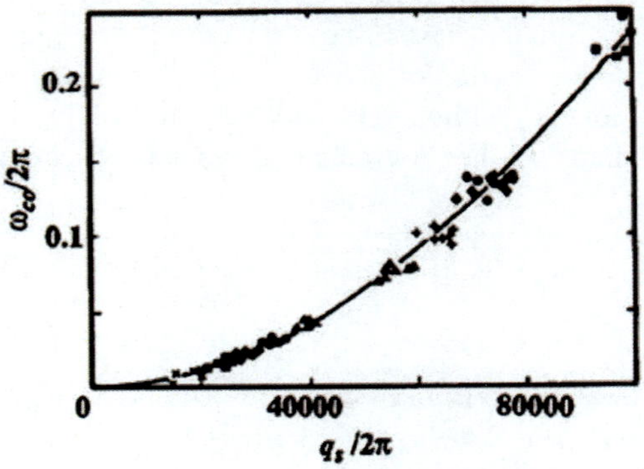

Fig. 14.6 The fracton dispersion curve obtained from the Brillouin determination of the cutoff frequency $\omega_{\rm co}$ and the scattering vector q. Symbols are the same as in Fig. 14.5 (after Courtens and Vacher 1989)

intensity was in agreement with simple phonon models. Fontana et al. (1990) reported on a study of low-frequency vibrational dynamics and electron-vibration coupling in AgI-doped glasses. By using both times-of-flight neutron scattering[4] and Raman scattering spectroscopies, they were able to determine the vibrational DOS and the frequency dependence of the electron-vibration coupling. The spectral dimension of the material was estimated to be about, $D_s \simeq 1.4$. Zemlyanov et al. (1992) employed inelastic neutron scattering measurements to study low-frequency vibrational excitations in polymethyl methacrylate. The measured DOS followed a power law in the energy, with a spectrum corresponding to a spectral dimension $D_s \simeq 1.8 \pm 0.05$. Other experimental efforts in this area were reviewed by Nakayama et al. (1994).

There have also been experimental studies of the dynamical properties of fractal materials without any scattering measurements. For example, Kopelman et al. (1986) measured the exciton recombination characteristics of naphthalene-doped microporous materials. The technique yields the spectral dimension of the embedded naphthalene structure, or, equivalently, the effective random walk fractal dimension D_w of the porous network. Kopelman et al., obtained $1 \leq D_s \leq 2$. Fischer et al. (1990) studied the trap-depth distribution of dibenzofuran singlet excitons and the temperature-dependent energy migration by time-resolved spectroscopy via synchrotron radiation and two-photon laser excitation and obtained $D_s \simeq 1.14$.

14.3 Vector Percolation and Vibrational Density of States

In this section, we describe and discuss the DOS in terms of the true vectorial nature of the vibrations. We consider heterogeneous materials with percolation-type disorder and focus our attention on the behavior of the DOS at the percolation threshold p_c, or above p_c but at length scales $L \ll \xi_p$, where ξ_p is the correlation length of percolation. For length scales $L \gg \xi_p$, a material is macroscopically homogeneous and, thus, its vibrational DOS is described by the classical laws in the phonon regime, Eq. (14.11).

14.3.1 *Scaling Theory*

Recall from Sect. 14.2 that the vibrational states in materials with fractal morphology, such as one at its percolation threshold p_c, or above p_c but over length scales $L \ll \xi_p$, are called fractons (phonons on fractals!). For macroscopically homogeneous materials, on the other hand, the vibrational DOS is given by Eq. (14.11) for low frequencies, i.e., long wavelengths or large length scales over which the material is homogeneous, such that $\omega < \omega_{co}$, where ω_{co} is a cutoff or crossover frequency. On

[4] A technique in which the initial position and velocity of a pulse of neutrons is fixed, and their final position and the time after the pulse that the neutrons are detected are measured.

the other hand, for materials with a fractal morphology, the vibrational DOS within the scalar percolation approximation is described by Eq. (14.17).

As discussed in Chap. 2, in the context of percolation models of disordered materials, the fractal dimension of the SSC representing a heterogeneous materials, is given by, $D_f = d - \beta/\nu$ and $D_w = 2 + (\mu - \beta)/\nu$, where β, μ and ν are the usual percolation and conductivity exponents, so that

$$D_s = 2\frac{\nu d - \beta}{2\nu + \mu - \beta} . \tag{14.45}$$

Equations (14.17) and (14.45) are valid as long as the scalar contributions to the elastic energy of the material dominate the vectorial contributions. An example of such a material is one that is described by the Born model (see Chap. 11). We do know, however, that the nature of the forces that are exerted in a material, or to the sites and bonds of the network representing it, and their contributions to the elastic energy are most important. Hence, to take into account the true vectorial nature of the vibrational DOS, we must introduce a new quantity D_s^e, called the *elastic* or *flexural spectral dimension* D_s^e, which is defined by replacing in Eq. (14.45) the conductivity exponent μ by f, the critical exponent of the elastic moduli near p_c, so that (Webman and Grest 1985)

$$D_s^e = 2\frac{\nu d - \beta}{2\nu + f - \beta} . \tag{14.46}$$

Accordingly, we must also define *two* distinct types of the DOS for vibrations of heterogeneous materials with fractal morphology. One is $\mathcal{N}(\omega)$, which is the DOS when the scalar contributions to the elastic energy of the material dominate those of all other factors for which Eq. (14.11) is valid. The second type is what we call the *vectorial* or *elastic* DOS $\mathcal{N}_e(\omega)$, which is for the case in which the vectorial contributions, such as those of the central and bond-bending (BB) forces, dominate the elastic energy. In this case, in analogy with Eq. (14.17), the vibrational DOS is given by

$$\mathcal{N}_e(\omega) \sim \omega^{D_s^e - 1} . \tag{14.47}$$

Equations (14.17) and (14.47) are both valid for high frequencies (i.e., short wavelengths or length scales) with $\omega > \omega_{\rm co}$. Such vibrational states are localized as long as D_s or D_s^e is less than 2 (Rammal and Toulouse 1983). As explained earlier in this chapter, $D_s \simeq 4/3$ for all $d \geq 2$, whereas D_s^e varies continuously with d. Since $f \geq \mu$, one must have $D_s^e \leq D_s$, and in fact D_s^e can be *less than one* (see below).

The difference between Eqs. (14.17) and (14.47) is striking. Since $D_s \simeq 4/3$, one has

$$\mathcal{N}(\omega) \sim \omega^{1/3} , \quad \text{scalar percolation approximation.} \tag{14.48}$$

If $\mathcal{N}(\omega)$ decreases with decreasing ω, then the material is mechanically stable. To understand this, recall that small frequencies imply large length scales, and if over such length scales $\mathcal{N}(\omega)$ decreases, the implication would be that there are fewer and fewer vibrational modes, i.e., the system is mechanically stable. Thus, materials with a fractal morphology in which the scalar percolation contributions to their elastic energy dominate all other contributions—those for which Eq. (14.48) is valid—are mechanically stable. On the other hand, if $f \simeq 3.75$ for 3D BB models (see Chap. 11) is used in Eq. (14.46), we obtain, $D_s^e \simeq 0.87$ for $d = 3$, so that Eq. (11.47) becomes

$$\mathcal{N}_e(\omega) \sim \omega^{-0.13}\ , \quad \text{bond-bending elasticity}\ , \tag{14.49}$$

That is, as ω decreases, there are larger and larger number of vibrating modes, implying that such materials *cannot become too large*, because otherwise they will lose their mechanical stability and, therefore, must rearrange themselves into more stable structures. This type of the DOS should be observed in materials with a fractal morphology in which the contributions of the BB forces dominate the elastic energy. On other hand, as discussed in Chap. 11, for gels whose elasticity is dominated by entropic effects, $f \simeq \nu d$, implying that, $D_s^e \approx 1.12$, so that

$$\mathcal{N}_e \sim \omega^{0.12}\ , \quad \text{materials with entropic elasticity}\ , \tag{14.50}$$

implying that such materials are also mechanically stable.

In analogy with the scalar percolation approximation, one can use Eq. (14.34) to define a dispersion relation:

$$\omega \sim [L(\omega)]^{-D_f/D_s^e}\ . \tag{14.51}$$

To derive Eq. (14.51), consider the vibrations of a material of linear size L. Although high-frequency modes of the vibrations are not affected by a change in the boundary conditions, the low-frequency modes will disappear from the spectrum. The crossover between the two regimes occurs at a frequency ω_L, such that

$$L \sim \Lambda(\omega_L)\ , \tag{14.52}$$

where $\Lambda(\omega)$ is the wavelength.

14.3.2 Crossover Between Scalar and Vector Models of Density of States

Feng (1985a, b) argued that in material with percolation disorder in which both the central and BB forces are present, another length scale $L_{\rm bb}$, in addition to the percolation correlation length ξ_p, is important. If $L_{\rm bb} \gg \xi_p$, then, at low frequencies Eq. (14.11) and at higher frequencies Eq. (14.17) govern the vibrational DOS, with a crossover at $\omega_{\rm co}$. If, however, $L_{\rm bb} \ll \xi_p$, then at low frequencies one is in the phonon regime until a characteristic frequency $\omega_{\rm bb}$ is reached at which there is a crossover from Eqs. (14.11) to (14.47). When the higher frequency $\omega_{\rm co}$ is reached, there is a second crossover from Eqs. (14.17) to (14.47). It is not difficult to guess that [see Eq. (14.51)]

$$\omega_{\rm bb} \sim \xi_p^{-D_f/D_s^e} . \tag{14.53}$$

According to Feng (1985b), the characteristic length scale $L_{\rm bb}$ is given by

$$L_{\rm bb} \sim \sqrt{\frac{\gamma}{\alpha}} , \tag{14.54}$$

where α and γ are, respectively, the usual central and BB force constants, defined by Eq. (11.41). In effect, $L_{\rm bb}$ is the length scale below which the motion, due to central forces, is energetically more favorable and above which the BB forces dominate. Large-scale computer simulations of Yakubo et al. (1990b), to be described below, confirmed such crossover behavior. Other efforts in this direction include those of Day et al. (1985), Cai and Thorpe (1989), and Monceau and Levy (1994).

14.3.3 Experimental and Numerical Confirmation of Vector Percolation Model of Density of States

The predictions of the vector percolation model for the vibrational DOS of heterogeneous materials have been supported by both large-scale numerical simulations and experimental data. In what follows we describe both.

14.3.3.1 Large-Scale Computer Simulation

Webman and Grest (1985) computed the vibrational DOS using vector percolation as described by the BB model, analyzing a lattice with $N \sim 10^3$ sites. They found, in agreement with Eq. (14.49), that the DOS was weakly divergent at low frequencies. Lam and Bao (1985) calculated the vibrational DOS of a triangular network with central forces and site percolation disorder, and found the DOS in the fracton regime is governed by Eq. (14.47) with $D_s^e \simeq 0.625$. Day et al. (1985) calculated the vibrational DOS of the same network but with central forces and 60% BB forces. They estimated the percolation threshold p_{ce} of the network and the elastic spectral dimension D_s^e in the ranges $0.4 \le p_{ce} \le 0.405$ and $1.25 \le D_s^e \le 1.3$, hence suggesting that the 60%—BB model does not belong to the universality class of a full BB model. Other efforts in this direction were reviewed by Nakayama et al. (1994).

Large-scale simulations of the vibrational DOS of the BB model were carried out by Yakubo et al. (1990b) who studied the crossover of the DOS from bending to stretching fractons at the percolation threshold of a 500×500 square network. They considered the limit in which the BB force constant γ was larger than the stretching constant α. The calculated DOS is shown in Fig. 14.7 in which $\gamma = 10\alpha = 0.133$ [see Chap. 11, Eq. (11.41)] . The network had a cutoff Debye frequency $\omega_D \approx 2.0784$. The steplike decrease of the states in the high-frequency regime in Fig. 14.7 indicates that the stretching motions are not excited above some frequency ω_0, whose value is determined by the stretching force constants α. For $\alpha = 0.0133$, ω_0 was estimated to be 0.2309 from the relation $\omega_0 = 2\sqrt{\alpha}$, coinciding with the computed value in Fig. 14.7.

To clarify the individual contributions by the bending and stretching forces, the ratio of the contribution of each to the overall elastic energy was calculated as a function of the frequency. The solid curve in Fig. 14.7 shows the ratio of the energy $\mathcal{H}_\alpha$ attributed to the stretching motion and the total energy $\mathcal{H}$. A crossover frequency $\omega_{\rm co}$ was defined based on the condition, $\mathcal{H}_\alpha(\omega_{\rm co})/\mathcal{H}(\omega_{\rm co}) = 1/2$, leading to $\omega_{\rm co} \sim 0.005$. The DOS in the vicinity of $\omega_{\rm co}$ is independent of ω. The crossover region from bending to stretching fractons extends over at least two orders of magnitude in frequency, because the ratio $\mathcal{H}_\alpha(\omega)/\mathcal{H}(\omega)$ of the energies increases logarithmically, as shown in Fig. 14.7. This is in contrast with the sharp crossover from phonons to fractons in scalar percolation approximation described above. The straight line through the solid circles in Fig. 14.7 has been drawn according to Eq. (14.48). Based on the data of Fig. 14.7 in the frequency range between $\mathcal{N}(\omega) \sim \omega_{\rm co}$ and ω_0, it is not clear that the $\omega^{1/3}$ law actually holds.

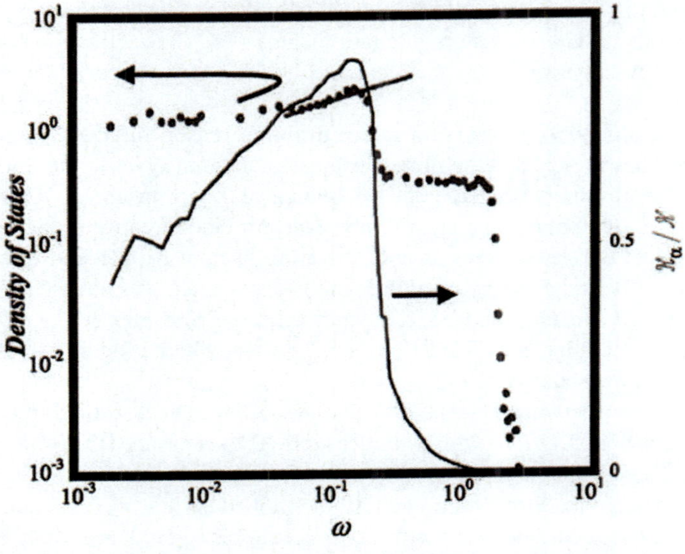

Fig. 14.7 Calculated density of states (solid circles) at the bond percolation threshold of a square network with stretching and bond-bending forces and the force constants $\gamma = 10\alpha = 0.133$. The straight line is guide to the eye for $\mathcal{N}(\omega) \sim \omega^{1/3}$, while solid curve shows the energy ratio $\mathcal{H}_\alpha/\mathcal{H}$ and its frequency dependence (after Yakubo et al. 1990b)

14.3.3.2 Experimental Data

Most of the experimental studies of vibrational DOS have involved aerogels, described earlier in this chapter. Consider the crossover or cutoff frequency ω_{co} as a function of q, given by Eq. (14.44). One must have, $\omega_{rmco} \sim q^{1.88}$, if silica aerogels are percolation fractals and $D_s \simeq 4/3$. Figure 14.6 presents such a plot for a series of aerogels with various densities. The data indicate that, $\omega_{co} \sim q^{1.9}$, which is in very good agreement with the theoretical prediction.

As described earlier, the vibrational DOS itself can be measured directly by incoherent neutron scattering. Vacher et al. (1989) used the technique to measure the DOS and reported that, $\mathcal{N}(\omega) \sim \omega^{0.85\pm0.15}$, which agrees neither with Eq. (14.17) nor with (14.47). This difference may be explained in various ways. One is that the material used in the study was not simply a sample-spanning percolation cluster, rather, in addition to the main gel network, many other smaller clusters (small branched polymers) could have been present, and when the solvent was removed, a *distribution* of clusters (rather than a single cluster) had been obtained. Thus, the incoherent scattering was not measured from just one percolation cluster, rather from a collection of them and, hence, the DOS should be written as

$$\mathcal{N}(\omega) \sim \omega^{2d/D_w-1} , \tag{14.55}$$

rather than Eq. (14.17) (i.e., D_f in the definition of D_s is replaced with d), because, as discussed in Chap. 2, the collection of all percolation clusters is *not* a fractal set, rather only the individual clusters at length scales $L \ll \xi_p$ are. Equation (14.55) then predicts that $\mathcal{N}(\omega) \sim \omega^{0.6}$, which is reasonably close to the experimental measurements.

Vacher et al. (1989) presented a rather different reason for the aforementioned difference. They argued that the finite clusters attach themselves to the SSC, making it denser, thereby increasing the *effective value* of D_s from about 4/3 to a larger value of 1.85. Such reasoning cannot, however, provide any information about the morphology of the dense network and, therefore, cannot be checked directly. One may also argue that the energy resolution in the experiment was not sufficiently high to yield enough accuracy. Indeed, the experiments of Schaefer et al. (1990), using more accurate techniques, yielded $D_s \simeq 1.22 \pm 0.14$, which is again consistent with the theoretical expectation of 1/3.

The crossover between Eqs. (14.17) and (14.47) was also studied in the experiments of Vacher et al. (1990) with silica aerogels. They measured the DOS of the gels and found their data that at low frequencies can be fitted with $D_s^e \simeq 0.9$, relatively close to the theoretical expectation of 1.12 mentioned earlier, whereas at higher frequencies the data could be fitted with $D_s \simeq 1.7 \pm 0.2$, again larger than $D_s \simeq 4/3$. These are shown in Fig. 14.8.

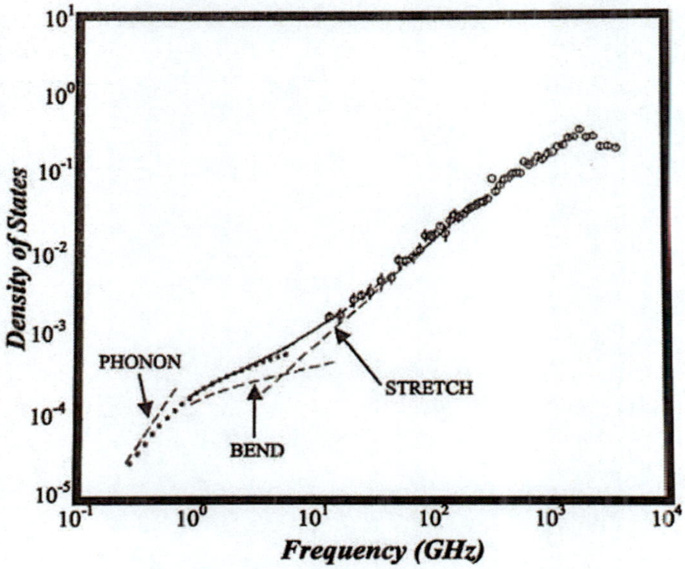

Fig. 14.8 Crossover in the scaling of the density of states of silica aerogels. Arrows indicate the locations of the crossover frequencies (after Vacher et al. 1990)

Boolchand et al. (1990) carried out experiments, including Mössbauer-Debye–Waller factores,[5] inelastic neutron scattering, Ramam scattering, and ultrasonic elastic moduli, using chalcogenide glasses that were of the general composition A_xB_{1-x}. As discussed earlier in Chap. 13, the average coordination number of such glasses is $\langle Z\rangle = 2(x+1)$, and according to the Phillips–Thorpe theory (see Chap. 13) and the vector percolation model, one should see a drastic difference in the vibrational DOS at $\langle Z\rangle \simeq 2.4$, corresponding to $x_c \simeq 0.2$. Indeed, Boolchand et al.'s experiments did indicate a threshold behavior at this composition. The same type of behavior was observed in the experiments of Kamitakahara et al. (1991), and Walter et al. (1988), although the threshold behavior in these experiments was not as strong as that in the experiments of Boolchand et al.

[5] The Lamb–Mössbauer factor, also called the elastic incoherent structure factor, is the ratio of elastic and total incoherent neutron scattering, or the ratio of recoil-free and the total nuclear resonant absorption in Mössbauer spectroscopy. The corresponding factor for coherent neutron or X-ray scattering is the aforementioned Debye–Waller factor.

Chapter 15
Hopping Conductivity of Heterogeneous Materials

15.1 Introduction

Hopping conduction in semiconductors was first associated by Hung and Gliessman (1950) with the observation that the activation energy of the conductivity in doped Ge exhibits a break at low temperatures T, and was attributed to a distinct mechanism of conduction. Mott[1] (1956) and Conwell (1956) proposed a model of conduction in which electrons conduct by thermally activated tunneling from a filled site to a vacant one, a process that is usually called *phonon-assisted hopping* (see also Chap. 14). The model was modified and extended by several researchers, the best known of which is perhaps the model of Miller and Abrahams (1960). They developed a model consisting of two parts, the quantum mechanical theory of the wave functions and of the transition rates W_{ij} from a localized state i to a localized state j, and a statistical mechanical theory of transport that employs such transition rates. Miller and Abrahams also showed how their model can be reduced to a random resistor network and utilized for computing the hopping conductivity of disordered solids.

It took researchers over a decade to discover certain deficiencies of the Miller-Abrahams resistor network model. It was also realized independently by Ambegaokar et al. (1971), Shklovskii and Efros (1971), Brenig et al. (1971), and Pollack (1972) that hopping conduction in semiconductors can be modeled successfully by invoking the concepts of critical-path analysis (CPA), described in detail in Chap. 7, and percolation theory (see below). Since their seminal papers, several electronic properties of semiconductors have been successfully predicted by percolation theory.

This chapter summarizes the important elements of the successful application of percolation to a technologically important problem. Our discussion is by no means

[1] Sir **Nevill Francis Mott** (1905–1996) was a British physicist who made fundamental contributions to many areas of physics, particularly the electronic structure of magnetic and disordered materials, and in particular amorphous semiconductors. He was awarded the 1977 Nobel Prize in Physics, which he shared with Philip W. Anderson and John Hasbrouck Van Vleck. Mott and Anderson had explained the reasons as to why magnetic or amorphous materials can sometimes be metallic and sometimes insulating.

M. Sahimi, *Applications of Percolation Theory*, Applied Mathematical Sciences 213,
https://doi.org/10.1007/978-3-031-20386-2_15

exhaustive as the number of published papers on the subject is too large. A thorough discussion can be found in the monograph of Shklovskii and Efros (1984), and more recently in Sahimi (2003a). An older account of some of such percolation predictions is given by Pollak (1978). In what follows we first describe the Miller-Abrahams resistor network, and then the application of percolation to predicting the hopping conductivity of disordered solids is explained, which utilizes a modification of the Miller-Abrahams network model.

15.2 The Miller-Abrahams Network

The starting point is Boltzmann's equation

$$\frac{\partial P_i}{\partial t} = \sum_j [W_{ji} P_j (1 - P_i) - W_{ij} P_i (1 - P_j)] \, , \tag{15.1}$$

where P_i is the probability that site i is occupied (by an electron). It is implicitly assumed that the occupation probabilities are uncorrelated. If repulsion can cause strong correlations, then the exclusion factor $(1 - P_i)$ should be omitted. In the linear (Ohmic) regime, the current is proportional to the applied field, and we linearize Eq. (15.1) by writing

$$P_i = P_i^0 + \Delta P_i \, , \tag{15.2}$$

$$W_{ij} = W_{ij}^0 + \Delta W_{ij} \, , \tag{15.3}$$

where superscript 0 denotes the equilibrium value, and Δ an increment proportional to the applied field. It is clear that the linearization implies that $\Delta W_{ij} = -\Delta W_{ji}$. We, thus, obtain the linearized version of Eq. (15.1):

$$\frac{\partial \Delta P_i}{\partial t} + \sum_j A_{ij} \Delta P_i - \sum_j A_{ji} \Delta P_j = \sum_j B_{ji} \Delta W_{ji} \, , \tag{15.4}$$

where

$$A_{ij} = W_{ij}^0 (1 - P_j^0) + W_{ji} P_j^0 \, , \tag{15.5}$$

$$B_{ij} = P_i^0 (1 - P_j^0) + P_j^0 (1 - P_i^0) \, . \tag{15.6}$$

Equation (15.4) is a set of linear equations for the unknowns ΔP_i. The equilibrium values P_i^0 are given by the Fermi distribution

$$P_i^0 = \frac{1}{\exp(E_i / k_B T) + 1} \, , \tag{15.7}$$

where E_i is the energy of a carrier on site i, as measured from the Fermi level,[2] and k_B is Boltzmann's constant. The equilibrium values W_{ij}^0 are given by

$$W_{ij}^0 = \frac{u_{ij}}{|\exp[(E_j - E_i)/k_b T] - 1|} , \tag{15.8}$$

with

$$u_{ij} = u_{ji} = \frac{1}{\tau_0}\exp(-2r_{ij}/a) . \tag{15.9}$$

Here, $1/\tau_0$ is on the order of a phonon frequency, r_{ij} is the distance between i and j, and a is a Bohr radiu.[3] It is assumed that τ_0 depends only weakly on r_{ij} and temperature T. It should be pointed out that the linearization is based on the non-trivial assumption that the site occupation number, which is either 0 or 1, can be replaced by a continuous variable, so that the governing equation can be written down for it.

Suppose that $\mathbf{F}$ is the intensity of an applied field, and $\mathbf{r}_i$ is the radius-vector of site i. The applied field changes the difference Δ_{ij} between the energies of sites i and j. Thus, in a linearized theory we should have

$$\Delta W_{ij} = \frac{dW_{ij}}{d\Delta_{ij}} e\mathbf{F}\cdot(\mathbf{r}_i - \mathbf{r}_j) = \frac{e\mathbf{F}\cdot(\mathbf{r}_i - \mathbf{r}_j)}{\sinh^2(\Delta_{ij}/k_B T)} u_{ij} , \tag{15.10}$$

where e is the charge of an electron. Miller and Abrahams defined a new variable V_i such that

$$P_i = P_i^0 + \Delta P_i \equiv \frac{1}{|\exp[(E_i - eV_i)/k_B T] + 1|} , \tag{15.11}$$

so that to first order

$$\Delta P_i = \frac{dP_i^0}{dE_i} eV_i = \frac{eV_i}{4k_B T \cosh^2(E_i/2k_B T)} , \tag{15.12}$$

The variable V_i is, in the linear regime, proportional to $\mathbf{F}$. Thus, one can transform the set of linear equations for P_i to another set for V_i, with the result being,

$$D_i \frac{\partial V_i}{\partial t} = \sum_j D_{ji} V_j - \sum_j D_{ij} V_i + \sum_j G_{ij} \mathbf{F}\cdot\mathbf{r}_{ij} , \tag{15.13}$$

where $D_i = P_i^0(1 - P_i^0)$, $D_{ij} = D_i A_{ij}$, and $G_{ij} = B_{ij} W_{ij}^0 W_{ji}^0 / u_{ij}$.

[2] The Fermi level of a solid material is the thermodynamic work needed to add one electron to the solid, and does not include the work required to remove the electron from wherever it came from. Understanding of how the Fermi level relates to electronic band structure in determining electrical properties, and how it relates to the voltage and flow of charge in an electronic circuit, is essential to an understanding of physics of solid materials.

[3] Recall that the Bohr radius is approximately equal to the most probable distance between the nucleus and the electron in a hydrogen atom in its ground state.

Consider, first, the steady-state case. We define a temperature-dependent conductance G_{ij} by

$$\frac{k_B T G_{ij}}{e^2} = P_i^0(1 - P_j^0)W_{ij}^0 = P_j^0(1 - P_i^0)W_{ji}^0 . \tag{15.14}$$

If we substitute Eq. (15.14) into the steady-state limit of Eq. (15.13), we obtain

$$\sum_j \left\{ \left[V_i - \frac{\mathbf{F} \cdot \mathbf{r}_i (W_{ij}^0 + W_{ji}^0)}{u_{ij}} \right] - \left[V_j - \frac{\mathbf{F} \cdot \mathbf{r}_j (W_{ij}^0 + W_{ji}^0)}{u_{ij}} \right] \right\} G_{ij} = 0 , \tag{15.15}$$

where $(W_{ij}^0 + W_{ji}^0)/u_{ij} = \coth(|\Delta_{ij}|/2k_B T)$. We are mainly interested in the regime for which $\coth(|\Delta_{ij}|/2k_B T) \sim 1$, in which case Eq. (15.15) becomes

$$\sum_j [(V_i - \mathbf{F} \cdot \mathbf{r}_i) - (V_j - \mathbf{F} \cdot \mathbf{r}_j)] G_{ij} = 0 . \tag{15.16}$$

Equation (15.16) represents a network of resistors, if we view $(V_i - \mathbf{F} \cdot \mathbf{r}_i)$ as the potential at site i. Then, $Z_{ij} = 1/G_{ij}$ is the resistance between sites i and j, and Eq. (15.16) is simply Kirchhoff's equation for site j. Miller and Abrahams treated Z_{ij} more generally and considered it as an impedance.

Under unsteady-state condition, the time-dependent term of Eq. (15.13) does not vanish, and Eq. (15.16) must be rewritten as

$$\frac{P_i^0 e^2 (1 - P_i^0)}{k_B T} \frac{\partial V_i}{\partial t} = \sum_j [(V_i - \mathbf{F} \cdot \mathbf{r}_i) - (V_j - \mathbf{F} \cdot \mathbf{r}_j)] G_{ij} . \tag{15.17}$$

To construct a more general network for the unsteady-state condition, we define a capacitance $C = P_i^0 e^2 (1 - P_i^0)/k_B T$ with a potential V_i across it, and refer all the potentials to the "ground" potential, which is zero. Because $\mathbf{F} \cdot \mathbf{r}_i$ is the applied potential at i, it is represented as an output from a generator connected in series with C between the ground and site i. There is an impedance Z_{ij} connected between any two junctions i and j, as well as a capacitor C_i in series with a generator connected to the ground. Using the expressions for P_i^0 and W_{ij}^0, and restricting our attention to the case when various site energies are of the order or larger than $k_B T$, we obtain

$$Z_{ij} = k_B T \frac{\exp[(|E_i| + |E_j| + |E_i - E_j|)/2k_B T]}{e^2 u_{ij}} , \tag{15.18}$$

$$C_i = \frac{e^2}{k_B T} \exp(-E_i/k_B T) . \tag{15.19}$$

Using Eq. (15.9), Eq. (15.18) is rewritten as

$$Z_{ij} = \frac{k_B T}{e^2} \exp(E_{ij}/k_B T + 2r_{ij}/a)\tau_0 \; , \tag{15.20}$$

where E_{ij} is either the energy of the site farther from the Fermi energy, or $E_{ij} = (|E_i| + |E_j| + |E_i - E_j|)/2$.

Equations (15.18) and (15.19) have an important implication: Even if the site energies are moderately distributed, the exponential dependence of Z_{ij} and C_i on these energies implies that they are enormously-broad distributed. This fact is used to reduce the computations of the effective properties of the network, since the broadness of the distribution of Z_{ij} implies that there are many small conductances that can be removed from the network. The resulting network is called the *reduced* network. Taking advantage of multiresolution wavelet transformations, Sahimi et al. (2005) and Pazhooesh et al. (2006) devised efficient numerical schemes to carry out simulation of hopping transport in such networks.

The procedure for constructing the reduced network is discussed in great details by Pollak (1978). Here, we present a summary of his discussion.

(i) One selects only the largest capacitances (for example, those within a numerical range) that exist in the network. All such capacitances are then considered as equal to the common value C. All other capacitances and their associated sites are then deleted from the network. Next, a resistance $Z_\ell = 2/(\omega C)$ is determined, where ω is the frequency at which the properties of the network are to be calculated.

(ii) One then discards all resistances that are larger than Z_ℓ, and replaces all resistances that are smaller than Z_ℓ by shorts. At high frequencies Z_ℓ is very small and, therefore, there are very few resistances smaller than or equal to Z_ℓ. But as the frequency is lowered, clusters of connected resistors appear.

(iii) The clusters merge together and form a sample-spanning cluster (SSC) when a critical frequency ω_c is reached. This is, of course, similar to the formation of the SSCs in percolation networks. Thus, one has a critical resistance $Z_c = 2/(\omega_c C)$ at which a SSC is formed for the first time. At a frequency smaller than ω_c, the reduction procedure becomes ineffective, since replacing the resistances by shorts would produce a macroscopic short throughout the network.

Miller and Abrahams were the first to calculate the hopping conductivity G of semiconductors using such reduced networks. They assumed that the statistical distribution of the resistances depends only on r_{ij}, but not the site energies. This was justified based on the experimental data for some semiconductors that indicated that impurity conduction exhibits a well-defined activation energy.

15.3 The Symmetric Hopping Model

It is clear that, if in Eq. (15.17) we assume, $\mathbf{F} = \mathbf{0}$, we obtain a symmetric hopping model in which [aside from the constant factor $P_i^0 e^2(1 - P_i^0)/(k_B T)$] V_i plays the same role as that of P_i in the master equation,

$$\frac{\partial P_i}{\partial t} = \sum_j W_{ij} \left[P_j(t) - P_i(t) \right] , \tag{15.21}$$

where W_{ij} is the transition rate between sites i and j, and $P_i(t)$ is the probability of being at site i at time t. Therefore, the formulation and analysis that have been developed for analyzing the master equation and its solutions under a variety of conditions can be utilized for computing the hopping conductivity of a disordered solid by the symmetric hopping model.

Equation (15.21) is, however, applicable when there is no external electric field in the system. If the field is non-zero, then hopping in the field's direction is more likely, and as a result one has a net current in the material, which may seem to make the problem very complex. The *fluctuation-dissipation theorem*[4] relates frequency-dependent conductivity $\hat{g}_e(\omega)$ in terms of the equilibrium, i.e., zero-field, current autocorrelation function (Kubo 1957). According to the theorem, if n is the density of the charge carriers, then, for a system of non-interacting carriers one has

$$\hat{g}_e(\omega) = \frac{ne^2}{k_B T} \hat{D}_e(\omega) , \tag{15.22}$$

where the frequency-dependent diffusivity $\hat{D}_e(\omega)$ is defined by

$$\hat{D}_e(\omega) = \int_0^\infty \langle v(0) v(t) \rangle e^{-i\omega t} dt . \tag{15.23}$$

Here, $v(t)$ is the velocity projected onto a fixed direction in space. At zero frequency Eq. (15.23) reduces to the well-known identity for the static diffusivity, in which case Eq. (15.22) reduces to the Nernst–Einstein relation.[5]

15.3.1 Exact Solution for One-Dimensional Materials

The exact solution for the symmetric hopping model in a linear chain with percolation disorder was derived by Odagaki and Lax (1980). Consider a linear chain in which

---

[4] Given a system that satisfies detailed balance, the fluctuation–dissipation theorem proves that thermodynamic fluctuations in a physical variable predict the response, quantified by the (general) admittance or impedance of the same physical variable (such as voltage, temperature difference, etc.), and vice versa, and applicable both to classical and quantum mechanical systems.

[5] **Walther Hermann Nernst** (1864–1941) was a German chemist who made fundamental contributions to thermodynamics, physical chemistry, electrochemistry, and solid state physics. He formulated the Nernst heat theorem that paved the way for the third law of thermodynamics, for which he won the 1920 Nobel Prize in Chemistry. He developed the Nernst equation that relates the reduction potential of a reaction to the standard electrode potential, temperature, and activities of the chemical species undergoing reduction and oxidation.

Albert Einstein needs no brief biography here.

each bond is present with probability p and disconnected with probability $(1-p)$. The transition rate along the active bonds is the same W_0 everywhere. The frequency-dependent effective diffusivity of the chain is written as

$$\hat{D}_e(\omega) = \sum_{n=1}^{\infty} n p_n \hat{D}_n(\omega) , \tag{15.24}$$

where n is the size of a cluster of active bonds, i.e., $n-1$ active bonds terminated by an inactive bond at each end of the cluster, p_n is the probability of finding such a cluster, and $\hat{D}_n(\omega)$ is the frequency-dependent diffusivity for a cluster of size n, given by

$$\hat{D}_n(\omega) = -\frac{\omega^2}{2n} \sum_{j,j_0}^{n} (j-j_0)^2 \tilde{P}(j, i\omega | j_0) . \tag{15.25}$$

Here, j_0 is the point at which the hopper begins its motion, and $\tilde{P}(j, i\omega|, j_0)$ is the Laplace transform of the probability that the hopper is at point j, given that it began its motion at j_0. The probability of finding a cluster of $n-1$ active bonds, terminated at its two ends by two inactive bonds, is given by

$$p_n = (1-p)^2 \, p^{n-1} . \tag{15.26}$$

It remains to calculate $\tilde{P}(j, \lambda | j_0)$. If we take the Laplace transform of the master equation, Eq. (15.21),

$$\lambda \tilde{P}_i(\lambda) - \delta_{j_0} = \sum_j W_{ij} [\tilde{P}_j(\lambda) - \tilde{P}_i(\lambda)] , \tag{15.27}$$

where λ is the Laplace transform variable conjugate to t, and the initial condition $P_i(0) = \delta_{j_0}$ has been used. Equation (15.27) can be written in a matrix form, $\mathbf{A}\tilde{\mathbf{P}}(\lambda, j_0) = \boldsymbol{\delta}(j_0)$, where $\mathbf{A}$, for a 1D system, is a $n \times n$ tridiagonal matrix. Since the transition rates for all the active bonds are assumed to be W_0, calculating the eigenvalues of $\mathbf{A}$ is not difficult. Suppose that $\boldsymbol{\Gamma} = (\Gamma_m, \Gamma_{3m}, \cdots, \Gamma_{(2n-1)m})$ is the eigenvector of $\mathbf{A}$. Then, it is not difficult to show (see for example, Noble and Daniel 1977) that, $\Gamma_m = \cos(m\pi/2n)$ (with $m = 0, 1, 2, \cdots, n-1$), and that the eigenvalue corresponding to $\boldsymbol{\Gamma}$ is simply $\lambda + 2W_0(1-\Gamma_{2m})$. Therefore,

$$\tilde{P}(j, \lambda | j_0) = \frac{1}{n} \left\{ \frac{1}{\lambda} + 2 \sum_{m=1}^{n-1} \frac{\Gamma_{(2j-1)m} \Gamma_{(2j_0-1)m}}{\lambda + 2W_0(1-\Gamma_{2m})} \right\} . \tag{15.28}$$

Replacing λ with $i\omega$ and using Eqs. (15.26) and (15.28) in (15.25), we obtain the following equation for the frequency-dependent diffusivity of a cluster (finite chain) of size n (Odagaki and Lax 1980):

$$\hat{D}_n(\omega) = 1 + \frac{1}{n}\sqrt{1 + 4/i\hat{\omega}}\left(\frac{1}{z_+^{2n} + 1} - \frac{1}{z_-^{2n} + 1}\right), \tag{15.29}$$

where $z_\pm = [(i\hat{\omega})^{1/2} \pm (4 + i\hat{\omega})^{1/2}]/2$, and $\hat{\omega} = \omega/W_0$. If we substitute Eq. (15.29) into (15.24), we obtain the desired expression for the frequency-dependent effective diffusivity. In particular, we find that in the low-frequency limit,

$$\hat{D}_e(\omega) \sim \frac{p(1+p)^2}{4(1-p)^4}\hat{\omega}^2 + i\frac{p}{2(1-p)^2}\hat{\omega}, \tag{15.30}$$

while at high frequencies

$$\hat{D}_e(\omega) \sim p - \frac{2p(1-p)}{\hat{\omega}^2} + i\frac{2p(1-p)}{\hat{\omega}}. \tag{15.31}$$

15.3.2 Exact Solution for Bethe Lattices

An exact solution of the symmetric hopping model can be derived, if one assumes that the disordered solid can be represented by a Bethe lattice of coordination number Z (Sahimi 1980). To begin with, note that Eq. (15.27) is equivalent to an electrical network, with $\tilde{P}_j(\lambda)$ corresponding to the voltage at site j and W_{ij} the conductivity of the bond joining sites i and j. The terms $-\delta_{j0}$ and $\lambda\tilde{P}_i(\lambda)$ represent, respectively, a current source and an electrical connection to the ground, with λ being a conductivity if it is real, or more generally an admittance if it is complex.

Consider an infinite Bethe lattice of random transition rates (conductances) with a statistical distribution $f(W)$ and a grounded conductance (admittance) connected to each site of the lattice, as well as a site O, and let G_i be the total conductance looking outward into the network from any site A, which is a nearest-neighbor of O (with conductance W_i). The conductance of the branch that starts at O is in parallel with the grounded conductance λ. Therefore, the conductance G, as seen from site O, is given by

$$G = \sum_{i=1}^{Z-1}\left(\frac{1}{G_i} + \frac{1}{W_i}\right)^{-1} + \lambda = \sum_{i=1}^{Z-1}\left(\frac{G_i W_i}{G_i + W_i} + \frac{\lambda}{Z-1}\right). \tag{15.32}$$

G and G_i are, however, statistically equivalent and, therefore, the problem is simply to determine the probability distribution of G_i and G self-consistently. That is, the distribution $H(G_i, \lambda)$ of G_i, when substituted into the right side of Eq. (15.32), must yield the same distribution for G on the left side of (15.32), implying that

$$H(G,\lambda)=\int\cdots\int\delta\left\{G-\left[\sum_{i=1}^{Z-1}\left(\frac{G_iW_i}{G_i+W_i}+\frac{\lambda}{Z-1}\right)\right]\right\}\prod_{i=1}^{Z-1}f(W_i)H(G_i,\lambda)dW_idG_i\ . \tag{15.33}$$

If we now take the Laplace transform of Eq. (15.33), exploit the facts that G_i and W_i are statistically independent; that all the W_i and G_i are identically distributed, and replace λ by $i\omega$, we obtain

$$\tilde{H}(s,i\omega)=\int_0^\infty \exp(-sG)H(G,i\omega)dG$$

$$=\left\{\int\int\exp\left[-s\left(\frac{GW}{G+W}+\frac{i\omega}{Z-1}\right)\right]f(W)H(G,i\omega)dGdW\right\}^{Z-1}\ . \tag{15.34}$$

The frequency-dependent effective conductivity $\hat{g}_e(\omega)$ of a 3D lattice can be accurately approximated by the microscopic conductivity $\hat{g}_m(\omega)$ of a Bethe lattice of coordination number Z, provided that Z is selected in such a way that the bond percolation threshold of the lattice is the same as that of the Bethe lattice. $\hat{g}_m$ is essentially the average $\langle G\rangle$ of the distribution $H(G,i\omega)$ and is given by, $\hat{g}_m=Z\langle G\rangle/(Z-1)$. Using the properties of the Laplace transform, it is straightforward to show that (Sahimi 1980) the frequency-dependent effective hopping conductivity, in the Bethe lattice approximation, is given by

$$\hat{g}_e(\omega)=Z\left[\int\int f(W)H(G,i\omega)\left(\frac{GW}{G+W}+\frac{i\omega}{Z-1}\right)dWdG\right]^{Z-2}\ , \tag{15.35}$$

which provides accurate estimates of $\hat{g}_e(\omega)$. Movaghar et al. (1980a, b) derived the same results as presented here using an entirely different approach. We will compare the predictions of the model of hopping conduction on a Bethe lattice with the experimental data latter in this chapter.

15.3.3 *Poor Man's Percolation: Effective-Medium Approximation*

We introduce a *uniform* lattice in which all the transition rates are equal to $\tilde{W}_e(\lambda)$, with site occupation probabilities being $\tilde{P}_i^0(\lambda)$, so that Eq. (15.27) becomes

$$\lambda\tilde{P}_i^0(\lambda)-\delta_{j_0}=\sum_j\tilde{W}_e(\lambda)[\tilde{P}_j^0(\lambda)-\tilde{P}_i^0(\lambda)]\ , \tag{15.36}$$

where the effective transition rate $\tilde{W}_e$ is yet to be determined. If one subtracts Eq. (15.27) from (15.36), after some rearrangements one obtains

$$(Z_i + \epsilon)[\tilde{P}_i(\lambda) - \tilde{P}_i^0(\lambda)] - \sum_j \left[\tilde{P}_j(\lambda) - \tilde{P}_j^0(\lambda)\right] = -\sum_j \Delta_{ij}\left[\tilde{P}_i(\lambda) - \tilde{P}_j(\lambda)\right] , \tag{15.37}$$

where Z_i is the coordination number of site i, $\epsilon = \lambda/\tilde{W}_e$, and $\Delta_{ij} = (W_{ij} - \tilde{W}_e)/\tilde{W}_e$. A *site-site* Green function G_{ik} is now introduced by the equation

$$(Z_i + \epsilon)G_{ik} - \sum_j G_{jk} = -\delta_{ik} . \tag{15.38}$$

G_{ik} is the response of the system at k if a unit current is injected into the network at i. In terms of G_{ik}, Eq. (15.37) is rewritten as

$$\tilde{P}_i(\lambda) = \tilde{P}_i^0(\lambda) + \sum_j \sum_k G_{ij}\Delta_{jk}[\tilde{P}_j(\lambda) - \tilde{P}_k(\lambda)] . \tag{15.39}$$

Equation (15.39) is an exact, albeit implicit solution of Eq. (15.21), applicable to *any* lattice, irrespective of its dimensionality or topological structure, which can also be rewritten in terms of the "flux," $\tilde{Q}_{ij}(\lambda) = \tilde{P}_i(\lambda) - \tilde{P}_j(\lambda)$. We denote bonds with Greek letters, assign directions to them, and let $\gamma_{\alpha\beta} = (G_{il} + G_{jk}) - (G_{jl} + G_{ik})$ be a *bond-bond* Green function, where i and l (j and k) are the network sites with tails (heads) of arrows on bonds α and β, respectively. Similar to G_{ij}, the bond-bond Green function $\gamma_{\alpha\beta}$ is the response of the system in bond β if a unit current is injected into the bond α. In terms of $\gamma_{\alpha\beta}$, Eq. (15.39) is rewritten as

$$\tilde{Q}_\alpha(\lambda) = \tilde{Q}_\alpha^0(\lambda) + \sum_\alpha \Delta_\beta \gamma_{\alpha\beta} \tilde{Q}_\beta(\lambda) . \tag{15.40}$$

Similar to Eqs. (15.39), (15.40) is also exact and expresses the flux $\tilde{Q}_\alpha$ in the bond α in the heterogeneous lattice as the sum of the flux $\tilde{Q}_\alpha^0$ in the same bond but in the uniform or effective-medium and the fluctuations in the flux through the bond that arises as the result of the heterogeneity of the network.

In practice, one cannot solve Eq. (15.40) for an arbitrary cluster of bonds and, thus, suitable approximate schemes must be developed. As described in Chaps. 2 and 7, in the effective-medium approximation (EMA) one assigns to all but a finite cluster of bonds in the network the effective transition rate $\tilde{W}_e(\lambda)$ (so that $\Delta_\beta \neq 0$ only for a finite set of bonds), and proceeds as above, now averaging over the transition rates of the bonds in the cluster in order to determine $\tilde{W}_e(\lambda)$. In the lowest-order approximation, only a single bond α has its transition rate W differing from $\tilde{W}_e(\lambda)$, in which case Eq. (15.40) yields

$$\left\langle \frac{1}{1 - \gamma_{\alpha\alpha}\Delta_\alpha} \right\rangle = 1 . \tag{15.41}$$

If $f(W)$ represents the statistical distribution of the transition rates, one obtains

$$\int_0^\infty \frac{f(W)}{1-\gamma_{\alpha\alpha}\Delta_\alpha}\mathrm{d}W = 1 \,. \tag{15.42}$$

Thus, $\tilde{W}_e(\lambda)$ is computed by requiring that the average of the excess flux, $\langle \tilde{Q}_\alpha - \tilde{Q}^0_\alpha \rangle$, or excess "potential," $\langle \tilde{P}_i - \tilde{P}^0_i \rangle$, vanish. It is not difficult to show that, for periodic lattices,

$$\gamma_{\alpha\alpha} = -2/Z + (2\epsilon/Z)G(\epsilon) = -p_c + (2\epsilon/Z)G(\epsilon) \,, \tag{15.43}$$

where $G(\epsilon) = -G_{ii}(\epsilon) > 0$, and $p_c = 2/Z$ is the EMA prediction of the bond percolation threshold of the periodic lattices already derived in Sect. 5.6. Equation (15.42) is general and can be used with an arbitrary $f(W)$. Let us note that the EMA is equivalent to the *coherent-potential approximation* (see for example, Yonezawa and Morigaki 1973; Faulkner and Stocks 1980).

The properties of diffusion in a heterogeneous material, modeled by a disordered lattice, can be characterized in terms of the mean-squared displacement $\langle R^2(t)\rangle$ of a particle, diffusing in the lattice. It is straightforward to show that,

$$\tilde{R^2}(\lambda) = \sum_{\mathbf{l}} l^2 \tilde{P}(\mathbf{l}, \lambda) = -\nabla^2_\omega \hat{P}(\omega, \lambda)|_{\omega=\mathbf{0}} \,, \tag{15.44}$$

where l is the length of a bond. Assuming that the length is unity for all the bonds of the lattice, we obtain

$$\langle R^2(t)\rangle = \mathcal{L}^{-1}\left[\frac{Z}{\lambda^2}\tilde{W}_e(\lambda)\right] \,, \tag{15.45}$$

where $\mathcal{L}^{-1}$ denotes an inverse Laplace transform.

Given Eqs. (15.22) and (15.23), it is clear how to estimate the frequency-dependent hopping conductivity by an EMA. All one must do is replacing in Eqs. (15.42) and (15.43) the Laplace transform variable λ by $i\omega$. Certain subtleties that arise from such a replacement must be carefully addressed.

We consider $d-$dimensional lattices of coordination number Z with percolation-type disorder, i.e., with a distribution of transition rates given by, $f(W) = (1-p)\delta_+(W) + p\delta(W - W_0)$, and summarize the main predictions of the EMA for frequency-dependent hopping conductivity $\hat{g}_e(\omega)$ of such models that were derived by, among others, Sahimi et al. (1983b) and Odagaki et al. (1983). In general, in the low-frequency regime, the leading terms of the frequency-dependent hopping conductivity $\hat{g}_e(\omega)$ is written as

$$\hat{g}_e(\omega) = \hat{g}_e(0) + \mathcal{R}(Z, d, p)h_r(\omega) + i\mathcal{I}(Z, d, p)h_i(\omega) \,, \tag{15.46}$$

where $\mathcal{R}$, $\mathcal{I}$, h_r and h_i are functions that depend on d and p. For $p < p_c$ one has $\hat{g}_e(0) = 0$. The EMA predicts that, for $p < p_c$, and regardless of d, one has

$$h_r(\omega) = \omega^2 \,, \quad h_i(\omega) = \omega \,. \tag{15.47}$$

Table 15.1 The EMA predictions for asymptotic form of $\mathcal{R}(Z,d,p)$ and $\mathcal{I}(Z,d,p)$ for $p < p_c$, where d and Z are the lattice dimensionality and coordination number, respectively

d	$\mathcal{R}(Z,d,p)$	$\mathcal{I}(Z,d,p)$
$1 < d < 2$	$\frac{2}{dZ^2}\left[\frac{c\Gamma(1-d/2)}{Z}\right]^{4/d}\frac{1-p}{(p_c-p)^{1+4/d}}$	$\frac{1}{Z}\left[\frac{2c\Gamma(d/2)\Gamma(1-d/2)}{Z(p_c-p)}\right]^{2/d}$
$d = 2$	$\frac{4c^2[\ln(p_c-p)]^2}{4(p_c-p)^3}$	$\frac{-2c\ln(p_c-p)}{Z^2(p_c-p)}$
$d > 2$	$\frac{4I_w^2(1-p)}{Z^4(p_c-p)^3}$	$\frac{2I_w}{Z^2(p_c-p)}$

Table 15.2 The EMA predictions for the frequency-dependent hopping conductivity at the percolation threshold p_c of a $d-$dimensional lattice of coordination number Z

d	$\hat{g}_e(\omega) - \hat{g}_e(\omega = 0)$
$1<d<2$	$\frac{1}{Z}\left[\frac{2Z}{Z-2}c\Gamma(d/2)\Gamma(1-d/2)\right]^{2/(d+2)}[\cos(d\pi/d+2)+i\sin(d\pi/d+2)]\,\omega^{d/(d+2)}$
$d = 2$	$\left[\frac{c}{2Z(Z-2)}\right]^{1/2}(1+i)(-\omega\ln\omega)^{1/2}$
$d > 2$	$\left[\frac{c}{Z(Z-2)}\right]^{1/2}(1+i)\omega^{1/2}$

Table 15.1 summarizes the EMA predictions for $p < p_c$. In this table, $p_c = 2/Z$ is the EMA prediction of the bond percolation threshold (see also Chap. 2), c is a constant of order 1, and I_w is called the Watson integral (Hughes 1995) which represents the value of the Green functions G at the origin in the limit $\lambda = 0$. The Watson integrals for the three cubic lattices are given by Eqs. (14.14)–(14.16).

Table 15.2 summarizes the EMA predictions for frequency dependence of the hopping conductivity at $p = p_c$, while Table 15.3 lists the same but for $p > p_c$, in which m_1 is a constant given by

$$m_1 = \int_{-\infty}^{1} \frac{\mathcal{N}(x)}{(1-x)^2}dx\,, \tag{15.48}$$

where $\mathcal{N}(x)$ is the density of states of the homogeneous (fully-connected) lattice at $p = 1$. For example, for the simple-cubic lattice in d dimensions, one has

$$\mathcal{N}(x) = \frac{\Gamma(d+1)}{\sqrt{\pi}\Gamma(d/2)}(1-x^2)^{d/2-1}\,. \tag{15.49}$$

The density of states for other regular 3D homogeneous lattices was calculated by Lax (1955). According to Table 15.3, the EMA predicts that, above the percolation threshold p_c and a certain dimensionality of the system, the real and imaginary part of $\hat{g}_e(\omega)$ depend on the frequency as ω^2 and ω, respectively, and that this “upper dimensionality” is $d = 4$.

Table 15.3 The EMA predictions for the frequency-dependent hopping conductivity of a d–dimensional lattice of coordination number Z above the percolation threshold p_c in the low-frequency limit

d	$\hat{g}_e(\omega) - \hat{g}_e(\omega = 0)$
$1 < d < 2$	$\dfrac{2c\Gamma(d/2)\Gamma(1-d/2)(Z-2)^{2-d/2}(1-p)}{Z^{d-1}(p-p_c)^{d/2}}[\cos(d\pi/4) + i\sin(d\pi/4)]\omega^{d/2}$
$d = 2$	$\dfrac{2c(1-p)}{Z(Z-2)(p-p_c)}(-\omega\ln\omega)i + \dfrac{\pi c(1-p)}{(Z-2)(p-p_c)}\omega$
$2 < d < 4$	$\dfrac{2I_w(1-p)}{Z(Z-2)(p-p_c)}\omega i + \dfrac{2c\Gamma(d/2-1)\Gamma(2-d/2)(1-p)}{Z^{d-1}(Z-2)^{2-d/2}(p-p_c)^{d/2}}\cos(d\pi/4)\omega^{d/2}$
$d = 4$	$\dfrac{2I_w(1-p)}{Z(Z-2)(p-p_c)}\omega i - \dfrac{2c(1-p)}{Z^3(p-p_c)^2}\omega^2\ln\omega$
$d > 4$	$\dfrac{2I_w(1-p)}{Z(Z-2)(p-p_c)}\omega i + \dfrac{2(1-p)}{Z^3(p-p_c)^2}\left[\dfrac{2I_w^2}{Z(p-p_c)} + m_1\right]\omega^2$

Unlike the percolation state of the system for which there are three distinct regimes—below, at, and above p_c—the EMA predicts that above p_c, and depending on the dimensionality d of the material, there are five distinct functions that yield the frequency dependence of the hopping conductivity.

For comparison with the exact solution of the problem for 1D materials, we list the EMA predictions for the same. In the low-frequency limit,

$$\hat{g}_e(\omega) = \frac{p(2-p)}{8(1-p)^4}\hat{\omega}^2 + i\frac{p(2-p)}{4(1-p)^2}\hat{\omega}\,, \tag{15.50}$$

whereas in the high-frequency limit,

$$\hat{g}_e(\omega) = p - \frac{2p(1-p)(2-p)}{\hat{\omega}^2} + i\frac{2p(1-p)}{\hat{\omega}}\,. \tag{15.51}$$

Therefore, once again, the EMA predictions are qualitatively, but not quantitatively, correct. Other types of EMAs, as well as mean-field analysis of asymmetric hopping have also been developed. The interested reader is referred to Sahimi (2003a) for complete details and comprehensive discussions.

15.4 Poor Man's Percolation: EMA Derivation of Mott's Formula

Mott (1968) pointed out that the exponential dependence of the resistances on the site energies cannot be ignored in most cases, because if the activation energy of a nearest-neighbor site is large, a hop to a distant site whose energy is lower may be easier than one to a nearest-neighbor site. How far the hopper can go depending on the ease of activation to higher energies and, therefore, the hopping distance and the resistance depend on the temperature. Such mechanism of hopping conduction is usually referred to as the *variable—range hopping* (VRH), in contrast with the original work of Miller and Abrahams that was restricted to nearest-neighbor hopping.

It is now generally believed that the VRH is the appropriate conduction mechanism at low temperatures, whereas nearest-neighbor hopping may be appropriate at high temperatures. Moreover, Mott showed that at low temperatures

$$g_e = g_0 \exp[-(T_0/T)^\alpha] , \tag{15.52}$$

which is an important characteristic of variable-range hopping conductivity, and is now one of the most famous results for the hopping conductivity of semiconductors. In general, α depends on the density of states near the Fermi level (see below). In Mott's theory the density of states was assumed to be constant, which results in $\alpha = 1/(d+1)$ for a $d-$dimensional material, G_0 and T_0 are some constants, and

$$T_0 = \frac{\lambda a^3}{k_B \mathcal{N}_F} , \tag{15.53}$$

where $\mathcal{N}_F$ is the density of states at the Fermi level E_F, and λ is a dimensionless parameter.

Equation (15.53) is particularly accurate for amorphous Ge in the range 60 K $\leq T \leq$300 K, with $T_0 \simeq 7 \times 10^7$ K. Similar temperature dependences have also been found for amorphous silicon and carbon, and vanadium oxide. As Pollak (1978) pointed out, however, unless g_e is measured over several orders of magnitudes, a $T^{-1/4}$ behavior should be viewed with caution, and should not be automatically interpreted as the evidence for variable-range hopping conductivity. Hill (1976), on the other hand, analyzed most of the published experimental data and showed that most of them do follow Eq. (15.53) with $\alpha = 1/4$. The conditions under which the behavior of g_e might deviate from $\alpha = 1/4$ will be discussed shortly.

We now describe the derivation of Eq. (15.52) by an EMA, if we assume that

$$W_{ij} = \nu_0 \exp(-2\alpha|R_{ij}|) \exp(-|E|/k_B T) , \tag{15.54}$$

where α is a constant, with

$$h(E) = \begin{cases} W_0^{-1} , & 0 \leq W \leq W_0 . \\ 0 , & \text{otherwise} \end{cases} \tag{15.55}$$

Movaghar and Schirmacher (1981) (see Sahimi 2003a, for complete details) developed what they called *two-site EMA*, a version of the EMA that is slightly more complex and accurate than the standard EMA described above, and derived the following expression for frequency-dependent conductivity,

$$\hat{g}_e(\omega) = \frac{e^2}{k_B T} \frac{N_c(1-n_c)}{\Omega} \hat{D}_e(\omega) , \tag{15.56}$$

where $n_c = N_c/N$ is number of hopping particles per sites, N_c is the total number of particles, while N is the total number of sites, and Ω is the total volume of the system. Here,

$$\hat{D}_e(\omega) = \frac{\langle \hat{\mathbf{R}}^2(\omega) \rangle}{6a_p} \hat{g}_1(\omega) , \tag{15.57}$$

with $a_p = e$, and e being the natural logarithm base. As shown by Movaghar and Schirmacher (1981), the quantity $\hat{g}_1(\omega)$ satisfies the following self-consistency equation

$$\hat{g}_1(\omega) = na_p[\hat{g}_1(\omega) + i\omega] \int\int \frac{h(E)C_{ij}(\mathbf{R}_{ij})W_{ij}}{i\omega + \hat{g}_1(\omega) + W_{ij}} d\mathbf{R}_{ij} dE , \tag{15.58}$$

where $\mathbf{R}_{ij}$ is the distance between sites i and j, and $C_{ij}(\mathbf{R}_{ij})$ is the correlation function between i and j. Moreover,

$$\langle \mathbf{R}^2(\omega) \rangle = \left\langle \mathbf{R}_{ij}^2 \frac{W_{ij}}{i\omega + W_{ij} + \hat{g}_1(\omega)} \right\rangle \left\langle \frac{W_{ij}}{i\omega + W_{ij} + \hat{g}_1(\omega)} \right\rangle^{-1} . \tag{15.59}$$

with $\langle \cdot \rangle$ denoting a configurational average.

We assume that the heterogeneity of the material is random, so that, $C_{ij}(\mathbf{R}_{ij}) = 1$, and substitute these in Eq. (15.58) in the limit $\omega = 0$, we obtain

$$\frac{4\pi k_B n a_p T}{8\alpha^3 W_0} \int_0^{W_0/k_B T} \int_0^{\infty} \frac{x^2}{(g_1/\nu_0)\exp(x+y) + 1} dx dy = 1 . \tag{15.60}$$

In the low-temperature limit—the regime of validity of the VRH model—$W_0/k_B T \to \infty$, we solve for g_1 to obtain

$$g_1 = \nu_0 \exp[-(T/T_0)^{1/4}] , \tag{15.61}$$

$$T_0 = \frac{24\alpha^3 a_p W_0}{\pi k_B n} . \tag{15.62}$$

Equation (15.61) is in agreement with (15.52). This example demonstrates the predictive power of the approximations that were derived and discussed above.

Note that Huinink et al. (2006) carried out extensive numerical calculations in order to study the structure and conductivity of the VRH percolation clusters in two dimensions. They showed that the VRH percolation clusters behave as standard percolation clusters, with the main difference between them being the existence of a temperature-dependent effective lattice constant.

15.5 Critical-Path Analysis

Over a decade after the original work of Miller and Abrahams (1960), Ambegaokar et al. (1971), Shklovskii and Efros (1971), Brenig et al. (1971), and Pollak (1972) realized that the transport paths that Miller and Abrahams had thought to be carrying most of the current in the network do not in fact carry any current in most situations. The reason is that, if one always proceeds through the nearest-neighbor sites, as in the Miller–Abrahams theory, one is certain to arrive at a site where a nearest-neighbor site is a large distance away and, therefore, it may be more efficient to go through farther neighbor sites. Ambegaokar et al. (1971) and Pollak (1972) invoked percolation in order to identify the correct current-carrying paths. To describe the models, we follow Ambegaokar et al. (1971) whose work is very elegant and conceptually simple. We already described the method in Chap. 7, and reiterate it here.

Ambegaokar et al. (1971) proposed that an accurate estimate of g_e is the critical percolation conductance g_c, which is the largest value of the conductance such that the subset of the network with $g_{ij} > g_c$ still contains a conducting SSC. Thus, they divided the network into three parts:

(i) A set of isolated clusters of high conductivity, with each cluster consisting of a group of sites connected together by conductances $g_{ij} \gg g_c$.

(ii) A small number of resistors with g_{ij} on the order of g_c, which connect together a subset of high conductance clusters to form the SSC. This was called *the critical subnetwork*, which is essentially the same as the static limit of the reduced network described above.

(iii) The remaining resistors with $g_{ij} \ll g_c$.

It should be clear that the resistors in group (ii) dominate the overall conductance of the network. Note that, as described in Chap. 7, the same ideas were used by Katz and Thompson (1986, 1987) to estimate the permeability and electrical conductivity of a porous medium with a broad pore-size distribution. Thus, the condition that $g_{ij} > g_c$, together with Eq. (15.20), is expressed by

$$\frac{r_{ij}}{r_m} + \frac{|E_i| + |E_j| + |E_i - E_j|}{2E_m} \leq 1 , \tag{15.63}$$

where $r_m = a \ln(g_0/g_c)/2$ is the maximum distance between any two sites between which a hop can occur, and $E_m = k_B T \ln(g_0/g_c)$ is the maximum energy that any initial or final state can have.

To construct the critical subnetwork, Ambegaokar et al. considered an empty network and, starting with the smallest resistors (largest conductor), inserted them in the network one by one. As more resistors are inserted, they connect and form clusters until the critical resistance, $Z_c = 1/g_c$, is reached at which a SSC is formed. To calculate Z_c, Abmegaokar et al. used Eq. (15.20), assuming that τ_0 is constant. Moreover, it was also assumed that the density of states $\mathcal{N}$ is constant near the Fermi level, $\mathcal{N}_F(E) = \mathcal{N}(E = E_F)$. Thus, Z_{ij} is a monotonic function of the random variable $\zeta = E_{ij}/k_BT + 2r_{ij}/a$, and the critical resistance Z_c defines a corresponding critical value of ζ, ζ_c. Around each site i such that $E_i < \zeta k_BT/2$, a sphere of radius

$$r_i = \frac{a}{2}(\frac{\zeta}{2} - \frac{E_i}{k_BT})\ , \tag{15.64}$$

is drawn, which increases with ζ. When two spheres overlap, a bond is inserted between the two sites with overlapping spheres. This happens only if inequality (15.63) is satisfied. Percolation occurs at ζ_c, corresponding to a critical radius r_c.

It should be clear that the problem, as formulated by Ambegaoker et al., is a site percolation process. Pollak (1972) analyzed it as a bond percolation phenomenon that allowed him to include the effect of short-range correlations, but his basic results are the same as those of Ambegaokar et al. Equation (15.64) indicates that there is a maximum radius $r_m = a\zeta_c/4$ and a maximum energy $E_m = \zeta_c k_BT/2$, with $\zeta_c = 2\ln(g_0/g_c)$. Thus, if two sites are separated by a distance larger than r_m, or farther from the energy level E_m, they will not contribute significantly to the conductivity. The volume of the sphere defined by Eq. (15.64) is $(\pi/6)a^3(\zeta_c/2 - E_i/k_BT)^3$, and the average volume $\langle V\rangle$, with averaging taken over all the sites with a sphere of non-zero radius, i.e., those for which $E_i < \zeta k_BT/2$, is given by

$$\langle V\rangle = \frac{\pi}{384}a^3\zeta_c^3 = \frac{\pi}{48}r_m^3\ . \tag{15.65}$$

On the other hand, ϕ_c, the volume fraction of the spheres at the percolation threshold, is given by, $\phi_c \simeq n\langle V\rangle$, where, assuming that the density of states is constant, $n = \mathcal{N}\zeta_c k_BT$ is the fraction of the sites with a sphere, i.e., those with an energy in the interval $(-E_m, E_m)$. Ambegaokar et al., estimated that $\phi_c \simeq 1/4$, somewhat larger than $\phi_c \simeq 0.15 - 0.17$, estimated by Scher and Zallen (1970) for 3D continuum percolation (see Chap. 3). One may also calculate the number of bonds per sites B_c of the network. We learned in Chap. 2 that for bond percolation, $B_c \simeq d/(d-1)$. For 3D amorphous materials—or, equivalently, continuum percolation—$B_c = 4\pi n r_c^3/3$, and computer simulations of Pike and Seager (1974) indicated that $B_c \simeq 2.8$ for 3D systems. On the other hand, B_c is related to the density of states $\mathcal{N}(E)$ by

$$B_c = \frac{4\pi n}{3}\frac{\int_{-E_m}^{+E_m}\mathcal{N}(E_i)dE_i\int_{-E_m}^{+E_m}(r_m^3 + 3r_m^2D_m + 3r_mD_m^2)\mathcal{N}(E_j)dE_j}{\int_{-E_m}^{+E_m}\mathcal{N}(E_i)dE_i}\ , \tag{15.66}$$

where D_m is the mean size of the sites. Combining $\phi_c = n < V > \simeq 1/4$ with Eq. (15.65), or using Eq. (15.66) together with the lattice or continuum value of B_c and the appropriate expression for r_m, one finally obtains

$$g_e \simeq g_c = g_0 \exp\left[-(T_0/T)^{1/4}\right]\ , \tag{15.67}$$

and

$$T_0 = \frac{16a^3}{k_b N} . \tag{15.68}$$

A comparison between Eqs. (15.68) and (15.62) indicates that the percolation model of Ambegaokar et al. predicts that $\lambda \simeq 16$. Thus, their model not only predicts the $T^{-1/4}$ behavior proposed by Mott, it also provides an estimate of the temperature T_0 defined by Eq. (15.61).

The pre-exponential factor g_0 has also been calculated by several research groups, since the quantitative prediction of g_c requires an accurate value of g_0. For example, using the work of Kurkijärvi (1974), Shklovskii and Efros (1975) proposed that $Z_0 = 1/g_0 = r_c(2r_c/a)^\nu R_0$, where ν is the correlation length exponent of 3D percolation, and R_0 is the resistance for $\zeta = \zeta_c$. Kirkpatrick (1974) suggested the same expression, except that he replaced ν by $\mu - 1$, where μ is the critical exponent of the conductivity of 3D percolation. Pollak (1972), Butcher and McInnes (1978), Butcher (1980), Movaghar et al. (1980a, b), and Movaghar and Schirmacher (1981) also calculated the pre-exponential factor g_0, although their results did not involve any critical exponent of percolation. The predictions of Butcher, Movaghar, and co-workers are particularly accurate. Note that, by calculating the effective conductivity of the VRH clusters, Huinink et al. (2006) showed that the central idea behind the CPA models is correct and that the aforementioned critical subnetworks do exist.

15.5.1 Effect of a Variable Density of States

In deriving the percolation model described above, it was assumed that the density of states $\mathcal{N}$ is constant near the Fermi level. Although the basic $T^{-1/4}$ law has been observed in many materials—see for example, Knotek et al. (1973), and Visčor and Yoffe (1982), among others—criticism of the formulation was raised by many (see e.g., Szpilka and Visčor 1982), mainly because the predicted and measured values of g_0 differed by several orders of magnitude. It was suggested that the assumption of a constant density of states in the percolation model may not be justified, and that one must use a variable density of states, in which case one would have to use Eq. (15.66) with a variable density of states.

Ortuno and Pollak (1983) studied the problem and proposed that, if the density of states is concave, then an appropriately modified percolation model can explain the data and remove the disagreement between the predictions and the experimental data. They used Eq. (15.66) with $\mathcal{N}_F(E) = \mathcal{N}(E_F)\exp(E/E_0)$, where E_0 is the exponential decay rate, and analyzed the problem in details. The predictions were then compared with the experimental data for amorphous Si and amorphous Ge; good agreement was found between the two. Moreover, the predicted value of g_0 was on the order of the experimental data.

Conduction in disordered granular materials occurs by tunneling of electrons and holes from charged grains to neutral ones, implying that electrons must be transferred from one neutral grain to another. Thus, each grain must be characterized by a charging energy $E_c = e^2/(\varepsilon D)$, where ε is the dielectric constant, and D is the grain size. As a result, a disordered granular material is characterized by a distribution $h(E_c)$, which is, of course, related to the grain-size distribution, and a density of states related to this distribution given by

$$\mathcal{N}(E) = \frac{1}{\langle \Delta E \rangle} \int_0^E h(E_c) dE_c \,, \tag{15.69}$$

where $\langle \Delta E \rangle$ is the average electronic-level separation inside the conducting grains. Thus, any distribution of energies $h(E_c)$ can be immediately transformed into one for the density of states. Normally, one expects that $\mathcal{N}(0) = 0$. In any composite material, however, one can have energy states other than those in the conducting grains, because there can be impurities that can contribute a non-zero $\mathcal{N}(E)$ at $E = 0$, for which there is in fact some experimental evidence.

To address this problem, Sheng and Klafter (1983) assumed that $\mathcal{N}(E) = \mathcal{N}_F + 1/\langle \Delta E \rangle$ and a log-normal distribution for $h(E_c)$, and calculated the hopping conductivity of the granular material, using Eq. (15.66). A two-parameter model was used to fit the results, with the fitting parameters being the width of the distribution $h(E)$, and $x = \pi n D_m/6$, and showed that varying the two parameters enables one to obtain a variety of temperature dependence of the conductivity, ranging from $T^{-1/4}$ to $T^{-1/2}$.

15.5.2 *Effect of Coulomb Interactions*

Hopping conductivity of several different amorphous materials exhibit a $T^{-1/2}$ behavior. As discussed in the preceding section, although a variable density of states, such as $\mathcal{N}(E) \sim E^\beta$ or $\mathcal{N}(E) = \mathcal{N}_F + cE^\beta$, can explain such data, the origin of such power laws was not clear for some time until it was explained by Efros and Shklovskii (1975), who suggested that long-range Coulomb interactions between localized electrons can generate a soft gap, called the Coulomb gap, in the density of states near the Fermi level. This implies that in a narrow gap centered around the Fermi level, the density of states cannot be constant and must vary with the energy, whereas outside the gap the density of states vanishes. If δ is the width of the gap, Efros and Shklovskii (1975) showed that $\mathcal{N}(\delta) \sim \delta^{d-1}$ in d dimensions. As a result, they suggested that there must be a crossover in 3D materials from the $T^{-1/4}$ behavior at relatively high temperatures, where the Coulomb gap is not effective, to $T^{-1/2}$ at lower temperatures.

Although a $T^{-1/2}$ behavior had been reported by several groups, the temperature below which the gap could be detected by conductivity measurements is usually too low in amorphous semiconductors. Generally speaking, the Coulomb gap does not affect the hopping conductivity of amorphous semiconductors and, moreover, it cannot be found in good metals. It can only affect those materials that have localized electronic states. For example, the Coulomb gap affects the hopping conductivity of doped crystallic semiconductors that provide experimental evidence for the crossover from $T^{-1/4}$ to $T^{-1/2}$ behavior.

15.5.3 Comparison with Experimental Data

We should keep in mind that Eq. (15.67) represents a lower bound to the true hopping conductivity of a network whose individual conductances vary over a broad range, because the critical subnetwork corresponds to replacing all $g_{ij} < g_c$ by 0, and all $g_{ij} \geq g_c$ by g_c in the original network. Equation (15.67) is exact only in the limit $T \rightarrow 0$. If $T > 0$, hops with conductance less than g_c also contribute to the macroscopic conductivity, implying that the optimal cutoff should be somewhat larger than g_c. Moreover, the percolation approach of Ambegaokar et al. cannot be used for 1D, or quasi-1D materials, because percolation disorder divides a linear chain into finite segments, making the problem meaningless.

The physical systems to which the 1D limit may be relevant are two classes of compounds that consist of weakly coupled parallel chains of strongly coupled atoms or molecules. The conductivity of such systems is highly anisotropic and, therefore, they may be treated as essentially 1D conductors. Well-known examples are salts of the organic ion-radical tetracyanoquinodimethane (TCNQ) and the square planar complexes of transition metals such as platinum and iridium.

Shante (1977) proposed a modification of the percolation model of Ambegaokar et al. that could take into account the effect of such complexities. His model was a bundle of chains in which hopping can occur along the chains and between them. Interchain hopping was assumed, however, to be much more difficult than intrachain hopping, but was still allowed. Shante's model also allowed the possibility of the intrachain hoppings in either 1D or 2D. At low temperatures Shante's model corresponded to 2D and 3D percolation, respectively, and consequently $T^{-1/3}$ and $T^{-1/4}$ behaviors were obtained. At high temperatures, the percolation model is no longer applicable, and Shante derived a T^{-1} behavior. Figure 15.1 compares the predictions of his model with the experimental data for N-methylphenazinium-TCNQ compounds; the agreement is excellent.

The predictions of the CPA of Ambegaokar et al. (1971) can also be obtained by the perturbation expansion, and EMA-type analysis, developed by Movaghar and co-workers, which were mentioned above. If one assumes that the transition rate between i and j is given by,

$$W_{ij} = \nu_0 \exp[-(|E_i| + |E_j| + |E_i - E_j|)/2k_B T - 2\alpha|\mathbf{R}_{ij}|] , \tag{15.70}$$

and uses the approach that resulted in Eqs. (15.60)–(15.62), assuming that the density of state is constant, then, in the low-temperature limit of hopping near E_F—the energy at the Fermi level—combining Eqs. (15.56) and (15.57) yields the following result (Movaghar et al. 1980a):

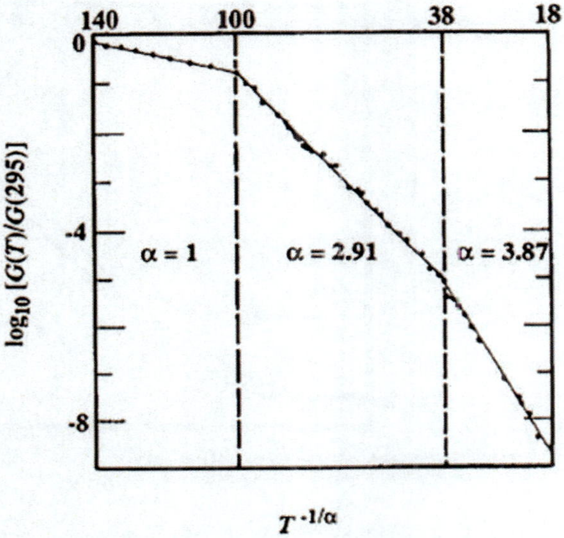

Fig. 15.1 Temperature dependence of hopping conductivity of NMP-TCNQ. Temperature is in Kelvin, and the straight lines are the best fits of the data (after Shante 1977)

$$\hat{g}_e(\omega) = \frac{\langle \mathbf{R}^2 \rangle}{6} \frac{e^2}{k_B T} \left[n N_F k_B T \nu_0 \hat{g}_1(\omega) \right] , \tag{15.71}$$

where

$$\hat{g}_1(\omega) = \frac{4 T a_p}{T_0} \left[i\omega + \hat{g}_1(\omega) \right] \int_0^\infty \frac{x^3 \exp(-x)}{i\omega + \hat{g}_1(\omega) + \exp(-x)} dx . \tag{15.72}$$

The experimental value, $T_0 = 3.6 \times 10^8$K for germanium was used in place of T_0/a_p, ν_0 was taken to be 10^{21} sec^{-1}, and Eq. (15.59) was used for computing $\mathbf{R}^2$.

Figure 15.2 compares the predicted real part of the frequency-dependent effective diffusivity at two temperatures with the experimental data for Ge. The agreement between the predictions and the data is excellent. In general, as described in Chap. 13, the real part of $\hat{g}_e(\omega)$, $\mathcal{R}(\omega)$ follows a power law

$$\mathcal{R}(\omega) \sim \omega^x , \tag{15.73}$$

where x is a temperature-dependent exponent, and $10^2 \leq \omega \leq 10^5$Hz. This is similar to what the EMA predicted; see Eq. (14.47). Movaghar et al.'s model predicted the temperature dependence of the exponent x to be linear, which is consistent with many experimental data.

The aforementioned model of Ortuno and Pollak was utilized by Maloufi et al. (1988), who used a variable density of states in order to fit their conductivity data for amorphous Si_ySn_{1-y} with $y = 0.47 - 1$. Figure 15.3 presents the fits of their data by the model, with the agreement being excellent over much of the temperature range.

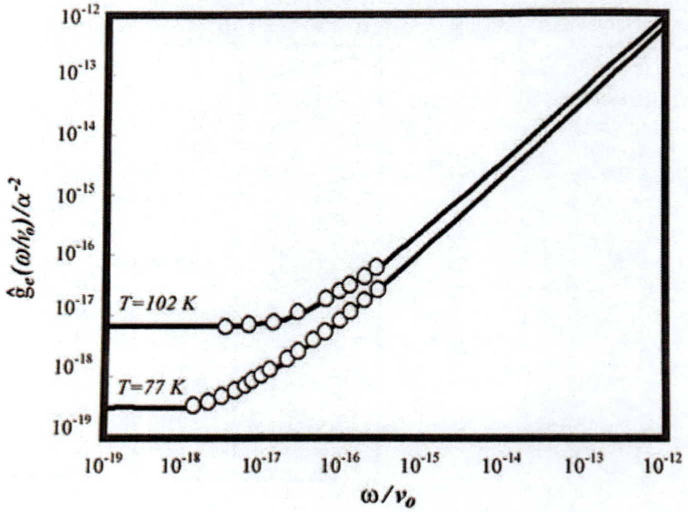

Fig. 15.2 Real part of the effective (frequency-dependent) hopping conductivity for two temperatures. Symbols are the experimental data, while the curves are the theoretical predictions. The absolute magnitude of the DC conductivity has been fitted to one curve (after Movaghar et al. 1980a)

As another example of a variable density of states, we consider $\mathcal{N}(E) \sim |E|^{\beta}$, where β is a positive constant, which was analyzed by Hamilton (1972) and Pollak (1972). Using Eq. (15.66), one obtains an equation similar to (15.56), but with $\alpha = (\beta+1)/(\beta+4)$, which does reduce to Eq. (15.66) when $\beta = 0$. The limit $\beta \to \infty$ corresponds to a system with a sudden onset of states away from the Fermi level, hence yielding a T^{-1} behavior for such materials.

An experimental realization of a power-law density of states was reported by Redfield (1973), who carried out a careful study of the hopping conductivity of heavily-doped and strongly-compensated GaAs. His data indicated a $T^{-1/2}$ behavior, rather than $T^{-1/4}$. Redfield showed that, although a $T^{-1/4}$ might look plausible, his data could be fitted extremely accurately by a $T^{-1/2}$ law. This is demonstrated in Figure 15.4, indicating that the data can be easily explained with a variable density of states, $\mathcal{N} \sim |E|^2$.

If the density of states is taken to be, $\mathcal{N}(E) = \mathcal{N}_F + cE^{\beta}$, where c is a constant, we obtain Eq. (15.56) with $\alpha = (\beta+1)/(\beta+2)$, which may correspond to a granular material with a broad particle-size distribution. Mehbod et al. (1987) showed that this type of density of states and the resulting exponent α can fit very well their experimental data for hopping conductivity of polymer–conducting–carbon black composites, which contain randomly dispersed particles in the matrix with a broad size distribution. In Mehbod et al.'s experiments the polymeric matrices were polystyrene, polyethylene, ethylene-propylene copolymer, and styrene-butadiene copolymer.

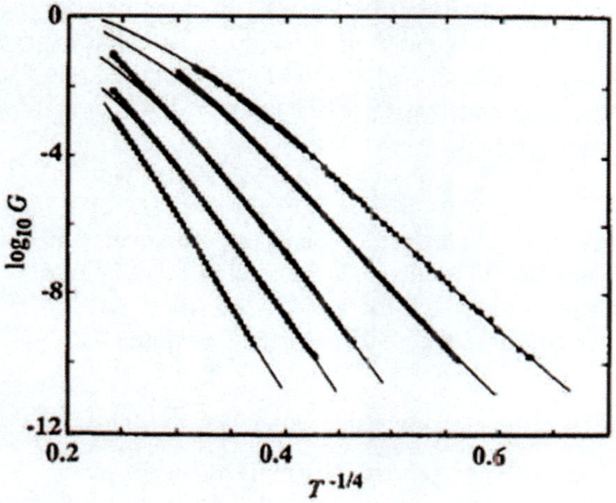

Fig. 15.3 Comparison of the predicted temperature dependence of hopping conductivity g_e of $a{-}\mathrm{Si}_y\mathrm{Si}_{1-y}$ (the lines), obtained with an exponential density of states, with the experimental data (symbols). The data are, from left, $y = 0.47, 0.62, 0.77, 0.9$, and 1. T is in Kelvin (after Maloufi et al. 1988)

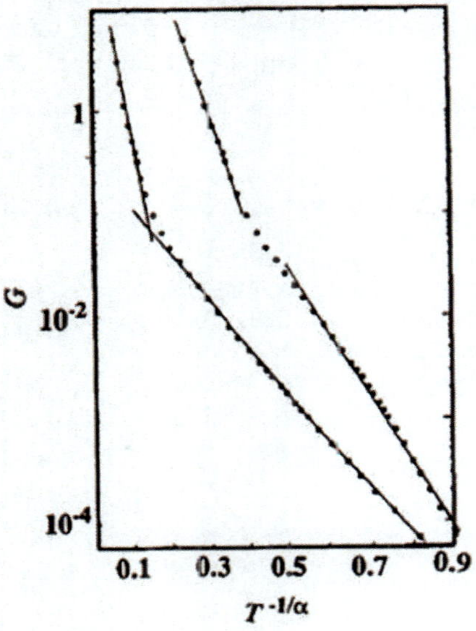

Fig. 15.4 Temperature dependence of hopping conductivity g_e of doped and compensated GaAs. Symbols represent the experimental data, the line on the left is their $T^{-1/2}$ fit, while that on the right is their $T^{-1/4}$ fit (after Redfield 1973)

The first convincing experimental evidence for the crossover, predicted by Efros and Shklovskii (1975) as the result of Coulomb interactions, was probably provided by Zhang et al. (1990). They measured the resistivity of five insulating compensated n-type CdSe samples, which are doped semiconductors. Their data clearly indicated a crossover from $T^{-1/4}$ behavior to $T^{-1/2}$ as the temperature was lowered by nearly three orders of magnitude from 15 to 0.04 K, over which the hopping energy becomes comparable with, and then smaller than, the energy gap introduced by Efros and Shklovskii (1975). The crossover temperature $T_{\rm co}$ was found to decrease with increasing concentration of the donor, which is expected.

Aharony et al. (1992) and Meir (1996) proposed a scaling theory for the crossover by suggesting that the effective resistivity $Z_e = 1/g_e$ of the material follows the following scaling law

$$\ln(Z_e/Z_0) = Ah(T/T_{\rm co}) , \tag{15.74}$$

where the scale factor A and the crossover temperature $T_{\rm co}$ both depend on the material's properties, but the scaling function $h(x)$ is universal with the properties that, $h(x) \sim x^{-1/4}$ for $x \gg 1$, and $h(x) \sim x^{-1/2}$ for $x \ll 1$. Moreover, using the percolation model, Aharony et al., proposed that

$$h(x) = \frac{x + [\sqrt{x+1} - 1]}{x[\sqrt{1+x} - 1]^{1/2}} , \tag{15.75}$$

and provided expressions for A and $T_{\rm co}$. The implication of Eqs. (15.74) and (15.75) is that, if $\ln(Z_e/Z_0)$ for many materials is plotted as a function of $T/T_{\rm co}$, then all the data will collapse onto a single universal curve. Figure 15.5 presents the collapse for the data reported by Zhang et al., indicating that the data collapse is essentially complete.

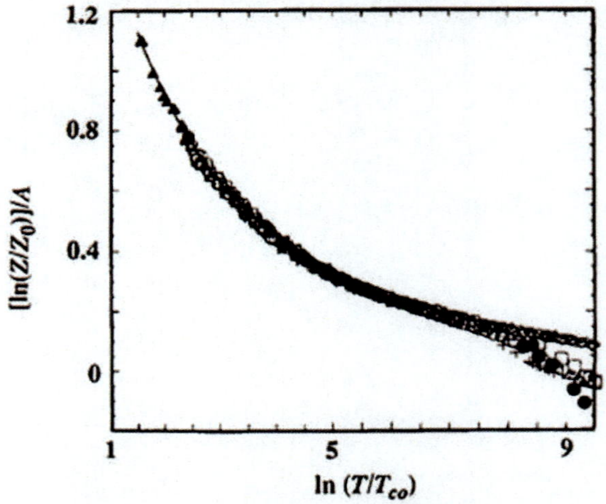

Fig. 15.5 Temperature dependence of hopping conductivity g_e of carbon black polymer composite. Temperature is in Kelvin (the data are from Zhang et al. 1990)

15.6 Effect of Fractal Morphology on Hopping Conductivity

As the preceding sections indicated, the $T^{-1/2}$ behavior of hopping conductivity may be explained in terms of the Coulomb interactions and the gap, which, however, do not play any role in many semiconductors that do exhibit the $T^{-1/2}$ behavior. There has been some experimental evidence that, if the material has a fractal morphology, then its hopping conductivity may exhibit the $T^{-1/2}$ behavior.

Lévy and Souillard (1987) suggested that in a material with fractal morphology impurity quantum states are *superlocalized*, i.e., their wave functions decay with the distance r as $\exp(-r^{\gamma})$, with $\gamma > 1$ (recall that in the classical Anderson localization, $\gamma = 1$). Thus, the possibility of superlocalization of electrons has been extensively studied. For example, Harris and Aharony (1987) and Aharony and Harris (1990) argued that one must distinguish between the behavior of a *typical* geometry, and that of a system with the *average* geometry, where the averaging is taken with respect to all the possible geometries of the system. Moreover, they argued that the latter yields $\gamma = 1$, i.e., there is no superlocalization of electrons. For the typical geometries, Aharony and Harris suggested that, proposed that

$$\max(\nu^{-1}, 1) \leq \gamma \leq D_{\min} , \tag{15.76}$$

where ν the critical exponent of percolation correlation length, and $D_{\min}$ is the fractal dimension of the minimum or chemical path described in Chap. 2. Thus, using the numerical values of ν and $D_{\min}$, one obtains $1 \leq \gamma(d = 2) \leq 1.13$ and $1.14 \leq \gamma(d = 3) \leq 1.34$.

Through extensive numerical simulations, Lambert and Hughes (1991) estimated that, $\gamma \simeq 1.13 \pm 0.06$ and 1.39 ± 0.07 for 2D and 3D percolation networks, respectively. Thus, the upper bound of Harris and Aharony may in fact be an exact result, i.e., $\gamma = D_{min}$. Earlier, and presumably less accurate, simulations of de Vries et al. (1989) had yielded $\gamma(d = 2) \simeq 1.0$, i.e., one does *not* have superlocalization in 2D. Harris and Aharony also suggested that for hopping conductivity of a percolation system in the fractal regime, i.e., either at p_c, or over length scales $L \ll \xi_p$ (where ξ_p is the percolation correlation length),

$$\alpha \simeq \frac{D_{\min}}{D_f + D_{\min}} , \tag{15.77}$$

consistent with $\gamma = D_{min}$. Equation (15.77) then yields $\alpha = 0.37$ and 0.35 for $d = 2$ and 3, respectively, indicating that the existence of a fractal morphology *cannot by itself* explain the observed $T^{-1/2}$ behavior of hopping conductivity of semiconductors in which Coulomb interactions are unimportant.

As mentioned above, early numerical studies of the problem were carried out by Lambert and Hughes (1991), but the most definitive study that the author is aware of was probably carried out by Kantelhardt and Bunde (1997), who studied localization

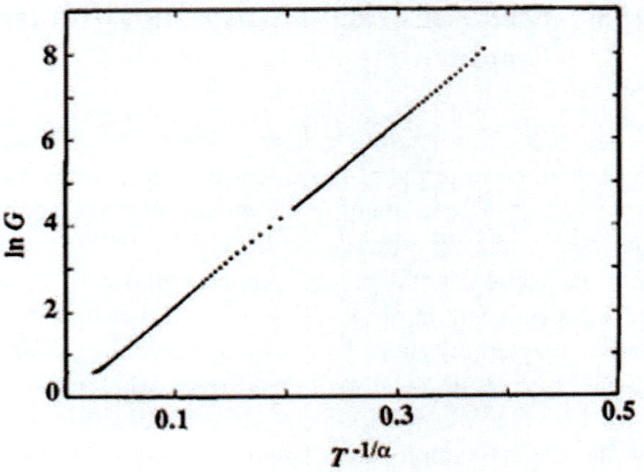

Fig. 15.6 Temperature dependence of hopping conductivity of carbon black polymer composites (after van der Putten et al. 1992)

of electronic eigenfunctions and vibrational excitations—the fractons described in Chap. 14—on percolation clusters, both in the topological (r_t) and Euclidean (r_e) space. They identified three regimes:

(i) The first regime corresponded to small r_t and r_e, in which electrons and fractons are superlocalized in the r_t space, but the localization is not exponential in the r_e space.

(ii) In the intermediate regime, stretched exponential *sublocalization* was found in both spaces with $\gamma \simeq 0.6$.

(iii) The third regime corresponded to large r_t and r_e in which the average behavior was found to depend strongly on the number of percolation configurations used in the averaging, but the asymptotic regime was a simple exponential decay with $\gamma = 1$, in agreement with the argument of Aharony and Harris.

None of these results supports, however, the view that a material with a fractal morphology can by itself explain the observed $T^{-1/2}$ behavior of hopping conductivity of semiconductors in which Coulomb interactions are unimportant.

van der Putten et al. (1992) measured the conductivity of carbon black polymer composites as a function of carbon black concentration, from a point in the vicinity of the percolation threshold p_c up to $33p_c$, and in the temperature range 4 K–300 K. Their data, shown in Fig. 15.6, indicated that α, the exponent defined by Eq. (15.56), is about 2/3. To interpret this value, they assumed that, $\mathcal{N}(E) \sim E^{\beta}$, and proposed that

$$\alpha = \frac{\gamma(\beta+1)}{D_f + \gamma(\beta+1)}, \tag{15.78}$$

where D_f is the fractal dimension of the 3D SSC at p_c. Therefore, with $\alpha = 2/3$ and $D_f \simeq 2.53$ one obtains $\beta \simeq 2.75$. Thus, their data can be best explained by a *combination* of a percolating fractal morphology—one at p_c—*and* a power-law density of states. Moreover, if we take, $\beta = 1$, then Eq. (15.78) predicts that $\alpha \simeq 1/2$, indicating that a linear density of states *plus* a percolating fractal morphology may be responsible for the reported $T^{-1/2}$ behavior of hopping the conductivity of semiconductors in which Coulomb interactions are unimportant.

Aharony et al. (1993) argued, however, that there is more than one way of interpreting the data of van der Putten et al. (1992) and, therefore, their interpretation may not provide the definitive solution to the problem. See also the experimental data for the conductivity of carbon black polymer composites reported by Mandal et al. (1997) who, however, claimed that their data can be understood on the basis of superlocalization.

15.7 Universality of Frequency-Dependent Hopping Conductivity

Many disordered solids have very similar frequency-dependent hopping conductivity. They include ion conducting glasses (Roling 1998), amorphous semiconductors (Long 1991), polycrystalline semiconductors (Kunar and Srivastava 1994), electron-conducting polymers (Jastrzebska et al. 1998), ion conducting polymers (Rozanski et al. 1995), transition metal oxides (Suzuki 1980), metal cluster compounds (van Staveren et al. 1991), and organic-inorganic composites (Bianchi et al. 1999). The universality was first pointed out by Taylor (1956) for ion conducting glasses. He showed that the dielectric loss for various glasses fall on a single curve, if the data are plotted versus scaled frequency, and that the activation energy of the DC conductivity was the same as that of the frequency that marks the onset of AC conduction. Isard (1961) noted that if one plots the normalized conductivity versus $C\omega/g_e(0)$, it exhibits universal behavior, expressed by

$$\hat{g}_e = \frac{g_e(\omega)}{g_e(0)} = F\left[C\frac{\omega}{g_e(0)}\right], \tag{15.79}$$

where $F(x)$ is the apparent universal function, and C is called the Taylor-Isard scaling constant, which is proportional to the inverse of the temperature T. Figure 15.7 presents an example of the Taylor-Isard universal scaling function for eight different ion conducting glasses (Roling 1998).

To explain the origin of the apparent universality of the AC conductivity, Dyre and Schr⊘der (2000) listed sixteen common characteristics of the vast majority of disordered solids, such as, for example, the facts that, (a) $g_e'(\omega)$, the real part of the AC conductivity, increases with ω, while $g_e''(\omega) \geq 0$, and (b) the real part of the conductivity depends on the frequency ω by a power law, $g_e'(\omega) \sim \omega^x$, with the exponent x being typically between 0.6 and 1.0; a decreasing function of the temperature; $x \simeq 1$ when the DC conductivity is essentially zero; when $x \simeq 1$, $g_e'(\omega)$ is almost independent of the temperature, and other common properties.

It should, however, be noted that there are also some "dissenting" views, emphasizing the *differences* between various disordered solids and their AC conductivity

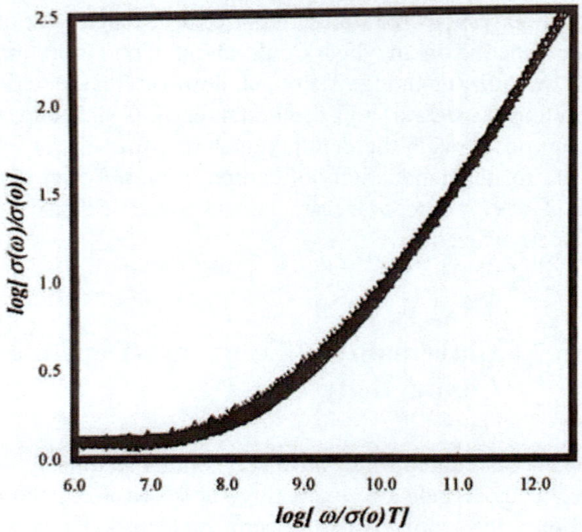

Fig. 15.7 Universal conductivity function for eight distinct ion conducting oxide glasses (after Roling 1998)

(see for example, Ngai and Moynihan 1998), because for some disordered solids the exponent x is slightly *larger* than 1.0 (Cramer and Buscher 1998). But, the vast majority of disordered solids do seem to exhibit universal AC conductivity.

The sixteen common AC characteristics of disordered solid materials provide strong empirical support for its universality, but there is also strong theoretical evidence for the universality. Sidebottom (1999) proposed the following universal scaling representation of the AC conductivity:

$$\hat{g}_e = \frac{g_e(\omega)}{g_e(0)} = F\left[\epsilon_0 \Delta\epsilon \frac{\omega}{g_e(0)}\right] , \tag{15.80}$$

so that the Taylor-Isard constant C is, $C = \epsilon_0 \Delta\epsilon$. The proof of Eq. (12.25) was provided by Schr⊘der and Dyre (2000).

Using the EMA, Dyre (1993) demonstrated that the rescaled AC conductivity does exhibit universal scaling as a function of the rescaled frequency. The activation energy $E_{\rm DC}$ of the DC conductivity, defined by, $g_e(0) \propto \exp(-\alpha E_{\rm DC})$ for large values of $\alpha = 1/(k_B T)$ (k_B is the Boltzmann's constant) can be determined by percolation. If $h(E)$ is the statistical distribution of the activation energies, then, based on the critical-path analysis $E_{\rm DC}$ is obtained as the solution of the following equation,

$$\int_0^{E_{\rm DC}} h(E) dE = p_c , \tag{15.81}$$

where p_c is the bond percolation threshold of the system. For a comprehensive the discussion of the interpretation of experimental data for the AC conductivity sees Léon et al. (2001).

Schrøder and Dyre (SD) (2008) studied the problem of frequency-dependent hopping conductivity in extremely disordered solids, and addressed two important issues: (a) experimental data (Dyre and Schrøder 2000) and numerical simulations (Sahimi et al. 2005; Pazhoohesh et al. 2006) indicated that the rescaled conductivity, $\tilde{g}_e = g_e(\omega)/g_e(0)$, may be a *universal* function of the rescaled frequency, $\tilde{\omega} = a\omega/\sigma(0)$, where a is a constant. (b) The universal conductivity curve is expressed by,

$$\ln \tilde{g}_e = (i\tilde{\omega}/\tilde{g}_e)^{D_f/2} , \tag{15.82}$$

where D_f is a fractal dimension associated with the cluster in which conduction occurs. In the EMA, one has (Sahimi 1984), $D_f = 2$. SD assumed that conduction occurs in the SSC at the percolation threshold p_c, and suggested, $D_f = D_s$, where D_s is the fracton dimension (see Chap. 14) that has the same approximate value of 4/3 in both 2D and 3D SSC. Numerical evidence presented by SD indicated that Eq. (15.82) with $D_f = D_s \approx 4/3$ provides accurate representation of the ac conductivity of a simple-cubic lattice.

On the other hand, Sahimi argued[6] (2022a) that the conduction cluster is *not* the SSC and, thus, D_f cannot be equal to its D_s. Sahimi (2022a) pointed out that SD had studied conduction by the random barrier (RB) model (Dyre 1988) in which transport takes place by hopping, with the charge carriers being subject to spatially random energy barriers. It is well known that in a system in which the hopping rates vary over many orders of magnitude, conduction occurs only in a very small portion of the system.

Sahimi (2022a) argued that, in fact, in the RB model conduction occurs in the *minimum spanning tree* (MST). His argument was as follows. Let E_{ij} be the energy, or barrier, associated with bond ij of a lattice. The MST is a cluster that visits every site in the lattice such that the total energy, $E = \sum_{ij} E_{ij}$, is *minimum*, with the constraint that visit to any site cannot create a closed loop. To construct the tree, one begins at a site i and selects the bond b connected to i with the lowest E_{ij}. Then, among all the *unvisited* bonds connected to b, the one with the lowest E_{ij} is selected, and so on. This is not only the physical basis of the RB model, but also for the *bond invasion percolation cluster* (BIPC) described in Chaps. 3 and 9, if invasion is from a single site, with the role of E_{ij} played by the capillary pressure needed to enter a bond. The MST, or the BIPC, is a fractal object with, $D_f \simeq 1.22$ and 1.37 in 2D and 3D, respectively, compared with $D_f \simeq 1.89$ and 2.53 for the SSC. Thus, the question is, what is the appropriate D_f in Eq. (15.82)?

Following SD's arguments, the correct D_f is the fracton dimension of the MSP or BIPC, not that of the SSC. Then, the relevant D_s will not take on the same value of 4/3 in both 2D and 3D. Indeed, for the MST and BIPC D_s takes on values very close to their D_f mentioned above.

[6] The original arguments were submitted to *Physical Review Letters* as a Comment on the paper by Schrøder and Dyre (2008). After exchanges with the authors and a referee, and the suggestion that the Comment should be published in *Physical Review B*, the author moved on. The analysis was finally archived as a preprint in July 2022.

The structure of the MST or BIPC is universal, because only the *order* of the energies E_{ij} matters, not their numerical values or their statistical distribution. Therefore, the universality of $\tilde{g}_e$ is due to the universality of the BIPC or the MST.

It is also interesting to note that the rescaled frequency-dependent permeability of porous media is also a universal function of a suitably rescaled frequency (Sahimi 2022b).

Chapter 16
Applications of Invasion Percolation

16.1 Introduction

A highly important problem in engineering, physics, and chemistry is developing efficient computational algorithms for studying complex phenomena. There is, of course, no "universal" algorithm that could be used for every computational problem. Each problem, or at least each class of problems that share similar traits, needs its own computational algorithm. Therefore, many ideas and concepts, as well as numerous algorithms have been put forward.

In the present chapter we describe one such algorithm, invasion percolation (IP) algorithm or, more generally, invasion algorithms, which has found numerous applications in a variety of contexts. We are already familiar with the IP. Chapter 3 provided a brief introduction to the subject; in Chap. 6 we described an application of the IP to modeling of regional seismicity (Rundle et al. 2020); in Chap. 9 we described and discussed the application of the IP to simulating two-phase flow in heterogeneous porous media; in Chap. 15 we mentioned its possible application to hopping conduction in extremely heterogeneous materials, while in Chap. 19 we will describe its application to problems at the interface of ecology, hydrology, and geochemistry. From a historical perspective, it was precisely the need for realistic, physics-based simulation of slow flow of two immiscible fluids in complex porous media that motivated the development of the IP. Various IP models were proposed (see Chap. 9; see also below) in order to incorporate the effects of the morphology of a porous medium, as well as the properties of the fluids at the microscopic level, into the dynamics of the fluid flow, simulated by the IP algorithm.

It was soon after its invention in early 1980s, however, that the IP was transformed into a key computational algorithm for gaining an understanding of a large and important class of non-equilibrium and irreversible processes in complex media. Thus, the static and dynamic characteristics of the IP models, which by themselves have proven to be fascinating, have been the subject of intense research. The goal of this chapter is to provide a review of other various applications of the IP, and its

M. Sahimi, *Applications of Percolation Theory*, Applied Mathematical Sciences 213,
https://doi.org/10.1007/978-3-031-20386-2_16

connection to other problems in statistical physics of complex media. A pedagogical review of the subject was given by Ebrahimi[1] (2010), which we follow and expand on.

16.2 Invasion Percolation

As already described in Chaps. 3 and 9, the IP is a growth process in which a cluster of the sites of a lattice that have been invaded by an agent, such as a fluid, grows into a stochastic field by selecting the path of the least resistance. The model was originally developed in the early 1980s by Lenormand and Bories (1980) and Chandler et al. (1982), but the name was coined by Wilkinson and Willemsen (1983) in order to emphasize that the IP is a modified form of percolation theory with a well-defined sequence of invasion events. The IP has many similarities, but also important differences, with the standard random percolation (RP) described in Chap. 2 and used throughout this book. For example, similar to the RP, one may define two versions of the IP: site IP and bond IP. One important difference is that, contrary to the RP, the IP cluster always percolates in the lattice and, therefore, there is no fine-tuned parameter, such as the percolation threshold p_c in the IP, although the breakthrough point, i.e., the point at which the invading agent first forms a sample-spanning cluster (SSC) and the invasion process reaches steady state, is the analog of p_c in the RP. This feature of the IP relates it to the paradigm of self-organized criticality (Stark 1991), i.e., a system that drives itself toward a critical state without any external driving force.

It is difficult to review the extensive literature on the IP. Our main goal is to describe the model's applicability to the study of some of the most interesting phenomena in statistical physics on the one hand, and some important problems in practice, on the other hand. To do so, we must understand the IP model's properties, as well as its intrinsic potential for being tailored for application to different problems.

In a simple initial configuration, such as the spherical geometry, the invasion starts from a single site at the center of a two-dimensional (2D) $L \times L$ lattice. But, a more appropriate geometry, at least as far as the problem of fluid displacement in a porous medium (see Chap. 9), as well as many other applications, is concerned the rectangular geometry where the invader invades from one face: Then, the medium is a $L \times 2L$ rectangle, or a $L \times L \times 2L$ lattice in 3D. In this case the macroscopic direction of the invasion is in direction of the longer edge, and the initially invaded cluster contains all the elements on the first row on one side of the lattice. The invasion is terminated when the cluster of the invaded sites (bonds) touches the opposite side. At this point—the aforementioned breakthrough point—the cluster spans the lattice.

As described in Chap. 9, during the invasion process, it may happen that the portion of the cluster boundary completely encircles a region of the lattice containing the

---

[1] The author is grateful to Professor Fatemeh Ebrahimi, University of Birjand, Iran, for her invaluable help.

defender. If the encircled region cannot be invaded, then one has a more complex variant of the IP, namely, IP with *trapping*, or TIP. Identifying the trapped sites or bonds and removing them from the list of the potential invasion domain is complex and time consuming. In practice, the simulation time grows as a power of N, the total number of the invaded sites. In the early versions of the TIP algorithms, the search for the trapped regions (if any) was done by scanning the entire lattice using the Hoshen–Kopleman algorithm (see Chap. 2) that identifies and labels all the connected regions filled by the defender that is disconnected from the outlet. Thus, the growth sites (bonds) are updated and sorted after each trapping event. Since the execution time for each call of the Hoshen–Kopelman algorithm scales as $\mathcal{O}(N)$, we should apply the algorithm only when the last growth events could potentially lead to the formation of the traps.

But for every lattice geometry the number of configurations that may lead to the trap formation is limited and, therefore, it is possible to diagnose them by developing relatively simple, accurate, and very fast routines, hence reducing the execution time significantly. For example, in the square lattice almost half of the invasion is unproductive regarding the formation of the traps. The method is precise, but not very efficient, as the computation time scales as $\mathcal{O}(N^2)$ for each lattice realization. But, because the trap formation is a local process and the interface is quasi-static, except in a small region around the last growth sites, Sheppard et al. (1999) could develop a highly efficient algorithm whose execution time scales as $\mathcal{O}(N \ln N)$; see Chap. 10.

To compare with the RP, as well as for the discussions in the rest of this chapter, three essential fractal dimensions of the IP models, namely, the fractal dimension $D_{\rm min}$ of the minimum path, that of the invasion cluster, D_f, and its backbone, $D_{\rm bb}$, are listed in Table 16.1. For the definitions of the three fractal dimensions, see Chap. 2.

Table 16.1 Numerical estimates of the fractal dimensions of the non-trapping IP (NTIP), trapping IP (TIP) models for both site and bond IP models, and their comparison with those of random percolation and optimal paths

Two-dimensional lattices			
Model	D_f	$D_{\rm bb}$	$D_{\rm min}$
NTIP	1.89	1.64	1.13
Site TIP	1.82	1.21	1.21
Bond TIP	1.82	1.21	1.21
Random percolation 1.89	1.64	1.13	
Optimal paths		1.21	1.21
Three-dimensional lattices			
NTIP	2.52	1.87	1.37
Site TIP	2.52	1.86	1.37
Bond TIP	2.52	1.46	1.46
Random percolation	2.53	1.87	1.37
Optimal paths		1.42	1.43

The shapes that IP clusters taken on in random and correlated disordered media were studied by Ebrahimi (2008).

When the correlations and heterogeneities in the structural properties of a disordered stochastic field cannot be ignored, very interesting properties emerge. For example, Knackstedt et al. (2000) simulated the bond and site TIP processes in correlated fields with long-range correlations, and reported non-universal behavior in scaling exponents of the model (see below). Their model can be classified as a "parametric" IP, in the sense that the statistical properties of the fields are characterized by a parameter, the Hurst exponent H that characterizes the extent and nature of the correlations in the model.

16.3 Extensions of Invasion Percolation

In addition to the well-known variants of the basic IP that we described in Chap. 9 and the above, several other relatively simple but useful extensions of the IP model have been developed.

16.3.1 *Self-Avoiding and Loopless Compressible Invasion Percolation*

Stark (1991) proposed a variant of the IP that he referred to as the *self-avoiding invasion percolation* (SWIP). Cieplak et al. (1994) and Barabási (1996) developed what they called loopless compressible IP (LCIP). In both models, the growth process is the same as the bond IP, but with the constraint that closed loops of the invaded sites or bonds are forbidden. The difference between the two is that, in the SWIP the growth rule is that no site may be invaded, if any of its neighbor sites—except the source site—are already occupied, whereas in the LCIP no bond may be invaded if two adjacent sites are already invaded. The SWIP generates drainage networks—usually referred to as the river networks—whereas in the LCIP the resulting configuration is a spanning tree and, if started from a central seed, it is the union of all the optimal polymer configurations from the center to each of the lattice sites. Note that the site-bond IP algorithm developed by Sahimi et al. (1998) and described in Chap. 9 also prevents loop formation.

16.3.2 *Gradient Invasion Percolation*

Gradient invasion percolation (GIP) (Birovljev et al. 1991; Meakin et al. 1992) is another variant of the parametric IP that extends the applicability of the model to the cases in which there exists a static gradient G in the original stochastic field. Percolation in a gradient was already described in Chap. 9. For the nth site in the GIP one has

$$r^n(\mathbf{x}) = r(\mathbf{x}) + Gx_\alpha \,, \tag{16.1}$$

where $r(\mathbf{x})$ is the original random resistance at location $\mathbf{x}$, G is the gradient, and α the spatial direction. Similar to the standard IP, the GIP has a static counterpart, the aforementioned gradient percolation (Chap. 9). Both models have applications to two-phase fluid flow in 3D porous media, when the presence of buoyancy forces adds a pressure gradient to the process.

16.3.3 Thermal Invasion Percolation

Another variant of the IP is to introduce thermal noise into the model. This version is called invasion percolation with temperature (Gabriell et al. 2000), or thermal IP (TeIP), and changes the dynamics from deterministic to stochastic such that the larger is the temperature, the stronger is the stochasticity. As the name suggests, the growth rule that was deterministic in the original IP versions is modified by introducing a Boltzmann factor,

$$P(\mathbf{x}, t) = \frac{\exp[-\beta r(\mathbf{x})]}{\sum_{\mathbf{x}'} \exp[-\beta r(\mathbf{x}')]} \tag{16.2}$$

as the probability of invading a lattice site at $\mathbf{x}$ at time t. The parameter β is inversely proportional to the temperature T, and the sum is over all the cluster's next neighbors $\mathbf{x}'$, which should be updated as the invasion proceeds. In the limit, $T \to 0$, we recover the usual IP. In the opposite limit, $T \to \infty$, the model produces the Eden growth, a growth process in which at each time step one of the un-invaded sites next to the invasion cluster is selected *at random* and added to the cluster.

16.4 Dynamics of Invasion Percolation: The Relation with Self-Organized Criticality

Experiments and numerical simulations have revealed the interesting nature of the dynamics of the IP: the advance of the front—the external perimeter—occurs with pockets connected to the invaded cluster by a narrow neck. This phenomenon, called *burst* (Roux and Guyon 1989) or *avalanche*, is in fact similar to the Haines jumps described in Chap. 4, leads the IP to a critical self-organized state in which avalanches of all sizes occur with a power-law distribution without any external driving force, which is why it is called self-organized. Furuberg et al. (1988) studied the dynamic pair correlation function $C(x, t)$ that represents the probability that a site at a distance x from a reference site is invaded at a time t later than the reference site, and found that it follows a dynamic scaling:

$$C(x, t) = x^{-1} f(x^z/t) . \tag{16.3}$$

It was shown that the dynamic exponent z is equal to D_f, the fractal dimension of the invading cluster. Such a dynamics brings the system into a situation whereby only a restricted range of the bonds or sites resistances is chosen as the IP cluster percolates indefinitely throughout the medium (Stark 1991). In other words, when the steady state is reached, the probability of invading a site above the percolation threshold p_c of the underlying lattice in the RP approaches zero; see Fig. 16.1. Therefore, Monte Carlo simulation of the IP generates accurate estimates of p_c of the lattice without any need to fine-tuning a parameter (Wilkinson and Willemsen 1983). It has been argued (Ebrahimi 2008) that this result may be responsible for the observed difference in the shape of the IP clusters, when they are compared with that of the clusters in the RP.

Technically, the IP belongs to the class of model with deterministic external dynamics in which the disorder is quenched, i.e., it is frozen in time. In other words, both the rules and the underlying stochastic field (such as a porous medium in which the IP occurs) are fixed and do not change during the process. If we apply the IP rule to a system with annealed disorder—one in which the disorder changes with the time (by, for example, the temperature)—with all the bonds (sites) at the interface having equal chances of being invaded, we recover the Eden model (see above), which produces compact clusters with $D_f = d$, with d being the spatial dimension, rather than a fractal cluster.

As mentioned above, Gabrielli et al. (2000) introduced the TeIP model by introducing thermal noise. In this context the IP and the Eden growth are two limiting cases that, as mentioned earlier, correspond to $T \to 0$ and $T \to \infty$, respectively. Gabrielli et al. also observed that for the intermediate, finite temperatures, the correlation length is finite and the criticality, characterized by the formation of self-similar

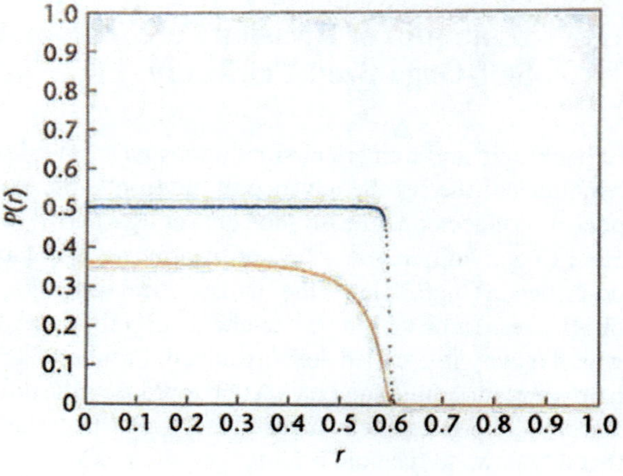

Fig. 16.1 $P(r)$, the distribution of the invaded sites at the breakthrough, for 1024×512 square lattices in the non-trapping IP (blue) and trapping IP (green) processes, averaged over 2000 lattice realizations (after Ebrahimi 2010; courtesy of Professor Fatemeh Ebrahimi)

fractal structure, is destroyed. This allows one to parameterize the degree of the criticality (fractality) of the system.

16.5 Applications

As mentioned in the Introduction, historically, the development of the IP model was motivated by the modeling of slow, capillarity-dominated fluid flow and two-phase displacement in porous media. This was described in detail in Chap. 9 and, therefore, it will not be considered in this chapter. But, over the last 40 years, the range of the applicability of the IP models has been expanded by developing new variations of the models, or linking them with seemingly unrelated phenomena. In what follows we describe some of the best-known applications of the IP models.

16.5.1 Modeling of Stream Networks

Due to the importance of drainage basins to the hydrologic cycle, and to the diversity and maintenance of life on Earth, they have been studied for a long time. In particular, a long-standing problem of characterizing drainage basins has been studied intensively, as it is important to many environmental issues, ranging from water management, to landslides, flood prevention, and aquatic dead zones. Drainage basins are simply land areas that slope toward a single outlet, such as a river mouth, or points of higher infiltration, or evaporation rates. The evolution of river networks is a consequence of headward growth and branching away from escarpments in a substrate. The presence of variations in the lithology, structure, soil developments, and topography makes the local substrate strength a stochastic field.

Scheidegger (1967) was probably one of the first investigators who proposed a quantitative model for the planar development of river networks. In his model that used a hexagonal lattice, channels were developed from an edge of the lattice that formed a line of basin outlets. At each "time step," an entire row of adjacent neighbors was simultaneously considered, with each of these nodes assigned at random a flow direction toward one of the two neighbors that were within the basin. The open boundary served as the location of the seed points, and the clusters of river basins were grown by a pixel layer at each time step. Although the model did produce patterns that were qualitatively similar to some of the patterns in actual river networks, it produced uniform headward growth, with the frontier of the cluster remaining smooth during development. In addition, each channel tended to follow a more or less straight path between its source and outlet.

Hack (1957) had discovered that the principal stream length L_p scales with the drainage basin area S as

$$L_p \sim S^{D_H} , \tag{16.4}$$

where D_H is a scaling exponent. On the other hand, if L is the relevant length of the system, then, $S \propto L^2$, and $L_p \sim L^{2D_H}$. Since, $L_p \sim L^{D_{\min}}$, where $D_{\min}$ is the fractal dimension of the minimum path (see Chap. 2), we obtain

$$D_H = \frac{1}{2} D_{\min} \, . \tag{16.5}$$

Analyzing the available data, Hack estimated that, $D_H \approx 0.6$. Hack's law was explained in terms of fractal concepts (Tarboton et al. 1988; Rinaldo et al. 1993; Maritan et al. 1996), and percolation (Stark 1991; Hunt 2016). Attempt was also made (Reis 2006) to explain Hack's law in terms of the constructal law (Bejan and Lorente 2013) that suggests that the natural tendency of all flow systems is to form configurations that, over time, generate progressively greater flow.

Stark (1991; see also Niemann et al. 2001) assumed that at any time the next failure takes place at the weakest point. On a lattice this is tantamount to the IP model. He also considered the physical fact that streams rarely bifurcate downstream and, therefore, introduced the additional constraint that closed loops cannot be formed, hence leading to the aforementioned SWIP, and used the model to simulate the evolution of stream (river, drainage) networks. Figure 16.2 presents an example of a drainage network that one obtains by simulating the SWIP model. According to Table 16.1, for TIP in 2D, $D_{\min} \approx 1.21$, so that $D_H \approx 0.605$, in excellent agreement with Hack's original estimate. Using the estimate of $D_{\min}$ for the NTIP, listed in Table 16.1, would not make any significant difference.

Fig. 16.2 An example of simulated drainage network grown by the SWIP model. The simulations were carried out in a 512 × 512 square lattice in which seeding was allowed from every site on one face of the lattice. The two parts show two different stages of the cluster growth. The heavy lines show the shortest path with a fractal dimension of 1.130, compared with that of the entire cluster, 1.896 (after Stark 1991)

Following the IP-based model of Stark (1991), Fehr et al. (2009; see also Fehr et al. 2011a, b), proposed another IP-based model for extracting watersheds from landscapes. In their model, simulated on a $L \times L$ square lattice with fixed boundary conditions in the vertical (growth of the network) direction and periodic boundary conditions in the horizontal direction, the upper and lower sides of the lattice represented the sinks of two basins, one in the north (N) and the other in the south (S). The height of each site i was represented by h_i. For each site i, the IP rule was applied, such that the basin (N or S) to which site i belongs was one that the IP-invaded cluster reached first. Therefore, all sites of the lattice belonged to one of the two basins, and the interface between them defined the watershed. To speed up the computations of identifying the interface, an efficient sweeping strategy was used: (a) Initially, the sites were selected along a straight line that connects the sinks. Therefore, when the IP algorithm from two neighboring sites evolved to different sinks, a segment of the watershed was between them. (b) From then on, the sites were selected only in the neighborhood of the already known watershed segments in order to reveal more segments of the watershed, resulting at the end in the complete watershed.

Oliveira et al. (2019) further refined the model by Fehr et al. Consider a rectangular lattice of size $L_x \times L_y$ in which, similar to the Fehr et al.'s model, the height of each site i is h_i. A height threshold height h^* is introduced such that, if $h_i > h^*$, then the ith site belongs to a cluster, called the height cluster that is composed of all connected sites with height larger than h^*. Otherwise, site i does not belong to any cluster. Oliveira et al. (2019) used $h^* = 0$, which for Earth corresponds to the sea level, and the height clusters define continents and islands on Earth. Thus, it is known a priori that they define drainage basins separated by several interface lines, whose specific sizes and shapes must be determined. The rules in Oliveira et al.'s model were similar to those proposed by Fehr et al.

Oliveira et al. (2019) studied two versions of the model. One was with the traditional periodic boundary conditions in the horizontal direction and unconventional periodic boundary conditions in the vertical direction, in order to simulate real landscapes. In the unconventional periodic boundary conditions, each site at the top (bottom) row is neighbor of every other site in the top (bottom) row and represents a mapping of a sphere onto a lattice. In the second model fixed boundary conditions in both directions were utilized for synthetic landscapes.

Simulations of Oliveira et al. indicated that the perimeter and area distributions of the basins display long tails, extending over several orders of magnitude and following approximately power laws, with exponents that depend on the spatial correlations and are invariant under the landscape orientation, not only for terrestrial, but also lunar and Martian landscapes. The terrestrial and Martian results were statistically identical, hence suggesting that a hypothetical Martian river would be similar to the terrestrial rivers. The simulations suggested that Hack's law may have its origin purely in the maximum and minimum lines of the landscapes.

Thus, IP-based models are able to explain the observed self-similarity in river networks and predicted structural quantities, such as D_H, as well as the network's fractal dimension, $D_f \simeq 1.82$ and other important quantities, such as the bifurcation, length and area ratios. All of Stark's results were consistent with the data for natural stream networks.

Stark's model produces tree-like structures. Although such structures are abundant in Nature, there are also many cases that contain closed loops. One example is

16.5.2 Healthy and Cancerous Vascular Networks

An important problem in medical science is the architecture of vascular networks in tumors. Compared to normal tissues, tumors are known to have many tortuous vessels, shunts, vascular loops, widely varying intervascular distances, and large avascular areas. Baish et al. (1996) examined the morphology of 2D vascular networks in the murine dorsal skinfold chamber preparation and could differentiate between distinct architectural features of normal and tumor vascular networks. Such differences also give rise to a fundamentally distinct flow of blood and transport paths. Baish et al. showed that normal capillaries, which are relatively straight and regularly spaced, are represented accurately by the so-called Krogh cylinder model, widely used in such studies. The model, named after August Krogh,[2] which is a simplified model of the tissue surrounding the capillary, and assumes that a cylindrical layer of tissue surrounds each capillary with the solute transferred only from that capillary, which is assumed to be a cylindrical tube of constant radius.

In contrast, when Baish et al. examined the morphology of tumor vascular networks, it turned out to be a self-similar fractal. They suggested that the tortuous vessels and wide range of avascular spaces found in tumors are better represented by the IP, and proposed the following model for it. One begins with a square lattice representing the potential paths of vascular growth, which occurs according to the IP rule, namely, by first assigning uniformly-distributed random strengths to each node on the lattice and, starting from an arbitrary node, occupying the lattice node adjacent to the current network that has the lowest strength, and continuing the process until the desired lattice occupancy is achieved. Blood vessels are assumed to connect all adjacent occupied lattice sites, with blood being supplied at the starting point and withdrawn from the point nearest to the opposite corner.

To retain only those parts of the network with non-zero flow, the backbone of the cluster was identified. Baish et al. used square lattices with relatively modest sizes, because they are better suited for vascular transport studies. With a lattice constant of 30–50 mm, the networks simulated represented a circulatory unit measuring roughly

[2] **Schack August Steenberg Krogh** (1874–1949) was a Danish professor at the department of zoophysiology at the University of Copenhagen from 1916 to 1945. He made important contributions to several fundamental discoveries in physiology, and developed the *Krogh's principle* according to which, "For such a large number of problems [in physiology] there will be some animal of choice, or a few such animals, on which it can be most conveniently studied," a principle that is essential to those branches of biology that rely on the comparative method, such as neuroethology, comparative physiology, and more recently functional genomics. He received the Nobel Prize in Physiology or Medicine for the discovery of the mechanism of regulation of the capillaries in skeletal muscle.

1–1.5 mm in width. In addition, the selected lattice size matched the known upper limit of fractal scaling observed in the tumor vasculature, as reported by Gazit et al. (1995; see also Naeim et al. 1996; Motamed et al. 1968), although the existence of larger avascular areas in tumors was already known (Ahlstrom et al. 1988).

The model was then used for predicting the effect of network morphology on flow and transport. Interesting results emerged. For example, the number of avascular spaces in tumors scaled with the size of the space. Thus, there is a few large avascular spaces and many smaller avascular ones between the vessels. The tortuosity of the vessels, represented by the elevated minimum path dimension, produces regions of locally flow-limited transport, and reduces flow through the tumor as a whole. The IP model provided a explanation for the long-standing paradox that tumor vasculature has a higher geometrical resistance than normal vasculature, even though it has a larger vessel diameter. The model could produce rate of oxygenation in normal and tumor tissues, clarifying the role of architectural obstacles to transport in tumors, and performed better than the standard Krogh cylinder model.

16.5.3 Simulating Spin Systems Near and Away from Criticality

Cluster algorithms, such as what was proposed by Swendsen and Wang[3] (1987), are among the most successful approaches to the development of highly efficient Markov processes that equilibrate successive phase space configurations. For a large class of spin models, the mapping proved by Fortuin and Kasteleyn (1972) demonstrated that any state of a lattice of spins can be described in terms of bond percolation clusters of, say, "down spins," in a surrounding background of "up spins" (Binder 1997). In this algorithm at each Monte Carlo step the spin configuration is updated by generating a new bond configuration from the current spin configuration by occupying some of the satisfied bonds, i.e., those across which the spins are aligned, of the lattice with a probability that depends on the temperature. Clusters of the sites connected by occupied bonds are locked into the same spin-type, and all the clusters are independently flipped with probability 1/2.

Machta et al. (1995) proposed a new algorithm based on the IP. In their approach the cluster growth is controlled by a modified bond IP rule, until the bond configuration fulfills a stopping condition. Two modifications were made in the usual bond IP. First, cluster growth was initiated at all lattice site. To understand this better, consider the standard bond IP. Every bond is initially on the perimeter of the cluster, and the invasion consists of adding bonds in a given random order. Initially, every site is a cluster, and in most steps of the cluster growth process two smaller clusters

[3] The algorithm is the first cluster scheme for Monte Carlo simulation of large systems near their critical point. It was designed for the Ising and Potts models and was generalized later to other systems. The algorithm's key ingredient is the random cluster model through percolation models of connecting bonds, due to Fortuin and Kasteleyn (1972).

are combined into a single larger cluster. The growth stops when a cluster becomes sample spanning. Thus, suppose that, $p = p(L)$ is the fraction of bonds accepted during the cluster growth in a system of linear size L. As Hammersley (1963) first pointed out, as $L \to \infty$, $p \to p_c$, the percolation threshold of the lattice. The second modification correlated the IP to an underlying spin configuration, which was done by allowing cluster growth along only satisfied bonds.

Thus, the way the invasion algorithm of Machta et al. (1995) is implemented is as follows. One begins with an Ising spin configuration S, attributes to the bonds of the lattice a random order. Correlated IP clusters are grown as described above until one of the clusters spans the system, which happens when the maximum separation in one of the d directions for a pair of nodes in the cluster is the system size L. After the growth process is terminated, each cluster is flipped with probability 1/2, which yields a new spin configuration S'. The bonds are then randomly reordered and the process starts anew.

A significant aspect of the algorithm is that it moves the spin configuration toward critical temperature T_c, regardless of whether the simulations are started in the high- or low-temperature phase. In other words, the transition temperature is an output of the algorithm, just as the percolation threshold p_c is an output of the IP. For example, if the initial spin configuration is in the low-temperature phase, then, the portion of satisfied bonds is greater than the critical value and, therefore, due to the relative abundance, a smaller fraction of bonds is needed to generate a spanning cluster than in the case of a critical spin configuration.

The computation time per Monte Carlo step of the method scales linearly with the number of spins, which is highly efficient, but with a prefactor somewhat larger than that for the Swendsen–Wang algorithm. The validity of the algorithm was demonstrated via Monte Carlo simulations of the Ising-Potts models in two and three dimensions (Machta et al. 1995, 1996).

Liverpool and Glotzer (1996) generalized the algorithm of Machta et al., in order to generate equilibrium spin configurations at $T \neq T_c$ by the following algorithm: One begins with all the spins up—although any initial configuration will work—throw bonds between nearest-neighbor parallel (satisfied) spins anywhere on the lattice with probability one. If any cluster of bonded spins reached a preselected length ℓ (with the length defined as the maximum linear extension of the lattice), each cluster is flipped with probability 1/2, and the procedure is repeated.

16.5.4 Optimization Problems

Computing the minimum spanning tree (MST), alluded to in Chap. 15, on a random graph is a fundamental problem in combinatorial optimization, and has been studied by computer scientists and applied mathematicians, as well as physicists and applied physicists, for decades. The solution of the problem has several important applications, such as image analysis, large-scale communication and transportation networks, to name a few (Barabási 1996; Dobrin and Duxbury 2001). The problem has close connections with the IP, with one caveat: The IP is a dynamic process that generates minimum spanning trees, whereas the MST itself is the global minimum of a cost function, which must visit every site in a graph. Thus, the MST corresponds to IP process that visits every site in a finite graph, well beyond the breakthrough point, i.e., the point at which a SSC of the invader is formed.

Consider a non-directed graph with N_n nodes that are interconnected through N_e edges with (a random) energy barrier ϵ_{ij} to move from node i to node j. The energy $E_{\rm MST}$ of the MST is the sum of the energies of the bonds that belong to the minimum spanning tree

$$E_{\rm MEST} = \sum_{(i,j)\,\in\,\text{tree}} \epsilon_{ij}\ . \tag{16.6}$$

Dobrin and Duxbury (2001) proved that the geometry of the MST on random graphs is fractal and universal, i.e., independent of the disorder, even if the disorder is strong. Furthermore, they showed that the paths on the MST are those on which the bond of maximum energy is minimal. Thus, the MST paths are those on which the energy barrier $\epsilon_{\rm barrier}$ is minimal. The barrier on such a path is the bond with the largest cost that lies on that path. In the steady-state limit, one has, $\epsilon_{\rm barrier} \rightarrow p_c$, implying that the barrier on the IP paths in a graph with edge costs drawn from a uniform distribution, takes on a value equal to the percolation threshold on that graph.

Prim (1957) introduced an efficient algorithm that determines the MEST among $N_n^{N_n-2}$ possible spanning trees and is as follows: (a) One selects an arbitrary vertex x_i. (b) Among all the vertices connected to x_i one identifies the one for which p_{ij}, a random number attributed to bond ij, is the smallest, and join x_i and x_j. (c) At the subsequent steps a new vertex is appended to the tree by searching for the bond that has the smallest weight p_{ik}, with x_i belonging to the tree, but not x_k. This implies that bonds that connect already occupied vertices are not eligible for growth. Prim (1957) showed that the tree generated by the algorithm is the smallest energy spanning tree for the graph.

Barabási (1996) noted the similarity between the Prim algorithm and the invasion bond percolation (IBP), and introduced a new version of the IBP that he dubbed the IBPO model, which grows the MEST. At every time step of his model the bonds invaded form the MEST are vertices of the IBP cluster, with energy defined as the sum of the invaded random bonds. The clusters generated by the IBPO model have the same scaling and dynamic properties as those in the standard IBP model, implying that the two models, as well as Prim's, belong to the same universality class.

The IBPO cluster is loopless, i.e., it is a tree, implying that loopless IBP belongs to the same universality class as the IBP. The cluster that is formed by the invaded bonds is the configuration that provides the solution of the global optimization problem by which one determines the smallest energy branching self-avoiding walk, connecting all vertices of a finite lattice. Moreover, the IBPO model is computationally more efficient than the IBP model.

The IBP and IBPO models differ in an additional trapping rule. In the latter only bonds that connect vertices of the cluster to empty vertices are eligible for growth. In addition, there may be bonds in the IBP model that are eligible for growth, which connect two already occupied vertices, and are trapped, since they are between two occupied vertices. Such trapped bonds in the IBPO model are not eligible for growth.

The IP model is also relevant to the problem of optimal paths in the strong disorder limit, i.e., when the bond energies are very broadly distributed. The optimal path is the best path that connects two given sites of a physical region with random landscape. It is a basic problem relevant to a host of interesting fields. In particular, Hansen and Kertész (2004) showed that, provided that distributions of the randomness are selected appropriately, the search for optimal paths links many problems in disordered media, including the IP, directed percolation, and directed and non-directed spanning polymers (see Chaps. 3 and 21). Porto et al. (1997) presented analytical analysis and the results of numerical simulations that suggested that the optimal path in an energy landscape in the strong disorder limit is in the universality class of the shortest path in IP with trapping. We will return to this important subject in Chap. 19.

A well-known example is the geometry of a domain wall, i.e., the path that separates up and down spin domains in 2D disordered ferromagnets. Cieplak et al. (1994) developed a new algorithm to determine the optimal paths in the limit of strong disorder and mapped it onto the strands of loopless compressible (non-trapping) IP. The mapping led to a definition of a new universal property of the IP clusters, the fractal dimension of the strands.

16.5.5 The Queuing Problem

Queuing theory deals with modeling and studying task processing in human dynamics, and high-technology communications. Barabási (2005; see also Oliveira and Barabási 2005) introduced a model of queuing in which one starts with a list, or queue of $n_0 \geq 2$ tasks. At every time step t one of the tasks is executed, which is replaced by $m(t)$ new tasks. If $m(t) = 1$, the queue length remains constant over time. The execution rule at each time step is given by fixing a random priority index $x_i \in [0, 1]$ for each task in the queue, and executing the task with the highest priority with a probability p. A randomly selected task is also executed with the probability $(1 - p)$.

Vázquez (2005) and Gabrielli and Caldarelli (2007) solved the model for $m = 1$ and $0 \leq p \leq 1$. Vaźquez (2005) showed that if $p \approx 1$, after a characteristic time $T(p)$, the system reaches a stationary state characterized by a waiting time distribution (WTD) $P_w(\tau)$, given by

$$P_w(\tau) = \frac{A(p)}{\tau} \exp[-\tau/\tau_0(p)] , \tag{16.7}$$

such that $\tau_0(p) \to \infty$ and $A(p) \to 0$ as $p \to 1$. In the limit $p = 1$ the time $T(p)$ diverges. Cafiero et al. (1996) derived exactly the complete WTD $P_w(\tau, t_0)$ out of stationarity, which depends on both the entrance time t_0 of a task and on its waiting time. This was done by mapping the problem onto the IP in $d = 1$.

Gabrielli and Caldarelli (2009) considered a particular version of the queueing model and showed that it can be mapped onto the IP model on a Cayley tree, i.e., a branching structure without closed loops (see Chap. 2). In their model, at each time step the task with the highest random priority is executed with a probability $p_0 = 1$ and is replaced in the queue by a constant number $m \geq 2$ of new tasks with random priorities, implying that the queue length grows linearly with time, and that at every time step $(m - 1)$ new tasks enter the queue. The IP on a Cayley tree was solved exactly by Nickel and Wilkinson (1983), and is as follows.

Consider a Cayley tree with branching ratio m or, equivalently, a Bethe lattice with coordination number $Z = m + 1$. At time $t = 0$ only the top vertex site of the tree is occupied (the injection site). To each site of the tree a random number $x \in [0, 1]$, representing *fitness*, and drawn from a uniform probability density function, is attributed. At each time step the site of the growth interface ∂C_g, defined at each time step t as the set of un-invaded sites connected by a first nearest-neighbor site to the growing cluster C_g of occupied sites up to that time, with the highest fitness (highest random number x) is occupied, and the cluster grows by one site. Because for each occupied site other new m sites enter the growth interface, the number of sites in, respectively, C_g and ∂C_g grows in time as t and $|\partial C_g| = m + (m - 1)t$.

The exact mapping of the queuing model onto the IP in a Cayley tree was achieved by identifying sites of the tree with tasks, fitness x with priority index, the growth interface ∂C_g in IP with the task list (the queue), and the growing IP cluster C_g with the set of executed tasks up to time t. The exact solution of the IP in the Cayley tree was then combined with the theory of random walks, in order to derive the exact solution of the queueing problem. Interesting insights were obtained. For example, some tasks seemed to remain indefinitely in the queue, before being carried out. In real world, the execution of a task often has the effect of generating an avalanche of new tasks, and executed tasks wait in the queue a broadly distributed waiting times, before their execution was done, which was reproduced by the model.

16.5.6 Social Dynamics

Social dynamics focuses on such phenomena as competition between two different opinions in a network of human beings, or agents, as a result of interaction among them. Several models for such phenomena have been proposed in the past, such as the Sznajd model (Sznajd-Weron and Sznajd 2000), the voter model proposed by Liggett (1999), the majority-rule model (Galam 2002), and the social impact model (Latané 1981). Such models incorporate the evolution of two competing states and can be mapped onto spin models. They have found applications in a broad range of disciplines. The spin-based models with short-range interactions produce, however, a steady state with either consensus of a single opinion, or equal concentrations of the two opinions. In real life, however, a stable coexistence with unequal concentrations of two opinions is commonly seen.

Shao et al. (2009) proposed a spin-type nonconsensus opinion (NCO) model, which produced non-trivial stable states in which stable coexistence of minority and majority opinions occurred. The stable state was reached, starting from a random initial configuration, after a dynamical process in a relatively short time. When the population of one opinion was above a certain critical threshold, even still minority, a large spanning cluster of a size proportional to the total population formed.

The basic assumption of the NCO model is that the formation of an opinion in a population is influenced both by one's own current opinion and that of his/her friends, which are represented as nearest neighbors in a lattice, representing the population or the society. Thus, a person is influenced by the majority opinion of the group, which includes his friends and himself. The two opinions are denoted by σ_- and σ_+, initially distributed at random in a network with the probabilities, respectively, of f and $(1 - f)$. A site i is selected at random that, together with its nearest-neighbor sites, forms a set of nodes N_i. At each time step, node i is converted to its opposite opinion, if it is in the local minority opinion. If the two opinions are equally present in the set N_i, with probability p, i is converted to its opposite opinion and remains unchange with probability $(1 - p)$. At each step of the simulation, every node is examined to see whether its opinion must change, with all the nodal updates made simultaneously at each simulation step. The system has reached a stable state, if no more changes occur.

Shao et al. (2009) simulated their model in a variety of networks, such as scale-free networks, which are characterized by a power-law degree distribution, $P(k) \sim k^{-\lambda}$ with $2 \leq k \leq k_{\max}$ and $\lambda = 5/2$ (see Chap. 17). They demonstrated the emergence of a phase transition in the stable state as the fraction of the sites with minority opinion was increased and the opinion became increasingly influential, even though the fraction was well below the value that would make the minority view the majority. The transition occurred at a non-trivial critical value f_c, which was dependent only upon the type of the network. The phase transition was of second-order (continuous) type, characterized by the existence of a large spanning cluster of the minority opinion with a size proportional to the total population.

Most importantly, Shao et al. (2009) demonstrated that their NCO model was in the same universality class as the TIP. Recall from Chap. 2 that the percolation cluster-size distribution n_s scales with s as, $n_s \sim s^{-\tau}$, with $\tau = 187/91 \approx 2.05$ in 2D, $\tau \approx 2.18$ in 3D, and $\tau = 5/2$ for three-like structures. In the 2D TIP, the cumulative distribution function has the form, $n_s(s' > s) \sim s^{1-\tau}$ with $\tau \approx 1.9 \pm 0.01$, which was also found for the cluster-size distribution of the Shao et al., model in 2D. Moreover, the fractal dimension D_f of the SSC in the 2D TIP has a fractal dimension of (see Table 16.1) $D_f \simeq 1.82$, while the stable clusters of the NCO model at f_c were also fractals with $D_f \approx 1.84$.

Chapter 17
Percolation in Random Graphs and Complex Networks

17.1 Introduction

Every lattice and network described so far in this book, which have been used to model disordered media, has been either one with a fixed coordination number in which each site is connected to only its nearest neighbors, or a stochastic one, such as the Voronoi network described in Chap. 3. Real world networks, such as the internet, and social, biological, and ecological networks are very different from the networks and lattices considered so far. Such structures constitute *complex networks* whose development was pioneered by Rapoport (1957), Gilbert (1959), and Erdös[1] and Rényi[2] (ER) (1959; 1960; 1961). Such networks contain N nods—sites or entities—and M edges—bonds or links—that are distributed randomly between the nodes. The *degree k of a node* is defined as the number of links emanating from it, i.e., the number of "neighbors" it has in the network, which is similar to the coordination number in Euclidean lattices described in Chap. 2. Alternatively, one may build up such networks by connecting every pair of the nodes (anywhere in the network) with equal probability, which generates a complex network essentially identical with the ER network.

Since complex networks play a fundamental role in many real-world problems, ranging from biological systems to the Internet, and because percolation and connectivity properties of such networks are essential to characterizing their properties, this chapter provides a brief summary of some of the most important properties of complex networks. Their applications will be described in the subsequent chapters.

---

[1] **Paul Erdös** (1913–1996) was a renowned Hungarian mathematician who made fundamental contributions to discrete mathematics, graph theory, number theory, mathematical analysis, approximation theory, set theory, and probability theory. He was a most prolific mathematicians, who proposed a large number of mathematical conjectures, and published 1500 papers with 500 collaborators.

[2] **Alfréd Rényi** (1921–1970) was a Hungarian mathematician, who made contributions to probability theory, combinatorics, graph theory, and number theory. A famous quote, "If I feel unhappy, I do mathematics to become happy. If I am happy, I do mathematics to keep happy" is attributed to him.

M. Sahimi, *Applications of Percolation Theory*, Applied Mathematical Sciences 213,
https://doi.org/10.1007/978-3-031-20386-2_17

Let us point out that Shirazi et al. (2009) developed a method by which stochastic processes, such as time series, are mapped onto complex networks. As examples, they constructed the networks for time series for free-jet and low-temperature helium turbulence, the German stock market index (the DAX), and white noise. They compared the network properties of the original time series that they had investigated with those for the shuffled and surrogate series[3], in order to quantify the effect of the long-range correlations and the fatness of the probability distribution functions of the series on the networks constructed. Shirazi et al. (2009) also demonstrated that the time series can be *reconstructed* with high precision by means of a simple random walk on their corresponding networks.

17.2 Erdös-Rényi Graph

There are actually two versions of the ER graph. In one version, denoted by $G(N, L)$, which was introduced by Erdös and Réyni, one selects a graph randomly and uniformly from the set of all graphs that have N nodes and L links or edges. The nodes have labels, which means that the graphs obtained from each other by permuting the nodes are viewed as distinct. For example, with $N = 2$ and $L = 3$ there are three two-edge graphs on three labeled nodes, one for each choice of the middle node in a two-edge path, with each of the three graphs included with probability $1/3$.

In the second model, denoted by $G(N, p)$ and introduced by Edward Gilbert, a graph is constructed by connecting labeled vertices randomly. Each edge is included in the graph with probability p, independently from all other edges. Thus, the probability p_G of generating a graph that has N nodes and L links is given by

$$p_G = p^L(1-p)^{L'-L} , \tag{17.1}$$

$$L' = \binom{N}{2} . \tag{17.2}$$

One may view p as a weight function. As p increases from 0 to 1, the model becomes increasingly likely to generate graphs with more edges, but increasingly less likely to produce graphs with fewer edges.

17.2.1 Properties of Erdös-Réyni Graph

The distribution of the degree k of a note in the ER graph is given by

[3] A surrogate model, also called a metamodel or an emulator, mimics the behavior of a simulated model as closely as possible, but is computationally cheap or cheaper to evaluate. They are constructed using a data-driven, bottom-up approach; see, for example, Queipo et al. (2005).

$$P(k) = \binom{N-1}{k} p^k (1-p)^{N-1-k} \tag{17.3}$$

which in the limit $N \to \infty$ and $Np =$ constant reduces to a Poisson's distribution

$$P(k) = \frac{(Np)^k \exp(-Np)}{k!}\,, \tag{17.4}$$

In their 1960 paper, Erdös and Réyni showed that, (a) if $Np < 1$, then they will almost surely have no connected components of size larger than $\mathcal{O}(\log N)$; (b) if $Np = 1$, then the graph will almost surely have a largest cluster whose size is on the order of $N^{2/3}$, and (c) if $Np \to c > 1$ with c being a constant, then the graph will almost surely have a unique giant cluster that contains a fraction of the nodes, and no other cluster will contain more than $\mathcal{O}(\log n)$ nodes. Erdös and Réyni also studied the connectivity of their graph and showed that $\ln N/N$ represents a sharp the threshold for the connectivity of the graph.

17.2.2 Percolation in Erdös-Réyni Graph

The Erdös-Rényi process is in fact unweighted bond percolation on the complete graph, which is a simple undirected graph in which every pair of distinct nodes is connected by a unique link, edge, or, to use the language of classical percolation, a bond. The transition at $Np = 1$ from a giant cluster to a small one is the obvious analogs of percolation on Euclidean lattices, as described in Chap. 2.

17.3 Small-World and Scale-Free Networks

Though very popular, it was shown that the ER networks do *not* describe the aforementioned real networks (Albert and Barabási 2002; Dorogovtsev and Mendes 2003; Newman 2003; Pastor-Satorras and Vespignani 2003). For example, Newman (2003) noted that the real-world complex networks tend to have high clustering in which neighbors of a given node tend to share a link or edge. In addition, it was shown that the degree sequence in the ER networks does not follow the expectation of random graph theory. If $n(k)$ is the number of nodes with degree k, then, the probability $P(k)$ that a randomly selected node has degree k is simply, $P(k) = n(k)/N$. On the other hand, in a ER network the expected degree sequence is a Poisson's distribution, given by Eq. (17.4) in which $Np = \langle k \rangle$, where $\langle k \rangle$ is the average degree of the network (Bollobás 1985). The mean distance $\langle \ell \rangle$ between two nodes is given by

$$\langle \ell \rangle = \frac{\ln N}{\ln\langle k \rangle}\,. \tag{17.5}$$

Such observations led to the development of *small-world network* (SWN) by Watts and Strogatz (1998). Consider, for example, a one-dimensional ring or a $d-$dimensional lattice ($d > 1$) with *rewiring*, which means removing a small fraction q of the links in the lattice, but adding new links at random between the nodes. If q is relatively small, the structure will have high clustering. Such structures are referred to as the SWNs because the distance between the nodes is small, on the order of $\log N$, whereas the same distance in the original $d-$dimensional grid is on the order of $N^{1/d}$.

The most important limitation of the Watts–Strogatz model is that it produces an unrealistic degree distribution, whereas networks in practice are often *scale-free* (SF) networks (see below) and inhomogeneous in degree, with hubs—nodes that are connected to a large number of other nodes—and a SF (power law) degree distribution, which we describe shortly. In this respect, real networks are perhaps better described by the Barabási-Albert (BA) model (Albert and Barabási 2002). But the BA model also fails to produce the high levels of clustering observed in many real networks. Thus, neither the Watts–Strogatz model nor the BA model should be viewed as fully realistic.

Let us describe the BA model. One begins with an initial connected network of N_0 nodes. Then, new nodes are added to the network one at a time, with each new node being connected to $n \leq N_0$ existing nodes with a probability proportional to the number of links that the existing nodes already have. Thus, the probability p_i that the new node is connected to node i is given by

$$p_i = \frac{k_i}{\sum_j k_j} , \tag{17.6}$$

where k_i is the degree of node i, and the sum is made over all pre-existing nodes j. The hubs tend to quickly connect with more links, whereas nodes with only a few edges are unlikely to be selected as the destination for a new link.

Analyzing the vast amount of data, it was shown that many real networks, such as the Internet and the World Wide Web (WWW) are characterized by a scale-free (SF) degree distribution, i.e., a power-law distribution given by

$$P(k) = (m-1)k_m^{m-1}k^{-m} , \quad k_m \leq k \leq k_M , \tag{17.7}$$

with k_m and k_M being, respectively, the minimum and maximum values of the degree k. For the BA graph, one has $m = 3$. The Internet corresponds to a SF network with $m \approx 2.5$. In fact, analysis of the data indicated (Albert and Barabási 2002; Dorogovtsev and Mendes 2003; Newman 2003; Pastor-Satorras and Vespignani 2003) that in many cases, $2 < m < 3$. Complex networks that are characterized by distribution (17.7) are referred to as the SF networks. Note that k_m must be provided by the user, since its value must be such that the distribution is normalizable, i.e., $\int_{k_m}^{k_M} P(k)dk = 1$. k_M, on the other hand, takes on a value that has been determined

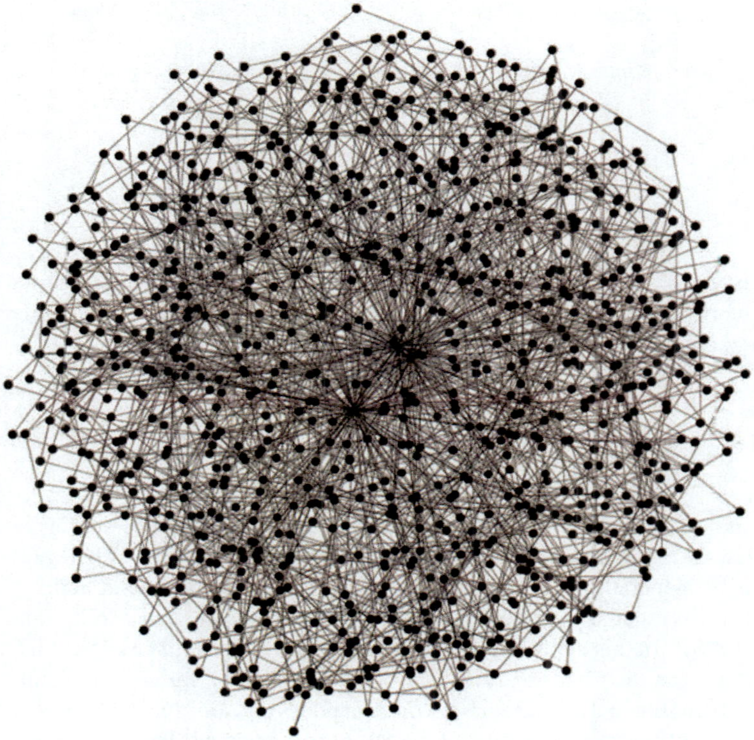

Fig. 17.1 A picture of a scale-free network

(Cohen et al. 2000) as the extreme value statistics of N variables, and is given by, $k_M = k_m N^{1/(m-1)}$. Figure 17.1 shows examples of the SF networks.

The first SF network was discovered by de Solla Price (1965). He studied the networks of citations between scientific papers, and showed that the number of links to papers, i.e., the number of citations they receive, had a heavy-tailed distribution following a Pareto distribution or power law and, thus, the citation network is a SF network. de Solla Price did not, however, use the term "scale-free network," until it was rediscovered by Albert and Barabási. In addition to the Internet and the web graph of the World Wide Web, other real-world examples of SF networks include some social networks, including collaboration networks, such as the collaboration of movie actors in films, and the co-authorship by mathematicians of papers; financial networks, such as interbank payment networks, and airline networks.

Note that SF networks with $m < 2$ were studied by Seyed-allaei et al. (2006). These are extremely skewed networks—which occur very frequently in systems of virtually or logically connected units—to have properties that are different from those of the SF networks with $m > 2$. The number of their links grows faster than the number of nodes, and they possess the small-world property, as their diameter increases by the logarithm of the size of the network and the clustering coefficient is finite.

The mean distance $\langle \ell \rangle$ between two nodes in a SF network is given by

$$\langle \ell \rangle = \begin{cases} \text{constant} & m = 2 \\ \ln(\ln N) & 2 < m < 3 \\ \ln N / \ln(\ln N) & m = 3 \\ \ln N & m > 3 \,. \end{cases} \tag{17.8}$$

Since for $2 < m < 3$ the mean distance increases only as $\ln(\ln N)$, as opposed to the ER networks in which $\langle \ell \rangle \ln N$, the SF networks for such values of m are referred to as *ultra-small networks*.

17.4 Generation of Scale-Free Networks

Let us first mention that Voitalov et al. (2019) described an approach for detecting power laws in empirical degree distributions in real-world networks. There are several methods for generating the SF networks, two of which are described here.

(i) The first method we describe was proposed by (see also Molloy and Reed (1995, 1998)). In this approach, for a given m, each node is first assigned its degree that follows the distribution (17.7), and is represented by stubs or half-links. Two stubs are then selected at random from distinct nodes and are connected by an edge or link. The process is repeated until all the stubs are connected. This method generates a SF network without any degree–degree correlation.

(ii) In the second approach (Goh et al. 2001) one starts with N nodes. A weight $w_i = i^{-\mu}$ is then attributed to node i ($i = 1, 2, \cdot, N$), where μ is a parameter in [0,1). A pair of nodes (i, j) is then selected with probabilities p_i and p_j, with $p_i = w_i / \sum_n w_n$, and similarly for p_j, and are connected with an edge, unless they are already connected. The process continues until NK edges are put in the system, where K is a parameter of the model, so that the mean degree of the distribution $\angle k \rangle = 2K$. The mean degree of node i is given by

$$\langle k_i \rangle = \left[\frac{2NK(1-\mu)}{N^{1-\mu}} \right] i^{-\mu} \,. \tag{17.9}$$

Equation (17.9) arises because the edges are connected to a node with a frequency proportional to the weight of that node. The degree distribution is still given by Eq. (17.7), except that, $m = 1 + 1/\mu$. The advantage of the approach is that many properties of the network can be determined analytically. The method generates, however, SF networks with degree–degree correlations, if $2 < m < 3$.

17.5 Percolation Properties

The percolation threshold of the BA network is $p_c = 0$. This implies that if we remove randomly nodes in the network, any fraction of nodes will not break the network. If, however, we remove only a relatively small fraction of nodes with the highest degrees, i.e., hubs, the network will collapse. More generally, the percolation threshold of a SF network of degree k is given by

$$p_c = \frac{1}{k-1} , \tag{17.10}$$

which is completely similar to the percolation threshold of a Bethe lattice, or Cayley tree, of coordination number Z, $p_c = 1/(Z-1)$, mentioned in Chap. 2. The reason is easily understood: the local structure of random graphs resembles a tree; that is, the number of nodes for closing a loop for a set of nodes with fewer nodes than order N is negligible. That is to say, even if there are closed loops, but their number is too small to affect the tree-like structure of the graph.

Since complex networks have a tree-like structure, it follows that the critical exponents that characterize various percolation quantities of interest on complex graphs correspond to mean-field exponents that are valid for spatial dimension $d = 6$ and higher, which are obtained by solving the percolation problem on a Bethe lattice. These exponents are listed in Table 2.3.

As discussed in Chap. 2, the largest cluster at the percolation threshold— what is normally referred to as the *giant cluster* in network theory— is a fractal object with a fractal dimension of D_f. For spatial dimensions $d \geq 6$, one has, $D_f = 4$ (see Table 2.3). In other words, the size S of the giant cluster scales as, $S \propto L^4$, where L is the linear dimension of the tree. On the other hand, the number of nodes N in the tree scales as, $N \propto L^6$, implying that, $S \propto N^{2/3}$.

The above simple analysis is correct, however, only if the degree distribution $P(k)$ decays with k quite fast, such as an exponential distribution. For the SF networks this is not the case, and the percolation critical exponents do depend on the exponent m. In fact, it has been shown that

$$S \propto \begin{cases} N^{(m-2)/(m-1)} & 3 < m < 4 \\ N^{2/3} & m > 4 . \end{cases} \tag{17.11}$$

In a complex network no embedding space is possible and, therefore, L is not well-defined. Thus, the most appropriate and well-defined distance is the chemical distance, defined as the shortest distance ℓ_c measured along the fractal structure of the clusters between two sites. Then, one has, $S \propto \ell_c^{D_c}$, where D_c is the *chemical dimension*. It has been shown that

$$D_c = \begin{cases} (m-2)/(\gamma-3) & 3 < m < 4 \\ 2 & m > 4\,. \end{cases} \tag{17.12}$$

Other percolation exponents have also been determined, and they all depend on m. For example, recall from Chap. 2 that the percolation probability $P(p)$, the probability that, when the fraction of uncut links is p, a given site belongs to the infinite, or the giant, cluster follows a power law near p_c given by, $P(p) \propto (p - p_c)^\beta$. For the SF networks,

$$\beta = \begin{cases} 1/(m-3) & 3 < m < 4 \\ 1 & m > 4\,, \end{cases} \tag{17.13}$$

where the second value represents the mean-field value for $d \geq 6$, listed in Table 2.3. Similarly, n_s, the number of clusters having s nodes, follows, $n_s \sim s^{-\tau}$. For the SF networks, one has,

$$\tau = \begin{cases} (2m-3)/(m-2) & 3 < m < 4 \\ 5/2 & m > 4\,, \end{cases} \tag{17.14}$$

and the second value is the mean-field value listed in Table 2.3. We should also point out that the percolation exponents in the SF networks that follow the statistics described by Eq. (17.9) have also been determined; see Lee et al. (2018).

That the percolation exponents depend on the exponent m that characterizes the power-law distribution of the node degrees in a SF network is somewhat similar to transport in continuum percolation. Recall from Chap. 3 that when the conductance distribution is of power-law type, the critical exponents that characterize the power-law behavior of the transport properties near p_c depend on the exponent of the conductance distribution.

Chapter 18
Percolation in Biological Systems

18.1 Introduction

Most biological systems are so complex that they preclude any reasonable description in terms of the exact interactions among their fundamental constituents. We do know, however, that many of most fascinating biological phenomena, and in particular those involving life, are the result of interactions between many elements. For example (Mora and Bialek 2011), many amino acids determine the structure of a single protein; the fate of a cell is determined by many genes, and shaping our thoughts and memories involve many neurons. But, for quite some time, application of statistical mechanics of disordered media, and in particular critical phenomena and percolation theory, to biological materials and phenomena seemed intractable, even though, due to their nature and the fact that they are driven by collective behaviors, the application seemed natural. But, considerable progress has been made over the past two decades, as it has been shown, through experiments and simulations, that a surprising number of such phenomena manifest behavior and properties that are akin to systems near a critical point. For example, as described by Krause and Ruxton (2002), groups of animals, such as flocking birds, move with fascinating coordination in that, instead of being dictated by a leader or responding to a common stimulus, the collective patterns of flock dynamics is self-organized, arising from local interactions between individuals who propagate information throughout the entire group. Flocks of birds are also highly responsive and cohesive in the face of predatory threat. The balance between order and high susceptibility indicates that a flock of birds may represent a critical state. Indeed, this has been demonstrated (Bialek et al. 2012).

Percolation or connectivity threshold also represents a critical point. There are many biological systems that are seemingly stochastic in nature and in which the role of connectivity of the microscopic elements or constituents is prominent. Examples include self-assembly of tobacco mosaic and other simple viruses (see, for example, Hohn and Hohn 1970), actin filaments (Poglazov et al. 1967) and flagella (Asakura et al. 1968), lymphocyte patch and cap formation (Karnovsky et al. 1972), many precipitation and agglutination phenomena, and many other biological phenomena

M. Sahimi, *Applications of Percolation Theory*, Applied Mathematical Sciences 213,
https://doi.org/10.1007/978-3-031-20386-2_18

(see below). Some of such phenomena, such as precipitation, occur spontaneously if the functional groups are sufficiently reactive. Thus, they depend on their level of chemical complexity and that of the solvent in which they occur. Other factors are not directly related to the solvent, but have a great influence on the outcome of biological processes. For example, in antigen–antibody reactions, clusters of all sizes react with one another, forming complex branched networks that grow in size as time progresses. We may also have reactions that can proceed by rapid addition of monomers to growing chains after a slow initiating event. Such phenomena are, therefore, similar to percolation processes. This chapter discusses applications of percolation concepts to such phenomena. We start with a discussion of antigen–antibody reactions and aggregations, and then describe several other biological processes to which percolation may be relevant.

18.2 Antigen–Antibody Reactions and Aggregation

Under suitable conditions, mixtures of antigen and antibodies—a system comprised of bifunctional and multifunctional monomers—form large networks or aggregates that contain many molecules of both. In general, the networks or branched structures react reversibly with one another. They are often insoluble and, hence, will precipitate. Since we are interested in the concentrations of antibodies and antigens as a function of time, it is important to examine the conditions under which precipitation may be expected to occur.

The first theory of antigen–antibody aggregation was developed by Goldberg (1952) who used combinatorial mathematics to derive results that are equivalent to those obtained for various percolation properties on a Bethe lattice. His work was extended by several others, including Aladjem and Palmiter (1965), Bell (1971), and Delisi (1974). Delisi and Perelson (1976) analyzed the problem extensively and derived several analytical results for the various properties of interest.

Consider aggregation of two types of monomers, S and L, of functionality 2 and $Z \geq 2$, respectively. The S monomers can be any of a number of substrates, such as immunoglobulin G (IgG), while the L monomers might be some appropriate ligand or antigen. The functional units on the monomers are equivalent to the sites in percolation on random networks described and utilized so far in this book. To analyze the problem, Goldberg (1952) made three basic assumptions: (a) all the free sites are equivalent; (b) no intramolecular reactions can occur; and (c) during the aggregation process cyclization reactions do not occur. Since closed loops are not formed, it immediately points to the similarity with percolation on Bethe lattices. Delisi and Perelson (1976) relaxed the second assumption to allow for intramolecular reactions. In the immunological literature, such reactions are often referred to as monogamous bivalency.

Several parameters were then introduced: (a) p_b, the probability that an antibody site picked at random is bound; (b) p_g, the probability that an antigen site picked at random is bound; (c) p_{br}, the probability that an antibody site is bound and has not reacted intramolecularly; and (d) p_{gr}, the probability that an antigen site is bound and has not reacted intramolecularly. Consider, then, an aggregate composed of x Z-functional antigens and $(x - 1 + n)$ bifunctional antibodies, where n is the number of antibodies that do not serve as connectors. Of these antibodies, n_1 are bound univalently, and n_2 bivalently. An aggregate with compositions x, n_1, and n_2 is called an (x, n_1, n_2)-mer.

The main quantity of interest is $C_{n_1,n_2}(a, p)$, the concentration of such aggregates. To determine it, we define $P(x, n_1, n_2)$ as the probability that a free site picked at random is part of an (x, n, n_2)-mer. If closed loops are not formed, then

$$P(x, n_1, n_2) = \frac{\text{the number of free site on } (x, n_1, n_2)\text{-mers}}{\text{Total number of free sites}}, \tag{18.1}$$

which implies that

$$P(x, n_1, n_2) = \frac{(Zx - 2x + 2 - 2n_2)C_{n_1,n_2}(x, p)}{L + S}. \tag{18.2}$$

We write $P(x, n_1, n_2)$ as

$$P(x, n_1, n_2) = \rho_l \omega_l \Omega_l + \rho_s \Omega_s \Omega_s , \tag{18.3}$$

where l and s label, respectively, free antigens and antibody sites. Here, ρ_l is the probability that the root is a free antigen site, ω_l the probability that the root is on an (x, n_1, n_2)-mer, given that it is a free L site, and Ω_l is the number of ways an (x, n_1, n_2)-mer can form, given that the root is a free L site, with analogous definitions for ρ_s, ω_s, and Ω_s. Each of these quantities can be calculated using the methods developed by Flory and Stockmayer for their theory of gelation described in Chap. 13, or by the methods developed for percolation on Bethe lattices. If closed loops are forbidden, then Goldberg (1952) showed that for

$$p = p_g = \frac{1}{Z - 1}, \tag{18.4}$$

there is a finite probability that an infinitely large aggregate forms. This is exactly the same as the site (or bond) percolation threshold of a Bethe lattice of coordination number Z (see Chap. 2).

Similar to the gelation and polymerization, the problem with such formulation is that the formation of closed loops is forbidden. Thus, the results that were derived represent the mean-field limit of percolation, i.e., a system whose spatial dimension is six or higher. In reality, antigen–antibody aggregation is a three-dimensional (3D) phenomenon and, therefore, we must use percolation on finite-dimensional lattices, and in particular 3D ones, in order to explain antigen–antibody aggregation phenomena.

18.3 Network Formation on Lymphocyte Membranes

According to modern theory of the clonal selection, the cells of the immune systems arise from the division and differentiation of the stem cells in the bone marrow (Burnet 1959). The cells that are potentially able to secrete antibodies are known as B-lymphocytes, which insert a homogeneous set of antibody-like receptor molecules on their membrane's surface, where they are used for recognition of complementary antigens. The population of B-cells is itself heterogeneous, i.e., different cells may have different types of membrane-bound receptor molecules. Under appropriate conditions, binding of complementary antigen to the receptors on a B-cell activates that cell to differentiate and/or proliferate into antibody-secreting cells. Moreover, the antibody secreted by the progeny of a particular B-cell is assumed to have specificity for the antigen, identical to that of the receptor molecules on the B-cell progenitor. Therefore, the role of antigen is cellular selection and amplification, and its mission is mediated by interaction with cellular receptors.

The process of proliferation and secreting large amounts of antibody is preceded by cross-linking of antigens to the receptors until a macroscopically large patch is first formed on the surface. The patch or network is the 2D analogue of 3D aggregation, described in the preceding section, a process called network or lattice formation. Antigen-stimulated lymphocyte lattice formation may play a key role in triggering immunocompetent cells. For immunogens with repeating arrays of antigenic determinants, e.g., polysaccharide flagellin, triggering is possible without the aid of helper cells, whereas other immunogens require the presence of T-cells or their products. It has been suggested that the T-cell requirement results from having to present the immunogen to the B-cells in an aggregated or network form, thus increasing its valence.

The relationship between network formation and biological activity is not completely clear yet. It is presumed that lattice formation provides some sort of stimulatory signal, but this may not be enough to trigger a cell. In any event, it is not unreasonable to assume that the triggering is a function of the strength of antibody–receptor interaction. Delisi and Perelson (1976) assumed that lattice formation is the 2D analogue of the 3D precipitation reaction described above and analyzed the problem in a similar way. To simplify the problem, they assumed that the effect of intramolecular reactions is unimportant. A well-understood characteristic of immune response to antigens is an increase in antigen–antibody affinity with time after immunization. The implication is that some aspects of the antigen–receptor interaction are equilibrium-controlled. Since antigens bound by more than one receptor dissociate very slowly, however, such binding may be effectively irreversible on the time scale of lattice formation. Delisi and Perelson (1976) derived several analytical results for the problem, using (implicitly) percolation on Bethe lattices, by a method similar to that of Fisher and Essam (1961) for percolation on the same lattice, including the cluster-size distribution, the behavior of the system near the point at which a large lattice is formed, i.e., near the percolation threshold, and membrane transport. Their results cannot be directly compared with any experimental information, however, as they were obtained with a Bethe lattice, which may not be directly relevant to 2D systems. Their work did indicate definitely the relevance of percolation to such biological phenomena.

18.4 Percolation in Immunological Systems

A relatively old, but still quite useful review of the relationship between immunology and critical phenomena was given by Perelson and Weisbuch (1997). Although immunology is an old research field, the application of percolation theory to it is relatively new, and some of our discussions here may not be upheld by new research in the years to come.[1]

The deadly disease of smallpox has been eradicated from the Earth for many years, but against AIDS (acquired immunodeficiency syndrome) no cure has been found yet, although considerable progress has been made for controlling it and reducing its devastating effects, as of the time of writing this book (2021). The two examples are perhaps the extreme cases success, and the lack of it, in application of immunology. So, how does the immune system work?

If we get the flu in winter, we will not normally get the same thickness shortly thereafter. Our body's white blood cells or lymphocytes produce antibodies or other cells that are able to neutralize the virus or other foreign antigens. Such antibodies fit the antigens the way a lock fits a key. Apart from small inaccuracies, one type of antigen fits only specific antibodies and vice versa. If we get a new virus, some antibodies accidentally—through mutations—may fit this antigen, our immune system notices this fit, produces more of such specific antibodies, and in this way combats the disease. When we get healthy again, a few very long-lived memory cells survive and allow a quicker response from our body, if the same virus returns. Vaccination produces a controlled amount of sickness so that antibodies and memory cells are formed for the specific disease. The immunodeficiency virus HIV seems to escape destruction by the immune system, and instead destroys slowly the T4 white blood cells, which are crucial as "helper" cells for the functioning of the immune response.

[1] The original version of this section, published in the first edition of this book, was written by the late Professor Dietrich Stauffer. The present section represents an updated version of it.

According to the now widely believed ideas of Jerne[2] (1974) antibodies can be treated in turn as antigens by the immune system, which then produces antiantibodies, antiantiantibodies, etc. Moreover, the inaccuracies of the lock-and-key relation between the various types of molecules let the same antibody fit slightly different antigens. In this way, all possible antibody shapes are connected, directly or indirectly, by antibody–antigen relations in a Jerne network, spanning the entire shape space.

What does this have to do with percolation? When we are ill, we take a medicine specific for that illness; not all possible types of medicines have been invented yet. Similarly, our immune system would destroy us if a flu infection could trigger all possible shapes of antibodies. Thus, the immune response should be very specific, i.e., the antibodies triggered by one specific antigen should form a finite cluster in the topological network of possible shapes. They should not percolate throughout this shape space. In other words,

Warning: Immunology has determined that percolation is dangerous to your health!!

de Boer and co-workers (1989, 1992) and Stewart and Varela (1991) investigated the conditions under which the immune response remains limited to a few types of antibodies, once a foreign antigen enters our body, i.e., the conditions under which percolation does *not* occur. Inherent in percolation theory, as described throughout this book, is the clear distinction between occupied and empty sites. Thus, percolation is the easiest to apply to immunology, if antibody molecules are either there or not there, without a more quantitative and more realistic distinction between such molecules according to their concentration. Indeed, "cellular automata" approximations—discrete systems in which each site can be in one of only two states, sick or healthy, and nothing in between—were developed in immunology by Kaufman et al. (1985) before percolation aspects were investigated. The combination of percolation and cellular automata methods then allowed computer simulation of five- to ten-dimensional shape spaces (Stauffer and Sahimi 1993, 1994; Sahimi and Stauffer 1993), as required for natural immune systems according to Perelson and Oster (1979).

Such studies of localization—not percolating—versus percolation can be done in two ways. We may look at the immune response, if no element of the immune

[2] **Niels Kaj Jerne** (1911–1994) was a Danish immunologist who is known for three important contributions to biological processes. (1) Instead of the body producing antibodies in response to an antigen, Jerne postulated that the immune system already has the specific antibodies it needs to fight antigens. (2) It was known that the immune system learns to be tolerant to the individual's own self, and Jerne postulated that the learning occurs in the thymus. (3) It was known that the T and B cells communicate with each other, and Jerne's network theory proposed that the active sites of antibodies are attracted to both specific antigens and to other antibodies that bind to the same site. Jerne shared the Nobel Prize in Physiology or Medicine in 1984 with **Georges Jean Franz Köhler** (1946–1995) and **César Milstein** (1927–2002) for "theories concerning the specificity in development and control of the immune system and the discovery of the principle for production of monoclonal antibodies."

system has yet been triggered. Then, our clusters are simply the sets of activated antibody types connected by the Jerne network bonds (Neumann and Weisbuch 1992). Alternatively, we may look at an immune system that has already evolved into some stationary equilibrium of present and absent antibody types and check for the changes made to this dynamic equilibrium by one specific type of antigen (Stauffer and Weisbuch 1992, Stauffer and Sahimi 1993, Stauffer and Sahimi 1994). This is analogous to the studies of chaos in a dynamical system, and has been used before in genetics (Kauffman 1969) and Ising magnets (Creutz 1986). Physicists often call it *damage spreading*, whereas experts on pattern recognition and neural networks refer to it as the *Hamming distance*.[3] Damage is the number of sites that differ in their spin or other characteristic values in a site-by-site comparison of two lattices, if the initial configurations of the two lattices differ by only one site. Hamming distance is used in neural networks to describe the number of pixels that differ in a comparison of two pictures. In particular, if one picture is the original aim and the other picture is an attempt to restore that original picture from a blurred or noisy version of it. Thus, in this sense damage spreading and the Hamming distance are the same.

Various studies have indicated that in the immunological models investigated, the question of whether the immune response remains localized or percolates through the entire system depends on the parameters that one selects, which is hardly a surprise to anyone familiar with the concept of a percolation threshold.

18.5 Percolation Conductivity in Biological Materials

Interactions between water and globular proteins influence their folding, enzymatic activity, and other properties. One way of studying various properties of such biological systems is through measuring their hydration-dependent dielectric losses and electrical conductivity in the hydration range critical for the onset of enzymatic functions. Behi et al. (1982) showed that hydrated protein powders exhibit dielectric dispersion at three distinct frequencies. The first one, $\Delta\epsilon_1$, occurred at frequencies near 1 HZ, with a corresponding change in the DC conductivity of the system, which was attributed to an isotope effect. The second and third dispersions, $\Delta\epsilon_2$ and $\Delta\epsilon_3$, occurred at much higher frequencies, 10^5 and 10^{10} Hz, respectively. Bone et al. (1981) and Behi et al. (1982) proposed that these two dielectric dispersions are due to the Debye relaxation of water bound on the surface of the macromolecule.

Careri et al. (1985) made careful measurements of dielectric losses of lysozyme powders of various hydration levels in the frequency range 10 kHz–10 MHz. The powders were prepared from the Worthington thrice crystalized and salt-free proteins. The dielectric permittivity of the protein is not measured directly, because the sample is only one part of the composite condenser that consists of a layer of the powder included between layers of dry air and glass. In the frequency range of interest to Careri et al., however, the glass did not display appreciable dielectric loss. Thus, their

[3] Named after mathematician **Richard Wesley Hamming** (1915–1998).

system was essentially a capacitor consisting of two layers of similar thicknesses, one of which had a vanishingly small conductivity and a dielectric constant close to that of the vacuum. The dielectric relaxation time t_d is predicted to be given by

$$t_d = \left(\frac{\epsilon_0}{g_e}\right)\sqrt{1+\epsilon}\,, \tag{18.5}$$

where ϵ_0 is the vacuum permittivity, ϵ is the relative dielectric constant, and g_e is the effective conductivity of the hydrated protein. Careri et al. confirmed that their system was insensitive to slight changes in the thickness of air or powder layers. In the frequency range of interest, the hydration dependence of ϵ was much weaker than that of g_e. Thus, a plot of t_d versus h, the hydration level, is essentially equivalent to a plot of g_e versus h.

Figure 18.1 displays the measured conductivity as a function of the hydration level h. To remove from the data the nearly negligible contribution of the non-percolative processes and systematic errors in the evaluation of capacitor geometry to the effective conductivity, the value of g_e at the percolation threshold was subtracted from g_e at all other points for any hydration level h. Careri et al. (1985) established that protonic conduction was the dominating contribution to the dielectric relaxation $\Delta\epsilon_2$ in the frequency range in which the data were collected. The relaxation was attributed to proton displacement on a single macromolecule. Since h is proportional to p, the usual occupation probability in percolation, we can write

$$g_e(h) - g_e(h_c) \sim (h - h_c)^{\mu}\,, \tag{18.6}$$

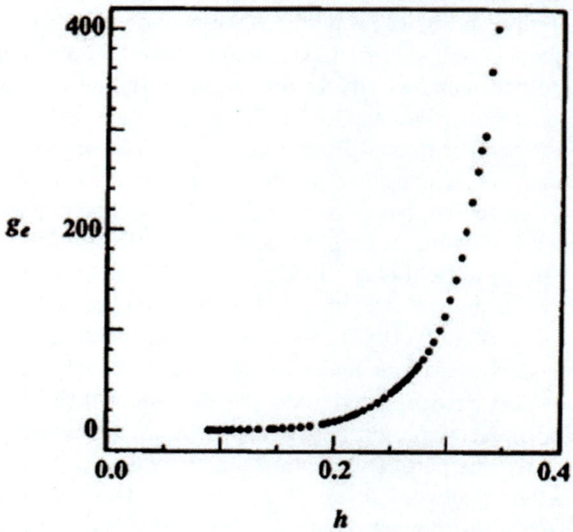

Fig. 18.1 Dependence of the effective conductivity g_e on the hydration level h in lysozyme powder at pH = 7 and 301 K. g_e has been normalized with the conductivity of the dry sample (after Careri et al. 1988)

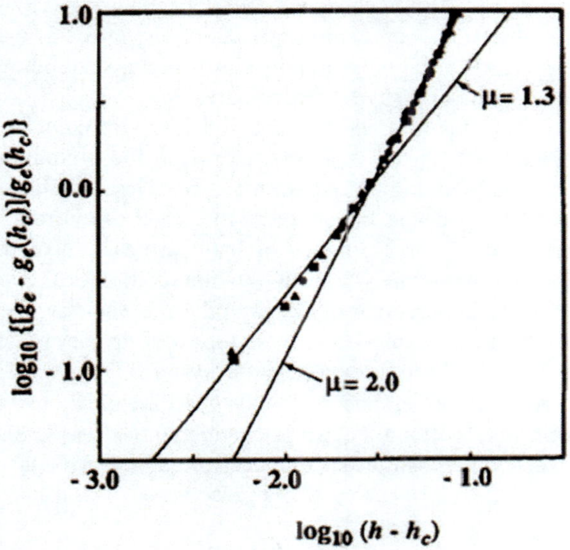

Fig. 18.2 Estimating the critical exponent μ of the effective conductivity of lysozyme powder at pH = 7. Symbols represent the measured data for native lysozyme hydrated with H_2O and D_2O, and a sample of 1 : 1 complex of lysozyme with $(GlcNAc)_4$, hydrated with H_2O (after Careri et al. 1988)

where h_c is the hydration level at the percolation threshold. The critical exponent μ was evaluated by Careri et al. for three distinct samples, namely, native lysozyme hydrated with H_2O, native lysozyme hydrated with D_2O, and 1 : 1 complex with $(GlcNAC)_4$ hydrated with H_2O. The results are shown in Fig. 18.2. There are two distinct regions: (a) a region for which we obtain a critical exponent $\mu \approx 1.3$ and (b) a region that yields $\mu \approx 2.0$. The first region is interpreted as being an essentially 2D system, as it is relatively thin. In the second region at higher hydration levels, intermolecular water bridges are established, giving rise to a higher dimensional system. Both exponents are in perfect agreement with the values of μ for 2D and 3D percolation conductivity (see Table 2.3).

What is the mechanism for the observed percolative conduction? Careri et al. (1988) argued that the observed percolation process consisted of proton transfer along a thread of hydrogen-bonded water molecules adsorbed on the protein surface. In their interpretation, water molecules are equivalent to the conducting elements in percolating system. The mean-free path of the protons at or above h_c is the distance between the poles of the macromolecules set by the boundaries of the molecule. The local structure of the protein itself is not important, only the structure of the clusters made of water molecules acting as interconnected conducting elements is significant.

The data have important biological consequences. The protein system used in the studies, with its surface sparsely covered by water or conducting elements, is similar to a protein membrane whose internal surface is sparsely populated, although the membrane itself is immersed in a solvent of near-unit water activity. Percolation theory suggests that membrane conduction is possible with channels only partly filled with conducting elements. Moreover, conduction can be turned off or on by adding or subtracting a few elements, without changing the basic structure of the membrane. Finally, since percolation focuses on the effect of disorder on conduction, we can bypass the need for a high level of structure extending over the full thickness of the membrane or its hydrocarbon core.

Rupley et al. (1988) extended the above studies to lightly hydrated purple membranes, which are more complex than the protein system. The conduction paths may be predominantly within one of several regions of the membrane, such as the lipid surface, the lipid-protein interface, or entirely within the protein. The existence of such preferred paths may give a 2D character to the conduction process in such membranes. Indeed, the measured critical exponent, $\mu \approx 1.23$ was close to that of 2D percolation, although the morphology of the system may be 3D. Bruni et al. (1989) extended such studies to a dry tissue where the conductivity process is integrated into complex, living systems. They used tissues of maize seeds, where water-induced effects lead to the onset integrated metabolism and, thus, to germination. The protonic nature of conduction process in such tissues was established by deuterium substitution. The critical exponent for the conduction was found to be, $\mu \approx 1.23$, close to that of 2D percolation.

18.6 Neuromorphic Computing

Neuromorphic computing is a computing paradigm that utilizes very large-scale integration (VLSI) systems with electronic analog circuits, in order to mimic neurobiological architectures in the nerve system. More specifically, “neuromorphic” is used to describe analog, digital, and mixed-mode analog/digital VLSI, as well as softwares that implement models of neural systems. Although the research field has at least a 30-year history (see, for example, Mead 1990), it was perhaps fabrication of atomic switches (Terabe et al. 2005) and memristors (Strukov et al. 2008) that led to a new paradigm for constructing architectures that make it possible to carry out neuromorphic computations.

Atomic switches (for a review see Hino et al. 2011) are nanoionic devices that control diffusion of metal cations and their reduction/oxidation reactions in the switching in order to either form or annihilate a metal atomic bridge, which is a conductive path between two electrodes in the on-state. Unlike conventional semiconductor devices, atomic switches provide highly conductive channels, even if their size is on the order of nanometer, which together with their low on-resistance and nonvolatility has made it possible to develop new types of programmable devices.

The existence of memristors, on the other hand, was suggested by Chua (1971) who, based on symmetry considerations, advanced the idea that in addition to the standard resistors R, capacitors C, and inductors L, there must be a fourth fundamental element in electrical circuits. Chua reasoned that six basic relations connect pairs of four fundamental variables in a circuit, namely, the current I, voltage V, charge q, and magnetic flux J. One relation is determined from the definition, $q = \int I(t)dt$, and another from Faraday’s law of induction, hence reducing the number of independent relations to four. Thus, there must also be a fourth circuit element. Chua called the new hypothetical element memory resistor, or memristor for short.

A memristor with memristance M is an element of an electrical circuit characterized by a link between the charge and magnetic flux, $dJ = Mdq$. If M is constant, one recovers the standard linear relation between J and q. If, however, M depends on q, no combination of nonlinear R, C, and L mimics the behavior of such a memristor. Then, a current-controlled memristor for circuit analysis is expressed by

$$V = \mathcal{R}(w)I \ , \tag{18.7}$$

$$\frac{dw}{dt} = I \ , \tag{18.8}$$

where w is the state variable of the device (in this case coinciding with the charge q) and $\mathcal{R}$ a generalized resistance that depends on the system's internal state. Chua and Kang (1976) considered a general *memristive* system described by

$$V = \mathcal{R}(I, w, t)I \ , \tag{18.9}$$

$$\frac{dw}{dt} = f(I, w, t) \ , \tag{18.10}$$

and, thus, the flux is no longer defined uniquely by q. Equation (18.9) implies that a memristive system is a particular nonlinear dynamical system, since there would be no current in the memristive system, if the voltage drop across it is zero. Therefore, although a memristor is a passive two-terminal element that is similar to a resistor, it is a particular type of nonlinear resistor that provides controllable resistance. As an electrical device, it is not yet available commercially.

Chua's proposal was not tested experimentally until Strukov et al. (2008) showed how memristive behavior arises in physical systems. They considered a thin semiconducting film of thickness ℓ, sandwiched between two metal contacts and divided into two parts: one with a high concentration of dopants (positive ions) and a low resistance $\mathcal{R}_{\mathrm{ON}}$, in series with an undoped material with a high resistance $\mathcal{R}_{\mathrm{OFF}}$. If an external voltage $V(t)$ is applied to the film, due to the dopants' drift the boundary between the doped and undoped parts moves forward. Thus, even in the simplest case—an ohmic conductor (OC) and linear ionic drift with average ionic mobility μ—one obtains

$$V(t) = \left[\mathcal{R}_{\mathrm{ON}}\frac{w(t)}{\ell} + \mathcal{R}_{\mathrm{OFF}}\left(1 - \frac{w(t)}{\ell}\right)\right] I(t) \ , \tag{18.11}$$

$$\frac{dw(t)}{dt} = \mu\frac{\mathcal{R}_{\mathrm{ON}}}{\ell} I(t) \ , \tag{18.12}$$

$$w(t) = \mu\frac{\mathcal{R}_{\mathrm{ON}}}{\ell} q(t) \ . \tag{18.13}$$

In this case, $w(t)$ represents the doped area's width. Combining Eqs. (18.11) and (18.13) yields an expression for the system's memristance $M(t)$. Since, $\mathcal{R}_{\rm ON} \ll \mathcal{R}_{\rm OFF}$, one obtains

$$M(t) = \mathcal{R}_{\rm OFF}\left[1 - \frac{\mu \mathcal{R}_{\rm ON}}{\ell^2} q(t)\right] . \tag{18.14}$$

The second term in the brackets is responsible for the memristive effect, since it is proportional to μ and $1/l^2$, which is very important at nanoscale.

Just as an Ohmic heterogeneous material with constant linear conductivity is too complex to be represented by resistors in series and/or parallel configurations, so also is a memristive material and, therefore, cannot be represented as memristors in parallel or series. Random resistor networks and their generalizations, described throughout this book, were developed precisely for this reason. Likewise, a network of interconnected memristive elements, or one that contains a mixture of memristors and OCs, is more realistic and versatile, and was fabricated by Borghetti et al. (2009) to which transistors were connected, and was capable of self-programming. The first neuromorphic computations involving a network of interconnected memristors were carried out by Nedaaee Oskoee and Sahimi (2011). In practice, neuromorphic computing on the hardware level can be realized by oxide-based memristors, spintronic memories, threshold switches, and transistors.

Using 2D models, simulations, and experiments, Brown and co-workers (Fostner et al. 2014; Fostner and Brown 2015; Mallinson et al. 2019; Pike et al. 2020) showed that one can obtain neuromorphic behavior in thin percolating films of nanoparticles. Experimentally, Pike et al. (2020) formed percolating networks of nanoparticles by depositing conducting nanoparticles onto silicon nitride substrates, which brought the particles into contact and formed interconnected clusters, made of particles in Ohmic contact, which are separated by tunnel gaps that have a distribution of sizes with fractal geometries. The tunnel gaps act as switching sites such that upon application of an external voltage stimulus (see below), atomic-scale filaments can be formed, and subsequently broken, in the tunnel gaps, changing the network conductance G. Deposition was terminated when the fraction p of the surface area covered with conducting particles approached the percolation threshold p_c. After deposition the overall structure of the network was held fixed.

To simulate the same system, Fostner and Brown (2015) used a model consisting of overlapping disks with 200×200 particle diameters, which resembled a continuum percolation model of the type described in Chap. 3 with a percolation threshold, $p_c \approx 0.676$. Tunneling was allowed explicitly to occur in the system. It was shown (Ambrosetti et al. 2010) that, with tunneling present, the effective conductivity of the system below p_c depends exponentially on surface coverage p. Simulating the system below p_c, where the conductance G of the system is due to tunneling across gaps between groups of connected particles, Fostner and Brown (2015) assumed that the connections between overlapping particles to have negligible resistance, and that each gap has a conductance, $G_g = A \exp(-\beta L)$, with A and β being two constants, and L the size of the gap, measured in units of the disks' diameter.

To study the properties of the network, the voltage across the system was increased in a triangular ramp from $V = 0$ to $V_{\max}$ and back to $V = 0$, with a constant step size of 0.025 V, and a number N_V of voltage steps was used to characterize both the voltage at any time and the time since the start of the ramp. Voltage ramping was done for consistency with the experimental protocol used in Ref. Fostner et al. (2014). Thus, all the quantities of interest may be studied as a function of N_V.

The switching process was simulated by two models, one stochastic, and one deterministic. First, the smallest gaps between each pair of clusters were identified, and the gaps with electric fields larger than a preset threshold E_t were replaced with a large conductance, $G_O = 10\ \Omega^{-1}$ with probability p_s, which accounted for the stochastic nature of the switching process. Pike et al. (2020) refined the model by allowing for two switching probabilities, p_s^+ and p_s^-, such that when the electric field in a gap or the current in a filament was greater than E_t, the switch was allowed to change state—switch on with probability p_s^+, or off with probability p_s^-. Doing so simulated the formation in the tunnel gap of an atomic wire, generated by surface diffusion or evaporation induced by the electric field. The effective conductance G of the network was then recalculated. The process, as well as the voltage ramps, continued until the system conductance converged to within 0.1% of the value on the previous step. For all the simulations, the cumulative number N_R of the replacements, or the switching events, as well as G and the electric current I, were recoded.

Figure 18.3 presents the macroscopic current for disk coverages $p = 0.55$ and 0.65, and the corresponding voltage ramp V, versus the number of voltage steps N_V. They indicate that, for $p = 0.55 < p_c$, the initial conductance is small, because the current flows through small tunneling conductors. It then increases in response to the increasing applied voltage during the second voltage cycle, because the tunneling

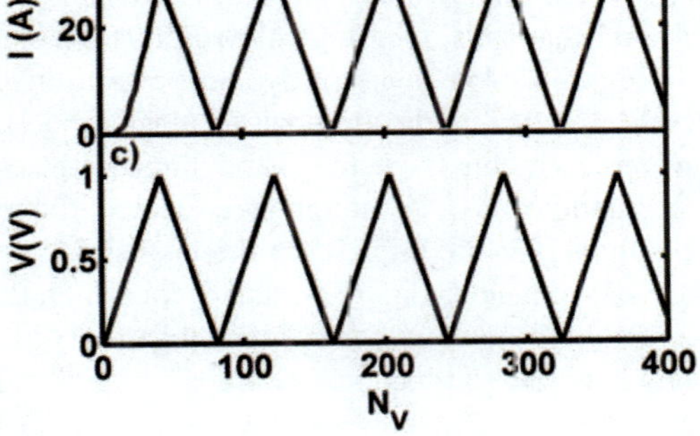

Fig. 18.3 Current I flowing the film as a function of time—the number of voltage steps N_V—for $p_s = 0.1$, and **a** $p = 0.55$; **b** $p = 0.65$, and **c** corresponding voltage ramp with $V_{\max} = 1$ V and voltage step $\Delta V = 0.025$ V (after Fostner and Brown 2015; courtesy of Professor Simon Brown)

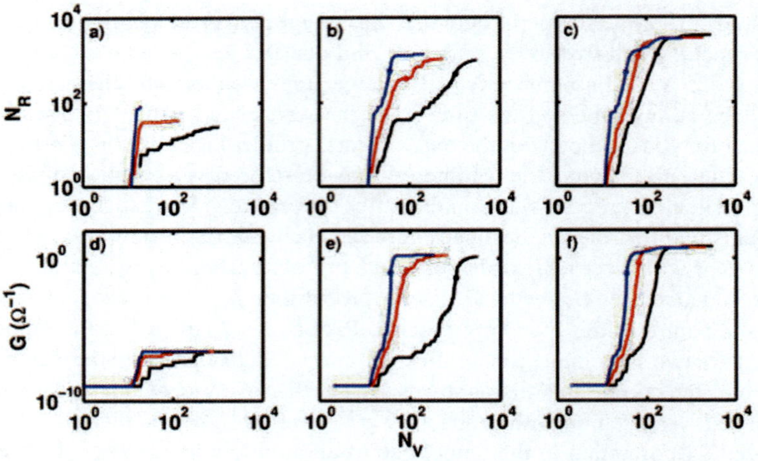

Fig. 18.4 Evolution of the number of tunnel junctions replaced N_R and the conductance G as a function of the number of voltage steps N_V (time) for $p = 0.55$ and $p_s = 0.01, 0.1$, and 0.8 (*black*, *red*, and *blue curves*, respectively, from right to left) for **a** and **d** $V_{\max} = 0.5$ V; **b** and **e** $V_{\max} = 1$ V, and **c** and **f** $V_{\max} = 5$ V. *Colored dots* mark the points at which the last tunneling conductor on the primary conduction path between the contacts is replaced by an Ohmic conductor G_O (after Fostner and Brown (2015); courtesy of Professor Simon Brown)

conductors are replaced (switched) by large conductances that represent atomic-scale wires. The switching event also reduces the potential difference between the two clusters of nanoparticles. Further increase in the current is caused by the subsequent cycles of the voltage ramp. After the fifth cycle, however, the network conductance is dominated by the high conductances G_O and saturates.

In contrast, there is large initial current for $p = 0.65$, and the conductance G saturates much more rapidly, since the tunneling gaps are both fewer and smaller and, therefore, the applied voltage causes a stronger increase in the electric field, increasing the number of switching events. Note that if the switching probability p_s is small, a larger number of voltage cycles would be needed, but eventually the current saturates again. The same type of patterns were observed in the experiments (Pike et al. 2020). These processes are qualitatively similar to leaky integration and fire mechanisms in biological neurons (see, for example, Burkitt 2006).

Figure 18.4 presents the dynamic evolution of N_R and G with N_V (time) for a range of $V_{\max}$ and p_s, indicating that increasing p_s^+ leads to a faster increase in the number of replacements (switching) and, thus, in the conductance. But, it also increases the fluctuations in G. Although, as expected, the maximum conductance for each set of parameters is the highest for the largest $V_{\max}$, the effect of $V_{\max}$ is more subtle. To see this, consider Fig. 18.4a and d, for which the threshold E_t was exceeded by only a relatively small number of tunnel gaps. Thus, the number of replacements was not large and the conductance saturated at small values, but still in the tunneling regime. For large $V_{\max}$, once E_t was exceeded, both N_R and G increased nearly exponentially until the last tunneling conductance on the conduction path between the contacts was replaced (marked by colored dots), hence defining a critical number of voltage steps. N_R eventually saturates because most of the relevant tunnel gaps are replaced, and the existence of an Ohmic conductance path implies that the voltage distribution is relatively uniform, so that the number of places that can exceed the threshold F_t is very few.

Some of such features were qualitatively similar to what happens in biological systems. Moreover, potentiation of the film happens for all surface coverages p, where potentiation refers to the sensitization of a conduction pathway that corresponds to a memory response in the system of interest, and results from an avalanche of neuronal connections. The size of the neuromorphic effect—the change in the conductance G—is more striking for systems with low particle coverages, because the films are more complex. As $p \to p_c$, the size of the connected groups of particles increases and approaches that of the simulated system, implying that the number of tunneling connections that must be traversed in order to span the system becomes smaller. This implies that the film contains fewer locations at which atomic-scale wires could be formed and, hence, that potentiation involves fewer switching events.

To obtain closer agreement with the experiments, Pike et al. (2020) developed a second model. They replaced the stochastic switching with a deterministic one in which the size d_i of each tunnel gap was changed in response to the electric field E_i in the gap according to

$$\Delta d_i = \begin{cases} r_d(E_i - E_t)\ , & \text{if } E_i \geq E_t \\ 0\ , & \text{otherwise} \end{cases} \tag{18.15}$$

The current flow I_j in each existing filament causes electromigration effects that decrease its width w_j according to

$$\Delta w_j = \begin{cases} r_w(I_j - I_t)\ , & \text{if } I_j \geq I_t \\ 0\ , & \text{otherwise} \end{cases} \tag{18.16}$$

Here, r_d and r_w are parameters that control the rates at which d and w change when, respectively, threshold fields E_t and currents I_t are exceeded. Experimentally, $E_t = 10$ V and $I_t = 0.01$ A, which were also used in the simulations.

The simulation results with the deterministic model and their comparison with the experimental data are shown in Fig. 18.5. There is a qualitative agreement between the data and the simulation results. Most important is the fact that the switching produces avalanches in the system as a result of the change in the macroscopic conductance G, which are qualitatively similar to observed neuronal avalanches in cortical tissue (see, for example, Fontenele et al. 2019). Both experiments and simulations indicated that the probability density functions of the size s and duration T_a of the avalanches follow power laws (Mallinson et al. 2019)

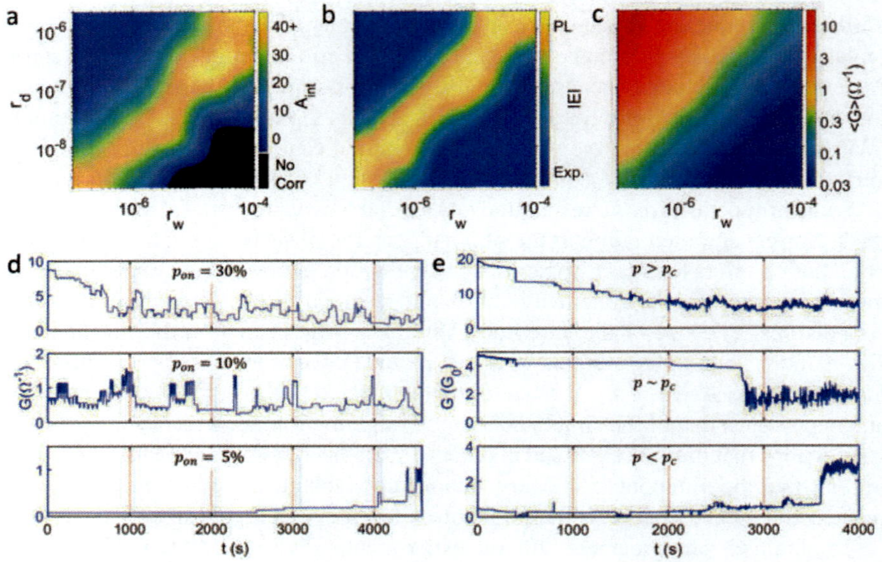

Fig. 18.5 Initial states in which 5, 10, and 30% of switches are "on" all self-tune toward $G \sim 0.5\ \Omega^{-1}$. Right: the Corresponding experimental data showing that under voltage stimulus devices with surface coverages self-tune toward critical states with $1 \leq G \leq 6G_0$. The difference between optimum values of G in the experiment and simulation is due to the choice of simulation parameters (after Pike et al. 2020; courtesy of Professor Simon Brown)

$$f(s) \sim s^{-\tau} , \tag{18.17}$$

$$h(T_a) \sim T_a^{-\alpha} , \tag{18.18}$$

both with a long tail at large times. Note the similarity between Eq. (18.17) and the percolation cluster-size distribution, described in Chap. 2. Equations (18.17) and (18.18) are the signature of a scale-free critical state. Experiments yielded, $\tau \simeq 1.6 \pm 0.1$ and $\alpha \simeq 2.0 \pm 0.1$. The deterministic model described above produced, $\tau \simeq 2.1 \pm 0.1$, and $\alpha \simeq 2.6 \pm 0.1$, in qualitative agreement with experiment. In addition, the autocorrelation function also followed a power law

The similarities and differences with a network of interconnected memristors should also be pointed out. The tunnel junctions, formed at gaps in the percolating film, which switch to a highly conducting state on the formation of an atomic scale wire, behave differently than memristors. But the tunnel gaps are memristive in the sense that, similar to a memristor, it is the device history that determines its state. Switching occurs at a well-defined threshold value of the electric field, but the memristors in the junction cannot be switched back to the low conduction state by simply reversing the polarity of the bias voltage.

18.7 Sensory Transmission and Loss of Consciousness

How dynamic communications between neurons lead to consciousness, or its loss, is a fundamental question that has been studied for decades. In addition to clinical studies, many attempts have been made to model individual electroencephalographic (EEG) features that are associated with loss of consciousness during general anesthesia (see, for example, Ching et al. 2012; Wang et al. 2014, and references therein), including the development of a stochastic model to describe general anesthesia as a thermodynamic phase transition (Steyn-Ross et al. 2001).

Zhou et al. (2015) developed a percolation model to determine the flow of information between neurons. The model produced typical EEG features under general anesthesia, as well as dose–response characteristics for loss of consciousness. Their model was based on a layered hierarchical fractal structure, from an input node to multiple output ones. The layers, meant to represent laminar and divergent organization seen (Sherman 2012) in mammalian thalamocortical structures, were generated by scale-invariant fractal expansion. Within each layer small-world (SW) network (see Chap. 17) properties among the nodes were generated using the Watts and Strogatz (1998) algorithm. As described in Chap. 17, an SW network is a graph in which most nodes are not neighbors of one another, but any node's neighbors are likely to be neighbors of each other, and most nodes can be reached from every other node by a small number of steps. Then, the typical distance L between two randomly selected nodes scales as, $L \propto \ln N$, where N is the number of nodes in the network. Such networks have found a large number of diverse applications, from geoscience (Yang 2001) to brain (Sporns et al. 2004; see also Chap. 17).

A directional weight W_{ij} from node j to node i was attributed to each edge of the network, with $W_{ij} \neq W_{ji}$ to represent the counterstream architecture of human brain. The anterior nodes were differentiated from the posterior ones by attributing different edge weights in the feedforward and feedback directions. Suppose that $A_i(t)$ and $P_j(t)$ are, respectively, the neural activity of node i and the preceding input from node j at time t. $A_i(t)$ is simply the weighted average of activities from all input nodes,

$$A_i(t) = \frac{\sum_i W_{ij} P_j(t)}{\sum_j W_{ij}} = \frac{W_{ii} P_i(t) + \sum_{j \neq i} W_{ij} P_j(t)}{W_{ii} + \sum_{j \neq j} W_{ij}} \tag{18.19}$$

where the input function $P_j(t)$ represents the accumulated history of neural activity from the preceding m time steps, weighted by exponentially decaying memory, i.e.,

$$P_j(t) = \frac{\sum_{\tau=1}^{m} \exp(-\tau) A_j(t-\tau)}{\sum_{\tau=1}^{m} \exp(-\tau)}, \tag{18.20}$$

Percolation is then used to assign randomly the weights W_{ij} to the edges from a sampling function, with probability p that represented the likelihood of activity transmission, according to

$$W_{ij} = \begin{cases} \mathcal{G}_{\text{CDF}}(U[0, 1]) & i \neq j \\ c \exp(-\lambda \sum_{k \neq i} W_{ki}) & i = j \end{cases} \tag{18.21}$$

Here, $\mathcal{G}_{\text{CDF}}$ is the Gaussian cumulative distribution function with its center at $(1 - p)$ and a standard deviation of 0.05, $U[0, 1]$ represents the uniform distribution, and c and λ are constant.

With decreasing p the probability of activity transmission along individual edges also decreases, representing the inhibition of information flow under anesthesia. Although various anesthetic classes act differently at molecular and cellular levels, the net effect is represented as a global inhibition of arousal (Brown et al. 2011). Sampling W_{ij} is done independently by p, when $i \neq j$, whereas for $j = j$ W_{ii} represents the memory of the past activity of the same node, which is influenced by the incoming connection strength. When all non-self-connections diminish, i.e., when $\sum_{k \neq i} W_{ki} \to 0$, self-connection is the dominating factor. Neuronal transmission, including axonal propagation and synaptic events, involves cycles of receptor inactivation, activation, and deactivation. To account for such dynamic behavior, edge weights were periodically resampled using the above sampling algorithm, with a periodicity proportional to $\exp(-ap)$, where a is a constant.

The percolation model was then validated by reproducing four key features of global EEG responses: (a) characteristic EEG waveforms, including burst suppression under deep anesthesia. A burst–suppression pattern is a discontinuous EEG that has periods of marked suppression or isoelectric intervals, alternating with "bursts" of activity, with or without embedded epileptiform features, and is commonly seen following hypoxic-ischemic injury. (b) An EEG power shift to lower frequencies with increasing anesthetic concentrations. (c) Synchronization of cortical nodes and (d) shift of α and δ rithms to the anterior, often called anteriorization. An α rithm is a uniform rhythm of waves in the normal EEG, exhibiting an average frequency of 10 per second, which is typical of a normal person awake in a quiet resting state. A δ rithm is an EEG wave with a frequency less than 3.5 per second, which is typical in deep sleep, in infancy, and in serious brain disorders.

As documented by Zhou et al. (2015), their percolation model reproduced all the four key features. The model indicated that as p decreases, a steep divergence of the transmitted signal from the original is developed, together with a loss of signal synchrony, as well as a sharp increase in information entropy in a critical manner, which resembles the precipitous loss of consciousness during anesthesia.

18.8 Actomyosin Networks

Actomyosin is the actin–myosin complex that forms within the cytoskeleton, and is inherently contractile, with the myosin motor protein able to pull on actin filaments, giving rise to contractile fibers that form the basis of skeletal muscle, and enabling cell motility and force generation at the sub-cellular level. An actomyosin network consists of self-assembled actin filaments—two-stranded helical polymers of the protein actin, with a diameter of around 6 nm—which have variable lengths, and are connected by cross-linkers and myosins.

Alvarado et al. (2013) (see also Sect. 18.15 below) fabricated a quasi-2D network of actin filaments with the size $2.5 \times 2.5\ \mathrm{mm}^2$, which were connected by fascin cross-linkers. Myosin motors pull the filaments together and exert contractile forces on the network, leading to cross-linker unbinding and actin filament movement that fundamentally altered the underlying actin network. Alvarado et al. traced the trajectories of the actin filaments during the contraction, in order to reconstruct the connectivity of the actin network throughout the dynamical process. Assuming that the number of cross-linkers in the system was M_c, three distinct regimes were observed: (a) For small M_c, the actin network was contracted into many foci, when the myosins were activated. Thus, if one retraces the actin molecules' position back to time $t = 0$, one observes that the resulting in foci originated from actin clusters with areas of similar sizes. (b) For intermediate values of M_c, the retracing to the original areas indicated a power-law distribution in sizes with an exponent of -1.91, which was similar to the exponent τ of the cluster-size distribution at the percolation threshold (see Chap. 2), $\tau = 187/91 \simeq 2.05$. The power-law cluster size distribution existed for a wide range of cross-linker concentrations, hence indicating that a critical state can be reached without fine-tuning, which is the hallmark of self-organized criticality. (c) For large M_c, the actin network contracted to a single cluster.

Lee and Pruessner (2016) proposed a 2D percolation model for the experimental observations that reproduced their essential features. They used a $L \times L$ square network with $N = L^2$ nodes. A strength s_k was attributed to a site k, selected from the Poisson's distribution with a mean value, $\langle s\rangle = M_c/N$, intended to mimic the random number of cross-linkers connected at each node. It was then assumed that at $t = 0$ the pulling force f_k at site k, which are generated by the molecular motor myosins, were also distributed according to Poisson's distribution with a mean value, $\langle f\rangle = M_m/N$, with M_m being the number of myosin motors in the system. In addition to f_k, adhesion connects the actin network to the boundary, and to mimic that Lee and Pruessner assumed that the strength of the boundary adhesion was s_A, and that the actin networks adhered to the top and bottom boundaries only.

Next, one must introduce the myosin-induced dynamics in the actin network. Lee and Pruessner assumed that when the pulling force f_k begins to act on nodes k, the sites rip and, therefore, disappear sequentially according to the reverse order of their unbinding propensity function $B(s_k, f_k)$, with B decreasing with s_k, but increasing with f_k. The disappearance of a node simulates unbinding of the cross-linkers from the actin filament due to loading. When a node disappeared, the pulling forces were adjusted to achieve force balance in the network, which was tantamount to assuming that the time scale t_u for unbinding was much larger than the corresponding time t_b to reach force balance.

In addition to the myosin unbindings, connected clusters of actin filaments may also contract. To include this in the model, Lee and Pruessner assumed that they occur if a connected cluster is detached from either boundary, and that the contraction time scale t_c is small compared to t_u, so that when a cluster is free to contract, it contracts to a focal point before the next unbinding event occurs. When a cluster was detached from the boundary, it could freely contract, and all forces due to mechanical tension ceased. Thus, the contracted cluster was removed from dynamics of the network and, thus, no node was deleted from the cluster.

It should be clear to the reader that the model resembles an invasion percolation (IP) process with trapping. The sequential node disappearance according to the reverse order of their unbinding propensity is an IP algorithm. That no node was removed from a detached cluster implies that such nods were trapped and could not be affected by the dynamics of the network.

The percolation model could also explain the aforementioned three regimes in the experiments of Alvarado et al. (2013). If M_c is small, many nodes will be empty and, thus, the actin network could not percolate, and fragments into many small foci. If, on the other hand, M_c is large, then the average strength of a node may be larger than the adhesion strength, $\langle s\rangle > s_A$, resulting in the detachment of the entire network from the adhesion boundary due to internal contraction. In between, one has networks with intermediate values of M_c, which is the regime in which the actin network percolates and the cross-linker unbinding occurs predominately in the interior.

The simulation of Lee and Pruessner confirmed the picture. However, the exponent τ of the cluster-size distribution was determined to be very close to $\tau = 187/91 \simeq 2.05$, indicating that the trapping effect was minimal and did not influence the results significantly. As a result, the model is close to the standard percolation model.

Although they did not mention percolation explicitly, a model that was developed by Wang and Wolynes (2012) for active contractility in actomyosin networks represents a percolation model in high-dimensional space, or on random graphs described in Chap. 17. Their model accounted for two key features of actomyosin self-organization, namely, the asymmetric load response of individual actin filaments, and the correlated motor-driven events that mimicked myosin-induced filament sliding. Wang and Wolynes (2012) studied the effect of the concentration and susceptibility of the motors on their collective behavior, as well as their interplay with the network connectivity, and how they regulate macroscopic contractility. Their work demonstrated that cooperative action of load-resisting motors in a force-percolating structure integrates local contraction/buckling events into a global contractile state through an active coarsening process.

18.9 Percolation Transition in the Mutation-Propagating Module Cluster in Colorectal Tumorigenesis

Tumorigenesis, the initial formation of a tumor in the body, involves epistatic interactions between genes and multiple somatic mutations. Certain types of genes do not mutate randomly, but tend to co-occur in cancer patients. In particular, colorectal cancer develops through the sequential accumulation of driver mutations, including adenomatous polyposis coli (APC), Kirsten rat sarcoma viral oncogene homolog (KRAS; pronounced K-raz), phosphatidylinositol 3-kinase (PI3K), and tumor protein p53 (TP53), hence suggesting that an important factor for the development and spread of cancer may be the cooperativity of driver mutations. The mechanism underlying such cooperative multiple mutations during tumorigenesis is not well understood, however.

Shin et al. (2017) proposed a percolation model in order to understand the cooperative phenomenon of somatic mutations in tumorigenesis in colorectal cancer. They first obtained somatic mutation data for colorectal cancer patients. Then, in order to use only high-confidence pathogenic variants, Shin et al. filtered the mutations based on their predicted pathogenicity, derived from five functional mutation prediction tools, which were then assigned to tumor samples by abstracting binary event calls, such that a genomic event either occurred—represented by "1"—or did not occur - denoted by "0"—in a gene for a given sample. Only patients with less than 300 mutations from the 223 colorectal cancer patients in The Cancer Genome Atlas (TCGA) were used, reducing the total number of patients to 198, for whom the expression data were obtained from the Firehose website.

The protein–protein interaction (PPI) network had been previously constructed by Hofree et al. (2013), of which Shin et al. considered only the largest connected subnetwork, because the network propagation was not available between a pair of nodes that did not connect with each other, either directly or indirectly. Shin et al. then integrated the expression and mutation profiles of the cancer patients with the PPI network, in order to obtain the adjacency matrix $\mathbf{A}$ of the network, as well as a patient-by-gene matrix that displayed the mutation profiles of the aforementioned binary (1, 0) states on $N = 10,968$ genes for the patients.

The entry A_{ij} of $\mathbf{A}$ represents the probability of interaction between nodes i and j along which the mutation influence propagates, which, for a given sample, was assumed to be proportional to the product of expression values of both genes in that sample, i.e., $A_{ij} \to A_{ij}E_iE_j$, with E_i indicating the expression value of gene i. Thus, patient-specific PPI networks were obtained, enabling the implementation of more realistic propagation of somatic mutations on the network (see below). Shin et al. used only samples that had both RNA-seq and somatic mutation data, obtained from the Dana-Farber Cancer Institute, and downloaded from the cBioPortal website.

To simulate the propagation of mutation effects in the PPI network, Shin et al. utilized an algorithm based on a random walk with restart on a network, developed by Vanunu et al. (2010). When a mutation occurred on a gene in the PPI network, a value of 1 was initially attributed to the mutated gene, which then propagated along the network neighborhood, such that higher values were assigned to non-mutated genes that were closer to the mutated one, according to, $\mathbf{F}_{t+1} = \alpha\mathbf{A}'\mathbf{F}_t + (1-\alpha)\mathbf{F}_0$ at step or time t, with $\mathbf{F}_0$ being a matrix of binary (1,0) states on genes such that 1 or 0 indicated whether a gene was mutated or not in the corresponding patient; $\mathbf{A}'$ a degree-normalized adjacency matrix of the PPI network, and α is the degree of diffusion of a mutation influence in the network. $\mathbf{A}'$ was set to be, $\mathbf{A}' = \mathbf{D}^{-1/2}\mathbf{A}\mathbf{D}^{1/2}$, in which $\mathbf{D}$ is a diagonal matrix such that its entries D_{ii} were the sum of row i of $\mathbf{A}$

A mutation cannot successfully affect all the nodes in the PPI network, but covers at most a few layers of nearest neighbors of the nodes. In this way, the effective boundary of mutation influences centered on the mutated node is predicted, inside which nodes have influence scores above a given threshold, and form a sub-network called a *mutation-propagating module* (MPM). The MPMs of individual mutations are distributed throughout the network, forming connected modules, the largest one of which, referred to as the giant cluster (GC; see Chap. 17)—the analog of the largest percolation cluster on lattices—could cover the network on the order of network size N. The GC can, therefore, impact significantly the properties of the network.

The threshold of mutation influences was set high enough that each MPM approximately included only nearest neighbors. Simulation of Shin et al. indicated that, on average, the GC covered about 20–40% of the entire network, even though patients only had an average of 20–40 somatic mutations, hence indicating that multiple somatic mutations interact cooperatively to expand the effective boundary of mutation influences. To check whether the cooperative effects in the GC in the network for each individual patient was due to nonrandom mutation profile, Shin et al. compared the size $S_{\rm GC}$ of the GC with the expected size $S^r_{\rm GC}$ of the same number of mutations that occurred randomly on the network. The former turned out to be much larger than the latter, with the differences being more striking when the cancer-related or driver genes were considered, rather than all the mutations. Thus, the cooperative effect among somatic mutations in cancer cannot be due to a random selection of mutations, but is determined by topological— connectivity—properties of somatic mutations in the PPI network. That is, of course, the signature of the percolation process.

Other aspects were examined as well. For example, if a GC forms based on nonrandom mutation profiles of a cancer patient, genes within it must be correlated with relevant biological and clinical characteristics of the patient. To test this and identify the biological processes that are inherent in the cluster, Shin et al. (2017) used a set of 50 "hallmark genes," and and identified hallmark gene sets enriched in the GC. The hallmark gene sets summarize and represent specific well-defined biological states or processes, display coherent expression, and were originally generated by a computational methodology based on identifying gene set overlaps and retaining those that display coordinate expression. They reduce noise and redundancy, providing a better-delineated biological space for GSEA (gene set enrichment analysis). It was hypothesized that the cooperation of multiple somatic mutations in a GC activates multiple cancer-related hallmarks that cooperatively enhance tumorigenesis.

Two tests were carried out to check the hypothesis. First, Shin et al. examined how closely the GC-based patient classification matches the previous consensus molecular subtypes (CMS) classification, proposed by Guinney (2015), which is considered a robust molecular classification for colorectal cancer. They also investigated which key biological features the subtypes have. The result was that the analysis of the hallmark gene set based on the giant percolating cluster (GPC) extracts all the features of the CMS groups, even though factor analysis was used, which is a type of unsupervised learning for feature selection or for data reduction. Moreover, the patient classification with the selected features exhibited very strong correlations with the CMS groups.

Second, Shin et al. showed that the hallmark gene sets enriched in cancer patients correlate with clinical tumor stages. To do so, they carried out statistical clustering analysis of the enrichment score of the hallmark gene sets for cancer patients according to the significant hallmark gene sets, of which 12 were selected to distinguish the tumor stages of cancer patients. The result indicated that the patient population may be divided into five clusters, of which three are strongly correlated with various tumor stages.

According to the analysis of Shin et al., scattered MPMs formed a GC in a PPI network and, more importantly, genes within the GC represented the phenotypic properties of cancer, including cell proliferation and metastasis. Therefore, Shin et al. proposed that the GC that results from somatic mutations of a cancer patient should be viewed as a GPC, i.e., the largest percolating cluster that integrates the influences of scattered somatic mutations, so that it confers phenotypic changes corresponding to cancer hallmarks. The question then is whether in practice the GC would also undergo a percolation transition on the accumulation of somatic mutations during cancer progression. The analysis of Shin et al. indicated that the answer is affirmative.

Thus, a GPC—cooperative mutation effects represented by a large connected cluster in a PPI network—can undergo a percolation transition during tumorigenesis. In the development of normal tissue, a number of somatic mutations can occur at various places in the PPI network, as a result of which the MPMs form a connected module, or a GC along with the accumulation of somatic mutations. Thus, tumorigenesis can be initiated by a certain driver mutation that connects scattered clusters into one, leading to the formation of a GPC that represents cancer hallmark. This is shown schematically in Fig. 18.6. Shin et al. also found that the most frequently observed

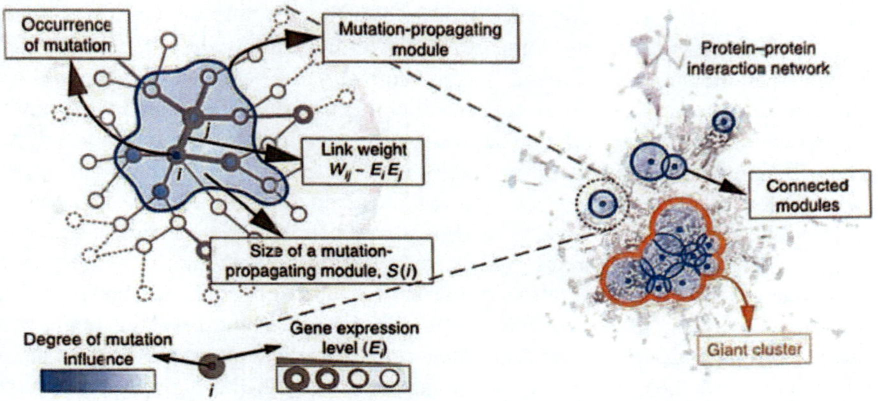

Fig. 18.6 Schematic of the percolation transition of cooperative mutational effects during tumorigenesis, showing formation of a giant cluster upon a protein–protein interaction network by the propagation of mutation influences. A mutation influence propagates along a PPI network and forms a mutation-propagating module, a subnetwork that is effectively influenced by the mutation. The link weight is determined by the product of the expression levels of its end genes, $W_{ij} \sim E_iE_j$. Several mutation-propagating modules occasionally form connected modules or a giant cluster, the largest connected module. The intensity of the color indicates the degree of mutation influence, while the width of a circle indicates the expression level of the corresponding gene (after Shin et al. 2017)

sequence of driver mutations that characterize colorectal cancer development might have been optimized in order to maximize the size of the GPC, hence providing a intriguing insight into the relationship between cancer development and percolation transition, which can be useful for understanding the fundamental mechanism of tumorigenesis.

18.10 Essential Nodes in Brain: Optimal Percolation

A most important problem in neuroscience is how the brain integrates distributed and specialized networks into a coherent information processing system. It is generally believed (see, for example, Sporns 2013) that integrating specialized modules in the brain are facilitated by a set of what is referred to as the *essential nodes*. Integrated brain networks exhibit long-range correlations in their activity, with the correlation structure often referred to as *functional connectivity.* Any perturbation in the essential nodes results in large disturbances in functional connectivity that affect global integration. Therefore, a fundamental step toward understanding how information is processed in brain circuits entails identifying the essential nodes, which in turn may contribute to the design of targeted interventions to restore or compensate dysfunctional correlation patterns in disease states of the brain.

There have been many studies utilizing various network centrality measures in order to identify the essential nodes. Such measures include hubs, nodes that have many connections to other nodes; betweeness centrality, a measure of centrality in a connected graph since there exists at least one shortest path between every pair of nodes; closeness centrality, a measure of centrality of a node in a network and calculated as the sum of the length of the shortest paths between the node and all other nodes in the graph; eigenvector centrality that measures the influence of a node in a network by assigning relative scores to all nodes in the network based on the fact that connections to high-scoring nodes contribute more to the score of a given node than equal connections to low-scoring ones, and other measures.

Each of such measures may be used as a way of ranking the nodes in order to determine the most influential nodes, with the nodes having the highest ranking viewed as essential for integration in the brain. Although each measure provides a distinct aspect of influence, they all predict that when the essential nodes are inactivated in a targeted intervention, integration in the overall network is to a large extent prevented. In other words, inactivated nodes with the highest rank inflict the largest damage to the long-range correlations. Therefore, the optimal centrality measure would be one that prevents integration of the network by inactivating the fewest number of nodes.

Del Ferraro et al. (2018) suggested that the problem can be addressed using percolation theory. One must identify the smallest set of nodes that, when inactivated, destroy the integration of the network, which is an NP-hard (non-deterministic polynomial-time hardness) problem. Del Ferraro et al. suggested that the problem can be mapped onto the *optimal percolation* (OP), first described by Morone and Makse

(2015). To understand the OP, first recall that the most influential nodes in a network are those that form a minimal set that guarantees global connection of the network, and are referred to as the *optimal influencers*. To be concrete and to see the relation with the percolation problem, consider a network of N nodes that are connected by M bonds or links, with an arbitrary degree distribution (i.e., arbitrary distribution of the nodal connectivity). As the reader knows, if the nodes are removed at random, then at a critical fraction q_c of the removed nodes—the percolation threshold—the network undergoes the percolation transition and its structure collapses. Therefore, the optimal influence problem is equivalent to estimating the minimum q_c of the influencers to fragment the network, so that the probability of having the GC would vanish.

This was the problem that was addressed by Morone and Makse (2015). They reformulated the problem as follows so that it can be addressed in a new framework. Consider a vector $\mathbf{n} = (n_1, \cdots, n_N)$ in a network of N nodes, such that $n_i = 0$ for the removed influencers, and $n_i = 1$, otherwise, with the fraction q of the removed influencers given by, $q = 1 - (N \sum_i n_i)^{-1}$. Suppose that $p_{i\to j}$ is the probability that a node i belongs to the GC in a network in which j has been removed. If there is no GC, then $p_{i\to j} = 0$ for all $i \to j$. We construct a $2M \times 2M$ matrix, defined by

$$\mathcal{M}_{k\to\ell,i\to j} \equiv \left.\frac{\partial p_{i\to j}}{\partial p_{k\to\ell}}\right|_{\{p_{i\to j}=0\}} .$$

Morone and Makse (2015) showed that for locally tree-like random graphs one has, $\mathcal{M}_{k\to\ell,i\to j} = n_i \mathcal{B}_{k\to\ell,i\to j}$. Here, $\mathcal{B}_{k\to\ell,i\to j}$ is the *non-backtracking matrix* of the network, a binary matricial representation of the topology of a network whose entries represent the presence of non-backtracking paths between pairs of different nodes, traversing a third intermediate one (Pastor-Satorras and Castellano, 2020). The entries of the matrix are non-zero, taking up a value of 1, only when $(k \to \ell, i \to j)$ form a pair of consecutive non-backtracking directed edges; i.e., $(k \to \ell, \ell \to j)$ with $k \neq j$.

One obtains a stable solution for the problem $\{p_{i\to j} = 0\}$, if the eigenvalues of $\mathcal{M}$, $\lambda(\mathbf{n}; q) \leq 1$. Thus, for a given $q \geq q_c$, the optimal influence problem may be reformulated as one of identifying the optimal configuration $\mathbf{n}$ that minimizes the largest eigenvalue $\lambda(\mathbf{n}; q)$ such that the optimal set $\mathbf{n}^*$ of Nq_c influencers is obtained when the minimum of the largest eigenvalue is at its critical threshold,

$$\lambda(\mathbf{n}^*; q_c) = 1 . \tag{18.22}$$

Morone and Makse then devised a numerical scheme to determine the configuration for which the largest eigenvalue approaches one. To do so, they recast the problem as one of optimization, whereby one minimizes the "energy" (or the cost function) of a many-body system (i.e., many interacting nodes in the network), where the form of the interactions was fixed by the non-backtracking matrix of the network, defined above, and minimizing the energy is tantamount to determining the configuration $\mathbf{n}^*$ with minimum q_c at which the GC collapses. To do so, they defined a ball of radius ℓ around every node, considered the nodes that belonged to the frontier $\partial\text{Ball}(i, \ell)$ and assigned to i a *collective influence* (CI) strength at level ℓ, defined by

$$\text{CI}_\ell(i) = (k_i - 1) \sum_{j \in \partial\text{Ball}(i,\ell)} (k_j - 1) , \tag{18.23}$$

where k_i is the degree of node i.

Thus, the computational algorithm based on Eq. (18.23) is as follows. Initially, all the nodes are present and, therefore, $n_i = 1$ for all of them. One node i^* with the largest CI_ℓ is removed, n_{i^*} is set to zero, and the degree of neighbors of i^* is reduced by one. The procedure is repeated to identify a new node with the largest CI_ℓ for removal. The computation is terminated when the GC does not exist anymore. The algorithm was implemented with three large networks, namely, the Erdös–Rényi graph (see Chap. 17), the web of Twitter users, and an immunization scheme on a personal contact network built by the phone calls by 14 million people in Mexico. In the last case, the method saved a large number of vaccines or, equivalently, identified the smallest possible set of people to quarantine.

Del Ferraro et al. (2018) used the OP algorithm of Morone and Makse, together with pharmacogenetic interventions in vivo, to predict and target nodes that are essential for global integration of a memory network in rodents. The computations predicted that integration in the memory network is mediated by a set of low-degree (low connectivity) nodes that are in the nucleus accumbens, which was confirmed with pharmacogenetic inactivation of the nucleus accumbens that eliminates the formation of the memory network, while inactivations of other brain areas leave the network intact.

18.11 Flexibility of Thought in High Creative Individuals

Flexibility of thought is considered a hallmark of creativity, because it indicates how an individual can process new data and ideas and expand his/her thought process. But, how can one quantify the flexibility of thought in a person? One approach is based on network science that represents semantic memory structure as a network, and was proposed by Borge–Holthoefer and Arenas (2010) and Baronchelli et al. (2013). Once the network representation is set up, percolation analysis can be used to study semantic memory and how it is affected by failures, i.e., cutting links in the network, and how it facilitates dynamical processes that act upon it. The former was used to study cognitive phenomena in patients with Alzheimer's disease (Borge-Holthoefer et al. 2011), while the latter was investigated in the context of memory retrieval (Arenas et al. 2012).

Kenett et al. (2018) proposed a percolation approach to study high-level cognition, and to test the hypothesis that the semantic network of highly creative people is more flexible than that of low creative individuals and, therefore, more robust. Semantic networks of low semantic creative (LSC) and high semantic creative (HSC) persons have been reported by Kenett et al. (2014), which were utilized by Kenett et al. (2018). The two networks, each with about 4,000 links, consisted of 96 cue words that were divided into groups of 4 concrete words from 24 categories, including fruits, musical instruments, vehicles, and others, and representing a priori components of the two networks. Kenett et al. (2018) modified the two networks by taking into account the number of participants who generated the associative features that determine the strength of semantic similarity between the cue words. The more similar associative response generated and the larger number of participants who generated the association responses to a pair of cue words, the stronger the link between this pair of cue words is.

Kenett et al. (2018) normalized the links in each network according to the mean number of associations per cue word, so as to remove the effect of the HSC individuals who generate a higher number of associations per cue words, when compared with the LSC individuals. The links between all pairs of cue words define a symmetric correlation matrix whose (i, j) entry represents the semantic similarity between cue words i and j, which can be studied in terms of an adjacency matrix of a weighted, undirected network in which each cue word is represented by a node, and a link between two cue words represents the semantic weight between them.

As usual, percolation analysis of the two networks was carried out by removing their links according to a pre-set threshold $\mathcal{T}$, such that all links with a weight smaller than the threshold were removed, the size (number of nodes) of the giant component in the network was determined, and the detached components (clusters) were identified. $\mathcal{T}$ was varied from the smallest weight in the network, the initial threshold $\mathcal{T}_i$, to a weight strength in which the giant component was smaller than three nodes, the final threshold $\mathcal{T}_f$, with its resolution selected according to the smallest difference between the sorted weights. Thus, one has a variety of step sizes that are the number of nodes of a component that disconnects from the GC in each step, most of which were groups of several nodes.

An illuminating quantity that was calculated is the percolation integral I_p, defined by

$$I_p = \int_{\mathcal{T}_i}^{\mathcal{T}_f} C_G(x)dx = \sum_{\mathcal{T}_i}^{\mathcal{T}_f} C_G(\mathcal{T})\mathcal{T}_r . \tag{18.24}$$

where C_G is the size of the GC at a threshold and $\mathcal{T}_r$ is the threshold resolution. I_p measures the speed by which the GC breaks. If the network breaks at a low value of the thresholds and has a steep percolation curve (see below), it will have a smaller I_p than a network with the same size of the GC that breaks with large thresholds and a flat percolation curve.

The effect of noise is important and must be studied. If one adds noise to the weights of the links and, for example, makes weaker weights stronger, and as a result obtains a value of I_p that is dramatically different from its pre-noise value, it implies that the percolation analysis has no physical significance. To test this, Kenett et al.

(2018) computed I_p over 500 realizations of the networks before carrying out the percolation analysis. The realizations were generated by adding a Gaussian noise to the network with a mean value of zero and a standard deviation σ that varied between 10^{-4} and 10^{-2}.

In addition to studying the effect of noise, analysis of the networks whose links were shuffled was also carried out in order to understand the effect of the structure of the network on the percolation analysis. In the shuffling, two links from two pairs of nodes are selected at random and are exchanged. For example, nodes n_1 and n_2 with link strength 0.5 and nodes n_3 and n_4 with link strength 0.7 are exchanged, so that the new network topology would be one in which n_1 and n_3 are connected with strength 0.5 and n_2 and n_4 with strength 0.7. This process was reiterated ten times the number of links in the network (i.e., about $\sim 40,000$ times) to ensure that most of the links in the network were replaced, and was also reiterated to generate 500 realizations, for each of which I_p was computed.

The hypothesis is that the semantic network of HSC individuals is more flexible than the LSC network and, therefore, their network will be more robust to link removal, hence breaking apart slower than the LSC network. The analysis of Kenett et al. (2018) revealed that the GC of the LSC network breaks apart faster at lower thresholds than the HSC network. This is shown in Fig. 18.7A. For the same threshold values, the LSC giant cluster is mostly smaller than its counterpart in the HSC network, indicating that the latter is more robust and its components are better connected. The percolation integral I_p turned out to be 14.76 for the LSC network, and 15.64 for the HSC one, computed for the type of curves shown in Fig. 18.7A.

The effect of noise is shown in Fig. 18.7B, indicating that both the LSC and HSC networks have a stable percolation process with significantly different integrals. The effect of link shuffling is demonstrated in Fig. 18.7C. Independent analysis of samples with the t test, carried out for the distribution of I_p values of the LSC networks and its comparison with the HSC ones, indicated that the percolation shuffled integral of the former is even slightly larger than that of the latter, implying that the robustness of the HSC network is more influenced by the shuffling of its structure. This suggests strongly that the difference between the empirical I_p of the two groups is driven by the structure of the networks, rather than by the link weights.

18.12 Cardiac Fibrosis and Arrhythmia

Cardiac contraction is controlled by abnormal propagation of nonlinear excitation waves in the heart, which can create life-threatening cardiac arrhythmias. Typically, such arrhythmias develop when the waves form turbulent rotational activity, or *reentry*. A well-known example of a pathology that leads to reentry formation in cardiac tissues is fibrosis (see, for example, Nguyen et al. 2014), which is the presence of a large number of nonexcitable cells that are scattered in the cardiac tissue, formed by myocytes that are excitable cells, inhomogeneously distributed, and are connected by gap junctions. The extracellular matrix, non-conducting gap junctions, blood ves-

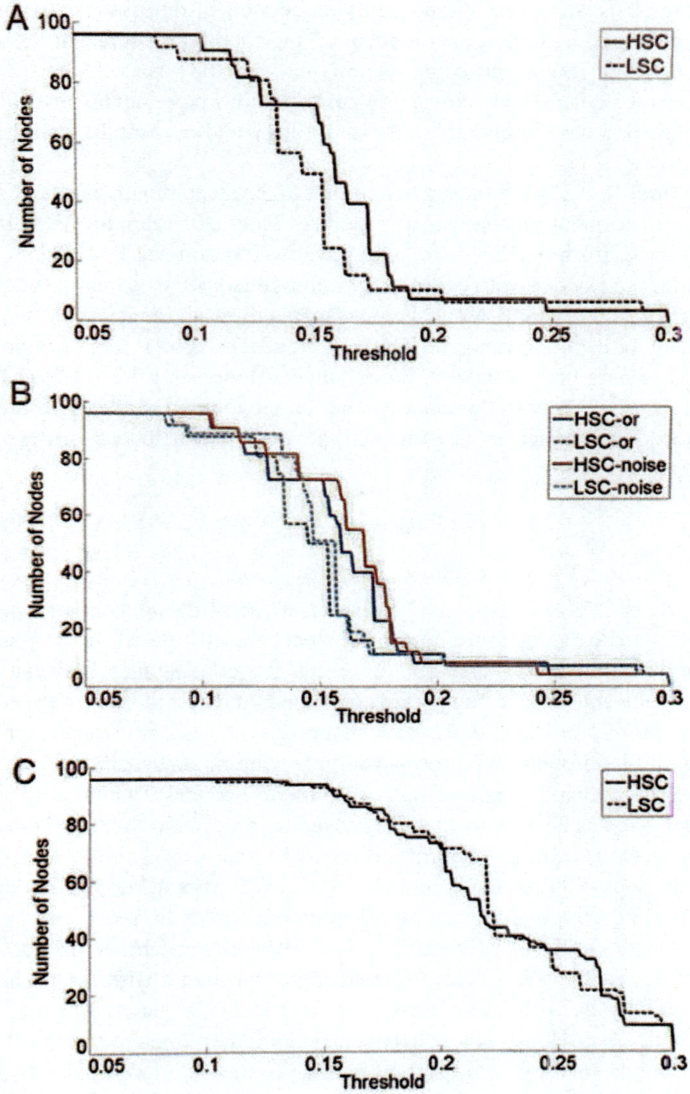

Fig. 18.7 **A** Percolation analysis of the LSC and HSC networks. **B** Effect of noise on the percolation analysis of the LSC and HSC networks, with addition of noise with standard deviations of 0.01 (LSC/HSC-noise) or without the noise (LSC/HSC-or). **C** An example of typical iteration of the percolation analysis on the shuffled links analysis in both networks (after Kenett et al. 2018)

sels or fibroblasts that nonexcitable cells embedded in the tissue, can disturb the gap-junction coupling among the myocytes. Non-conducting connections in cardiac tissue strongly affect wave propagation and increases the possibility of arrhythmias. An important issue is the link between the pattern of fibrosis and the onset of arrhythmia, as it has practical implications for cardiology and development of arrhythmia treatment.

Alonso and Bär (2013) first suggested that heterogeneities in cardiac tissue may provide an anatomical mechanism for reentry. They considered a 2D cellular network with some link being non-conducting whose fraction was ϕ. Thus, $\phi = 0$ and 1 corresponded, respectively, to completely homogeneous tissue, and a set of disconnected myocytes. For small ϕ a wave propagates through the network, whereas for large ϕ the wave cannot propagate. Thus, as the reader knows, there is a percolation threshold ϕ_c where both behaviors interchange. Alonso and Bär (2013) carried out numerical simulations with the model in which they varied ϕ. The governing equation for propagation of action potential at node i (or a cell with its center at i) is given by

$$\frac{\partial V_i}{\partial t} = -(I_{fi} + I_{so} + I_{si}) + \sum_{j=1}^{Z} g_{ij}(V_j - V_i) , \tag{18.25}$$

where I_{fi}, I_{so} and I_{si} are, respectively, fast inward sodium, slow outward potassium, and slow inward calcium ionic currents, which are all related to the potential V_i (Fenton and Karma 1998), and g_{ij} represents the conductance between i and j, taking on a constant value if i and j are connected, and zero if i and j are not linked.

The simulations indicated that a wave that propagates in homogeneous tissues and interacts with a heterogenous region—a damaged region on the cardiac muscle—can generate reexcitations through a small-scale reentry process, which may emerge as ectopic beats—extra heartbeats that are caused by a signal to the upper chambers of the heart (the atria) from an abnormal electrical focus—or as reentries. Generating such reexcitations requires the combination of two factors inside the damaged area, i.e., a fraction of non-conducting links close to but above the percolation threshold, and a change of the local properties toward weaker excitability. In addition, the probability of reentry was shown to be strongly correlated with the probability of formation of clusters with a size larger than a typical size that emerge near ϕ_c.

In another paper, Alonso et al. (2016) carried out the same type of simulation in a 3D model, represented by a thin slab of cardiac tissue, in order to take into account the fact that atrial tissue is much thinner than ventricular one. Their simulations indicated the strong and non-trivial effect of the thickness, even for thin tissues, on the probability of micro-reentries and ectopic beat generation, and a strong correlation of the occurrence of micro-reentry with ϕ_c. There was also a qualitative agreement between the 3D simulated electrograms in the fibrotic region with the experimentally observed complex fractional atrial electrograms (CFASs).

Vigmond et al. (2016) used an approach similar to Alonso and co-workers, except that they used 2D synthetic fibrosis patterns, discretized like a square network. The pattern was made heterogeneous by a percolation algorithm, namely, by generating a random number R for each elementary block and removing it if R was less than a threshold set for the element, representing the fibrotic density. Isotropic fibrosis was generated by using a uniform threshold for the removal of the elements, whereas anisotropic fibrosis was produced by setting the threshold of removal for even rows much higher than for the odd ones and, therefore, even rows represented disruptions to lateral coupling. The simulations indicated that reentry may be obtained with a two-phase behavior (conducting and non-conducting) and depends on fibrotic density. The CFAEs were recorded above fibrotic regions and, consistent with clinical data, electrogram duration and fractionation increased with more rapid pacing.

Since the early twentieth Century, two conditions have been widely recognized for the formation of reentry (Mines,[4] 1913), namely, the existence of a critical wavelength, and the so-called *unidirectional blocks*. In the case of the former, in order for a rotation to continue, its period must be longer than what is referred to as the *refractory time* t_r, which is the period of time in which an organ cannot repeat a particular action. In other words, t_r is the amount of time that an excitable membrane requires to become ready for a second stimulus, once it returns to its resting state following an excitation. Thus, the existence of such a temporal condition suggests that the wavelength of the rotational activity must be larger than a critical value, namely, vt_r, where v_r is the wave velocity. Structural heterogeneity, such as clusters of fibrosis, provides a natural morphology around which the rotation can occur.

The second condition, the existence of unidirectional blocks, is often necessary for the initiation of reentry (Kléber and Rudy 2004). The wave is locally blocked from propagating in one direction, but can do so in the opposite direction. Morphological heterogeneity created by fibrosis contributes to the existence of unidirectional block. As pointed out above, Alonso and Bär (2012) also showed that the probability of reentry in fibrotic tissue is strongly correlated with the probability of the formation of fibrotic clusters with a size larger than the typical size.

Pashakhanloo and Panfilov (2021) introduced a model for predicting the probability of reentry formation based on the size distribution of what they referred to as *minimal functional clusters* in the tissue, which is essentially a percolation model and represents a refinement of the earlier model by Alonso and co-workers. In their model, one begins with a $L \times L$ square lattice in which fibrosis is modeled by randomly removing elements from the lattice with probability p. For any given p, the simulations are carried out using the Beeler–Reuter–Drouhard–Roberge ionic model, which is similar to Eq. (18.25). The stimulus is applied from the lattice's boundary in order to generate planar waves. Reentry was considered successful if the activation in the tissue lasted longer than a certain time, following application of the stimulus. The model successfully simulated the formation of reentry, when macroscopic

[4] **George Ralph Mines** (1886–1914) was a pioneering English cardiac electrophysiologist who made two fundamental contributions to cardiac electrophysiology. One was his proposal that an action potential wavefront could propagate in a circle, repeatedly activating the tissue, while his second contribution was the discovery of the vulnerable period of the heart—the time during the cardiac cycle when a single stimulus can induce ventricular fibrillation. On 7 November 1914, Mines was found unconscious in his laboratory, and died tragically later that day, cutting short a brilliant career.

rotational activity was initiated by a microscopic unidirectional block, such that the wave was blocked from propagating from the left to right direction, but could do so in the opposite direction.

A cluster is defined (Alonso and Bär 2013) as a set of connected inexcitable elements that form an obstacle, a purely topological characteristic of the texture that, without any conduction blocks, would solely determine the path of the propagating wave. Such blocks, in the presence of fibrosis-induced conduction blocks, form at the junction of, or close proximity to, two or more regions of fibrosis that connect the fibrotic clusters to form a larger cluster of unexcitable tissue, which is what Pashakhanloo and Panfilov referred to as the functional cluster, which is a more refined concept of fibrotic cluster, as it takes into account not only the morphology of the texture but also the properties of propagating waves, i.e., conduction blocks.

Pashakhanloo and Panfilov studied the size distribution of the functional clusters. A "correlation length" ξ was defined in a manner similar to that for percolation clusters (see Chap. 2),

$$\xi^2 = \frac{2\sum_s R_s^2 s^2 n_s}{\sum_s s^2 n_s}, \tag{18.26}$$

where n_s is the number of clusters of size s, and R_s is the radius of gyration of the cluster. ξ depends not only p but also on excitability e. The important point is that ξ for the functional clusters is not equal to that of the fibrotic clusters, and is strongly affected by excitability. Moreover, the percolation threshold at which a spanning functional or fibrotic cluster forms was found to agree with the wave percolation threshold, i.e., the probability of formation of a total wave block in the lattice. The functional clusters manifest the dependence of percolation on the excitability, such that lower excitability resulted in a lower threshold, because a larger number of blocks existed.

Dynamically, after the occurrence of a unidirectional block, the tissue behind it can be re-excited from the opposite direction, if the propagation delay, caused by wave propagation around clusters to reach the conduction block from the opposite side, is longer than the refractory period t_r. But because there are multiple paths for the wavefront to reach there, the delay should be determined by the shortest path. This gives rise (Pashakhanloo and Panfilov 2021) to a *minimal functional cluster* associated with a unidirectional block, which is constructed by disconnecting the encompassing functional cluster at the location of the block, and identifying among the resulting subclusters the one with the smallest *perimeter*. The physical significance of the cluster with minimum perimeter is that the wave must—partially or fully—travel along its perimeter in order to reach the other side of the block, and possibly cause reexcitation.

Finally, we note that Rabinovitch et al. (2021) utilized a 2D cellular automata model to simulation transport, percolation properties, and tortuosity in heart, and Falkenberg et al. (2019) used a similar, but more complex, 3D model to study mechanisms of atrial fibrillation (AF). Their simulations produced spontaneous emergence of the AF that was in agreement with the clinically observed diversity of activation.

Moreover, the simulations indicated that the difference in surface activation patterns is a direct consequence of the thickness of the discrete network of heart muscle cells in which electrical signals flow and percolate to reach the imaged surface.

18.13 Connectivity of Temporal Hierarchical Mobility Networks During COVID-19

The coronavirus disease 2019 (COVID-19), an unprecedented pandemic, affected more than 200 countries. As of the time of the final editing of this section (12 August 2022), the total number of infected people has been 585,086,861, with 6,422,914 having lost their lives (COVID-19, as reported by the World Health Organization). In order to understand how the infectious disease spreads, one must study its relation with the mobility pattern of a population. For example, some studies indicated (see, for example, Kraemer et al. 2020) that the early spatial patterns of COVID-19 infection in China correlated with the population mobility fluxes, and that the correlation decreased as some local control policies for limiting mobility were put in place.

Deng et al. (2021) demonstrated the relevance of percolation process and phase transition to human mobility networks on the county level. Their study indicated the possibility of developing effective strategies for controlling the mobility flow at critical bridges [connecting urban areas] and containing the transmission of COVID-19. To gain a deeper understanding of the percolation effect, one must construct mobility-based networks in order to accurately model the transmission of COVID-19. To do so, one must consider some key properties: (a) The networks must have proper resolution in order to account for both the spatial and temporal dynamics of mobility networks. For example, some studies aggregated human mobility on the census block group or county levels, whereas aggregation removes details and information on other levels, resulting in inaccurate prediction for the transmission patterns. (b) Some studies pointed to the effectiveness of inter-city travel restrictions for reducing the imported COVID-19 incidence rate, but one must gain a better understanding of how local measures and responses could change.

He et al. (2022) studied the percolation of the human mobility network that included over 175,000 of the census block groups in the United States. In their network model each bond between two sites, which represent the blocks, is given a weight. Two block groups have a connection (a bond between them) on any day in which people travel between them. If more people travel between them, the weight increases. To analyze the properties of the network, He et al. (2022) proposed a hierarchical structure of the mobility network to understand human dynamics during the pandemic, in which the network of inter-Metropolitan Statistical Area (MSA) was composed of 378 MSAs, with each comprising of many census block groups, which was at the intra-MSA level. To implement the percolation process in the network, He et al. (2022) deleted a fraction q of the weakest links (lowest weights) and determined the giant component of the network, since it is easier to break the weak links than the strong ones. They demonstrated the existence of a critical value q_c of q, such that when $q > q_c$, the network breaks down abruptly because the large enough *second largest component* of the network becomes disconnected from the giant cluster. The value of q_c is dynamic, and He et al. (2022) computed its daily value for each intra-MSA and inter-MSA.

The study of He et al. (2022) pointed to three crucial features of percolation in dynamic mobility networks. One is the hierarchical structure of the temporal mobility networks based on the concept of cluster formation in the classic percolation theory. The second feature is the universality of phase transition in mobility networks at q_c, a concept that by now is very familiar to the reader of this book, demonstrating that it is the percolation process that controls the universality of transitions in such networks on various levels, from block groups to inter-MSAs. The third feature is the association between critical threshold q_c and the key characteristics of local MSAs, which improves the ability for assessing and predicting the vulnerability of mobility networks.

A related problem is vaccinating people and developing a strategy for maximum efficiency so that the society can eventually reach herd immunity. One such strategy is vaccinating those that have more exposure to others, which has been shown theoretically to be highly effective. The problem with this strategy is that identifying such individuals—sometimes referred to as "hot spotting"—is difficult, hence preventing its large-scale implementation. Penney et al. (2021) proposed an approach for using the technology underlying digital contact tracing in order to boost vaccine coverage. To do so, Penney et al. (2021) modeled the spread of disease using percolation in random graphs, described in Chap. 17. They defined a quantitative measure, the efficiency, which they defined as the percentage decrease in the reproduction number per percentage of the population vaccinated. Penney et al. (2021) demonstrated that optimal implementations of their percolation model can achieve herd immunity with as little as half as many vaccine doses as a non-targeted strategy.

18.14 Protein Structure

Theoretically, there are 10^{400} possible sequences for a protein of medium size. But, the sequences of only about 10^8 proteins are currently known, whereas the number of extant sequences is estimated to be 10^{34}, and it is believed that over the past four billion years of evolution up to 10^{43} different protein sequences may have been explored. Such statistics imply that the Darwinian protein evolution, based on mutation of the genotype and subsequent natural selection of the phenotype, excludes the possibility that extant sequences are randomly scattered in the theoretical sequence space. Instead, they are expected to form a connected network in which functional sequences and mutations form the nodes and links or edges (Smith 1970). Then, the fundamental questions (Smith 1970) are whether all existing proteins are part of a single network with a single starting point, and what fraction of the functional sequence space has been explored. One is also interested to learn how large is the space of functional, but never-born, proteins.

The 10^8 known protein sequences are not equally distributed, but cluster into homologous families (Rappoport et al. 2012). Due to their sparsity (separation between the clusters), however, most clusters have neighboring nodes that differ by multiple mutations. The sparsity may be explained by the extinction of ancestor sequences (Thornton 2004). Alternatively, one may predict that as our knowledge of known sequences of proteins expands, the sparsity decreases.

Buchholz et al. (2017) used percolation theory to analyze the sequence space of known proteins. In particular, they utilized the power-law cluster-size distribution of percolation, $n_s \sim s^{-\tau}$ (see Chap. 2), to study the clusters in the network of the extant sequences. They began by updating the databases on α/β hydrolases (abH, 395,000 sequences), cytochrome P450 monooxygenases (CYP, 53,000 sequences), thiamine diphosphate-dependent decarboxylases (DC, 39,000 sequences), and β-hydroxyacid dehydrogenases/imine reductases (bHAD, 31,000 sequences) (see Buchholz et al. 2017 for references to the databases)). Representative sequences for each homologous family were selected as seed sequences, after which family databases for short-chain dehydrogenases/reductases (SDR, 141,000 sequences) and ω-transaminases (oTA, 121,000 sequences) were established based on seed sequences derived from literature. Then, sequence identities of high-scoring sequence pairs were calculated for each protein database, and sequence pairs with a distinct sequence identity cutoff were clustered.

The cluster-size distributions n_s for the six datasets were then constructed. To do so, n_s was analyzed for cluster sizes, $s = 1, 2, \ldots, 1000$ and, because for large s the data become increasingly sparse, the corresponding histogram distributions were constructed by counting the number of clusters $n_{i,j} = \sum_{s=i}^{j} n_s$ for cluster sizes $i \leq s \leq j$. Three theoretical models were then tested against the data: (a) a Gaussian distribution; (b) the exponential distribution, $n_s \sim \exp(-bs)$, where b is a fitting parameter, and (c) the power-law distribution, as predicted by percolation theory.

When the data were plotted in log-log fashion, they differed considerably. $\log n_s$ according to the Gaussian distribution increased gradually with $\log s$ and decayed rapidly for s larger than the mean value of the distribution. The exponential distribution decays rapidly, of course, for all s, whereas for the power-law distribution $\log n_s$ depends linearly on $\log s$. Figure 18.8 presents the results for the six sets of the data that follow the power-law distribution. The exponent τ was found to be between 2.4 and 3.3. As described in Chap. 17, biological networks represent structures with high dimensionality. Therefore, the lower value of 2.4 is consistent with the mean-field value, $\tau = 5/2$ (see Table 2.3). As described in Chap. 17, for scale-free networks described by the degree distribution (17.7), the exponent τ is given by, $\tau = (2\gamma - 3)/(\gamma - 2)$, where γ represents the exponent of the power law (17.7). Thus, if we use $\tau \approx 3.3$, we obtain $\gamma \approx 2.77$. In other words, the protein sequences may be represented by a scale-free network with the exponent $\gamma \approx 2.77$.

For related work see Deb (2009), Weber and Pande (2015), and Peng et al. (2015).

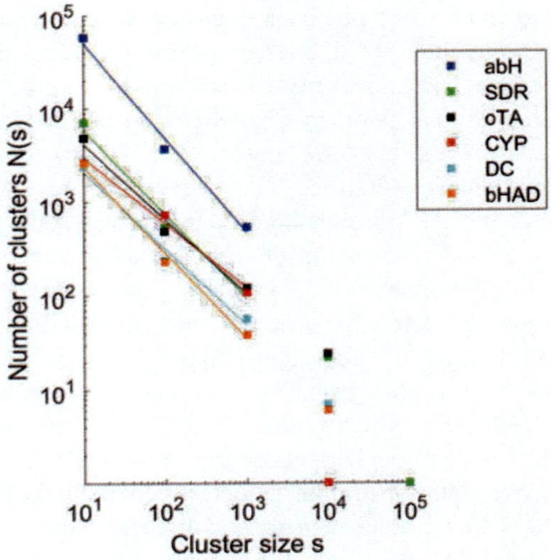

Fig. 18.8 Cluster-size distribution of α/β hydrolases (abH), short-chain dehydrogenases/reductases (SDR), ω-transaminases (oTA), cytochrome P450 monooxygenases (CYP), thiamine diphosphate-dependent decarboxylases (DC), and β-hydroxyacid dehydrogenases/imine reductases (bHAD). They follow the power-law distribution (after Buchholz et al. 2017; courtesy of Professor Jürgen Pleiss)

18.15 Molecular Motors and Mechanical Properties of Cells and Active Gels

Molecular motors in the cytoskeletal filamentous actin (F-actin) network generate internal stresses that determine the mechanical properties of cells. Some aspects of this were already described in 18.8. It is due to motor activity that the cells have the ability to contract the surrounding extracellular matrix (ECM), which is essentially a biopolymer network whose elastic properties are strongly influenced by the active contractility. This happens in both reconstituted intracellular F-actin networks with myosin motors (see, for example, Mizuno et al. 2007; Gordon et al. 2012) and in the ECM with contractile cells (Lam et al. 2011). Active polymer networks are heterogeneous and contain soft or floppy modes of deformation (see Chap. 11). Motor-induced contractile stresses are coupled to the soft modes (Broedersz and MacKintosh 2011), which produce a nonlinear elastic response, which is different from the nonlinearities arising from single fiber elasticity. In addition, the coupling casts doubts on the presumed equivalence of internal—motor—and external stress. As a result, as has been emphasized throughout this book, their properties cannot be quantitatively described by the classical linear continuum theory of elasticity.

As described in Chap. 11, one can form elastic networks by interconnected straight fibers (bonds) with linear stretching and bending elasticity. If a fraction of the fibers is removed, one has an elastic percolation model. Sheinman et al. (2012; see also Sheinman et al. 2015) proposed such a model to study the effect of motor-generated stresses in percolating fiber networks. They used an FCC network that allowed them to study the effect of the fiber connectivity over an extended range, since the coordination number of the intact FCC network is $Z = 12$. To introduce motor activity, they incorporated contractile, static, and strain-independent force dipoles between neighboring nodes of the network, inserted randomly with a probability q, while the fibers were modeled as linear elastic bonds with a stretching modulus α_1 and bending rigidity α_2 (see Chap. 11, where the two were denoted, respectively, by α and γ). Assuming that the equilibrium length of the fibers is unity, the total energy of the network is given by (see also Chap. 11)

$$\mathcal{H} = \frac{1}{2}\alpha_1 \sum_{\langle ij \rangle} e_{ij}(|\mathbf{r}_{ij}| - 1)^2 + \frac{1}{2}\alpha_2 \sum_{\langle ijk \rangle} e_{ij}e_{jk}\left(\frac{\mathbf{r}_{ij} \times \mathbf{r}_{jk}}{|\mathbf{r}_{ij}||\mathbf{r}_{jk}|}\right) + f\sum_{\langle ij \rangle} q_{ij}|\mathbf{r}_{ij}| \ . \tag{18.27}$$

Here, $\mathbf{r}_{ij} = \mathbf{u}_i - \mathbf{u}_j$, with $\mathbf{u}_i$ denoting displacement) of node i, $e_{ij} = 1$ or 0 if the fiber is, respectively, present or cut, f is the dipole rigidity, and $q_{ij} = 1$, if a motor acts between nodes i and j, or $q_{ij} = 0$, otherwise.

The model was then studied by an effective-medium approximation (EMA)—poor man's percolation—and numerical simulation. To develop the EMA, Sherman (2012) set $\alpha_2 = 0$, which reduced their network to one somewhat similar to a central-force spring network described in Chap. 11. The development of their EMA paralleled what was described in Chap. 11, but somewhat more complex, due to the additional term in the total energy $\mathcal{H}$, Eq. (18.27), representing the contribution of the motors. Their final EMA equation for the effective stretching modulus α_e is given by

$$\int_0^\infty \frac{\alpha_{ij} - \alpha_{1e}(\sigma_M)}{1/u_{\rm EM} + \alpha_{ij} - \alpha_{1e}(\sigma_M)} h(\alpha_{ij})d\alpha_{ij} = 0 \ , \tag{18.28}$$

where $u_{\rm EM}$ is the displacement of a bond in the unperturbed effective medium due to a unit force acting on a bond, α_{ij} is the stretching modulus of fiber ij, $h(\alpha_{ij})$ represents the probability density function of α_{ij}, and $\sigma_M = \sqrt{8}\,qf$ within the EMA. Thus, with a percolation distribution, $h(\alpha_{ij}) = p\delta(\alpha_{ij} - 1) + (1-p)\delta(\alpha_{ij})$, where $\alpha_{ij} = \alpha_1 e_{ij}$, one obtains the following expression for the shear modulus of the network:

$$\mu_{\rm EM} = \frac{5\sqrt{2}}{72}\alpha_{1e} + \frac{5}{6}\sigma_M \ . \tag{18.29}$$

Recall from Chap. 11 that a central-force FCC network at the percolation threshold p_c has a mean connectivity of $Z_{\rm CF} = 6$ [see Eq. (11.40)], and that for coordination numbers $Z < Z_{\rm CF}$ the network is not rigid. In the present case, however, the presence of the motors means that even below $Z_{\rm CF}$ the network is rigid with a finite effective shear modulus. Thus, one may view motor stress as an external field that stabilizes floppy networks. Numerical simulations of Sherman (2012) indicated that close to $Z_{\rm CF}$ one has the anomalous scaling,

$$\mu_e \sim \alpha_e^{1-x}\sigma_M^x \ , \tag{18.30}$$

with $x \approx 0.4$. The EMA predicts that, $x = 1/2$. If the bending rigidity is not ignored, i.e., if $\alpha_2 \neq 0$ in Eq. (27), then, Sherman (2012) showed that

$$\mu_e \sim \alpha_{2e}^{1-y} \sigma_M^y , \tag{18.31}$$

with their numerical simulations yielding $y \approx 0.6$, where α_{2e} is the effective bending rigidity.

In another paper from the same group (Alvarado et al. 2013), extensive experimental data were analyzed to show that myosin motors contract crosslinks actin polymer networks with a scale-free size distribution (see Chap. 17), and that the effect occurs over a broad range of crosslink concentrations. Their analysis indicated that the motors reduce connectivity of the networks by unbinding the crosslinks, and that, in order to coordinate global contractions, motor activity should be low. Otherwise, they force initially well-connected networks to a critical state at which rupture forms across the entire network.

To understand the interplay between motor activity and the connectivity of active cytoskeletal network, Alvarado et al. (2013) varied the density of myosin motors and fascin crosslinks, given by the molar ratios R_M and R_C, and prepared a series of active networks with constant myosin activity, $R_M = 0.01$, and gradually increased R_C. The experiments indicated that even at low R_C, the motors contract actin networks, albeit on small length scales. As R_C was increased, contraction occurred on larger length scales. Eventually, the motors break up the network into multiple disjoint clusters. At still higher R_C, motor activity contracted the entire network into a single dense cluster. The transition from local to macroscopic contraction represented a percolation transition: Below the transition point, the network is only locally correlated without long-range connectivity, whereas above a well-defined critical connectivity the system establish global correlations and corrections.

To demonstrate the connection to the percolation transition, Alvarado et al. (2013) noted that, according to percolation theory, the largest cluster of size ξ_1 (which is the percolation correlation length ξ_p; see Chap. 2) increases monotonically with p, the probability of making connections between the nodes, whereas the second-largest cluster with a size ξ_2 should exhibit a peak right at the connectivity threshold, with ξ_1 and ξ_2 both approaching the system's linear size L. The data of Alvarado et al. (2013) was in agreement with this picture; see Fig. 18.9.

Another test was carried out by Alvarado et al. (2013) to further establish the connection of the problem to percolation theory. It was based on the cluster-size distribution of percolation networks. As described in Chap. 2, close to the connectivity threshold, one has a power-law size distribution, with an exponent (in 2D) of, $\tau = 187/91 \approx 2.05$ [see Eq. (2.14) and Table 2.3]. To see whether this prediction is borne out by their data, Alvarado et al. (2013) looked for networks that satisfied the constraints, $\xi_1 \sim \xi_2 \sim L$, and replotted all the data separately in the $\xi_1 - \xi_2$-space. This is shown in Fig. 18.10a. Since by definition, $\xi_2 < \xi_1$, all samples are within a triangle in $\xi_1 - \xi_2$-space. Figure 18.10a identifies clearly the samples at the triangle's peak, where $\xi_1 \sim \xi_2 \sim L$, and represents the critically connected network. Samples to the left of the peak are for low R_C, which is identified as the local contraction

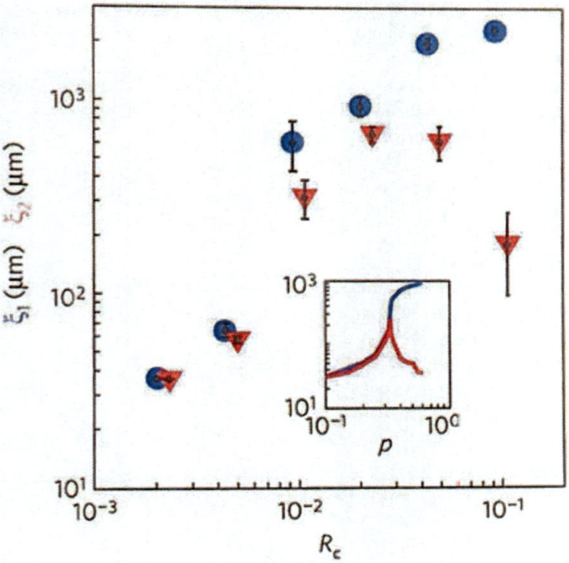

Fig. 18.9 Dependence of ξ_1, the size of the largest cluster (*blue circles*) and ξ_2, the linear dimension of the second largest cluster (*pink triangles*) on R_C. Error bars denote standard errors of the mean for repeat experiments. Inset: Predicted dependence of ξ_1 (*blue*) and ξ_2 (*pink*) on connection probability p according to percolation theory (after Alvarado et al. 2013)

regime, whereas those to the right represent samples with high R_C and therefore, the global contraction regime.

Next, the entire distribution of cluster sizes was plotted; see Fig. 18.10b. The data are again consistent with percolation theory, since the critically connected regime exhibits a power-law cluster-size distribution over more than two orders of magnitude in measured area. The power-law exponent is $\tau \approx 1.9$. Although the estimate is only about 5% smaller than the predicted value of 2.05 in 2D, τ for the percolation clusters cannot be less than 2.

If one reverses the roles of occupied and vacant sites or bonds in a percolation lattice, the problem can be thought of as finding the distribution of hole sizes within a single large cluster (Hu et al. 2016). Thus, the holes studied are not simple percolation holes, but are within the surrounding backbone. Using scaling analysis, Hu et al. (2016) suggested that, $\tau = 1 + D_{\rm BB}/2$ for the backbone holes, where $D_{\rm EB} \simeq 1.64$ is the fractal dimension of percolation backbone in 2D (see Chap. 2). Thus, one obtains, $\tau \simeq 1.82$. If, however, one considers a model of simple holes within a percolation cluster, one obtains (Hu et al. 2016), $\tau = 1 + D_f/2$, where $D_f = 91/48 \simeq 1.895$ is the fractal dimension of 2D sample-spanning percolation cluster. Therefore, one obtains $\tau \simeq 1.94$, consistent with the experimental estimate (see Fig. 18.10).

Note also that the distributions of the other two regimes also agree with percolation theory, because the local contraction regime has a short-tail distribution with a sharp cutoff. On the other hand, the global contraction regime exhibits a bimodal distribution with two well-separated length scales, one with the percolating cluster with a size $\xi_1 \sim L$, and a second one for small disjointed clusters with a typical size, $\xi_2 \ll LL$.

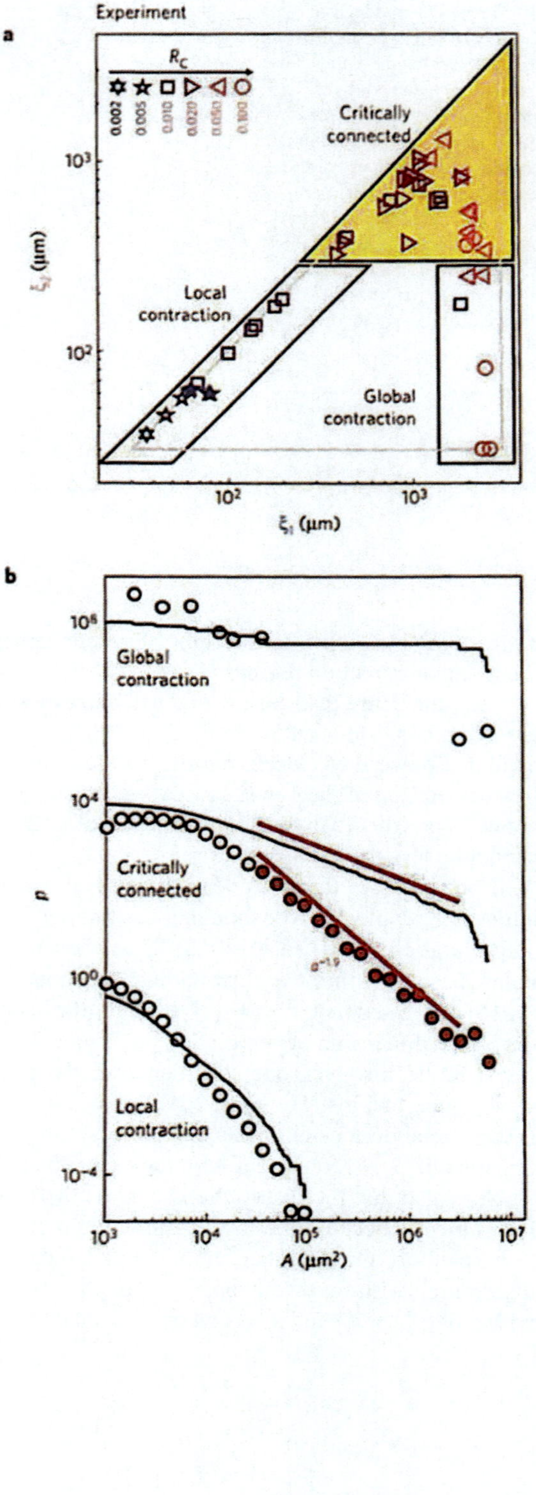

Fig. 18.10 **a** Scatter plot of 48 samples for six values of R_C in $\xi_1 - \xi_2$-space. Boxes delimit three distinct regimes: local contraction ($\xi_1 < 300$ μm), critically connected ($\xi_1 \geq 300$ μm and $\xi_2 \geq 300$ μm), and global contraction ($\xi_1 \geq 1,500$ μm and $\xi_2 < 300$ μm). **b** Histogram (*circles*) and complementary cumulative probability distribution (*solid lines*) of cluster areas A for the three regimes in (**a**). For the critically connected regime, the data (*red circles*) are consistent with a power-law distribution (*solid red lines*) with an exponent of -1.91 ± 06, $p = 0.52$. The visual form of the complementary cumulative probability distribution does not depend on bin size. The slope of the complementary cumulative probability distribution is equal to one plus the slope of the histogram, because the latter is the absolute value of the derivative of the complementary cumulative probability distribution (after Alvarado et al. 2013)

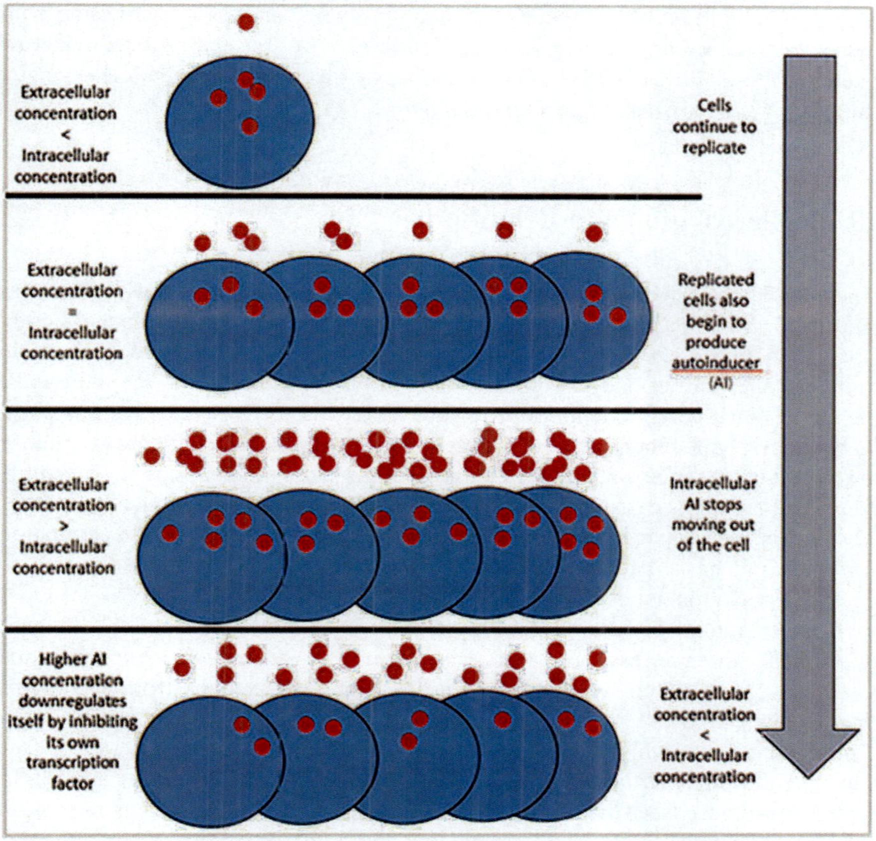

Fig. 18.11 Overview of how quorum sensing works in bacteria (from https://asm.org/Articles/2020/June/How-Quorum-Sensing-Works)

In addition, Alvarado et al. (2013) demonstrated that restructuring the network breaks it into clusters because motor stresses unbind crosslinks. Thus, one may predict that increased motor activity—increased R_M or motor force—should generate smaller clusters. The simulations and experiments of Alvarado et al. (2013) confirmed the predictions. If, however, motor forces become larger than the unbinding threshold of the crosslinks that maintain network stability, crosslink constraints may fail across the system, similar to fracture of heterogeneous networks (Sahimi and Goddard 1986; de Arcangelis et al. 1989; see Sahimi 2003b for a comprehensive review) and motors are free to slide actin filaments. Actin–myosin gels that are well-connected respond to motor sliding by undergoing a global contraction in which the meshwork collapses to a tightly packed cluster with a density higher than that of the the surrounding fluid.

For discussions of active stress in fiber networks in the context of biological functions see also Dasanayake et al. (2011), Lenz (2014), and Ronceray et al. (2016).

A comprehensive review of these fascinating problems and their relationship with percolation theory is given by Alvarado et al. (2017). The percolation model of Sheinman et al. (2012, 2015) has been studied by several groups; see, for example, Böttcher et al. (2016), and Lee and Pruessner (2016) described earlier.

18.16 Percolation and Evolution

One may describe evolution as a percolation problem in a multidimensional fitness landscape (Gavrilets and Gravner 1997; Gravner et al. 2007; Mustonen and Lassig 2009; Catalan et al. 2017; see also Katsnelson et al. 2018). In this model selectable traits or genes correspond to "dimensions," and their number is in the thousands, the model corresponds to a high-dimensional percolation. Thus, one may imagine a landscape with mountain ridges and plateaus that are partially covered with water. If the paths that go under water never continue, in the sense of being sample-spanning, whereas there are accessible paths (in the percolation sense) above the water level whose existence depends on the evolutionary temperature, or effective population size, then, one has a percolation model.

The percolation model becomes clearer if we consider the critical (sample-spanning) percolation cluster consisting of paths that follow the "shores"—the lines of minimal fitness necessary for survival. There is considerable evidence that, due to the cost of selection, actual evolution does not deviate much from such paths (Barton 1995; Bell 2013). Since the percolation model of evolution is one in a high-dimensional space, which is well represented by a Bethe lattice or Cayley tree (see Chap. 2), the implication is that the SSC—or the GC since the model is in high-dimensional space (see Chap. 17)—has a tree structure without double paths or dead ends and, as such, is an "optimal" simple line. Moreover, recall from Chap. 2 that the tree–like structure of the critical percolation cluster (in a high-dimensional space) has only one route between any two points on such a landscape.

Using such concepts, Gravner et al. (2007) studied two such percolation models of evolution. One represented a phenotype space in which individuals were characterized by a large number n of continuously varying traits. If the fitnesses are randomly assigned, viable phenotypes form a giant connected cluster that percolates throughout the phenotype space, if the viability probability is larger than $1/2^n$. In the second model, genotype-to-phenotype and phenotype-to-fitness maps were explicitly accounted for, allowing for neutrality at both phenotype and fitness levels. This generated a fitness landscape with tunable correlation length. Phenotypic neutrality and correlation between fitnesses may reduce the percolation threshold (see Chap. 3), and correlations at the phase transition point between local and global are most conducive to the formation of the giant cluster.

18.17 Microbial Communications

It was assumed for a long time that bacteria are "loners" that act on their own as organisms without any interactions with others. Accumulated experimental evidence over the years forced researchers to rethink this view. Not only do bacteria communicate with each other through signal transmission but also do so over length scales that are much larger than their own individual cell sizes. Such communications result in the formation of biolfilms and producing antibiotics, as well as pathogenesis.

Communication between bacteria is based on chemical signaling molecules that are called *autoinducers*. Such molecules regulate bacterial gene expression in a process that is referred to as *quorum sensing*. Similar to various languages, the signals also vary between species, with some bacteria being able to interpret many different signals, whereas others could respond to only a select few. It is through quorum sensing that individual bacteria within colonies coordinate and carry out colony-wide functions, such as sporulation, bioluminescence, and biofilm formation.

The way quorum sensing works is as follows (American Society for Microbiology, 2020). Individual bacterium synthesizes autoinducers. Acyl-homoserine lactone autoinducers, produced by Gram-negative bacteria, diffuse passively (no reaction) through their thin cell wall, whereas Gram-positive bacterial autoinducers that consist of peptide must be actively transported through their peptidoglycan cell wall using a transporting system called the adenosine triphosphate (ATP)-binding cassette. As the bacteria keep reproducing, more individual cells also produce autoinducers that move out of the cells, hence increasing their concentration in the extracellular matrix, eventually reaching a critical threshold for their mass that makes it energetically unfavorable for intracellular autoinducers to keep leaving the cell, thus increasing their intracellular concentration. As their concentration in intracellular increases, autoinducers bind to their receptors and trigger signaling cascades that alter transcription factor activity and, therefore, gene expression, which for many bacteria, includes downregulating autoinducer synthesis in a negative feedback loop.

For example, the Gram-negative pathogenic bacteria *Vibrio cholerae* (*V. cholerae*) uses quorum sensing for virulence during a cholera infection by building biofilms to help transport nutrients between colonies, while protecting them at the same time. It is due to their ability of forming biofilms within the host that bacteria's reproduction cycle and eventual secretion of cholera toxin are successful, one of two virulence factors that contribute to between 21,000 and 143,000 cholera deaths worldwide each year.

To transmit signals through, for example, electrical signals in the nervous system, and coordinate cellular actions at long distances, bacterial communities must overcome two major challenges. One is how to achieve reliable signal transmission to desired target sites. The second challenge is due to the fact that bacterial communities manifest significant cell-to-cell heterogeneity, which may act as obstacles for long-range signal propagation (Raj and van Oudenaarden 2008; Symmons and Raj 2016). Thus, if, for example, only a fraction of the cells contribute to signal transmission, the cell-to-cell heterogeneity could result in the propagating signal to dieing out before reaching its target (Cao et al. 1999; Alonso et al. 2016). Therefore, the question of the effect of heterogeneity on bacterial communities in long-range signal transmission is important. Then, once one talks about heterogeneity and only a fraction of the cells contributing to signal propagation, the relevance of percolation theory becomes immediately clear.

Larkin et al. (2018) developed a percolation model to describe the connectivity and clustering statistics of firing cells in the biofilm. They used a triangular lattice, since cells in their experimental biofilms had a modal value of six neighbors. Thus, the cells had six nearest neighbors to which they connect, forming triangles. If a fraction ϕ of the cells are firing, then, at the percolation threshold ϕ_c a SSC of such cells forms. This corresponds to site percolation. Larking et al. utilized a $L \times W$ lattice with $L = 30$ and $W = 200$ cells, roughly corresponding to the size of their experimental systems, with the signal propagating in the L direction. Due to the asymmetric dimensions of the lattice, they estimated that $\phi_c \approx 0.45$, about 10% less than the exact site percolation threshold of the triangular lattice, $\phi_c = 1/2$ (see Chap. 2). To describe the dynamics of signal propagation between firing cells, Larkin et al. used the FitzHugh–Nagumo model,[5] which is a slightly more complex model than what is described by Eq. (18.25).

Larking et al. also carried out experiments to compare with their simulations. The bacteria used was *Bacillus (B.) subtilis* NCIB 3610. *B. subtilis*, commonly found in soil, is a gram-positive model organism for basic research and an industrial workhorse, while *B. subtilis* NCIB 3610 is an undomesticated strain that exhibits phenotypes lost from the more common domesticated laboratory strains. In order to measure membrane potential of individual bacteria within the biofilms during electrochemical signaling, Larkin et al. (2018) used the fluorescent reporter thioflavin-T (ThT), which acts as a Nernstian membrane potential indicator, such that the higher the membrane potential of the cell, the larger is the amplitude of the fluorescent ThT signal. Measurements at the scale of a single cell in the biofilm indicated that only some cells exhibited pulses in electrical activity, with the rest not participating in signaling. This is shown in Fig. 18.12a and b. The distribution of membrane potential amplitudes developed during signal propagation was bimodal, with the fraction of signaling cells being, $\phi = 0.43 \pm 0.02$, which is in agreement with the percolation threshold obtained from their simulation.

Also measured was the spatial distribution of signaling, or firing, cells that formed clusters of various sizes; see Fig. 18.12c. The distribution of cluster sizes followed a power law with an exponent, $\tau \simeq 2.0$, in excellent agreement with 2D percolation, $\tau = 187/91 \simeq 2.05$; see Fig. 18.12d. Thus, the spatial organization of firing cells within the bacterial community may be organized near the percolation threshold.

Silva et al. (2019) took advantage of the percolation effect in order to study how to interrupt bacterial communications, both experimentally and with computer sim-

---

[5] Named after Richard FitzHugh who suggested the model (FitzHugh 1961), and Jin-Ichi Nagumo et al. (1962) who developed its equivalent circuit, the model describes a prototype of an excitable system, such as a neuron. It is an example of a relaxation oscillator because, if the external stimulus I_{ext} exceeds a threshold, the system will exhibit a characteristic excursion in phase space, before the variables relax back to their rest values.

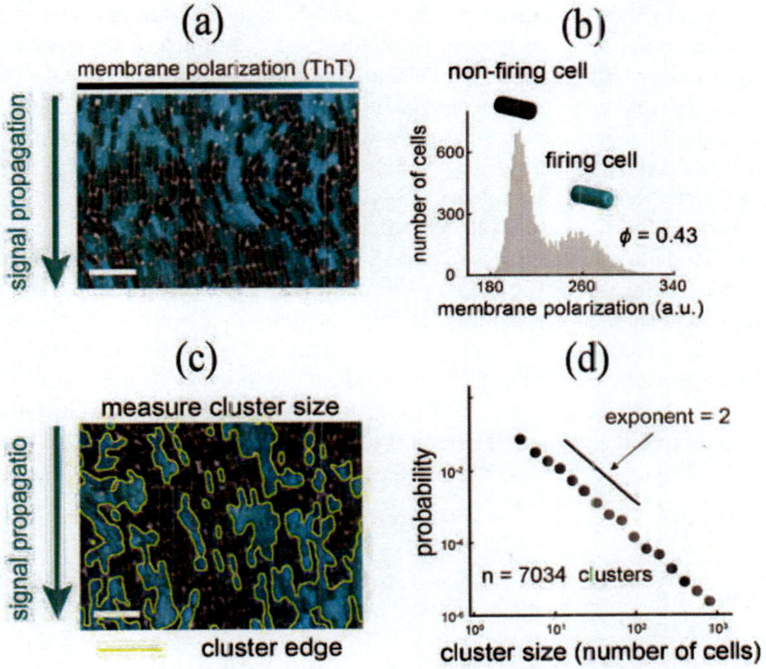

Fig. 18.12 **a** Membrane polarization is heterogeneous at the single-cell level within the biofilms, with cyan overlay indicating fluorescence of ThT. Scale bar is 10 μm. **b** Histogram of individual cell ThT intensity with 14, 936 cells, during a signal pulse. The bimodal histogram indicates that only a fraction of firing cells (*cyan*) participate in signaling. **c** Firing cells are spatially clustered within biofilms. *Yellow outlines* indicate cluster edges identified by image analysis based on ThT fluorescence. Scale bar is 10 μm. **d** The cluster-size distribution, computed for 7,034 clusters, follows a power law with an exponent of $\tau \simeq 2$ (After Larkin et al. 2018; courtesy of Professor Gürol Süel of the University of California, San Diego)

ulation. Robustness of the communication networks to interference was studied by analyzing spatial patterns of activation in a model bacterial community that consisted of a *sender cell*, which produced and responded to autoinducer (AI) signals, and a *degrader cell* that produced an enzyme capable of destroying the signal. Silva et al. used *B. subtilis* that secretes the degradative enzyme AiiA. *B. subtilis* interacts with many soil strains that use acyl-homoserine lactone (AHL)-based quorum sensing. Thus, by varying the ratio of the two strains one can study how they influence the ability of the senders to communicate over long distances.

Silva et al. (2019) mixed sender and degrader cells with a specified ratio and applied it to the top of an LB-agar plate. Cells emit signals that diffuse and spread out via cell division, but are not motile. Sender cells are a strain of *Escherichia (E.) coli* that contain the LuxRI quorum sensing system and a green fluorescent protein (GFP) reporter gene for quorum sensing activation. As a result of increasing the concentration of the AHL, the signaling molecule increased the expression of the GFP reporter gene, with the response function being nonlinear, exhibiting a sharp increase in cellular fluorescence when the amount of signal exceeded a threshold concentration. Degrader cells, on the other hand, are a strain of *E. coli* that express constitutively the lactonase enzyme AiiA, which remained inside the degrader cells and degraded AHL quorum sensing signals that diffuse into the cell. The complete details of the experiments are given by Silva et al. (2019).

Experiments by Silva et al. (2019) indicated the existence of a percolation transition that leads to a sudden breakdown in long-range communication at a critical strain ratio. The scaling exponents at the percolation transition point were found to be independent of the specific properties of individual strains. Three percolation exponents, namely, β, γ, and ν that characterize, respectively, the power-law behavior of percolation probability, mean cluster size, and the correlation length near the percolation threshold—see Eqs. (2.8), (2.11), and (2.12)—were estimated based on the experimental data. The results were, $\beta \simeq 0.17 \pm 0.06$, $\gamma \simeq 2.28 \pm 0.36$, and $\nu \simeq 1.36 \pm 0.12$, which are consistent with their exact values in 2D, namely, $\beta = 5/36 \simeq 0.139$, $\gamma = 43/18 \simeq 2.39$, and $\nu = 4/3 \simeq 1.33$.

Chapter 19
Percolation Theory, Ecology, Hydrology, and Geochemistry

19.1 Introduction

Three important scientific fields are ecology, hydrology, and geochemistry. In ecology one studies the relationships between living organisms, including humans and their physical environment. The organisms that are considered by ecologists can be at distinct and vastly disparate length scales, ranging from the individuals, population, and community, to ecosystems and biosphere. Dynamically interacting systems of organisms, the communities they make up, and the non-living components in their environment are what constitute an ecosystem in which such processes as primary production, nutrient cycling, and niche construction occur and regulate the flux of energy and matter through an environment.

Hydrology is the study of flow, distribution, and management of water on Earth. This includes the water cycle, water resources, and environmental watershed sustainability. It focuses on how water circulates throughout the Earth through various pathways and at different rates, and factors that contribute to the phenomenon. It includes *groundwater hydrology*—usually called *hydrogeology*—which quantifies groundwater flow and solute transport studied in Chaps. 7–9; *infiltration*, the process by which water enters the soil, some of which is absorbed, while the rest percolates down to the water table; *surface water flow*, which quantifies flow of surface water and solute transport in it, and *precipitation and evaporation*, both of which are important elements of water cycle.

Geochemistry is the science of the composition, structure, processes, and other physical aspects of Earth. The focus of this branch of science is on examining the distribution of chemical elements in rocks and minerals, and the movement of such elements in soil and water systems. An important part of geochemistry is *chemical weathering*, caused by chemical reactions between rain water and the mineral grains in rocks, particularly when the water is slightly acidic.

M. Sahimi, *Applications of Percolation Theory*, Applied Mathematical Sciences 213,
https://doi.org/10.1007/978-3-031-20386-2_19

19.2 Chemical Weathering

Chemical weathering, especially hydrolysis and oxidation, represents the first step in soil production, since the results of such reactions are typically new minerals, such as clays and soluble salts. The processes associated with chemical weathering cannot occur without water and take place more rapidly at higher temperatures. Thus, warm and damp climates provide the best conditions for such reactions.

In general, there are three types of chemical weathering: (a) carbonation or solution, which removes rock in solution by acidic rain water. An important example is weathering of limestone by rain water that contains dissolved CO_2, a process that is sometimes called carbonation. (b) Oxidation, by which rock is broken down by oxygen and water. For example, if rock is iron rich, the process often gives it a rusty-colored weathered surface. (c) Hydrolysis, the process of breakdown of rock by acidic water to produce clay and soluble salts, as mentioned above.

Although, at first glance, the three aforementioned scientific disciplines may seem to have very little to do with percolation theory, a deeper study of them reveals that percolation is certainly relevant to describing the phenomena that the scientists of three fields study. In Chap. 4, we learned how to characterize porous media, and in particular rock, where many ecological, hydrological, and geochemical processes occur. Part of water cycle is through soil and rock, and the presence of fractures in rock and flow of water in highly heterogeneous porous formations create complexities in the cycle; we learned this aspect in Chap. 5. Solute transport and its relation with percolation theory was studied in Chap. 8. Thus, we are already familiar with many aspects of the relevance of percolation theory to the three aforementioned scientific disciplines. What this chapter intends to do is focusing on the intersection of the three disciplines, and demonstrate how percolation theory helps one to gain a much deeper understanding of the phenomena involved. A good review of the subject has been given by Hunt et al. (2021c), and parts of this chapter represent an updated and rewritten version of the review.[1]

As the discussions so far should have made it clear, water plays a fundamental role in all the three scientific fields. In particular, water cycle is perhaps the most important to life, food production, vegetation, and a healthy environment. Thus, we begin the discussion of this chapter with water cycle and how it relates to percolation theory.

19.3 The Water Cycle

Water in rivers, seas, oceans, and other large aqueous bodies vaporizes and enters the atmosphere, but returns to Earth through precipitation. Some of the water penetrates soil and other porous formations and flows as groundwater, but is eventually extracted and brought back to the surface. This completes the water cycle. The partitioning

[1] I would like to thank Professor Allen Hunt of Wright State University for his invaluable help.

of water into various systems has some complications, however, since some water falls directly back into the oceans and seas before reaching the continents, and the water that rains onto the continents may be recycled through the atmosphere, most notably in tropical rainforests where it can evaporate and fall back as precipitation many times before reaching the sea. Such a cycle is of key importance to climate change, understanding, and management of which is vital to life on Earth.

The continental water cycle is also linked fundamentally to the carbon cycle through various venues. One route is through water-limited chemical weathering of silicate minerals, the surface portion of the rock cycle, which is responsible for sequestering over 99% of Earth's carbon pool in the form of carbonate minerals, such as limestone. The second venue is plant growth, which is responsible for fossil energy resources, such as coal and oil, as well as the food supply.

It has been shown that both aspects of the water cycle—chemical weathering and plant growth—are best described by spatiotemporal scaling relationships developed based on percolation theory. Moreover, the proportionality of rates of plant growth and mineral weathering to the associated water fluxes links profoundly the carbon and water cycles. The link makes it possible to predict the basic partitioning of water at the Earth's terrestrial surface, known as the water balance, in terms of the critical exponents of percolation theory, by applying an ecological premise according to which ecosystems that dominate are optimally successful at drawing down atmospheric carbon in the process of reproducing themselves.

Therefore, the carbon cycle is linked critically with the rock cycle, since long-term silicate weathering rates and the formation of carbonate minerals tend to balance the outgassing of carbon dioxide from volcanoes, although small changes in the rates over geologic time scales are associated with major changes in Earth's past climate, and may be, for example, responsible for the most profound extinction event in the geologic record, the Permian–Triassic Extinction event.[2]

Thus, let us consider the water cycle in greater detail. Precipitation on the terrestrial Earth may have various potential fates. If it is in the liquid state, it can evaporate directly from the surface or vegetation canopy; it can run off along the surface, or it can penetrate the surface. The water that penetrates the surface is used by plants and transpired, or can continue flowing slowly downward to the water table and wind up in streams. If it arrives as snow, it can melt and then follow in any of the same pathways. Otherwise, it accumulates or sublimes directly to the atmosphere. The combination of evaporation and transpiration is called *evapotranspiration* (ET, denoted by E_T). Surface and subsurface run-off together are simply called run-off, Q. If the change ΔS in subsurface water storage S is negligible, then $Q + E_T = P$, where P is the total precipitation. If climate does not change and human beings do

[2] Also known as the *End-Permian Extinction*, the event formed the boundary between the Permian and Triassic geologic periods, and occurred approximately 252 million years ago. It is considered to be the Earth's most severe known extinction event. 57% of biological families, 83% of genera, 81% of marine species, and 70% of terrestrial vertebrate species became extinct. It is also the largest known mass extinction of insects. The consensus is that the causes of extinction were (a) elevated temperatures and (b) widespread oceanic anoxia and ocean acidification due to the large amounts of CO_2 that were emitted by the eruption of the Siberian Traps.

not mine water from the subsurface, then ΔS approaches zero over sufficiently long times. Since water storage is the key to agriculture, stream flow, forest health, and many other aspects of life on Earth, the continued extraction of water makes full understanding of the interrelationship of aforementioned water fluxes and resources an imperative goal.

The water taken up by the plants contributes to carbon capture in plant growth and reproduction, as well as mineral and nutrient adsorption. The water that penetrates the soil to reach the water table facilitates weathering of bedrock to produce soil and, in the process, draws carbon from the atmosphere and sequesters it in the soil and, ultimately, in sediments in the oceans' floors. People compete along such paths for water and the products that result from carbon capture.

The soil and the bedrock may be represented by random networks, and plants as directed graphs or trees. As water flows along such networks, diffusion and advection in the flowing water of dissolved solids and gases through the soil and the plants represent transport processes. Growth of plants is limited by their ability to access water and nutrients through the soil, and chemical weathering is limited by the ability of flowing water to transport the products of the weathering away from the reaction front. Even the agricultural products that reach human transportation networks may be limited by interruption of the supply chain, limiting them in reaching their destinations.

The system that we have described is called the *food–energy–water nexus* (FEW nexus), as solar energy is converted to bioenergy at its center. The nexus links the global water and carbon cycles, both of fundamental importance to human needs, and of central significance in the geologic evolution of Earth. It is then clear that the properties of transport processes in the aforementioned networks, which are fundamentally heterogeneous, are relevant to all aspects of the FEW nexus. Research over the past 40 years has demonstrated that determining such properties is perhaps most accurate when the concepts of percolation theory are involved. In some cases, an effective-medium approximation—poor man's percolation—is more straightforward to use. We now describe the progress that has been made and we are aware of.

19.4 Scaling Theory of Low-Temperature Geochemistry

Geochemical reactions near the surface or within water bodies occur at relatively low temperatures. The primary interests are the closely related processes of soil formation and chemical weathering, including the contribution by acids, released through the metabolism or decay of plants. Nevertheless, the results that we describe here have a wider range of applicability. To further complicate the discussion, the primary lines of investigation into chemical weathering have utilized continuum modeling that is based on principles of mass, momentum, and energy conservation. In particular, in the field of geochemistry, modeling surface reactions in partially or fully saturated media almost always employs some version of the convective–diffusion equation (CDE; see Chap. 8), a partial differential equation that governs the change in solute concentration as a result of molecular diffusion and a nonuniform flow field, subject to a concentration gradient. Although use of the CDE for atmospheric applications is appropriate, its application to phenomena that occur in porous media is problematic, and its use is motivated mainly by convention and convenience. The main issue is the flow and transport coefficients that appear in the equation. Unless they are evaluated accurately by taking into account the heterogeneity of the system in which the phenomenon occurs, such equations not particularly suited to complex networks representing soil, rock, and plants.

The parameters of the CDE, particularly for use over large length scale, are known to have non-trivial scale and velocity dependence. The implications of not including such scale-dependent coefficients go well beyond the difficulties of making accurate predictions, to the tendency to misinterpret the relevance of convection and diffusion to surface reactions in geomedia. Other consequences include attributing the slow-down in such reactions over time to changing conditions on the surface, even though the dependence of the reaction rates on both time and flow rate appears to be rather universal across mineral composition. The most important problems with the use of the CDE are in its difficulty in predicting solute transport over length scales much larger than a single pore, as well as a relatively homogeneous medium, and time scales that exceed typical laboratory measurements. The discrepancies between the predictions and experiment are, therefore, related to the difficulty in treating, within a continuum framework, the scale dependence of the effects of heterogeneity and connectivity. It is the effect of such factors that necessitate invoking the concepts of percolation theory.

As described in Chap. 8, the CDE relates the temporal/spatial changes in solute concentration C to the flow field and effective diffusion coefficients D_L and D_T representing, respectively, the dispersion coefficients in the longitudinal direction—parallel to the direction of macroscopic flow—and the transverse direction, perpendicular to the flow direction:

$$\frac{\partial C}{\partial t} + \mathbf{v} \cdot \nabla C = D_L \frac{\partial^2 C}{\partial x^2} + D_T \nabla_T^2 C \ , \tag{19.1}$$

where x is the direction of the macroscopic flow and ∇_T^2 denotes the Laplacian in the transverse directions. Although the dispersion coefficients are roughly proportional to the fluid's velocity (see Chap. 8), their proportionality to the scale of the experiment over 10 orders of magnitude variations in the length scale, from microns to 10 km, is not predicted or supported by the CDE, as the scale dependence of the coefficients requires taking into account the medium morphological characteristics.

More importantly, however, the solution of Eq. (19.1) generates Gaussian concentration fields, superimposed on a linear time dependence of the position of the centroid of the solute distribution. Depending on which term of Eq. (19.1) dominates, the chief effect on solute transport can be either diffusion or convection. Thus, a dimensionless number, the Péclet number Pe, which is the ratio of a microscopic diffusion time t_d and a microscopic convection time t_c, is introduced (see Chap. 8). Pe is large when convection is rapid and diffusion is slow, and small under the opposite conditions. Calculating Pe is simple: given a pore scale ℓ and a typical fluid velocity, such as the mean flow v_m, Pe is given by $\mathrm{Pe} = t_d/t_c = (\ell^2/D)/(\ell/v_m) = \ell v_m/D$, where D is the molecular diffusivity. Although many aspects of such a calculation is subject to debate, including whether the ℓ and v_m should be considered scale dependent and/or time dependent, or whether ℓ should be different for convection and diffusion, such potential complications are not addressed further here. If the velocity is large, convective effect is much stronger than diffusion, and the solute distribution will spread minimally *transversely* and is carried along with the flow, such that a characteristic solute transport distance is proportional to the time. If, on the other hand, the fluid velocity is small, the effective solute spreading is due to diffusion that diminishes with increasing spread, but there will be no dependence on the flow velocity. The conclusion is, therefore, that one can generate either a proportionality of solute transport rate to fluid flow, or its proportionality with a diminishing power of the time, but not both *simultaneously* (see also Chap. 8).

Experimental data for chemical weathering and soil formation, however, indicate both a time dependence similar to diffusion and proportionality to flow velocity. Although power laws are prevalent in spatial ecological systems (see, for example, Pascual and Guichard 2005), such experimental data are perhaps best understood within the context of convective solute transport as described by percolation theory and its power laws near the percolation threshold, first pointed out by Sahimi et al. (1982, 1983a). In particular, the three-dimensional (3D) scaling relationship for the mean distance x for transport of a solute in terms of the time t is given by

$$x \sim t^{1/D_{\mathrm{bb}}}, \tag{19.2}$$

with D_{bb} being the fractal dimension of the percolation backbone (see Chap. 2). One then has $v = dx/dt \sim t^{1/D_{\mathrm{bb}}-1}$. The numerical value of $1/D_{\mathrm{bb}} - 1$ in 3D for saturated conditions, corresponding to random percolation, is -0.47 (with $D_{\mathrm{bb}} \simeq 1.87$ in 3D; see Table 2.3). In order to make proportionality in Eq. (19.2) an equality, one requires two parameters, namely, a spatial scale x_0, such as a typical pore length ℓ, or the median particle size, and a time scale, $t_0 = x_0/v_0$, where v_0 is a typical pore fluid velocity, such as v_m. Then,

$$x = x_0 \left(\frac{t}{t_0}\right)^{1/D_{\mathrm{bb}}}. \tag{19.3}$$

Hunt et al. (2021a) showed that, for chemical weathering limited by the rate of solute transport, the solute velocity can be transformed directly to the reaction rate by multiplying it by a mean molar density of the reaction products. The proportionality of the two, while not universal, will nonetheless generate very similar reaction rates for a wide variety of minerals under the conditions that reactions are transport limited. The relevance of dx/dt to solute transport velocities implies that Eq. (19.3) itself should be applicable for predicting the weathered or soil depth. Since the two quantities are usually employed to mean different depths, Eq. (19.3) can at most give a well-defined equality for one of the two quantities.

It should be pointed out that the long tails in the distribution of the solute arrival times, which is often seen in the experimental data and are associated with the retardation of the solute transport, can be destroyed by diffusion, since a solute can diffuse into and out of the more tortuous paths. Taking into account only the second possibility, Hunt and Skinner (2010) showed that, for any given value of Pe, there exists a length scale beyond which the long-tailed distributions disappear, and that this important crossover point occurs at smaller scales with decreasing Pe. While a solute can also diffuse into such paths, the implied probability that the solute concentration is higher on such paths, because it has advanced more rapidly along such paths, will tend to accentuate the effects of diffusion in destroying the long tails of the distribution, justifying the given approach to lowest order.

White and Brantley (2003) demonstrated that field-scale chemical weathering rates for each of four distinct silicate minerals decline over time according to the same power law, with an exponent near -0.6, for periods of under a year to 6 million years. In a 2-year-long laboratory experiment, the weathering rate of one of the four minerals—plagioclase—decayed according to a power law with an exponent, -0.47. Similarly, calculation of the velocity of solute transport based on the centroid of its spatial distribution yields, over 11 decades of time, an approximate power law with exponent -0.6, although both at shorter time periods and the longer ones, the fundamental scaling of the velocity with the aforementioned exponent of -0.47 is obtained. Thus, the value of the exponent between -0.47 and -0.6 is in agreement with both laboratory- and field-derived weathering rates (i.e., -0.47 and -0.6, respectively) over time periods from minutes up to 6 million years. Others have reported that weathering rates are linear in the water flow rate or, equivalently, inversely proportional to a water residence time, both in the laboratory and in the field, extending over 10 orders of magnitude variations in flow speeds. We may, therefore, conclude that Eq. (19.3) has predictive power, not only for solute transport, but also for transport-limited surface reaction rates in porous media, including chemical weathering. Thus, the relevant question is, when should one expect surface reaction rates in porous media to be transport limited?

The answer to the question is formulated by comparing the rate of reaction when fresh surfaces are exposed and reaction products are removed instantaneously, with the rate by which convective solute transport removes reaction products when the reaction is effectively instantaneous. The reaction time is given in terms of reaction volume per unit area, with a naïve linear dependence on a length scale. One must also determine the appropriate input surface area over which the reaction occurs, as direct data are seldom available on the appropriate surface area in natural porous media, and how the area may have evolved over time, which could be over time scales up to 100 million years. This leads to another dimensionless group in terms of the ratio of a convective and reaction time scale, known as the Damköhler number, Da.[3] When the convection time is much larger than the reaction time, the reaction is transport limited; otherwise, it is limited by the kinetics of the reaction.

[3] Named after **Gerhard Damköhler** (1908–1944), a German chemist and the Director of Chemical Reaction Engineering at the Institute of Physical Chemistry in Göttingen, Germany

It is here that the advantage of gaining understanding of solute convection based on percolation theory manifests itself: solute convection time increases more rapidly with distance than linearly, making Da scale dependent, with a tendency for transport limitations to dominate as the scale of the experiment, natural or laboratory, increases. Thus, at short time scales, reaction rates vary widely with the mineral types and temperature, but over long time scales the temperature dependence disappears and universal time and fluid flow rate are set. Expressing Da in terms of the solute velocity, a relevant length scale, and the initial reaction kinetics makes it, in principle, possible to formulate the reaction rate in terms of its initial value and Da.

Note that if it happens that diffusion is a more effective means for removing reaction products, one can substitute the convective time scale with x^2/D, generating a slightly different expression for Da. Since the ratio of the two expressions for Da is just the Péclet number Pe, one can address the relevance of the two mechanisms by examining Pe. In that case, however, the proposed physical mechanism is distinct from what is described here.

The expression for estimating soil depth is often expressed in an abstract way, namely, that the soil depth z is a function of climate (cl), organisms (o), relief (r), parent material (p), and time (t), i.e., $z = f(cl, o, r, p, t)$. In the early times of studying the problem, no details were even suggested as to how such a functional dependence can be evaluated and established. Only recently have some possibilities for a specific functional form begun to be proposed, and even then its testing is somewhat limited.

Recall the earlier assertion that the fundamental length scale of the pore network should be given by the median particle size, which clearly represents a clear role for the parent material. A second role exists, however, because under near-surface conditions the reaction kinetics of carbonate minerals is about five orders of magnitude larger than that of silicate minerals and, therefore, the importance of solute transport limitations extends down to much smaller length and time scales for carbonates than for silicates. The role of chemical weathering, which is plant and climate driven, in this discussion is to fix the functional form of Eq. (19.3). Next, consider the flux of organic and carbonic acids into and through the soil. Such plant and atmospheric reagents are transported by water into the soil and, therefore, water flow rate is also important to soil depth, and is given by the difference between precipitation and absolute evapotranspiration, divided by the soil porosity ϕ, in order to convert to a pore velocity.

Thus, the question is, do plants and climate have two distinct effects on soil depth? Time is explicit in Eq. (19.3). Relief becomes more important when erosional effects are significant, in which case Eq. (19.3) would not be appropriate for the soil depth. One should instead set the time derivative of Eq. (19.3), which is equal to the soil formation rate, to be equal to the erosion rate E in order to determine a steady-state soil depth. Then,

$$z = x_0[(P - E_T)/D_{\mathrm{bb}}\phi E]^{1/(D_{\mathrm{bb}}-1)} \,. \tag{19.4}$$

Note that, for 3D media, the exponent in Eq. (19.4) is about 1.15, which, as shown below, has consequences in other contexts as well. Thus, the erosion rate E is represented explicitly. If z is known for a given slope, then the soil depth can be expressed as a function of the slope angle. Equation (19.4) is appropriate in regions with more active tectonics, whereas Eq. (19.3) is more suitable for more stable continental interiors, since there erosion rates are typically smaller.

For a range of conditions, however, neither of the two limits represented by Eqs. (19.3) and (19.4) may be operative, as erosion is significant, but not dominant. Then, by differentiating Eq. (19.3) one determines the rate of soil production, expressed as a function of soil depth and, therefore, the change in the soil depth is given by

$$\frac{dz}{dt} = \left(\frac{1}{D_{bb}}\right)\left(\frac{x_0}{t_0}\right)\left(\frac{x_0}{x}\right)^{D_{bb}-1} - E\,. \tag{19.5}$$

Equation (19.4) is obtained by setting $dz/dt = 0$. Otherwise, Eq. (19.5) may be integrated numerically in order to obtain the soil depth z as a function of time. For early times, Eq. (19.3) is recovered, whereas in the limit of large time, Eq. (19.4) is obtained. Thus, the general soil depth interpolates continuously from Eq. (19.3) to (19.4) as the time scale increases.

The validity of the proposed scaling relationships has been assessed in a variety of studies, some of which have been direct comparisons of the predictions of Eq. (19.3) with experimental data for surface reaction rates, while many other cases have been related to predicting soil depth and soil production. That the validity of the predictions can be tested by comparing them with experimental data for soil production is due to the fact that soil production is ordinarily limited by chemical weathering.

To use Eq. (19.3) to make predictions, only two parameters are needed, namely, a fundamental length scale, which we take it to be the median grain size d_{50}, and a time scale, which is clearly d_{50}/v_0, where v_0 can be taken to be $(P - E_T)/\phi$. To use Eq. (19.4) or (19.5), an additional parameter E is required. If we use the experimental values of the three parameters, Eq. (19.5) can be used with no adjustable parameters to predict soil depths as a function of time, for up to millions of year in time scale, in the world's largest suitable database (Egli et al. 2018), and for a range of additional soils in China, California, and Peru (Yu and Hunt 2017a), soil depths as a function of slope angle (Yu et al. 2019), soil production rates as a function of time and space, and of climate (Hunt and Ghanbarian 2016).

Yu and Hunt (2017b) showed that a typical value of the particle size, about 30 μm, and a typical precipitation value of approximately 1 m/yr provide predictions with high accuracy to within 10% deviation from the data for depth of deep tropical weathering layers over time scales up to 50 million years. Other data that related weathering depths of as much as 250 m to time scales of up to 150 million years were also shown to conform with the predictions of Eq. (19.5), implying that this equation can forecast the depth of the critical zone on all time scales between the break-up of the supercontinent Pangaea and the present.

19.5 Weathering Rinds

Weathering rinds describe the depths of the weathered layer of rocks, located at or near the surface. One of their original uses was for dating surface deposits, such as alluvium or lava flows. Their thicknesses are typically measured in millimeters or centimeters, rather than meters, even over time scales up to millions of years. Interestingly, however, their thicknesses do not usually increase with time as a power law with an exponent of 0.53, but much more often with an exponent close to 0.8. Both exponents can, however, be explained within the framework of percolation theory.

Weathering requires water, which does not penetrate rock as easily as in unconsolidated media. Instead, microfractures, common to most surfaces except for some volcanic glasses, such as obsidian for which diffusion does indeed appear to be the dominant process that triggers weathering, help deliver water to the interior of the rocks, known as *clasts*. But the flux of water that enters, for example, a buried clast, is roughly equal to the value of $(P - E_T)$ multiplied by the ratio of the hydraulic conductivity of the rock and the soil. Since the near-surface clasts are the products of the breaking off of pieces of rock by fracturing, their chief remaining fractures are of sizes much smaller than the original ones, and their hydraulic conductivities are usually orders of magnitude smaller than the surrounding soil. Thus, the time scale that enters into the scaling relationship is much larger than what determines soil formation. Since, however, the objects near the surface are ordinarily well above the water table, the water that enters usually occurs under wetting conditions associated with rainfalls. Thus, the appropriate values of the percolation exponents that enter the calculation of solute transport are those of invasion percolation with trapping (TIP) under wetting conditions for 2D microfractures, which, as described in Chaps. 9 and 16, is (see Table 16.1) $D_{\rm bb} \simeq 1.217$, and, therefore, the exponent that appears in Eq. (19.2) or (19.3) is $1/D_{\rm bb} \simeq 0.82$. Most of the experimental data have also reported the value of the exponent to be very close to 0.82.

One well-known exception to the above discussion is in weathering of pillow lava clasts, emplaced below sea level, which proceeds under saturated conditions with a time dependence more nearly in accord with Eq. (19.3), but with $D_{\rm bb} \simeq 1.87$. Exclusion of the potential role of diffusion is, however, also more difficult in that case.

19.6 Vegetation Growth

The ability for predicting the growth rates of vegetation may aid in predicting characteristics of carbon cycle. The most comprehensive description of vegetation growth is within the framework of *allometric scaling* (West et al. 1997; West and Brown 2005). Allometric scaling relations are characteristic of all organisms. They have been observed in numerous experiments, and have also been derived based on general models that describe how essential materials are transported through space-filling fractal networks of branching tubes. The model assumes that the energy dissipated is minimized, and that the terminal tubes do not vary with body size. We summarize some important results of allometric scaling that are connected to the percolation scaling described below.

The single most important prediction of allometric scaling is that the metabolism of plants is proportional to the plant mass to the power of 3/4. Other important related predictions include tree height being proportional to the trunk diameter to the 2/3 power, and height growth rates being constant in time. Even as an approximation, the prediction that the growth rates of tree height should be constant at young stand ages is clearly inaccurate.

In fact, it is well known that the rate of growth of tree height increases with time after germination, since trees use the majority of their resources for the initial development of a root system, rather than utilizing them for growth of their heights. However, trees typically do achieve their greatest height growth rates during the first growing season, if for no reason other than the fact that they have not experienced a period of little to no growth yet. Thus, there is a rapid drop in growth rates toward the end of the growing season. Moreover, at greater ages, which is typically when the canopy height is reached, tree growth rates decrease markedly. Neglecting the initial allocation to the roots still leaves a sigmoidal-shaped growth curve, a result familiar from site index curves in forestry. What is important in the present context is, however, that even in the middle of the growth curve, typically from months to decades, the growth rate is not constant, but continuously declining. From the perspective of percolation theory, this is a result of the constraints developed in root–soil interactions.

The growth of roots in soils has been logically considered to be driven by gradients of nutrients and water needed for photosynthesis, plant function, growth, and reproduction. Nutrients are concentrated near the surface. The search for nutrients and water should generate an optimal success/cost ratio. Such a strategy can be approximately generated by an IP algorithm that was described in Chap. 16, which in 2D heterogeneous media develops optimal paths which have a fractal dimension of 1.21 (Cieplak et al. 1994; see also Chap. 16).

The model described here for tree growth assumes that the rate of root tip extension is constant through time, but that the rate at which root radial extent R_E can increase is a decreasing function of time. This particular assumption follows from the relationship between total root length ℓ_r, and R_E, which, according to the IP percolation model of optimal paths in 2D, is given by

$$\ell_r \sim R_E^{D_{\text{op}}} , \tag{19.6}$$

where D_{op} is the fractal dimension of the optimal paths in strongly heterogeneous media. Since the root length increases linearly in time, its root radial extent should increase as power law in time with an exponent that is the inverse of the fractal dimension of the optimal paths. Thus,

$$R_E = x_0 \left(\frac{t}{t_0}\right)^{1/D_{\text{op}}} . \tag{19.7}$$

It should be pointed out that the value of x_0 in Eq. (19.7) is not necessarily the same for all soils, and has been suggested to be the smaller of the median grain size d_{50} and a characteristic plant xylem diameter. Since plant xylem tends to have values between 1 and 100 μm, however, a geometric mean value of 10 μm is probably a reasonable estimate of a typical value of a xylem diameter, which is not greatly different from the characteristic particle size selected above, namely, 30 μm. In fact, a proportionality factor of 0.3 that relates pore sizes to particle sizes and was suggested by Gvirtzman and Roberts (1991) leads to nearly equivalent values for pore and xylem diameters. Since, as will be shown below, a little less than two-third of precipitation P may return to the atmosphere as E_T, and a little more than one-third (more than half as large as the E_T) should penetrate the soil for both cases of soil formation and vegetation growth, the fundamental time and space scales should be very nearly the same. Consequently, on a logarithmic plot of x and t, the predicted values of R_E and soil depth form straight lines of distinct slopes, radiating (nearly) from a common point. Experimental data conform to this prediction, a strong argument in favor of underlying convective mechanism for both processes.

Although two independent scale parameters, namely, x_0 and t_0 are required in Eq. (19.7), in practice, they may be combined into one. The ratio x_0/t_0 is a pore-scale velocity, but the ratio $x_0/t_0^{0.83}$, while originally defined in terms of pore-scale quantities may be equally expressed in terms of seasonal fluxes. Thus, one has equivalently,

$$R_E = T_g \left(\frac{t}{t_g}\right)^{1/D_{\text{bb}}} , \tag{19.8}$$

provided that, $T_g/t_g^{1/D_{\text{bb}}} = x_0/t_0^{1/D_{\text{bb}}}$, at some value of t. Here, T_g is the growing season transpiration, while t_g represents the growing season length. The same x_0 and t_0 that bound soil depth as a function of age −30 μm and t_0 estimated from pore-scale flow rates ranging from 0.025 to 25 m/yr—also provide bounds on vegetation height as a function of time. With the aid of the Biometric and Allometric Database (BAAD), Hunt et al. (2017) showed that for a growing season of 6 months, the observed maximum and minimum values of ecosystem transpiration, namely, 1.65 and 0.025 m, bound the existing data of woody plant height equally accurately.

Hunt and Manzoni (2016) discussed a large number of trees and some other plant species that conform to the above growth model. Since then, it has also been possible to show that the proportionality of R_E to T_g explains the observed intra-specific variation of tree height with slope curvature, but minimal dependence on slope gradient, the substrate hydraulic characteristics, soil chemistry, plant hydraulic conductivity, and climate.

Note that Hunt and Manzoni (2016) also showed that the sublinear time dependence of Eq. (19.7) leads to a modification of the allometric scaling relationship between tree height and tree diameter, namely,

$$h \propto d^{2/(3D_{op})} = d^{0.551} , \tag{19.9}$$

if we use the 2D estimate of $D_{op} \simeq 1.21$. Two meta-studies confirmed that Eq. (19.9) is a more accurate summary of the world's known data, including ca. 100,000 individual tree measurements with emphasis on tropical forests than the original allometric scaling prediction.

19.7 River Drainage

Hack (1957) proposed an empirical relationship between the length L of the longest streams (measured from any site to the edge of the sub-basin) and the area A of their basins, given by (see also Chap. 16)

$$L = cA^h , \tag{19.10}$$

where c is a constant, and the exponent h is about 0.6 for most basins, and varies slightly from region to region. It slightly decreases for basins larger than 8,000 mile^2 or 20,720 km^2. In addition to the catchment scales, Hack's law is also followed by unchanneled small-scale surfaces, if the morphology measured at high resolutions (Cheraghi et al. 2018). Maritan et al. (1996) showed by energy minimization of both homogeneous and heterogeneous river networks that, over a range of parameter values, there are three distinct universality classes, and determined the scaling exponents of all three, including Hack's law. Notably, their theory also predicts the fractal dimension of river sinuosities—the tendency to meander back and forth across its floodplain in an S-shaped pattern over time—in accord with the fractal dimension of D_{min} of the *minimum path* (also called *chemical path*) of random percolation (see Chap. 2) whose value in 2D is 1.13 (see Table 2.3).

In actual rivers, the relationship between Hack's law and the sinuosity is simple, since the length of the drainage, not the stream, is proportional to the square root of the drainage area. Thus, the exponent h in Hack's law is exactly half the sinuosity exponent, i.e., $h = D_{min}/2 \simeq 0.57$, in close agreement with $h \simeq 0.6$ that is widely quoted. Hu et al. (2016) interpreted this in terms of the dominant character of the substrate: a highly heterogeneous substrate requires the optimal paths fractal dimension, $D_{op} \simeq 1.21$; thus, $h = D_{op}/2 \simeq 0.61$ for the sinuosity. On the other hand, a less heterogeneous substrate should be consistent with a fractal dimension of D_{min}, i.e., 1.13 and, $h = D_{min}/2 \simeq 0.57$. Thus, with this interpretation of the observed range of Hack's law exponent h of 0.57–0.61 is very nicely mapped out by the percolation exponents.

Hunt et al. (2021a) showed that Eq. (19.7), with the same d_{50} and nearly the same range of flow rates, can map out the observed range of integrated river basin lengths as a function of time. A fit of the equation to the data yielded an exponent $1/1.19$, very close to $1/1.21 = 1/D_{op}$. In fact, the appropriate flow rates were about a factor of five larger than those that were appropriate for vertical flow in the vadose zone, and exactly in agreement with data given by the U.S. Geological Survey for regional groundwater velocity v_g, 0.3 m/yr to 100 m/yr. Thus, with L being the length of the river, one has

$$L = d_{50}\left(\frac{t}{t_g}\right)^{1/D_r}, \tag{19.11}$$

where $D_r \approx 1.19$. Thus, as proposed above, one is tempted to conjecture that the exponent D_r in Eq. (19.11) is the same as the fractal dimension D_{op} of optimal paths in 2D. However, the river already represents the optimal path according to Maritan et al. (1996; see also Rinaldoa et al. 2014 for a comprehensive review) and the above discussion of Hack's law. While this is a conceivable interpretation, it would not be analogous to the growth of roots, since it is the straight line distance covered by roots that grow more slowly than linearly in time, whereas the roots appear to grow at a constant rate.

One mechanism by which an equation similar to Eq. (19.11) can be derived is for the organization of the rivers to be a 3D structure, for which the fractal dimension of the optimal paths is $D_{\mathrm{op}} \simeq 1.43$. In that case, if the river fractal dimension of sinuosity is between 1.13 and 1.21, the optimal flow paths in the subsurface would be longer than those along the surface by an exponent that lies between $1.43/1.13 = 1.27$ and $1.43/1.21 = 1.18$. Another possibility is to argue that only the organization of subsurface flow path in the direction of the river flow should be relevant, which is a 2D configuration that would directly produce the fractal dimension of 1.21. In either interpretation, the 2D tortuosity of the surface flow paths could be considered in some sense as a projection of a greater tortuosity in the subsurface. It should also be pointed out that river tortuosities are known to diminish when they are near continental scales.

Considerable evidence has already linked processes associated with groundwater flow rates in integrating river drainages, including a range of physical characteristics, such as landscape relief, bifurcation angles, etc., providing justification for linking river integration times to groundwater flow rates. The values of regional groundwater flow rates are also in accord with the result that flow is proportional to $P - E_T$. Overall, the upper limit of observed vadose zone flow rates is closer to the middle of the observed groundwater flow rates. As seen in Fig. 19.1, this is very nearly what is observed. Note that the limit of (re)integration of the largest continental drainage systems, such as the Amazon and Yangtze River, which began its reorganizing about 60–90 million years ago, and ended within approximately the last 10 million years, connects the time scale of about 80 million years with the length scale of 7,000 km. Such limits represent, within a factor of two, the time since the break-up of Pangaea—Earth's supercontinent that consisted of all the present continents—about 150–200 million years ago, and its size, or that of Eurasia today, which is about 12,000 km.

In another investigation (Hunt et al. 2021b), actual values of current groundwater flow rates for the Mojave Desert in California, which is about 2 m/yr through the Pleistocene era, and the Sonoran Desert in Arizona, which is 1 m/yr in the San Pedro

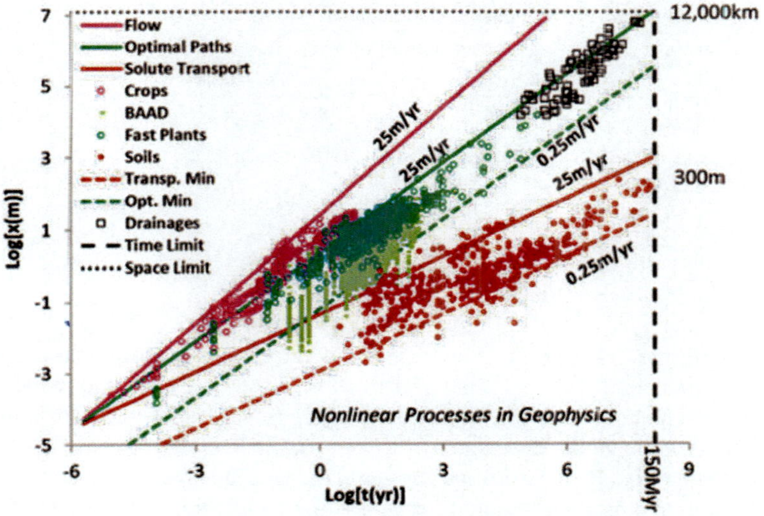

Fig. 19.1 Simultaneous test of scaling relationships between size and age in organizing 346 pairs of data for the depth of soil (brown), 1000 pairs for the linear dimension of natural vegetation and fungi (green), 1093 pairs, including Eucalyptus and other plantations, for crops (magenta), and 5538 (lighter green) pairs from the BAAD compilation. The common origin at 1 Î¼m and 1 s relates to typical fluid velocities and pore sizes of a fine-textured soil; the different slopes for the green and brown lines arise from the distinct topologies of solute transport through soils and roots, while the magenta line represents a scale-independent fluid velocity (Courtesy of Professor Allen Hunt)

River Basin, 5 m/yr along the Gila mainstem, and closer to 1 m/yr for its northern tributaries, the Verde and Salt Rivers, were used to predict the lengths of the Mojave River, since its initiation about 3.8 million years ago, and the Gila River and its tributaries since its initiation about 13 million years ago. Notably, the predictions for each case were very accurate, even though the Mojave River integration developed from its headwaters and proceeded downstream, whereas the Gila River integration began at the present-day Colorado River and proceeded upstream, though its northern tributaries, likely integrated top down, are similar to the Mojave.

The success of Eq. (19.11), when relevant flow rates from the vicinity of the drainages are used in it, motivates some further hypotheses. Consider extensional tectonic rates that may produce regional extension rates from about 0.1 to 2 cm/yr. If the rate of extension exceeds that of integration, one should expect failure of river basins to integrate. Given that river integration rates decrease with increasing length scales, one should expect a maximum river length to be associated with a given climate and extensional velocity. With increasing climatic aridity that maximum river length should decrease. Setting Eq. (19.11) for a river length equal to the product of a tectonic velocity, v_T and the time yield an expression for a time scale, t_I, which, when resubstituted into $t_I v_T$, yields a length scale L, which is the longest length river expected under the given climatic and tectonic conditions and is given by

Table 19.1 Typical values for the three important fluxes

Extension (cm/yr)	Groundwater flow (m/yr)	P (m/yr)
0.5	.5	0.05
1	1	0.1
2	2	0.2

$$L = d_{50}\left(\frac{v_g}{v_T}\right)^{1/(D_r-1)} \tag{19.12}$$

with v_g being the groundwater flow rate (flow velocity). The combination of the exponents produces $1/(D_r - 1)$, a power that exceeds 5, is a large enough value to approximate threshold behavior, i.e., either v_g is large enough to produce the drainage or there is no drainage of that length.

Suppose that Eq. (19.12) is utilized by setting $L = 1000$ km in order to determine the minimum precipitation required for generating an integrated drainage of length 1000 km. Thus, use $P - E_T P - 0.6P = 0.4P$, which is close to the global average. One also needs to relate $P - E_T$ to a pore-scale velocity, which, as discussed above, is achieved by dividing it by a typical porosity, 0.4. Then, the precipitation would be given quite simply by roughly 10% of the necessary groundwater flow rate, as shown in Table 19.1.

For western Arizona, extension rates are slightly less than 0.5 cm/yr (Brady 2002) and P is on the order of 10 cm/yr, but higher in neighboring mountain ranges. The Gila River drainage, extending from New Mexico to Colorado River on the Arizona–California border, exceeds 1000 km, although it has not yet, after 13 million years, succeeded in integrating disconnected basins along and near the continental divide near its headwaters in New Mexico. Moreover, through flow from New Mexico to Colorado River is a rare occurrence. Incorporating other surface transport phenomena, such as gravitational and aeolian processes, which tend more to fill in topographic lows than excavate them, would raise the precipitation requirements for drainage integration to some, as yet unknown, degree. The existence of such complicating factors is not in dispute as, for example, in the Sahara Desert, continental-scale paleodrainage systems can be imaged from space, and are now undetectable on the surface, having been filled in with mobile sand, or otherwise obscured.

19.8 Water Balance

It is important, particularly in the present context, that the exponents governing the power-law dependence of soil development and chemical weathering on time, as well as vegetation growth rates, are given by percolation theory. However, the dependence of soil formation on the downward water flux (infiltration) and those of vegetation growth rates R_E and tree height on the upward water flux (transpiration) are of even greater importance for the scientific disciplines of ecology, hydrology, shallow geochemistry, and geomorphology. Such dependence would have been more difficult to identify without the ability to remove the time variable from the analysis of climatic dependence. The importance of the percolation variables to the context of these separate domains is even greater when we consider the water balance.

Consider, again, Eq. (19.8). For the case that $t = t_g$, a growing season, the equation yields T_g for a given R_E, or tree height, which is very nearly the same as R_E between heights of 0.5 and 40 m. For a root system confined to grow along optimal paths in two dimensions, a reasonable assumption is that its mass is equal to its R_E raised to a power equal to the mass fractal dimensionality of 2D percolation, which is $D_f \simeq 1.9$, and thus $T_g^{D_f}$. Thus, if below-ground biomass is a relatively constant fraction of the total biomass of a plant (or ecosystem) biomass, one should expect that the net primary productivity (NPP, denoted by N_P) of vegetation should be proportional to $T_g^{D_f}$. Ecosystem levels of the NPP measure the yearly net drawdown of CO_2 from the atmosphere in producing biomass. This prediction has been confirmed to a high degree of accuracy by comparing the results of 20 experimental studies over climates, ranging from the Namibian Desert to the tropical Savannahs and rainforests. In particular, in these cases, yearly transpiration ranged from about 2 cm/yr to 1650 cm/yr, and produced a variation in N_P of about a factor of 6400, from 0.5 gC/m^{2}yr to 3500 gC/m^{2}yr.

The dependence of the NPP on transpiration does not describe all of its variability, which for any particular value of E_T can vary by as much as a factor of 20. The actual dependence of the NPP on transpiration may generate less variability, since the fraction of E_T that is used as transpiration by the plants can vary. On average, however, transpiration is nearly 70% of E_T. Studies that relate the NPP to transpiration alone are, however, very difficult to perform and evidence on this particular dependence is virtually non-existent in the literature. Since subsurface run-off is also in most cases the majority of the run-off, approximation of transpiration by E_T and subsurface run-off by Q does not involve a change in the fundamental ratio of E_T and Q from the ratio of transpiration to subsurface run-off, but can add a significant contribution to the variability of the ratio of E_T and Q.

Roots are mostly confined within the top 1 to 2 meters of soil, sometimes even as little as half a meter. Fungal and bacterial symbionts are associated with the root nutrient and water search, and receive energy (sugar) produced by the trees in return. Thus, it is logical to take the product of $T_g^{1.9}$ and the soil depth in order to express the 3D structure of the below-ground productivity. According to Eq. (19.4), however, the soil depth under steady-state conditions is proportional to $(P - E_T)$. Thus, water used to drive plant growth is not available to produce soil. By approximating the total run-off as subsurface run-off Q and substituting it for $(P - E_T)$ in Eq. (19.4), and by approximating E_T by T, one can write for the NPP $(P - Q)^{D_f}$. Then,

$$N_P \propto (P - Q)^{D_f} Q^{1/(D_{bb}-1)} \ . \tag{19.13}$$

Note that the value of the power of Q in 3D is 1.15. At this point, it becomes possible to formulate the hypothesis that the ecosystem that is most successful at reproducing itself through the process of drawing down atmospheric carbon and converting it to biomass will be one that dominates under the associated climatic condition. Such an optimization can be performed easily on Eq. (19.13) by setting $dN_P/dQ = 0$ with the result (when numerical values of D_f and $D_{\rm bb}$ are used) that $Q = 0.377P$ and $E_T = 0.623P$. Considering the complexity of Earth's terrestrial biosphere, the fact that such a simple result is within only 1.5% of the global average of E_T, whose estimates range from about $0.59P$ to $0.67P$, is truly remarkable.

19.9 Variability and Climate

On global scale, it may be permissible to approximate the biosphere as largely 2D with a thickness of about 1 m, given by the most fertile portion of soil. It may also be sufficient to ignore regional climate variability, where in some places the solar radiation is more than enough to evaporate all the water arriving as rain, but in other places there is not nearly enough solar energy to evaporate the precipitation, and most of P simply runs off. These local complications are of great interest to climate scientists and water resources stewards, as well as to agricultural planners.

We address the variability of the evapotranspiration between watersheds or drainages. If water is not transported to drainage through the subsurface, or extracted from the subsurface, or replaced in the subsurface, then one has the quantity $PE_T + Q$, as described above. The question then is what kind of limits can we impose on E_T? It is necessary in the following analysis to define the potential evapotranspiration P_E as the depth of water that the available solar energy could evaporate, if it was all used to evaporate water. Then, it is clear that E_T cannot exceed P_E. It is equally apparent that within such a closed basin, E_T cannot exceed P. Budyko[4] (1958) argued that the important limits are as follows. If $P_E < P$, actual evapotranspiration cannot exceed the potential evapotranspiration, but if $P_E > P$, the relevant limit is that $E_T < P$. When data do not follow such bounds, it must be assumed that one of the above assumptions has been violated. Most commonly, the problem is in the mining of groundwater for irrigation, which amounts to an extra source of water not measured by precipitation. In fact, such mining is common throughout the western United States, and amounts to as much as 4% of P per year throughout the Columbia River Basin.

[4] **Mikhail Ivanovich Budyko** (1920–2001) was a Soviet and Russian climatologist and one of the founders of physical climatology. His pioneering studies on global climate, calculating temperature of Earth by considering simple physical model of equilibrium in which the incoming solar radiation absorbed by the Earth's system is balanced by the energy re-radiated to space as thermal energy, were truly groundbreaking. His book (Budyko, 1956, in Russian; published in English in 1958) transformed climatology from a qualitative into a quantitative physical science. He directed the compilation of an atlas illustrating the components of the Earth's heat balance in 1963.

For $P_E > P$, the assumption was that vegetation covered only a portion of the surface equal to P/P_E, and that water arriving elsewhere evaporated. The optimization procedure that led to Eq. (19.13) was then performed for the fraction of the ground covered by vegetation. It was assumed in the same regime that the fraction of P that is equal to $(P - P_E)$ simply runs off, and the optimization procedure described that led to Eq. (19.13) could be applied to the remaining $P = P_E$.

19.10 Plants

Some plants, particularly in more arid regions, but also Eucalypts in southeastern Australia's temperate rainforests, have more nearly isotropic root systems. Such forests have higher transpiration demands, and the growth rates of the Eucalyptus species tend to decay more rapidly than the species in tropical rainforests, as well as in the temperate rainforests of the Pacific Northwest in the United States, for example. In fact, their transpiration demands and growth rates, though initially extraordinarily high, follow the prediction that uses a power law with an exponent that is the fractal dimension of 3D optimal paths (see above), $D_{\mathrm{op}} \simeq 1.46$. Thus, one should also not use the 2D mass fractal dimensionality to predict the dependence of N_P on transpiration, but, rather, the 3D value, $D_f \simeq 2.53$. Dimensional constraints prevent multiplying by the soil depth to a power higher than 0.5 in this case. Then, one can optimize the expression

$$N_P = (P - Q)^{2.5} \left(Q^{1.15}\right)^{0.5} . \quad (19.14)$$

The result is $E_T \simeq 0.813P$. In fact, one can replace the power 2.53 by the appropriate fractal dimension of an observed root system and use observed variability in plant root fractal dimensionality to predict E_T. The largest study available gives fractal dimensionalities for a collection of 55 species of plants common to the Great Plains ecosystem of the United States, 38 forbs and 17 grasses. The grasses make up the dominant biomass. The given fractal dimensionalities of the forbs conformed to 2.5; those of the grasses fell into two groups, one with fractal dimensionality 1.9 and the second with fractal dimensionality 2.65. The three values are still very close to those predicted by percolation theory, but it is also possible to obtain some estimates of uncertainty in the root fractal dimension by addressing the variability in D_f among individual species.

19.11 Improving Production of Plants with Plague Susceptibility

Propagation of a disease in plants and vegetables is an important problem, not only from a scientific perspective, but also for its social and economical consequences.

Plagues of insects or gastropods, on one hand, and the spread of diseases caused by bacteria, fungi, and oomycetes, on the other, pose a great threat to the production of fruits and vegetables, since they can reduce their production, and even destroy them completely. They can also transmit the disease to other plantations that share the irrigation system. Several models of disease propagation in plants have been proposed (see, for example, Shaw 1994; Bailey et al. 2000; Mundt and Sackett 2012).

One such model, based on percolation, was proposed by Ramírez et al. (2018) in which the sites in a lattice represent two different types of plants, say A and B, which are growing on a given type of soil. Each tree type has a particular pathogen susceptibility, defined as the probability of being infected by a specific pathogen, which we denote by χ_A and χ_B for the two types. Next, consider a percolation lattice in which a fraction p of the sites are occupied by the two types of tree. Suppose that p_A is the probability that a site is occupied by tree A, so that $p_B = 1 - p_A$ is the probability that an available site is occupied by a plant of variety B.

The average number of sites available for spreading a disease is $N_d = \langle N_A \rangle + \langle N_B \rangle$, with, for example, $\langle N_A \rangle$ being the average number of susceptible sites of type A. Thus, $\langle N_A \rangle = N p_A \chi_A p$, and $\langle N_B \rangle = N(1 - p_A)\chi_B p$, where N is the total number of sites. One must also take into account the fraction of sites that are inoculated with the pathogen, which are located in the occupied sites or in resistant plants. Suppose that p_I is the probability that a site in the lattice is inoculated. Then, the average number $\mathcal{N}$ of sites through which a disease can propagate is given by

$$\mathcal{N} = N p_e = N_d + (N - N_d) p_I \ , \tag{19.15}$$

where the second term on the right side accounts for the inoculated sites that conform to the situation described above. Equation (19.15) implies that one has a percolation process on the lattice with an effective probability p_e given by

$$p_e = p_I + (1 - p_I)[p_A \chi_A + (1 - p_A)\chi_B] p \ . \tag{19.16}$$

The effective percolation threshold p_c^e of the process is obtained from Eq. (19.16) by setting $p_e = p_c$ and $p = p_c^e$, which yields

$$p_c^e = \frac{p_c - p_I}{(1 - p_I)[p_A \chi_A + (1 - p_A)\chi_B]} \ , \tag{19.17}$$

where p_c is the percolation threshold of the original lattice.

As discussed by Ramírez et al. (2018), one must also consider the extent of disease incidence on the sown plants, since the pathogen can spread on plants that are susceptible and belong to the same cluster. Some pathogens may exhibit latency stages when they are in an adverse environment and, thus, any point in the lattice can be the source of infection, including a site with no plant or one with a plant that resists the pathogen. Therefore, if the initial point of infection is an empty site or one occupied by a plant resistant to the pathogen, then the disease may be transmitted to more than one adjacent cluster, even if they are disjoint. Thus, the average number of sites at which the pathogen causes damage is slightly larger than the average cluster size for $p_c^e \sim p_c$, as a result of spreading to disjoint clusters by the initial infection point. For $p_e < p_c$, on the other hand, the contribution of the initial infection point is through finite clusters, including isolated sites, whereas for $p_e > p_c$ the initial infection site belongs to the sample-spanning cluster. Therefore, if there is more than one initial point of infection in the system, one expect the appearance of sites that connect two adjacent disjoint clusters to be amplified.

Ramírez et al. (2018) carried out simulations on a square lattice, since crops are planted in parallel rows on the soil, implying that the seeds are sown in a square lattice arrangement. The distance between the neighboring site was selected according to the maximum length that the pathogen can travel. Since one does not know a priori whether a seed was sown at a given site, one assign 1 to each entry in the matrix based on the occupation probability p. Given a value of p_A, one assigns randomly to each occupied site—to each entry with a value of 1—a plant of type A or B by generating for each site a random number $0 < R < 1$ such that if $R < p_A$, a plant of type A is sown there. Otherwise, a plant of type B is sown.

The inoculated cells in the initial configuration were taken in randomly and uniformly in the lattice with probability p_I, which represents the fraction of inoculated sites in the lattice, which may or may not propagate the pathogen to neighboring plants, depending on the pathogen susceptibility of each type of plant. Thus, a plant of type A (B) is infected and develops the disease if a random number R, selected uniformly, is less χ_A (χ_B); otherwise, the plant remains healthy. If the plant is infected and develops the disease, then its site value is changed from 1 to 0.

Ramírez et al. (2018) also carried out experimental study of the same problem in a system that mimicked their simulation model. They prepared the substrate by mixing peat moss and sieved soil (2-mm mesh) in a 1:2 volume–volume mixture, which was placed in plastic double bags of 6 kg of high-density polyethylene, and was sterilized in an electric autoclave. Three types of chili seed were used, from which groups of 100 seeds were selected for carrying out the experiments. For each bioassay, aluminum trays were used, and of sterile substrate to each was added, after which the seed was spread homogeneously on the trays and covered with more of substrate, moistened with enough water. The trays were watered daily to maintain the humidity until the buds of seedlings emerged. It took 8 days after sowing to observe the initial growth, after which it was fertilized every 7 days with a fertilizer.

The microorganisms were taken from a phytopathogenic oomycete strain. Each oomycete was inoculated in the same sterile substrate used for the preparation of the trays. Segments of the growths were inoculated in plastic bags containing the substrate and were mixed by shaking them for 3 weeks to ensure the growth of the oomycete throughout the substrate, after which they were incubated at room temperature. Each tray planted contained, on average, about 80 seedlings, and each of the oomycetes was inoculated into the three trays corresponding to the three types of chili. The fertilization of the plants was stopped after the microorganisms were inoculated, but the humidity was maintained at field capacity during the entire experiments. Thirty five days after sowing, live plants were counted in each tray and the survival percentage was calculated.

The survival rate was determined experimentally by exposing a number of plants to the pathogen and counting the number of alive plants after 5 months. If $\mathcal{S}$ (as a percentage) is the survival rate of a plant type exposed to a pathogen, then the pathogen susceptibility is $\chi = 1 - \mathcal{S}/100$. Depending on the pathogen susceptibility of each plant, the probability p_A for which the pathogen will only spread on finite clusters was determined, even if all the sites were sown. This is shown in Fig. 19.2, where the combinations of pathogen susceptibility χ_B and mixture proportion (probability) p_A that prevent the formation of the spanning cluster of diseased trees for fixed $\chi_A = 1.0,\ 0.75,\ 0.50$, and 0.25 in various regular lattices in the limit $p_I \to 0$, corresponding to a single initial inoculation point, is shown.

The simulations of Ramírez et al. were in agreement with their experiments. Thus, using a percolation-based model, one can devise a strategy for preventing spread of a disease in plants, hence increasing their production of fruits. One strategy consists of sowing two types of plants with different susceptibilities to a specific pathogen, and organized as a percolating lattice in order to maximize the number of plants that survive an infestation. The lattice spacing coincides with the maximum distance

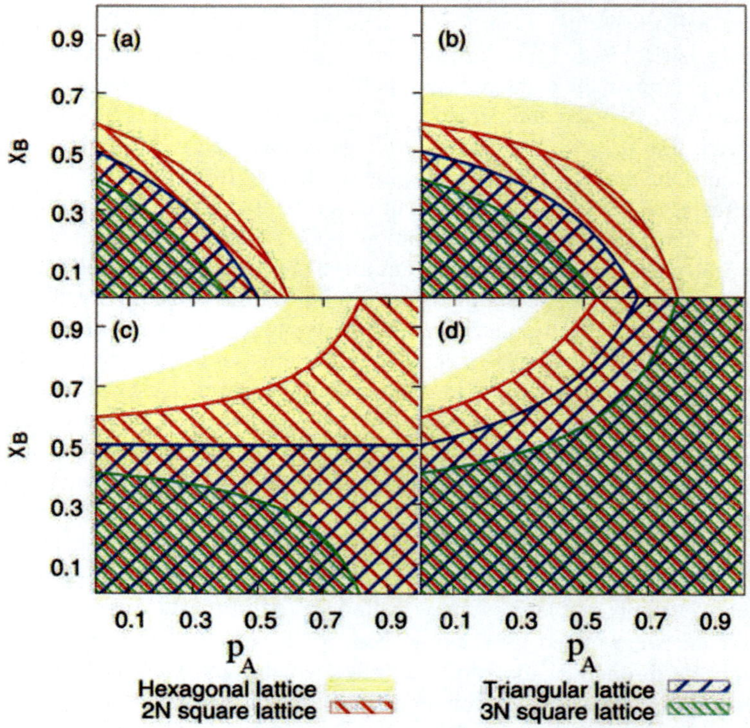

Fig. 19.2 The conditions for no percolation in regular lattices of trees when the density of inoculated cells p_I is small, $p_I \to 0$ for susceptibilities **a** $\chi_A = 1.0$; **b** 0.75; **c** 0.5, and **d** 0.25 (after Ramírez et al. 2018)

that the pathogen can travel before entering a state of dormancy or dying due to starvation. Under particular conditions of pathogen susceptibility, there are values of p_A for which the disease will only propagate on finite clusters, even if the entire soil is sown.

19.12 Spread of Fungi in Soil

Bailey et al. (2000) used percolation theory to study invasion and persistence of fungal parasites and saprotrophs in soil. The important factor in the phenomenon is the ability of the fungus to spread by mycelial growth and expansion of fungal colonies. The lateral spread of parasites depends on the endogenous supply and movement of nutrients within the fungal colony, the growth habit of the colony, and the distances between susceptible host roots or other organs. Clearly, if susceptible hosts are too far apart, local invasion ceases as the fungus exhausts its nutrient supply before infecting a new host. This brings out the issue of connectivity and, therefore, the relevance of percolation theory. Indeed, Bailey et al. (2000) visualized spread as occurring through a population of discrete sites on a lattice, whose sites represent discrete nutrient sources, consisting of susceptible roots or discrete fragments of organic matter, with the entire lattice representing these oil in which disease spread occurs.

The lattice can be 3D for nutrient fragments, such as comminuted straw, leaf, or root tissue. For lateral spread of disease, however, it can be restricted to a horizontal plane that passes through a population of seeds, roots, hypocotyls, or other organs. Therefore, as Bailey et al. (2000) discussed, local spread and, hence, progressive invasion by colony expansion depend on the spatial distribution of uncolonized sites. The fungus will spread as long as it can make contact with these sites, in which case it creates an expanding patch whose size is limited only by that of the system in which the fungus grows. The fungus will stop spreading, however, if it cannot make contact with new sources of uncolonized hosts or organic fragments, which results in a patch (cluster) of finite size. The crucial role of "making contacts"—making connected path—is described by percolation theory. In other words, there is a critical distance r_c such that for distances $r < r_c$ invasive spread of the pathogen among a population of nutrient sites can occur, whereas for distances $r > r_c$ the spread of the pathogen is finite. Clearly, the critical distance r_c corresponds to a percolation threshold.

Thus, Bailey et al. (2000) used percolation theory to describe saprotrophic growth of a soil-borne fungal plant pathogen, *Rhizoctonia (R.) solani*, which spreads in a finite system in a population of nutrient sites organized on a triangular lattice. They could predict the existence of a threshold probability of colonization between sites. If the probability of colonizing a neighboring site is above the threshold, the fungus can spread invasively and creates large patches (clusters), but the growth below the threshold is finite and is restricted to small patches.

To produce experimental data, small spots of agar of 1% (low nutrient) or 10% (high nutrient) (weight/volume) potato dextrose agar (PDA) were used to provide reproducible substrates at each site on a structure corresponding to a triangular lattice, with no significant decay during the course of the experiment. Pairs of agar sites that consisted of a donor and recipient were positioned at 4, 6, 8, 10, 12, 14, 16, 18, 20, or 22 mm apart (from center to center) in Petri plates. Fifteen replica of each distance were generated for completely randomized design, and the experiment was repeated for low- and high-nutrient sites. The donor site of each agar pair was inoculated with a single hyphal strand, *c.*, 1 mm in length, removed from the growing edge of a 4-day-old colony of *R. solani* Kühn, grown on water agar. The plates were assessed daily for 20 days for colonization of the recipient sites.

Using their experimental data, Bailey et al. (2000) constructed a probability profile, which describes the change in probability of colonization with distance between a donor (colonized) and a recipient (uncolonized) site, and used percolation threshold p_c to estimate the critical distance r_c. One example is shown in Fig. 19.3. To test the model, they also used the known r_c from their experiments to estimate p_c. As Fig. 19.3 indicates, the probability p of colonization decays sigmoidally as the distance r between donor and recipient sites increases. Threshold distances r_c for bond percolation on the triangular lattice (for which, $p_c \simeq 0.35$; see Table 2.1) were obtained by inverse prediction by fitting the data to a functional form, which was done by maximum likelihood with the assumption that the errors were binomial. As Fig. 19.3 also indicates, increasing the nutrient status of the agar resulted in a steeper profile, shifted to the right and an increase in the threshold distance between sites from $r_c = 8.1 \pm 0.49$ mm for low nutrient sites, to $r_c = 11.8 \pm 0.93$ mm for sites with a high-nutrient status.

Perhaps, the most important insight that percolation theory provided in this case was the fact that one can distinguish between invasive (sample spanning) and noninvasive (finite) saprotrophic spread of *R. solani*. Moreover, percolation theory provided the tool to predict how changing the distance between uncolonized sites affects the colonization of organic matter. Colonization profiles that summarize the probability of transmission of the fungus with distance between donor and recipient sites identified a threshold distance r_c for either invasive or noninvasive spread of the fungus in a population of agar sites. Invasive spread did not depend on the furthest extent of mycelial growth, evident in the tails of colonization profiles from the placement experiments, but, instead, it was associated with a threshold distance r_c that was different for low- and high-nutrient sites, but corresponded to a common percolation threshold, $p_c \simeq 0.35$ for the triangular lattice.

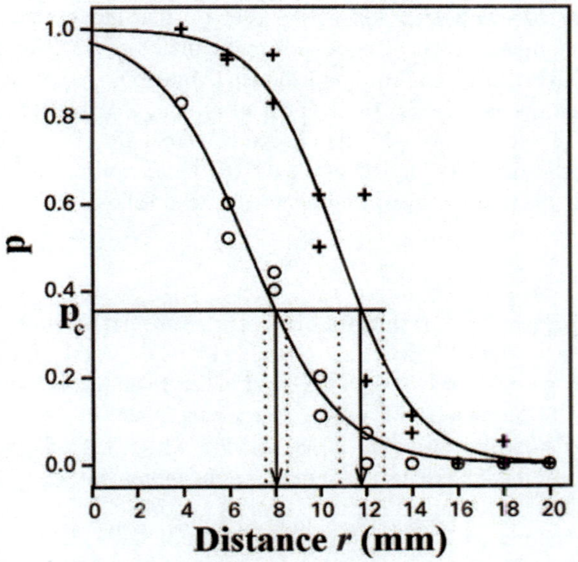

Fig. 19.3 Dependence of the probability p of colonization of *Rhizoctonia solani* on the distance r between donor and recipient sites with low (1% potato dextrose agar (PDA), open circles), and high (10% PDA, crosses) nutrient agar. The experimental data (symbols) were fitted with the functions, $p = 1/\{\exp[0.51(r - 6.85)]\}$ and $p = 1/\{\exp[0.61(r - 10.83)]\}$, respectively, in order to provide profiles for the probability p of colonization with distance. *Dotted vertical lines* represent 95% confidence intervals about the estimated threshold distance r_c for invasion (after Bailey et al. 2000)

19.13 Pattern of Tropical Deforestation

The significance of tropical forests to the global carbon cycle cannot be overstated. More than half of the known species worldwide reside in these forests and have been studied intensively for decades. The rapid increase in agriculture, logging, and urban growth, as well as climate change that interrupts the water cycle brought about by severe draught have, however, led to unprecedented losses of tropical forest, with the average annual deforestation rates since the 1990s being around 0.5%. Deforestation rates are not the same around the globe, with Asia and particularly Brazil being currently the hot spots. As the extent of forests decreases, it gives rise to fragmentation in which patches of forest are split into several smaller ones, somewhat similar to the phenomenon of fragmentation described in Chap. 10. The sizes of the fragments are not obviously the same and, therefore, there is a distribution of fragment sizes.

Taubert et al. (2018) used percolation theory to analyze the current forest fragmentation structures in tropical and subtropical regions of the Americas, Africa, Asia, and Australia (denoted here as Asia-Australia) on a high-resolution forest cover map with approximately 21 billion pixels, each $30 \times 30\ m^2$. For each continent, they used a clustering algorithm to count and analyze the size and perimeter distribution of all detectable forest fragments separately. Taubert et al. (2018) identified more than 130 million forest fragments across all continents, with their sizes ranging over 11 orders of magnitude and reaching 427 million hectares (Mha). The largest forest fragment in south America, located in the Amazon, spans around 45% of its total forest area, whereas the largest fragment on Borneo in Asia covers only 18% of the forest.

Analysis of the data by Taubert et al. (2018) indicated that the fragment-size distribution, i.e., the number n_s of fragments of size s, follows Eq. (2.14),

$$n_s \sim s^{-\tau} . \tag{19.18}$$

Figure 19.4 presents their results for three regions around the globe. As Table 2.3 indicates, the exact value of τ in 2D is $\tau = 187/91 \simeq 2.05$. Analysis of Taubert et al. yielded $1.9 \leq \tau \leq 1.98$, very close to the exact value. In addition, the fractal dimension D_f of the 2D fragments is given by (see Chap. 2), $D_f = 2/(\tau - 1) = 91/48$, while the analysis by Taubert et al. (2018) yielded, $D_f \simeq 1.87$, for Asia-Australia, and $D_f \simeq 1.92$ for the Americas, again in excellent agreement with 2D percolation. Finally, the exponent τ_p (the analog of τ) for the perimeter of the fragments (Ziff 1986) was also estimated. The results were $\tau_p \simeq 2.16$, 2.21, and 2.23 for, respectively, the Americas, Asia-Australia, and Africa, to be compared with the exact value in 2D, $\tau_p = 15/7 \simeq 2.14$.

These results suggest that the tropical forest fragmentation is a critical phenomenon, occurring at or near a percolation threshold. In order to explain why this should be, Taubert et al. (2018) developed a forest fragmentation model, dubbed FRAG. A landscape with $C_{\max}$ cells of size s (30×30 m^2) was used. The cells were

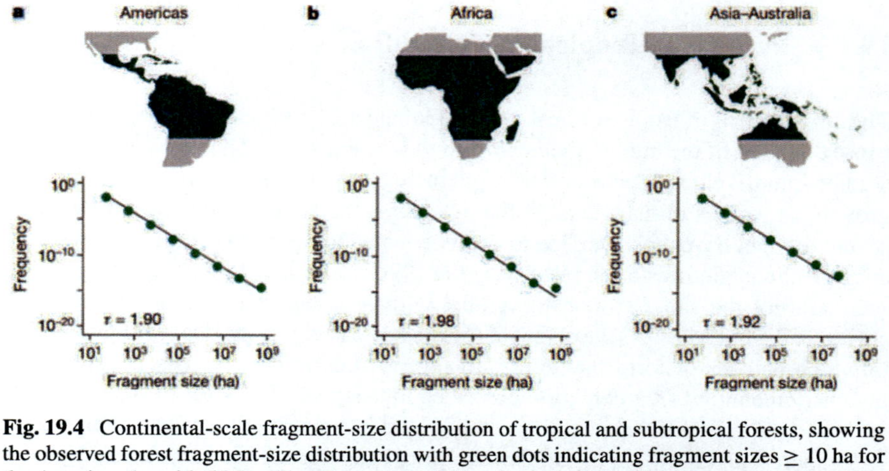

Fig. 19.4 Continental-scale fragment-size distribution of tropical and subtropical forests, showing the observed forest fragment-size distribution with green dots indicating fragment sizes ≥ 10 ha for the Americas (**a**, with 55.5 million fragments), Africa (**b**, with 44.8 million fragments), and Asia-Australia (**c**, with 30.5 million fragments). *Solid lines* represent the fits of power-law distributions with exponent τ. The maps show the entire land area of the tropical belt and indicate the selected tropical regions in green (after Taubert et al. 2018)

either forested or deforested. The simulation would begin with a fully forested area, i.e., all the cells were in the "forest" state, with the total forest area (landscape area) being $A_{\max} = C_{\max}s$. In each time step, taken to be 1 year, some forest area was cleared by a constant deforestation rate d_F in units of percentage per year, so that a certain number of forest cells, $d_F C_{\max}$, were selected randomly and converted to a deforested area.

For simplicity, cells could not grow trees and go back to a forest state, so that the forest area was successively reduced over time, until it was cleared in its entirety. Note that deforestation rates used in the model could also be interpreted as net deforestation rates that result from reforestation and gross deforestation that occur at random sites. The results were then analyzed using the same methods that were used for the analysis of the high-resolution forest cover map in terms of remaining forest area, fragment numbers, and their mean size, as well as the fragment-size distribution. Dynamics of fragment numbers scaled with landscape size $C_{\max}$. Thus, normalized fragment numbers were computed by dividing the absolute number of fragments by $C_{\max}$. The results reproduced those of the empirical data.

Chapter 20
Explosive Percolation and Its Applications

20.1 Introduction

In Chap. 3, we described many variants of the classical random bond or site percolation. Such variants were motivated by a particular application, or for explaining certain phenomena in which the connectivity of the elements of a system appears to play a prominent role in its macroscopic properties. As discussed in Chap. 13, in almost all the models that were described in Chap. 3, the percolation transition was continuous and second order: it was characterized by a divergent susceptibility or mean cluster size S (see Chap. 2), an infinite correlation length ξ_p, and power-law decay of the correlation function near the percolation threshold. In addition, the sample-spanning cluster (SSC) at p_c, or above p_c but over length scales $L \ll \xi_p$, has a fractal and statistically self-similar structure with a fractal dimension D_f defined and given in Chap. 2.

But, the percolation transition does not always have to be continuous and second or higher order. The connectivity of the system can *abruptly* change at p_c and, with it, other properties of interest. One example is certain versions of bootstrap percolation described in Chap. 3. Another good example, described in Chap. 11, is the rigidity transition in BCC lattice (Arbabi and Sahimi 2021). A third example is invasion percolation with long-range, non-decaying correlations (Knackstedt et al. 2000). An important new percolation model that can lead not only to a unusual phase transition, but also to many novel phenomena, not seen in the classical percolation and its variants described in Chap. 3, is *explosive percolation* (EP), first introduced by Achlioptas et al. (2009). In addition, EP has found several important applications. Thus, the purpose of this chapter is to describe the essential features of EP, and discuss its potential and realized applications. A good elementary introduction of EP is given by D'Souza (2021), while comprehensive reviews of EP and its properties are given by D'Souza and Nagler (2015), Boccaletti et al. (2016) and D'Souza et al. (2019). This chapter relies heavily on these review.

M. Sahimi, *Applications of Percolation Theory*, Applied Mathematical Sciences 213,
https://doi.org/10.1007/978-3-031-20386-2_20

20.2 Explosive Percolation

It is now known that EP can be the result of many types of graph evolution processes that delay the formation of large clusters and break the multiplicative coalescence rule of standard models of percolation described in Chap. 2. Because of this, many unique phenomena can occur during the emergence of large-scale connectivity.

20.2.1 Explosive Percolation Through the Achlioptas Process

What should one do to delay or enhance the percolation transition? This question was apparently first addressed by Dimitris Achlioptas (a Professor at the University of Athens at the time of writing this chapter) (D'Souza and Nagler 2015). In his algorithm, one begins with N isolated nodes and, instead of choosing one edge in each discrete time t, one selects two candidate edges, say $\{e_1, e_2\}$. Given a set of pre-determined selection criteria, one then selects the edge that best satisfies the criteria and adds it to the graph, while discarding the second edge for the current time step. This is what is usually referred to as the *power of two choices* in randomized algorithms (see, for example, Azar et al. 1999; Mitzenmacher 2001). If the selection criteria include keeping clusters small, then it will delay the onset of percolation, whereas growing a large cluster as quickly as possible enhances percolation. Note that "time" in such models is defined as the ratio of the total number L of the added edges and the total number nodes, namely, $t = L/N$.

Bohman and Frieze (2001), who in fact first dubbed the above algorithm as the *Achlioptas process* (AP), were interested in delaying the formation of a giant cluster in a graph evolution process similar to the Erdös–Rényi (ER) graph (see Chap. 17). In order to analyze the AP process analytically, they considered "bounded-size" rules in which all components of size K or greater are treated equivalently. This is now referred to as the Bohman and Frieze (BF) process in which the edge e_1 is accepted if it joins two isolated nodes, and e_2 is rejected; otherwise, the reverse is accepted, so that only clusters of size one—isolated nodes—are distinguished, and all clusters of size $K \geq 2$ are treated equivalently.

The process just described is similar to aggregation process that also begins with a collection of N isolated particles, monomers, or nodes. Thus, the first edge joins together two distinct components—two isolated particles or nodes—and forms a cluster of size two. If the clusters' sizes are sufficiently small, the probability that a randomly selected edge is within a cluster that is already formed is negligibly small and, therefore, the probability that each edge that one adds to the growing graph joins two disjoint clusters (a single node or a small cluster) is high. When the process of forming the graph is posed in this context and with such assumptions, it becomes completely similar to the aggregation processes (Ben-Naim and Krapivsky 2005) and can be modeled by a mean-field approach, i.e., the well-known Smoluchowski[1] rate equation (Smoluchowski 1916),

$$\frac{dn_s}{dt} = \sum_{i+j=s} K_{ij} n_i n_j - 2n_s \sum_j K_{sj} n_j \ , \tag{20.1}$$

with n_i being the density of clusters of size i, i.e., the number of clusters of size i, divided by the total number of clusters.

Note that the BF process may also be analyzed by the rate equations. At early times, long before the giant cluster emerges, all clusters are of size at most $\mathcal{O}(\log N)$ and, as discussed earlier, the edges that are added to the growing graph join previously disjoint clusters or nodes with high probability. In the critical region, however, in which clusters grow sufficiently large, edges within the clusters are more likely to be selected and added to the growing graph, which is why the Smoluchowski equation breaks down near the percolation threshold.

Note also that one may design a process by which percolation on a lattice is avoided. This is connected to studies of highly optimized tolerance (HOT) (Carlson and Doyle 1999) in which one uses optimization, by either natural selection or engineering design, in order to generate power-law distributions of cluster sizes, and places the clusters strategically in locations that minimize "damage" to particular classes of random fluctuations. The configurations that are generated on a lattice by the HOT process are "extreme" in that, they have a compact structure (non-fractal), hence avoiding a spanning cluster even at the highest densities.

20.2.2 Explosive Percolation Through Achlioptas Process and the Product Rule

Achlioptas et al. (2009) studied the AP with unbounded-size selection rules in which clusters with distinct sizes were treated uniquely. They carried out simulation of the process in a random graph by introducing what they called the *product rule* (PR), which was as follows and is shown schematically in Fig. 20.1a. Consider N isolated nodes, and select two candidate edges $\{e_1, e_2\}$ uniformly at random at each discrete time step t. Suppose that a and b are the nodes linked by e_1, c, and d are those connected by e_2, and $|S_i|$ is the size of the cluster that contains node i. Then, if $|S_a||S_b| < |S_c||S_d|$, edge e_1 is added to the graph; otherwise, e_2 is added. Therefore, only the edge that minimizes the product of the two sizes of the two clusters joined by e_1 is retained. Since the probability that an edge inside a cluster of size $|S_i|$ is

[1] Polish physicist (although he was born near Vienna, the present Austria) **Marian Smoluchowski** (1872–1917) was a pioneer of statistical physics. He made seminal contributions to kinetics of matter; discovered density fluctuations in a gas phase; explained, independently of Albert Einstein, the Brownian motion; and proposed the governing equation for diffusion in an external potential field.

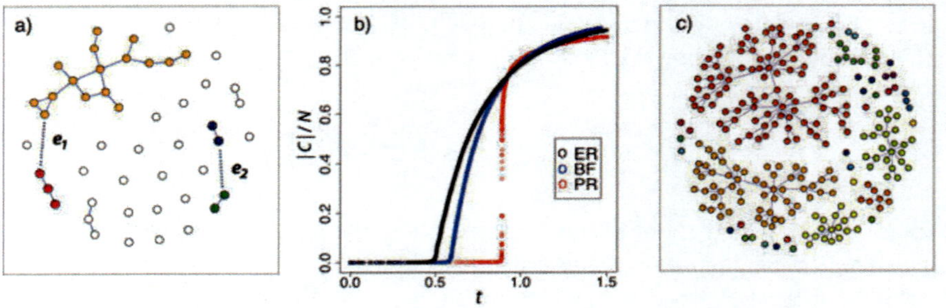

Fig. 20.1 **a** Schematic of the evolution of the graph with the PR. At each time step, two edges e_1 and e_2 compete for being added to the graph. In this example, the product of the clusters merged by e_1 is $3 \times 16 = 48$, whereas for e_2 is $2 \times 2 = 4$ and, thus, e_2 is accepted and e_1 rejected. **b** Typical evolution of fractional size of the largest cluster, $|C|/N$ as a function of edge density t for the ER, BF, and PR processes on a system with $N = 10^6$ nodes. **c** An example of a network with $N = 250$ nodes, grown by the PR, with the nodes in each distinct cluster rendered in the same color, with the largest cluster of size $|C|$ being indicated in red (after D'Souza et al. 2019)

proportional to $(|S_i|/N)^2$, which vanishes as $N \to \infty$ in the subcritical regime in which clusters are sublinear in the system size N. Thus, the PR selection criterion is used if nodes $a = b$ and/or $c = d$ without affecting the percolation transition.

Figure 20.1b compares a typical realization of the PR, ER, and BF processes, starting with a system of $N = 10^6$ nodes, which indicates that the onset of large-scale connectivity is considerably delayed for the PR process. Moreover, the large-scale connectivity for the PR process emerges drastically, going from sublinear to a level roughly equal to the ER and BF processes and happening during an almost imperceptible change in the edge density.

Figure 20.1b also indicates that the transition in EP with the PR rule is abrupt and resembles a first-order phase transition. Achlioptas et al. (2009) analyzed the transition by considering a scaling window $\Delta_N(\lambda, A)$ as a function of the system size N, which measures the number of edges needed for $|S|$ to transition from being smaller than N^λ to being larger than AN. If $A = \lambda = 1/2$, then Δ_N is the number of edges that must be added in order for the order parameter, the mean cluster size, to transition from $|S| \leq N^{1/2}$ to $|S| \geq N/2$. Simulation of Achlioptas et al. (2009) with $N = 6 \times 10^7$ nodes indicated that, $\Delta_N(1/2, 1/2) \propto N^{2/3}$, so that $\Delta_N(1/2, 1/2)/N \to N^{-1/2}$, which vanishes as $N \to \infty$, occurring at a percolation threshold $t_c \approx 0.888$. At first, this indicates a first-order discontinuous transition, which Fig. 20.1b also indicates, but later works showed (see below) that the transition is, in fact, continuous, although its universality class was unusual.

Others have studied EP generated with PR on the scale-free networks, described in Chap. 17 (Cho et al. 2009; Radicchi and Fortunato 2009) and on 2D regular lattices (Ziff 2009). While both cases manifested behavior similar to what was described above, the case of scale-free networks proved to be very interesting. Recall from Chap. 17 and Eq. (17.7) that the connectivity distribution in scale-free networks follows a power law, characterized by an exponent m. Numerical simulations (Cho et al. 2009; Radicchi and Fortunato 2009) provided evidence that there is a critical value m_c above which the system undergoes a discontinuous transition, with m_c being a *tricritical point.*[2] Moreover, it was shown that all the various models exhibit a scaling behavior with characteristics of continuous, second-order phase transitions, including a power-law cluster-size distribution with an exponent whose values for $m = 2.5$, 2.8, and 3.5 were, respectively, $\tau \approx 2.15$, 2.13, and 2.94, as well as a diverging mean cluster size that followed a power law (Cho et al. 2009; Ziff 2010; Radicchi and Fortunato 2010) with the corresponding exponents (Radicchi and Fortunato 2010) of $\gamma \approx 0.24$, 0.42, and 0.15, which are highly unusual when we compare them with those of percolation on regular 2D and 3D lattices, as well as the Bethe lattices.

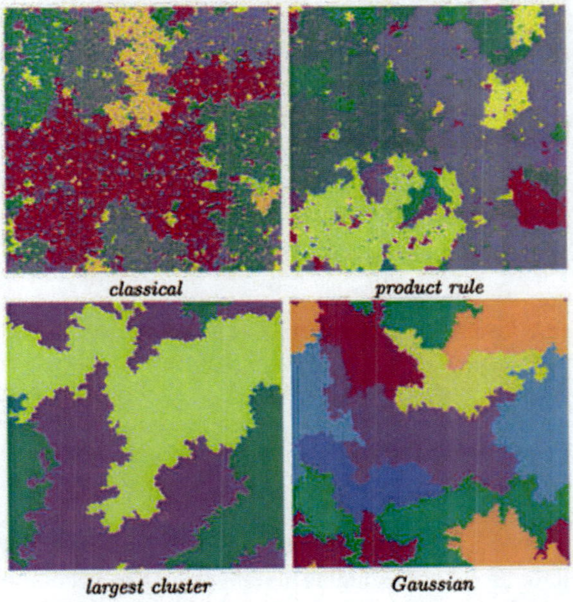

Fig. 20.2 Snapshots of the system for four percolation models, obtained on a square lattice of size 1024^2 sites at the percolation threshold (after Araújo and Herrmann 2010; courtesy of Dr. Nuno Araújo)

[2] A point at which three phases coexist is called a triple point for a one-component system, and a tricritical point is one in the phase diagram at which three-phase coexistence terminates.

20.2.3 Explosive Percolation Through the BFW Model

Bohman, Frieze, and Wormald (BFW) (2004) introduced the following model that leads to EP. One begins with N isolated nodes, and at each time step adds one edge to the growing network. Suppose that u is the total number of selected bonds, L is the number of occupied bonds, and k is the stage of the process; initially, $k = 2$. The first bond is occupied at random; $L = u = 1$. Then, one selects at random an edge out of all the possible edges, or the unoccupied ones, after which the maximum cluster size $S_{\max}$ is measured, if the selected edge is occupied. If $S_{\max} < k$, then the edge is occupied (added to the network). If, on the other hand, $L/u \geq g(k) = 1/2 + \sqrt{1/(2k)}$, then u is increased by 1. Otherwise, k is increased by one. One checks again whether $S_{\max} < k$, and so on. The procedure is iterated from $L = 0$, which corresponds to a bond occupation fraction $p = L/(2N) = 0$, to $L = 2N$ where all the bonds in the system are occupied, $p = 1$. Chen and D'ouza (2011) showed that the BFW model produces a discontinuous phase transition, and that multiple stable giant clusters can coexist. Schneider et al. (2012) demonstrated that the same model, when implemented on a regular lattice, also produces discontinuous phase transition. The model has been studied on other networks, such as scale-free and ER graphs.

20.2.4 Explosive Percolation by the n-Edge Model

One may generalize the process described above and consider $n \geq 2$ candidate edges at each time step, with n being a fixed constant. Thus, n nodes are selected at random at each time step, and two of them are connected, given certain rules. The limit $n = 2$ corresponds to the ER model. Explosive percolation can also be generated by the model introduced by Friedman and Landsberg (2009) in which the "triangle rule" was invoked such that three possible edges between three randomly selected nodes were considered as candidate edges. D'Souza and Mitzenmacher (2010) analyzed the model in detail. In addition, an EP model with the *sum rule* was introduced in which, instead of the product, one minimizes the sum of the clusters' sizes that would be joined (Cho and Kahng 2011a, b; Riordan and Warnke, 2011); see Bastas et al. (2014) for a comprehensive review.

20.2.5 Explosive Percolation by the Spanning Cluster—Avoiding Rule

Cho et al. (2013) and Ziff (2013) developed another EP model on regular lattices, which was based on avoiding the SSC. The model is now referred to as the spanning cluster-avoiding (SCA) model. In their model at each time step t, one selects at random n unoccupied edges and classifies into two groups, bridge and non-bridge edges. Bridge (non-bridge) edges are those that would (would not) form a SSC when occupied and, thus, the model strictly avoids occupation of bridge edges and, therefore, one of the non-bridge edges is randomly selected and occupied. If the n potential edges are all bridge edges, one of them is selected at random and labeled as occupied. As soon as a SSC is formed, the restrictions on the occupation of the edges are lifted. The procedure continues until all edges are occupied.

Numerical simulations and theoretical analyses showed that the EP in the SCA model may be either discontinuous or continuous for $d < d_u = 6$, depending on the number of eligible edges n. For $d \geq d_u$, the transition was shown to be continuous for any finite and fixed value of n. If, however, n is varied with the system size N, a discontinuous transition can also occur for $d \geq d_u$.

Note that in the SCA model on regular Euclidean lattices with an n–edge rule (see above), the percolation transition for emergence of a SSC is discontinuous for a lattice with dimension $d < d_u = 6$, so long as $n \geq n_c = d/(d - D_{\mathrm{bb}})$, where D_{bb} is the fractal dimension of the backbone (see Chap. 2), which has been calculated analytically and estimated numerically by Cho et al. (2013). Therefore, for two-dimensional lattice, $n_c \approx 2.554$ and, hence, setting n as low as 3 is sufficient to produce a discontinuous transition. There is, however, a remarkable distinction between the case with $n = n_c$ and one with $n > n_c$. In the former case, the discontinuous percolation transition occurs at some intermediate value of the threshold t_c. On the other hand, for $n > n_c$ the process is global. Therefore (D'Souza and Nagler 2015), when the SSC emerges, it encompasses the entire system, which also happens for an n–edge AP on a random graph in the limit $n \to \infty$. This occurs because (Rozenfeld et al. 2010), instead of metric or geometrical confinements, the rule makes it possible for the process to have unrestricted access to the entire collection of clusters generated on the graph. Thus, a giant percolating cluster is formed only in the final step of the process when only one cluster has survived.

20.3 Continuous Versus Discontinuous Phase Transition: Powder Keg and Unusual Scaling

What is the condition for having a discontinuous phase transition in EP? Friedman and Landsberg (2009) proposed a condition under which *any* percolation transition may be discontinuous. According to their criterion, for such a transition to occur, there must exist a *powder keg*—a collection of a sub-extensive number of clusters that together contain cN nodes, with c being a positive constant, implying that the number of nodes in such clusters should diverge as $N \to \infty$. We already mention that the BFW model leads to a discontinuous phase transition.

Araújo and Herrmann (2010) proposed two models for EP that led to discontinuous transition. In one model, called "largest cluster model," an edge is randomly selected among the empty ones. If its occupation would not lead to the formation or growth of the largest cluster, it is always occupied; otherwise, it is occupied with probability

$$\min\left\{1, \exp\left[-\alpha\left(\frac{s-\langle s\rangle}{\langle s\rangle}\right)^2\right]\right\}, \tag{20.2}$$

where s is the size of the cluster that would be formed by occupying the bond being considered currently, and $\langle s\rangle$ is the average cluster size after occupying the bond, with the parameter α controlling the broadness of size distribution. If $\alpha \leq 0$, one recovers the usual percolation, since the size of the largest cluster is always greater than, or equal to, the average cluster size and, thus, all new bonds are occupied. On the other hand, if $\alpha > 0$, the occupation probability given by Eq. (20.2) suppresses the formation of a cluster significantly larger than the average, inducing a homogenization of cluster sizes.

In the second model, a new bond is selected from the list of empty ones and occupied with probability given by Eq. (20.2). For internal connections, one considers an s twice the cluster size. Because Eq. (20.2) is a Gaussian with average size $\langle s\rangle$ and size dispersion $\langle s\rangle\sqrt{2\alpha}$, the model is referred to as the Gaussian model. In this model, the occupation probability is assigned to all the new bonds, even when they are not related to the largest cluster. This not only provides complete control over clusters greater than the average size, but also over the smaller ones. For $\alpha = 0$, all bonds have the same probability and, therefore, the model reduces to classical bond percolation. For $\alpha < 0$, the growth of larger clusters is favored in two distinct ways: they differ more from the average value and have more empty bonds than the smaller ones. Yet, for all $\alpha < 0$ the model recovers the classical universality class of percolation.

Figure 20.2 presents snapshots for four models of bond percolation, namely, classical, PR, largest cluster, and Gaussian, all of which were obtained at their respective percolation thresholds. As the figure indicates, clusters of very different sizes are obtained for classical percolation and the PR and, in fact, the cluster-size distribution is characterized by a power law. For the two models proposed by Araújo and Herrmann (2010), however, a characteristic cluster size is seen, and both models lead to a localized cluster-size distribution. Araújo and Herrmann (2010) showed that both models produce Gaussian cluster-size distributions and compact (non-fractal) clusters as one would expect in first-order transitions. Moreover, they showed that cluster's perimeters are fractal at the transition point, with a fractal dimension of 1.23 ± 0.03. Note that although Araújo and Herrmann (2010) considered the Gaussian probability, since this is what one expects for the cluster-size distribution at a first-order transition point, any other function could be used, so long as it constrains the largest cluster differing significantly in size from the average cluster.

On the other hand, using extensive numerical simulations, as well as rate equations (see above), several research groups (da Costa et al. 2010; Grassberger et al. 2011; Lee et al. 2011; Tian and Shi 2012) showed the EP as modeled by AP rule leads to a continuous phase transition. To demonstrate this, da Costa et al. (2010) considered an EP model, which they called the dCDGM model,[3] which was as follows. Starting with N initial nodes, one selects at random two nodes and compares the clusters to

[3] The model's name consists of the initials of the authors' last names.

which they belong. One selects a node between the two that belongs to the smallest cluster. Next, a second pair of nodes is selected and, as in the previous step, one chooses the node that belongs to the smallest of the two clusters, and then adds a link between the two selected nodes, which merges the two smallest clusters. This procedure is repeated a large number of time steps in order to grow a graph and study the EP process on it.

It should be pointed out that a cluster can be selected several times, which is similar to percolation clusters. The model invokes the key element of other EP models, including that of Achlioptas et al. (2009), described earlier, since to link nodes one selects the minimal clusters from a few possibilities, but the selection process is more stringent than that of Achlioptas et al. (2009), because it guarantees that the product of the sizes of two merging clusters is *always* the smallest of the other possibilities. This is easily seen by recognizing that each of the first pair of selected nodes in any step may connect with any node of the subsequent second pair, which is in contrast to selection from only two possibilities in the EP model of Achlioptas et al. (2009). Therefore, if the transition in the dCDGM model is continuous, then so also is the transition in the EP model of Achlioptas et al. (2009).

To study the model numerically, da Costa et al. (2010) defined two key quantities. One was $R(s) = sn_s/\langle s\rangle$, with $\langle s\rangle$ being the average cluster size, and n_s the usual percolation quantity, namely, the number of clusters of size s. Then, $\sum_s R(s) = 1 - S$, with S being the mean size of all the clusters. The second quantity was $Q(s)$, the probability that if one selects at random two nodes, then the smaller of the two clusters to which the two nodes belong is of size s. It is straightforward to see that $\sum_s Q(s) = 1 - S^2$. da Costa et al. (2010) then introduced the cumulative distributions, $R_{\text{cumu}}(s) = \sum_{u=s}^{\infty} R(s)$, and $Q_{\text{cumu}} = \sum_{u=s}^{\infty} Q(s)$. Then,

$$Q(s) = [R_{\text{cumu}}(s) + R_{\text{cumu}}(s+1) + 2S]\, R(s) = [2 - 2R(1) - 2R(2) - \cdots - 2P(s-1) - P(s)]P(s)\,, \tag{20.3}$$

and the rate equation governing $R(s)$ is given by

$$\frac{\partial R(s;t)}{\partial t} = s \sum_{u+v=s} Q(u;t)Q(v;t) - 2sQ(s;t)\,, \tag{20.4}$$

which is a generalization of the Smoluchowski rate equation, Eq. (20.1). Here, t_c is defined as the point at which $R(s)$ follows a power law over the full range of s. Note that $R(s)$ follows the power law, $R(s) \sim s^{1-\tau}$, while one has $Q(s) \sim s^{3-2\tau}$, with τ being the usual percolation exponent for the cluster-size distribution (see Chap. 2).

The results are shown in Fig. 20.3. Extensive numerical simulations of da Costa et al. (2010) yielded, $t_c \approx 0.923$ (note the high value of t_c). The usual percolation exponents (defined in Chap. 2) were estimated to be $\tau \approx 2.047$ (5/2), $\beta \approx 0.055$ (1.0), and $\gamma \approx 0.857$ (1.0), where numbers in the parentheses indicate the corresponding mean-field values of the exponent (see Table 2.3). In addition, the fractal dimension of the largest cluster was estimated to be $D_f \approx 2.333$ (4.0), while the upper critical

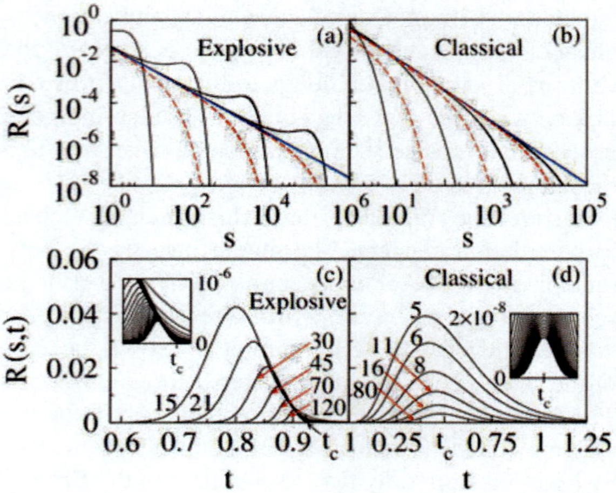

Fig. 20.3 **a** The evolution of the distribution $R(s)$ below and at (solid lines) and above (dashed lines) the percolation threshold of EP. **b** Same as in **a**, but for the classical percolation (corresponding to the mean-field limit). **c** Dependence of $R(s; t)$ on t for a set of cluster sizes s for EP. Numbers on curves indicate s. **d** Same as in **c**, but for classical percolation. The insets show the results for large s (after da Costa et al. 2010)

dimension (at which the mean-field theory becomes exact) turned out to be $d_u = D_f + 2\beta \approx 2.445$ (which for classical percolation on regular lattices is $d_u = 6$; see Chap. 2). In addition, the finite-size effects are characterized by the shift,

$$t_c(\infty) - t_c(N) \sim N^{-2/d_u} \sim N^{-0.818} , \tag{20.5}$$

which should be compared with Eq. (2.42). That the upper critical dimension is not even 3 is truly remarkable, given that the graphs have infinite dimensions. Work of da Costa et al. was followed by that of Riordan and Warnke (2011) who provided evidence that all AP models on ER graphs give rise to continuous percolation transitions.

20.3.1 *The Importance of Selecting the Correct Order Parameter*

As described in Chap. 2, the order parameter in classical percolation is $P(p)$, the fraction of bonds or sites in the SSC, which is unique, with the phase transition being continuous. The literature on EP also adopted $P(p)$ as the order parameter, even though in at least some cases many macroscopic clusters emerged. Thus, the question of whether EP represents a first- or second-order transition should be viewed within this context: is $P(p)$ the right order parameter? The standard practice in (canonical) statistical mechanics is to select the order parameters, as well as the control parameter (which in classical percolation is p, the fraction of active sites or bonds; see Chap. 2) such that they are conjugate to each other. On the other hand, a canonical approach to percolation is possible only for very specific models, such as Hamiltonian graph models (Bizhani et al. 2012).

As da Costa et al. (2014) argued, "for continuous phase transitions the order parameter cannot be chosen in an arbitrary way, by demanding only that it is 0 in the normal phase and non-zero in the ordered phase. For these transitions the order parameter must satisfy several strict conditions that are well known in the theory of phase transitions," such as hyperscaling relations mentioned in Chap. 2, i.e., the relations between the critical exponents as a function of the space dimensionality d. Thus, for the order parameter to satisfy a number of constraints, it needs to be based on the details of the model, such as the choice for the parameter n in the aforementioned $n-$edge processes.

On the other hand, if the order parameter is assumed to be the total relative size of *all* the macroscopic clusters, models of EP with an initial continuous transition, but with multiple giant components, may be globally continuous. Although such a perspective neglects mesoscopic structures, it does show that many of the observed discontinuities, as well as non-self-averaging behaviors (see below) may in fact be a consequence of the choice of the order parameter.

Just as important as the order parameter is the control parameter, which, in classical percolation, as already mentioned, is p. The control parameter is important because its choice crucially determines which features are or should be measured. If, instead of p, time is used as the control parameter, the fraction of links p and time t will be equivalent for reversible percolation. In particular, if time is a discrete variable, then a time step forward implies adding a link or edge, whereas a step backward means removing a link. As links are selected completely randomly, the standard percolation process is reversible and can be described as a reversible kinetic model.

But the same issues are not as clear in the EP. In fact, the issue is intriguing because the EP models, regardless of whether they are continuous or discontinuous, are usually irreversible. For example, the diffusion-limited cluster aggregation model of the EP (see below) can exhibit a discontinuous phase transition, if the order parameter is studied as a function of p, the relative number of aggregation events, but not as a function of t, defined as the physical time in the Smoluchowski equation, described above. Thus, if we approach the problem from the perspective of statistical mechanics, selecting a typical control parameter can be a delicate issue and crucially alter the critical behavior and the type of phase transition.

20.3.2 *Reversible Versus Irreversible Phase Transition*

As already mentioned, the classical percolation phase transition, which is continuous and second order, is reversible. When phase transitions are reversible, regardless of whether the control parameter is tuned forward or backward, the order parameter always exhibits the same continuous transition at the same critical point—in this case the percolation threshold. But, generally speaking and importantly, first-order phase transitions emerge from irreversible processes. But it must be emphasized that if a phase transition is continuous, it does not necessarily imply it is also reversible. The best evidence for this is the fact that, as discussed above, the great majority of the EP models represent continuous transitions, but are irreversible because (D'Souza et al. 2019) it is typically impossible to construct the reverse of an EP process on random networks from the forward link-addition process. This is due to the fact that the microscopic rule is based on cluster information (their sizes and structure, for example), which is lost during the process.

Discontinuous phase transitions are usually hysteretic, i.e., a given state depends on the history of the process through which the current state was arrived at.[4] This feature will be discussed further later on, when we describe an important application of EP.

20.4 Applications of Explosive Percolation

Many potential applications of EP have been suggested, with some being speculation, and others being more concrete. In what follows, such applications are described.

20.4.1 Electric Breakdown

Electric breakdown is an important and costly problem in the production of integrated circuits (see Sahimi 2003b, for a comprehensive review and discussion). The manufacturing of such circuits involves millions of nanoscale metallic and semiconducting devices that are placed on a substrate. The system can, however, be polluted with metallic particles that can induce electric breakdown, malfunctioning, and a shorter lifetime of the circuits. Therefore, one would like to understand the conditions under which and the speed by which such a system collapses. To address the problem, one must clearly take into account the fact that deposition of the metallic particles is hindered by the local electric field. To understand the complexities of the problem, Oliveira et al. (2014) proposed the following model that can lead to explosive phase transition.

Consider an $L \times L$ tilted square lattice with periodic boundary conditions in one direction and a voltage drop $V_0 = 1$ applied in the second direction. Assume, for simplicity, that the bonds of the lattice are either associated with highly resistive elements, or metallic particles, with resistances $R = 1$ and 0, respectively. Nodes that are connected by metallic bonds form metallic clusters and, thus, have the same electric potential. Resistive bonds that connect sites of the same metallic cluster are referred to as the *internal bonds*, whereas all other resistive bonds are called *merging*

[4] We are already familiar with this concept. In Chap. 9, we saw that the way the relative permeabilities of multiphase fluid flow are reached at any saturation depends on the process, i.e., the history; see also the discussions later in this chapter.

bonds. There is no electric field across internal bonds and, therefore, metal dust can be deposited on them with a probability q times larger than those deposited on merging bonds, with the factor $q > 1$ describing the relative deposition disadvantage due to the presence of the electric field. Moreover, the probability of adsorbing a metallic particle decreases monotonically with the strength ΔV of the gradient of the local electric field. Thus, Oliveira et al. (2014) assumed that the deposition probability decays as $1 - (\Delta V/V)^{\chi}$, with χ being an adjustable parameter. The probability p_d for metallic bonds replacing a resistive one—i.e., the deposition process—is given by

$$p_d = \left(\frac{p_0}{q}\right)\left\{1 + (q-1)\left[1 - \left(\frac{\Delta V}{V_0}\right)^{\chi}\right]\right\}, \tag{20.6}$$

where p_0 is the deposition probability in the absence of a voltage drop. Note that if a bond is of internal type, we have $p_d = p_0$, since $\Delta V = 0$. Clearly, the probability of metallic particles leaving the surface is simply $1 - p_d$. The three parameters of the model have clear physical meaning. p_0 is a measure of the amount of metallic dust; q represents the enhancement of deposition in the absence of a local electric field, while χ signifies the dependence on the strength of the local field. If $\chi = \infty$, all resistive bonds, both merging and internal, can be replaced with the same probability p_0, so that this limit represents bond percolation, which is also the case if $q = 1$ for *any* value of χ. Oliveira et al. (2014) showed that one recovers the $q-$state Potts model, a generalization of the Ising model, for $\chi = 0$ (see Wu 1982, for a comprehensive discussion), to which we shall return shortly.

Simulation of Oliveira et al. (2014) indicated that, irrespective of the initial configuration of the lattice, a steady state is reached after a certain number of iterations n_i. For fixed q and χ, increasing p_0 leads one to reach a value $p_c(q, \chi)$ at which the system electrically breaks down, and a spanning metallic cluster may emerge after steady state, making the system fluctuate strongly between resistive and metallic configurations. In particular, $p_c(q, \infty) = 1/2$, since one recovers bond percolation in this limit. For $\chi = 0.1$, for example, the fraction p of nodes in the largest metallic cluster oscillates around two well-defined values, $p \approx 0.05$ and 0.45, yielding a bimodal distribution. In addition, the height of the peaks in the fluctuations grows with $L^{d/2}$ and the distance between them does not vanish with L, which represent signatures of a discontinuous transition. On the other hand, although the variable p for $\chi = 1$ also fluctuates around an average value, its distribution is unimodal with no sign of metastability emerging, indicating a continuous transition in this limit.

Figure 20.4 presents snapshots of the steady-state configurations for both cases at and around the threshold p_c. While for $\chi = 0.1$ a gigantic compact metallic cluster abruptly emerges, suggesting a discontinuous transition, the largest cluster that emerges for $\chi = 1$ appears to be fractal, hence indicating a continuous transition. As a result, the electric breakdown is either smooth or explosive, depending on the values of χ and q.

Note that, as mentioned above, in the limit $\chi = 0$, merging bonds become metallic with probability p_0/q and internal bonds with probability p_0, which is precisely the Monte Carlo procedure to obtain the Coniglio–Klein clusters (Coniglio and

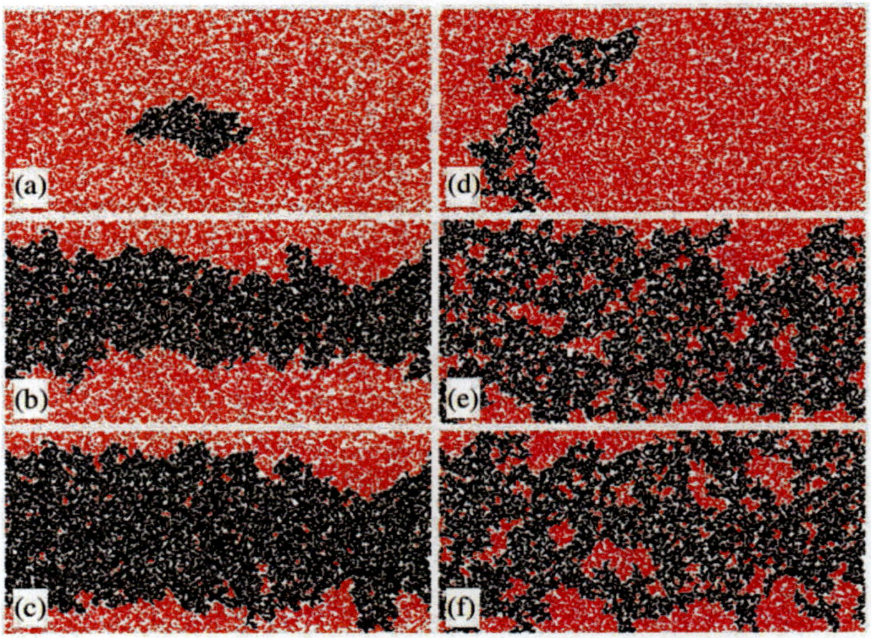

Fig. 20.4 Snapshots of steady-state configurations for $q = 10$, $L = 128$, and $\chi = 0.1$, with **a** $p_0 = 0.57$, **b** 0.58, and **c** 0.59, and for $\chi = 1$ with **d** $p_0 = 0.49$, **e** 0.5, and **f** 0.51. Metallic bonds that belong to the largest cluster are colored in black, while the bonds that are in the other metallic cluster appear in red. For $\chi = 0.1$, the largest metallic cluster is compact and grows abruptly at the threshold, $p_c \approx 0.58$, while the cluster for $\chi = 1$ is fractal and grows rather smoothly with p_0 (after Oliveira et al. 2014; courtesy of Dr. Nuno Araújo)

Klein 1980), which represent correlated clusters (for a comprehensive discussion see Coniglio and Fierro 2021) for the $q-$state Potts model, as was shown by Gliozzi (2002). Thus, the exact transition point for the square lattice in the limit $\chi = 0$ is given by (Wu 1982),

$$p_c(q, \chi = 0) = \frac{\sqrt{q}}{1 + \sqrt{q}} . \tag{20.7}$$

It should also be pointed out that the nature of phase transition in 2D in the Potts model changes at $q = 4$ from continuous and second order for $q < 4$, to first order and discontinuous for $q > 4$. Thus, the model of Oliveira et al. (2014) not only provides a physical realization of the EP and classical percolation, but is also linked with many important models in statistical physics.

20.4.2 Formation of Bundles of Nanotubes with Uniform Size

Nanotubes, both of carbon (Oberlin et al. 1976; Iijima 1991) and silicon-carbite (Malek and Sahimi 2010; Khademi and Sahimi 2011; Barghi et al. 2014) type, have numerous applications, ranging from gas storage, drug delivery, and separation processes, to electronics, optoelectronics, sensors, fabrication of nanodevices, and composite materials. As described in Chap. 12, due to their electrical properties, adding a small number of nanotubes to nanodevices and composite materials can modify significantly their transport properties (see, for example, Choi et al. 2001; Kim et al. 2010), particularly when it comes to a possible transition from an insulator to a conductor. As the discussion so far throughout this book should have made it clear, the transition is characterized and modeled by percolation theory. The percolating path is usually formed through the nanotubes.

Experiments indicate (see, for example, Homma et al. 2002; Postma et al. 2005) that nanotubes with large aspect ratios (ratio of their length and diameter) play a significant role in designing nanoscale devices. As we discussed in Chaps. 3 and 5, such objects may be represented by long sticks with essentially zero width, so that a random network of such nanotubes with large aspect ratios can be mapped onto a continuum percolation system (see Chap. 3; for a comprehensive discussion of continuum percolation see Balberg 2021). Such a model for a bundle of nanotubes was first proposed by Du et al. (2005) and Kyrylyuk and van der Schoot (2008).

Two experimental methods produce bundles of nanotubes, namely, laser-vaporized and electric-arc techniques (Thess et al. 1996; Journet et al. 1997). Indeed, they have produced close-packed single-walled nanotubes (SWNTs) with uniform diameters. On the other hand, percolation theory tells us that the bundle's size cannot be uniform, unless growth of large bundles is suppressed, since in the classic percolation clusters of all sizes are produced. This implies that the effective growth mechanism for formation of SWNTs bundle with uniform diameters is not ordinary percolation, since it does not have any mechanism to suppress aggregation between large bundles to produce a uniform size. But, as we have learned in this chapter, the suppression can be achieved by the mechanism that drives the EP and, thus, may be exploited in fabrication of nanotube bundles with uniform sizes. This important insight was first pointed out by Kim et al. (2010), who developed a continuum EP model to demonstrate the utility of their idea.

In their model, one begins by placing two types of sticks, α and β, on a plane at random positions with random orientations. Suppose that $\{s_{\alpha_1}, s_{\alpha_2}, \cdots, s_{\alpha_n}\}$ ($\{s_{\beta_1}, \cdots, s_{\beta_m}\}$) represent the sizes of the clusters that form a new large cluster with the size $\sum_{k=1}^{n} s_{\alpha_k} + 1$ ($\sum_{k=1}^{m} s_{\beta_k} + 1$) by placing the α (β) sticks on the 2D plane. As in the EP with the product rule (see above), one computes

$$\pi_\alpha = \Pi_{i=1}^{n} s_{\alpha_i} , \quad \text{and} \quad \pi_\beta = \Pi_{i=1}^{m} s_{\beta_i} . \tag{20.8}$$

Then, if $\pi_\alpha \leq \pi_\beta$ ($\pi_\alpha \geq \pi_\beta$), stick β (α) is removed from the system. This process prefers the connection between small clusters, causing cluster repulsion or suppress-

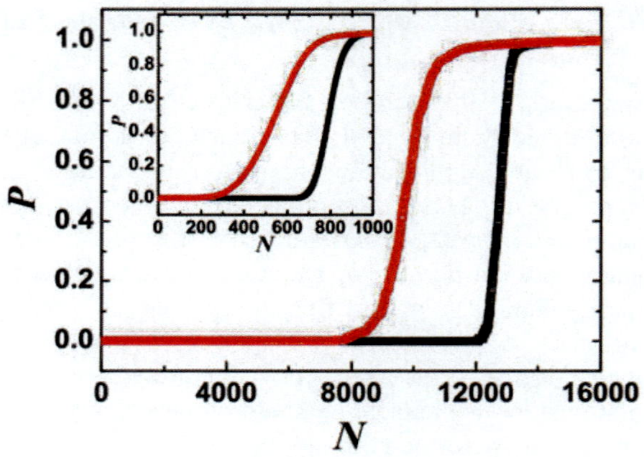

Fig. 20.5 Comparison of the computed percolation fraction P as a function of N, the total number of sticks placed in a given square, for ordinary percolation (OP, squares) and explosive percolation generated by the Achlioptas process (AP) with the product rule (circles) (after Kim et al. 2010)

ing the growth of large clusters, which is what is needed for producing bundle of SWNTs with uniform sizes.

The order parameter is simply the percolation fraction (Chap. 2) $P = n_c/N$, where n_c is the number of sticks in the SSC and N is the total number of sticks placed in a given square. Figure 20.5 compares the order parameter between ordinary percolation (OP) and the AP with a system size of $L = 40$, indicating that (a) the AP delays the transition, which is what one observes in complex networks (as well as in 2D loopless percolation, since the original AP does not allow formation of loops) and (b) the transition under AP is more abrupt than in the OP (although the transition is still continuous; see above).

Perhaps a more accurate way of checking whether a transition is abrupt is through measuring hysteresis, a history-dependent property of a system that, due to the existence of metastable states, usually—but not always—occurs in first-order phase transitions (Landau and Binder 2000). In other words, a given state of the system depends on how that state was arrived at. Two-phase flow in porous media, studied in Chap. 9, is one example in which the relative permeabilities and capillary pressure are history dependent, but vanishing of the relative permeabilities at the residual saturation (percolation threshold) still represents a continuous phase transition.

Thus, if there is hysteresis in the problem of fabricating bundles of nanotubes, then the route for changes in P during adding the sticks should be different from removing them. Figure 20.6 compares the results for when one adds the sticks and when one removes them for $L = 40$ and 10. There is clearly hysteresis. Therefore, because the formation of nanotube bundles and stick percolation with the AP share the same growth mechanism that suppresses coagulation of different bundles or clusters, the EP with PR rule should provide a framework for fabrication of nanotube bundles.

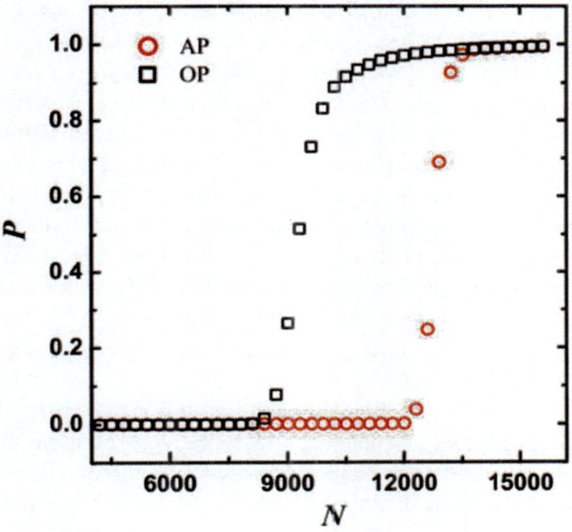

Fig. 20.6 Hysteresis in the percolation fraction P for network size $L = 40$. Black and red squares represent the results when, respectively, one adds the sticks and removes them. s. Inset shows the results for $L = 10$ (after Kim et al. 2010)

20.4.3 Crackling Noise in Fractional Percolation

Fragmentation is an important process that is encountered in everyday life, ranging from shattering of glass and breaking combustion fuel (see Chap. 10), to cracking of sidewalk. An important aspect of fragmentation is that the sizes of the fragments are of the same order of magnitude as the parent pieces, which explains why the case in which one fragment is much smaller than the other fragment has not been studied much. At the same time, fragmentation belongs to a class phenomena and systems that when pushed slowly, they crackle such that one may even be able to hear, such as when coal is combusted.

Schneider et al. (2013) modeled the phenomenon when one fragment is small, while the size of the other fragments is significantly larger, and demonstrated that crackling noise emerges from the process that they modeled. Their model begins with N nodes and L edges or links. Initially, the nodes are not connected and $L = 0$. At each step, one selects at random three different nodes or vertices v_1, v_2, and v_3 that belong to three clusters (which could be the same) with sizes S_1, S_2, and S_3 with $S_1 \geq S_2 \geq S_3$. One then connects two of the nodes v_i and v_j for which $\Delta_{ij} = fs_i - s_j$ $(1 \leq i < j \leq 3)$ is the smallest, where $0 < f \leq 1$ is fixed. If there are multiple minima, one selects one at random from among them. Thus, the process favors connections between two clusters that are of similar sizes, since the size of the larger cluster is rescaled by a factor f, which in essence represents the target fraction. Note that since $\Delta_{12} \leq \Delta_{13}$, one considers only Δ_{12} and Δ_{23}. Note that a fragmentation process in which $s \to [gs := s_i,\ (1-g)s := s_j]$ is the inverse process for which f determines the magnitude of the discontinuities in the order parameter, where $g = (1+f)/(1+2f)$. The density of edges in the network is $p = L/N$, and p_c is the critical link or edge density at which the (first) connectivity phase transition occurs. It should already be clear that the process has similarities to the EP, generated by the AP. Schneider et al. (2013) dubbed their model *fractional percolation*, since coalescence of clusters that differ substantially in their sizes is systematically suppressed.

Next, the sizes S_i are normalized by N, the total number of nodes in the system, so that $s_1 = S_1/N \geq s_2 = S_2/N \geq \cdots$. Schneider et al. (2013) showed that the largest cluster cannot merge with those smaller than $fS_1/(1+f)$. If fewer than three distinct clusters are selected, either an intracluster link is added and the size of largest cluster is necessarily unchanged, or the size of the largest cluster doubles. This implies that as soon as the size of the second largest cluster exceeds $s_1 f/(1+f)$ in an infinite system, it merges with the largest one, $s_{\rightarrow} s_1 + s_2$. A third macroscopic cluster is, with high probability, impossible and, therefore, the second largest cluster is also resetted, $s_2 \rightarrow 0$.

The result is that after the first-phase transition, for $p > p_c$ and $N \rightarrow \infty$, the size of the largest cluster either remains constant or jumps discontinuously. Since, as we now know, the first transition is continuous, the process exhibits infinitely many discontinuous transitions that are arbitrarily close to the first transition at $p = p_c$. In addition, Scheidegger (1967) showed that the process is non-self-averaging, i.e., the (relative) variance defined by $\sigma^2 = (\langle s_1^2 \rangle - \langle s \rangle^2)/\langle s_2 \rangle^2$ remains finite for $p \geq p_c$ as $N \rightarrow \infty$.

Schneider et al. (2013) also demonstrated that their process reproduces the main features of the so-called Barkhausen noise, which is the noise in the magnetic output of a ferromagnet when a magnetizing force applied to it is changed. The noise is caused by rapid changes of size of magnetic domains, which are similarly magnetically oriented atoms in ferromagnetic materials. This is similar to the way the sizes of the clusters change in the model of Schneider et al. (2013). Thus, one has crackling noise in the type of percolation process modeled with fractional growth rules, implying randomly distributed jumps in the order parameter. Such jumps represent discontinuous phase transitions. If, however, such mechanisms are mixed, even weakly, with mechanisms that merge clusters purely at random, the transitions vanish or become at most weakly discontinuous.

20.4.4 Diffusion-Limited Cluster Aggregation

Diffusion-limited cluster aggregation (DLCA), first proposed by Meakin (1983) and Kolb et al. (1983), is a generalization of the diffusion-limited aggregation (DLA) model developed by Witten and Sander (1981). Both are fundamental processes that have a very large number of applications, ranging from colloidal systems and polymerization, to fluid flow in porous media (for a comprehensive review of this aspect, see Sahimi 2011), and practical, real-world problems, such as the structure of asphaltene aggregates (Rassamdana et al. 1996; Rassamdana and Sahimi 1996); see Meakin (1998) for an excellent review.

In the DLA model, the site at the center of a lattice is occupied by a stationary particle. A particle is then injected into the lattice, far from the center, which diffuses on the lattice. If it reaches a surface site, i.e., an empty site, which is a nearest neighbor to the stationary particle, it sticks to that site and remains there permanently. Another particle is then injected into the lattice, which diffuses on the lattice until it reaches another surface (empty) site and sticks to it, and so on. If this process is continued for a long time, a large aggregate is formed. Computer simulations (see Meakin 1998 for a comprehensive review) indicated that the DLAs are fractal objects with fractal dimensions of 1.7 and 2.5, in 2D and 3D, respectively.

In the DLCC aggregation model (Meakin 1983; Kolb et al. 1983), one begins with an empty lattice. At time $t = 0$, lattice sites are selected at random and occupied by particles, until a small fraction p_0 of the sites are occupied by the particles. Each occupied site can contain only one particle. A randomly selected cluster of occupied sites, including a single site, is then selected and moved (diffuses) in a randomly selected direction. Then, the perimeter sites of the cluster—the set of sites that are adjacent to the cluster—are examined to see whether they are occupied by other particles or clusters. If so, the perimeter particles or clusters are added to the cluster that was moved to form a larger cluster. Once a larger cluster is formed, it is not allowed to break up again. Another randomly selected cluster is moved again, its perimeter is examined for possible formation of a larger cluster, and so on. The DLCC aggregates are also fractal objects with fractal dimensions of 1.45 and 1.8 in 2D and 3D, respectively.

Cho and Kahng (2011a, b) showed that DLCA can be mapped onto the framework of EP, by showing that Brownian motion of the clusters suppresses the mobility of the largest clusters, hence impeding their growth, leading to the discontinuous emergence of a giant cluster as a function of the number of aggregation events, or "time." Naturally, the diffusivity or speed of movement of a cluster should be inversely proportional to some power of their mass. Cho and Kahng (2011a, b) also considered the case in which a cluster's velocity was inversely proportional to the cluster's mass to a power η. They mapped out the tricritical point that separates discontinuous from continuous phase transitions as a function of η.

Cho et al. (2016) also showed that if the growth of the largest cluster in the DLCA is somehow controlled by, for example, a mechanism akin to the AP, the resulting gelation transition—formation of the largest cluster during the aggregation (see Chap. 13)—may exhibit very diverse patterns as a function of the mechanism of the growth control or the control parameter of the mechanism. The patterns included multiple stochastic discontinuous transitions, genuine non-self-averaging, and ultra-slow convergence of the transition point.

20.4.5 Emergence of Molecular Life

Every living being grows by self-replication. The process involves some sort of autocatalysis that is a necessary ingredient for the emergence of molecular life.

Zimmer et al. (2018) argued that early molecular life represents sustained emergence of large, but finite-size molecules, with its dynamics governed by finite-size effects of the type that have been described in this book (see Chap. 2). To demonstrate the correctness of their claim, they simulated the following model, which was originally developed by Worst et al. (2016).

Zimmer et al. considered a polymerization process in which polymers consisting of indistinguishable monomers concatenate either spontaneously or autocatalytically, which is similar to an actual biological system where the elementary building blocks are formed by DNA strands with a length of ten base pairs, with autocatalytic reproduction being realized by template-based ligation. In the model, reactions occur spontaneously with a rate constant k_2 or autocatalytically with a rate constant $k_1[A_n + m]$, so that the governing equation for the densities of the polymers, represented by $[\cdot]$, at time t is given by

$$\frac{d[A_{n+m}]}{dt} = (k_1[A_{n+m}] + k_2)\,[A_n][A_m] - \sum_{k\geq 1}\left(1 + \delta_{n+m,k}\right)(k_1[A_{n+m+k}] + k_2)\,[A_{n+m}][A_k]\,, \tag{20.9}$$

where $\delta_{i,j}$ is the Kronecker delta. Zimmer et al. carried out their simulation, beginning with $N = 2^K$ monomers for some constant K, and selected their time step such that $k_2 = 1$. Two quantities were monitored. One was the normalized number of reactions, $p =$ (number of reactions)/N, while the second one was the order parameter P, the normalized size of the largest polymer, $P = S_1/N$. Note that p denotes both the probability of having a link (as in percolation) and the relative number of reactions, since every reaction introduces exactly one new link.

The limit $k_1 = 0$ describes spontaneous assembly of polymers with two reactive ends, for which it is known that as $N \to \infty$, the order parameter P has a deterministic jump from 0 to 1 at the end of the reaction process, when $p = 1$. The end of the assembly process at which a single macromolecule emerges signifies gelation, and is described by the (trivial) 1D percolation, since it exhibits a simple discontinuous percolation transition at $p = p_c = 1$.

The opposite limit, $k_1 \to \infty$, defines a system in which spontaneous concatenation is completely outcompeted by autocatalytic concatenation and, therefore, it is restricted to phases in which the latter process cannot occur due to lack of templates. Thus, larger polymers emerge only by merging of two polymers of *exactly* the same size, leading to a cascade of events in which the largest polymer doubles in size. In the last step, two equally large clusters merge, $A_{N/2} + A_{N/2} \to AN$, so that it represents a special case of growth of networks (see Chap. 17 and discussions earlier in this chapter) in which the two smallest clusters always combine, hence representing the deterministic limit of growth of some random networks. The last merger occurs at exactly $p = p_c = 1$. Close to $p = 1$, the reaction rate is extremely low since the system contains only a few clusters, implying that the last steps of the polymerization take up most of the physical time t, with the reactions ending at about $t \approx p_c = 1$.

The conclusion is that in the limits $k_1 = 0$ and ∞ the gelation or percolation transition is discontinuous as $N \to \infty$. As Zimmer et al. (2018) argued, the transition

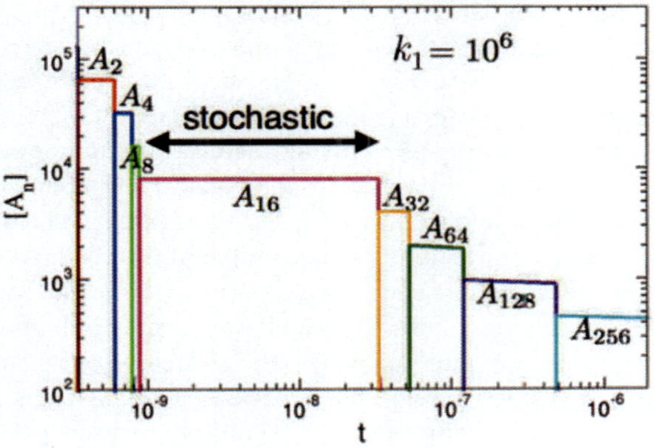

Fig. 20.7 Dynamic evolution of molecules' densities, shown for a single realization with $k_2 = 1$ and $N = 2^{17}$ (after Zimmer et al. 2018)

is, in fact, discontinuous for any value of k_1 (and k_2), because a continuous transition would require acceleration of the growth of large polymers through joining small clusters, which cannot happen in this process. Compared to the limit $k_1 \to \infty$, however, all other values of k_1 result in a slower autocatalytic merging of smaller molecules. Moreover, compared to the limit $k_1 = 0$, any other value of k_1 is equivalent to reducing k_2 from the limit $k_2 \to \infty$ to a finite value and, thus, a lower rate of spontaneous concatenation.

Another interesting feature of the model is that there are abrupt transitions between phases in which molecules of only one size are present, if the autocatalytic rate constants are very large, $k_1 \gg k_2$. This is shown in Fig. 20.7, which indicates that the sizes are in powers of 2, giving rise to a doubling cascade. In other words, after a molecule of size 2^n is spontaneously formed, the fast autocatalysis leads to an almost instantaneous transformation of all molecules of size 2^{n-1} into pairs and, hence, into molecules of size 2^n. Simulation of Zimmer et al. (2018) indicated that there is a critical value k_{1c} of k_1, such that for $k_1 > k_{1c}$ molecules of length 2^n are the dominant length in the system. k_{1c} depends on the initial monomer number N, and the simulations indicated that $k_{1c} \propto 1/N$.

Finally, it was shown that any finite-size system of the type studied is non-self-averaging. This was demonstrated by computing the relative variance, defined by $\sigma^2 = \langle G - \langle G \rangle \rangle^2 / \langle G \rangle^2$, and showing that it remains finite for any finite system, and vanishes only in the limit, $N \to \infty$, and that such a limit can be approached very slowly.

20.4.6 *Evolutionary Process in Protein Networks*

To describe the evolution of the structure of protein networks, Rozenfeld et al. (2010) built the Human Protein Homology Network (HPHN), a weighted network in which the weights represent the degree of homology—similarity—between two proteins.

As shown by Medini et al. (2006), homologous proteins organize themselves in network modules. Thus, a large number of proteins may have evolved by duplication-divergence events from a single ancestral protein, and have preserved the phylogenetic relationships in the network representation. This can lead to clustering of homologous proteins in distinct families, leading eventually to formation of giant protein clusters. The data for building the HPHN was obtained from the similarity matrix of proteins (SIMAP) project (at http://boinc.bio.wzw.tum.de/boincsimap/), which is composed of proteins in the human cell, where two proteins are linked if they are homologous.

Since any two proteins may have a degree of homology, one begins from a fully connected weighted network. The similarity between two proteins was calculated by the SIMAP through the expected value (E-value) method, which quantifies the level of statistically significant similarity between proteins. More precisely, E-value is calculating as the number of sequences in a database that are expected in a random search to align equally or more significantly with the input sequence than the hit that was found, and may be determined by the Smith–Waterman algorithm (Smith and Waterman 1981). The algorithm carries out *local* sequence alignment in order to determine similar regions between two strings of nucleic acid or protein sequences and, instead of examining the entire sequence, it compares segments of all possible lengths, and optimizes the similarity measure. Note that the E-value ranges anywhere from 0 to infinity, with 0 indicating a perfect alignment. Since an E-value that is exactly zero cannot be really found, one must set a threshold for it. Thus, Rozenfeld et al. (2010) considered links between proteins with an E-value of up to 10^{-10}. The resulting network had $N = 21,709$ nodes and $L = 1,289,345$ links.

Rozenfeld et al. (2010) hypothesized that there exists a spanning skeleton of the network that connects all the different network areas. There are, however, local modules of well-connected proteins, implying that it is much more probable that the links of new proteins have a much larger weight within a module, rather than with proteins that are further away. The simulation began with an empty network of all proteins in the HPHN with no links between them. The genetic distance between proteins was computed through the score ratio (SR) that, for two proteins 1 and 2, is defined as

$$s_{ij} = \frac{\text{alignment score between } i \text{ and } j}{\text{score of self-aligning protein } i} ,$$

where the SR is given by $\text{SR} = \min(s_{12}, s_{21})$. Then, four strategies were used to build up the network: (a) starting with the largest weight link, Rozenfeld et al. (2010) added one link or edge at a time in decreasing order of the weight corresponding to the SR; (b) the same as in (a), but in increasing order of the weights; (c) using the PR (product rule) described earlier; and (d) adding edges at random.

Figure 20.8 summarizes the results, presenting the fraction S/N of the nodes in the largest cluster as a function of p, the fraction of added links, $p = L_t/L$, where L_t is the total number of links added to the network up to time (step) t, and S is the size of the cluster. The fastest growth of the protein network is one in which links are added according to the rule (b). The PR algorithm delays the formation of the

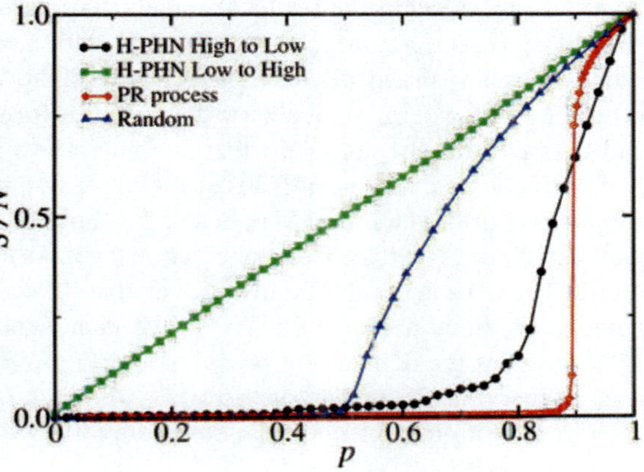

Fig. 20.8 Fraction S of nodes in the largest cluster of HPHN as a function of p of links according to the four different rules (after Rozenfeld et al. 2010)

SSC and the emergence of the EP longer than the other cases. The protein network that is grown by adding the edges according to rule (a) above is very close to the EP with the PR algorithm, case (c), and exhibits its main features. Both (a) and (c) favor growth of small clusters whose sizes grow slowly until the explosion stage where the SSC emerges.

20.4.7 Global Trade Network and Transportation

The discussion throughout this book, and particularly in this chapter, should have made it clear that the essential aspects of the emergence of large-scale connectivity have been explained by two important theoretical approaches. One is percolation theory that has provided the key insight that network formation processes are driven by adding new links or edges stochastically to the network according to some local rule. The important theoretical framework is global optimization methods that tell us network formation is controlled by a central authority or driven by a single global objective function (see Sahimi and Tahmasebi, 2021, for a comprehensive review of optimization methods). The two theoretical frameworks have armed us with enough knowledge and insight to understand the formation of a wide variety of networks.

Socioeconomic networks are, however, different from many other types of networks described in this chapter and Chap. 17 in that they are driven by "local agents" who make individual decisions based on optimizing their own goals. The result is a network constrained by many individual who, at the same time, interact with each other through optimization problems.

Schiulaz et al. (2018) investigated how large-scale networks are formed based on decisions made by rational agents. Each agent minimizes individually the costs for production, transaction, and transportation in an underlying transport network. Schiulaz et al. (2018) developed an exact mapping of the resulting nonlinear trading optimization problem onto a local model that leads to the EP via suppression of large clusters. Because of the cost of transportation, individual nodes—the agents—satisfy their demand locally as long as it is feasible. When the percolation threshold is reached, i.e., when a giant cluster or the SSC is formed, the products are purchased and shipped from the producers that do the test best globally.

Specifically, Schiulaz et al. (2018) analyzed a model for formation of fundamental supply network, which has N nodes and L links. The nodes represent agents who must satisfy their own local demand, while the links form the potential transportation routes. Thus, the network of trades that evolves between the agents is similar to bond percolation. Each node $i \in \{1, 2, \cdots, N\}$ is an agent with a fixed demand D_i. An agent satisfies the demand by buying supplies $S_{ki} \geq 0$ from any node k (including itself) satisfying the constraint that $\sum_{k=1}^{n} S_{ki} = D_i$. To achieve reasonable profit, each tries to minimize its cost C_i, defined by

$$C_i = \sum_k \left[C_{ki}^P + C_{ki}^T\right] , \tag{20.10}$$

where superscripts P and T denote production and transportation. In general, C_{ki}^P depends nonlinearly on purchases S_{ki}. Thus, we write

$$C_{ki}^P = p_k(S_k)S_{ki} , \tag{20.11}$$

where p_k, the cost per unit, depends on the total production cost S_k for agent k, given by $S_k = \sum_j S_{kj}$. The cost K_{ki}^T associated with any transaction is proportional to the amount of goods S_{ki} transported, as well as the effective distance ℓ_{ki} between agents k and i, so that

$$C_{ki}^T = p_T S_{ki} \ell_{ki} , \tag{20.12}$$

where $\ell_{ki} = \sum_e \ell_e$, with ℓ_e being the length of the edges along the shortest path between k and i, and p_T is the cost of transaction per unit good and unit distance (e.g., fuel costs, number of hours for transportation, etc.)

Equation (20.10) defines the optimization problem that all agents must solve simultaneously, giving rise to the network of optimal purchases S_{ki}. The resulting state of the network then corresponds to a Nash equilibrium. Nash equilibrium, named in honor of John Nash[5] is the most common way to define the solution of a non-cooperative game that involves two or more players in which each player knows the equilibrium strategies of the other players, and no player gains anything by changing only their own strategy. Note that if the demand of all agents is the same, $D_i \equiv D$,

[5] **John Forbes Nash, Jr.** (1928–2015), was an American mathematician who made fundamental contributions to game theory, differential geometry, and the study of partial differential equations. His work has provided insight into the factors that govern chance and decision-making in complex systems found in everyday life, and are widely used in economics. Together with **John Charles Harsanyi** (1920–2000), the Hungarian economist, and **Reinhard Justus Reginald Selten** (1930–2016), a German economist, Nash received the 1994 Nobel Prize in Economic Sciences. The 2002 movie, *A Beautiful Mind*, was roughly based on his life.

the optimal supplier is identified *locally*, since an agent i only has to ask its direct neighbors about their current suppliers to find its optimal supplier.

To demonstrate the utility of their ideas, Schiulaz et al. (2018) analyzed several problems. Consider, first, the case with large transaction costs, $p_T \to \infty$, and only internal production $i^* = i$ and $S_{ii} = D_i$, where i^* represents the optimal supplier. As p_T decreases, so also do the transaction costs, and agents minimize their total costs by making external purchases S_{ki} from agents with lower production costs. Transaction costs eventually disappear at $p_T = 0$, with all agents having the same supplier minimizing their production costs. To quantify such effects, the size $S(i^*)$ of the connected clusters in the network defined by these purchases was computed, where S is the number of agents $\{i_1, i_2, \cdots\}$ with the same supplier i^*. In particular, the size $S_1(p_T)$ of the currently largest cluster was computed. Schiulaz et al. (2018) analyzed a case in which the production cost per unit p_k decreased linearly, $p_k(S_k) = b_k - aS_k$, where $a \geq 0$ directly quantifies the strength of the economies of scale. The results were qualitatively unchanged for all forms of decreasing p_k.

The existence of discontinuous or continuous percolation transition in this model can be demonstrated by studying a random network embedded in a unit square, which reveals the significance of economies of scale. Thus, weak economies of scale (small a) lead to a continuous growth of the largest cluster, but sufficiently strong economies of scale lead to a discontinuous evolution of the size of the largest cluster in the network. A local decrease in the transaction costs triggers a cascade of decisions by agents: as the cluster size increases, the production costs of its supplier decrease and a large fraction of all agents join the connected cluster. Thus, a giant connected cluster emerges in a continuous phase transition with weak economies of scale, or a

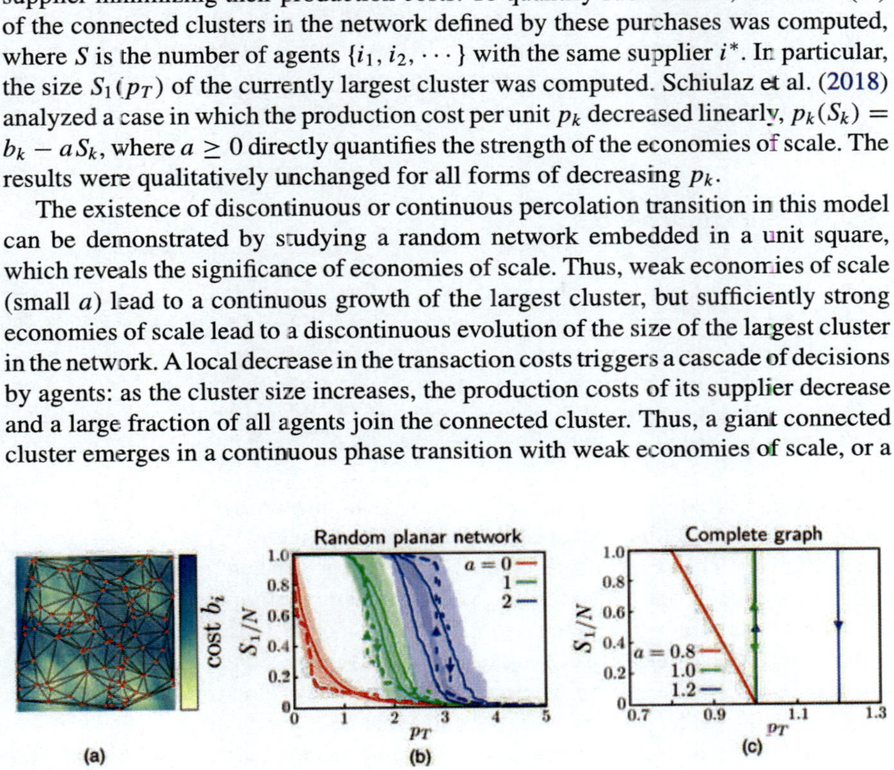

Fig. 20.9 **a** A small random network embedded in a unit square. The parameters $b_i = i/N \in [0, 1]$ are the real part of a smooth random function $b(x, y)$, generated from $2^{10} \times 2^{10}$ discrete Fourier modes with zero mean amplitude and variance, $S(\omega_x, \omega_y) = 1/(\omega_x^2 + \omega_y^2)^2$ (corresponding to a fractional Brownian motion with the Hurst exponent $H = 1/2$). **b** Single realizations (dashed lines) of the evolution of the relative size of the largest cluster for a random 2D network with $N = 10^4$ nodes, averaged (solid lines and shading) over 100 realizations of b_i. **c** The predicted behavior for a completely connected network in the mean-field limit, $N \to \infty$. Weak economies of scale $a < a_c$ lead to a continuous growth of the largest cluster. Sufficiently strong economies of scale $a > a_c$ generate a discontinuous transition, where a_c is the critical value of a. Reversing the process by increasing the transaction costs leads to a direct reversal for weak economies of scale, but strong economies of scale experience hysteresis (after Schiulaz et al. 2018)

discontinuous phase transition with strong economies of scale. These are shown in Fig. 20.9.

A realistic network topology, an elementary model of a world transport network, was also analyzed by Schiulaz et al. (2018). In this case, the nodes represent individual countries or regions, while the links make the transport routes via land between neighboring countries, or shipping routes through sea. Different network structures were analyzed by varying the costs for various modes (countries) of transportation, modifying the effective distances of transport via land and sea. If sea transport is expensive, the network has a large diameter, and multiple large clusters appear in various regions and merge when p_T becomes small. This is shown in Figs. 20.10a–20.10d. Conversely, if sea transport is cheap, the network becomes densely connected with a small diameter, similar to the complete graph, and the single largest cluster grows in a sudden cascade, as Figs. 20.10e–20.10g indicate.

20.4.8 Social Networks

Pan et al. (2011) analyzed empirical data from two distinct real-world social networks. One was a mobile phone call network, while the second one was a co-authorship networks of scientists. In each case, they built up the aggregate interaction network from the data and used it as the underlying substrate for a percolation process. In their analysis, all the edges were initially considered as "unoccupied," and an AP was used to sequentially "occupy" edges (or insert them into the growing network).

Pan et al. (2011) used a variant of the minimum cluster (MC) rule (see above), where the number of links that are compared during the addition process was treated as a parameter. They show that both the network structure and the number of links compared influence the universality class of the percolation process, whether it is ordinary or the EP, of the percolation transition. To demonstrate this, Pan et al. (2011) carried out finite-size scaling (see Chap. 2) using subnetworks, which they selected based on the known external properties of the empirical networks. The resulting estimates for the critical exponents were consistent with the view that the EP transition is second order (see above). Moreover, Pan et al. (2011) illustrated a connection between links selected by the MC rule and community structure by showing that at the percolation threshold the cluster structure that arises from the application of the MC rule contains information about the community structure of the network.

20.4.9 Explosive Immunization

Immunization of networks against epidemic spreading of a virus, diseases, and other types of destructive agents is a highly important problem. At the time of writing

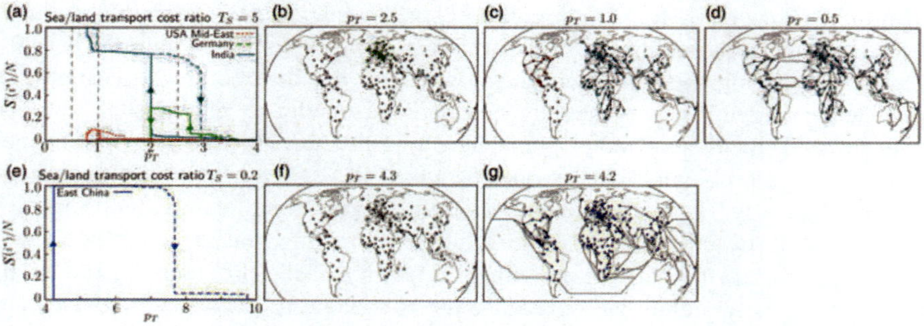

Fig. 20.10 Global connectivity induced by locally optimal decisions in a model of world transport network. **a** Evolution of the size $S_i(i^*)$ of the clusters identified by a supplier i^* when sea transport is more expensive (by a factor $T_S = 5$) than through land. The network diameter is large and multiple large clusters emerge in various parts of the world. **b–d** Network structure and active trade links for various values of the transaction costs p_T, when land routes are preferred to sea transport. **e** Evolution of the size of the emerging cluster for small sea transport costs ($T_S = 0.2$). Since sea routes connect most countries cheaply, the network diameter is small and only one large cluster emerges in a single large cascade. The transition occurs for larger p_T as the overall diameter of the network is much smaller. **f** and **g** The state of the network immediately before and after the percolation transition (after Schröder et al. 2018)

this chapter, the coronavirus pandemic was still going on strong and, thus, efficient immunization of large communities was of utmost importance. One of the most important aspects of this problems is that, COVID-19 or any other infectious disease spreads in a population by using the network of contacts between hosts (individuals). Thus, immunization corresponds to an attack that destroys the network on which the disease can spread. Hence, vaccination of network nodes (the people) is often the most effective way of preventing large epidemics. Although several strategies have been proposed for this purpose, including, for example, manipulating the network topology (Schiff 2011; Schneider et al. 2013), the approach proposed by Clusella et al. (2016) is interesting, practicable, and highly efficient, and exploits the concepts of the EP.

The most important task is to identify the nodes—blockers—that if removed, the network connectivity is destroyed, the most important of which—dubbed *superblockers*—are assumed to be equivalent to *superspreaders*, which are the most efficient infection-spreading nodes. Thus, identifying the superblockers and vaccinating them is a most efficient strategy to cut into pieces the network and stop the spread of a disease. In the algorithm proposed by Clusella et al. (2016), in a network of N a fraction qN is vaccinated, while the remaining unvaccinated nodes would be susceptible to the infection. The size of an invaded cluster depends on the fraction q of vaccinated nodes, the type of epidemic, and its virulence. The maximum fraction of nodes infected at any time is, however, bounded by the relative size $S_1(q)$ of the largest cluster of susceptible nodes $\mathcal{G}(q)$. If $S_1(q)$ is kept as small as possible, it will ensure that epidemic outbreaks of any type are as small as possible for a vaccination level q (Schneider et al. 2013). When the network is very large, $N \rightarrow \infty$, immunization is achieved by fragmenting the network so that $S_1(q) = 0$. Thus, one may define an immunization threshold q_c as the smallest value of q at which $S_1(q) = 0$. Generally speaking, the smaller q_c, the more effective is the immunization strategy, because the epidemic is prevented by vaccinating a smaller set of nodes.

In general, identifying superspreaders is a non-deterministic polynomial-time (NP) complete problem, which is one for which the correctness of each solution is verified quickly, i.e., in polynomial time, and a brute-force search algorithm can find a solution by trying all possible solutions, implying that the concept of the NP completeness concerns the worst-case difficulty of the problem. Therefore, it is computationally prohibitive. To reduce the computations, Clusella et al. used an "inverse strategy," first suggested by Schneider et al. (2012) in which one begins with a network where all the nodes are potentially dangerous and thereby vaccinated, i.e., at time $t = 0$, $q = 1$, and then the increasingly dangerous nodes are progressively unvaccinated, that is, they are made susceptible to infection. This is, of course, closely related to the EP. In addition, Clusella et al. (2016) used different schemes for $q > q_c$ and $q < q_c$, and utilized the Neumann–Ziff algorithm (see Chap. 2) for identifying susceptible clusters.

Given a network configuration with a mixture of vaccinated and susceptible nodes, Clusella et al. (2016) selected at random n candidates among the vaccinated nodes and unvaccinated the least dangerous, i.e., the weakest blocker. They based the selection on a node score that quantifies its blocking ability, with the idea that harmless nodes should be identified on the basis of the size of the cluster of susceptible nodes they would join if unvaccinated, keeping in mind that such clusters should be kept small, as well as the local effective connectivity that measures their potential danger if made susceptible. Because the relative importance of these two ingredients is significantly different below and above the immunization threshold q_c, Clusella et al. (2016) used two different scores.

The first score was defined by

$$\sigma_i^{(1)} = k_i^{\text{eff}} + \sum_{\mathcal{C}\in\mathcal{N}_i} (\sqrt{S} - 1) , \tag{20.13}$$

and was used for large q. In Eq. (20.13) k_i^{eff}, which quantifies the potential danger to the effective local connectivity, is determined from

$$k_i^{\text{eff}} = k_i - L_i - M_i(\{k_i^{\text{eff}}\}) , \tag{20.14}$$

where k_i is the connectivity at node i, L_i is the number of leaves—dead ends—that do not lead to anywhere and, therefore, are deducted. M_i is the number of strong hubs. It is subtracted because in the inverse protocol they will likely be vaccinated in case of an epidemic. Clusella et al. (2016) analyzed several networks, leading them to identify strong hubs as those nodes that satisfy $k_i^{\text{eff}} \geq K$, with $K \approx 6$. Equation (20.14) implies that, compared to nodes without hub neighbors, nodes that are surrounded by hubs may play only a minor blocking role for spread and can be left unvaccinated.

In Eq. (20.13), S is the size of a cluster $\mathcal{C}$, which depends on q, and $\mathcal{N}_i$ is the set of all clusters connected to node i. Using $\sigma_i^{(1)}$ yields small values of q_c.

But for $q < q_c$ one should not keep a small $S_1(q)$, since in that region it leads to big jumps in $S_i(q)$ when two large clusters join, as a result of which many nodes that are at the interface between the two clusters suddenly become harmless without actually being treated as such. Thus, to prevent cluster merging in that region, Clusella et al. (2016) used a second score given by

$$\sigma_i^{(2)} = \begin{cases} \infty & \text{if } \mathcal{G}(q) \text{ is not in the set } \mathcal{N}_i \ , \\ |\mathcal{N}_i| & \text{else, if arg min}_i |\mathcal{N}_i| \text{ is unique} \ , \\ |\mathcal{N}_i| + \epsilon S_2 & \text{else} \ , \end{cases} \tag{20.15}$$

where $|\mathcal{N}_i|$ is the number of clusters in the neighborhood of node i, S_2 is the size of the second largest cluster, and $\epsilon \ll 1/N$ is a small positive number. Thus, in this region, only candidates that are neighbors to the giant cluster are selected, among which the candidate with the smallest number of neighboring clusters is selected. If this is not unique, the candidate for which the second largest neighboring cluster is the smallest is picked.

Clusella et al. (2016) used their algorithms with many networks, including the Erdös–Rényi (ER) and scale-free (SF) networks (see Chap. 17), and a real-world network, a network of Scottish cattle movements, which turns out to be an SF network

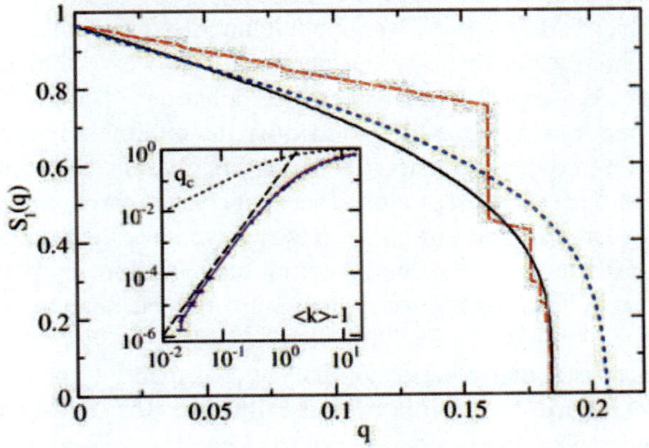

Fig. 20.11 Relative size $S_1(q)$ of the largest clusters (i.e., one normalized by the total number of nodes N) as a function of q for the Erdös–Rényi networks with $N = 10^6$ and mean connectivity $\langle k \rangle = 3.5$. The dashed curve with jumps is for the case in which the algorithm is used with score $\sigma_i^{(1)}$ is used for all q with 2000 candidates, and $K = 6$ (K is the lower bound for k_i^{eff}). The continuous curve is obtained with $\sigma_i^{(2)}$ for $q < q_c$, where $S_1(q) = 1/500$. The dotted lines are the results of optimal percolation of Morone and Maske. The inset shows a logarithmic plot of q_c (straight line) against $\langle k_i \rangle - 1$, the power law (20.16), while the dotted curve shows the result for random immunization (after Clusella et al. 2016)

with an exponent $m = 2.7 \pm 0.06$ for its power-law connectivity distribution; see Eq. (17.7). Figure 20.11 shows the results for the ER network with a mean connectivity $\langle k \rangle = 3.5$. The best results, in terms of preventing the spread of an infection as represented by the decay of the size $S_1(q)$ of the largest cluster, were obtained by using the scores $\sigma_i^{(1)}$ and $\sigma_i^{(2)}$ for $q > q_c$ and $q < q_c$, respectively, and are shown as solid curve in the main plot of Fig. 20.11, which should be compared with the dashed line obtained by using $\sigma_i^{(1)}$ for all values of q.

Note that the big jumps correspond to joining of big clusters. Also shown in Fig. 20.11 are the results reported by Morone and Makse (2015) using optimal percolation algorithm, which was described in Chap. 18. Clusella et al. (2016) estimated that $q_c \leq 0.1838$, which is also smaller than the best estimate, $q_c \approx 0.192$ by Morone and Makse (2015); see, however, the next section on optimal dismantling. Clusella et al. (2016) also carried out extensive simulation with several ER networks with various mean connectivities $\langle k \rangle$ and estimated that for small $\langle k \rangle$,

$$q_c \sim (\langle k \rangle - 1)^{2.6} . \tag{20.16}$$

This is shown in the inset of Fig. 20.11.

20.4.10 *Optimal Dismantling and Explosive "Death"*

Dismantling a network, such as the Twitter and Instagram networks, refers to removal of nodes from a network in order to fragment it into disconnected clusters. Thus, the problem is essentially the reverse of a percolation process whereby one tries to connect the isolated clusters in order to create a giant one. The problem of *optimal* dismantling is one in which one tries to identify the minimum number of nodes in a network whose removal fragments it, which is clearly related closely to explosive immunization studied in the last section. The giant cluster of a network can propagate an epidemic to a large fraction of nodes. Thus, if we view the removal of nodes as vaccination of individuals who cannot transmit the epidemic anymore, dismantling the giant cluster may be viewed as a way of organizing a vaccination campaign, albeit an extreme one.

Interestingly, it is actually easier to analyze a dismantling problem (Bradonjic et al. 2013) if we fragment a network by optimally removing edges, rather than nods. In this case, the problem is usually referred to as an *interdiction problem* (see, e.g., Zenklusen 2014). Optimal dismantling, which was presumably studied for the first time by Janson and Thomason (2008) who were interested in optimal fragmentation of random graphs (see Chap. 17), is also closely related to optimal percolation, described in Chap. 18 and mentioned in the last section above; see also Albert et al. (2000), Callaway et al. (2000), and Cohen et al. (2001) for earlier works. The problem is essentially identical with *explosive death*, explosive breakdown of a system. For example, Zhao et al. (2018) studied the emergence of explosive death in networked van der Pol oscillators with conjugate variables coupling, and demonstrated that the network structures play a critical role in identifying the types of explosive death behaviors. A van der Pol oscillator is a non-conservative one with nonlinear damping, which evolves at time t according to the following equation:

$$\frac{d^2x}{dt^2} - \mu(1 - x^2)\frac{dx}{dt} + x = 0 , \qquad (20.17)$$

with x being the position coordinate and μ a scalar parameter indicating the nonlinearity and the strength of the damping. Zhao et al. used a reverse EP process in order to cause explosive death in the network. In principle, both optimal dismantling and interdiction problem result in a discontinuous first-order percolation transition.

To define the optimal dismantling problem, we define more concretely a *C-dismantling set* of a graph as a cluster of nodes whose removal yields another graph with its largest cluster having a size (number of nodes) of *at most* C. The $C-$dismantling number of a graph is the minimal size of such a set. Determining if the $C-$dismantling number of a graph is smaller than some constant is an NP-complete decision problem (see above for the definition of such a problem). Braunstein et al. (2016) developed an algorithm for optimal dismantling by formulating it as a problem of cost minimization in which the cost function is minimized over the feasible configurations of the activation times $\{t_i\} \in \{0, \cdot, t\}^{\mathrm{T}}$, where feasible means that for all vertices i, either $t_i = 0$ in which case i is included in the cluster whose size is decreasing, or $t_i > 0$, in which case it satisfies a certain constraint, and T denotes the transpose operation. Braunstein et al. (2016) called their approach the *Min-Sum* algorithm, since the cost function to be minimized is a sum over the various configurations at various times.

Braunstein et al. (2016) showed that their algorithm produces very efficient dismantling. In particular, using the ER graph with $N = 10^7$ nodes and a mean connectivity $\langle k\rangle = 3.5$, the algorithm identified $C = 10^3$ dismantling sets (clusters), removing 17.8 percent of the nodes, which should be compared with the optimal percolation algorithm of Morone and Makse (2015) that removes 20.6 percent of the nodes (see above and Chap. 18). Moreover, the algorithm dismantled the Twitter network with 532,000 nodes into components smaller than $C = 10^3$ using only 3.4 percent of the nodes, which should be compared with that of optimal percolation that needs 5.6 percent of the nodes to dismantle the same network. This should also be compared with the algorithm described in the last section. See also Zdeborová and Zhou (2016) for another highly efficient dismantling algorithm based essentially on reverse EP.

20.4.11 Spreading of Information

An important problem in the age of the Internet, Twitter, and other social networks is the manner and speed by which any type of information is spread throughout a society. Zhu et al. (2018) used the EP to model outbreaks of information sharing in social networks, arguing that abrupt and large-scale bursts of information sharing, particularly when it comes to sensitive issues, may lead to social instability, particularly if the spreading information is false. Zhu et al. (2018) combined models of epidemic spreading on an $L \times L$ lattice with certain selection rules in order to determine to which neighboring node a node spreads its information, hypothesizing that when the authenticity of the information cannot be verified, a node spreads the information or rumor to the smallest group it has access to. To implement this, Zhu et al. (2018) used the EP with the Sum Rule (see above) (rather than the product rule) in their simulations. They demonstrated that their algorithm increases the latency period when the information stays localized in small groups. At a critical stage in the evolution, however, a global outbreak abruptly occurs, and the information is globally shared.

We note that, much earlier, Stauffer and Sahimi (2006) studied a model of how opinions about a given "extreme" subject, about which various groups of a population have different degrees of enthusiasm for or susceptibility to, such as fanaticism, extreme social and political positions, and terrorism, may spread. Interestingly, their model was, in a certain limit, the discrete analog of a deterministic continuum model suggested by others. Stauffer and Sahimi simulated their model on a scale-free network, as well as lattices with infinite- or short-range interactions.

In another paper, Stauffer and Sahimi (2007) proposed a model for spreading information that took into account the nature of the interactions network, and provided some key insights into the phenomenon. Using simulations on scale-free networks, Stauffer and Sahimi (2007) showed (a) the existence of a fundamental difference between a hierarchical network whereby people are influenced by those that are higher in the hierarchy but not by those below them, and a symmetrical network where person-on-person influence works mutually and (b) that a few "fanatics" can actually influence a large fraction of the population, either temporarily in the hierarchical networks or permanently in symmetrical networks. In addition they showed that even if the "fanatics" disappear, the population may still remain susceptible to the positions originally advocated by them. The model was, however, general and applicable to any phenomenon for which there is a degree of susceptibility to in the population.

Chapter 21
Directed Percolation and Its Applications

21.1 Introduction

As described briefly in Chap. 3, a variant of the classical percolation called *directed percolation* (DP), has been shown to have fascinating relations with a whole host of complex dynamical phenomena. When the DP was original proposed in the late 1970s (see below), it was mostly a mathematical curiosity, and perhaps a good "toy" model for learning computer simulation of a problem in statistical physics. But it was shown over time that the DP does have many important applications, which have actually been established, not only by precise computer simulations but also experiments.

Consider, for example, one of the most important applications of the DP, namely, turbulent flow, which is one of the most complex hydrodynamic phenomena, characterized by chaotic variations in the pressure and fluid velocity fields. It is the chaotic variations that make turbulent flow such a complex problem, so much so that Nobel Laureate Richard Feynman[1] called it the most important unsolved problem in classical physics. Many fluid flows that occur in nature, or are generated in engineering applications, are in the turbulent regime. In addition, we see turbulence in everyday phenomena, ranging from external flow over all types of vehicles, such as cars and airplanes, to terrestrial atmospheric circulation, fast river flows, surf, and smoke from a chimney.

At first, it may seem strange that there is a relation between percolation and turbulent flow, since the former is a purely geometrical model, whereas the latter is a chaotic dynamical phenomenon. Chorin and Akao (1991) suggested, however, that

[1] **Richard Phillips Feynman** (1918–1988) was an American theoretical physicist, who is best known for his fundamental contributions to path integral formulation of quantum mechanics, the theory of quantum electrodynamics, the physics of the superfluidity of supercooled liquid helium, and particle physics, for which he proposed the parton model, a model of hadrons (composite subatomic particles), such as protons and neutrons. In recognition of his contributions to the development of quantum electrodynamics, Feynman received the Nobel Prize in Physics in 1965, jointly with **Julian Seymour Schwinger** (1918–1994) and **Shinichiro Tomonaga** (1906–1979).

M. Sahimi, *Applications of Percolation Theory*, Applied Mathematical Sciences 213,
https://doi.org/10.1007/978-3-031-20386-2_21

some properties of turbulent flow can be studied through the analysis of the statistical equilibria of vortex filaments, which are imaginary spatial curves that induce rotary flow in the space through which they pass. For example, one may think of a vortex filament as the center of a tornado with the associated circulatory flow around its core. By considering a single filament and calculating its energy and entropy as a function of the temperature and the vortex length, Chorin (1991) discovered an interesting percolation property of vortex filaments near the maximum entropy state that consists of "polymeric" configurations, which explained the mechanics that create the maximum entropy state.

The connection that Chorin suggested between turbulence and percolation was qualitative and based on random percolation. More recent works have revealed a deep direct relation between turbulence and the DP. Therefore, this chapter will first describe the early contributions to the relation between turbulence and percolation theory by Chorin and others, and then turns its attention to the link between the DP and several of its important applications, including the transition between laminar and turbulent flows.

21.2 Statistics of Vortex Filaments and Percolation

As pointed out above, although the latest research indicates that it is the DP that is most closely linked with the transition from laminar flow to turbulence, earlier research had sought to connect the classical percolation to the transition. In this vein, Chorin (1988, 1991, 1992) considered a vortex filament that was made of N oriented links coinciding with the edges of a regular cubic lattice of bond length h in the three-dimensional (3D) space. The filament is not closed, but is self-avoiding, i.e., no point of it is the end point of more than two links, implying that no two links coincide. Chorin assumed that the probability of having a given vortex configuration is proportional to the Gibbs weight, $\exp(-E/T)$, with E and T being the energy and temperature, and both positive and negative T were allowed, which was already allowed in statistical mechanics of vortices in 2D studied by Onsager[2] (1949). Chorin defined a quantity,

$$\nu_{1,N} = \frac{\log\langle r_N\rangle}{\log N}\,, \tag{21.1}$$

where r_N is the end-to-end length of the vortex, and $\langle\cdot\rangle$ indicates an average over the Gibbs weight, and we define, $\nu_1 = \lim_{N\to\infty} \nu_{1,N}$. If $T = \pm\infty$, then, all config-

[2] **Lars Onsager** (1903–1976) was a Norwegian physical chemist and theoretical physicist. He derived the exact solution of the two-dimensional Ising model in zero external field, as well as the Onsager reciprocal relations that express the equality of certain ratios between flows and forces in thermodynamic systems that are out of equilibrium in which the notion of local equilibrium exists, for which he received the Nobel Prize in Chemistry in 1968.

urations of equal length are equally probable. Such self-avoiding configurations of equal weight are a standard model of linear polymers, with ν_1 being called the Flory exponent defined by, $\langle r_N\rangle \sim N^{\nu_1}$; in 3D, $\nu_1 \approx 0.59$. Therefore, $D_1 = 1/\nu_1$ is the fractal dimension of the filaments.

Chorin also defined a function $\phi(r)$ by

$$\phi(r) = \sum_{|i-j|\le r} \boldsymbol{\xi}_i \cdot \boldsymbol{\xi}_j \,, \tag{21.2}$$

where $\boldsymbol{\xi}_i$ is the vorticity vector, with i being a multi-index that denotes the location of a link. Let

$$\nu_{2,N} = \frac{\log\langle r_N\rangle}{\log\langle \phi(\langle r_N\rangle)\rangle} \,. \tag{21.3}$$

Note that $\phi(r)$ is the integral (sum) of the two-point correlation function of ξ over a sphere of radius r, so that $(1/S)d\phi/dr$ is the two-point correlation function, with S being the surface of the sphere. Fourier transform of the trace of the velocity correlation tensor yields an energy spectrum for the velocity field given by, $E(k) \sim k^{-1/\nu_2}$ for large wave numbers k conjugate to r, with $\nu_2 = \lim \nu_{2,N}$ as $N \to \infty$. More generally, if the support of the vorticity has a fractal dimension D_0, and $E(k) \sim k^{-\zeta}$, then, $\zeta = D_0 - D_1 + D_2$, with $D_2 = 1/\nu_2$. For a finite N, we write, $\zeta_N = D_0 - D_{1,N} + D_{2,N}$, with $D_{1,N} = 1/\nu_{1,N}$ and $D_{2,N} = 1/\nu_{2,N}$. ζ is called the Kolmogorov exponent.

Using numerical simulations, Chorin (1991) showed that

$$\lim_{N\to\infty} \nu_{1,N} = \begin{cases} 1 & T < 0 \,, \\ \nu_1 \approx 0.592 & T = \infty \,, \\ 0 & T > 0 \,, \end{cases} \tag{21.4}$$

and

$$\lim_{N\to\infty} \nu_{2,N} = \begin{cases} 1 & T < 0 \,, \\ \nu_2 \approx 2.7 & T = \infty \,, \\ \infty & T > 0 \,, \end{cases} \tag{21.5}$$

Moreover,

$$\lim_{N\to\infty} \zeta_N = \begin{cases} 3 & T < 0 \,, \\ \zeta_0 \approx 1.68 & T = \infty \,, \\ \infty & T > 0 \,. \end{cases} \tag{21.6}$$

These results can now be interpreted in terms of percolation concepts. If a vortex at negative temperature has a high energy, i.e., when it is very long, it becomes smooth, since $\nu_1 = 1$, and its spectrum corresponds to that of a smooth flow with $\zeta \ge 3$, as shown by Chorin (1988). On the other hand, if $\nu_1 = 0$, then, $r \sim N^0$, implying that the end-to-end length of the vortex does not increase with N, and it rolls up into a tight "ball," not spanning large distances. Thus, $T = \infty$ represents a percolation threshold, because infinitely extended vortices can exist on one side, but not on the other side. The threshold corresponds to a maximum entropy state and polymeric configurations

Bershadskii (1994) made the connection between percolation and turbulent flow more explicit. He began his analysis by making the well-known observation that turbulent diffusion produces highly twisted and distorted layers, in which the concentration of a passive scalar varies strongly, and that it is the layers with high concentrations that are important. He dubbed such systems *scalar fractal*, which is not the same as the fractal sets that characterize other fluid dynamical properties of turbulent flows, such as those described above. A particle of a passive scalar executes a random walk on a passive fractal so that its trajectory is linked with the motion of a scalar fractal. Thus, for an external observer, the motion of the particle resembles a random walk of the passive scalar.

Bershadskii (1994) divided the motion of such particles into three groups: (a) One in the turbulent diffusion proper in which the particle executes a random walk on the scalar fractal, including its dead ends. (b) The motion of the particle on the turbulent fractal that bypasses the dead ends. (c) The motion of the particle in the turbulent fractal, taking the shortest paths that connect two widely separated points. The second type is reminiscent of random walk on the backbone of percolation clusters (see Chap. 2), since motion into the dead ends is not allowed. The third type is similar to a random walk on the *elastic backbone*, i.e., the set of all shortest paths on the original backbone that, when elongated, would give the first contribution to a restoring force (Herrmann et al. 1984; Sampaio Filho et al. 2018).

The dynamic evolution of the scalar fractal is self-similar, which implies that the percolation time, which is on the order of turnover time, should be much smaller than the characteristic time scale of the self-similar evolution of the scalar fractal, which is of the order of molecular diffusion time. According to Monin and Yaglom (1975), the turnover time scale τ_t is given by

$$\tau_t \sim \left[k^2 E_v(k)\right]^{-1/2} , \tag{21.7}$$

where $E_v(k)$ is the energy spectrum of the fluctuations of the velocity field v. The characteristic time scale τ_c is given by

$$\tau_c \sim \left[k^3 \frac{\langle v^2 \rangle}{C_s^2} E_c(k)\right]^{-1/2} , \tag{21.8}$$

where C_s is the concentration of the passive scalar. Now, assume that, initially, large-scale distribution of a passive scalar fluctuates over a single length scale ℓ_0. As the system evolves, one has a cascade process that gives rise to the formation of a hierarchy of fluctuations over length scales $\ell_n \sim q^{-n}\ell_0$, where q is the multiplicity of the breakdown process of larger scales into smaller ones. If n is large enough, the characteristic time scale $\tau_c^{(n)}$ of the nth order is given by

$$\tau_c^{(n)} \sim \left[k_n^3 \frac{\langle v^2 \rangle}{\langle C_s^2 \rangle} E_c(k_n) \right]^{-1/2} , \tag{21.9}$$

with $k_c \sim \ell_n$. Equation (21.9) is based on Eq. (21.8). Thus, the formation of the cascade takes a time, $\tau_\infty \sim \sum_n^\infty \tau_c^{(n)}$, which is a convergent series provided that, $E_c(k) \sim k^{-\zeta}$ and $\zeta < 3$ (see above). The most important quantity is $\tau_\infty - \tau_N \sim \sum_{n=N}^\infty \tau_c^{(n)}$. For large N one has,

$$\tau_\infty - \tau_N \sim \sum_{n=N}^{\infty} \ell_n^{(3-\zeta)/2} \sim \sum_{n=N}^{\infty} q^{-n(3-\zeta)/2} \sim q^{N(\zeta-3)/2} . \tag{21.10}$$

One now needs to quantify the cascade process of scale splitting in turbulent flow. One way of doing so is by representing large-scale two-dimensional turbulence as a collection of wave packets (Bershadskii et al. 1993), according to which after N break-up $M(\tau_N) \sim q^N$ fluctuations are formed over length scales $\ell_N \sim q^{-N}\ell_0$. The cluster of such wave packets occupies a volume with an effective length scale ℓ_e and a fractal dimension D_t, so that

$$M(\tau_N) \sim \ell_e^{D_t} . \tag{21.11}$$

Since,

$$M(\tau_N) \sim (\tau_\infty - \tau_N)^{2/(\zeta-3)} , \tag{21.12}$$

it follows that

$$\ell_e(\tau_N) \sim (\tau_\infty - \tau_N)^{-2/D_t(\zeta-3)} , \tag{21.13}$$

Because $\zeta < 3$, ℓ_e represents a diverging length scale. As described in chapter two, the diverging length scale in percolation is the correlation length ξ, whose power-law divergence near the percolation threshold is characterized by the exponent ν. Therefore,

$$\nu = \frac{2}{D_t(3-\zeta)} . \tag{21.14}$$

It is straightforward to show that $\zeta = D_t$, implying that Eq. (21.14) is a quadratic equation for ζ with the solutions

$$\zeta = 3/2 \pm (9/4 - 2/\nu)^{1/2} . \tag{21.15}$$

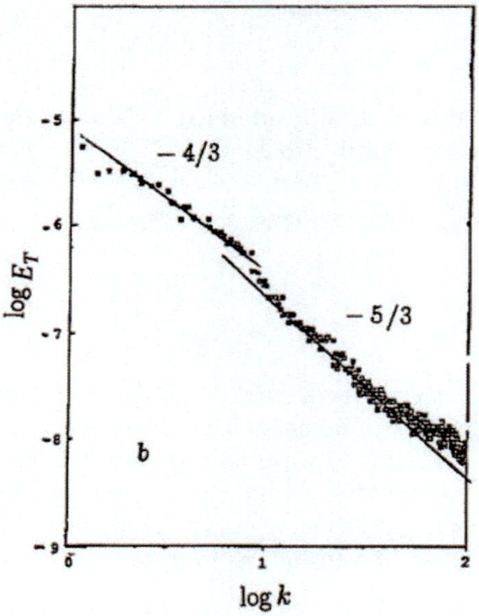

Fig. 21.1 Spectrum of temperature fluctuations in turbulent flow in an external magnetic field (after Bershadskii and Branover 1994)

If one now uses the well-known Corrsin[3]–Obukhov[4] value, $\zeta = 5/3$ (Obukhov 1949; Corrsin 1951), Eq. (21.15) yields, $\nu \approx 0.9$, in perfect agreement with 3D percolation (see Chap. 2), but not $\nu = 4/3$ for 2D percolation. In other words, the turbulence is *quasi*-2D, not strictly 2D. One also obtains $\zeta = 4/3$ for the same value, $\nu \approx 0.9$. One may also determine the fractal dimension D_h of a 2D section of the cluster of passive scalar, i.e., the fractal dimension of the intersection of the perimeter and the inner past of the cluster. It is straightforward to show that, $D_t + D_h = 3$, and therefore, $D_h = 4/3$ and 5/3 for $\zeta = 5/3$ and 4/3. Experimental evidence for a crossover between energy spectra with $\zeta = 4/3$ and 5/3 was reported by Nastrom et al. (1986). Bershadski and Branover (1994) also presented experimental evidence for the same, in terms of temperature fluctuations of a passive scalar during the flow of mercury, to which an external magnetic field was applied. This is shown in Fig. 21.1.

Bershadskii and Branover (1994) refined their values of the fractal dimension D_h of the perimeter. Experimental data for isotropic growth of clouds (Branover et al. 1993) to growth restricted in a layer (see above) indicate that one has, $D_h = 4/3$ at small scales, whereas $D_h = 7/4$ at much larger scales. The latter value is identical

[3] **Stanley Corrsin** (1920–1986) was an American physicist, fluid dynamicist, and Professor of Engineering at Johns Hopkins University. He made fundamental contributions to fluid dynamics, and in particular turbulence. He was awarded the Fluid Dynamics Prize of the American Physical Society 1983, and the Theodore von Kármán Medal of the American Society of Civil Engineers.

[4] **Alexander Mikhailovich Obukhov** (1918–1989) was a Russian physicist and applied mathematician known for his important contributions to statistical theory of turbulence and atmospheric physics. He is considered as one of the founders of modern boundary layer meteorology.

to the fractal dimension of the *hull* (i.e., perimeter) of 2D percolation clusters whose value is, $D_h = 1 + 1/\nu = 7/4$ (Saleur and Duplantier 1987).

21.3 Directed Percolation

The link between percolation and turbulent flow that was established by Chorin, as well as Bershadskii, is interesting, but is concerned with the geometrical features of turbulence. The link between the dynamics of the two is through the DP, and not the classical random percolation. To the author's knowledge, the DP was introduced for the first time by Arrowsmith and Essam (1977, 1979) as a problem on directed graphs. Obukhov (1980) was the first to study it in regular lattices. Thus, we first describe the essential features of the DP and then explain its relation with turbulent flow.

Consider a 2D or 3D lattice in which a fraction p of the bonds are present, or active, or occupied, but the rest are not. Next, pick a direction, say from between two opposite faces of a 3D lattice or between the "top" and the "bottom" of a 2D lattice and attribute orientation to the bonds in that direction, so that one can travel along the orientation, but not in the opposite. It is clear that the cluster of connected sites in the DP is a subset of the corresponding cluster in isotropic percolation. An example is shown in Fig. 21.2, which also compares it with isotropic percolation. A physical realization of the DP is a network of diodes that allow electrical current only in one direction along the bonds. Another example is water spreading in a piece of paper on a flat surface as being an example of isotropic percolation, but if we turn the paper vertically and ignore capillary action, gravity changes the behavior of water flow to the DP.

It should then be clear that the system is anisotropic and, therefore, it has two, not one, correlation lengths. One is $\xi_{\parallel}$, the longitudinal correlation length in the direction

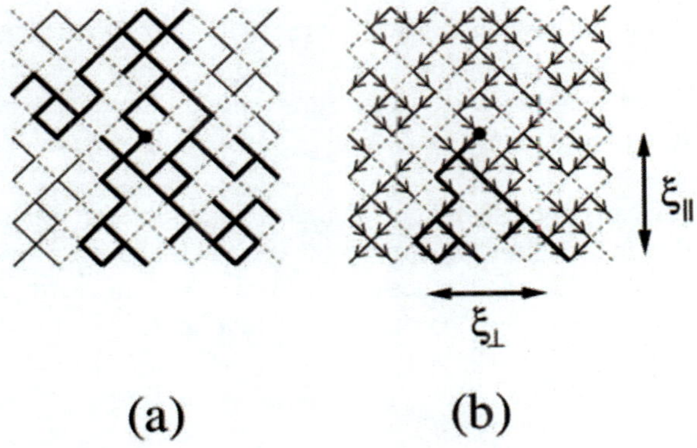

Fig. 21.2 Isotropic (**a**) and directed percolation (**b**) on the square lattice. $\xi_{\parallel}$ and $\xi_{\perp}$ are, respectively, the longitudinal and transverse correlation lengths

Table 21.1 Estimates of site and bond percolation thresholds, p_{cs}^D and p_{cb}^D, of some two- and three-dimensional lattices

Lattice	p_{cb}^D	p_{cs}^D
Square	0.6447	0.7054
Triangular	0.4780	0.5956
Simple cubic lattice	0.3821	0.4352
Body-centered cubic lattice	0.2873	0.3445
Face-centered cubic lattice	0.199	–

of macroscopic orientation of the lattice, while the second one is $\xi_\perp$, the transverse correlation length in the direction perpendicular to the main orientation. These are also shown in Fig. 21.2. Thus, the DP transition at the percolation threshold p_c^D of a lattice is characterized by two diverging length scales:

$$\xi_\parallel \sim (p - p_c^D)^{-\nu_\parallel} , \tag{21.16}$$

$$\xi_\perp \sim (p - p_c^D)^{-\nu_\perp} , \tag{21.17}$$

where $\nu_\parallel$ and $\nu_\perp$ are the critical exponents analogous to ν, the critical exponent of correlation length in isotropic percolation (see Chap. 2). Due to the constraint that directionality imposes on the lattice, the percolation threshold p_c^D of a lattice in the DP is much larger than the corresponding threshold in the isotropic lattice. Table 21.1 lists the percolation thresholds of some 2D and 3D lattices, reported by Essam et al. (1988) and Grassberger (1989).

Directed percolation can also be viewed as a dynamical system, if the preferred direction is associated with time, instead of space. Thus, 2D DP becomes (1+1) DP, and 3D DP is referred to as (2+1) DP. The upper critical dimension of the DP is, $d_u = 5$, as opposed to $d_u = 6$ for isotropic percolation (see Chap. 2). Then, in the dynamical interpretation of the DP, the upper critical dimension is 4.

In a manner similar to isotropic percolation described in Chap. 2, the critical exponents β and γ may be defined. Recall from Chap. 2 the scaling and hyperscaling relations between the various critical exponents, such as $\nu d = 2\beta + \gamma = \beta + 1/\sigma$. They can be generalized for the DP. Thus, for example, one has, $\gamma + 2\beta = (d - 1)\nu_\perp + \nu_\parallel$. Estimates of the critical exponents for 2D and 3D DP, as well as their exact mean-field values that are valid at the upper critical dimension $d_u = (4 + 1)$, are listed in Table 21.2. Also included is the "dynamical" exponent, $z = \nu_\parallel/\nu_\perp$ (see below).

One may also define an angle ϕ as the average opening of a cluster below p_c^D, or that of the sample-spanning cluster (SSC) above p_c^D in which case $\tan\phi$ is simply the reaction front in a reaction model in which the active or present sites of the cluster that react. Clearly, $\tan\phi = \xi_\perp/\xi_\parallel$, and therefore,

$$\tan\phi \sim (p - p_c^D)^{\nu_\parallel - \nu_\perp} . \tag{21.18}$$

Table 21.2 Estimates of the critical exponents of directed percolation in d−dimensional systems, as well as their exact (mean-field) values at the upper critical dimension $(4+1)$

Exponent	$(1+1)$	$(2+1)$	Mean field (4+1)
β	0.276	0.584	1
$\nu_\parallel$	1.733	1.286	1
$\nu_\perp$	1.096	0.729	1/2
γ	2.277	1.60	1
σ	2.554	2.18	2
z	1.581	1.764	2

Another exponent, called the *meandering exponent* ζ, is defined as follows. Suppose that $\mathbf{r}(t)$ is the d projection of the $(d+1)$ position vector $[\mathbf{r}(t), t]$, which describes the directed polymer in a random medium. A directed polymer (Kardar 1987) is defined as follows: For each bond ij in a regular lattice, one selects a random energy ϵ_{ij} according to a probability distribution, and then considers all directed random walks of length L, starting at a fixed origin on the lattice. By definition, then, a directed walk is one that is stretched along a single longitudinal direction, with fluctuations in the transverse directions only. Physically, a directed polymer is a stretched polymer that does not contain loops or overhangs. Then, $\langle \mathbf{r}^2(t) \sim t^{2\zeta}$. Then, it can be shown that (Hinrichsen 2000), $\zeta = \nu_\perp/\nu_\parallel = 1/z$

Directed percolation may be viewed as a reaction–diffusion system (Hinrichsen 2000). Suppose that the active sites represent particles A, while the inactive ones are simply vacancies V. The particles can either destroy themselves, or produce an offspring A. If two particles arrive at the same site, they combine to produce a single A. Thus, the DP may be interpreted as a reaction-diffusion process described by

$$\begin{aligned} \text{diffusion:} \quad & V + A \to A + V \\ \text{self-destruction:} \quad & A \to V \\ \text{offspring production:} \quad & A \to A + A \\ \text{combination:} \quad & A + A \to A \end{aligned}$$

Depending on the ratio between offspring production and self-destruction, the reaction–diffusion process may either remain active or reach an empty state from where it cannot escape, which is the *absorbing* state of the DP.

21.4 General Model of Directed Percolation and Random Resistor–Diode Networks

Just as percolation networks of resistors have been studied, and used widely to model a variety of phenomena, one can also consider a network of resistors and diodes. The most general form of such models was introduced by Redner (1983), which we follow closely to describe. Consider, for example, a square lattice whose bonds could be one of the following types: (a) *positive* diodes that point upward or from left to right, with occupation probability p_+; (b) negative diodes that, with occupation probability p_-, act in the direction opposite to the positive ones; and (c) resistors that occupy the bonds with probability p. The fraction of the vacant bonds is therefore, $q = 1 - (p_+ + p_- + p)$. Then, the following limits emerge:

(i) If $p \neq 0$ and $q \neq 0$, but $p_\pm = 0$, one recovers the standard isotropic percolation.

(ii) If $p_+ \neq 0$ and $q \neq 0$, but $p = p_- = 0$, one has the DP.

(iii) If $p_+ \neq 0$ and $q \neq 0$, but $p_- = q = 0$, one has *reverse percolation*.

(iv) if $p_+ \neq 0$, $q \neq 0$, and $p \neq 0$, but $p_- = 0$, one has an *oriented* random resistor–diode network.

(v) If $p_\pm \neq 0$, $p \neq 0$ and $q \neq 0$, one has a random resistor–diode network in which the diodes have random orientations.

In reverse percolation, there is a transition from one-way to two-way percolation as p reaches the percolation threshold p_c. Two-dimensional DP and reverse percolation are related by duality (Dhar et al. 1981; Redner 1982a). Therefore, the manner in which the angle $\phi \to \pi$ in the reverse transition is identical to the way in which $\phi \to 0$ in the DP.

For oriented resistor–diode networks, one has the phase diagram shown in Fig. 21.3, which represents the intersection of the half-spaces $p \leq 1, q \leq 1$, and $p_+ \leq 1$ with the plane $p + q + p_+ = 1$. Each corner of the triangle represents a pure state, i.e., a network filled by the bonds corresponding to the indicated probability, whereas an interior corresponds to a lattice in which all three types of bonds are present. Note that the point denoted by "isotropic" is a tricritical point, one at which all three phases, i.e., the DP, and reverse and isotropic percolation, are *simultaneously* critical (see in this chapter). If negative diodes are also present in the network, then, as described above, one has a random resistor–diode network for which the phase diagram is a tetrahedron (Redner 1983), since there are four components in the network and, therefore, one also has a *multicritical point* (Pfeuty and Toulous 1977). For both models, namely, oriented and random resistor–diode networks, one can define a macroscopic conductivity and the associated critical exponent (Redner 1982b), similar to μ, the critical exponent for the conductivity of classical percolation defined in Chap. 2.

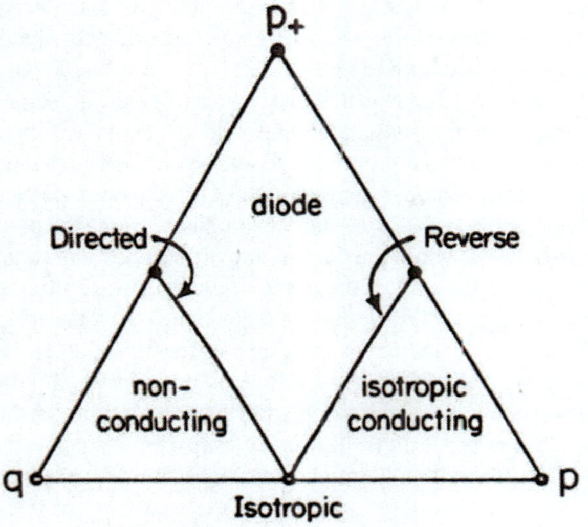

Fig. 21.3 Phase diagram of the oriented resistor–diode network (after Redner 1983)

21.5 Directed Percolation and Transition Between Laminar and Turbulent Flows

Let us first describe briefly the problem. Osborne Reynolds[5] was presumably the first who studied the transition from laminar to turbulent flow in pipes (Reynolds 1883). The Reynolds number is defined by, $\mathrm{Re} = \rho\langle v\rangle L/\mu$, where $\langle v\rangle$ and L are the average or characteristic fluid velocity (which, in pipe flow, are the cross-sectional averaged fluid velocity) and length scale of the system (the diameter of the pipe), and ρ and μ are the fluid's density and viscosity. Reynolds observed localized clusters of turbulence, which are now called "puffs," that split or decay spontaneously.

Precise experiments by Wygnanski and Champagne (1973) described systematically the phase diagram of the transition from laminar to turbulent flow as a function of the Reynolds number. It is well known that laminar flow in pipes is linearly stable for all Reynolds numbers, but small disturbances may trigger the transition to the turbulent state. If Re is sufficiently low, the fluid flow is always laminar, as any turbulent disturbances decay immediately. When, however, $1650 < \mathrm{Re} < 2050$, the puffs are metastable and, as shown by Hof et al. (2008), their lifetime grows very strongly with increasing Reynolds number (see below). For $\mathrm{Re} > 2050$, the characteristic lifetime of the puffs increases, and they begin to split and exhibit complex spatiotemporal behavior. When Re exceeds a critical value, $\mathrm{Re}_c \approx 2500$, a sufficiently large inlet disturbance generates a uniform state of turbulence. Under this condition, a "slug"

[5] **Osborne Reynolds** (1842–1912) was an Irish-born British mathematician who made fundamental contributions to fluid dynamics, as well as heat transfer between solids and fluids. The dimensionless Reynolds number Re, as well as Reynolds stresses, are named in his honor.

grows with a clearly defined turbulent–laminar interface and a velocity that scales roughly with $\sqrt{\mathrm{Re}-\mathrm{Re_c}}$ (Sreenivasan and Ramshankar 1986). It is this transition that is linked with the DP.

Next, we describe the analogy between the DP and the transition from turbulent to laminar flow (Sipos and Goldenfeld 2011). Active states in a 3D lattice correspond to coarse-grained regions of viscous scale ℓ_μ in which the turbulence intensity exceeds a threshold, whereas inactive states correspond to patches of the fluid that are laminar. As described above, the lattice dimension along the percolating direction is associated with time t, while the percolating probability p is analogous to the Reynolds number Re in the vicinity of the percolation transition, although the mapping $p \leftrightarrow \mathrm{Re}$ does not have to be linear. For the metastable puffs, the region with $\mathrm{Re} < 2050$ is mapped onto $p < p_c^D$, whereas the growing fronts with $\mathrm{Re} > 2500$ is analogous to $p > p_c^D$. The critical region in the DP maps onto the spatiotemporal regime, but one should keep in mind that this region and the percolation threshold p_c^D are not strictly defined, except in the limit of infinite system size.

To this author's knowledge, the first time that a possible link between the DP and turbulent flow, and in particular the transition from laminar to turbulent flow, was suggested was in a paper by Pomeau (1986). He studied the behavior of a front that separates a turbulent region from a laminar one, particularly at the onset of propagation when the velocity of the front is very small. This is a difficult problem since turbulent fluctuations are stochastic and, therefore, one must focus on the stochasticity, ignoring other features. Experimental data for the transition was reported by Roshko[6] (1961) for the transition to turbulence in the wake of a cylinder, indicating that the transition occurs by bursts.

Thus, Pomeau (1986) envisioned a model of an oscillation coherent parallel to one spatial direction as the vortex shedding[7] in the wake of a cylinder. Vortices are generated at the back of the body and detach periodically from either side of it, forming a what is called a *Von Kármán vortex street*, which is a repeating pattern of swirling vortices. The flow past the object generates alternating low-pressure vortices on its downstream side, with the object tending to move toward the low-pressure zone. Pomeau (1986) assumed that the coherence is lost by a phenomenon similar to the intermittent transition of a single oscillation, since intermittency coherent parallel to the cylinder is not possible because synchronization of turbulence would imply that the interaction of neighboring turbulent oscillators is strong enough to prevent exponential divergence of their trajectories. However, the interaction becomes weak at large distances.

---

[6] **Anatol Roshko** (1923–2017) was a Canadian-born physicist and engineer, and the Theodore von Kármán Professor of Aeronautics at the California Institute of Technology. A member of the National Academy of Engineering, he made significant contributions to turbulent flow and gas dynamics. Roshko was a fellow of the American Physical Society, and the American Academy of Arts and Sciences, and received the Fluid Dynamics Prize of the American Physical Society and the Timoshenko Medal of the American Society of Mechanical Engineers.

[7] Vortex shedding is an oscillatory flow that develops when a fluid flows past a bluff body at certain velocities, which depend on the size and shape of the body.

Therefore, in Pomeau's hypothetical model, a turbulent state is similar to what is induced by the intermittent transition in oscillations, and the interaction between neighboring oscillators is viewed as a type of "contamination" process, i.e., a process in which the local state of an oscillator may jump from regular to intermittent, with the interaction being equivalent to a change of the control parameter from its value in the stable-to-the-intermittent range. This implies that one must consider the fact that in the absence of triggering by a turbulent neighbor, a local oscillator has a finite life time in its turbulent state, if it is below the threshold of spontaneous firing, as assumed. Thus, consider a string of such oscillators. If an oscillator is in a turbulent state, it may either relax spontaneously toward its quiescent state, or contaminate (propagate to) its neighbors, keeping in mind that if the neighbors are already in the turbulent state, the contamination does not change anything.

But, as explained above, such a process is precisely the way the DP is defined, which is characterized by a well-defined percolation threshold that, in the present problem, is a function of the ratio of two probabilities, i.e., the probabilities of going back to regular oscillations before or after contaminating the neighbors, and staying put. As Pomeau (1986) pointed out, this link can be tested by experiments, because the critical exponents of the DP depend only on the spatial dimension, and not the details of the flow system.

Therefore, suppose that g is the control parameter for which the velocity of the front that separates turbulent from laminar regions vanishes. Near the critical value g_c of g, the front velocity v_f should follow a power law, $v_f \sim |g - g_c|^\alpha$, with α being a universal critical exponent. As described above, one dimension of the DP may be thought of as the time and, therefore, to compare with the data for the wake of the cylinder one must do so with the exponents of the DP in a 2D system, with the velocity measured spanwise in the direction of "time." Because each oscillator is linearly stable, subcritical turbulence in the percolation model would depend on random triggering by external perturbations, which represent the root, say, at the top pf the lattice in the percolation model. To avoid that, one should have a localized and permanent source of turbulence in the flow. Then, below the onset of extended turbulence, the permanent source of fluctuations would trigger turbulence over a domain of spanwise diverging extent near the onset that should follow a universal power law.

21.5.1 Numerical Simulation

Numerical simulations of Chaté and Manneville (1989) and Houlrik et al. (1990) did not support Pomeau's conjecture regarding the link between the DP and turbulence and, in fact, the computed critical exponents turned out to be nonuniversal. It was, however, suggested by Grassberger and Schreiber (1991) that the apparent nonuniversality is due to traveling solitary excitations with long lifetimes, and that one should, in principle, recover the DP universality class only for very long time scales. Rolf et al. (1998) took the next step by studying a coupled-map lattice (CML) model that exhibits spatiotemporal intermittency, which was introduced by Chaté and Manneville (1989). The CML utilizes individual maps that can be in two distinct states, namely, a chaotic or "turbulent" state, or in a "laminar" one, with the latter being an *absorbing* state for a single map, meaning that once the motion is in the laminar state, it cannot escape it. Interesting dynamical patterns emerge in such coupled systems, including penetration of turbulent patches into a laminar state that, due to the strong fluctuations, it has been dubbed *spatiotemporal intermittency*.

The original CML with one spatial and one dynamical dimension is written in terms of a field $u_i(t)$ at site i and time t as

$$u_i(t+1) = f[u_i(t)] + \frac{1}{2}\epsilon\Delta_f u_i(t) , \tag{21.19}$$

where f is the local map, and $\Delta_f u_i(t) = f[u_{i-1}(t)] - 2f[u_i(t)] + f[u_{i+1}(t)]$, which is essentially a finite-difference approximation to second-order derivative of $f(x)$. Here, ϵ is the strength of the coupling between site i and its two neighbors. The local map is given by

$$f(x) = \begin{cases} rx\ , & x \in [0, 1/2] \\ r(1-x)\ , & x \in [1/2, 1] \\ x\ , & x \in [1, r/2]\ . \end{cases} \tag{21.20}$$

Thus, for $x \leq 1$, the chaotic motion of f is governed by a standard tent map of slope r (so-called due to the tent-like shape of the graph of f). When $r > 2$, however, the trajectory may escape to a laminar state with $x > 1$, a state that is marginally stable because the slope in the line of fixed points is 1. Although as mentioned above, the laminar state is absorbing, it is no longer possible to have an absorbing state if the maps are coupled, since the interactions with its neighbors can pull a laminar site back into chaotic motion.

In this model, the dynamics is parallel or *synchronous*, i.e., all sites i are updated *simultaneously*. Marcq et al. (1996) suggested that the scaling properties of the CML, using coupled logistic maps, depend on whether synchronous or *asynchronous* updating is used, with the latter meaning that in each step a random site on the CML is selected and is iterated forward. When the updating is synchronous, the critical exponents are nonuniversal.

The nonuniversality motivated Rolf et al. (1998) to consider a CML with asynchronous updating, meaning that at time t a site i_r is selected at random and is iterated according to Eq. (21.19), while all the other variables keep their values, so that

$$u_{i_r}(t + 1/L) = f[u_{i_r}(t)] + \frac{1}{2}\epsilon\Delta_f u_{i_r}(t) , \tag{21.21}$$

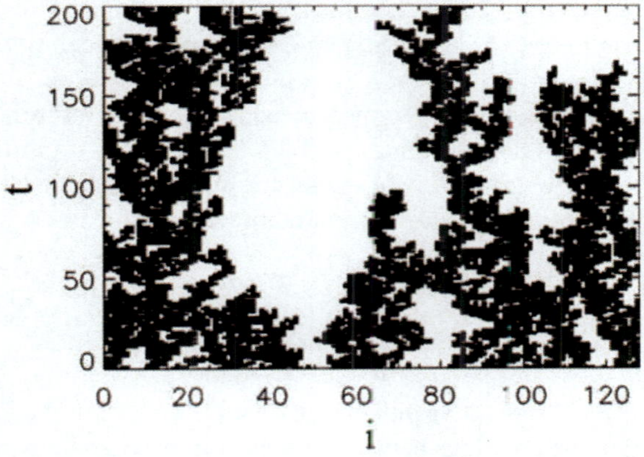

Fig. 21.4 Time evolution of the asynchronous CML with the parameters, $r = 3.0$, $\epsilon = 0.58$ and system size $L = 128$. The turbulent sites with $u \leq 1$ are black, while the laminar sites with $u > 1$ are white (after Rolf et al. 1998)

with L being the linear size of the lattice. For $i \neq i_r, u_i(t + 1/L) = u_i(t)$. Figure 21.4 shows a pattern generated by the asynchronous updating of the sites (the system size was 128×128). Because only one site was updated at a time, a new horizontal line in the time axis was added only after $L = 128$ time steps ($t \to t + 1$). As Fig. 21.4 indicates, the turbulent sites, shown by black, percolate through the system, sometimes ending in a dead-end bond. Next, the critical exponents were estimated. To do so, Rolf et al. (1998) located the critical line in the parameter plane (ϵ, r) by measuring the absorption time $\tau(r, \epsilon, L)$—the time it takes the system, starting from a random initial state, to reach the absorbing state—averaged over an ensemble of initial conditions. τ diverges at the critical point $\epsilon = \epsilon_c(L)$ according to,

$$\tau(\epsilon_c, L) \sim L^z \ , \tag{21.22}$$

where z is the aforementioned dynamical exponent. The order parameter $P(\epsilon, L, t)$, defined as the fraction of turbulent sites in the lattice (see also Chap. 2), averaged over many initial states, follows a power law as the critical line is approached from the above, $P \sim (\epsilon - \epsilon_c)^\beta$, which as the discussion in Chap. 2 showed, can be written in terms of a scaling function:

$$P(\epsilon_c, L, t) \sim L^{-\beta/\nu_\perp} h(t/L^z) \ , \tag{21.23}$$

where the scaling function $h(x)$ has the property that at times much smaller than the absorption time, one expects a power-law decay in time due to critical correlations and, thus, the $L-$dependent prefactor in Eq. (21.23) must drop out. Therefore, for $t \ll L^z$, one must have

$$P(\epsilon_c, L, t) \sim t^{-\beta/\nu_\parallel} \ . \tag{21.24}$$

On the other hand, for times much larger than the absorption time, one has uncorrelated decay of $P(\epsilon, L, t)$ and, therefore, the scaling function $h(x)$ will decay exponentially.

Rolf et al. (1998) also considered spatial correlations in the model. Note that although any dynamics of the CML will end up in the absorbing or laminar state, there are still nontrivial spatial correlations in a long lasting quasi-stationary situation. The correlations are computed based on the pair correlation function, defined by

$$C_j(t) = \frac{1}{L}\sum_{i=1}^{L}\langle u_i(t)u_{i+j}(t)\rangle - \langle u_i(t)\rangle^2 , \tag{21.25}$$

with $\langle\cdot\rangle$ indicating an average over different initial conditions. If ϵ is small, implying that the coupling between the sites is weak, one expects the spatial correlations to decay exponentially with distance, but, at $\epsilon = \epsilon_c$, one expects an algebraic decay of correlations, that is,

$$C_j(t) = j^{1-\eta}\psi[j/\xi_{\rm CML}(t)] . \tag{21.26}$$

Here, $\xi_{\rm CML} \sim t^{1/z}$ is the correlation length as the CML moves toward a steady state.

Thus, all the important critical exponents were computed by finite-size scaling for several values of r, the important parameter of Eq. (21.20). While ϵ_c did depend on r, the critical exponents did not. The results were, $z \approx 1.59$, $\beta \approx 0.28$, $\beta/\nu_\parallel \approx 0.16$, and $\eta \approx 1.50$. These are all in excellent agreement with the estimates of the same exponents for the DP, listed in Table 21.2.

Sipos and Goldenfeld (2011) simulated the bond DP in $(3 + 1)$ dimensions, using the traveling puff as the reference frame, which usually travels more slowly than the laminar mean flow velocity $\langle v\rangle$. The disturbance in the inlet of pipe flow experiments was modeled as the initial region of active—turbulent—sites. At each time step, each turbulent site of the lattice stayed turbulent with probability p, or decayed to the laminar state with probability $(1 - p)$. The turbulent lattice sites (x, y, z) could also activate adjacent laminar sites with probability p. The adjacent sites were selected differently for odd and even time steps, since the bond DP that Sipos and Goldenfeld studied was on the diagonal lattice (to simulate time forward). Therefore, for even time steps, adjacent sites were $(x + 1, y, z)$, $(x, y + 1, z)$, and $(x, y, z + 1)$, whereas, for odd time steps, they were $(x - 1, y, z)$, $(x, y - 1, z)$, and $(x, y, z - 1)$. The sites were updated sequentially.

Sipos and Goldenfeld (2011) studied two key quantities. One was the survival probability p_s of turbulent puffs in pipe flow, which is known to be given by (Faisst and Eckhardt 2004)

$$p_s(\mathrm{Re}, t) = \exp\left[\frac{t - t_0}{\tau(\mathrm{Re})}\right] . \tag{21.27}$$

Here, p_s is the probability that the turbulent puff survives after flowing for a time t, t_0 is the time for formation of the puffs, while τ is a characteristic lifetime, which of course depends on the Reynolds number. Experimentally, Hof et al. (2008) found that

$$\tau(\text{Re}) = \tau_0 \exp[\exp(c_1\text{Re} + c_2)] \ , \tag{21.28}$$

with $\tau_0 \sim L/\langle R\rangle$, and c_1 and c_2 being two fitting parameters. In the simulations, $p_s(p,t)$ is the probability that there will be active sites in the lattice after t DP steps, from which τ can be estimated. Note that, in the simulation of DP, t_0 is the time over which the initial state is remembered by the system, but is not a constant, and depends on p. Thus, one must compute the survival lifetimes of many initializations of the DP, and then construct the cumulative distribution function (CDF) of the results, which approximates $p_s(p,t)$. From the fit of the CDF to Eq. (21.27) one obtains $\tau(p)$ and $t_0(o)$. The computed results agreed with Eq. (21.28).

The second quantity that Sipos and Goldenfeld (2011) computed was the growth rate of the active DP clusters for $p > p_c$, which is related to the growth rate of turbulent slugs. The speed v_f at which the front of the percolating clusters propagate into the neighboring inactive region is given by, $v_f \sim \xi_\perp/\xi_\| \sim (p - p_c^D)^{\nu_\| - \nu_\perp}$. Thus, according to Table 21.2, the exponent for v_f should be 0.557, which is consistent with experiments (de Lozar and Hof, 2022). The computed growth rates by Sipos and Goldenfeld (2011) were in good agreement with the experiments.

Perhaps most interestingly, Doostmohammadi et al. (2017) showed, using numerical simulations, that for a distinct class of turbulence at *low* Re, the transition to turbulence in a channel is also characterized by the emergence of puffs (see above) that are generated by the microscopic activity of biological fluids, even thought, unlike the previous cases, the low Reynolds number turbulent-like flows that they studied apparently lacked a perfectly unique absorbing state. Doostmohammadi et al. showed that the critical exponents that they computed are in the universality class of the DP.

In their study, Doostmohammadi et al. (2017) solved the continuum equations of active nematics in micro-channels. The transition to turbulence occurs when one increases the amount of local energy injection (i.e., activity) in the living fluids, which in a confined channel leads to spontaneous symmetry breaking,[8] hence generating unidirectional flow, followed by an oscillatory regime that is characterized by distorted streamlines. If the activity is increased further, a stable lattice of vortices throughout the channel emerges, which transitions to meso-scale turbulence at still higher activities. The intermediate vortex lattice in active matter has been observed experimentally in motility assays of microtubules (Sumino et al. 2012) and in bacterial suspension in a confined channel (Wioland et al., 2016). These are shown in Fig. 21.5.

[8] Symmetry breaking refers to a phenomenon in which infinitesimally small fluctuations that act on a system crossing a critical point decide the system's fate by determining which branch of a bifurcation should be taken. Although looking from outside the system, the choice will seem arbitrary, it is not. The "breaking" refers to the fact that such transitions usually bring the system from a symmetric but disorderly state into one or more definite states.

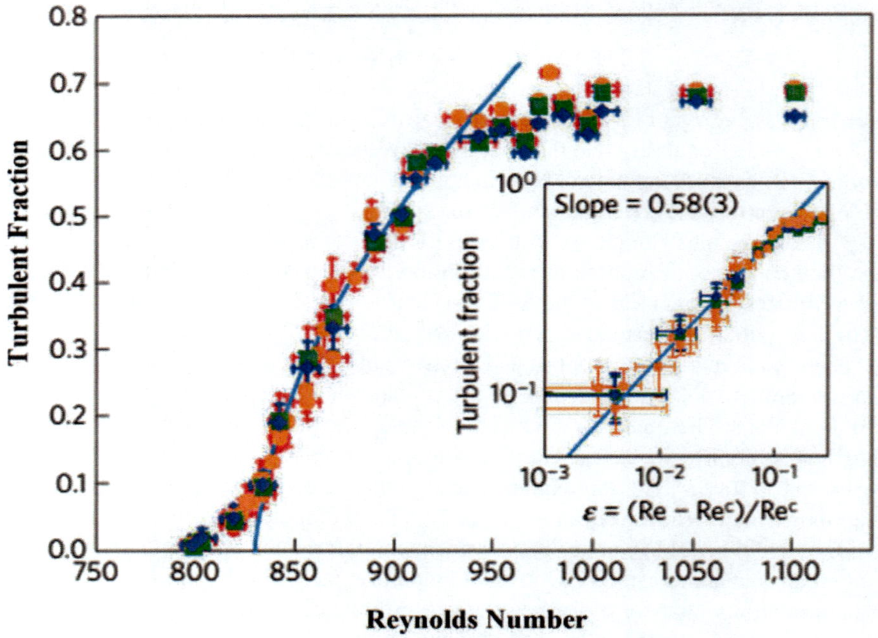

Fig. 21.5 **a** A highly ordered flow vortex lattice formed at lower activities, and **b** active turbulence, fully established at higher activities. Lower panels in (**a**) and **b** show the height-averaged enstrophy signal along the channel. **c** Coexistence of the vortex lattice and meso-scale turbulence close to the transition point. The zoomed-in panel in **d** illustrates the formation of active puffs from the vortex lattice. Color maps show vorticity contours with blue and red corresponding to clockwise and anti-clockwise vortices, respectively. The average radius of vortices is $0.32h$, with h being the channel height. Solid black lines show streamlines of the flow (after Doostmohammadi et al. 2017)

The governing equations for the velocity field $\mathbf{v}$ are given by

$$\nabla \cdot \mathbf{v} = 0\,, \tag{21.29}$$

$$\rho\left(\frac{\partial}{\partial t} + \mathbf{v}\cdot\nabla\right)\mathbf{v} = \nabla \cdot \mathbf{\Pi}\,, \tag{21.30}$$

with $\mathbf{\Pi}$ being the stress tensor. In addition to the usual order parameter, one also needs an additional order parameter for the orientational order of active fluids. To account for it, the nematic tensor, $\mathbf{Q} = \frac{3}{2}q(\mathbf{nn} - \mathbf{I})/3$, was considered, where q is the coarse-grained magnitude of the orientational order, $\mathbf{n}$ is the director—a unit vector parallel to the average molecular orientation—and $\mathbf{I}$ is the identity tensor. The governing equation for $\mathbf{Q}$ is given by

$$\left(\frac{\partial}{\partial t} + \mathbf{v}\cdot\nabla\right)\mathbf{Q} - \mathbf{S} = \Gamma\mathbf{H}\,, \tag{21.31}$$

where Γ is the rotational diffusivity. Let, $\mathbf{V} = \mathbf{Q} + \mathbf{I}/3$. Then, the co-rotation $\mathbf{S}$ is given by

$$\mathbf{S} = (\lambda\boldsymbol{\sigma} + \boldsymbol{\Omega}) \cdot \mathbf{V} + \mathbf{V} \cdot (\lambda\boldsymbol{\sigma} - \boldsymbol{\Omega}) - 2\lambda\mathbf{V}(\mathbf{Q} : \nabla\mathbf{v}) \; , \tag{21.32}$$

which represents the response of the orientation field to the extensional and rotational components of the velocity gradients, represented by the strain rate tensor, $\boldsymbol{\sigma} = \frac{1}{2}(\nabla\mathbf{v} + \nabla\mathbf{v}^{\mathrm{T}})$, and the vorticity, $\boldsymbol{\Omega} = -\frac{1}{2}(\nabla\mathbf{v} - \nabla\mathbf{v}^{\mathrm{T}})$, with T denoting the transpose operation, and λ being the tumbling parameter. If θ is the angle that the director $\mathbf{n}$ makes with respect to the flow direction, λ is defined by, $\lambda = (\cos 2\theta)^{-1}$. With E being the total free energy, the relaxation of the orientational order is determined by the molecular field $\mathbf{H}$, given by

$$\mathbf{H} = -\frac{\partial E}{\partial \mathbf{Q}} + \nabla \cdot \left[\frac{\partial E}{\partial(\nabla\mathbf{Q})}\right] . \tag{21.33}$$

The total energy E is given by

$$E = \frac{1}{2}A\mathbf{Q}^2 + \frac{1}{3}B\mathbf{Q}^3 + \frac{1}{4}C\mathbf{Q}^4 + \frac{1}{2}K(\nabla\mathbf{Q})^2 \; , \tag{21.34}$$

with K being the elastic constant. The stress tensor, $\boldsymbol{\Pi}$ is written as, $\boldsymbol{\Pi} = \boldsymbol{\Pi}^{\mathrm{visc}} + \boldsymbol{\Pi}^{\mathrm{elastic}} + \boldsymbol{\Pi}^{\mathrm{act}}$. Here, $\boldsymbol{\Pi}^{\mathrm{visc}} = 2\mu\boldsymbol{\sigma}$ is the usual viscous term; $\boldsymbol{\Pi}^{\mathrm{elastic}}$ is the contribution of the nematic elasticity, given by

$$\boldsymbol{\Pi}^{\mathrm{elastic}} = -\mathcal{P}\mathbf{I} + 2\lambda\mathbf{V}(\mathbf{Q} : \mathbf{H}) - \lambda\mathbf{H} \cdot \mathbf{V} - \lambda\mathbf{V} \cdot \mathbf{H} - \nabla\mathbf{Q} : \left[\frac{\partial E}{\partial(\nabla\mathbf{Q})}\right] + \mathbf{Q} \cdot \mathbf{H} - \mathbf{H} \cdot \mathbf{Q} \, , \tag{21.35}$$

where $\mathcal{P}$ is the pressure, and, $\boldsymbol{\Pi}^{\mathrm{act}} = -\zeta_a\mathbf{Q}$, with ζ_a being the activity coefficient, is the active contribution to the stress tensor.

Doostmohammadi et al. (2017) solved the governing equations for a 2D velocity field $\mathbf{v}$ and nematic director $\mathbf{n}$ by a hybrid of lattice Boltzmann and finite-difference methods (see, for example, Thampi et al. 2014) with the parameters $A = 0$, $B = 0.3$, $C = -0.3$, $\Gamma = 0.34$, $K = 0.04$, $\lambda = 0.3$, $\rho = 1$ and $\mu = 2/3$, in lattice Boltzmann units. A channel with a height $h = 25$ and length $L = 3,000$, with no-slip boundary conditions at the walls and periodic boundary conditions applied to its extremities, were used.

To link the simulations to the DP, Doostmohammadi et al. (2015) considered site percolation in $(1 + 1)$ dimensions on a diagonal square lattice. In nematic flows through the channel, the inactive (unoccupied in percolation) state corresponds to the vortex lattice state, and the activated (occupied) phase represents the active puffs of meso-scale turbulence. At time t, each site is occupied with some probability p_2 if both backward sites (at time $t - 1$) are occupied, and with probability p_1 if only one backward site is occupied. The DP corresponds to $p = p_1 = p_2$. When p reaches the percolation threshold p_c, the system transitions from the absorbing phase of entirely inactive states to one in which the stationary density of active sites is non-zero, which, as already described, belongs to the DP universality class.

The turbulence fraction, the area fraction, or the order parameter $P(\mathcal{A})$ occupied by active puffs in the channel, was then computed as a function of $(\mathcal{A} - \mathcal{A}_c)/\mathcal{A}_c$. Well below the critical point $\mathcal{A}_c$ (corresponding to p_c^D), active puffs have a short lifetime and rarely split, implying a negligible turbulence fraction in the steady state. As $\mathcal{A}_c$ is approached, however, puff decay becomes less likely and splitting time decreases significantly. For $\mathcal{A} > \mathcal{A}_c$, the puff population does not die out, producing a steady-state, non-zero turbulence fraction $P(\mathcal{A})$. When $P(\mathcal{A})$ was fitted to a power law, $P(\mathcal{A}) = [(\mathcal{A} - \mathcal{A}_c)/\mathcal{A}_c]^\beta$, Doostmohammadi et al. (2017) obtained, $\beta \simeq 0.275 \pm 0.043$, in excellent agreement with that of the DP in $(1 + 1)$ dimension, listed in Table 21.2. Two other exponents were also computed and were found to be in agreement with those of the DP.

The agreement is remarkable because, as Doostmohammadi et al. (2017) pointed out, it points to a parallelism between the low Reynolds number meso-scale turbulence in living fluids with a characteristic vorticity length scale, and high Reynolds number inertial turbulence, which is scale-invariant. Moreover, active puffs are generated at a small activity-independent rate. The agreement can be understood by recognizing that generating the puffs corresponds to a weak, non-ordering field conjugated to the turbulence fraction, which is known to have no detectable effect on the exponents sufficiently close to the critical point.

21.5.2 Experimental Confirmation

To this author's knowledge, the first attempt for testing by experiments Pomeau's conjecture for the link between the transition to turbulence and the DP was made by Daviaud et al. (1990; see also Daviaud et al. 1992). They studied Rayleigh–Bénard convection[9] in quasi-1D geometries, including a rectangular and an annular cell. To study the destabilization process of the homogeneous convective pattern, they carried out a series of experiments for increasing Rayleigh number Ra, the ratio of the time scales for conductive and convective thermal transport. For larger Ra, a spatiotemporal intermittent (STI) regime emerged, indicated by a strong decrease of the spatial coherence, leading to a mixing of turbulent patches within laminar domains, signaling a second-order phase transition. The critical exponents for the transition were estimated and compared with those of the DP. But, except for the exponent β, the estimates did not agree with those of the DP. Other efforts (see, for example, Michalland et al. 1993; Degen et al. 1996; Vallette et al. 1997) also failed to obtain exponents that were in agreement with those of the DP.

[9] The Rayleigh–Bénard convection is a natural phenomenon that occurs in a planar horizontal layer of fluid heated from below, with the fluid forming a regular pattern of convection cells known as Bénard cells.

Rupp et al. (2003), whose idea was to drive the fluid system into the chaotic phase with a large excitation, and then quench it toward the STI regime, took a significant step toward confirming the link between the DP and turbulent transition. To do so, they exploited the fact that applying a sufficiently strong magnetic field to a flat ferrofluid surface generates a pattern of liquid spikes. Rupp et al. (2003) made a ring of such spikes around the pole shoe of an electromagnet. The spikes, when they are under the influence of an additional oscillating current, moved, combined, and split, hence generating a chaotic phase. A charged-coupled device (CCD) camera recorded the motion of the spikes, which were then digitized and analyzed, producing data for determining the scaling exponents of the mean chaotic fraction of the system, and the correlation lengths and the inactive intervals in both space and time. Except for the critical exponent $\nu_{\parallel}$ associated with the longitudinal (time) correlation length $\xi_{|}$, the results were in reasonable agreement with those of the DP: $\beta \approx 0.3 \pm 0.05$, $\nu_{\parallel} \approx 0.7 \pm 0.05$, $\nu_{\perp} \approx 1.2 \pm 0.05$, and $\eta \approx 1.7 \pm 0.05$. Note, however, that the statistical errors were rather large, and the estimate for $\nu_{\parallel}$ is way different from the corresponding value for the DP (see Table 21.2).

Takeuchi et al. (2007; with full details given in Takeuch et al. 2009) reported consistency of the DP's critical behavior with the transition of two different turbulent states of a driven liquid crystal system. They studied an electrohydrodynamic phase transition in a thin layer of nematic liquid crystals, materials that are made up of long rod-like molecules that spatially order, such that the rods' long axes generally point in the same direction, making them anisotropic. The direction of average molecular orientation defines what is called the director field that can be subject to various defects. If one applies an increasing external voltage V to the system, the quasi-2D system moves from laminar through a succession of turbulent phases with the STI regimes in between. Takeuchi et al. focused on the transition between two topologically different turbulent states, which they referred to as the dynamic scattering modes 1 and 2 (DSM1 and DSM2). They were observed successively as the amplitude of the voltage V was increased. The difference between the two states was in their density of topological defects in the director field. Such defects in the DSM2 state, called disclinations, are generated and elongated considerably due to shear, causing loss of macroscopic nematic order and a lower light transmittance. In the DSM1 state, on the other hand, the defects are hardly elongated and their density remains very low. Therefore, the two phases can be distinguished experimentally simply by shining a light through the system and keeping track of dark and light patches, which was accomplished with a CCD.

Near and above the threshold voltage V_c, which signals the appearance of the DSM2, a STI was observed, with the DSM2 patches moving around in a DSM1 background, suggesting the existence of an absorbing phase transition with the DSM1 playing the role of the absorbing state. The key ingredient is active DSM2 patches that evolve in space-time, essentially by contamination of the DSM1 absorbing state, and recession from the DSM2 state. Thus, the order parameter $P(V)$ is just the fraction of the entire surface occupied by active DSM2 regions. Therefore, near V_c, one has, $P(V) \sim (V^2 - V_c^2)^{\beta}$. Note that deviations from criticality are measured in terms of V^2, rather than V, because the dielectric torque that drives the convection is proportional to V^2.

Takeuchi et al. carried out a wide variety of experiments, including steady-state experiments by pushing the system into the critical regime and measuring such quantities as the correlation length and order parameter, and their power-law behavior near the transition point. They employed two types of experiments. One was critical-quench experiments that characterize the system as it decays from fully active initial conditions, while the second one was critical-spreading experiments that measure the opposite behavior, namely, propagation of active regions from a single activated seed.

Takeuchi et al. measured twelve critical exponents, five scaling functions, and eight scaling relations. Their estimates for the exponents were $\beta \approx 0.59$, $\nu_\parallel \approx 1.29$, and $\nu_\perp \approx 0.75$, all of which are in good agreement with the most accurate estimates of the critical exponents of the $(2+1)$ DP listed in Table 21.2. In addition, they reported several other critical exponents that were in excellent agreement with those of the DP. For example, they measured the distributions $N(\ell)$ and $N(\tau)$ of the sizes ℓ and durations τ of the inactive DSM1 regions. The distribution $N(\ell_x)$ in the x-direction was obtained by detecting all the inactive segments in the x-direction for all y and times t. These distributions decay as power laws at the critical voltage V_c up to a finite-size exponential cutoff, with

$$N(\ell_x) \sim \ell_x^{-\mu_x} , \tag{21.36}$$

$$N(\ell_y) \sim \ell_y^{-\mu_y} , \tag{21.37}$$

$$N(\tau) \sim \tau^{-\mu_\parallel} , \tag{21.38}$$

with μ_x and μ_y indicating that the exponent $\mu_\perp$ was measured in the x- and y-directions. The measured exponents were $\mu_x \approx 1.10$, $\mu_y \approx 1.23$, and $\mu_\parallel \approx 1.61$, to be compared with those of the DP, $\mu_\perp \approx 1.204$ and $\mu_\parallel \approx 1.549$.

Since, as discussed above, the previous experimental attempts had mostly failed, the question is why was the critical behavior of the DP observed in this set of experiments. Takeuchi et al. stated that the key to their success was finding a suitable absorbing state, bringing quenched disorder under control, and having sufficient averaging, but the most important factor is clearly the absorbing state. Whereas in the previous experiments the absorbing states were theoretically laminar or fluctuation-free, in the real world long-range effects due to laminar rigidity or propagating fluctuations can break the DP scaling, or reduce the number of degrees of freedom. In the experiment of Takeuchi et al., however, the absorbing state was also turbulent and fluctuating, hence reducing long-range effects and ensuring that fluctuations do not destroy the absorbing state.

Sano and Tamai (2016) carried out an experimental investigation of channel flow in order to identify a distinctive transition to turbulence with critical scaling properties. Water at 25 °C was used as the fluid, and the channel's walls consisted of plates of 25-mm-thick polymethyl methacrylate (PMMA) with optical surface quality. It was 6 m long and consisted of three pieces with $1,960 \times 1,000$ mm^2 slots, with the ends of each slot being reinforced by welding. The side walls were made of PMMA strips of size $50 \times 5 \times 1,000$ mm^3, with the precision of the depth being ± 0.1 mm. Cross-braces were placed at 425-mm intervals along the channel, in order to avoid further deflection due to static pressure load. The channel inlet was connected to a buffering box through a smoothly curved contracting joint with an area contraction ratio of 1:20. In order to generate a turbulent boundary condition, a grid was placed near the inlet. Note that when the grid was covered with seven layers of mesh screens, the flow remained laminar in the entire channel, at least up to Reynolds number of 1,400, but because the covering by mesh screen was not sufficient at the edge, turbulent flow did not decay near either end of the buffering box located at axial distance $z = 0$ mm and $z = 900$ mm when $\mathrm{Re} = 1,400$. Thus, turbulent flows were injected from the inlet gradually, which grew and spread. Flow velocity was controlled by controlling the speed of the pump and the opening of the valve. The flow rate was measured by a flow meter.

A visualization technique with tracer particles was used. The particles were metal-coated mica platelets, 10–20 μm in diameter and 3 μm in thickness, which were added to water for visualization, with their concentration being low enough to keep the change of viscosity of water negligible. The platelets tended to align perpendicular to shear stress, implying that they were parallel to the $x-z$ surface in laminar flow, whereas they rotated in turbulent spots. The grazing angle illumination provided moderate light reflections from the laminar regions towards the front, whereas scattering from the turbulent spots was omni-directional and its intensity deviated significantly from that of the laminar regions. Six projectors were attached 250 mm apart and 300 mm above the (x, z) surfaces in order to illuminate the channel surface at a grazing angle and to attain a reasonably uniform intensity of illumination. Three CCD cameras facing the center of the $x-z$ plane of each slot synchronously captured movies of the spatiotemporal dynamics of the flow of each slot. The turbulent fraction $P(x)$—the area fraction of the turbulent region—was measured at six positions where the incidence angles from each of the six light sources to each measuring position in the channel were almost equal and, at the same time, the reflection angles from the measuring position to each of the three CCD cameras were almost equivalent. The results were fitted to

$$P(\mathrm{Re}) = P_0 \left(\frac{\mathrm{Re} - \mathrm{Re}_c}{\mathrm{Re}_c} \right)^{\beta}, \tag{21.39}$$

where Re_c is the critical Reynolds number. The results are shown in Fig. 21.6. Measurements of Sano and Tamai (2016) yielded, $\beta \approx 0.58 \pm 0.08$ and $\mathrm{Re}_c \approx 830 \pm 4$. The estimate for β is in excellent agreement with that of $(2 + 1)$ DP, shown in Table 21.2. Sano and Tamai (2016) also measured three other critical exponents, defined by Eqs. (21.36)–(21.38), and found them to be in good agreement with those of $(2 + 1)$ DP.

Further confirmation of the link between the transition to turbulence and the DP was reported by Lemoult et al. (2016). They carried out experiments with Couette

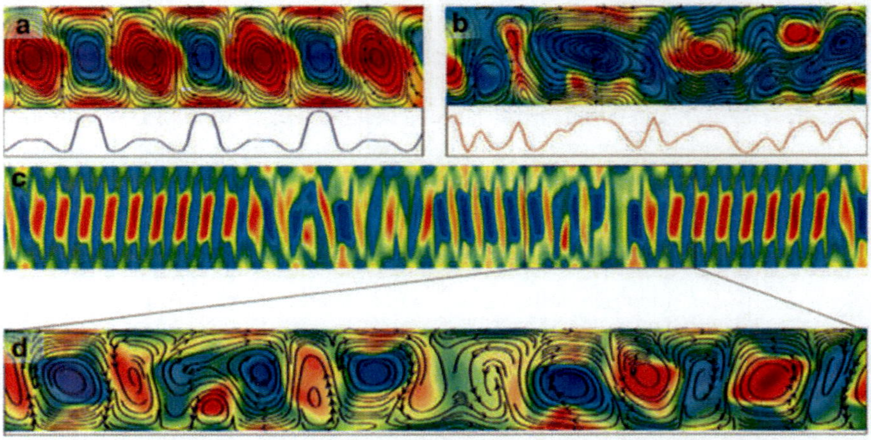

Fig. 21.6 The turbulent fraction P versus the Reynolds number Re at various downstream locations: $x/h = 1,292$ (orange square), $x/h = 1,880$ (blue diamond), and $x/h = 2,096$ (green square). Error bars are one standard deviation. Inset: a log-log plot of P as a function of reduced Reynolds number with $\mathrm{Re}_c = 830$. The solid blue lines are the best fit. The data below Re_c were removed for fitting (after Sano and Tamai 2016)

flow, which is when a fluid is sheared between two surfaces, at least one of which moves with a constant speed. Earlier experimental work to link the DP to turbulent transition (Bottin et al. 1998b; Bottin and Chaté 1998a) using Couette flow had failed, and in fact had reported a first-order phase transition, rather than a second-order one. The experiments of Lemoult et al. (2016) were carried out in a circular Couette geometry, in which water was held in the space between two concentric cylinders. The inner diameter of the outer cylinder, 220 mm, was 1 mm larger than that of the outside diameter of the inner cylinder and, therefore, there was a $2h = 500 \pm 10\mu$m gap, and an azimuthal aspect ratio, circumference/$h = 2750$. The fluid in the axial direction was confined by a top and a bottom end wall, attached to the inner cylinder, and the axial aspect ratio was, cylinder height/$h = 16$. Because of the confinement effect, turbulence was observed only at high Re (see below). The experiments were carried out at a constant temperature.

In the experiments turbulent flow was generated at sufficiently large Reynolds number, Re = 2000. Then, Re was reduced to a value close to the critical point, allowing the flow to evolve. After initial transients decayed, the percentage P of the domain that was in turbulent motion, i.e., the turbulent fraction, was measured and averaged over long times. For each Re, the turbulent fraction was averaged over 105 or more (dimensionless) advective time units, $t\langle v\rangle/h$, and Re was varied between 1600 and 1800. The experiments indicated that by decreasing Re the turbulent fraction also decreased continuously, signaling a second-order phase transition. When the data were fitted to power law (21.32), the estimate, $\beta \simeq 0.25 \pm 0.04$ was obtained, ex good agreement with that of (1 + 1) DP, $\beta \simeq 0.276$ (see Table 21.2). The exponents

$\mu_{\parallel} \simeq 1.84 \pm 0.02$ and $\mu_{\perp} \simeq 1.75 \pm 0.02$ were obtained from the data, in excellent agreement with, respectively, 1.84 and 1.748 for (1+!) DP.

21.6 Turbulent Flow and Ecological Collapse

Shih et al. (2016) carried out the extensive simulation of flow in pipes, demonstrating that, at a large scale, a zonal flow (see below) emerges that is activated by anisotropic turbulent fluctuations. Small-scale turbulence was then suppressed by the zonal flow, generating the predator–prey dynamics, which is a model of the dynamics of ecological systems. They solved the Navier–Stokes equations in a pipe of length $L = 10D$, where D is the diameter The Reynolds number at which transitional turbulence occurs is higher for short pipes and, therefore, the simulations were carried out at Re $= 2,600$, based on estimates of when puff decay transitions to its splitting. Since only transitional behavior was of interest in this problem, decomposition of large-scale modes that would indicate collective behavior, and small-scale modes that would be representative of turbulent dynamics were studied. Thus, Shih et al. (2016) computed $(\langle v_\theta\rangle, \langle v_z\rangle)$, the velocity components in cylindrical coordinates averaged over angular and axial directions θ and z. Note that $\langle v_r\rangle = 0$. This represents the aforementioned zonal flow. The numerical results suggested that the large-scale zonal flow and small-scale turbulence are necessary, and perhaps sufficient, which are components of an effective coarse-grained description of transitional turbulence.

To link the flow simulations to predator-prey model, Shih et al. (2016) simulated a three-component model, namely, nutrient (E), prey (B), and predator (A), representing, respectively, laminar flow, turbulence, and zonal flow. The model was simulated in a 2D strip represented by a 401×11 lattice. Clearly, the control parameter is the prey birth rate b. If b is small enough, the population is metastable, and cannot sustain itself, decaying with a finite lifetime $\tau_d(b)$. With increasing b the lifetime of the population, and in particular that of the prey, increases rapidly. When b is large enough, the decay of the initial population is not observed. Instead, the initially localized population splits after a time $\tau_s(b)$, spreading outward and spontaneously splitting into multiple clusters.

To show that the model is in the universality class of the DP, Shih et al. (2016) considered the limit of the transition to prey extinction, so that its population is very small and, thus, no predator can survive. Then, the prey–predator model simplifies to

$$B_i \to E_i\ , \quad B_i + E_j \to B_i + B_j\ , \quad B_i + E_j \to E_i + B_j\ .$$

where the second and third reactions occur between nearest neighbors i and j on the lattice. This model is, however, exactly the same as the reaction-diffusion model for the DP, as shown by Ódor (2004), and described earlier in this chapter.

Thus, a most interesting result emerges a fluid on the edge of turbulence has the same dynamic scaling properties that a predator–prey ecosystem has on the edge of extinction, both of which are in the universality class of the DP.

21.7 Catalytic Reactions

Ziff, Gulari, and Barshad (ZGB) (1986) developed a model for describing the catalytic reaction, $CO + O \to CO_2$ on a platinum surface, represented by a square lattice. CO and O_2 with mole fractions y and $(1 - y)$ are brought into contact with the catalytic surface. Each site on the surface is either vacant, or occupied by one CO or one O atom. CO molecules can occupy vacant sites with a probability y, whereas the oxygen molecules can split into two O atoms and occupy two neighboring sites with probability $(1 - y)$. When CO and O atoms are neighbors, they react to produce a CO_2 molecule, which leaves the surface. This simulates the Langmuir–Hinshelwood reaction.

If the surface is entirely covered by CO or O, no reaction would be possible, and the surface is deactivated and poisoned, representing the absorbing states of the model. The phase diagram of the model is shown in Fig. 21.7. If $y < y_1 \simeq 0.389$, the poisoned state with O atom emerges, but for $y > y_2 \simeq 0.525$, the surface always evolves to a CO-poisoned state. It is for $y_1 < y < y_2$ that the surface is catalytically active. But the transitions at the two critical points have different characteristics. Whereas the transition at y_2 is believed to be first order and discontinuous, the one

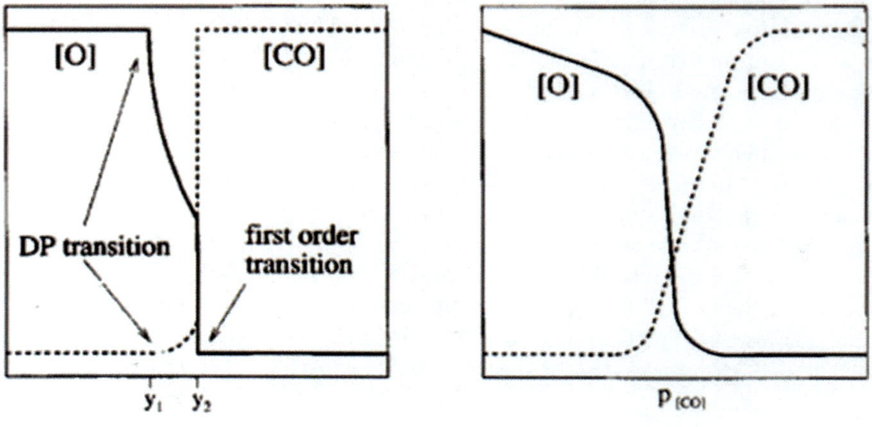

Fig. 21.7 Catalytic reaction of $CO + O \to CO_2$ according to the ZGB model (left) and experiments of Ehsasi et al. (right)

at y_1 is continuous, and is in fact conjectured to belong to the universality class of the DP.

The ZGB model attracted considered attention. Several extensive sets of numerical simulations were carried out (Grinstein et al. 1989; Meakin 1990; Jensen et al. 1990; Aukrust et al. 1990; Voigt and Ziff 1997) in order to check the question of the link between the ZGB model with the DP. These simulations did indeed confirm that the transition at $y = y_1$ is continuous and belongs to the universality class of the DP.

Ehsasi et al. (1989) measured the rate of oxidation of CO over (210) and (111) single-crystal platinum surfaces as a function of reactant pressures, P_{O} and P_{CO}, and the temperature. Their data indicated qualitative agreement with the ZGB model. The agreement improved when Ehsasi et al. (1989) modified the ZGB model and incorporated the possibility of desorption of CO from the surface according to some probability.

21.8 Depinning of Interface in Two-Phase Flow in Porous Media

The structure of a fluid interface in a porous medium during imbibition, i.e., when a wetting fluid displaces a non-wetting fluid, is very interesting. The cluster of the invading fluid during imbibition is compact, but capillary forces (see Chap. 4) lead to random local pinning of the interface that results in an interface with a rough self-affine structure. This was demonstrated by the experiments of Rubio et al. (1989). They performed their experiments in a thin (2D) porous medium made of tightly packed clean glass beads of various diameters. Water was injected into the porous medium to displace the air in the system. The motion of the interface was recorded and digitized with high resolution.

Thus, let us review briefly the dynamics of rough surfaces and interfaces, since rough self-affine surfaces are encountered in numerous systems, and occurs over a broad range of length scales, from nanometers in biological growth (Barabási and Stanley 1995) and fabrication of thin films, to meters or larger in fluid flow in porous media (Sahimi 2011). In particular, the structure of such surfaces and the dynamics of their growth have been studied with both discrete and continuum models (see Meakin 1998, for a comprehensive review).

Scaling analysis of such phenomena is also very useful, because their dynamics may be characterized by power laws and universality classes. In particular, one of the most important properties of a growing surface is the interface width $w(t)$, defined by

$$w(t) = \left(\langle h^2\rangle - \langle h\rangle^2\right)^{\frac{1}{2}} . \tag{21.40}$$

where $h(\mathbf{x}, t)$ is the height of the surface at position $\mathbf{x}$ at time t, and $\langle\cdot\rangle$ is its average over various positions. For most growth phenomena that start from a flat substrate, the width $w(t)$ follows the scaling law proposed by Family and Vicsek (1985):

$$w(t) \sim t^{\beta} F(L/t^{1/z}) \sim \begin{cases} t^{\beta} \ , \ t < t_{\times} \\ L^{\alpha} \ , \ t > t_{\times} \end{cases} \tag{21.41}$$

in which α and β are, respectively, the roughness and growth exponents, $z = \alpha/\beta$ is the dynamic exponent, $F(x)$ is the scaling function, and $t_{\times}$ is a crossover time (corresponding to a crossover length scale $L_{\times} \sim t^{1/z}$), the time at which a transition occurs from growth to one in which $w(t)$ is saturated and no longer grows with time.

One of the most successful continuum models for describing a wide range of interface growth is the model proposed by Kardar, Parisi, and Zhang (KPZ) (1986)

$$\frac{\partial h(x,t)}{\partial t} = \nu \nabla_T^2 h(x,t) + \frac{v}{2} [\nabla h(x,t)]^2 + \mathcal{A}\eta(x,t) \ , \tag{21.42}$$

in which ν is the surface tension, v is the growth velocity perpendicular to the interface, ∇_T^2 denotes the Laplacian in the transverse direction, and η represents a random noise term, with $\mathcal{A}$ being its amplitude, with the following statistical properties:

$$\langle \eta(x,t) \rangle = 0 \ , \tag{21.43}$$

$$\langle \eta(x,t)\eta(x',t') \rangle = 2\mathcal{A}\delta(x-x')\delta(t-t') \ . \tag{21.44}$$

Due to its nonlinearity, the exact solution of the KPZ equation has not been derived in any dimension. In (1+1)-growth on a line, the exact values, $\alpha_{\mathrm{KPZ}} = 1/2$ and $\beta_{\mathrm{KPZ}} = 1/3$ have been derived through scaling analysis and renormalization group method, and by random-matrix theory. In higher dimensions, however, the exponents cannot be determined analytically and, therefore, numerical simulation has been used.

It was the experimental work of Takeuchi and Sano (2010) that confirmed the predictions of the KPZ equation for the growth of rough, self-affine surfaces. They studied growing interfaces of turbulent liquid crystals (see above), focusing on interfaces between two topologically different turbulent states, the aforementioned dynamic scattering modes 1 and 2 (DSM1 and DSM2), the differences between which were already described above. The experiments were similar to what was already described. Takeuchi and Sano (2010) measured the interface width $w(L,t)$ and the correlation function, $C(L,t) = \langle [h(x+\ell,t) - h(x,t)]^2 \rangle$, with $\langle \cdot \rangle$ denoting an average over the ensembles, and confirmed the KPZ predictions.

Now that we have equipped ourselves with the necessary tools for analyzing rough self-affine surfaces, let us go back to the two-phase flow problem in porous media. Koplik and Levine (1985) proposed a linear version of the KPZ equation, given by

$$\frac{\partial h(x,t)}{\partial t} = \nu \nabla_T^2 h(x,t) + v + \mathcal{A}\eta(x,t) \ . \tag{21.45}$$

Consider the Koplik–Levine equation in zero transverse dimension,

$$\frac{\partial h}{\partial t} = v + \mathcal{A}\eta(h) \ . \tag{21.46}$$

It is then straightforward to see why pinning of the fluid interface may occur. If $v > \mathcal{A}$, where $\mathcal{N}_{\max}$, then $\partial h/\partial t > 0$, and the interface always moves with a velocity fluctuating around v. If, however, $v < \mathcal{A}$, the interface will eventually arrive at a point where $v + \mathcal{A} = 0$, and is pinned there. Therefore, there must be a pinning transition at some finite value of $\mathcal{A}$.

The opposite of pinning an interface, is its depinning, i.e., applying a force F large enough that can get the interface moving again. If F is low enough, the interface will remain pinned in certain pores. Above a critical threshold F_c, however, the interface begins moving through the medium with an average velocity v. Close to the transition point, v is expected to scale as

$$v \sim \left(\frac{F - F_c}{F_c}\right)^{\theta} \ . \tag{21.47}$$

Large values of α are due to heterogeneity of porous medium, because of which the interface does not move according to the KPZ equation, but propagates by avalanches.

Two universality classes for this type of interfacial growth have been proposed. For linear growth, the interface is described by a random field Ising model with the exponents $\alpha = 1$, $z = 4/3$, and $\theta = 1/3$ in $(1 + 1)$ dimensions. If a KPZ-type nonlinearity is present, however, the roughening process should exhibit a depinning transition, which is related to the DP, as argued by Tang and Leschhorn (1992) and Makse et al. (1998). We should, however, keep in mind that the underlying DP mechanism of depinning transitions differs significantly from an ordinary DP process in a porous medium subjected to a gravitational or another external field, since, in this case, the moving interface must percolate along a given direction so that the flow is unidirectional. In depinning transition, however, water may flow both along the external force, as well as in the opposite direction.

A simple model that exhibits a depinning transition was suggested by Hinrichsen (2000), in which the pores are represented by cells of a diagonal square lattice, and a (wetting) fluid flows to neighboring cells by crossing the edges of the cell. The edges, depending on the direction of the flow, can either be open or closed (in the percolation sense). Assume that all edges are permeable in downward direction, but they can only be crossed in upward direction with the probability p. Hence, starting with a horizontal row of wet cells at the bottom, one obtains a compact cluster of wet cells. The size of the cluster (and, therefore, the penetration depth of the fluid) depends on p, which, if it is large enough, the cluster is "infinitely" large (sample-spanning), corresponding to a moving interface. But if p is sufficiently small, the cluster is bound from above, implying that the interface becomes pinned.

Due to the structure of the pore space, a pinned interface may be interpreted as a directed path along impermeable edges, from one boundary of the system to the opposite. Thus, the interface becomes pinned only if there exists a directed path of impermeable bonds that connect the boundaries of the system, implying that the depinning transition is related to the DP process that runs perpendicular to the direction of growth. The interface becomes pinned at the lowest lying branch of the DP backbone (see Chap. 2). Therefore, the roughening exponent α is the same as what is called the meandering exponent ζ of the DP, described earlier in this chapter:

$$\alpha = \zeta = \frac{\nu_\perp}{\nu_\parallel} , \tag{21.48}$$

and the velocity exponent θ is, $\theta = \alpha$. Buldyrev et al. (1992a) carried out experiments in which a water-based ink penetrated a paper. Analyzing the surface of the ink front, they determined α to be, $\alpha \simeq 0.63 \pm 0.04$. According to Table 21.2, $\alpha = 0.729/1.286 \simeq 0.57$ close to the experimental value. In another set of experiments with several types of pore space, Buldyrev et al. (1992b) measured the roughness exponent to be, $\alpha \simeq 0.5 \pm 0.05$, which is also close to the theoretical prediction.

21.9 Flowing Sand

Daerr and Douady (1999) reported on a series of elegant experiments. In their experiments, glass beads, representing sand, with diameters of 250–425 μm were poured uniformly on a plane of size 1 m at an angle of inclination of ϕ_0, starting at the top of the surface that was covered by a rough velvet cloth. As the beads flowed downward, a thin layer of thickness $h = h_d(\phi_0)$ that consisted of several monolayers settled and remained stationary. The sand layer of thickness h_d was dynamically stable, with h_d decreases with increasing ϕ_0

For each ϕ_0 there existed, however, another thickness $h_s > h_d$, beyond which a static sand layer became unstable, implying that there existed a region in the (ϕ, h) plane in which a static layer was stable, but a flowing one was unstable. Now, take the settled system at thickness $h_d(\phi_0)$ and increase its angle of inclination to $\phi = \phi_0 + \Delta\phi$, making sure that the system stays within the region of mixed stability. Although the layer did not flow spontaneously, disturbing it generated flow at the top, and an avalanche propagated, leaving behind a layer of thickness $h_d(\phi)$. The avalanches were a more or less regular triangle with an opening angle θ, which decreased if $\Delta\phi$ did so and vanished as $\Delta\phi \to 0$.

On the other hand, when the plane was lowered so that $\Delta\phi < 0$, the system's thickness $h_d(\phi_0)$ was less than its current thickness of dynamic stability $h_d(\phi)$. In this case, an initial perturbation should not propagate, but rather die out after a certain time, or beyond a certain size $\xi_\parallel$. Thus, one may expect

$$\xi_\parallel \sim (-\Delta\phi)^{-\nu_\parallel} . \tag{21.49}$$

One elegant aspect of these experiments was the fact that pouring sand at inclination ϕ_0 produced a critical system precisely at the borderline, with respect to changing the angle, between a stable system with $\phi < \phi_0$ in which perturbations died out, and an unstable regime with $\phi > \phi_0$ in which perturbations persisted and spread. The results of the experiments reported by Daerr and Douady (1999) were puzzling, because the aforementioned threshold phenomenon produced clusters whose shapes differed from those seen in the DP, as they were much more compact.

There is, however, a corresponding model called compact-directed percolation (CDP), which is unstable against perturbations toward the standard DP behavior, first studied by Domany and Kinzel (1984). Because there was no fine-tuning of the experimental system, the results of Daerr and Douady (1999) were surprising, since the DP is a generic system and, therefore, one would expect to find non-compact active regions with the DP exponents. Janssen (1981) and Grassberger (1982) conjectured that any continuous spreading transition from a fluctuating active phase into a single frozen one should belong to the universality class of the DP, if there are only short-range interactions and no exceptional properties, such as higher symmetries or quenched randomness.

Hinrichsen et al. (1999) associated the threshold phenomenon with the DP, assuming that near the angle of preparation ϕ_0 the behavior of the sand system is related to a DP problem with $\Delta p \propto p - p_c^D$. Recall that near the DP transition point, the angle of opening θ is given by Eq. (21.18). Thus, we expect to have

$$\tan\theta \sim (\Delta\phi)^{\nu_\parallel - \nu_\perp} . \tag{21.50}$$

Therefore, one can determine $\nu_\parallel$ and $\nu_\perp$ in the experiments, and compare their values with those associated with the DP and CDP. For the CDP, one has (Dickman and Tretyakov 1995), $\nu_\parallel = 2$, $\nu_\perp = 1$, and $\beta = 0$. They proposed a model that could explain qualitatively the crossover from the DP to CDP. For the CDP, their model produced, $\nu_\parallel - \nu_\perp \simeq 0.98 \pm 0.05$, in agreement with the theoretical result.

21.10 Neuronal Avalanche in Brain

As is well known, living neurons exhibit diverse patterns of activity. As discussed in the previous chapters, many phenomena, such as earthquakes and forest fire, propagate in systems that are in a critical state in which the sizes of the events exhibit no characteristic scale and are described by power laws. Inspired by this fact, Beggs and Plenz (2003) hypothesized that a similar type of activity with complex emergent properties may exist in networks of cortical neurons. To check the validity of their hypothesis, Beggs and Plenz (2003) studied mature organotypic cultures and acute slices of rat cortex by recording continuously spontaneous local field potentials using a 60-channel multielectrode array, and showed that propagation of spontaneous activity in cortical networks is described by equations that govern avalanches. In particular, Beggs and Plenz (2003) reported occurrences of intermittent bursts of local field potentials in in vitro multielectrode recordings of cultured and acute slices of the rat brain, which occurred with a clear separation of time scales, dubbed *neuronal avalanches*.

A neuronal avalanche is characterized by its size S—the total number of significant voltage deflections recorded by electrodes between periods of silence—as well as by its duration T, which is the number of consecutive time bins spanned by an avalanche. The data collected by Beggs and Plenz (2003) could be well represented by power–law distributions for both the sizes of avalanches and their durations

$$P_s(S) \sim S^{-\tau} , \tag{21.51}$$

$$P_d(T) \sim T^{-\tau_d} . \tag{21.52}$$

The two exponents were reported to be, $\tau \simeq 3/2$ and $\tau_d \simeq 2$. Due to the scale-invariance of the two distributions, it was immediately speculated (Beggs and Plenz 2003; Chialvo 2004, among others) that brain may operate in some sort of a critical state, undergoing a second–order phase transition. Beggs and Plenz (2003) themselves proposed that the observations and data can be explained based on a critical branching process (Harris 1963), since the values of the two exponents were compatible with that process. The proposition was further supported by the fact that the branching parameter close to the critical value of 1 was also obtained from the data.

But, the exponents are also compatible with the mean–field values of the same in the DP, as shown by Muñoz et al. (1999). Moreover, as already discussed above, stochastic models that have a unique absorbing state, which in the present context implies that all neurons are quiescent, and have no other symmetries, usually belong to the DP universality class. Therefore, it is only natural to ask whether this is also the case in the present problem. Carvalho et al. (2021) addressed this problem.

But, before describing Carvalho et al.'s work, it should be pointed out that other experimental data indicated deviations from the aforementioned values of the two exponents, τ and τ_d. For example, as pointed out by Carvalho et al., ex vivo recordings of the turtle visual cortex (Shew et al. 2015), spike avalanches of rats under ketamine–xylazine anesthesia (Ribeiro et al. 2010), and electroencephalography (EEG) or magnetoencephalography (MEG), or M/EEG, avalanches in resting or behaving humans (Zhigalov et al. 2015) exhibited significant deviations from the values of the two exponents. Moreover, it is well known that non-critical systems can also exhibit power-law distributions for their properties. Thus, one needs an extra test to check the relevance of critical phenomena, and by extension of the DP, to the present problem. It was suggested by Touboul and Destexhe (2017) that since at criticality the average avalanche size $\langle S\rangle$ for a given duration T follows the power law,

$$\langle S\rangle \sim T^{1/(\sigma\nu z)} , \tag{21.53}$$

where σ, ν and z are the usual critical exponents of dynamic processes, such as the DP (see Table 21.2), if the phenomenon is at criticality, the three exponents should satisfy the following scaling relations (Muñoz et al. 1999, Friedman et al. 2012):

$$\frac{1}{\sigma \nu z} = \frac{\tau_d - 1}{\tau - 1} . \tag{21.54}$$

Note that this scaling relation is satisfied by the mean-field values of the exponents of the DP, as both sides of the equations are equal to 2.

Fontenele et al. (2019) studied cortical spike avalanches of urethane-anesthetized rats with an experimental setup that is known to yield highly variably spiking activity, ranging from very asynchronous to very synchronous population activity. The two regimes are characterized by distinct ranges of the coefficient of variation (CV) of the population firing rate, which is believed to be a simple marker of cortical states. Fontenele et al. (2019) reported, $\tau = 1.52 \pm 0.09$ and $\tau_d = 1.7 \pm 0.1$, so that, $(\tau_d - 1)/(\tau - 1) = 1.4 \pm 0.1$, and $1/(\sigma \nu z) = 1.28 \pm 0.03$ and, therefore, the scaling relation (21.54) is approximately satisfied with an intermediate value of CV. This suggests a phase transition away from both the synchronized and desynchronized ends of the spectrum of the spiking variability. Note also that the aforementioned values of the avalanche exponents are not all compatible (within error bars) with their mean-field values for the DP.

Carvalho et al. attributed such inconsistencies to finite-size effects. That is, they proposed that the measured activity is based on a very small fraction of the total number of neurons in a given area of the brain. In fact, it has been shown (see, for example, Priesemann et al. 2014; Wilting and Priesemann 2019) that such finite-size effect do indeed affect the apparent distribution of avalanches. To demonstrate the finite-size effects, Carvalho et al. (2021) developed a model and carried out numerical simulations with a large number of neurons, on the order of 10^5, but intentionally restricted their analysis to a small number of cells, $\sim 10^2$, to mimic the fact that, in practice, one can only record and analyze a few hundred neurons among the millions in a rat's brain

Carvalho et al. carried out their simulations using two distinct discrete models. One was the so-called excitatory/inhibitory network of Girardi-Schappo et al. (2016) in which each neuron is a stochastic leaky integrate-and-fire unit with discrete time step equal to 1 ms, connected in an all-to-all graph. The details of the model, which is quite complex, are given in the original reference, and will not be repeated here. The second simpler model consisted of a network of probabilistic excitable cellular automata in a random graph (Kinouchi and Copelli 2006), which is very similar to a branching process, and is known to mimic the changing inhibition-excitation levels of cortical cultures. The details of the model are as follows.

Each site i ($i = 1, \cdots, N$) has five states that are the silent state, $s_i = 0$; the active state, $s_i = 1$ corresponding to a spike, and the remaining three states with $s_i = 2, 3$, and 4 that will not respond to incoming stimuli. Each site receives input from K presynaptic neighbors that are selected randomly at the start ($K = 10$ was used) and kept fixed throughout the simulations. A quiescent site i becomes excited, i.e.,

$s_i(t) = 0 \to s_i(t+1) = 1$, with probability p_{ij}, if a presynaptic neighbor j is active at time t. At each time step, all presynaptic neighbors are swept and independently considered so that its probability is given by

$$P_s[s_i(t+1) = 1 | s_i(t) = 0] = 1 - (1 - h_i) \prod_{j \in \mathcal{N}(i)}^{K} [1 - p_{ij} s_j(t)] , \qquad (21.55)$$

with h_i being the probability of site i spiking due to an external stimulus and, $\mathcal{N}(i)$ the set of presynaptic neighbors of i. The remaining transitions happen with probability 1, including the transition $4 \to 0$ that returns i to its initial quiescent state. The time step of the model was equivalent to 1 ms. The probabilities p_{ij} were selected from a uniform distribution in an interval $[0, 2\lambda_n/K]$, with $\lambda = K\langle p_{ij}\rangle$ being the branching ratio, which is the control parameter of the model. It is known (Kinouchi and Copelli 2006) that the model undergoes a phase transition of the mean-field DP type at a critical value, $\lambda = \lambda_c = 1$. For $\lambda < 1$, the system is in the subcritical phase and eventually reaches the absorbing state in which $s_i = 0$, $\forall i$. For $\lambda > 1$, on the other hand, the system is in the supercritical phase, exhibiting self-sustained activity with a nonzero stationary density of population firings. A single random neuron was stimulated ($h_i = 1$) only when the system reached the absorbing state, sparking the network activity and subsequently being set back to $h_i = 0$. The system was initially set with a single randomly selected site being active and the others in the silent state. Carvalho et al. also presented experimental data for five rats.

From their experimental data, Carvalho et al. (2021) identified a CV value of 1.46 ± 0.08 that indicated the phase transition point, at which they estimated that, $\tau = 1.54 \pm 0.12$, $\tau_d = 1.73 \pm 0.18$, and $1/(\sigma\nu z) = 1.3 \pm 0.02$, which are not consistent with the mean-field values for the DP. When all the simulation data were analyzed, the exponents, $\tau = 3/2$, $\tau_d = 2$, and $1/(\sigma\nu z) = 2$ were obtained. When the data for only one hundred neurons, the same as in the experimental data, were analyzed, the results were different. Carvalho et al. (2021) obtained a CV value of 1.41 ± 0.05, signaling a phase transition at which the *effective values*, $\tau = 1.65 \pm 0.02$, $\tau_d = 1.87 \pm 0.03$, and $1/(\sigma\nu z) = 1.34 \pm 0.02$ were obtained. Thus, the mean-field DP phase transition with subsampling does reproduce an entire range of experimentally observed avalanches across a range of the CV values.

21.11 Satellite Networks

Another important application of the DP is to Low Earth Orbit (LEO) satellites. We only describe briefly the main ideas of the application, and refer the interested readers to the original references. Since thousands of LEO satellites cover Earth, their networks can provide coverage and services that cannot be provided by terrestrial communication systems. But, they also suffer from long propagation delay, redundancy overhead, and uncertainty of their communication links. On the other hand, the regular and dense LEO satellite constellation also provides motivation for designing new network architecture and protocol.

Hu et al. (2020) used the DP to design a new routing system for LEO satellites. In their design, each satellite routes a packet over several Inter-Satellite-Links (ISLs) towards the intended destination, without relying on link-layer retransmissions. Their algorithm overcomes the link redundancy overhead and delay/reliability tradeoff, because it can control the size of percolation cluster for routing. More precisely, consider a $m \times n$ network. Each satellite has four ISLs and receives at most two copies of the same packet [since it is directed], and it will also send out at most two copies. In the DP routing, each satellite, which acts as a router in the LEO backbone network, will relay any packet to the neighboring satellite which is closer to the destination of the packet. In other words, similar to the DP, the packet always moves in one direction.

Thus, in the DP routing, the total transmission cost of a packet is the number of links, $2mn + m + n$, which is much more efficient the standard flooding. At the same time, the load is also evenly distributed to all the links in the network. Since the satellites move, their motion may lead to changes in the topology of the grid or network, but the location-based DP routing decides based on the *current* locations of neighboring satellites. An important benefit of DP routing is that, given the redundant but evenly distributed transmissions, the end-to-end reliability is substantially improved without relying on link-layer retransmission.

Using the Starlink[10] as an example, Hu et al. (2020) demonstrated that with their DP routing, the intercontinent propagation delay is reduced significantly, while the reliability is several orders higher than single-path optimal routing.

---

10 Starlink is a satellite internet constellation operated by Space Exploration Technologies Corporation, known as SpaceX, which provides satellite Internet access coverage to 36 countries.

Chapter 22
Percolation in Large-Scale Systems

22.1 Introduction

Many of the applications of percolation theory that were described in the previous chapters are now well established in the sense that percolation is invoked to explain the phenomena and the related problems. Even the most recent applications are those associated with biological systems, and those involving general random networks and graphs of the type described in Chap. 17, are also well established.

In this chapter, several lesser-known applications, including some that are seemingly unusual, or to some extent "surprising," are described. Such applications only go to show the power of percolation theory, and the importance of connectivity and its role in the macroscopic properties of heterogeneous systems.

22.2 Mobile Ad Hoc Networks

In wireless ad hoc networks, a set of mobile hosts with wireless interfaces form a temporarily connected structure without the aid of any fixed infrastructure of a centralized administration. The nodes of such networks communicate directly with each other within their wireless transmission ranges. Outside nodes must, however, rely on some intermediaries to relay messages. Dynamic paths are used by the nodes to transmit packets (messages). Thus, a multihop process takes place whereby several intermediate hosts relay the packets sent by the source host before they reach their destination. As a result, every node functions as a router. Furthermore, the success of the communication depends strongly on other nodes' cooperation.

Such systems have been dubbed *mobile ad hoc networks* (MANETs), which represent random, self-configurable, and rapidly deployable networks. The main goal of developing the MANETs is not only obtaining better service but also having networks that can serve in situations in which no other means of communications can operate. Examples include networks that are used in battlefields, in search-and-rescue

M. Sahimi, *Applications of Percolation Theory*, Applied Mathematical Sciences 213,
https://doi.org/10.1007/978-3-031-20386-2_22

operations, and in networks of sensors. The emergence of the MANETs is not only due to their capability for serving in hazardous conditions but also their low cost, as well as their fast and simple deployment.

The major challenges to address in the development of ad hoc networks are developing efficient routing, *naming*, *service discovery*, and *security provisions*. Naming is the automatic assignment and manipulation of names—the internet protocol (IP) addresses in the present context—to the communicating nodes/networks that may join, leave, merge, or even partition from a MANET. Service discovery refers to algorithms and protocols that are used to locate, and sometimes advertise, services provided by some mobile nodes by other requesting nodes. Security provisions refer to structural considerations made in the design of the networks' connectivity, protocols, and other subsystems, and aim to secure communication in terms of robustness, privacy, and correctness. Such challenges may be overcome, if one develops a suitable model for the MANETs, which can help the development of a better protocol efficiency.

As in any large-scale network—whether it is a network of pores in a porous medium or a communication network—estimation of any global or macroscopic property based on the local observations or data poses difficulty to computing in general, and ad hoc networking in particular. This implies, in the context of the MANETs, that the mobile nodes are supposed to "think" globally while "observing" only the local environment, such as their connectivity and range of transmission.

One of the most important features of any wireless network is the ability to provide some means of complete or partial connectivity among the communicating nodes. The network is considered *connected* if for every pair of nodes a communication path between them can be identified. Consequently, a *partially connected* network provides the communication means only for a certain fraction of the nodes. As the discussions throughout this book should have made it clear, connectivity of a network influences its operations. For example, a reliable multicast, a concurrent resource management, and many other applications and services require complete connectivity of the network, whereas the anycast-based ones require only partial connectivity. Anycasting is sending a message to at least one member of a certain group. As an illustrative example in the military domain, it is considered a common achievement for the enemy to cut off the communication connectivity of its foe. As another example, in a civil search-and-rescue operation after an earthquake, the search teams should remain connected even if subsequent earthquakes intercept the communication lines. Therefore, it is wise, and indeed practical, to proactively monitor the network and provide enough connectivity.

Sometimes, the goal of the network is not necessarily interconnecting the nodes, but to convey some control commands to a few particular actors. For example, consider a military network of sensors guiding an unmanned autonomous vehicle (UAV), or a civil network of monitoring potential fires, using mobile sensors. The common feature of all such cases is the need to convey a message from one side of the network to the other side, which is usually called the *command chain*. In such a problem, the cluster of the nodes making the command chain is precisely the same as the giant cluster (GC), introduced in Chap. 17 and used in the subsequent chapters.

Characterizing the effect of the local connectivity on the behavior of communication network is not, of course, a new problem. Glauche et al. (2003) focused on identifying a physical local coordination rule for the transmission power, with the goal of guaranteeing the global connectivity of the entire network. To analyze the properties of their model, Glauche et al. allocated a power P to each node based on some locally observable parameter, such as the number of its links, which is equal to the number of its one-hop neighbors, i.e., those that are directly linked to a node. Franceschetti et al. (2005, 2007) invoked percolation to study the effect of unreliable communication on the connectivity of wireless ad hoc networks. The imperfect links were modeled by considering a random connection model in which each pair of points was connected according to some probability; see also Koushanfar et al. (2007).

In addition, graph theory (see Chap. 17) has been used in the past to model and describe ad hoc communication networks and their connectivity; see, for example, Dousse and Thiran (2004), Iyer and Manjunath (2006), and Kumar et al. (2008). But these works considered simple structures, such as one-dimensional (1D) models, and attempted to derive the necessary and sufficient conditions for the connectivity of the graph. Much of what they presented, were, however, known, in one form or another, in the literature on the classical percolation problem.

Mohammadi et al. (2009) proposed a percolation model for the MANETs, and characterized the connectivity and structure of the command chain, the GC. To see the link between percolation theory and the MANETS, one may view the active or present sites in a percolation system as the nodes in a MANET, with the difference that in the latter the nodes are randomly distributed in space and, thus, cannot be part of any regular lattice with nearest-neighbor connections. Moreover, two nodes are connected if they are within their respective transmission range, implying that a node can be connected to a large number of other nodes, some of which may be at very long distances. These are, of course, the basic features of complex networks described in Chap. 17. Because of the transmission range of each node, as well as its number of nearby nodes, there is randomness in the degree of the connectivity of each node. The two quantities, namely, the transmission range and the density of the nodes, are the important parameters that determine whether a given link between two nodes is "alive" or "dead." A link between two nodes may die by, for example, their mobility that takes them out of the transmission range. A node may die because, for example, the battery that it uses is drained and can no longer supply the required power.

To carry out the simulations, Mohammadi et al. (2009) used a 2D system of size 1000 m $\times$ 1000 m. A number of nodes were inserted in the system at random, after which their transmission ranges were set, which may or may not be equal. Two nodes A and B are connected if and only if the distance r_{AB} between them satisfies, $r_{AB} \leq \min(R_A, R_B)$, where R_A and R_B are the transmission ranges of the nodes. Thus, for every node A, all other nodes B that are within the transmission range are identified and linked to A, if the criterion is satisfied.

The model has some similarities with the variable-ranging hopping model that was developed in the 1970s, and described in Chap. 15, for estimating the hopping conductivity of amorphous semiconductors. Recall that in that model a hopper could jump from a site to any other site, if the energetics of the system were favorable for the hop, hence acting as if the site is connected to far away nodes, not only its nearest neighbor ones. In the present model, however, there are "physical" links between any node and many other nodes in the network, provided that they are all within each other's transmission range. Note that the networks that Mohammadi et al. (2009) developed and simulated were *not* of the scale-free or small-world type of networks described in Chap. 17.

The percolation behavior of the wireless networks is evaluated based on the two measures of strength defined above. Consider, for example, a network in which the GC percolates for a certain transmission range, implying that the command chain connectivity is established. To understand how the wireless network loses its percolation connectivity, one simply decreases the strengths by removing the nodes and/or cutting the links, and estimates the number of nodes and links that can be removed while keeping the network connected.

To evaluate the performance of a MANET, one must compute several important characteristic quantities. They are, (a) the connectivity, quantified by the connectivity ratio, defined as the fraction of the size of the largest completely connected part of the network and the total number of the nodes. A network is connected when the connectivity ratio is 1; (b) coverage, since an ad hoc network is not only intended for providing connectivity between the nodes but also for its ability for covering an area, a certain distance, etc., with sustainable connections, with a good example being sensor fields for both military and civil applications that must cover an area to provide sustainable chains of data flow among the sensors, and (c) the shortest path count between two nodes, because if the network loses one (or more) nodes and/or links, and does so repeatedly such that the shortest path between two nodes fails for some reason, one can identify the next shortest path between the same two nodes, given their fixed transmission ranges. From a user's perspective, the number of routes, starting from the shorter ones, can be determined after a path loss due to, for example, the mobility (link loss), power drain (node loss), or a combination of the two. This would then be an important factor for measuring the networking quality.

Two other important properties are the connectivity/command-chain strength, and power consumption. Mobility is an inherent characteristic of the MANETs, which imposes frequent link loss on the network, hence threatening its connectivity, coverage, and the pre-established routes. On the other hand, in some cases, such as hand-held devices, the battery power drain may also cause a serious node loss, which in turn might hurt the vital networking parameters, which is why the connectivity strength is defined. Generally speaking, power is a major challenge in wireless networking because the more "dead" nodes one has, the less powerful networking structure one achieves. Transmission power consumption follows the transmission range as a power law with a certain exponent.

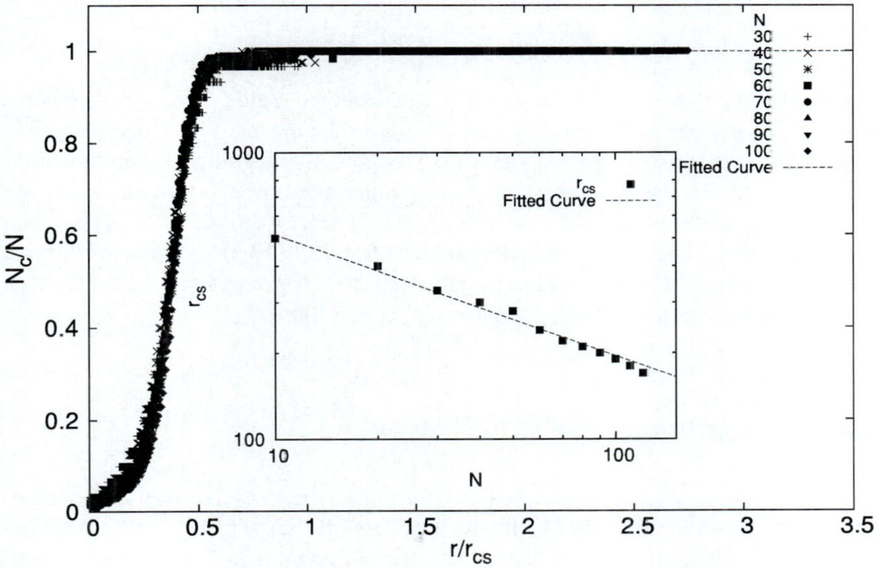

Fig. 22.1 Scaled maximum cluster size N_c as a function of the scaled transmission range r_{cs} for several total number of nodes N. The inset shows the logarithmic plot of r_{cs} as a function of N (after Mohammadi et al. 2009)

Mohammadi et al. (2009) studied the percolation properties of their MANET network, and carried out a scaling analysis, in order to identify their universal features. For example, they simulated networks with various numbers of nodes and transmission ranges, assumed to be the same for all the nodes, and determined the size S_c of the giant cluster. Their simulations indicated that

$$\frac{S_c}{N} \propto f(r/r_{cs}) \ , \tag{22.1}$$

where $f(x)$ is the scaling function. This is shown in Fig. 22.1. Note the sharp transition, which resembles a first-order discontinuous transition. The inset of the figure presents a logarithmic plot of the critical transmission range r_{cs} for the formation of the GC as a function of the total number of nodes N, with the dashed line representing a power law,

$$r_{cs} \sim N^{-\beta} \ , \tag{22.2}$$

with $\beta \approx 0.5$, implying that if the numerical value of the node density is known, each node can "guess" its own r_{cs}. Similarly, r_{cg}, the critical transmission range for forming a GC, was found to follow power law (22.2). In addition, the scaling function was found to be well approximated by

$$f(x) = \frac{1 + c_1 \exp(-c_2 x)}{1 + c_3 \exp(-c_2 x)}. \quad (22.3)$$

with $c_1 \approx 104$, $c_2 \approx 21$, and $c_3 \approx 3151$. Note that, in the region $r \approx r_{cs}$ one presumably has a power law and the data do indicate it, but the extent of that region appears to be small. Thus, each node can "guess" its local density by a single communication with its neighbors. Then, it can, making a simple approximation, assume that the global density of the nodes is the same as the local one so that it can also "guess" the numerical value of r_{cg}. The results are then used in Eq. (22.3) in order to estimate the size of the GC. Other percolation and connectivity properties of the MANET networks were also computed by Mohammad et al. (2009).

22.3 Topography of Earth and Moon

Topography, the study of land surfaces, lays the underlying foundation of a landscape. It refers to, for example, valleys, rivers, or craters on the surface. The distinguishing factor for topographic maps is the elevation. In geographic information system (GIS) mapping, one uses digital elevation models for a terrain. The vast majority of topographic maps show contour lines, which are simply lines of equal elevation.

Let us first define the power spectrum $\mathcal{S}(\omega)$ of a set of data. It is defined as the square of the coefficients in a Fourier series representation of the data (time or spatial series), which measures the average variation of the data at different wavelengths ω. Over a wide range of length scales, the power spectrum $\mathcal{S}(\omega)$ of linear transects of Earth's topography has a remarkable characteristic scaling relation with the wave number ω as

$$\mathcal{S} \sim \omega^{-\alpha}. \quad (22.4)$$

If adjacent data points in a series are completely uncorrelated, then $\mathcal{S}$ will be a constant, whereas, for strongly correlated ones relative to points that are far apart, it will be large (small) at small ω (large) wavelengths. Earlier works (Vening Meinesz[1] 1951; Sayles and Thomas 1978; Newman and Turcotte 1990) indicated that $\alpha \simeq 2.0$. A more careful study of the spectral analysis of Earth's data, averaged over all latitudes, indicated, however, that, $\alpha \simeq 1.88 \pm 0.10$, which was supported by cross-check analysis (Fallah et al. 2016). This is shown in Fig. 22.2. Similar scaling relations have been identified in Earth's bathymetry—the underwater equivalent of topography (Bell 1975)—the topography of natural rock surfaces (Brown and Scholz 1985), and that of Venus (Kucinskas et al. 1992).

Stochastic processes that are characterized by the power-law power spectrum given by Eq. (22.4) are usually referred to as the fractional Brownian motion. In that case the exponent α is usually written as, $\alpha = 2H + 1$, with H being the Hurst

[1] **Felix Andries Vening Meinesz** (1887–1966) was a Dutch geophysicist and geodesist, known for his invention of a precise method for gravimetry, and discovering anomalies in continental drift. The crater Vening Meinesz on the Moon is named in his honor.

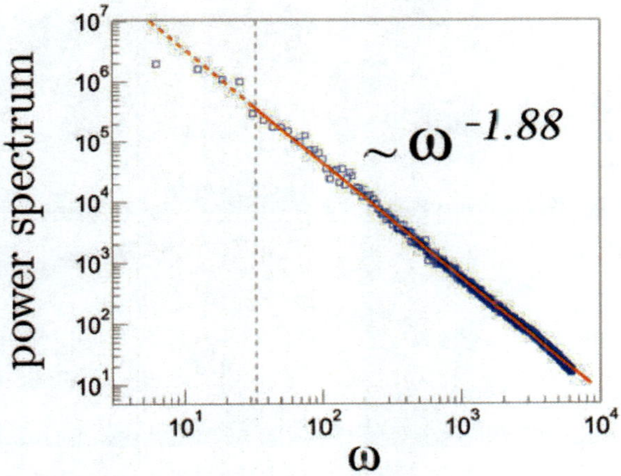

Fig. 22.2 The power spectrum of Earth's data versus the wave number ω. Solid line indicates the power-law fit to the data in the scaling region, followed by the extrapolated dashed line for the large length scales (after Saberi 2021; courtesy of Dr. Abbas Ali Saberi)

exponent such that $0.5 < H \leq 1$ indicates positive correlations between successive data points, whereas $0 < H < 0.5$ is representative of negative correlations, with $H = 1/2$ denoting random successive increments. Saberi (2021) provides a good review of such properties.

An excellent review of the relationship between percolation and statistical topography is given by Isichenko (1992). Saberi (2013) developed a percolation description of global topography of Earth by defining a geoid-like level as an equipotential spherical surface, acting as a counterpart to the percolation parameter p (see Chap. 2). When the hypothetical water level is decreased from the highest to lowest available heights on Earth, a geometrical phase transition occurs at a certain critical level h_c around which most parts of the landmass join together. Remarkably, the critical level h_c coincides with the current mean sea level, $h_c = 0$ on Earth. Saberti assumed the height relief $h(r, \theta, \phi)$ to be on a sphere of unit radius $r = 1$, which also coincides with the present mean sea level (as zero height level) on Earth, hence expressing all the corresponding lengths in units of the Earth's average radius.

Next, as Saberi (2013) suggested, one imagines that the global landscape is flooded such that the continental land masses are crisscrossed by a series of narrow channels, so that the resulting sea level everywhere on Earth would coincide with a spherical surface—the geoid. Then, all regions above the water level are colored differently as disjoint islands, while the rest is left white; see Fig. 22.3. If the water level is high, there will be small disconnected islands, whereas, if it is low, there will be disconnected lakes. Thus, there should be a critical sea level h_c at which a percolation transition takes place.

As usual, one must define an order parameter that, in this case, is defined as the probability of any site to be part of the largest island, completely similar to the classical percolation. As Fig. 22.4 shows, the order parameter for islands has a sharp decrease around the zero height level, right at the present mean sea level, which is indicative of a geometrical phase transition at this level. A similar analysis for the

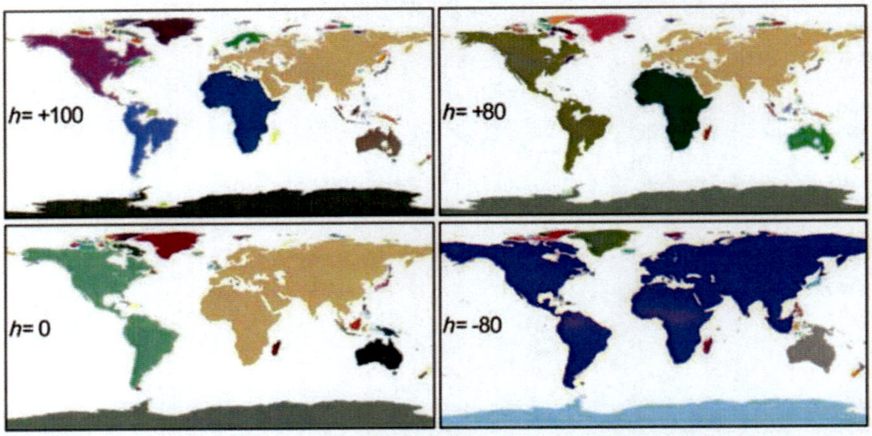

Fig. 22.3 Schematic illustration of the continental aggregation by decreasing the sea level, indicating a percolation transition at the present mean sea level around which the major parts of the landmass join together (after Saberi 2021; courtesy of Dr. Abbas Ali Saberi)

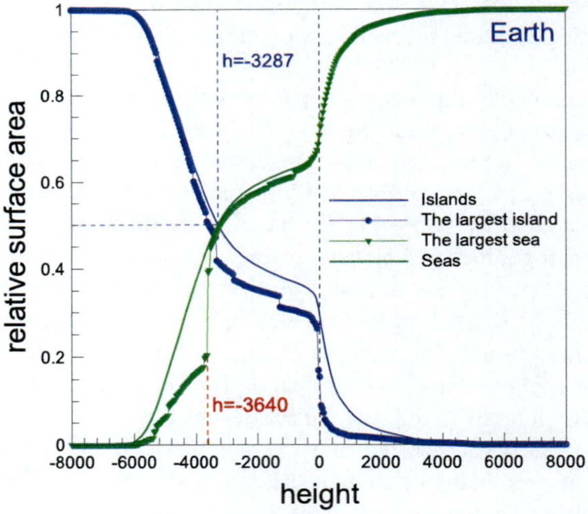

Fig. 22.4 Relative surface area of the largest island (circles) and the largest sea (triangles), followed by the total surface area of the islands and the oceans (solid curves) divided by the total area 4π of Earth, as a function of the sea level (in m). One critical level is distinguished by each order parameter. The oceanic critical level is close to the level $h = -3287$ m at which the total island and oceanic surface areas are equal (after Saberi 2021; courtesy of Dr. Abbas Ali Saberi)

oceanic clusters—where disjoint oceans at each level are differently colored, leaving islands white—gave rise to a discontinuous jump in the oceanic order parameter at around 3640 m. This is also shown in Fig. 22.4.

Saberi (2013) also computed the mean island size S, defined by

$$S = \frac{\sum_s s^2 n_s(h)}{\sum_s s n_s(h)}, \tag{22.5}$$

completely similar to the classical percolation (see Chap. 2), where $n_s(h)$ is the average number of islands of size s at level h, and the sum is over all islands, except the largest one. The correlation length ξ is also defined in a manner analogous to the classical percolation:

$$\xi^2 = \frac{2\sum_s R_s^2 s^2 n_s(h)}{\sum_s s^2 n_s(h)}, \tag{22.6}$$

with R_s being the radius of gyration of a given island of size s. Saberi (2013) showed that both quantities diverge at the percolation threshold h_c, as one expects for a second-order percolation transition.

In a subsequent work, Saberi and co-workers (Fan et al. 2019) developed a different percolation model for the topography of Earth. They used the topographic data of the ETOPO1 Global Relief Model, which is used to estimate the volumes of the world's oceans and to derive a hypsographic curve of Earth's surface (see, https://www.ngdc.noaa.gov/mgg/global/global.html). The resolution was 1 arc min, hence implying that the map contained $N = 10800 \times 21600$ grid points. The present mean sea level (zero height) was assumed as a vertical datum of the height relief $h(\phi_i, \theta_i)$, where ϕ_i and θ_i are, respectively, the corresponding latitude and longitude of grid point i.

As the first step, all the grid points in the ETOPO1 Global Relief Model were ranked according to their height $h(\phi_i, \theta_i)$, from the largest to the smallest. To each site of a 2D lattice, a number was attributed according to the position of its height in the rank. Then, starting from an unoccupied lattice, the sites were occupied one by one according to their ranking, i.e., one first selects the site with the highest height, then the second, and so on. At each step, the fraction of occupied sites p increases by the inverse of the total number of sites N in the Earth's relief landscape, which generates a configuration of occupied sites corresponding to every p. Each node has indeed four nearest neighbors.

The percolation clusters were then identified based on classical graph theory, namely, a cluster is a subset of nodes such that there exists at least one path from each node in the subset to another. Periodic and free boundary conditions were used, respectively, along the zonal and the meridional directions. A series of subnetworks, or subgraphs, were obtained, which were denoted by G_m. As in classical percolation (see Chap. 2), the order parameter was defined as the relative size of the largest cluster. Since Earth is spherical, the largest cluster in the spatial relief network is defined by

$$P(p) = \frac{\max \left[\sum_{i \in G_i} \cos(\phi_i), \cdots, \sum_{i \in G_m} \cos(\phi_i), \cdots \right]}{\sum_{i=1}^{N} \cos(\phi_i)} . \tag{22.7}$$

Fan et al. (2019) argued that Earth's relief network undergoes several abrupt and statistically significant phase transitions, i.e., it exhibits a significant discontinuity in the order parameter $P(p)$. The size of the ith gap $g_i(p)$ for each fraction p is defined by

$$g_i(p) = P(p) - P(p - 1/N) , \tag{22.8}$$

so that g_1 is the largest gap, g_2 is the second largest gap, and so on. The larger the gap g_i is, the larger are the two clusters before merging. Therefore, g_1 as the order parameter is a natural candidate for a possible percolation transition and formation of a GC of size $\mathcal{O}(N)$.

Fan et al. (2019) argued that their network, just before the jump in g_1, is characterized by four major communities, with the largest one being the Afro-Eurasia continental landmass, followed by the Americas, and the Antarctica, with the fourth largest being Oceania. A critical node (located at 64.458 333 °N, 171.141667 °W) connects the largest and second largest clusters at the percolation threshold $p_c \approx 0.321$, with altitude level at $h = -43$ m under the current sea level. It is, however, not clear yet whether the percolation transition is continuous or discontinuous at p_c.

Saberi (2013) also analyzed the topography of the Moon, as an example of the most heavily studied waterless body with completely different surface properties and interior mechanisms. In this case, the order parameter exhibits two rather small jumps at altitude levels around 960 and 1360 m, as do the correlation length and the mean island size. The Moon's height distribution features a single global peak at a level of about -950 m, quite close to one of the critical levels located at around -960 m. Measurement of the correlation length and the mean cluster size for the "oceanic" clusters indicates that they exhibit only one critical level very close to -960 m. This feature may imply that this critical level is more important for the description of the global topography of the Moon, and it is also quite close to the level $h = -1049$ m at which the total island and oceanic surface areas are equal, hence implying that the island and oceanic percolation thresholds coincide.

The critical level at $h = 1360$ m unravels a characteristic feature of the lunar farside. It is well known that Moon's nearside is low and flat, dominated by volcanic maria, whereas the farside is mountainous and deeply cratered. The maps of the elevation levels at $h = 1360$ and $h = 1340$ m, on both sides of the critical level indicated aggregation of two main mountainous islands that are separated by a very narrow passageway, perhaps the hallmark of a rather nonrandom origin of the formation of the lunar farside highlands.

22.4 Ancient Sea Level on Mars

Martian dichotomy refers to a major feature of the Martian crust, which is the sharp contrast between the northern and southern hemispheres of the planet. The dichotomy manifests itself topographically, since the presence of heavily cratered and elevated highlands in the southern hemisphere are distinctly different from the lowland plains in the northern hemisphere by a reduction in crustal thickness of about 30 km (see, for example, Zuber 2001). The origin of the dichotomy is currently unknown. Several theories have been proposed to explain the dichotomy, one of which is that the northern hemisphere was once covered by a vast ocean (see, for example, Perron et al. 2007), the best evidence for which is observation of several possible paleoshorelines that are distributed predominantly near the margins of the northern plains of Mars (see, for example, Clifford and Parker 2001). The existence of the ocean has been challenged (Malin and Edgett 1999) based on the observation of elevation variations of significant long-wavelength trends along the identified shorelines of up to two kilometers that do not follow surfaces of equal gravitational potential that should hold for a sea level.

Proceeding along the same lines as in the development of his percolation model of Earth's topography, Saberi (2020) applied the model to Mars, and provided evidence for the existence of ancient sea level on the planet. He argued that if a massive ancient ocean existed in the entire northern lowlands of Mars, it should have been surrounded by a relatively long gravitational equipotential contour line, implying that it should have left a clear trace on Mars' present topography. To detect it, he used the high-resolution and globally distributed topographic data of Mars, provided by the Mars Orbiter Laser Altimeter (Zuber et al. 1992).

Figure 22.5 plots the total length of the isoheight (or contour) lines on Mars as a function of height. The curve resembles the bimodal histogram of the Martian topography with two evident peaks at level heights $h = -2040$ m and $h = 1400$ m with a local minimum in between, with the zero-level height defined as the equipotential surface whose average value at the equator is 3396.2 km. This is similar to the same plot for Earth, also shown in Fig. 22.5, with two distinct maxima in the continental platforms and oceanic floors.

Saberi (2020) also computed the fractal dimension D_f of the longest isoheight line on Mars at various altitudes, which reflects the geometrical irregularity of Mars' topography with a significant response around the level height $h = 1400$ m, where it attains a value $D_f \approx 1.2$, very close to the overall value, which was also reported for Earth's shorelines (Baldassarri et al. 2008; Boffetta et al. 2008).

The correlation length ξ was also computed for both continental and oceanic clustering, using Eq. (22.6). It was found that ξ diverges at the same level height $h_c = 1400$ m, indicated by Fig. 22.5. This should be compared with the computed correlation length on Earth, described above, which also appears to diverge exactly at Earth's present mean sea level, which can be interpreted as implying that Mars's northern hemisphere was once covered by an ocean early in its geologic history, with the water level up to the height of about 1400 m.

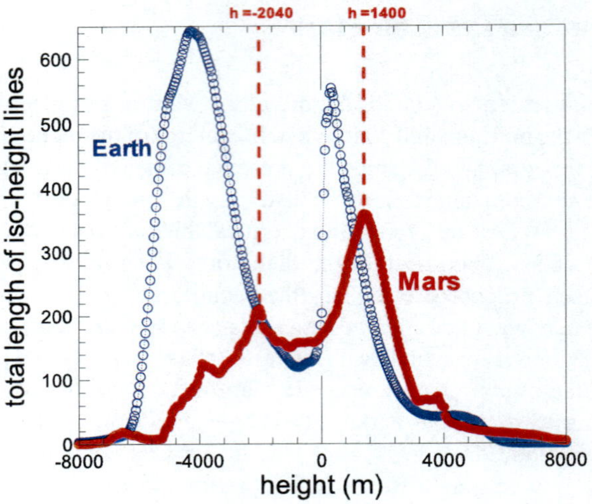

Fig. 22.5 Total length of the isoheight lines on Mars (solid circles) and Earth (open circles) as a function of altitude. The zero level height corresponds to the present mean sea level on Earth, but has a different meaning on Mars (see the text). The two vertical dashed lines specify two distinct level heights, $h = -2040$ m and $h = 1400$ m with a maximum length of the isoheight lines in the northern lowlands and southern highlands of Mars, respectively. In case of Earth the two oceanic and continental peaks unravel common similarities between the two planets. Note that the present mean sea level on Earth (vertical solid lines) is located just below the continental maximum. All length scales are expressed in units of the average radius of the corresponding planet. The resolutions of the analyzed data sets for Earth and Mars are, respectively, $10,800 times 21,600$ and 4000×8000 grid points. Spherical shape of the planets was taken into account when calculating the length of the isoheight lines (after Saberi 2020; courtesy of Dr. Abbas Ali Saberi)

Saberi (2020) defined the strength of the largest continent as the area of the largest island, divided by 4π, since he assumed that the Martian topography is on a sphere of unit radius (whose surface is, therefore, 4π). The calculated strength of the largest continent, when plotted versus the fraction f of total land area, exhibited a large jump at a certain fraction f_c, hence providing evidence for a percolation phase transition in the topography of Mars. This is shown in Fig. 22.6. A very interesting aspect of Fig. 22.6 is the comparison of the results with similar data for Earth, since in both planets the major geometrical phase transition occurs at the same fraction of the total landmass area, 29 percent, which is also consistent with the present mean sea level on Earth that makes up 71 percent of it is surface. This is remarkable since the same percentage of water on both Earth and Mars is predicted without using any adjustable parameter, hence providing strong support for the proposal that Mars had a huge ocean billions of years ago.

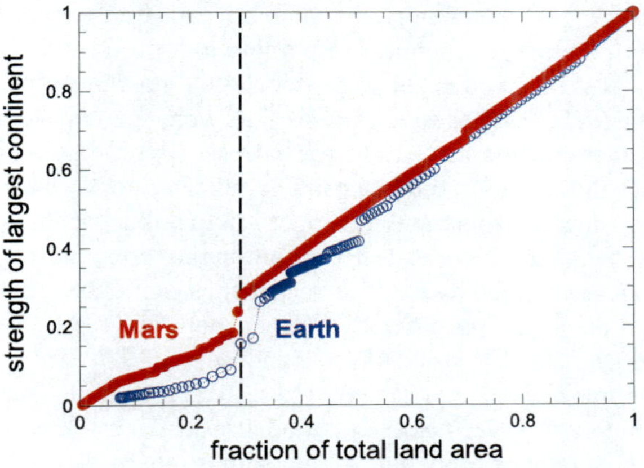

Fig. 22.6 Relative surface area of the largest island of Mars (solid circles) and Earth (open circles) as a function of the total fraction of the landmass area. The vertical dashed line marks the percentage of the present landmass area on Earth, which coincides with the level height $h = 1400$ m on Mars, including the maximum length of the isoheight lines (after Saberi 2020; courtesy of Dr. Abbas Ali Saberi)

22.5 Forest Fires

Forest fires are a worldwide phenomenon. In California, where the author lives, forest fires occur for a variety of reasons, ranging from sparks by defective electric transmission lines, to natural causes such as lightening, and man-made fires (see https://www.fire.ca.gov/stats-events/). Spread of fire in a forest belongs to a general class of problems in which a front propagates in a heterogeneous medium, similar to some of the phenomena associated with various diseases that were described in Chap. 18. Due to the destructive power of forest fires, their properties have been carefully monitored and recorded and, therefore, there are ample data in the literature.

The idea that if a tree in a forest is set on fire, it could spread to its nearest-neighbor trees and, then, depending on a variety of factors, spread throughout the forest, hints at the possibility of modeling forest fires as a critical phenomenon, and in particular percolation. Such factors include supply of adequate oxygen, wind velocity, land's topography, age and type of trees, and, of course, rainfalls. The first step toward such goal was taken, to our knowledge, by MacKay and Jan (1984). They represented a forest by a densely packed triangular lattice, i.e., one in which all the sites were occupied by identical trees, and the spread of fire was a localized surface phenomenon, implying that a burning tree was able to ignite only its nearest neighbors.

In the model, the fire begins with the central site ablaze at "time" $t = 0$. The propagation of the fire to neighboring "warm" trees would occur with probability p, leading to four types of sites at any given time t, namely, (a) sites with burnt trees; (b) sites with burning trees; (c) warm trees (unburnt trees that are nearest neighbors to burning trees); and (d) the remaining trees (sites). At each time step of the simulation, the propagation of the fire from ignited or burning trees to warm trees would occur with probability p. At the end of the time step, all the previously burning trees are considered burnt (dead) and the warm trees that were ignited would be the new burning centers that create a new set of warm trees, which may include unignited warm trees from the previous time steps. The process stops when the system is "exhausted," meaning that there are no further ignited trees, or the fire reaches the edge of the forest (lattice). Note that there is a trapping effect in the model in that, an unburnt region surrounded by burnt or vacant sites cannot be ignited.

Others refined the model proposed by MacKay and Jan (1984); see for example, von Niessen and Blumen (1986) and Ohtsuki and Keyes (1986) who, in the presence of a wind during fire, related the problem to directed percolation (see Chap. 21). In particular, Albinet et al. (1986) studied a model in which a fraction p of the sites of a lattice, either square or triangular, were combustible trees, while the rest were not. The combustible trees were characterized by two parameters, τ_i and τ_b, and by the type of interaction between the sites. τ_b was the number of time steps that an ignited site remained in the burning state and was, thus, a measure of the heat released by a site. τ_i, on the other hand, was a measure of the heat required to ignite a site and corresponded to the number of time steps required to ignite a site, if one site in interaction with it was burning. Every site could interact with up to its fifth neighbor.

At time $t = 0$, all the combustible sites in the first two rows were ignited. At each successive time step, the heat transferred from each burning site to its neighbors was calculated from the type of interaction (i.e., in terms of the number of neighbors interacting with a given neighbor). A search was then carried out to see if new sites received enough heat to ignite, and whether burning sites have burned long enough to become extinguished. The process was continued until either (a) a site on the last row was ignited, or (b) all sites were extinguished, and the system was exhausted. Therefore, the model was essentially a site percolation problem with three additional features, namely, the two parameters τ_i and τ_b, and the interaction of any site with other sites that are farther than nearest neighbors.

In the simulation, the lattice was scanned from left to right and row by row, and every time a burning site was identified, the heat transferred was computed and the affected sites were ignited, or extinguished, as necessary. One time step was equivalent to sweeping the lattice once, and in the limiting case of a densely populated lattice with $\tau_i = \tau_b = 1$, the fire front traversed the lattice in one time step. The critical densities, or percolation thresholds, were determined by using the algorithm. They depend, of course, on the range of interactions between a site and its neighbors.

As always, one must define an order parameter $P(p)$. Albinet et al. (1986) defined the order parameter by

$$P(n, t, p)) = \lim_{n,t\gg 1, n\leq L} \frac{\text{number of burnt trees in the } n\text{th row}}{pL}, \tag{22.9}$$

for a lattice of size $L \times L$. This definition is essentially the same as the order parameter in classic percolation theory, introduced in Chap. 2, which is defined by the fraction of sites belonging to the sample-spanning percolation cluster. If one defines a characteristic time, $\theta \propto [(p - p_c)/p_c]^{-\tau}$, then, it is straightforward to show that near the percolation threshold, the order parameter follows the following scaling law:

$$P(n, t, p) = n^{-\beta/\nu} f(n/\xi, t/\theta) , \quad (22.10)$$

where β and ν are the usual exponents for the order parameter and the correlation length ξ, and $f(x, y)$ is the scaling function. At p_c, one must have

$$P(n, t, p_c) = n^{-\beta/\nu} f_1(tn^{-\tau/\nu}) , \quad (22.11)$$

with $f_1(z)$ being another scaling function related to $f(x, y)$. This implies that the mean propagation speed n—the number of rows to which the fire has spread—must scale as

$$n \propto t^{\nu/\tau} . \quad (22.12)$$

Note that the exponent ν/τ is a measure of the acceleration, or deceleration, of the fire front. The total number $N_b(t, p)$ of burnt trees is given by

$$N_b(t, p) = pL \int_n P(n, t, p)dn \propto t^{(\nu-\beta)/\tau} . \quad (22.13)$$

The calculations by Albinet et al. yielded $1.2 < \nu < 1.4$; recall from Chap. 2 that in 2D, $\nu = 4/3$.

Study of the mean speed $n(t)$ of fire propagation at $p = p_c$ indicated that one can discern three regions. In region I, the system is in a transitory state where the ignition of the entire first two rows ignites quick advance of the fire front, as the system "remembers" the initial conditions. In region II, one has a steady propagation in which the above scaling laws are valid. In region III, the finite size of the system is the prevailing factor, so that fire propagation is restricted by the border of the forest, decreasing the speed of the mean position of the front.

Calculations of Albinet et al. (1986), yielded, $\tau/\nu \approx 1.15$, so that the mean propagation speed scales as, $n(t) \sim t^{0.87}$. The sample-spanning percolation cluster of the burnt trees also had a fractal structure with a fractal dimension $D_f \approx 1.89$, in agreement with the exact value (see Chap. 2), $D_f = 91/48$. The length of the fire front—the number F of burned sites that are in contact with unburnt sites when the fire has just penetrated the forest—was also found to scale with L as, $F \propto L^{D_p}$, with $D_p \approx 1.8$.

Beer and Enting (1990) carried out actual laboratory experiments in order to test the predictions of the model by Albinet et al. (1986). Their estimates for the exponents ν/τ and $(\nu - \beta)/\tau$ did not agree with what Albinet et al. (1986) had reported. In particular, they noted that the estimate of ν/τ increases as the size of the burnt area does. The reason, as discussed by Beer and Enting (1990), is that the geometry of flame radiation is such that ν/τ increases as the curvature increases. Thus, as the size of the forest increases, large clusters can exist, hence larger scales and larger curvatures come into play. The conclusion was that, while percolation theory is certainly relevant, a simple site or bond percolation model may not be adequate to explain all aspects of forest fires. Perhaps, given that even a quiescent fire generates its own winds, its presence affects the outcome of the model and, in particular, universality class of the exponents. The presence of wind implies that directionality is an important factor and, thus, requires the use of directed percolation (see Chap. 21), as attempted by Ohtsuki and Keyes (1986).

Rabinovich et al. (2002) suggested the conditions under which percolation theory is relevant to the combustion of a heterogeneous medium. If the characteristic scale of combustion wave is smaller than the sizes of structural elements of the reacting system, one will have a percolation combustion regime in which the process has a form of successive ignition and burning of connected combustible elements that constitute a cluster spanning the opposite boundaries of the system. Their study indicated that the percolation regimes are possible if (a) each element burns independently, meaning that the ratio of the combustion zone scale and the size of the burning elements should be less than unity, and the heat exchange between particles should be hampered; (b) the heat losses should be close to the extinction limit; and (c) the percentage of combustible elements should be in the vicinity of the percolation threshold. See also Grinchuck and Rabinovich (2004) who studied combustion of a heterogeneous mixture, and investigated the limit in which percolation theory is relevant, as well as Lam et al. (2020) who used numerical simulations to study flame propagation in discrete particulate clouds, and linked the phenomenon to percolation.

22.5.1 Self-Organized Critical Model of Forrest Fires

It should be pointed out that forest fires have also been modeled as a phenomenon that manifests self-organized critical (SOC) behavior. The notion of the SOC was first introduced by Bak, Tang, and Wiesenfeld (1987) as a mechanism by which complexity arises in nature. Self-organized criticality is a property of dynamical systems that have a critical point, which acts as an attractor. The macroscopic behavior of such systems exhibits the spatial and/or temporal scale-invariance that is characteristic of those undergoing a phase transition at a critical point, except that there is no need to fine-tune a control parameters to a particular value because, effectively, the system tunes itself as it evolves toward criticality.

Bak et al. (1990) introduced an SOC model for forest fires. In their model, the state of the "forest" was updated in parallel according to three rules: (a) a burning tree becomes an empty site of a lattice; (b) a green tree becomes a burning tree if at least one of its nearest neighbors is burning; and (c) at an empty site, a tree grows with probability p. Their work was followed by the well-known SOC model of Drossel and Schwabl (1992), and the work of Malamud et al. (1998). In the latter case, it was shown that a simple SOC model of forest fires reproduces the main feature of the phenomenon, namely, the fact that forest fires manifest power-law frequency-area statistics over many orders of magnitude. The practical implication of the model is that the frequency–area distribution of small and medium fires can be used to quantify the risk of large fires, which is routinely done for earthquakes.

A more realistic SOC model of forest fires was developed by Strocka et al. (1995). Their model included *both* trees and bushes. Thus, in addition to the usual nearest-neighbor fire spreading, both trees and bushes could ignite each other at the same lattice site. Numerical simulations of Strocka et al. in two dimensions indicated that the bushes with the higher growth rate remain critical, whereas the self-organization of the trees is destroyed by quick bushfires that ignite the tree clusters before they can become large.

Duarte (1997) introduced cellular automata for simulating heterogeneous bushfire propagation and computed their critical properties. A cellular automaton is a regular lattice of cells (or nodes representing the cells), each of which is in one of a finite number of states, such as "on" and "off." Each cell has a set of cells that represent its neighborhood. An initial state at time $t = 0$ is set by assigning a state for each cell. Time t is advanced by one after a new generation is generated according to some fixed rule that determines the new state of each cell in terms of its current state and those of its neighbors. The rule for updating the state of cells is usually the same for each cell, and does not change over time either. The updating of the states is carried out simultaneously (see, for example, Wolfram 1983; Toffoli and Margolus 1987; Schiff 2011).

In Duarte's model, the underlying model takes continuous fuel with variable moisture characterized by either a fixed value or a random choice of extinction determined by the moisture level. Elliptical (circular) propagation from a single ignition point was assumed. Several classes of universality were identified that were identical to be those of directed and usual (isotropic) percolation, even though a significant range of variation of the many parameters was simulated. Thus, Duarte's model linked several important classes of percolation problems.

Although Clar et al. (1995, 1997) did introduce a non-equilibrium percolation model with self-organized critical states that could perhaps be used for explaining certain aspects of forest fires, we do not describe such models here because their percolation aspects emerge only in certain limits. The interested reader can consult the original references.

22.6 Percolation in Sea Ice

A most complex heterogeneous material is sea ice, which consists of pure ice with brine and air inclusions in its microstructure. The size and geometry of sea ice depend on its crystal structure and, obviously, the temperature and bulk salinity. What makes sea ice different from many other types of disordered materials is that even a small change in the temperature has a dramatic effect on its morphology and effective properties. It is known that if the volume fraction p of brine is less than a critical value of $p_c \approx 0.05$, columnar sea ice is effectively impermeable to fluid flow, whereas, for $p > p_c$, brine flows through the ice. Based on the relation between brine volume, temperature T and salinity S, p_c corresponds to a critical temperature $T = -5$ °C for $S = 5$ ppt, which is referred to as the "law of fives." Direct observations indicate that, due to gravity drainage, the rate of change of sea ice salinity, dS/dt, vanishes for $p < 0.05$, and that the permeability of thin sea ice decreases by more than two orders of magnitude as the surface temperature is lowered in a small critical region around -5 °C (4).

Although flow of brine is fundamental to many processes, such as sea ice production through freezing of flooded ice surfaces, heat transport through the sea ice, replenishment of nutrients for sea ice algal communities, and remote sensing, the transition that controls brine transport received little attention until Golden et al. (1998) proposed a percolation model for the phenomenon, and showed that it provides a theoretical foundation for understanding the critical behavior of sea ice near the transition point. The model enabled Golden et al. (1998) to explain the aforementioned "law of fives," the behavior of the fluid permeability near the critical temperature, and the reported data for surface flooding on sea ice in the Weddell Sea, part of the Southern Ocean whose land boundaries are defined by the bay formed from the coasts of Coats Land and the Antarctic Peninsula, and East Antarctic region.

It should be clear that the standard lattice percolation model, described in Chap. 2, with the open and closed bonds representing, respectively, brine and ice, cannot explain the very low percolation threshold, $p_c \approx 0.05$ (recall from Chap. 2 that bond percolation threshold of the simple-cubic lattice, for example, is ≈ 0.25). Thus, to explain the low p_c, Golden et al. (1998; see also Golden 2001) considered the behavior of composite materials that consist of conducting particles embedded in an insulating matrix. As pointed out earlier in this book, it has been reported (Kusy and Turner 1971; McLachlan et al. 1990) that compacting large polymer particles of typical radius R_p with much smaller metal particles of radius R_m results in low percolation thresholds for the metal particles, and produces highly conducting composites. Golden et al. (1998) observed that the morphology of such compressed powders is very much similar to the cellular microstructure of columnar sea ice, with the key parameter being the ratio, $\mathcal{R} = R_p/R_m$. For large $\mathcal{R}$ the numerical value of p_c is not very sensitive to the exact value of $\mathcal{R}$. For example, for $17 \leq \mathcal{R} \leq 44$ one obtains $0.03 <\leq p_c \leq 0.7$.

Golden et al. (1998) analyzed photo-micrographs of morphology of sea ice and used typical brine inclusion sizes to estimate that for their sea ice composite $\mathcal{R} \approx 24$. Assuming that the metal particles are spherical, a good approximation for p_c for the compressed composites is given by (Kusy 1977)

$$p_c = \frac{4\phi_c}{4\phi_c + \mathcal{R}\epsilon}, \tag{22.14}$$

where ϵ is a reciprocal planar packing factor and ϕ_c is the critical surface area fraction of the larger particles, which, for percolation, must be covered by the smaller

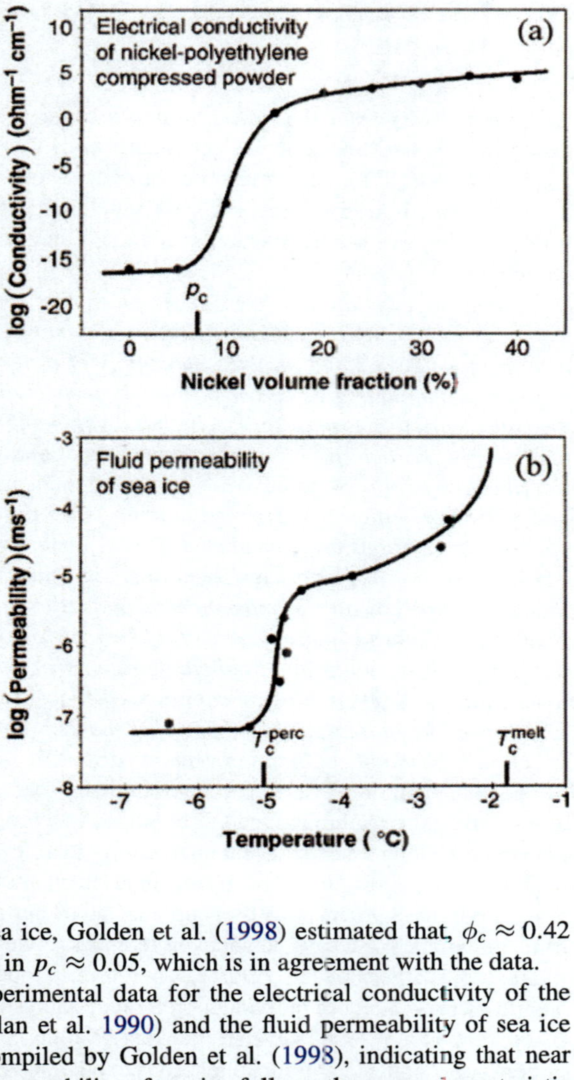

Fig. 22.7 Comparison of **a** the electrical conductivity of compressed powders of large polyethylene particles of radius R_p and small nickel particles of radius R_m, with $\mathcal{R} = R_p/R_m = 16$, and **b** the fluid permeability $K_e(T)$ of thin young sea ice as a function of surface temperature T. There seems to be a second transition for $K_e(T)$ at the melting point T_c^{melt}, where log(permeability) increases rapidly (after Golden et al. 1998)

(conducting) particles. For sea ice, Golden et al. (1998) estimated that, $\phi_c \approx 0.42$ and $\epsilon \approx 1.27$, hence resulting in $p_c \approx 0.05$, which is in agreement with the data.

Figure 22.7 compares experimental data for the electrical conductivity of the compressed powder (Mclachlan et al. 1990) and the fluid permeability of sea ice (Ono and Kasai 1985), as compiled by Golden et al. (1998), indicating that near the critical temperature, the permeability of sea ice follows the same characteristic behavior as that of the electrical conductivity of the composite of composite media near a percolation threshold. As described in Chap. 2, near the percolation threshold the fluid permeability K_e follows the power law (2.19). For the sea ice data Golden et al. estimated e to be $e \approx 2.5$, which is somewhat larger than $e \approx 2$ in random percolation (see Chap. 2), but not too far from it.

22.7 Lifecycle of Industrial Products and Consumer Demand

An important concept in the industry is *product lifecycle*, which is used in the study of the evolution of an industrial product. Such a study was first carried out by Abernathy and Utterback (1978), who developed the concept of a dominant design by studying the automobile industry. Generally speaking, three stages are distinguished, namely, (a) the explorative stage; (b) the development stage; and (c) the mature stage. To quote Klepper (1997):

"Three stages in an industry's development are commonly recognized: an early explorative stage, an intermediate development stage, and a mature stage. The first or early formative stage involves the supply of a new product of relatively primitive design, manufactured on comparatively unspecialized machinery, and marketed through a variety of exploratory techniques. Volume is typically low. A high degree of uncertainty characterizes business experience at this stage. The second stage is the intermediate development stage in which manufacturing techniques are more refined and market definition is sharpened; output grows rapidly in response to newly recognized applications and unsatisfied market demands. A high but somewhat lesser degree of uncertainty characterizes market outcomes at this stage. The third stage is that of a mature industry. Management, manufacturing, and marketing techniques all reach a relatively advanced degree of refinement. Markets may continue to grow, but do so at a more regular and predictable rate ... (e)stablished connections with customers and suppliers (including capital market access) all operate to buffer changes and thereby to limit large shifts in market shares. Significant innovations tend to be fewer and are mainly of an improvement variant."

Frenken et al. (2008) used percolation theory to develop a theoretical framework and simulate the aforementioned three stages of a product lifecycle. It has been recognized for a long time that agent interactions are relevant to understanding diffusion (spreading) of innovation. This means that the owner of a given product can induce similar purchasing decisions by people in his or her social circle. But, at the same time, individuals also have personal preferences. For example, sports cars are of little interest to a father of four children. Therefore, a model of adoption (of a product) should incorporate both individual preferences and social imitation.

Frenken et al. (2008; see also Frenken 2006) assumed that agents—the site of a lattice—are connected by social relations to other agents, and that the social network is a regular lattice, so that all agents are symmetrically connected to four neighbors, i.e., a square lattice. Application of percolation concepts to adoption implies that an agent becomes aware of a novel product only when a neighbor purchases it for the first time and, therefore, decides whether to adopt it, so that word-of-mouth is the only mechanism by which agents gain information about new products. The agent adopts the product based on its preferences, as indicated by its "minimal requirement," which is a level of value-for-money below which the agent does buy the product. Conversely, any product that an agent becomes aware of with a value-for-money index above the minimum threshold will be purchased. It is here that percolation enters the model, since in order to attribute preferences to agents, Frenken et al. assigned a random number, drawn from a uniform distribution in [0,1), which represented the minimal requirement demanded by an agent before adopting the product. As such, their model is similar to that of Solomon et al. (2000), which will be described in 22.9.

To simulate the model, one first initializes the system by assigning a product an exogenous value-for-money index level p in the interval [0,1). The system is then initialized by selecting at random a few agents and offering them the product. They will buy it only if the value-for-money index p of the product is above the agent's minimal requirement. In each subsequent simulation time step t, the agents that are neighbors to those who made a purchase at time $t-1$ will have the opportunity to buy a product, but it will be done only if the product exceeds the agent's minimal requirement.

It should be clear that all agents whose minimal requirements are higher than p will never buy the product. At the same time, however, it is also possible that agents with requirements lower than p will not adopt because, in the absence of previously adopted neighbors, they never get a chance to evaluate the product. Since preferences are randomly distributed according to a uniform distribution in [0,1), the maximum extent of spreading of a product with value-for-money index p is exactly p. For example, a product with $p = 0.8$ will be adopted by at most 80 percent of all agents, since 20 percent of the agents will have a minimum requirement exceeding 0.8.

What is the implication of the existence of a percolation threshold for this model? It implies that there exists a critical value-for-money index p_c such that for product with values $p > p_c$, the diffusion rate will be close to their value-for-money index, whereas, for $p < p_c$, the diffusion rate will be significantly lower than their value-for-money index. Thus, for $p > p_c$, the information about a new product will almost fully percolate through the social network, hence triggering nearly all potential customers to adopt the product, whereas, for $p < p_c$, the information will not percolate, implying that many potential adopters will not adopt the product, simply because they never become aware of it through contact with other adopters.

Figure 22.8 presents the spreading of a product in a market simulated on a 400×400 square lattice at the end of the simulation, when no more purchasing occurs. Figure 22.8a shows the state of the system for $p = 0.55$, whereas the results shown in Fig. 22.8b correspond to $p = 0.6$, just above the site percolation threshold, $p_c \approx 0.593$ (see Chap. 2). Dark cells are agents who did not buy the product, whereas brighter ones represent agents who did. In both cases, ten sites were initially selected at random and were offered the opportunity to buy the product, and then the model evolved iteratively according to the procedure described above.

When $p < p_c$, only some of the initial agents buy the product initially, and then "infect" their neighbors by spreading the information about the existence of the novel product. But, eventually, each of the cases encounters clusters of too demanding agents; i.e., agents with minimal requirements higher than the product's value-for-money index and, therefore, diffusion of the products is stopped before the vast majority of agents have the opportunity to consider it. On the other hand, when the product value-for-money index is above p_c—in this case, $p = 0.6$—the diffusion of the product can be slowed by groups of highly demanding consumers,

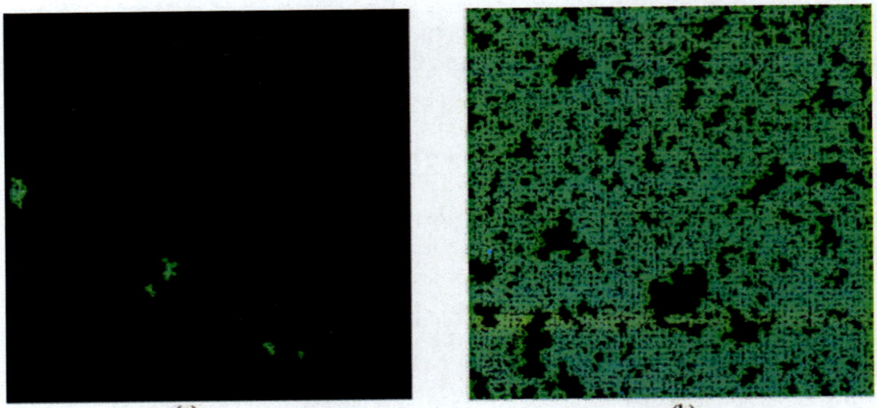

(a) (b)

Fig. 22.8 Spreading of a single product on a 400×400 square lattice and initialized with 10 randomly selected agents, for **a** $p = 0.55$ and **b** $p = 0.6$ (after Frenken et al. 2008)

but because a large majority of willing consumers can make their purchase, close to 60 percent of all consumers will adopt the new product. Therefore, the model and simulations demonstrate clearly that increasing the value-for-money index of the product only slightly from p, slightly below p_c to p, and slightly above p_c lead to a sudden increase in the rate of the product's diffusion, i.e., the familiar percolation transition.

Frenken et al. (2008) generalized their model to the case in which more than one product is offered for the adoption and purchase, and showed that the model reproduces qualitatively all features of product lifecycles observed in actual markets.

Zeppini and Frenken (2018) used the same type of model to study spreading ("diffusion") a new product in a society, represented by the type of social networks described in Chap. 17. Because consumers have different attitudes toward the price of a new product, a critical price exists that defines a phase transition from a phase in which spreading remains localized, to one in which the product spreads macroscopically. Just below the critical price, the consumer surplus is maximized and, therefore, one can systematically compare the economic efficiency and viability of network structures by investigating their critical price. Simulating the model on a variety of networks, including small-world and scale-free networks (see Chap. 17), Zeppini and Frenken (2018) showed that networks with low clustering are the most efficient, because clustering leads to redundant information flow, hence impeding effective spreading of a product. They also showed that the more equal a society, the more efficient the spreading process.

22.8 Spreading of Technological Innovations

It is well known that innovations emerge more highly clustered than if they were purely random. The distribution of their importance is highly skewed, which appears to follow a power law, which is, of course, a characteristic of percolation systems. This hints at the possibility that percolation concepts may be relevant to spreading of technological innovation, and that has been shown to be indeed the case.

Mort (1991; see also Mort 1994) studied how specific marketing approaches that have facilitated successful innovations, such as the Xerox 914 copier, can be linked to the concept of connectivity, the essential feature of percolation theory. He also described the essential role that connectivity plays in the spread of certain information, communication-based innovations, such as the FAX machines, and other network-based products, as well as the utility and value of such products to the customer. As he stated (Mort 1994).

A product's introduction to the marketplace is not the end of the innovation process, merely the end of the beginning. In other words, the spread of new products in the marketplace is a requisite of innovation, and one can gain some useful insights by viewing it as a percolation phenomenon involving diffusive flow with connectivity through a granular medium [i.e., a lattice of sites], identifying the market with the medium and its real and potential customers with occupied percolation sites.

Figure 22.9 presents the growth in the production of FAX machines from 1975 to 1989. It indicates that despite intensive marketing by Xerox Corporation of an affordable office FAX machine—the Telecopier—in 1966, the volume produced (and, hence, sold) remained essentially constant, until the abrupt rise in production that reflected much higher demand, starting around 1983–84. This is a particularly good example of the role connectivity—and, hence, percolation theory—in large-scale societal problems (some of which were also described in Chap. 18 in the context of biological systems), because it involves communication of information—the FAX machine—rather than its creation or reproduction, whose value to a customer is intrinsically linked to the number already sold, i.e., the extant connectivity. In other words, even though a FAX machine had essentially been invented and produced in 1966, due to the lack of connectivity for spreading the information about its use and potentials, its sales did not take off until around 1983, when it had been sold to enough customers to cause a rapid increase in the spread of information about it through the interconnectivity of national and international communities.

The inset of Fig. 22.9 presents the percolation model in which the market is represented by a square lattice, with its occupied sites indicating actual and potential customers. The basic features of the model are somewhat similar to those of the model by Frenken et al. (2008), described in the last section. Note the sharp rise in the number of FAX machines sold, beginning around 1983, which resembles completely the behavior and features of many percolation properties (see Chap. 2).

For a related, but more complex, percolation model of technological innovation see Silverberg and Verspagen (2005).

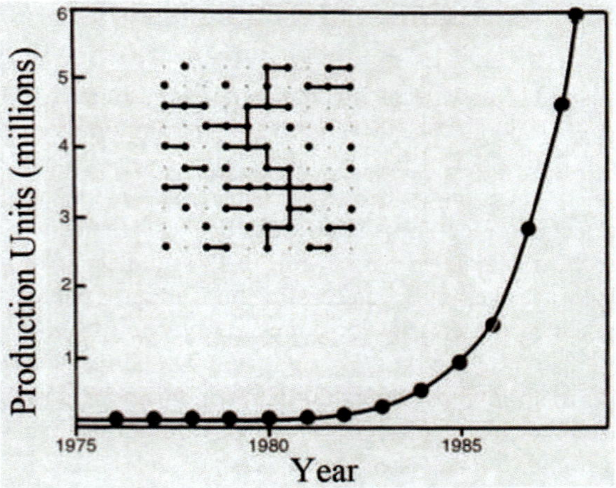

Fig. 22.9 Data for production of FAX machines in Japan that exhibit a sharp transition, characteristic of a percolation process. The inset shows a 2D percolation model that reproduces the same features (after Mort 1994)

22.9 Social Percolation

Solomon et al. (2000) addressed several questions related to collective phenomena in social systems. More specifically, they addressed the following questions:

(i) Why does one observe either hits or failures in not only certain markets, such as toys, gadgets, and movies but also in the adoption of technological changes that have political and economical measures, and in the political arena, rather than a featureless distribution of partial successes or failures?

As we have learned in this book, particularly based on some of the applications described so far in this chapter, percolation theory may provide the correct answer to the question. If so, the next question to address is:

(ii) Since economic agents are adaptive and profit-seeking, should we observe some meta-dynamics driving the market, either close to the percolation threshold or away from it?

To address these questions, Solomon et al. adopted the language of those who are interested to go and see a given film. They considered a lattice or a random graph (see Chap. 17) with percolation critical threshold p_c, in which each site contains an agent i—the "film-goer"—which communicates information about the film and its quality q to its nearest neighbors, i.e., the agents that are on sites to which site i is joined by a link. The dynamics of the model consists of having each agent i either viewing or not viewing a particular film according to a prescribed procedure. In the initial state of the system at "time" $t = 0$, a small number of agents are informed about the movie.

Each agent i, who is initially informed about the film, will decide to go to see it, if and only if the quality q—a number between zero and one—of the film is larger than his or her personal preference p_i, i.e., if $q > p_i$. The agents who decide not to see the film do not play any further role until the end of a given time step. One time step, $t \to t + 1$ of the model represents simulating the spread of film viewing across the entire lattice or graph. Note, however, that the initial agents who decided to go to the film become themselves sources of information about it. Thus, their nearest neighbors j are also informed about the film and, therefore, must decide, according to the condition $q > p_j$ criterion, whether to see the film. This process could continue until the procedure stops by itself, i.e., when all the neighbors of all the agents that saw the film up to now, either saw the film already, or decided already not to go to a cinema to see it.

The model, which is similar to what was described above for products lifecycles, is, of course, a standard percolation problem, because the agents' preferences are frozen and do not change in time. But, the model already makes predictions that are in sharp contrast with the standard economic models, since if the personal preferences p_i take independent random values distributed uniformly in [0,1), then the average probability of an agent seeing the film, once one of its neighbors did so, is the film quality q. Thus, if a film happens to have a quality $q < p_c$, then after a certain time its diffusion—or, more precisely, spreading of information about it—among the public will stop, the film becomes a flop, and will not reach any significant percentage of its potential public. But, if the film quality is $q > p_c$ and not too close to p_c, it will reach most of its potential viewers as islands of interested agents will percolate. In that case, the film will be viewed by roughly a fraction q of the entire viewers population. This is in sharp contrast with the prediction that is made in the full rational perspective of standard economics: since agents are supposed to have full knowledge about the film, they should go to see it in proportion to q without any threshold. But, the standard economics prediction is in contradiction with the observed distribution of hits and flops; see, for example, Farrell (1998) who studied the cinema industry.

The model so far is a static percolation system. But in social problems, other factors also play important roles. For example, the natural evolution of viewers' choice and taste, as well as the business interest of the film makers, might keep the system close to a critical point or percolation threshold. For example, after they saw the movies, the agents can become more demanding and, thus, increase their preferences p_i. Those who preferred not to see the movie, may lower their expectations and preferences, i.e., decrease their p_i, while uninformed agents keep their preferences intact. In addition, after hits (flops) the film producers will decrease (increase) the quality q of the film(s), in order to remain above the critical threshold, while minimizing their cost.

Solomon et al. (2000) simulated the simplest model on $L \times L$ square lattices that took such effects into accounts. The choice of a square lattice, a 2D system, was motivated by the fact that large cities represent essentially 2D systems. Solomon et al. (2000) simulated the simplest dynamics, one in which the quality q of the film increased by Δq, if no sample-spanning cluster was formed between top and bottom of the lattice, whereas it decreased by Δq, otherwise. The viewer's preference p_i, initially distributed randomly in [0,1), would change by $\pm\Delta p$, depending on whether or not agent (site) i watched the film. The Leath algorithm (see Chap. 2) was used to determine which sites were connected to the top line (assumed to be all occupied), and stopped once one site of the cluster reached the bottom line. Thus, one such

Leath cluster growth corresponded to one time step of the preference and quality dynamics, $t \rightarrow t + 1$, occurring on the occasion of the release of a new film.

Computer simulations of Solomon et al. (2000) indicated that,

(i) For adaptive film quality, which occurs for $\Delta p = 0$ and $\Delta q > 0$, the quality q moves to the percolation threshold p_c.

(ii) For $\Delta p > 0$ and $\Delta q = 0$ one has adaptive customer preferences so that the p_i distribution drifts toward a single peak centered on the fixed q value, taken equal to 0.5 (i.e., no sample-spanning cluster) or 0.593 (one or more spanning clusters).

(iii) When $\Delta p > 0$ and $\Delta q > 0$, both p_i and q move toward $p_c \approx 0.593$, even if initially, $q = 0.5$. Thus, this generalizes invasion percolation (see Chap. 16), exhibiting self-organized criticality (see above). Therefore, whereas standard percolation is observed at percolation threshold of 0.593, the social percolation model of Solomon et al. (2000) drifts toward p_c *automatically*.

Thus, one concludes that, under reasonable assumption about consumers behavior in certain commodities market, percolation is a relevant paradigm that explains the occurrence of double peaked market share distributions, observed in practice. Moreover, for certain types of preferences and quality adjustments, the market evolves "naturally" toward self-organized criticality in the neighborhood of p_c. One should, however, keep in mind that a percolation model is not the only way to explain the data. There are other models, such as that of Weisbuch and Boudjema (1999), which do not resort to percolation and indicate that when agents are influenced by the number of their neighbors, which take any of the two decisions, similar effects emerge due to the presence or absence of seed configurations from which cluster growth starts.

References

Abeles, B., H.L. Pinch, and J.I. Gittleman, Percolation conductivity in W-Al_2O_3 granular metal films, *Phys. Rev. Lett.* **35**, 247 (1975).

Abernathy, W.J., and J. Utterback, Patterns of industrial innovation, *Technol. Rev.* **50**, 41 (1978).

Abu-Hassanein, Z.S., C.H. Benson, and L.R. Blotz, Electrical resistivity of compacted clays, *J. Geotech. Eng.* **122**, 397 (1996).

Acharya, R.C., S.E.A.T.M. Van der Zee, and A. Leijnse, Approaches for modeling longitudinal dispersion in pore networks, *Adv. Water Resour.* **30**, 261 (2007).

Achlioptas, D., R.M. D'Souza, and J. Spencer, Explosive percolation in random networks, *Science* **323**, 1453 (2009).

Adam, M., M. Delsanti and D. Durand, Mechanical measurements in a reaction bath during the polycondensation reaction near the gelation threshold, *Macromolecules* **18**, 2285 (1985).

Adam, M., M. Delsanti, D. Durand, G. Hild, and J.P. Munch, Mechanical properties near gelation threshold, comparison with classical and 3d percolation theories, *Pure Appl. Chem.* **53**, 1489 (1981).

Adam, M., M. Delsanti, J.P. Munch, and D. Durand, Size and mass determination of clusters obtained by polycondensation near the gelation threshold, *J. Physique* **48**, 1809 (1987).

Adam, M., M. Delsanti, R. Okasha, and G. Hild, Viscosity study in the reaction bath of the radical copolymerisation of styrene divinylbenzene, *J. Physique Lett.* **40**, L539 (1979).

Adam, M., D. Lairez, M. Karpasas, and M. Gottlieb, Static and dynamic properties of cross-linked poly(dimethylsiloxane) pregel clusters, *Macromolecules* **30**, 5920 (1997).

Adam, S., S. Cho, M.S. Fuhrer, and S. Das Sarma, Density inhomogeneity driven percolation metal-insulator transition and dimensional crossover in graphene nanoribbons, *Phys. Rev. Lett.* **101**, 046404 (2008).

Adler, J., Bootstrap percolation, *Physica A* **171**, 453 (1991).

Adler, J., R.G. Palmer, and M.H. Meyer, Transmission of order in some unusual dilute systems, *Phys. Rev. Lett.* **58**, 882 (1987).

Adler, J., and D. Stauffer, Evidence for non-universal exponents in bootstrap percolation, *J. Phys. A* **23**, Lii19 (1990).

Adler, P.M., and J.-F. Thovert, *Fractures and Fracture Networks* (Kluwer, Dordrecht, 1999).

Adler, P.M., J.-F. Thovert, and V.V. Mourzenko, *Fractured Porous Media* (Oxford University Press, London, 2013).

Adolf, D., J.E. Martin, and J.P. Wilcoxon, Evolution of structure and viscoelasticity in an epoxy near the sol-gel transition, *Macromolecules* **23**, 527 (1990).

M. Sahimi, *Applications of Percolation Theory*, Applied Mathematical Sciences 213, https://doi.org/10.1007/978-3-031-20386-2

Aharony, A., S. Alexander, O. Entin-Wohlman, and R. Orbach, Scaling approach to phonon-fracton crossover, *Phys. Rev. B* **31**, 2565 (1985).

Aharony, A., and A.B. Harris, Superlocalization, correlations and random walks on fractals, *Physica A* **163**, 38 (1990).

Aharony, A., A.B. Harris, and O. Entin-Wohlman, Was superlocalization observed in carbonblack-polymer composites, *Phys. Rev. Lett.* **70**, 4160 (1993).

Aharony, A., Y. Zhang, and M.P. Sarachik, Universal crossover in variable range hopping with Coulomb interactions, *Phys. Rev. Lett.* **68**, 3900 (1992).

Ahlstrom, H., R. Christofferson, and L.E. Lorelius, Vascularization of the continuous human colonic cancer cell line LS 174 T deposited subcutaneously in nude rats, *APMIS* **96**, 701 (1988)

Aki, K., in *Earthquake Prediction: an International Review*, edited by D. Simpson and P. Richards (American Geophysical Union, Washington, 1981), p. 566.

Aladjem, R., and M.T. Palmiter, The antigen-antibody reaction: V. A quantitative theory of antigen-antibody reactions which allows for heterogeneity of antibodies, *J. Theor. Biol.* **8**, 8 (1965).

Albert R., and A.L. Barabási, Statistical mechanics of complex networks, *Rev. Mod. Phys.* **74**, 47 (2002).

Albert, R., H. Jeong, and A.L. Barabási, Error and attack tolerance of complex networks, *Nature* **406**, 378 (2000)

Albinet, G., G. Searby, and D. Stauffer, Fire propagation in a 2-D random medium, *J. Physique* **47**, 1 (1986).

Alexander, S., Is the elastic energy of amorphous materials rotationally invariant? *J. Physique* **45**, 1939 (1984).

Alexander, S., Bernasconi, J., and Orbach, R., Low energy density of states for disordered chains, *J. Physique Colloq.* **39**(C6), 706 (1978).

Alberts, B., A. Johnson, J. Lewis, M. Raff, K. Roberts, and P. Walter, Molecular Biology of the Cell, 4th ed. (Garland Science, New York, 1994).

Alexander, S. and Orbach, R., Density of states of fractals: 'fractons', *J. Physique Lett.* **43**, L625 (1982).

Alexandrowicz, Z., Critically branched chains and percolation clusters, *Phys. Lett.* **A80**, 284 (1980).

Al-Futaisi, A., and T.W. Patzek, Extension of Hoshen-Kopelman algorithm to non-lattice environments, *Physica A* **321**, 665 (2003).

Allain, C., L. Limat and L. Salomé, Description of the mechanical properties of gelling polymer solutions far from gelation threshold: generalized effective-medium calculations of the superconductor-conductor site percolation problem, *Phys. Rev. A* **43**, 5412 (1991).

Allain, C., and L. Salomé, Sol-gel transition of hydrolyzed polyacrylamide + chromium III: Rheological behavior versus cross-link concentration, *Macromolecules* **20**, 2957 (1987a).

Allain, C., and L. Salomé, Hydrolysed polyacrylamide/Cr^3 gelation: Critical behaviour of the rheological properties at the sol-gel transition, *Polym. Commun.* **28**, 109 (1987b).

Allain, C., and L. Salomé, Gelation of semidilute polymer solutions by ion complexation: Critical behavior of the rheological properties versus cross-link concentration, *Macromolecules* **23**, 981 (1990).

Allen, L.C., B. Golding, and W.H. Haemmerle, Dynamic shear modulus for two-dimensional bond percolation, *Phys. Rev. B* **37**, 913 (1988).

Alonso, S., and M. Bär, Reentry near the percolation threshold in a heterogeneous discrete model for cardiac tissue, *Phys. Rev. Lett.* **110**, 158101 (2013).

Alonso, S., R.W. dos Santos, and M. Bär, Reentry and ectopic pacemakers emerge in a three–dimensional model for a slab of cardiac tissue with diffuse microfibrosis near the percolation threshold, *PLoS One* **11**, e0166972 (2016).

Alvarado, J., M. Sheinman, A. Sharma, F.C. MacKintosh, and G.H. Koenderink, Molecular motors robustly drive active gels to a critically connected state, *Nat. Phys.* **9**, 591 (2013).

Alvarado, J., M. Sheinman, A. Sharma, F.C. MacKintosh, and G.H. Koenderink, Force percolation of contractile active gels, *Soft Matter* **13**, 5624 (2017).

Amaefule, J.O., and L.L. Handy, Thermal stability of surfactants for reservoir application, *Soc. Pet. Eng. J.* **22**, 371 (1982).

Amaral-Labat, G., M. Sahimi, A. Pizzi, V. Fierro, and A. Celzard, Mechanical properties of heat-treated organic foams, *Phys. Rev. E* **87**, 032156 (2013).

Ambegaokar, V.N., B.I. Halperin, and J.S. Langer, Hopping conductivity in disordered systems, *Phys. Rev. B* **4**, 2612 (1971).

Ambrosetti, G., C. Grimaldi, I. Balberg, T. Maeder, A. Danani, and P. Ryser, Solution of the tunneling-percolation problem in the nanocomposite regime, *Phys. Rev. B* **81** 155434 (2010).

American Society for Microbiology, How quorum sensing works; https://asm.org/Articles/2020/June/How-Quorum-Sensing-Works; accessed 20 January 2022.

Aminzadeh, F., T.A. Tafti, and D. Maity, An integrated methodology for sub-surface fracture characterization using microseismic data: A case study at the NW Geysers, *Comput. Geosci.* **54**, 39 (2013).

Anderson, P.W., Theory of dirty superconductors, *J. Phys. Chem. Solids* **11**, 26 (1959).

Andrade, J.S., A.M. Auto, Y. Kobayashi, Y. Shibusa, and K. Shirane, Percolation conduction in vapour grown carbon fibre, *Physica A* **248**, 227 (1998).

Andrade, J.S., S.V. Buldyrev, N.V. Dokholyan, S. Havin, P.R. King, Y. Lee, G. Paul, and H.E. Stanley, Flow between two sites on a percolation cluster, *Phys. Rev. E* **62**, 8270 (2000).

Androutsopoulos, G.P., and C.E. Salmas, Tomography of macro-meso-pore structure based on mercury porosimetry hysteresis loop scanning. Part II: MP hysteresis loop scanning along the overall retraction line, *Chem. Eng. Commun.* **181**, 179 (2000).

Araújo, N.A.M., and H.J. Herrmann, Explosive percolation via control of the largest cluster, *Phys. Rev. Lett.* **105**, 035701 (2010).

Arbabi, S., and M. Sahimi, Absence of universality in percolation models of disordered elastic media with central forces, *J. Phys. A* **21**, L863 (1988a).

Arbabi, S., and M. Sahimi, Elastic properties of three-dimensional percolation networks with stretching and bond-bending forces, *Phys. Rev. B* **38**, 7173 (1988b).

Arbabi, S., and M. Sahimi, On three-dimensional elastic percolation networks with bond-bending forces, *J. Phys. A* **23**, 2211 (1990a).

Arbabi, S., and M. Sahimi, Critical properties of viscoelasticity of gels and elastic percolation networks, *Phys. Rev. Lett.* **65**, 725 (1990b).

Arbabi, S., and M. Sahimi, Computer simulations of catalyst deactivation - I. Model formulation and validation, *Chem. Eng. Sci.* **46**, 1739 (1991a).

Arbabi, S., and M. Sahimi, Computer simulations of catalyst deactivation - II. The effect of morphological, transport and kinetic parameters on the performance of the catalyst, *Chem. Eng. Sci.* **46** 1749 (1991b).

Arbabi, S., and M. Sahimi, Mechanics of disordered solids. I. Percolation on elastic networks with central forces, *Phys. Rev. B* **47**, 695 (1993).

Arbabi, S., and M. Sahimi, Elastic moduli of body-centered cubic lattice near rigidity percolation threshold: Finite-size effects and evidence for first-order phase transition, *Phys. Rev. E* **103**, 042314 (2021).

Archie, G.E., The electrical resistivity log as an aid in determining some reservoir characteristics, *Trans. Am. Inst. Mech. Eng.* **146**, 54 (1942).

Arenas, A., J. Borge-Holthoefer, and Y. Moreno, Topological versus dynamical robustness in a lexical network, *Int. J. Bifurcat. Chaos* **22**, 1250157 (2012).

Aris, R., On the dispersion of a solute in a fluid through a tube, *Proc. Roy. Soc. Lond.* **A235**, 67 (1956).

Aris, R., The longitudinal diffusion in flow through a tube with stagnant pockets, *Chem. Eng. Sci.* **11**, 194 (1959).

Arns, C.H., M.A. Knackstedt, and N.S. Martys, Cross-property correlations and permeability estimation in sandstone, *Phys. Rev. E* **72**, 046304 (2005).

Arrowsmith, D.K., and J.W. Essam, Percolation theory on directed graphs, *J. Math. Phys.* **18**, 235 (1977).

Arrowsmith, D.K., and J.W. Essam, Percolation theory on multirooted directed graphs, *J. Math. Phys.* **20**, 101 (1979).

Asakura, S., G. Eguchi, and T. Lino, 1968, Unidirectional growth of Salmonella flagella in vitro, *J. Molec. Biol.* **35**, 327 (1968).

Attarimoghaddam, A., A. Kharaghani, E. Tsotsas, and M. Prat, Kinematics in a slowly drying porous medium: Reconciliation of pore network simulations and continuum modeling, *Phys. Fluids* **29**, 022102 (2017).

Aukrust, T., D.A. Browne, and I. Webman, Critical behavior of an autocatalytic reaction model, **41**, 5294 (1990).

Axelos, M.A.V., and M. Kolb, Crosslinked biopolymers: Experimental evidence for scalar percolation theory, *Phys. Rev. Lett.* **64**, 1457 (1990).

Aydin, M., T. Yano, F. Evrendilek, and V. Uygur, Implications of climate change for evaporation from bare soils in a Mediterranean environment, *Envir. Monitor. Assess.* **140**, 123 (2008).

Azar, Y., A.Z. Broder, A.R. Karlin, and E. Upfal, Balanced allocations, *SIAM J. Comput.* **29**, 180 (1999).

Babadagli, T., Fractal analysis of 2-D fracture networks of geothermal reservoirs in south-western Turkey, *J. Volcan. Geother. Res.* **112**, 83 (2001).

Bailey, D.J., W. Otten, and C.A. Gilligan, Saprotrophic invasion by the soil-borne fungal plant pathogen Rhizoctonia solani and percolation thresholds, *New Phytol.* **146**, 535 (2000).

Baish, J.W., Y. Gazit, D.A. Berk, M. Nozue, L.T. Baxter, and R.K. Jain, Role of tumor vascular architecture in nutrient and drug delivery: An invasion percolation-based network model, *Microvascular Research* **51**, 327 (1996).

Bak, P., *How Nature Works: The Science of Self-Organized Criticality* (Copernicus, New York, 1996).

Bak, P., K. Chen, and C. Tang, A forest-fire model and some thoughts on turbulence, *Phys. Lett. A* **147**, 297 (1990).

Bak, P., K. Christensen, L. Danon, and T. Scanlon, Unified scaling law for earthquakes, *Phys. Rev. Lett.* **88**, 178501 (2002).

Bak, P. and C. Tang, Earthquakes as a self-organized critical phenomenon, *J. Geophys. Res.* **B94**, 15635 (1989).

Bak, P., C. Tang, C., and K. Wiesenfeld, Self-organized criticality: an explanation of $1/f$ noise, *Phys. Rev. Lett.* **59**, 381 (1987).

Balberg, I., A comprehensive picture of the electrical phenomena in carbon black-polymer composites, *Carbon* **40**, 143 (2002).

Balberg, I., Continuum percolation, in *Encyclopedia of Complexity and Systems Science*, edited by R. A. Meyers, Vol. 2 (Springer, New York, 2009).

Balberg, I., Principles of the theory of continuum percolation, in *Complex Media and Percolation Theory*, edited by M. Sahimi and A.G. Hunt (Springer, New York, 2021), p. 89.

Balberg, I., C.H. Anderson, S. Alexander, and N. Wagner, Excluded volume and its relation to the onset of percolation, *Phys. Rev. B* **30**, 3933 (1984a).

Balberg, I., D. Azulay, D. Toker, and O. Millo, Percolation and Tunneling in composite materials, *Int. J. Mod. Phys. B* **18** 2091 (2004).

Balberg, I., and N. Binenbaum, Computer study of the percolation threshold in a two-dimensional anisotropic system of conducting sticks, *Phys. Rev. B* **28**, 3799 (1983).

Balberg, I., N. Binenbaum, and S. Bozowski, Anisotropic percolation in carbon black-polyvinylchloride composites, *Solid State Commun.* **47**, 989 (1983).

Balberg, I., N. Bienbaum, and N. Wagner, Percolation thresholds in the three-dimensional sticks system, *Phys. Rev. Lett.* **52**, 1465 (1984b).

Balberg, I., B. Berkowitz, and G.E. Drachsler, Application of a percolation model to flow in fractured hard rocks, *J. Geophys. Res.* **B98**, 10015 (1991).

Baldassarri, A., M. Montuori, O. Prieto-Ballesteros, and S.C. Manrubia, Fractal properties of isolines at varying altitude revealing different dominant geological processes on Earth, *J. Geophys. Res.* **113**, E09002 (2008).

Ball, P.C., and R. Evans, Temperature dependence of gas adsorption on a mesoporous solid: capillary criticality and hysteresis, *Langmuir* **5**, 714 (1989).
Balogh, J., and B.G. Pittel, Bootstrap percolation on the random regular graphs, *Random Struct. Algorithms* **30**, 257 (2007).
Banavar, J.R., and D.L. Johnson, Characteristic pore sizes and transport in porous media, *Phys. Rev. B*, **35**, 7283 (1987).
Barabási, A.-L., Invasion percolation and global optimizaion, *Phys. Rev. Lett.* **76**, 3750 (1996).
Barabási, A.-L., The origin of bursts and heavy tails in human dynamics, *Nature* **435**, 207 (2005).
Barabási, A.-L., and H.E. Stanley, *Fractal Concepts in Suface Growth* (Cambridge University Press, Cambridge, 1995).
Barghi, S.H., T.T. Tsotsis, and M. Sahimi, Hydrogen sorption hysteresis and superior storage capacity of silicon-carbide nanotubes over their carbon counterparts, *Int. J. Hydrogen Energy* **39**, 21107 (2014).
Baronchelli, A., R. Ferrer-i-Cancho, R. Pastor-Satorras, M. Chater, and M.H. Christiansen, Networks in cognitive science, *Trends Cogn. Sci.* **17**, 348 (2013).
Barton, C.C., Fractal analysis of the scaling and spatial clustering of fractures, in *Fractals and Their Use in the Earth Sciences*, edited by C.C. Barton and P.R. LaPointe (Geological Society of America, Washington, 1992).
Barton, C.C., Fractal analysis of scaling and spatial clustering of fractures, In, *Fractals in Earth Sciences*, edited by C.C. Barton anmd P.R. LaPointe (Plenum, New York, 1995a), p. 141.
Barton, C.C., Bedrock geological map of Hubbard Brooks experimental forest and maps of fracture and geology in road cuts along Interstate 93, Grafton County, New Hampshire. *Rep. I-2562, U.S. Geological Survey Misc. Invest. Ser.* **2**, 1 (1995b).
Barton, C.C., and P.A. Hsieh, *Physical and Hydrological-Flow Properties of Fractures*, Guidebook T385 (Las Vegas, American Geophysical Union, 1989).
Barton, C.C., and E. Larsen, in *Proceedings of the International Symposium on Fundamentals of Rock Joints*, edited by O. Stephansson (Bjorkliden, Sweden, 1985), p. 77.
Barton, C.C., T.A. Schutter, W.R. Page, and J.K. Samuel, Computer generation of fracture networks for hydrologic-flow modelling, *Trans. Amer. Geophys. Union* **68**, 1295 (1987).
Barton, N.H., Linkage and the limits to natural selection, *Genetics* **140**, 821 (1995).
Bastas, N., P. Giazitzidis, M. Maragakis, and K. Kosmidis, Explosive percolation: Unusual transitions of a simple model, *Physica A* **407**, 54 (2014).
Basu, U., and P.K. Mohanty, Self-organised criticality in stochastic sandpiles: Connection to directed percolation, *Euro Phys. Lett.* **108**, 60002 (2014).
Batrouni, G., B. Kahng, and S. Redner, Conductance and resistance jumps in finite-size random resistor networks, *J. Phys. A* **21**, L23 (1988).
Bauhofer, W., and J.Z. Kovacs, A review and analysis of electrical percolation in carbon nanotube polymer composites, *Compos. Sci. Technol.* **69**, 1486 (2009).
Bebbington,, M., D. Vere-Jones, and X. Zheng, Percolation theory: a model for rock fracture? *Geophys. J. Int.* **100**, 215 (1990).
Bednorz, J.G., and K.A. Müller, Possible high T_c supperconductivity in the Ba–La–CU–O system, *Z. Phys. B* **64**, 189 (1986).
Beeckman, J.W., I.S. Nam, and G.F. Froment, Stochastic modeling of catalyst deactivation by site coverage, *Stud. Surf. Sci. Catal.* **34**, 365 (1987).
Beer, T., and I.G. Enting, Fire spread and percolation modeling, *Math1. Comput. Modelling* **13** (11), 77 (1990).
Beggs, J.M., and D. Plenz, Neuronal avalanches in neocortical circuits, *J. Neurosci.* **23**, 11167 (2003).
Behi, Behi, J., S. Bone, H. Morgan, and R. Pethig, Effect of deuterium-hydrogen exchange on the electrical conduction in lysozyme, *Int. J. Quantum Chem. Biol. Symp.* **9**, 367 (1982).
Bejan, A., and S. Lorente, Constructal law of design and evolution: Physics, biology, technology, and society, *J. Appl. Phys.* **113**, 151301 (2013).

Bell, G.I., Evolutionary rescue and the limits of adaptation, *Philos. Trans. R. Soc. Lond. B* **368**, 20120080 (2013).
Bell, G.I., Mathematical model of clonal selection and antibody production, III. The cellular basis of immunological paralysis. *J. Theor. Biol.* **33**, 339 (1971).
Bell, Jr., T.H., Statistical features of sea-floor topography, *Deep-Sea Res.* **22**, 883 (1975).
Benguigui, L., Experimental study of the elastic properties of a percolating system, *Phys. Rev. Lett.* **53**, 2028 (1984).
Benguigui, L., Lattice and continuum percolation transport exponents: experiments in two-dimensions, *Phys. Rev. B* **34**, 8176 (1986).
Benguigui, L., and P. Ron, Experimental realization of superelasticity near the percolation threshold, *Phys. Rev. Lett.* **70**, 2423 (1993).
Beniot, C., E. Royer, and G. Poussigue, The spectral moments method, *J. Phys.: Condens. Matter* **4**, 3125 (1992).
Ben-Naim, E., and P.L. Krapivsky, Kinetic theory of random graphs: From paths to cycles, *Phys. Rev. E* **71**, 026129 (2005).
Ben-Noah, I., and S.P. Friedman, Review and evaluation of root respiration and of natural and agricultural processes of soil aeration, *Vadose Zone J.* **17**, 170119 (2018).
Bergstad, M., and N. Shokri, Evaporation of NaCl solution from porous media with mixed wettability, *Geophys. Res. Lett.* **43**, 4426 (2016).
Benzi, R., L. Biferale, S. Ciliberto, M.V. Struglia, and R. Tripiccione, Generalized scaling in fully developed turbulence, *Physica D* **96**, 162 (1996)
Bergman, D.J., Exact relations between critical exponents for elastic stiffness and electrical conductivity of two-dimensional percolating networks, *Phys. Rev. E* **65**, 026124 (2002).
Bergman, D.J., E. Duering, and M. Murat, Discrete network models for the low-field Hall effect near a percolation threshold: Theory and simulations, *J. Stat. Phys.* **58**, 1 (1990).
Bergman, D., and Y. Imry, Critical behavior of the complex dielectric constant near the percolation threshold of a heterogeneous material, *Phys. Rev. Lett.* **39**, 1222 (1977).
Bergman, D.J., and Y. Kantor, Critical properties of an elastic fractal, *Phys. Rev. Lett.* **53**, 511 (1984).
Bergman, D.J., and D. Stroud, Scaling theory of the low-field Hall effect near the percolation threshold, *Phys. Rev. B* **32**, 6097 (1985).
Berkowitz, B., Analysis of fracture network connectivity using percolation theory, *Math. Geol.* **27**, 467 (1995).
Berkowitz, B., and I. Balberg, Percolation theory and its application to groundwater hydrology, *Water Resour. Res.* **29**, 775 (1993).
Berkowitz, B., O. Bour, P. Davy, and N. Odling, Scaling of fracture connectivity in geological formations, *Geophys. Res. Lett.* **27**, 2061 (2000).
Berman, D., B.G. Orr, H.M. Jaeger, and A.M. Goldman, Conductances of filled two-dimensional networks, *Phys. Rev. B* **33**, 4301 (1986).
Bernard, R.A., and R.H. Wilhelm, Turbulent diffusion in fixed beds of packed solids, *Chem. Eng. Prog.* **46**, 233 (1950).
Bernasconi, J., Conduction in anisotropic disordered systems: Effective-medium theory, *Phys. Rev. B* **9**, 4575 (1974).
Bernasconi, J., Real-space renormalization of bond-disordered conductance lattices, *Phys. Rev. B* **18**, 2185 (1978).
Bershadskii, A., Large-scale percolation and diffusion in turbulent stratosphere, *Physica A* **206**, 120 (1994).
Bershadskii, A., and H. Branover, Free and layer turbulent percolation: topological instabilities and their suppression, *J. Phys. I France* **4**, 1115 (1994).
Bershadskii, A., E. Kit, and A. Tsinober, Spontaneous breaking of reflexional symmetry in real quasi-two-dimensional turbulence: stochastic travelling waves and helical solitons in atmosphere and laboratory, *Proc. R. Soc. Lond. A* **441**, 147 (1993).
Bershadskii, A., and K.R. Sreenivasan, Extended self-similarity of the small-scale cosmic microwave background anisotropy, *Phys. Lett. A* **319**, 21 (2003).

Beyne, A.O.E., and G.F. Froment, A percolation approach for the modeling of deactivation of zeolite catalysts by coke formation, *Chem. Eng. Sci.* **45**, 2089 (1990).
Beyne, A.O.E., and G.F. Froment, A percolation approach for the modeling of deactivation of zeolite catalysts by coke formation: diffusional limitations and finite rate of coke growth, *Chem. Eng. Sci.* **45**, 2089 (1990).
Beyne, A.O.E., and G.F. Froment, A percolation approach for the modeling of deactivation of zeolite catalysts by coke formation: diffusional limitations and finite rate of coke growth, *Chem. Eng. Sci.* **48**, 503 (1993).
Bialek, W., A. Cavagna, I. Giardina, T. Mora, E. Silvestri, M. Viale, and A.M. Walczak, Statistical mechanics for natural flocks of birds, *Proc. Natl. Acad. Sci. U.S.A.* **109**, 4786 (2012).
Bianchi, R.F., H.P. Souza, T.J. Bonagamba, H.C. Panepucci, and R.M. Faria, Ionic conduction and structural properties of organic-inorganic composite based on poly(propileneglycol), *Synth. Met.* **102**, 1186 (1999).
Bianco, V., and G. Franzese, Hydrogen bond correlated percolation in a supercooled water monolayer as a hallmark of the critical region, *J. Mol. Liquids* **285**, 727 (2019).
Biercuk, M.J., M.C. Llaguno, M. Radosavljevic, J.K. Hyun, A.T. Johnson, and J.E. Fischer, Carbon nanotube composites for thermal management, *Appl. Phys. Lett.* **80**, 2767 (2000).
Bijeljic, B., A.H. Muggeride, and M.J. Blunt, Pore-scale modeling of longitudinal dispersion, *Water Resour. Res.* **40**, W11501 (2004).
Bijeljic, B., and M.J. Blunt, Pore-scale modeling of transverse dispersion in porous media, *Water Resour. Res.* **43**, W12S11 (2007).
Billaux, D., J.P. Chiles, K. Hestir, and J.C.S. Long, Three-dimensional statistical modelling of a fractured rock mass. An example from the Fanay-Augeres mine, *Int. J. Rock Mech. Min. Sci. Geomech. Abstr.* **26**, 281 (1989).
Binder, K., Applications of Monte Carlo methods to statistical physics, *Rep. Prog. Phys.* **60**, 487 (1997).
Bird, R.B., W.E. Stewart, and E.N. Lightfoot, *Transport Phenomena*, 2nd corrected ed. (Wiley, New York, 2007).
Birovljev, A., L. Furuberg, J. Feder, T. Jøssang, K.J. Måløy, and A. Aharony, Gravity invasion percolation in two dimensions: Experiment and simulation, *Phys. Rev. Lett.* **67**, 584 (1991).
Bizhani, G., M. Paczuski, and P. Grassberger, Discontinuous percolation transitions in epidemic processes, surface depinning in random media, and Hamiltonian random graphs, *Phys. Rev. E* **86**, 011128 (2012).
Bizmark, N., J. Schneider, R.D. Priestley, and S.S. Datta, Multiscale dynamics of colloidal deposition and errosion in porous media, *Sci. Adv.* **6**, eabc2530 (2020).
Blunt, M.J., and H. Scher, Pore-level model of wetting, *Phys. Rev. E* **52**, 6387 (1995).
Boccaletti, S., J.A. Almendral, S. Guan, I. Leyva, Z. Liu, I. Sendiña-Nadal, Z. Wang, and Y. Zou, Explosive transitions in complex networks: structure and dynamics: percolation and synchronization, *Phys. Rep.* **660**, 1 (2016).
Boffetta, G., A. Celani, D. Dezzani, and A. Seminara, How winding is the coast of Britain? Conformal invariance of rocky shorelines, *Geophys. Res. Lett.* **35**, L03615 (2008)
Bohman, T., and A. Frieze, Avoiding a giant component, *Random Struct. Algorithms* **19**, 75 (2001).
Bohman, T., A. Frieze, and N.C. Wormald, Avoidance of a giant component in half the edge set of a random graph, *Random Struct. Algorithms* **25**, 432 (2004).
Böhmer, R., and C.A. Angell, Correlations of the nonexponentiality and state dependence of mechanical relaxations with bond connectivity in Ge-As-Se supercooled liquids, *Phys. Rev. B* **45**, 10091 (1992).
Bollobás, B., *Random Graphs* (Academic Press, London, 1985).
Bollobás, B., and O. Riordan, *Percolation* (Combridge University Press, Cambridge, 2006).
Bolton, F., and D. Weaire, Rigidity loss transition in a disordered 2D froth, *Phys. Rev. Lett.* **65**, 3449 (1990).
Bone, S., J. Eden, and R. Pethig, Electrical properties of proteins as a function of hydration and NaCl content, *Int. J. Quantum Chem. Biol. Symp.* **8**, 307 (1981).

Bonnet, E., O. Bour, N.E. Odling, P. Davy, I. Main, P. Cowie, and B. Berkowitz, Scaling of fracture systems in geological media, *Rev. Geophys.* **39**, 347 (2001).

Bonnet, P., D. Sireude, B. Garnier, and O. Chauvet, Thermal properties and percolation in carbon nanotube-polymer composites, *Appl. Phys. Lett.* **91**, 201910 (2007).

Boolchand, P., R.N. Enzweiler, R.L. Cappelletti, W.A. Kamitakahara, Y. Cai, and M.F. Thorpe, Vibrational thresholds in covalent networks, *Solid State Ionics* **39**, 81 (1990).

Boolchand, P., and M.F. Thorpe, Glass-forming tendency, percolation of rigidity, and onefold-coordinated atoms in covalent networks, *Phys. Rev. B* **50**, 10366 (1994).

Bordi, F., C. Cametti, J. Rouch, F. Sciortino, and P. Tartaglia, Cluster formation in water-in-oil microemulsions at percolation: evaluation of the electrical properties, *J. Phys.: Condens. Matter* **8**, A19 (1996).

Borge–Holthoefer, J., and A. Arenas, Semantic networks: Structure and dynamics, *Entropy* (Basel) **12**, 1264 (2010).

Borge–Holthoefer, J., Y. Moreno, and A. Arenas, Modeling abnormal priming in Alzheimer's patients with a free association network, *PLoS One* **6**, e22651 (2011).

Borgman, O., P. Fantinel, W. Lühder, L. Goehring, and R. Holtzman, Impact of spatially correlated pore-scale heterogeneity on drying porous media, *Water Resour. Res.* **53**, 5645 (2017).

Borghetti, J., Z. Li, J. Straznicky, X. Li, D.A.A. Ohlberg, W. Wu, D.R. Stewart, and R.S. Williams, A hybrid nanomemristor/transistor logic circuit capable of self-programming, Proc. Natl. Acad. Sci. U.S.A. **106**, 1699 (2009).

Böttcher, L., O. Woolley-Meza, E. Goles, D. Helbing, and H.J. Herrmann, Connectivity disruption sparks explosive epidemic spreading, *Phys. Rev. E* **93**, 042315 (2016).

Bottin, S., and H. Chaté, Statistical analysis of the transition to turbulence in plane Couette flow, *Eur. Phys. J. B* **6**, 143 (1998).

Bottin, S., F. Daviaud, P. Manneville, and O. Dauchot, Discontinuous transition to spatiotemporal intermittency in plane Couette flow, *Eur. Lett.* **43**, 171 (1998).

Bose, S., P. Raychaudhuri, R. Banerjee, P. Vasa, and P. Ayyub, Mechanism of the size dependence of the superconducting transition of nanostructured Nb, *Phys. Rev. Lett.* **95**, 147003 (2005).

Bouchaud, E., M. Delsanti, M. Adam, M. Daoud, and D. Durand, Gelation and percolation: swelling effect, *J. Physique Lett.* **47**, 539 (1986).

Bour, O., and P. Davy, Connectivity of random fault networks following a power-law fault length distribution, *Water Resour. Res.* **33**, 1567 (1997).

Bour, O., and P. Davy, On the connectivity of three-dimensional fault networks, *Water Resour. Res.* **34**, 2611 (1998).

Bourret, A., Low-density silica aerogels observed by high-resolution electron microscopy, *Europhys. Lett.* **6**, 731 (1988).

Bradonjic, M., M. Molloy, and G. Yan, Containing viral spread on sparse random graphs: Bounds, algorithms, and experiments, *Internet Math.* **9**, 406 (2013).

Brady, R., Very high slip rates on continental extensional faults: new evidence from (U-Th)/He thermochronometry of the Buckskin Mountains, Arizona, *Earth Planetary Sci. Lett.* **197**, 95 (2002).

Branover, H., A. Bershadskii, A. Eidelman, and M. Nagorny, Possibility of simulating geophysical flow phenomena by laboratory experiments, *Bound. Layer Met.* **62**, 117 (1993).

Braunstein, A., L. Dall'Asta, G. Semerjian, and L. Zdeborová, Network dismantling, *Proc. Natl. Acad. Sci. U.S.A.* **113**, 12368 (2016).

Bray, Y.L., and M. Prat, Three-dimensional pore network simulation of drying in capillary porous media, *Int. J. Heat Mass Transfer* **42**, 4207 (1999).

Brenner, H., and L.J. Gaydos, The constrained Brownian movement of spherical particles in cylindrical pores of comparable radius: Models of the diffusive and convective transport of solute molecules in membranes and porous media, *J. Colloid Interface Sci.* **58**, 312 (1977).

Bresser, W., P. Boolchand, and P. Suranyi, Rigidity percolation and molecular clustering in network glasses, *Phys. Rev. Lett.* **56**, 2493 (1986).

Brenig, W., G. Dï¿½hler, and P. Wï¿½lfle, Theory of thermally assisted electron hopping in amorphous solids, *Z. Phys.* **246**, 1 (1971).
Brinker, C.J., and G.W. Scherer, *The Physics and Chemistry of Sol-Gel Processing* (Academic, San Diego, 1990).
Broadbent, S.R., and J.M. Hammersley, Percolation processes. I. Crystals and mazes. *Proc. Camb. Philos. Soc.* **53**, 629 (1957).
Broedersz, C.P., and F.C. MacKintosh, Molecular motors stiffen non-affine semiflexible polymer networks, *Soft Matter* **7**, 3186 (2011).
Brown, E.N., P.L. Purdon, and C.J. Van Dort, General anesthesia and altered states of arousal: a systems neuroscience analysis, *Annu. Rev. Neurosci.* **34**, 601 (2011).
Brown, S.R., and C.H. Scholz, Broad bandwidth study of the topography of natural rock surfaces, *J. Geophys. Res.* **90**, 12575 (1985).
Bruderer, C., and Y. Bernabé, Network modeling of dispersion: Transition from Taylor dispersion in homogeneous networks to mechanical dispersion in very heterogeneous ones, *Water Resour. Res.* **37**, 897 (2001).
Bruni, F., G. Careri, and A.C. Leopold, Critical exponents of protonic percolation in maize seeds, *Phys. Rev. A* **40**, 2803 (1989).
Buchenau, U., M. Morkenbusch, G. Reichenauer, and B. Frick, Inelastic neutron scattering from virgin and densified aerogels, *J. Non-Crystl. Solids* **145**, 121 (1992).
Buchholz, P.C.F., S. Fademrecht, and J. Pleiss, Percolation in protein sequence space. *PLoS ONE* **12**, e0189646 (2017).
Bufe, C.G., and D.J. Varnes, Predictive model of seismic cycle of the Great San Francisco Bay Region, *J. Geophys. Res. B* **98**, 9871 (1993).
Bug, A.L.R., S.A. Safran, G.S. Grest, and I. Webman, Do interactions raise or lower a percolation threshold? *Phys. Rev. Lett.* **55**, 1896 (1985).
Buldyrev, S.V., A.-L. Barabási, F. Caserta, S. Havlin, H.E. Stanley, and T. Vicsek, Anomalous interface roughening in porous media: Experiment and model, *Phys. Rev. A* **45**, R8313 (1992a).
Buldyrev, S.V., A.-L. Barabási, S. Havlin, J. Kertész, H.E. Stanley, and H.S. Xenias. Anomalous interface roughening in 3D porous media: Experiment and model, *Physica A* **191**, 220 (1992b).
Bunde, A., W. Dieterich, and H.E. Roman, Monte Carlo studies of ionic conductors containing an insulating second phase, *Solid State Ionics* **18-19**, 147 (1986a).
Bunde, A., M.D. Ingram, P. Maass, and K.L. Ngai, Diffusion with memory: a model for mixed alkali effects in vitreous ionic conductors, *J. Phys. A* **24**, L881 (1991).
Burganos, V.N., C.A. Paraskeva, and A.C. Payatakes, Three-dimensional trajectory analysis and network simulation of deep bed filtration, *J. Colloid Interf. Ssi.* bf 48, 167 (1992).
Burger, N., A. Laachachi, M. Ferriol, M. Lutz, V. Toniazzo, and D. Ruch, Review of thermal conductivity in composites: mechanisms, parameters and theory, *Prog. Poly. Sci.* **61**, 1 (2016).
Burgess, C.G.V., D.H. Everett, and S. Nuttall, Adsorption hysteresis in porous materials, *Pure & Appl. Chem.* **61**, 1845 (1989).
Burkitt, A.N., A review of the integrate-and-fire neuron model: I. Homogeneous synaptic input, *Biol. Cybern* **95**, 1 (2006).
Burnet, M., *The Colonal Selection Theory* (Vanderbilt University Press, Nashville, 1959).
Butcher, P.N., Calculation of hopping transport coefficients (for impurity bands), *Philos. Mag. B* **42**, 799 (1980).
Butcher, P.N., and J.A. McInnes, Analytical formulae for DC hopping conductivity r-percolation in 3D, *Philos. Mag.* **32**, 249 (1978).
Cafiero, R., A. Gabrielli, M. Marsili, and L. Pietronero, Theory of extremal dynamics with quenched disorder: Invasion percolation and related models, *Phys. Rev. E* **54**, 1406 (1996).
Cai, J., W. Wei, X. Hua, and D.A. Wood, Electrical conductivity models in saturated porous media: A review, *Earth Sci. Rev.* **171**, 419 (2017).
Cai, M., B.F. Edwards, and H. Han, Exact and asymptotic scaling solutions for fragmentation with mass loss, *Phys. Rev. A* **43**, 656 (1991).
Cai, Y., and M.F. Thorpe, Floppy modes in network glasses, *Phys. Rev. B* **40**, 10535 (1989).

Callaghan, P.T., *Principles of Nuclear Magnetic Resonance Microscopy* (Oxford University Press, New York, 1991).

Callaway, D.S., M.E.J. Newman, S.H. Strogatz, and D.J. Watts, Network robustness and fragility: Percolation on random graphs, *Phys. Rev. Lett.* **85**, 5468 (2000).

Candau, S.J., Ankrim, M., Munch, J.P., Rempp, P., Hild, G., and Osaka, R., 1985, in *Physical Optics of Dynamical Phenomena in Macromolecular Systems*, edited by B. Sedlácek (Belin, De Gruyter), p. 145.

Cao, J.-M., Z. Qu, Y.-H. Kim, T.-J. Wu, A. Garfinkel, J.N. Weiss, H.S. Karagueuzian, and P.-S. Chen, Spatiotemporal heterogeneity in the induction of ventricular fibrillation by rapid pacing, *Circ. Res.* **84**, 1318 (1999).

Capuzzi, G., P. Baglioni, C.M. Gambi, and E.Y. Sheu, Percolation phenomenon of calcium bis(2-ethylhexyl) sulfosuccinate water-in-oil microemulsions by conductivity and dielectric spectroscopy measurements, *Phys. Rev. E* **60**, 792 (1999).

Careri, G., M. Geraci, A. Giansanti, and J.A. Rupley, Protonic conductivity of hydrated lysozyme powders at megahertz frequencies, *Proc. Natl. Acad. Sci. U.S.A.* **82**, 5342 (1985).

Careri, G., A., Giansanti, and J.A. Rupley, Critical exponents of protonic percolation in hydrated lysozyme powders, *Phys. Rev. A* **37**, 2703 (1988).

Carlson, J.M., and J. Doyle, Highly optimized tolerance: A mechanism for power laws in designed systems, *Phys. Rev. E* **60**, 1412 (1999).

Carmesin, I., and K. Kremer, The Bond fluctuation method : A new effective algorithm for the dynamics of polymers in all spatial dimensions, *Macromolecules* **21**, 2819 (1988).

Castner, T.G., N.K. Lee, G.S. Cieloszyk, and G.L. Salinger, Dielectric anomaly and the metal-insulator transition, *Phys. Rev. Lett.* **34**, 1627 (1975).

Catalan, P., C.F. Arias, J.A. Cuesta, and S. Manrubia, Adaptive multiscapes: an up-to-date metaphor to visualize molecular adaptation, *Biol. Direct.* **12**, 7 (2017).

Celzard, A., M. Krzesiñskab, J.F. Marêché, and S. Puricellia, Scalar and vectorial percolation in compressed expanded graphite, *Physica A* **294**, 283 (2001).

černý, J., Critical path analysis for continuum percolation, *Ann. I. H. Poincaré - PR* **40**, 661 (2004).

Chalopin, Y., S. Volz, and N. Mingo, Upper bound to the thermal conductivity of carbon nanotube pellets, *J. Appl. Phys.* **105**, 084301 (2009).

Chalupa, K., P.L. Leath, and G.R. Reich, Bootstrap percolation on a Bethe lattice, *J. Phys. C* **12**, L31 (1979).

Chandler, R., J. Koplik, K. Lerman, and J.F. Willemsen, Capillary displacement and percolation in porous media, *J. Fluid Mech.* **119**, 249 (1982).

Charlaix, E., E. Guyon, and N. River, Criterion for percolation threshold in a random array of plates, *Solid State Commun.* **50**, 999 (1984).

Charlaix, E., E. Guyon, and S. Roux, Permeability of a random array of fractures of widely varying apertures, *Trans. Porous Media* **2**, 31 (1987a).

Charlaix, E., J.-P. Hulin, and T.J. Plona, Experimental study of tracer dispersion in sintered glass porous materials of variable compaction, *Phys. Fluids* **30**, 1690 (1987b).

Charlaix, E., J.-P. Hulin, C. Leroy, and C. Zarcone, Experimental study of tracer dispersion in flow through two-dimensional networks of etched capillaries, *J. Phys. D* **21**, 1727 (1988a).

Carvalho, T.T.A., A.J. Fontenele, M. Girardi-Schappo, T. Feliciano, L.A.A. Aguiar, T.P.L. Silva, N.A.P. de Vasconcelos, P.V. Carelli, and M. Copelli, Subsampled directed-percolation models explain scaling relations experimentally observed in the brain, *Front. Neural Circuits* **14**, 576727 (2021).

Chaté, H., and P. Manneville, Role of defects in the transition to turbulence via spatiotemporal intermittency, *Physica D* **37**, 33 (1989).

Chatzis, I., and F.A.L. Dullien, Modeling pore structure by 2-D and 3-D networks with application to sandstones, *J. Can. Pet. Technol.* **16**, 97 (1977).

Chatzis, I., and F.A.L. Dullien, Application of the theory of percolation for a model of drainage in porous media and relative permeability of injected non-wetting liquids, *Rev. Inst. Franc. Pet.* **37**, 183 (1982).

Chatzis, I., and F.A.L. Dullien, The modeling of mercury porosimetry and the relative permeability of mercury in sandstones using percolation theory, *Int. Chem. Eng.* **25**, 47 (1985).
Chatzis, I., and N.R. Morrow, Correlation of capillary number relationships for sandstone, *Soc. Pet. Eng. J.* **24**, 555 (1984).
Chauvet, F., P. Duru, S. Geoffroy, and M. Prat, Three periods of drying of a single square capillary tube, *Phys. Rev. Lett.* **103**, 124502 (2009).
Chelidze, T.L., A percolation model of the solid body destruction and the forecast of earthquakes, *Doklady Akad. Nauk* **246**, 54 (1979).
Chelidze, T.L., Percolation and fracture, *Phys. Earth Planet. Int.* **28**, 93 (1982).
Chelidze, T.L., and Y. Guéguen, Evidence of fractal fracture, *Int. J. Rock. Mech. Min. Sci. Geomech. Abstr.* **27**, 223 (1990).
Chelidze, T.L., and Y.M. Kolesnikov, Modelling and forecasting the failure process in the framework of percolation theory, *Izvestiya Earth Physics* **19**, 347 (1983).
Chen, C.C., and Y.C. Chou, Electrical-conductivity fluctuations near the percolation threshold, *Phys. Rev. Lett.* **54**, 2529 (1985).
Chen, J.-D., and J. Koplik, Immiscible fluid displacement in small networks, *J. Colloid Interface Sci.* **108**, 304 (1985).
Chen, J.-D., and N. Wada, Visualization of immiscible displacement in a three-dimensional transparent porous medium, *Exp. Fluids* **4**, 336 (1986).
Chen, S.H., J. Roucht, F. Sciortino, and P. Tartaglia, Static and dynamic properties of water–in–oil microemulsions near the critical and percolation points, *J. Phys.: Condens. Matter* **6**, 10855 (1994).
Chen, W., and R.M. D'ouza, Explosive percolation with multiple giant components, *Phys. Rev. Lett.* **106**, 115701 (2011).
Cheraghi, M., A. Rinaldo, G.C. Sander, P. Perona, and D.A. Barry, Catchment drainage network scaling laws found experimentally in over–land flow morphologies, *Geophys. Res. Lett.* **45**, 9614 (2018).
Chialvo, D.R., Critical brain networks, *Physica A* **304**, 756 (2004).
Chigada, P.I., J. Wang, B. Al-Duri, J. Wood, and S.P. Rigby, Modelling of pore structure evolution during catalyst deactivation and comparison with experiment, *Chem. Eng. Sci.* **65**, 5550 (2010).
Ching, S.N., P.L. Purdon, S. Vijayan, N.J. Kopell, and E.N. Brown, A neurophysiological-metabolic model for burst suppression, *Proc. Natl. Acad. Sci. U.S.A.* **109**, 3095 (2012).
Cho, Y.S., S. Hwang, H.J. Herrmann, and B. Kahng, Avoiding a spanning cluster in percolation models, *Science* **339**, 1185 (2013).
Cho, Y.S., and B. Kahng, Discontinuous percolation transitions in real physical systems, *Phys. Rev. E* **84**, 050102 (2011a).
Cho, Y.S., and B. Kahng, Suppression effect on explosive percolation, *Phys. Rev. Lett* **107**, 275703 (2011b).
Cho, Y.S., J.S. Kim, J. Park, B. Kahng, and D. Kim, Percolation transitions in scale-free networks under the Achlioptas process, *Phys. Rev. Lett.* **103**, 135702 (2009).
Cho, Y.S., M.G. Mazza, B. Kahng, and J. Nagler, Genuine non-self-averaging and ultraslow convergence in gelation, *Phys. Rev. E* **94**, 022602 (2016).
Choi, S.U.S., Z.G. Zhang, W. Yu, F.E. Lockwood, and E.A. Grulke, Anomalous thermal conductivity enhancement in nanotube suspensions, *Appl. Phys. Lett.* **79**, 2252 (2001).
Chorin, A.J., Spectrum, dimension, and polymer analogies in fluid turbulence, *Phys. Rev. Lett.* **60**, 1947 (1988).
Chorin, A.J., Equilibrium statistics of a vortex filament with applications, *Commun. Math. Phys.* **141**, 619 (1991).
Chorin, A.J., A vortex model with superfluid and turbulent percolation, *J. Statist. Phys.* **69**, 67 (1992).
Chorin, A.J., and J. Akao, Vortex equilibria in turbulence theory and quantum analogues, *Physica D* **52**, 403 (1991)
Chua, L.O., Memristor - The missing circuit element, *IEEE Trans. Circuit Theory* **18**, 507 (1971).

Chua, L.O., and S.M. Kang, Memristive devices and systems, *Proc. IEEE* **64**, 209 (1976).
Chubynsky, M.V., and M.F. Thorpe, in *Physics and Applications of Disordered Materials*, edited by M. Popescu (INOE, Bucharest, Romania, 2002), p. 229.
Chubynsky, M.V., and M.F. Thorpe, Algorithms for three-dimensional rigidity analysis and a first-order percolation transition, *Phys. Rev. E* **76**, 041135 (2007).
Chui, S.T., Disappearance of the Coulomb charging energy and low-temperature resistivity of granular metals, **Phys. Rev. B 43**, R14274 (1991).
Cieplak, M., A. Maritan, and J.R. Banavar, Optimal path and domain walls in the strong disorder limit, *Phys. Rev. Lett.* **72**, 2310 (1994).
Clar, S., B. Drossel, and F. Schwabl, Self-organized critical and synchronized states in a nonequilibrium percolation model, *Phys. Rev. Lett.* **75**, 2722 (1995).
Clar, S., K. Schenk, and F. Schwabl, Phase transitions in a forest-fire model, *Phys. Rev. E* **55**, 2174 (1997).
Clarkson, M.T., and S.I. Smedley, Electrical conductivity and permittivity measurements near the percolation transition in a microemulsion. I. Experiment, *Phys. Rev. A* **37**, 2070 (1988).
Clerc, J.P., G. Giraud, S. Alexander, and E. Guyon, Conductivity of a mixture of conducting and insulating grains: Dimensionality effects, *Phys. Rev. B* **22**, 2489 (1980).
Clerc, J.P., G. Giraud, J.M. Laugier, J.M. and Luck, The electrical conductivity of binary disordered systems, percolation clusters, fractals and related models, *Adv. Phys.* **39**, 191 (1990).
Clerc, J.P., G. Giraud, and J. Roussenq, *C. R. Acad. Sci. Paris B* **281**, 227 (1975).
Clément, E., C. Baudet, E. Guyon, and J.-P. Hulin, Invasion front structure in a 3D model porous medium under a hydrostatic pressure gradient, *J. Phys. D* **20**, 608 (1987).
Clifford, S.M., and T.J. Parker, The evolution of the Martian hydrosphere: implications for the fate of a primordial ocean and the current state of the northern plains, *Icarus* **154**, 40 (2001)
Clusella, P., P. Grassberger, F.J. Pérez-Reche, and A. Politi, Immunization and targeted destruction of networks using explosive percolation, *Phys. Rev. Lett.* **117**, 208301 (2016).
Cohen, M.H., and G.S. Grest, Liquid-glass transition, a free volume approach, *Phys. Rev. B* **20**, 1077 (1979).
Coniglio, A., Thermal phase transition of dilute s-state Potts and n-vector models at the percolation threshold, *Phys. Rev. Lett.* **46**, 260 (1981).
Cohen, R., K. Erez, D. ben–Avraham, and S. Havlin, Breakdown of the internet under intentional attack, *Phys. Rev. Lett.* **86**, 3682 (2001).
Cohen, M.H., and J. Jortner, Effective medium theory for the Hall effect in disordered materials, *Phys. Rev. Lett.* **30**, 696 (1973).
Cohen, M.H., J. Jortner, and I. Webman, Percolation conductivity in granular metal films, *Phys. Rev. B* **17**, 4555 (1978).
Cohen, R., K. Erez, D. ben-Avraham, and S. Havlin, Resilence of the Internet to random breakdown, *Phys. Rev. Lett.* **85**, 4626 (2000).
Colby, R.H., J.R. Gilmor, and M. Rubinstein, Dynamics of near-critical polymer gels, *Phys. Rev. E* **48**, 3712 (1993).
Coleman, J.N., S. Curran, A.B. Dalton, A.P. Davey, B. McCarthy, W. Blau, and R.C. Barklie, Percolation-dominated conductivity in a conjugated-polymer-carbon-nanotube composite, *Phys. Rev. B* **58**, R7492 (1998).
Coniglio, A., U. De Angelis, and A. Forlani, Pair connectedness and cluster size, *J. Phys. A* **10**, 1123 (1977)
Coniglio, A., and A. Fierro, Correlated percolation, in *Complex Media and Percolation Theory*, edited by M. Sahimi and A.G. Hunt (Springer, New York, 2021).
Coniglio, A., and W. Klein, Clusters and Ising critical droplets: a renormalisation group approach, *J. Phys. A* **13**, 2775 (1980).
Coniglio, A., H.E. Stanley, and W. Klein, Site-bond correlated-percolation problem: A statistical mechanical model of polymer gelation, *Phys. Rev. Lett.* **42**, 518 (1979).
Coniglio, A., H.E. Stanley, and W. Klein, Solvent effects on polymer gels: A statistical-mechanical model, *Phys. Rev. B* **25**, 6805 (1982).

Conrad, H., U. Buchenau, R. Schatzler, G. Reichenauer, and J. Fricke, Crossover in the vibrational density of states of silica aerogels studied by high-resolution neutron spectroscopy, *Phys. Rev. B* **41**, 2753 (1990).

Constantinides, G.N., and A.C. Payatakes, Three dimensional network model for consolidated porous media. Basic studies, *Chem. Eng. Commun.* **81**, 55 (1989).

Conwell, E.M., Impurity band conduction in germanium and silicon, *Phys. Rev.* **103**, 51 (1956).

Côté, J. and J.-M. Konrad, A generalized thermal conductivity model for soils and construction materials, *Can. Geotech. J.* **42**, 443 (2005).

Cotton, J.P., *Thése, Université Paris, 6* (1974).

Courtens, E., and R. Vacher, Structure and dynamics of fractal aerogels, *Z. Phys. B* **68**, 355 (1987).

Courtens, E., Vacher, R., and Stoll, E., Fractons observed, *Physica D* **38**, 41 (1989).

COVID-19: https://covid19.who.int/

Craciun, F., C. Galassi, and E. Roncari, Experimental evidence for similar critical behavior of elastic modulus and electric conductivity in porous ceramic materials, *Europhys. Lett.* **41**, 55 (1998).

Cramer, C., and M. Buscher, Complete conductivity spectra of fast ion conducting silver iodide/silver selenate glasses, *Solid State Ionics* **105**, 109 (1998).

Crane, R.W., N.P. Armitage, A. Johansson, G. Sambandamurthy, D. Shahar, and G. Grüner, Survival of superconducting correlations across the two-dimensional superconductor-insulator transition: A finite-frequency study, *Phys. Rev. B* **75**, 184530 (2007).

Creutz, M., Deterministic Ising dynamics, *Ann. Phys. (NY)* **167**, 62 (1986).

Coupette, F., and T. Schilling, Exactly solvable percolation problems, *Phys. Rev. E* **105**, 044108 (2022).

Courtens, E., R. Vacher, and J. Pelous, in *Fractals: Physical Origin and Properties*, edited by L. Peitronero (Plenum, London, 1990), p. 285.

Cruden, D.M., Describing the size of discontinuities, *Int. J. Rock. Mech. Min. Sci. Geomech. Abtr.* **14**, 133 (1977).

Dabrowski, A., Adsorption - from theory to practice, *Adv. Colloid Interface Sci.* **93**, 135 (2001).

Corrsin, S., On the spectrum of isotropic temperature fluctuations in isotropic turbulence, *J. Appl. Phys.* **22**, 469 (1951).

da Costa, R.A., S.N. Dorogovtsev, A.V. Goltsev, and J.F.F. Mendes, Explosive percolation transition is actually continuous, *Phys. Rev. Lett.* **105**, 255701 (2010).

da Costa, R.A., S.N. Dorogovtsev, A.V. Goltsev, and J.F.F. Mendes, Solution of the explosive percolation quest: Scaling functions and critical exponents, *Phys. Rev. E* **90**, 022145 (2014).

Dabrowski, A., Adsorption - from theory to practice, *Adv. Colloid Interface Sci* **93**, 135 (2001).

Dadvar, M., M. Sohrabi, and M. Sahimi, Pore network model of deactivation of immobilized glucose isomerase in packed-bed reactors - I: Two-dimensional simulations at the particle level, *Chem. Eng. Sci.* **56**, 2803 (2001).

Dadvar, M., and M. Sahimi, Pore network model of deactivation of immobilized glucose isomerase in packed-bed reactors - II: Three-dimensional simulations at the particle level, *Chem. Eng. Sci.* **57**, 939 (2002).

Dadvar, M., and M. Sahimi, Pore network model of deactivation of immobilized glucose isomerase in packed-bed reactors - III: Multiscale modeling, *Chem. Eng. Sci.* **58**, 4935 (2003).

Daerr, A., and S. Douady, Two types of avalanche behaviour in granular media, *Nature* **399**, 241 (1999).

Dai, U., A. Palevski, and G. Deutscher, Hall effect in a three-dimensional percolation systems, *Phys. Rev. B* **36**, 790 (1987).

Daigle, H., Application of critical path analysis for permeability prediction in natural porous media, *Adv. Water Resour.* **96**, 43 (2016).

Daigle, H., B. Ghanbarian, P. Henry, and M. Conin, Universal scaling of the formation factor in clays: Example from the Nankai Trough, *J. Geophys. Res.: Solid Earth* **120**, 7361 (2015).

Darcel, C., O. Bour, P. Davy, and J.R. de Dreuzy, Connectivity properties of two-dimensional fracture networks with stochastic fractal correlation, *Water Resour. Res.* **39**, 1272 (2003).

Das, R.K., R.S. Schechter, and M.M. Sharma, The role of surface roughness and contact deformation on the hydrodynamic detachment of particles from surfaces, *J. Colloid Interf. Sci.* **164**, 63 (1994).
Dasanayake, N.L., P.J. Michalski, and A.E. Carlsson, General mechanism of actomyosin contractility, *Phys. Rev. Lett.* **107**, 118101 (2011).
Dashtian, H., N. Shokri, and M. Sahimi, Pore-network model of evaporation-induced salt precipitation in porous media: the effect of correlations and heterogeneity, *Adv. Water Resour.* **112**, 59 (2018).
Dashtian, H., N. Shokri, and M. Sahimi, Pore-network simulation of drying of heterogeneous and stratified porous media, in *Convective Heat Transfer in Porous Media*, edited by Y. Mahmoudi, K. Hooman, and K. Vafai (Taylor & Francis, London, 2019).
Dashtian, H., Y. Yang, and M. Sahimi, Non-universality of the Archie exponent due to multifractality of the resistivity well logs, *Geophys. Res. Lett.* **42**, 10655 (2015).
Das Sarma, S., M.P. Lilly, E.H. Hwang, L.N. Pfeiffer, K.W. West, and J.L. Reno, Two-dimensional metal-insulator transition as a percolation transition in a high-mobility electron system, *Phys. Rev. Lett.* **94**, 136401 (2005).
Daoud, M., Distribution of relaxation times near the gelation threshold, *J. Phys. A* **21**, L973 (1988).
Daoud, M., Viscoelasticity near the sol-gel transition, *Macromolecules* **33**, 3019 (2000).
Daoud, M., and A. Coniglio, Singular behaviour of the free energy in the sol-gel transition, *J. Phys. A* **14**, L301 (1981).
Daoud, M., F. Family, and G. Jannik, Dilution and polydispersity in branched polymers, *J. Physique Lett.* **45**, 199 (1984).
Daviaud, F., M. Bonetti, and M. Dubois, Transition to turbulence via spatiotemporal intermittency in one-dimensional Rayleigh-Bénard convection, *Phys. Rev. A* **42**, 3388 (1990).
Daviaud, F., J. Lega, P. Bergé, P.Coullet, and M. Dubois, Spatio-temporal intermittency in a 1D convective pattern: Theoretical model and experiments, *Physica D* **55**, 287 (1992).
David, C., Y. Guéguen, and G. Pampoukis, Effective medium theory and network theory applied to the transport properties of rock, *J. Geophys. Res.* 95(B5), 6993 (1990).
Davis, H.T., L.R. Valencourt, and C.E. Johnson, Transport processes in composite media, *J. Am. Ceram. Soc.* **58**, 446 (1975).
Davis, L.C., and B.E. Artz, Thermal conductivity of metal-matrix composites, *J. Appl. Phys.* **77**, 4954 (1995).
Day, A.R., R. Tremblay, and A.-M.S. Tremblay, Spectral properties of percolating central force elastic networks, *J. Non-Cryst. Solids* **75**, 245 (1985).
de Arcangelis, L., A. Hansen, H.J. Herrmann, and S. Roux, Scaling laws in fracture. *Phys. Rev. B* **40**, 877 (1989).
Deb, D., Saraswathi Vishveshwara, and Smitha Vishveshwara, Understanding protein structure from a percolation perspective, *Biophys. J.* **97**, 1787 (2009).
Debenedetti, P.G., and H.E. Stanley, Supercooled and glassy water, *Phys. Today* **56**(6), 40 (2003).
de Boer, R.J., and P. Hogeweg, Idiotypic networks incorporating T-B cell cooperation. The conditions for percolation, *J. Theor. Biol.* **139**, 17 (1989).
de Boer, R.J., L.A. Segel, and A.S. Perelson, Pattern formation in one- and two-dimensional shape-space models of the immune system, *J. Theor. Biol.* **155**, 295 (1992).
de Dreuzy, J.-R., P. Davy, and O. Bour, Percolation parameter and percolation-threshold estimates for three-dimensional random ellipses with widely scattered distributions of eccentricity and size, *Phys. Rev. E* **62**, 5948 (2000).
Degen, M., I. Mutabazi, and C.D. Andereck, Transition to weak turbulence via spatiotemporal intermittency in the Taylor-Dean system, *Phys. Rev. E* **53**, 3495 (1996).
de Gennes, P.-G., Exponents for excluded volume problem as derived by the Wilson method, *Phys. Lett. A* **38**, 339 (1972).
de Gennes, P.-G., On a relation between percolation theory and the elasticity of gels, *J. Physique Lett.* **37**, L1 (1976a).
de Gennes, P.-G., La percolation: Un concept unificateur, *La Recherche* **7**, 919 (1976b).
de Gennes, P.-G., Incoherent scattering near a sol gel transition, *J. Physique Lett.* **40**, L197 (1979).

de Gennes, P.G., Suspension colloïdales dans un mélange binaire critique, *C.R. Acad. Sci.* **292**, 701 (1981).

de Gennes, P.-G., Capture d'une "fourmi" par des piéges sur un amas de percolation, *C. R. Acad. Sci. Paris* **296 II**, 881 (1983a)

de Gennes, P.-G., Hydrodynamic dispersion in unsaturated porous media, *J. Fluid Mech.* **136**, 189 (1983b).

de Gennes, P.-G., and E. Guyon, Lois générales pour l'injection d'un fluide dans un milieu poreux alétoire, *J. Mech.* **17**, 403 (1978).

De Gregoria, P., A. Lawler, and K.A. Dawson, Bootstrap percolation, in *Complex Media and Percolation Theory*, edited by M. Sahimi and A.G. Hunt (Springer, New York, 2021), p. 149.

Del Ferraro, G., A. Moreno, B. Min, F. Morone, Ú. Pérez-Ramírez, L. Pérez-Cervera, L.C. Parra, A. Holodny, S. Canals, and H.A. Makse, Finding influential nodes for integration in brain networks using optimal percolation theory, *Nat. Commun.* **9**, 2274 (2018).

del Gado, E., L. de Arcangelis, and A. Coniglio, Elastic properties at the sol-gel transition, *Europhys. Lett.* **46**, 288 (1999).

Delisi, C., A theory of precipitation and agglutination reactions in immunological systems, *J. Theor. Biol.* **45**, 555 (1974).

Delisi, C., and A.S. Perelson, The kinetics of aggregation phenomena I. Minimal models for patch formation on lymphocyte membranes *J. Theor. Biol.* **62**, 159 (1976).

de Lozar, A., and B. Hof, Universality at the onset of turbulence in shear flows, e-print arXiv:1001.2481 @010); accessed 19 March 2022.

Dembovskii, S.A., V.V. Kirilenko, and Ju. A. Buslaev, *Izv. Akad. Nauk USSR, Neorg, Mat.* **7**, 328 (1971).

Deng, H., J. Du, J. Gao, and Q. Wang, Network percolation reveals adaptive bridges of the mobility network response to COVID-19, *PLoS One* **16**, e0258868 (2021).

Deprez, N., D.S. McLachlan, and I. Sigalas, The measurement and comparative analysis of the electrical and thermal conductivities, permeability and Young's modulus of sintered nickel, *Solid State Commun.* **66**, 869 (1988).

Deprez, N., D.S. McLachlan, and I. Sigalas, The measurement and comparative analysis of the electrical and thermal conductivities and permeability of sintered nickel, *Physica A* **157**, 181 (1989).

Deptuck, D., J.P. Harrison, and P. Zawadzki, Measurement of elasticity and conductivity of a three-dimensional percolation system, *Phys. Rev. Lett.* **54**, 913 (1985).

Derrida, B., R. Orbach, and K.W. Yu, Percolation in the effective-medium approximation: Crossover between phonon and fracton excitations, *Phys. Rev. B* **29**, 6645 (1984).

de Solla Price, D.J., Networks of scientific papers, *Science*, **149**, 510 (1965)

Deutscher, G., O. Entin-Wohlman, S. Fishman, and Y. Shapira, Percolation description of granular superconductors, *Phys. Rev. B* **21**, 5041 (1980).

Deutscher, G., and M.L. Rappaport, Critical currents of superconducting aluminium-germanium and lead-germanium thin film alloys near the metal-insulator transition, *J. Physique Lett.* **40**, L219 (1979).

Deutscher, G., M.L. Rappaport, and Z. Ovadyahu, Random perculation in metal-Ge mixtures, *Solid State Commun.* **28**, 593 (1978).

Devreux, F., J.P. Boilot, F. Chaput, L. Malier, and M.A.V. Axelos, Crossover from scalar to vectorial percolation in silica gelation, *Phys. Rev. E* **47**, 2689 (1993).

de Vries, P., H. De Raedt, and A. Lagendijk, Localization of waves in fractals: Spatial behavior *Phys. Rev. Lett.* **62**, 2515 (1989).

Dhar, D., M. Barma, and M.K. Phani, Duality transformations for two-dimensional directed percolation and resistance problems, *Phys. Rev. Lett.* **47**, 1238 (1981).

Dias, M.M., and A.C. Payatakes, Network models for two-phase flow in porous media. Part 1. Immiscible microdisplacement of non-wetting fluids, *J. Fluid Mech.* **164**, 305 (1986a).

Dias, M.M., and A.C. Payatakes, Network models for two-phase flow in porous media. Part 2. Motion of oil ganglia, *J. Fluid Mech.* **164**, 337 (1986b).

Dias, M.M., and D. Wilkinson, Percolation with trapping, *J. Phys. A* **19**, 3131 (1986).
Diaz, C.E., I. Chatzis, and F.A.L. Dullien, Simulation of capillary pressure curves using bond correlated site percolation on a simple cubic network, *Transp. Porous Media* **2**, 215 (1987).
Dicarlo, D.A., A. Sahni, and M.J. Blunt, The effect of wettability on three-phase relative permeability, *Transp. Porous Media* **39**, 347 (2000).
Dickman, R., and A. Yu. Tretyakov, Hyperscaling in the Domany–Kinzel cellular automaton, *Phys. Rev. E* **52**, 3218 (1995).
Dienes, J.K., Los Alamos Scientific Laboratory, Report LA-8553-PR (1980), p. 19.
Djabourov, M., J. Leblond, and P. Papon, Gelation of aqueous gelatin solutions. II. Rheology of the sol-gel transition, *J. Phys. France* **49**, 333 (1988).
Dobrin, R., and P.M. Duxbury, Minimum spanning trees on random networks, *Phys. Rev. Lett.* **86**, 5076 (2001).
Domany, E., and W. Kinzel, Equivalence of cellular automata to Ising models and directed percolation, *Phys. Rev. Lett.* **53**, 311 (1984).
Domb, C., and M.F. Sykes, Cluster size in random mixtures and percolation processes, *Phys. Rev.* **122**, 77 (1961).
Doostmohammadi, A., T.N. Shendruk, K. Thijssen, and J.M. Yeomans, Onset of meso–scale turbulence in active nematics, *Nat. Commun.* **8**, 15326 (2017).
Dorogovtsev, S.N., A.V. Goltsev, and J.F.F. Mendes, $k-$core organization of complex networks, *Phys. Rev. Lett.* **96**, 040601 (2006).
Dorogovtsev, S.N., and J.F.F. Mendes, *Evolution of Networks: from Biological Nets to the Internet and WWW* (Oxford University Press, Oxford, 2003).
Dousse, D., and P. Thiran, Connectivity vs. capacity in dense ad hoc networks, *Proc. IEEE INFORCOM* (2004), p. 486.
Doyen, P.M., Permeability, conductivity, and pore geometry of sandstone, *J. Geophys. Res.* **93**, 7729 (1988).
Dresselhaus, G., M.S. Dresselhaus, and R. Saito, *Physical Properties of Carbon Nanotubes* (World Scientific, Singapore, 1998).
Drossel, B., and F. Schwabl, Self-organized critical forest-fire model, *Phys. Rev. Lett.* **69**, 1629 (1992).
Druger, S.D., A. Nitzan, and M.A. Ratner, Dynamic bond percolation theory: A microscopic model for diffusion in dynamically disordered systems. I. Definition and one?dimensional case, *J. Chem. Phys.* **79**, 3133 (1983).
Druger, S.D., M.A. Ratner, and A. Nitzan, Generalized hopping model for frequency-dependent transport in a dynamically disordered medium, with application to polymer solid electrolytes, *Phys. Rev. B* **31**, 3939 (1985).
D'Souza, R.M., Explosive percolation processes, in *Complex Media and Percolation Theory*, edited by M. Sahimi and A.G. Hunt (Springer, New York, 2021), p. 405.
D'Souza, R.M., J. Gómez-Gardeñes, J. Nagler, and A. Arenas, Explosive phenomena in complex networks, *Adv. Phys.* **68**, 123 (2019).
D'Souza, R.M., and M. Mitzenmacher, Local cluster aggregation models of explosive percolation, *Phys. Rev. Lett.* **104**, 195702 (2010).
D'Souza, R.M., and J. Nagler, Anomalous critical and supercritical phenomena in explosive percolation, *Nature Phys.* **11**, 531 (2015).
Du, F., J.E. Fischer, and K.I. Winey, Effect of nanotube alignment on percolation conductivity in carbon nanotube/polymer composites, *Phys. Rev. B* **72**, 121404(R) (2005).
Duarte, J.A.M.S., Bushfire automata and their phase transitions, *Int. J. Mod. Phys. C* **8**, 171 (1997).
Dubi, Y., Y. Meir, Y. Avishai, Nature of the superconductor-insulator transition in disordered superconductorsature, *Nature* **449**, 876 (2007).
Dubois, M. and B. Cabane, Light-scattering study of the sol-gel transition in silicon tetraethoxide, *Macromolecules* **22**, 2526 (1989).
Dubrov, V.E., M.E. Levinshtein, and M.S. Shur, Anomaly in the dielectric permeability in metal-dielectric transitions. Theory and simulation, *Sov. Phys.-JETP* **43**, 2014 (1976).

Duda, A., Z. Koza, and M. Matyka, Hydraulic tortuosity in arbitrary porous media flow, *Phys. Rev. E* **84**, 036319 (2011).

Duering, E., and D.J. Bergman, Scaling properties of the elastic stiffness moduli of a random rigid-nonrigid network near the rigidity threshold: theory and simulations, *Phys. Rev. B* **37**, 9460 (1988).

Dullien, F.A.L., *Porous Media: Fluid Transport and Pore Structure*, 2nd ed. (Academic Press, New York, 1992)

Dunn-Rankin, D., and A.R. Kerstein, Numerical simulation of particle size distribution evolution during pulverized coal combustion, *Combust. Flames* **69**, 193 (1987).

Durand, D., M. Delsanti, M. Adam, and J.M. Luck, Frequency dependence of viscoelastic properties of branched polymers near gelation threshold, *Europhys. Lett.* **3**, 297 (1987).

Durian, D.J., Foam mechanics at the bubble scale, *Phys. Rev. Lett.* **75**, 4780 (1995)

Durian, D.J., Bubble-scale model of foam mechanics: Melting, nonlinear behavior, and avalanche, *Phys. Rev. E* **55**, 1739 (1997).

Duval, E., A. Boukenter, T. Achibat, B. Champagnon, J. Serugetti, and J. Dumas, Structure of silica aerogels and vibrational dynamics in fractal materials and glasses: electron microscopy and low-frequency Raman scattering, *Philos. Mag. B* **65**, 181 (1992).

Duxbury, P.M., D.J. Jacobs, M.F. Thorpe, and C. Moukarzel, Floppy modes and the free energy: Rigidity and connectivity percolation on Bethe lattices, *Phys. Rev. E* **59**, 2084 (1999).

Dynes, R.C., J.P. Garno, and J.M. Rowell, Two-dimensional electrical conductivity in quench-condensed metal films, *Phys. Rev. Lett.* **40**, 479 (1978).

Dyre, J.C., The random free?energy barrier model for ac conduction in disordered solids, *J. Appl. Phys.* **64**, 2456 (1988).

Dyre, J.C., Universal low-temperature ac conductivity of macroscopically disordered materials, *Phys. Rev. B* **48**, 12511 (1993).

Dyre, J.C., and T.B. Schr⊘der, Universality of ac conduction in disordered solids, *Rev. Mod. Phys.* **72**, 873 (2000).

Ebrahimi, F., The shape of invasion percolation clusters in random and correlated media, *J. Stat. Mech.*, P04005 (2008).

Ebrahimi, F., Invasion percolation: A computational algorithm for complex phenomena. *Comput. Sci. Eng.* **12** (No. 2), 84 (2010).

Efros, A.L., and B.I. Shklovskii, Coulomb gap and low temperature conductivity of disordered systems, *J. Phys. C* **8**, L49 (1975).

Efros, A.L., and B.I. Shklovskii, Critical behaviour of conductivity and dielectric constant near the metal-non-metal transition threshold, *Phys. Status Solidi B* **46**, 475 (1976).

Eggarter, T.P., and M.H. Cohen, Simple model for density of states and mobility of an electron in a gas of hard-core scatterers, *Phys. Rev. Lett.* **25**, 807 (1970).

Egli, M., A.G. Hunt, D. Dahms, G. Raab, C. Derungs, S. Raimondi, and F. Yu, Prediction of soil formation as a function of age the percolation theory approach, *Front. Emviron. Sci.* **6**, 108 (2018).

Ehsasi, M., M. Matloch, O. Frank, J.H. Block, K. Christmann, F.S. Rys, and W. Hirschwald, Steady and nonsteady rates of reaction in a heterogeneously catalyzed reaction: Oxidation of CO on platinum, experiments and simulations, *J. Chem. Phys.* **91**, 4949 (1989).

Elber, R., and M. Karplus, Low-frequency modes in proteins: Use of the effective-medium approximation to interpret the fractal dimension observed in electron-spin relaxation measurements, *Phys. Rev. Lett.* **56**, 394 (1986).

Ellenbroek, W.G., V.F. Hagh, A. Kumar, M.F. Thorpe, and M. van Hecke, Rigidity loss in disordered systems: Three scenarios, *Phys. Rev. Lett.* **114**, 135501 (2015).

Endo, H.K., J.C.S. Long, C.R. Wilson, and P.A. Witherspoon, Model for investigating mechanical transport in fracture networks, *Water Resour. Res.* **20**, 1390 (1984).

Erdös, P., and A. Réyni, On random graphs, *Publ. Math. Inst. Hung. Acad. Sci.* **6**, 290 (1959).

Erdös, P., and A. Réyni, On the evolution of random graphs, *Publ. Math. Inst. Hung. Acad. Sci.* **5**, 17 (1960).

Erdös, P., and A. Réyni, on the strength of connectedness of a random graph, *Acta. Math. Acad. Sci. Hung.* **12**, 261 (1961).
Essam, J.W., Percolation and cluster size, in Phase Transition and Critical Phenomena, edited by C. Domb and M.S. Green, Volume II (Academic Press, London, 1972), p. 197.
Englman, R., Y. Gur and Z. Jaeger, Fluid flow through a crack network in rocks, *J. Appl. Mech.* **50**, 707 (1983).
Englman, R., Z. Jaeger, and A. Levi, Percolation theoretical treatment of two-dimensional fragmentation in solids, *Phil. Mag. B* **50**, 307 (1984).
Ertel, W., K. Frobose, and J. Jackle, Constrained diffusion dynamics in the hard-square lattice gas at high density, *J. Chem. Phys.* **88**, 5027 (1988).
Essam, J.W., A.J. Guttmann, and K. De'Bell, On two-dimensional directed percolation, *J. Phys. A* **21**, 3815 (1988).
Everitt, B., *Cluster Analysis* (Wiley, West Sussex, 2011).
Ewing, R.P., and A.G. Hunt, Dependence of the electrical conductivity on saturation in real porous media, *Vadose Zone J.* **5**, 731 (2006).
Fadda, G.C., D. Lairez, and J. Pelta, Critical behavior of gelation probed by the dynamics of latex spheres, *Phys. Rev. E* **63**, 061405 (2001).
Faillettaz, J., A. Pralong, M. Funk, and N. Deichmann, Evidence of log-periodic oscillations and increasing icequake activity during the breaking-off of large ice masses, *J. Glaciology* **54**, 725 (2017).
Faisst, H., and B. Eckhardt, Sensitive dependence on initial conditions in transition to turbulence in pipe flow, *J. Fluid Mech.* **504**, 343 (2004).
Fallah, B., A.A. Saberi, and B. Sodoudi, Emergence of global scaling behaviour in the coupled earthatmosphere interaction, *Sci. Rep.* **6**, 34005 (2016).
Falkenberg, M., A.J. Ford, A.C. Li, R. Lawrence, A. Ciacci, N.S. Peters, and K. Christensen, Unified mechanism of local drivers in a percolation model of atrial fibrillation, *Phys. Rev. E* **100**, 062406 (2019).
Family, F., and A. Coniglio, Cross over from percolation to random animals and compact clusters, *J. Phys. A* **13**, L403 (1980).
Family, F., and T. Vicsek, Scaling of the Active zone in the Eden process on percolation networks and the ballistic deposition model, *J. Phys. A* **18**, L75 (1985).
Fan, J., J. Meng, and A.A. Saberi, Percolation framework of the Earth topography, *Phys. Rev. E* **99**, 022304 (2019).
Fantinel, P., O. Borgman, R. Holtzman, and L. Goehring, Drying in a microfluidic chip: experiments and simulations, *Sci. Rep.* **7**, 15572 (2017).
Farago, O., and Y. Kantor, Entropic elasticity of two-dimensional self-avoiding percolation systems, *Phys. Rev. Lett.* **85**, 2533 (2000).
Farrell, W., *How Hits Happen* (Harper-Collins, New York, 1998).
Faulkner, J.S., and G.M. Stocks, Calculating properties with the coherent-potential approximation, *Phys. REv. B* **21**, 3222 (1980).
Fehr, E., J.S. Andrade, Jr., S.D. da Cunha, L.R. da Silva, H.J. Herrmann, D. Kadau, C.F. Moukarzel1, and E.A. Oliveira, New efficient methods for calculating watersheds, *J. Stat. Mech.*, P09007 (2009).
Fehr, E., D. Kadau, J.S. Andrade, Jr., and H.J. Herrmann, Impact of perturbations on watersheds, *Phys. Rev. Lett.* **106**, 048501 (2011a).
Fehr, E., D. Kadau, N.A.M. Araújo, J.S. Andrade, Jr., and H.J. Herrmann, Scaling relations for watersheds, *Phys. Rev. E* **84**, 036116 (2011b).
Feng, Q., S. Li, X. Han, and S. Wang, Network simulation for formation impairment due to suspended particles in injected water, *J. Pet. Sci.Eng.* **133**, 384 (2015).
Feng, S., Percolation properties of granular elastic networks in two dimensions, *Phys. Rev. B* **32**, 510 (1985a).
Feng, S., Crossover in spectral dimensionality of elastic percolation systems, *Phys. Rev. B* **32**, 5793 (1985b).

Feng, S., B.I. Halperin, and P.N. Sen, Transport properties of continuum systems near the percolation threshold, *Phys. Rev. B* **35**, 197 (1987).

Feng, S., and P.N. Sen, Percolation on elastic networks: new exponent and threshold, *Phys. Rev. Lett.* **52**, 216 (1984).

Feng, S., M.F. Thorpe, and E. Garboczi, Effective-medium theory of percolation on central-force elastic networks, *Phys. Rev. B* **31**, 276 (1985).

Feng, X., W.J. Bresser, and P. Boolchand, Direct evidence for stiffness threshold in chalcogenide glasses, *Phys. Rev. Lett.* **78**, 4422 (1997).

Fenton, F.H., and A. Karma, Vortex dynamics in three-dimensional continuous myocardium with fiber rotation: Filament instability and fibrillation, *Chaos* **8**, 22 (1998).

Fernholz, D., and V. Ramachandran, the giant $k-$core of a random graph with a specific degree sequence (2003); https://www.cs.utexas.edu/users/vlr/papers/kcore03.pdf

Ferri, B., J. Frisken, and D.S. Cannell, Structure of silica gels, *Phys. Rev. Lett.* **67**, 3626 (1991).

Ferry, J.D., *Viscoelastic Properties of Polymers* (Wiley, New York, 1980).

Finney, J.L., Random packing and the structure of simple liquids. I. Geometry of random close packing, *Proc. Roy. Soc. London A* **319**, 479 (1970).

Fischer, U., C. von Borczyskowski, and N. Schwentner, Singlet-exciton transport and spatial and energetic disorder in dibenzofuran crystals, *Phys. Rev. B* **41**, 9126 (1990).

Fisher, M.E., The theory of critical point singularities, in *Critical Phenomena*, Proceedings of 1970 Enrico Fermi International School in Physics, Course No. 51, Varenna, Italy, edited by M.S. Green (Academic Press, New York, 1971), p. 1

Fisher, M.E., and J.W. Essam, Some cluster size and percolation problems, *J. Math. Phys.* **2**, 609 (1961).

FitzHugh, R., Impulses and physiological states in theoretical models of nerve membrane, *Biophysical J.* **1**, 445 (1961).

Fitzpatrick, J.P., R.B. Malt, and F. Spaepen, Percolation theory and the conductivity of random close packed mixtures of hard spheres, *Phys. Lett. A* **47**, 207 (1974).

Fletcher, T.H., A.R. Kerstein, R.J. Pugmire, and D.M. Grant, Chemical percolation model for devolatilization. 2. Temperature and heating rate effects on product yields, *Energy Fuels* **4**, 54 (1990).

Fletcher, T.H., A.R. Kerstein, R.J. Pugmire, M.S. Solum, and D.M. Grant, Chemical percolation model for devolatilization. 3. Direct use of ^{13}C NMR data to predict effects of coal type, *Energy Fuels* **6**, 414 (1992).

Flory, P.J., Molecular size distribution in three dimensional polymers. I. Gelation, *J. Am. Chem. Soc.* **63**, 3083 (1941).

Fontana, A., F. Tocca, M.P. Fontana, B. Rosi, and A.J. Dianoux, Low-frequency dynamics in superionic borate glasses by coupled Raman and inelastic neutron scattering, *Phys. Rev. B* **41**, 3778 (1990).

Fontenele, A.J., N.A.P. de Vasconcelos, T. Feliciano, L.A.A. Aguiar, C. Soares-Cunha, B. Coimbra, L. Dalla Porta, S. Ribeiro, A.J. Rodrigues, N. Sousa, P.V. Carelli, and M. Copelli, Criticality between cortical states, *Phys. Rev. Lett.* **122**, 208101 (2019).

Forero-Sandoval, I.Y., A.P. Franco-Bacca, F. Cervantes-Álvarez, C.L. Gómez-Heredia, J.A. Ramírez-Rincón, J. Ordonez-Miranda, and J.J. Alvarado-Gil, Electrical and thermal percolation in two-phase materials: A perspective, *J. Appl. Phys.* **131**, 230901 (2022).

Fortuin, C.M., and P.W. Kasteleyn, On the random-cluster model, *Physica* **57**, 536 (1972)

Fostner, S., R. Brown, J. Carr, and S.A. Brown, Continuum percolation with tunneling, *Phys. Rev. B* **89**, 075402 (2014).

Fostner, S., and S.A. Brown, Neuromorphic behavior in percolating nanoparticle films, *Phys. Rev. E* **92**, 052134 (2015).

Fox, T.G., and P.J. Flory, Further studies on the melt viscosity of polyisobutylene, *J. Phys. Chem.* **55**, 221 (1951).

Franceschetti, M., O. Dousse, D.N.C. Tse, and P. Thiran, Closing the gap in the capacity of wireless networks via percolation theory, *IEEE Trans. Information Theor.* **53**, 1009 (2007).

Franceschetti, M., L. Booth, M. Cook, R. Meester, and J. Bruck, Continuum percolation with unreliable and spread-out connections, *J. Statist. Phys.* **118**, 721 (2005).
Freltoft, T., J.K. Kjems, and D. Richter, Density of states in fractal silica smoke-particle aggregates, *Phys. Rev. Lett.* **58**, 1212 (1987).
Frenken, K., Technological innovation and Complexity theory, *Econ. Innov. New Techn.* **15**, 137 (2006).
Frenken, K., G. Silverberg, and M. Valente, A percolation model of the product lifecycle, Danish Research Unit for Industrial Dynamics Working Paper No. 08-20 (2008).
Frette, V., J. Feder, T. Jøssang, and P. Meakin, Buoyancy-driven fluid migration in porous media, *Phys. Rev. Lett.* **68**, 3164 (1992).
Freund, F., and D. Sornette, Electromagnetic earthquake bursts and critical rupture of peroxy bond networks in rocks, *Tectonophysics* **431**, 33 (2007).
Fricke, J., Aerogels, *Sci. Amer.* **258** (No. 5), 68 (1988).
Friedman, E.J., and A.S. Landsberg, Construction and analysis of random networks with explosive percolation, *Phys. Rev. Lett.* **103**, 255701 (2009).
Friedman, N., S. Ito, B.A.W. Brinkman, M. Shimono, R.E. Lee DeVille, K.A. Dahmen, J.M. Beggs, and T.C. Butler, Universal critical dynamics in high resolution neuronal avalanche data, *Phys. Rev. Lett.* **108**, 208102 (2012).
Friedman, S.P., and N.A. Seaton, Critical path analysis of the relationship between permeability and electrical conductivity of three-dimensional pore networks, *Water Resour. Res.* **34**, 1703 (1998).
Frohlich, C., and S.D. Davis, Teleseismic *b* values; or, much ado about 1.0, *J. Geophys. Res.* **98**, 631 (1993).
Froment, G.F., Modeling of catalyst deactivation, *Appl. Catal. A Gen.* **212**, 117 (2001).
Furuberg, L., J. Feder, A. Aharony, and T. Jøssang, Dynamics of invasion percolation, *Phys. Rev. Lett.* **61**, 2117 (1988).
Frydman, A., O. Naaman, and R.C. Dynes, Universal transport in two-dimensional granular superconductors, *Phys. Rev. B* **66**, 052509 (2002).
Gao, C., S. Zhang, F. Wang, B. Wen, C. Han, Y. Ding, and M. Yang, Graphene networks with low percolation threshold in ABS nanocomposites: selective localization and electrical and rheological properties, *ACS Appl. Mater. Interfaces* **6**, 12252 (2014).
Gao, J.-F., Z.-M. Li, Q.-j. Meng, and Q. Yang, CNTs/ UHMWPE composites with a two-dimensional conductive network, *Mater. Lett.* **62**, 3530 (2008).
Gabrielli, A., and G. Caldarelli, Invasion percolation and critical transient in the Barabśi model of human dynamics, *Phys. Rev. Lett.* **98**, 208701 (2007).
Gabrielli, A., and G. Caldarelli, Invasion percolation on a tree and queueing models, *Phys. Rev. E* **79** (2009) 041133.
Gabriell, A., G. Caldarelli, and L. Pietronero, Invasion percolation with tempreture and the nature of self-organized criticality in real systems, *Phys. Rev. E* **62**, 7638 (2000).
Galam, S., Minority opinion spreading in random geometry, *Eur. Phys. J. B* **25**, 403 (2002).
Garboczi, E.J., Effective force constant for a central-force random network, *Phys. Rev. B* **37**, 318 (1988).
Garboczi, E.J., and M.F. Thorpe, Effective-medium theory of percolation on central force elastic networks. II. Further results, *Phys. Rev. B* **31** 7276, (1985).
Garboczi, E.J., and M.F. Thorpe, Effective-medium theory of percolation on central-force elastic networks. III. The superelastic problem, *Phys. Rev. B*, **33**, 3289 (1986a).
Garboczi, E.J., and M.F. Thorpe, Density of states for random-central-force elastic networks, *Phys. Rev. B* **32**, 4513 (1986b)
Garboczi, E.J., and M.F. Thorpe, Density of states for random central-force elastic networks, *Phys. Rev. B* **32**, 4513 (1986).
Garfunkel, G.A., and M.B. Weissman, Noise scaling in continuum percolating films, *Phys. Rev. Lett.* **55**, 296 (1985).
Garland, J.C., Granular properties of high T_c superconductors, *Physica A* **157**, 111 (1989).

Gauthier-Manuel, B., and E. Guyon, Critical elasticity of Polyacrylamide above its gel point, *J. Physique Lett.* **41**, L503 (1980).

Gauthier-Manuel, B., E. Guyon, S. Roux, S. Gits, and F. Lefaucheux, Critical viscoelastic study of the gelation of silica particles, *J. Physique* **48**, 869 (1987).

Gavalas, G.R., Analysis of char combustion including the effect of pore enlargement, *Combust. Sci. Technol.* **24**, 197 (1981).

Gavrilets, S., and J. Gravner, Percolation on the fitness hypercube and the evolution of reproductive isolation, *J. Theor. Biol.* **184**, 51 (1997).

Gazit, Y., D. Berk, M. Leunig, L.T. Baxter, and R,K. Jain, Scale-invariant behavior and vascular network formation in normal and tumor tissue, *Phys. Rev. Lett.* **75**, 2428 (1995).

Gefen, Y., A. Aharony, and S. Alexander, Anomalous diffusion on percolation clusters, *Phys. Rev. Lett.* **50**, 77 (1983).

Geiger, A., F.H. Stillinger, and A. Rahman, Aspects of the percolation process for hydrogen-bond networks in water, *J. Chem. Phys.* **70**, 4185 (1979).

Gelves, G.A., B. Lin, U. Sundararaj, and J.A. Haber, Low electrical percolation threshold of silver and copper nanowires in polystyrene composites, *Adv. Funct. Mater.* **16**, 2423 (2006).

Georgiev, D.G., P. Boolchand, and M. Micoulaut, Rigidity transition and molecular structure of As_xSE_{1-x} glasses, *Phys. Rev. B* **62**, R9228 (2000).

Gerber, A., and G. Deutscher, Upper critical field of superconducting Pb films above and below the percolation threshold, *Phys. Rev. Lett.* **63**, 1184 (1989).

Ghanbarian, B., Applications of critical path analysis to uniform grain packings with narrow conductance distributions: I. Single-phase permeability, *Adv. Water Resour.* **137**, 103529 (2020a).

Ghanbarian, B., Applications of critical path analysis to uniform grain packings with narrow conductance distributions: II. Water relative permeability, *Adv. Water Resour.* **137**, 103524 (2020b).

Ghanbarian, B., Predicting single-phase permeability of porous media using critical-path analysis, in *Complex Systems and Percolation Theory*, edited by M. Sahimi and A.G. Hunt (Springer, New York, 2021).

Ghanbarian B., and H. Daigle, Thermal conductivity in porous media: Percolation-based effective-medium approximation, *Water Resour. Res.* **52**, 295 (2016a).

Ghanbarian, B., and H. Daigle, Permeability in two-component porous media: Effective-medium approximation compared with lattice-Boltzmann simulations, *Vadoze Zone J.* **15**, vzj2015.05.0071 (2016b).

Ghanbarian, B., H. Daigle, A.G. Hunt, R.P. Ewing, and M. Sahimi, Gas and solute diffusion in partially saturated porous media: Percolation theory and Effective Medium Approximation compared with lattice Boltzmann simulations, *J. Geophys. Res. Solid Earth* **120**, 182 (2015).

Ghanbarian, B., M. Esmaeilpour, R.M. Ziff, and M. Sahimi, Effect of pore-scale heterogeneity on scale-dependent permeability: Pore-network simulation and finite-size scaling analysis, *Water Resour. Res.* **57**, e2021WR030664 (2021).

Ghanbarian, B., and A.G. Hunt, Universal scaling of gas diffusion in porous media, *Water Resour. Res.* **50**, 2242 (2014).

Ghanbarian, B., A.G. Hunt, R.P. Ewing, and M. Sahimi, Tortuosity in porous media: A critical review, *Soil Sci. Soc. Am. J.* **77**, 1461 (2013a).

Ghanbarian, B. A.G. Hunt, R.P. Ewing, and T.E. Skinner, Universal scaling of the formation factor in porous media derived by combining percolation and effective medium theories, *Geophys. Res. Lett.* **41**, 3884 (2014).

Ghanbarian, B., A.G. Hunt, M. Sahimi, R.P. Ewing, and T.E. Skinner, Percolation theory generates a physically based description of tortuosity in saturated and unsaturated porous media, *Soil Sci. Soc. Am. J.* **77**, 1920 (2013b).

Ghanbarian, B., and F. Javadpour, Upscaling pore pressure-dependent gas permeability in shales, *J. Geophys. Res. Solid Earth* **122**, 2541 (2017).

Ghanbarian, B., and M. Sahimi, Electrical conductivity of partially-saturated packings of particles, *Transp. Porous Media* **118**, 1 (2017).

Ghanbarian, B., M. Sahimi, and H. Daigle, Modeling relative permeability of water in soil: Application of effective-medium approximation and percolation Theory, *Water Resour. Res.* **52**, 5025 (2016).

Ghanbarian, B., C. Torres-Verdín, L.W. Lake, and M. Marder, Gas permeability in unconventional tight sandstones: Scaling up from pore to core, *J. Pet. Sci. Eng.* **173**, 1153 (2019).

Ghanbarian, B., Torres-Verdín, C., Skaggs, T.H., Quantifying tight-gas sandstone permeability via critical path analysis, *Adv. Water Resour.* **92**, 316 (2016).

Ghanbarian-Alavijeh, B., and A.G. Hunt, Comparison of the predictions of universal scaling of the saturation dependence of the air permeability with experiment, *Water Resour. Res.* **48**, W08513 (2012a).

Ghanbarian-Alavijeh, B., and A.G. Hunt, Unsaturated hydraulic conductivity in porous media: percolation theory, *Geoderma* **187-188**, 77 (2012b).

Gilbert, E.N., Random graphs, *Annal. Math. Statist.* **30**, 1141 (1959).

Girardi-Schappo, M., G.S. Bortolotto, J.J. Gonsalves, L.T. Pinto, and M.H.R. Tragtenberg, Griffiths phase and long–range correlations in a biologically motivated visual cortex model, *Sci. Rep.* **6**, 29561 (2016).

Gist, G.A., A.H. Thompson, A.J. Katz, and R.L. Higgins, Hydrodynamic dispersion and pore geometry in consolidated rock, *Phys. Fluids* **A2**, 1533 (1990).

Glass, R.J., and M.J. Nicholl, Quantitative visualization of entrapped phase dissolution within a horizontal flowing fracture, *Geophys. Res. Lett.* **22**, 1413 (1995).

Glass, R.J., M.J. Nicholl, and V.C. Tidwell, Challenging models for flow in unsaturated, fractured rock through exploration of small scale processes, *Geophys. Res. Lett.* **22**, 1457 (1995).

Glass, R.J., and D.L. Norton, in *Proceedings of the Third Annual International Conference on High Level Radioactive Waste Management* (American Nuclear Society, Las Vegas, 1992), p. 717.

Glauche, I., W. Krause, R. Sollacher, and M. Greiner, Continuum percolation of wireless ad hoc communication networks, *Physica A* **325**, 577 (2003).

Gliozzi, F., Simulation of Potts models with real q and no critical slowing down, *Phys. Rev. E* **66**, 016115 (2002).

Goh, K.-I., B. Kahng, and D. Kim, Universal behavior of load distribution in scale-free networks, *Phys. Rev. Lett.* **87**, 278701 (2001).

Goldberg, R.J., A theory of antibody-antigen reactions. I. Theory for reactions of multivalent antigen with bivalent and univalent antibody, *J. Am. Chem. Soc.* **74**, 5715 (1952).

Golden, K.M., Brine percolation and the transport properties of sea ice, *Annal. Glaciol.* **33**, 28 (2001).

Golden, K.M., S.F. Ackley, V.I. Lytle, The percolation phase transition in sea ice, *Science* **282**, 2238 (1998).

Goldman, A.J., R.G. Cox, and H. Brenner, Slow viscous motion of a sphere parallel to a plane wall - I. Motion through a quiescent fluid, *Chem. Eng. Sci.* **22**, 653 (1967).

Goldman, A.M., and N. Markovic, Superconductor-insulator transitions in the two-dimensional limit, *Phys. Today* **51**, 39 (1998).

Good, I.J., The number of individuals in a cascade process, *Proc. Camb. Philos. Soc.* **45**, 360 (1949).

Gordeyev, S.A., F.J. MacEdo, J.A. Ferreira, F.W.J. Van Hattum, and C.A. Bernardo, Transport properties of polymer-vapour grown carbon fibre composites, *Physica B* **279**, 33 (2000).

Gordon, D., A. Bernheim-Groswasser, C. Keasar, and O. Farago, Hierarchical self-organization of cytoskeletal active networks, *Phys. Biol.* **9**, 026005 (2012).

Gordon, M., and J.A. Torkington, Scrutiny of the critical exponent paradigm, as exemplified by gelation, *Pure Appl. Chem.* **53**, 1461 (1981).

Gouyet, J.F., M. Rosso, and B. Sapoval, Fractal structure of diffusion and invasion fronts in three-dimensional lattices through the gradient percolation approach, *Phys. Rev. B* **37**, 1832 (1988).

Granek, R., and A. Nitzan, Correlated dynamic percolation: Many bond effective?medium theory, *J. Chem. Phys.* **90**, 3784 (1989)

Granek, R., and A. Nitzan, Dynamic percolation theory for diffusion of interacting particles, *J. Chem. Phys.* **92**, 1329 (1990).

Grannan, D.M., J.C. Garland, and D.B. Tanner, Critical behavior of the dielectric constant of a random composite near the percolation threshold, *Phys. Rev. Lett.* **46**, 375 (1981).

Grant, D.M., R.J. Pugmire, T.H. Fletcher and A.R. Kerstein, Chemical model of coal devolatilization using percolation lattice statistics, *Energy Fuels* **3**, 175 (1989).

Grassberger, P., On phase transitions in Schlögl's second model, *Z. Phys. B: Cond. Mat.* **47**, 365 (1982).

Grassberger, P., Directed percolation in 2 + 1 dimensions, *J. Phys. A* **22**, 3673 (1989).

Grassberger, P., C. Christensen, G. Bizhani, S.-W. Son, and M. Paczuski, Explosive percolation is continuous, but with unusual finite size behavior, *Phys. Rev. Lett.* **106**, 225701 (2011).

Grassberger, P., and T. Schreiber, Phase transitions in coupled map lattices, *Physica D* **50**, 177 (1991).

Gravner, J., D. Pitman, and S. Gavrilets, Percolation on fitness landscapes: effects of correlation, phenotype, and incompatibilities, *J. Theor. Biol.* **248**, 627 (2007).

Grest, G.S., and M.H. Cohen, Percolation and the glass transition, in *Percolation Structures and Processes*, edited by G. Deutscher, R. Zallen, and J. Adler, *Annals of the Israel Physical Society* **5** (Adam Hilger, Bristol, 1983), p. 187.

Grest, G.S., I. Webman, S.A. Safran, and A.L.R. Bug, Dynamic percolation in microemulsions, *Phys. Rev. A* **33**, 2842 (1986).

Griffiths, A., On the movement of a coloured index along a capillary tube, and its application to the measurement of the circulation of water in a closed circuit, *Proc. Phys. Soc. Lond.* **23**, 190 (1911).

Grimmett, G.R., *Percolation* (Grundlehren der mathematischen Wissenschaften, Springer, Berlin, 1999).

Grinchuk, P., and O. Rabinovich, Percolation phase transition in combustion of heterogeneous mixtures, *Combustion, Explosion and Shock Waves* **40**, 408 (2004).

Grinstein, G., Z.-W. Lai, and D.A. Browne, Critical phenomena in a nonequilibrium model of heterogeneous catalysis, *Phys. Rev. A* **40**, 4820(R) (1989).

Guéguen, Y., C. David, and P. Gavrilenko, Percolation and fluid transport in the crust, *Geophys. Res. Lett.* **18**, 931 (1991).

Guéguen, Y., and J.K. Dienes, Transport properties of rocks from statistics and percolation, *Math. Geol.* **21**, 1 (1989).

Guilfoyle, D.N., P. Mansfield, and K.J. Packer, Fluid-flow measurement in porous media by echo-planar imaging, *J. Magn. Reson.* **97**, 342 (1992).

Guinney, J., *et al.*, The consensus molecular subtypes of colorectal cancer, *Nat. Med.* **21**, 1350 (2015).

Gupta, P.K., and A.R. Cooper, Topologically disordered networks of rigid polytopes, *J. Non-Cryst. Solids* **123**, 14 (1990).

Gutenberg, B., and C.F. Richter, Magnitude and energy of earthquakes, *Annali di Geofisica* **9**, 1 (1956).

Gvirtzman, H., and P.V. Roberts, Pore scale spatial analysis of two immiscible fluids in porous media, *Water Resour. Res.* **27**, 1167 (1991).

Haan, S.W., and R. Zwanzig, Series expansion in a continuum percolation problem, *J. Phys. A* **10**, 1547 (1977).

Hack, J.T., Studies of longitudinal stream profiles in Virginia and Maryland, U.S. Geological Survey Professional Paper 294-B (1957).

Hagiwara, T., Archie's *m* for permeability, SPE Paper 13100 (1984).

Halfpap, B.L., and S.M. Lindsay, Rigidity percolation in the germanium-arsenic-selenium alloy system, *Phys. Rev. Lett.* **57**, 847 (1986).

Hajirezaie, S., X. Wu, and C.A. Peters, Scale formation in porous media and its impact on reservoir performance during water flooding, *J. Natural Gas Sci. Eng.* **39**, 188 (2017).

Halley, J.W., and W.K. Holcomb, Conductivity of a three–component "reactive" percolation model, *Phys. Rev. Lett.* **40** 1670 (1978).

Halperin, B.I., S. Feng, and P.N. Sen, Differences between lattice and continuum percolation transport exponents, *Phys. Rev. Lett.* **54**, 2391 (1985).

Hamilton, E.M., Variable range hopping in a non uniform density of states, *Philos. Mag.* **26**, 1043 (1972).

Hammersley, J.M., in *Methods in Computational Physics*, Volume 1, edited by B.J. Alder, S. Fernbach, and M. Rotenberg (Academic Press, New York, 1963).

Hammonds, K.D., H. Deng, V. Heine, and M.T. Dove, How floppy modes give rise to adsorption sites in zeolites, *Phys. Rev. Lett.* **78**, 3701 (1997).

Hamzehpour, H., S. Pazoki, M. Khazaei, and M. Sahimi, Dependence of percolation and flow properties of fracture networks on the morphology, *Physica A* **584**, 125361 (2021).

Hansen, A., and J. Kertész, Phase diagram of optimal psths, *Phys. Rev. Lett.* **93**, 040601 (2004).

Hansen, A., and S. Roux, Multifractality in elastic percolation, *J. Stat. Phys.* **53**, 759 (1988).

Hansen, A., and S. Roux, Universality class of central-force percolation, *Phys. Rev. B* **40**, 749 (1989).

Harris, A.B., and A. Aharony, Anomalous diffusion, superlocalization and hopping conductivity on fractal media, *Europhys. Lett.* **4**, 1355 (1987).

Harris, T.E., *The Theory of Branching Processes* (Springer, Berlin, 1963).

Harrison, A.K., and R. Zwanzig, Transport on a dynamically disordered lattice, *Phys. Rev. A* **32**, 1072 (1985).

Hasan, S., V. Joekar-Niasar, N.K. Karadimitriou, and M. Sahimi, Saturation-dependence of non-Fickian transport in porous media, *Water Resour. Res.* **55**, 1153 (2019).

Hashemi, M., B. Dabir, and M. Sahimi, Dynamics of two-phase flow in porous media: Simultaneous invasion of two fluids, *AIChE J.* **45**, 1365 (1999a).

Hashemi, M., M. Sahimi, and B. Dabir, Monte Carlo simulation of two-phase flow in porous media: Invasion with two invaders and two defenders, *Physica A* **267**, 1 (1999b).

Hashemi, M., M. Sahimi, and B. Dabir, Percolation with two invaders and two defenders: volatile clusters, oscillations, and scaling, *Phys. Rev. Lett.* **80**, 3548 (1998).

Hatton, C.G., I.G. Main, and P.G. Meredith, A comparison of seismic and structural measurements of fractal dimension during tensile subcritical crack growth, *J. Struct. Geol.* **15**, 1485 (1993).

Hazlett, R.D., and M.J. Furr, Percolation model for permeability reduction in porous media by continuous-gas foams, *Ind. Eng. Chem. Res.* **39**, 2709 (2000).

He, H., H. Deng, Q. Wang, and J. Gao, Percolation of temporal hierarchical mobility networks during COVID-19, *Phil. Trans. R. Soc. A* **380**, 20210116 (2022).

He, H., and M.F. Thorpe, Elastic properties of glasses, *Phys. Rev. Lett.* **54**, 2107 (1985).

Head, D.A., A.J. Levine, and F.C. MacKintosh, Deformation of cross-linked semiflexible polymer networks, *Phys. Rev. Lett.* **91**, 108102 (2003).

Heiba, A.A., H.T. Davis, and L.E. Scriven, Effect of wettability on two-phase relative permeabilities and capillary pressures, SPE Paper 12172 (1983).

Heiba, A.A., H.T. Davis, and L.E. Scriven, Statistical network theory of three-phase relative permeabilities, SPE Paper 12690 (1984).

Heiba, A.A., M. Sahimi, L.E. Scriven, and H.T. Davis, Percolation theory of two-phase relative permeability, SPE Paper 11015 (1982).

Heiba, A. A., M. Sahimi, L.E. Scriven, and H.T. Davis, Percolation theory of two-phase relative permeability, *SPE Reservoir Eng.*, **7**, 123 (1992).

Hendrickson, B., Conditions for unique graph realizations, *SIAM J. Comput.* **21**, 65 (1992).

Henkes, S., D.A. Quint, Y. Fily, and J.M. Schwarz, Rigid cluster decomposition reveals criticality in frictional jamming, *Phys. Rev. Lett.* **116**, 028301 (2016).

Herrmann, H.J., D.C. Hong, and H.E. Stanley, Backbone and elastic backbone of percolation clusters obtained by the new method of 'burning,' *J. Phys. A* **17**, L261 (1984).

Herrmann, H.J., and M. Sahimi, Fluid penetration through a crack in a pressure gradient, *J. Phys. A* **26**, L1145 (1993).

Hestir, K., and J.C.S. Long, Analytical expressions for the permeability of random two-dimensional Poisson fracture networks based on regular lattice percolation and equivalent media theories, *J. Geophys. Res.* **95**, 21565 (1990).

Hilfer, R., Geometric and dielectric characterization of porous media, **Phys. Rev. B 44**, 60 (1991).

Hill, R.M., On the observation of variable range hopping, *Phys. Stat. Sol.* **A35**, K29 (1976).

Hino, T., T. Hasegawa, K. Terabe, T. Tsuruoka, A. Nayak, T. Ohno, and M. Aono, Atomic switches: atomic-movement-controlled nanodevices for new types of computing, *Sci. Technol. Adv. Mater.* **12**, 1 (2011).

Hinrichsen, H., On possible experimental realizations of directed percolation, *Braz. J. Phys.* **30**, (2000); https://doi.org/10.1590/S0103-97332000000100007

Hinrichsen, H., A. Jiménez–Dalmaroni, Y. Rozov, and E. Domany, Flowing sand: A physical realization of directed percolation, *Phys. Rev. Lett.* **83**, 4999 (1999).

Hirata, T., Fractal dimension of fault systems in Japan: Fractal structure in rock fracture geometry at various scales, *Pure Appl. Geophys.* **131**, 157 (1989).

Hirata, T., T. Satoh, and K. Ito, Fractal structure of spatial distribution of microfracturing in rock, *Geophysics* **90**, 369 (1987).

Hodgson, D.F., and E.J. Amis, Dynamic viscoelastic characterization of sol-gel reactions, *Macromolecules* **23**, 2512 (1990).

Hof, B., A. de Lozar, D.J. Kuik, and J. Westerweel, Repeller or attractor? Selecting the dynamical model for the onset of turbulence in pipe flow, *Phys. Rev. Lett.* **101**, 214501 (2008).

Hofree, M., J.P. Shen, H. Carter, A. Gross, and T. Ideker, Network–based stratification of tumor mutations, *Nat. Methods* **10**, 1108 (2013).

Hohn, T., and B. Hohn, Structure and assembly of simple rna bacteriophages, *Adv. Virus. Res.* **16**, 43 (1970).

Holly, E.E., S.K. Venkataraman, F. Chambon, and H.H. Winter, H.H., Fourier transform mechanical spectroscopy of viscoelastic materials with transient structures, *J. Non-Newtonian Fluid Mech.* **27**, 17 (1988).

Homma, Y., Y. Kobayashi, T. Ogino, and T. Yamashita, Growth of suspended carbon nanotube networks on 100-nm-scale silicon pillars, *Appl. Phys. Lett*. **81**, 2261 (2002).

Horváth, V.K., F. Family, and T. Vicsek, Comment on self-affine fractal interfaces from immiscible displacement in porous media, *Phys. Rev. Lett.* **65**, 1388 (1990).

Horváth, V.K., F. Family, and T. Vicsek, Dynamic scaling of the interface in two-phase viscous flows in porous media, *J. Phys. A* **24**, L25 (1991a).

Horváth, V.K., F. Family, and T. Vicsek, Anomalous noise distribution of the interface in two-phase fluid flow, *Phys. Rev. Lett.* **67**, 3207 (1991b).

Hoshen, J., and R. Kopelman, Percolation and cluster distribution. I. Cluster multiple labelling technique, *Phys. Rev. B* **14**, 3438 (1976).

Hosseini, S.A., and M. Alfi, Time-lapse application of pressure transient analysis for monitoring compressible fluid leakage, *Greenhouse Gases: Sci. Technol.* **6**, 352 (2016).

Houlrik, J.M., I. Webman, and M.H. Jensen, Mean-field theory and critical behavior of coupled map lattices, *Phys. Rev. A* **41**, 4210 (1990).

Hsu, W.Y., and T. Berzins, Percolation and effective-medium theories for perfluorinated ionomers and polymer composites, *J. Polym. Sci. Polym. Phys. Ed.* **23**, 933 (1985).

Hu, H., R.M. Ziff, and Y. Deng, No-nnclave percolation corresponds to holes in the cluster backbone, *Phys. Rev. Lett.* **117**, 185701 (2016).

Hu, J., L. Cai, C. Zhao, and J. Pan, Directed percolation routing for ultra-reliable and low-latency services in Low Earth Orbit (LEO) satellite networks, in *Proceedings of IEEE 92nd Vehicular Technology Conference (VTC2020-Fall)* (2020).

Huang, X., and C. Zhi, *Polymer Nanocomposites: Electrical and Thermal Properties* (Springer, Berlin, 2016).

Huang, Y., C. Ellingford, C. Bowen, T. McNally, D. Wu, and C. Wan, Tailoring the electrical and thermal conductivity of multi-component and multi-phase polymer composites, *Int. Mater. Rev.* **65**,129 (2020).

Hughes, B.D., *Random Walks and Random Environments*, Volumes 1 & 2 (Oxford University Press, London, 1995).

Hughes, B.D., M. Sahimi, L.E. Scriven, and H.T. Davis, Transport and conduction in random systems, *Int. J. Eng. Sci.* **22**, 1083 (1984).

Hughes, R., *Catalyst Deactivation* (Academic Press, New York, 1984).

Huinink, H.P., P.A. Bobbert, W.F. Pasveer, and M.A.J. Michels, Structure and conductivity of clusters generated by variable-range hopping percolation, *Phys. Rev. B* **73**, 224204 (2006).

Hulin, J.-P., E. Charlaix, T.J. Plona, L. Oger, and E. Guyon, Tracer dispersion in sintered glass beads with a bidisperse size distribution, *AIChE J.* **34**, 610 (1988a).

Hulin, J.P., E. Clément, C. Baudet, J.F. Gouyet and M. Rosso, Quantitative analysis of an invading-fluid invasion front under gravity, *Phys. Rev. Lett.* **61**, 333 (1988b).

Hundley, M.F. and A. Zettl, Temperature-dependent ac conductivity of thin percolation films, *Phys. Rev. B* **38**, 10290 (1988).

Hung, C.S., and J.R. Gliessman, The resistivity and Hall effect of Germanium at low temperatures, *Phys. Rev.* **79**, 726 (1950).

Hunt, A.G., Applications of percolation theory to porous media with distributed local conductances, *Adv. Water Resour.* **24**, 279 (2001).

Hunt, A.G., Percolative transport and fractal porous media, *Chaos, Solitons, and Fractals* **19**, 309 (2004a).

Hunt, A.G., Continuum percolation theory and Archie's law, *Geophys. Res. Lett.* **31**, L19503 (2004b).

Hunt, A.G., Continuum percolation theory for saturation dependence of air permeability, *Vadose Zone J.* **4**, 134 (2005a).

Hunt, A.G., Possible explanation of the values of Hack's drainage basin, river length scaling exponent, *Nonlin. Processes Geophys.* **23**, 91 (2016).

Hunt, A.G., R. Ewing, and B. Ghanbarian, *Percolation Theory for Flow in Porous Media*, 3rd ed. (Springer, Heidelberg, 2014a).

Hunt, A.G., B. Faybishenko, and B. Ghanbarian, Predicting characteristics of the water cycle from scaling relationships, *Water Resour. Res.* **57**, e2021WR030808 (2021a).

Hunt, A.G., B. Faybishenko, and B. Ghanbarian, Non-linear hydrologic organization, *Nonlin. Processes Geophys.* **28**, 599 (2021b).

Hunt, A.G., and G.W. Gee, Water retention of fractal soil models using continuum percolation theory: Tests of Hanford site soils, *Vadose Zone J.* **1**, 252 (2002a).

Hunt, A.G., and G.W. Gee, Application of critical path analysis to fractal porous media: Comparison with examples from the Hanford Site, *Adv. Water Resour.*, **25** 129 (2002b).

Hunt, A.G., and B. Ghanbarian, Percolation theory for solute transport in porous media: geochemistry, geomorphology, and carbon cycling, *Water Resour. Res.* **52**, 7444 (2016).

Hunt, A.G., B. Ghanbarian, and R.P. Ewing, Saturation dependence of solute diffusion in porous media: universal scaling compared with experiments, *Vadose Zone J.* **13**, vzj2013.08.0204 (2014b).

Hunt, A.G., R. Holtzman, and B. Ghanbarian, A percolation–based approach to scaling infiltration and evapotranspiration, *Water* **92**, 104 (2017).

Hunt, A.G., and S. Manzoni, *Networks on Networks: The Physics of Geobiology and Geochemistry* (Institute of Physics, Bristol, 2016).

Hunt, A.G., T.E. Skinner, R.P. Ewing, and B. Ghanbarian-Alavijeh, Dispersion of solutes in porous media, *Eur. Phys. J. B* **80**, 411 (2011).

Hunt, A.G., and T.E. Skinner, Longitudinal dispersion of solutes in porous media solely by advection, *Phil. Mag.* **88**, 2921 (2008).

Hunt, A.G., and T.E. Skinner, Incorporation of effects of diffusion into advection-mediated dispersion in porous media, *J. Stat. Phys.* **140**, 544 (2010).

Hunt, A.G., F. Yu, and B. Ghanbarian, Application of percolation theory to reaction and flow in geochemical systems in soil and rock, in *Complex Media and Percolation Theory*, edited by M. Sahimi and A.G. Hunt (Springer, New York, 2021c).

Huseby, O., J.-F. Thovert, and P.M. Adler, Geometry and topology of fracture systems, *J. Phys. A* **30**, 1415 (1997).
Huseby, O., J.-F. Thovert, and P.M. Adler, Dispersion in three-dimensional fracture networks, *Phys. Fluids* **13**, 594 (2001).
Iijima, S., Helical microtubules of graphitic carbon, *Nature* **354**, 56 (1991).
Imdakm, A.O., and M. Sahimi, Transport of large particles in flow through porous media, *Phys. Rev. A* **36**, 5304 (1987).
Imdakm, A.O., and M. Sahimi, Computer simulation of particle transport processes in flow through porous media, *Chem. Eng. Sci.* **46**, 1977 (1991).
Isard, J.O., A study of the migration loss in glass and a generalized method of calculating the rise of dielectric loss with remperature, *Proc. Inst. Electr. Eng.* **109B** (Suppl. No. 22), 440 (1961).
Ishihara, A., Determination of molecular shape by osmotic measurements, *J. Chem. Phys.* **18**, 1446 (1950).
Isichenko, M.B., Percolation, statistical topography, and transport in random media, *Rev. Mod. Phys.* **64**, 961 (1992).
Iyer, S.K., and D. Manjunath, Topological properties of random wireless networks, *Sadhana* **31**, 117 (2006).
Izadbakhsh, A., and F. Khorasheh, Simulation of activity loss of fixed bed catalytic reactor of MTO conversion using percolation theory, *Chem. Eng. Sci.* **66**, 6199 (2011).
Jaeger, H.M., D.B. Haviland, B.G. Orr, and A.M. Goldman, Onset of superconductivity in ultrathin granular metal films, *Phys. Rev. B* **40**, 182 (1989).
Jaccarino, V., and L.R. Walker, Discontinuous occurence of localized moments in metals, *Phys. Rev. Lett.* **15**, 258 (1965).
Jackson, R., *Transport in Porous Catalysts* (Elsevier, Amsterdam, 1977).
Jacobs, D.J., and M.F. Thorpe, Generic rigidity percolation: The pebble game, *Phys. Rev. Lett.* **75**, 4051 (1995).
Jacobs, D.J., and M.F. Thorpe, Generic rigidity percolation in two dimensions, *Phys. Rev. E* **53**, 3682 (1996).
Jacquin, C., Caractere fractal des interfaces fluid-fluide en milieu poreu, *C. R. Acad. Sci. Paris* **B300**, 721 (1985).
Jambhekar, V.A., R. Helmig, N. Schröder, and N. Shokri, Free-flow-porous-media coupling for evaporation-driven transport and precipitation of salt in soil, *Transp. Porous Media* **110**, 251 (2015).
Janson, S., and A. Thomason, Dismantling sparse random graphs, *Combin. Probab. Comput.* **17**, 259 (2008).
Janssen, H.K., On the nonequilibrium phase transition in reaction-diffusion systems with an absorbing stationary state, *Z. Phys. B: Cond. Mat.* **42**, 151 (1981).
Jastrzebska, M.M., S. Jussila, and H. Isotalo, Dielectric response and a.c. conductivity of synthetic dopa-melanin polymer, *J. Mater. Sci.* **33**, 4023 (1998).
Jensen, I., H.C. Fogedby, and R. Dickman, Critical exponents for an irreversible surface reaction model, *Phys. Rev. A* **41**, 3411(R) (1990).
Jerauld, G.R., Ph.D. Thesis, University of Minnesota, Minneapolis (1985).
Jerauld, G.R., J.C. Hatfield, L.E. Scriven, and H.T. Davis, Percolation and conduction on Voronoi and triangular networks: a case study in topological disorder, *J. Phys. C* **17**, 1519 (1984a).
Jerauld, G.R., L.E. Scriven, and H.T. Davis, Percolation and conduction on the 3D Voronoi and regular networks: A second case study in topological disorder, *J. Phys. C* **16**, 3429 (1984b).
Johnson, D.L., J. Koplik, and L.M. Schwartz, New pore-size parameter characterizing transport in porous media, *Phys. Rev. Lett.*, **57**, 2564 (1986).
Jones, L.M., and P. Molnar, Some characteristics of foreshocks and their possible relationship to earthquake prediction and premonitory slip on faults, *J. Geophys. Res.* **84**, 3596 (1979).
Journet, C., W.K. Maser, P. Bernier, A. Loiseau, M. Lamy de la Chapelle, S. Lefrant, P. Deniard, R. Lee, and J.E. Fischer, Large-scale production of single–walled carbon nanotubes by the electric–arc technique, *Nature* **388**, 756 (1997).

Juretchke, H.J., R. Landauer, and J.A. Swanson, Hall effect and conductivity in porous media, *J. Appl. Phys.* **27**, 838 (1956).

Kagan, Y.Y., and L. Knopoff, Stochastic synthesis of earthquake catalogs, *J. Geophys. Res.* **86**, 2853 (1981).

Kamitakahara, W.A., R.L. Cappelletti, P. Boolchand, B. Halfpap, F. Gompf, D.A. Neumann, and H. Mutka, Vibrational densities of states and network rigidity in chalcogenide glasses, *Phys. Rev. B* **44**, 94 (1991).

Kanamori, H., and D. Anderson, Theoretical basis of some empirical relations in seismology, *Bull. Seism. Soc. Am.* **65**, 1073 (1975).

Kantelhardt, J.W., and A. Bunde, Electrons and fractons on percolation structures at iticality: Sublocalization and superlocalization, *Phys. Rev. E* **56**, 6693 (1997).

Kaliski, S., On a model of the continuum with essentialy non-symmetric tensor of mechanical stress, *Arch. Mech. Stosow* **15**, 33 (1963).

Kang, K., and S. Redner, Fluctuation-dominated kinetics in diffusion-controlled reactions, *Phys. Rev. A* **32**, 435 (1985)

Kantor, Y., and I. Webman, Elastic properties of random percolating systems, *Phys. Rev. Lett.* **52**, 1891 (1984).

Kantzas, A., and I. Chatzis, Network simulation of relative permeability curves using a bond correlated-site percolation model of pore structure, *Chem. Eng. Commun.* **69**, 191 (1988).

Kapitulnik, A., and G. Deutscher, Percolation characteristics in discontinuous thin films of Pb, *Phys. Rev. Lett.* **49**, 1444 (1982).

Kapitulnik, A., J.W.P. Hsu, and M.R. Hahn, M.R., in *Physical Phenomena in Granular Materials*, edited by G.D. Cody, T.H. Geballe, and P. Sheng (Pittsburgh, Materials Research Society, 1990), p. 153.

Kardar, M., Domain walls subject to quenched impurities, *J. Appl. Phys.* **61**, 3601 (1987).

Kardar, M., G. Parisi, and Y.C. Zhang, Dynamic scaling of growing interfaces, *Phys. Rev. Lett.* **56**, 889 (1986).

Kargar, F., Z. Barani, R. Salgado, B. Debnath, J.S. Lewis, E. Aytan, R.K. Lake, and A.A. Balandin, Thermal percolation threshold and thermal properties of composites with high loading of graphene and boron nitride fillers, *ACS Appl. Mater. Interfaces* **10**, 37555 (2018).

Karnovsky, M.J., E.R. Unanue, and M. Leventhal, Ligand-induced movement of lymphocyte membrane macromolecules. II. Mapping of surface moieties, *J. Exp. Med.* **136**, 807 (1972).

Katsnelson, M.I., Y.I. Wolf, and E.V. Koonin, Towards physical principles of biological evolution, *Phys. Script.* **93**, 043001 (2018).

Katz, A.J., and A.H. Thompson, Quantitative prediction of permeability in porous rock, *Phys. Rev. B* **34**, 8179 (1986).

Katz, A.J., and A.H. Thompson, Prediction of rock electrical conductivity from mercury injection measurements, *J. Geophys. Res.* **B92**, 599 (1987).

Katz, A.J., A.H. Thompson, and R.A. Rashke, Numerical simulation of resistance steps for mercury injection under the influence of gravity, *Phys. Rev. A* **38**, 4901 (1988).

Kauffman, S.A., Metabolic stability and epigenesis in randomly constructed genetic nets, *J. Theor. Biol.* **22**, 437 (1969).

Kaufman, M., J. Urbain, and R. Thomas, Towards a logical analysis of the immune response, *J. Theor. Biol.* **114**, 527 (1985).

Kazemi, F., Z. Mohammadpour, S.M. Naghib, Y. Zare, and K.Y. Rhee, Percolation onset and electrical conductivity for a multiphase system containing carbon nanotubes and nanoclay, *J. Mater. Res. Technol.* **15**, 1777 (2021).

Keating, P.N., Relationship between the macroscopic and microscopic theory of crystal elasticity. I. Primitive Crystals, *Phys. Rev.* **152**, 774 (1966).

Kechavarzi, C., and K. Soga, Determination of water saturation using miniature resistivity probes during intermediate scale and centrifuge multiphase flow laboratory experiments, *Geotech. Test. J.* **25**, 95 (2002).

Keilis-Borok, V.I., and L.N. Malinovskaya, One regularity in the occurrence of strong earthquakes, *J. Geophys. Res.* **69**, 3019 (1964).

Keilis-Borok, V.I., and A.A. Soloviev (eds.), *Nonlinear Dynamics of the Lithosphere and Earthquake Prediction* (Springer, Heidelberg, 2002).

Kenett, Y.N., D. Anaki, and M. Faust, Investigating the structure of semantic networks in low and high creative persons, *Front. Hum. Neurosci.* **8**, 407 (2014).

Kenett, Y.N., O. Levy, D.Y. Kenett, H.E. Stanley, M. Faust, and S. Havlin, Flexibility of thought in high creative individuals represented by percolation analysis, *Proc. Natl. Acad. Sci. U.S.A.* **115**, 867 (2018).

Kenkel, S.W., and J.P. Straley, Percolation theory of nonlinear circuit elements, *Phys. Rev. Lett.* **49**, 767 (1982).

Kerstein, A.R., Equivalence of the void percolation problem for overlapping spheres and a network problem, *J. Phys. A* **16**, 3071 (1983).

Kerstein, A.R., and A.L.R. Bug, Scaling theory of pore growth in a reactive solid, *Phys. Rev. B* **34**, 1754 (1986).

Kerstein, A.R., and B.F. Edwards, Percolation model for simulation of char oxidation and fragmentation time-histories, *Chem. Eng. Sci.* **42**, 1629 (1987).

Kerstein, A.R., and S. Niksa, Fragmentation during carbon conversion: predictions and measurements, in *Twentieth Symposium (International) on Combustion*, the Combustion Institute, Ann Arbor, Michigan (1984), p. 941.

Kesten, H., *Percolation for Mathematicians* (Birkhäuser, Boston, 1982).

Khademi, M., and M. Sahimi, Molecular dynamics simulation of pressure-driven water flow in silicon-carbide nanotubes, *J. Chem. Phys.* **135**, 204509 (2011).

Khan, S.R., E.M. Pedersen, B. Kain, A.J. Jordan, and R.P. Barber, Jr., Superconductor-insulator transition in granular Pb films near a superconducting ground plane, *Phys. Rev. B* **61**, 5909 (2000).

Kim, B.W., S.H. Park, R.S. Kapadia, and P.R. Bandaru, Evidence of percolation related power law behavior in the thermal conductivity of nanotube/polymer composites, *Appl. Phys. Lett.* **102**, 243105 (2013).

Kim, B.W., S.H. Park, and P.R. Bandaru, Anomalous decrease of the specific heat capacity at the electrical and thermal conductivity percolation threshold in nanocomposites, *Appl. Phys. Lett.* **105**, 253108 (2014).

Kim, H., S. Kobayashi, M.A. AbdurRahim, M.J. Zhang, A. Khusainova, M.A. Hillmyer, A.A. Abdala, and C.W. Macosko, Graphene/polyethylene nanocomposites: Effect of polyethylene functionalization and blending methods, *Polymer* **52** 1837 (2011).

Kim, Ki.H., and D. Stroud, Critical currents of composite superconductors: Model calculations, *Phys. Rev. B* **45**, 2417 (1992).

Kim, T., G. Kim, W.I. Cho, Y.-K. Kwon, and J.-M. Zuo, Electrical transport in small bundles of single-walled carbon nanotubes: Intertube interaction and effects of tube deformation, *Appl. Phys. Lett.* **96**, 173107 (2010).

Kim, Y., Y.-k. Yun, and S.-H. Yook, Explosive percolation in a nanotube–based system, *Phys. Rev. E* **82**, 061105 (2010).

King, P.R., and M. Masihi, *Percolation Theory in Reservoir Engineering* (World Scientific, Singapore, 2018).

Kinloch, I.A., J. Suhr, J. Lou, R.J. Young, and P.M. Ajayan, Composites with carbon nanotubes and graphene: An outlook, **Science 362**, 547 (2018).

Kinouchi, O., and M. Copelli, Optimal dynamical range of excitable networks at criticality, *Nat. Phys.* **2**, 348 (2006).

Kirkpatrick, S., Classical transport in disordered media: Scaling and effective-medium theories, *Phys. Rev. Lett.* **27**, 1722 (1971).

Kirkpatrick, S., Percolation and conduction, *Rev. Mod. Phys.* **45**, 574 (1973).

Kirkpatrick, S., in *Proceedings of 5th International Conference on Amorphous and Liquid Semiconductor*, edited by J. Stuke and W. Brenig (Taylor and Francis, London, 1974), p. 183.

Kirkpatrick, S., Percolation thresholds in granular films - non-universality and critical current, in *Inhomogeneous superconductors - 1979*, edited by D.U. Gubser, T.L. Francavilla, J.R. Leibowitz, and S.A. Wolf, *AIP Conference Proc.* **58** (1979), p. 79.

Kirkwood, J.G., The skeletal modes of vibration of long chain molecules, *J. Chem. Phys.* **7**, 506 (1939).

Kjems, J.K., Thermal transport in fractal systems, *Physica A* **191**, 328 (1993).

Kléber, A.G., and Y. Rudy, Basic mechanisms of cardiac impulse propagation and associated arrhythmias, *Physiol. Rev.* **84**, 431 (2004).

Klepper, S., Industry life-cycles, *Ind. Corporate Change* **6**, 145 (1997).

Knacksted, M.A., B.W. Ninham, and M. Monduzzi, Diffusion in model disordered media, *Phys. Rev. Lett.* **75**, 653 (1995).

Knackstedt, M.A., M. Sahimi, and A.P. Sheppard, Invasion percolation with long-range correlations: First-order phase transitions and nonuniversal scaling properties, *Phys. Rev. E* **61**, 4920 (2000).

Knopoff, L., T. Levshina, V.I. Keilis-Borok, and C. Mattoni, Increased long-range intermediate-magnitude earthquake activity prior to strong earthquakes in California, *J. Geophys. Res.* **101**, 5779 (1996).

Knotek, M.L., M. Pollak, T.M. Donovan, and H. Kurtzman, Thickness dependence of hopping transport in amorphous-Ge films, *Phys. Rev. Lett.* **30**, 853 (1973).

Koch, D.L., and J.F. Brady, Dispersion in fixed beds, *J. Fluid Mech.* **154**, 399(1985).

Koch, R.H., R.B. Laibowitz, E.I. Alessandrini, and J.M. Viggiano, Resistivity-noise measurements in thin gold films near the percolation threshold, *Phys. Rev. B* **32**, 6932 (1985).

Kogut, P.M., and P.L. Leath, Bootstrap percolation transitions on real lattices, *J. Phys. C* **14**, 3187 (1981).

Kogut, P.M., and J.P. Straley, Distribution-induced non-universality of the percolation conductivity exponents, *J. Phys. C.* **12**, 2151 (1979).

Kolb, M., R. Botet, and R. Jullien, Scaling of kinetically growing clusters, *Phys. Rev. Lett.* **51**, 1123 (1983).

Kopelman, R., Fractal reaction kinetics, *Science* **241**, 1620 (1988).

Kopelman, R., S. Parus, and J. Prasad, Fractal-like exciton kinetics in porous glasses, organic membranes, and filter papers, *Phys. Rev. Lett.* **56**, 1742 (1986).

Koplik, J., and T.J. Lasseter, One- and two-phase flow in network models of prous media, *Chem. Eng. Commun.* **26**, 285 (1984)

Koplik, J., and T.J. Lasseter, Two-phase flow in random network models of porous media, *Soc. Pet. Eng. J.* **25**, 89 (1985).

Koplik, J., and H. Levine, Interface motion through a random background, **Phys. Rev. B 32**, 280 (1985).

Koplik, J., C. Lin, and M. Vermette, Conductivity and permeability from microgeometry, *J. Appl. Phys.* **56**, 3127 (1984).

Koplik, J., S. Redner, and D. Wilkinson, Transport and dispersion in random networks with percolation disorder, *Phys. Rev. A* **37**, 2619 (1988).

Koponen, A., M. Kataja, and J. Timonen, Permeability and effective porosity of porous media, *Phys. Rev. E* **56**, 3319 (1997).

Korzhenevskii, A.L., and A.A. Luzhkov Density of phonon-fracton states of disordered solids in the vicinity of percolation phase transition, *Zh. Eksp. Teor. Fiz.* **99**, 530 (1991) [*Sov. Phys. JETP* **72**, 295 (1991)].

Kostek, S., L.M. Schwartz, and D.L. Johnson, Fluid permeability in porous media: Comparison of electrical estimates with hydrodynamical calculations, *Phys. Rev. B* **45**, 186 (1992).

Koudina, N., R. Gonzalez Garcia, J.-F. Thovert, and P.M. Adler, Permeability of three-dimensional fracture networks, *Phys. Rev. E* **57**, 4466 (1998).

Koushanfar, F., A. Davare, D.T. Nguyen, A. Sandiovanni-Vincentelli, and M. Potkonjak, Techniques for maintaining connectivity in wireless ad-hoc networks under energy constraints, *ACM Trans. Embedded Comput. Systems* **6**, Article 16 (2007).

Kowal, D., and Z. Ovadyahu, Disorder induced granularity in an amorphous superconductor, *Solid State Commun.* **90**, 783 (1994).
Kraemer, M.U.G., C.-H. Yang, B. Gutierrez, C.-H. Wu, B.Klein, D.M. Pigott, L. du Plessis, N.R. Faria, R. Li, W.P. Hanage, J.S. Brownstein, M. Layan, A. Vespignani, H. Tian, C. Dye, O.G. Pybus1, and S.V. Scarpino, The effect of human mobility and control measures on the COVID-19 epidemic in China, *Science* **368**, 493 (2020).
Krause, J., and G. Ruxton, *Living in Groups* (Oxford University Press, London, 2002).
Krishna, K., and J.A. Wesselingh, The Maxwell-Stefan approach to mass transfer, *Chem. Eng. Sci.* **52**, 861 (1997).
Kubo, R., Statistical-mechanical theory of irreversible processes. I. General theory and simple application to magnetic and conduction problems, *J. Phys. Soc. Jpn.* **12**, 570 (1957).
Kucinskas, A.B., D.L. Turcotte, J. Huang, and P.G. Ford, Fractal analysis of Venus topography in Tinatin Planatia and Ovda Regio, *J. Geophys. Res.* **97**, 13635 (1992).
Kumar, A., D. Manjunath, and J. Kuri, *Wireless Networking* (Elsevier, Amsterdam, 2008), Chap. 9.
Kunar, B.K., and G.P. Srivastava, Dispersion observed in electrical properties of titanium-substituted lithium ferrites, *J. Appl. Phys.* **75**, 6115 (1994).
Kurkijärvi, J., Conductivity in random systems. II. Finite-size-system percolation, *Phys. Rev. B* **9**, 770 (1974).
Kusy, R.P., and D.T. Turner, Electrical resistivity of a polymeric insulator containing segregated metallic particles, *Nature* **229**, 58 (1971)
Kusy, R.P., Influence of particle size ratio on the continuity of aggregates, *J Appl. Phys.* **48**, 5301 (1977).
Kwon, S., H.W. Cho, G. Gwon, H. Kim, and B.J. Sung, Effects of shape and flexibility of conductive fillers in nanocomposites on percolating network formation and electrical conductivity, *Phys. Rev. E* **93**, 032501 (2016).
Kyrylyuk, A.V., M.C. Hermant, T. Schilling, B. Klumperman, C.E. Koning, and P. van der Schoot, Controlling electrical percolation in multicomponent carbon nanotube dispersions, *Nat. Nanotechnol.* **6**, 364 (2017).
Kyrylyuk, A.V., and P. van der Schoot, Continuum percolation of carbon nanotubes in polymeric and colloidal media, *Proc. Natl. Acad. Sci. U.S.A.* **105**, 8221 (2008).
Laibowitz, R.B., and Y. Gefen, Dynamic scaling near the percolation threshold in thin Au films, *Phys. Rev. Lett.* **53**, 380 (1984).
Laman, G., On graphs and rigidity of plane skeletal structures, *J. Eng. Math.* **4**, 331 (1970).
Lamé, G., Examen des différentes méthodes employées pour Résoudre les Problémes de Géométrie (Vve Courcier, Paris, 1818).
Landau, D.P., and K. Binder, *A Guide to Monte Carlo Simulations in Statistical Physics* (Cambridge University Press, Cambridge, England, 2000).
Landauer, R., The electrical resistance of binary metallic mixtures, *J. Appl. Phys.* **23**, 779 (1952).
Lagar'kov, A.N., L.V. Panina, and A.K. Sarychev, Effective magnetic permeability of composite materials near the percolation threshold, *Sov. Phys. JETP* **66**, 123 (1987) [*Zh. Eskp. Teor. Fiz.* **93**, 215 (1987)].
Lam, F.Y.K., X.C. Mi, and A.J. Higgins, Dimensional scaling of flame propagation in discrete particulate clouds, *Combustion Theor. Model.* **24**, 486 (2020).
Lam, P.M., and W. Bao, Recursion method for the density of states and spectral dimension of percolation networks, *Z. Phys. B* **59**, 63 (1985).
Lam, W.A., O. Chaudhuri, A. Crow, K.D. Webster, T.-D. Li, A. Kita, J. Huang, and D.A. Fletcher, Mechanics and contraction dynamics of single platelets and implications for clot stiffening, *Nat. Mater.* **10**, 61 (2011).
Lambert, C.J., and G.D. Hughes, Localization properties of fractons in percolating structures, *Phys. Rev. Lett.* **66**, 1074 (1991).
Laplace, P.S., *Oeuvres completes de Laplace, t IV, Supplément au livre X du traité de la mécanique céleste* (Complete Work of Laplace, tome 4, Supplement to Book 10 of the Treatise on Celestial

Mechanics) (1806), p. 394. See also, *2eme supplément au livre X* (2nd Supplement to Book 10), p. 419.
LaPointe, P.R., A method to characterize fracture density and connectivity through fractal geometry, *Int. J. Rock Mech. Min. Sci. Geomech. Abstr.* **25**, 421 (1988).
Lapp, A., L. Leibler, F. Schosseler, and C. Strazielle, Scaling Behaviour of pregel sols obtained by end-linking if linear chains, *Macromolecules* **22**, 2871 (1989).
Larkin, J.W., X. Zhai, K. Kikuchi, S.E. Redford, A. Prindle, J. Liu, S. Greenfield, A.M. Walczak, J. Garcia-Ojalvo, A. Mugler, and G. M. Süel, Signal percolation within a bacterial community, *Cell Systems* **7**, 137 (2018).
Larson, R.G., Percolation in Porous Media with Application to Enhanced Oil Recovery, M.S. Thesis, the University of Minnesota, Minneapolis (1977).
Larson, R.G., and H.T. Davis, Conducting backbone in percolating Bethe lattice, *J. Phys. C* **15**, 2327 (1982).
Larson, R.G., H.T. Davis, and L.E. Scriven, Displacement of residual nonwetting fluid from porous media, *Chem. Eng. Sci.* **36**, 75 (1981a).
Larson, R.G., and N.R. Morrow, Effects of sample size on capillary pressure in porous media, *Powder Technol.* **30**, 123 (1981).
Larson, R.G., L.E. Scriven, and H.T. Davis, Percolation theory of residual phases in porous media, *Nature* **268**, 409 (1977).
Larson, R.G., L.E. Scriven, and H.T. Davis, Percolation theory of two phase flow in porous media, *Chem. Eng. Sci.* **36**, 57 (1981b).
Latané, B., The psychology of social impact, *Am. Psychol.* **36**, 343 (1981).
Laugier, J.M., J.P. Clerc, G. Giraud, and J.M. Luck, AC properties of 2D percolation networks: A transfer matrix approach, *J. Phys. A* **19**, 3153 (1986b).
Lax, M., Molecular field in spherical mode, *Phys. Rev.* **97**, 629 (1955).
Le, D., H. Hoang, and J. Mahadevan, Impact of capillary-driven liquid films on salt crystallization, *Transp. Porous Media* **80**, 229 (2009).
Le, K.H., A. Kharaghani, C. Kirsch, and E. Tsotsas, Discrete pore network modeling of superheated steam drying, *Drying Technol.* **35**, 1584 (2017).
Leath, P.L., Cluster size and boundary distribution near percolation threshold, *Phys. Rev. B* **14**, 5046 (1976).
Leclerc, D.F., and G.H. Neale, Monte Carlo simulations of radial displacement of oil from a wetted porous medium: fractals, viscous fingering and invasion percolation, *J. Phys. A* **21**, 2979 (1988).
Le Doussal, P., Percolationlike exponent for the conductivity of highly disordered resistor networks, *Phys. Rev. B* **39**, 881 (1989).
Lee, C.F., and G. Pruessner, Percolation mechanism drives actin gels to the critically connected state, *Phys. Rev. E* **93**, 052414 (2016).
Lee, D., B. Kahng, Y.S. Cho, K.-I. Goh, and D.-S. Lee, Recent advances of percolation theory in complex networks, *J. Korean Phys. Soc.* **73**, 152 (2018).
Lee, H.K., B.J. Kim, and H. Park, Continuity of the explosive percolation transition *Phys. Rev. E* **84**, 020101(R) (2011).
Lee, J.H., F. Trier, T. Cornelissen, D. Preziosi, K. Bouzehouane, S. Fusil, S. Valencia, and M. Bibes, Imaging and harnessing percolation at the metal-insulator transition of $NdNiO_3$ nanogaps, *Nano Lett.* **19**, 7801 (2019).
Lee, S.-I., Y. Song, T.W. Noh, X.-D. Chen, and J.R. Gaines, J.R., Experimental observation of nonuniversal behavior of the conductivity exponent for three-dimensional continuum percolation systems, *Phys. Rev. B* **34**, 6719 (1986).
Lee, Y., J.S. Andrade, S.V. Buldyrev, N.V. Dokholoyan, S. Havlin, P.R. King, G. Paul, and H.E. Stanley, Traveling time and traveling length in critical percolation clusters, *Phys. Rev. E* **60**, 3425 (1999).
Le Febvre du Prey, E.J., Factors affecting liquid-liquid relative permeabilities of a consolidated porous medium, *Soc. Pet. Eng. J.* **13**, 39 (1973).

Lehmann, P., and D. Or, Evaporation and capillary coupling across vertical textural contrasts in porous media, *Phys. Rev. E* **80**, 046318 (2009).

Leibler, L., and F. Schosseler, Gelation of polymer solutions: an experimental verification of the scaling behavior of the size distribution function, *Phys. Rev. Lett.* **55**, 1110 (1985).

Lemieux, M.A., P. Breton, and A.-M.S. Tremblay, Unified approach to numerical transfer matrix methods for disordered systems: application to mixed crystals and to elasticity percolation, *J. Physique Lett.* **46**, L1 (1985).

Lemoult, G., L. Shi, K. Avila, S.V. Jalikop, M. Avila, and B. Hof, Directed percolation phase transition to sustained turbulence in Couette flow, *Nat. Phys.* **12**, 254 (2016).

Lenormand, R., and S. Bories, Description d'un mécanisme de connexion de liaisons destiné a l'étude du drainage avec piégeage en milieu poreux, *C. R. Acad. Sci. Paris B* **291**, 279 (1980).

Lenormand, R., E. Toubol and C. Zarcone, Numerical models and experiments on immiscible displacements in porous media, *J. Fluid Mech.* **189**, 165 (1988).

Lenormand, R., and C. Zarcone, Invasion percolation in an etched network: measurement of a fractal dimension, *Phys. Rev. Lett.* **54**, 2226 (1985a).

Lenormand, R., and C. Zarcone, Two-phase flow experiments in a two-dimensional permeable medium, *Physicochem. Hydro.* **6**, 497 (1985b).

Lenormand, R., C. Zarcone, and A. Sarr, Mechanisms of the displacement of one fluid by another in a network of capillary ducts, *J. Fluid Mech.* **135**, 337 (1983).

Lenz, M., Geometrical origins of contractility in disordered actomyosin networks, *Phys. Rev. X* **4**, 041002 (2014).

Léon, C., A. Rivera, A. Várez, J. Sanz, J. Santamaria, and K.L. Ngai, Origin of constant loss in ionic conductors, *Phys. Rev. Lett.* **86**, 1279 (12001).

Leverett, M.C., Capillary behavior in porous solids, *Trans. AIME* **142**, 159 (1941).

Levine, S., and D.L. Cuthiell, Relative permeabilities in two-phase flow through porous media: An application of effective medium theory, *J. Can. Pet. Tech.* **25**, 74 (1986).

Levinshtein, M.E., M.S. Shur, and E.L. Efros, On the relation between critical indices and percolation theory, *Sov. Phys.-JETP* **41**, 386 (1975).

Levinshtein, M.E., M.S. Shur, and E.L. Efros, Galvanomagnetic phenomena in disordered systems. Theory and simulation, *Sov. Phys.-JETP* **42**, 1120 (1976).

Lévy, Y.-E., and B. Souillard, Superlocalization of electrons and waves in fractal media, *Europhys. Lett.* **4**, 233 (1987).

Lewinski, T., Physical correctness of Cosserat-type models of honeycomb grid plates, *J. Theor. Appl. Mech.* **23**, 53 (1985).

Li, J., and M. Östling, Percolation thresholds of two-dimensional continuum systems of rectangles, *Phys. Rev. E* **88**, 012101 (2013).

Li, Q., C.M. Soukoulis, and G.S. Grest, Vibrational properties of percolating clusters: Localization and density of states, *Phys. Rev. B* **41**, 11713 (1990).

Li, Y., W.G. Laidlaw, and N.C. Wardlaw, Sensitivity of drainage and imbibition to pore structures as revealed by computer simulation of the displacement process. *Adv. Colloid Interface Sci.* **26**, 1 (1986).

Li, Y.-J., M. Xu, and J.-Q. Feng, Dielectric behavior of a metal-polymer composite with low percolation threshold, *Appl. Phys. Lett.* **89**, 072902 (2006).

Liang, N.T., Y. Shan, and S.-Y. Wang, Electronic conductivity and percolation theory in aggregated films, *Phys. Rev. Lett.* **37**, 526 (1976).

Liggett, T.M., *Stochastic Interacting Systems: Contact, Voter, and Exclusion Processes* (Springer, Berlin, 1999)

Limat, L., Percolation and Cosserat elasticity: exact results on a deterministic fractal, *Phys. Rev. B* **37**, 672 (1988a).

Limat, L., Rotationally invariant elasticity in a planar fractal network, *Phys. Rev. B* **38**, 512 (1988b).

Limat, L., Micropolar elastic percolation: the superelastic problem, *Phys. Rev. B* **38**, 7219 (1988c).

Limat, L., Elastic and superelastic percolation networks: imperfect duality, critical Poisson ratios, and relations between microscopic models, *Phys. Rev. B* **40**, 9253 (1989).

Lin, Y.G., D.T. Mallin, J.C.W. Chien, and H.H. Winter, Dynamical mechanical measurement of crystallization-induced gelation in thermpplastic elastomeric poly(propylene), *Macromolecules* **24**, 850 (1991).

Liu, X., and H.v. Löhneysen, Low-temperature thermal properties of amorphous AS_xSe_{1-x}, *Phys. Rev. B* **48**, 13486 (1993).

Liu, H., L. Zhang, and N.A. Seaton, Analysis of sorption hysteresis in mesoporous solids using a pore network model, *J. Colloid Interface Sci.* **156**, 285 (1993).

Liu, K., S. Henkes, and J.M. Schwarz, Frictional rigidity percolation: A new universality class and its superuniversal connections through minimal rigidity proliferation, *Phys. Rev. X* **9**, 021006 (2019).

Liu, J., and K. Regenauer-Lieb, Application of percolation theory to microtomography of structured media: Percolation threshold, critical exponents, and upscaling, *Phys. Rev. E* **83**, 016106 (2011).

Liu, J., and K. Regenauer-Lieb, Application of percolation theory to microtomography of rocks, *Earth-Sci. Rev.* **214**, 103519 (2021).

Liu, H., L. Zhang, and N.A. Seaton, Determination of the connectivity of porous solids from nitrogen sorption measurements. II. Generalisation, *Chem. Eng. Sci.* **47**, 4393 (1992).

Liverpool, T.B., and S.C. Glotzer, Fixed-cluster acceleration algorithm for spin systems, *Phys. Rev. E* **53**, R4255 (1996).

Lobb, C.J., and M.G. Forrester, Measurement of nonuniversal critical behavior in a two-dimensional continuum percolating system, *Phys. Rev. B* **35**, 1899 (1987).

Lobb, C.J., and D.J. Frank, Percolation critical exponents for conductance and critical current in two dimensions, *AIP Conference Conference Proceedings* **58**, 308 (1979).

Lomnitz-Adler, J., Asperity models and characteristic earthquakes, *Geophys. 1. Roy. Astr. Soc.* **83**, 435 (1985).

Long, A.R., Hopping conductivity in the intermediate frequency regimes, in *Hopping Transport in Solids,* edited by M. Pollak and B. Shklovskii (Elsevier, Amsterdam, 1991), p. 207.

Long, J.C.S., and D. Billaux, From field data to fracture network modeling: An example incorporating spatial structure, *Water Resour. Res.* **23**, 1201 (1987).

Long, J.C.S., P. Gilmour, and P.A. Witherspoon, A model for steady state fluid flow in random, three-dimensional networks of disk-shaped fractures, *Water Resour. Res.* **21**, 1105 (1985).

Long, J.C.S., J.S. Remer, C.R. Wilson, and P.A. Witherspoon, Porous media equivalents for networks of discontinuous fractures. *Water Resour. Res.* **18**, 645 (1982).

Long, J.C.S., and P.A. Witherspoon, The relationship of the degree of interconnection to permeability in fracture networks, *J. Geophys. Res.* **90**, 3087 (1985).

Lorenz, B., I. Orgzall, and H.O. Heuer, Universality and cluster structures in continuum models of percolation with two different radius distributions, *J. Phys. A* **26**, 4711 (1993).

Loring, R.F., Configurational relaxation and diffusion of a flexible polymer in a dynamically disordered medium, *J. Chem. Phys.* **94**, 1505 (1991).

Loring, R.F., and S. Mukamel, Phonon and fracton vibrational modes in disordered harmonic structures: a self-consistent theory, *Phys. Rev. B* **34**, 6582 (1986).

Lovász, L., and Y. Yemini, On generic rigidity in the plane, *SIAM J. Alg. Disc. Meth.* **3**, 91 (1982).

Lu, J., R. Darvari, J.P. Nicot, P. Mickler, and S.A. Hosseini, Geochemical impact of injection of Eagle Ford brine on Hosston sandstone formation: Observations of autoclave water-rock interaction experiments, *Applied Geochem.* **84**, 26 (2017).

Lubensky, T.C., and J. Isaacson, Field Theory for the statistics of branched polymers, gelation and vulcanization, *Phys. Rev. Lett.* **41**, 829 (1978).

Machta, J., Y.S. Choi, A. Lucke, T. Schweizer, and L.M. Chayes, Invaded cluster algorithm for equilibrium critical points, *Phys. Rev. Lett.* **75**, 2792 (1995).

Machta, J., Y.S. Choi, A. Lucke, T. Schweizer, and L.M. Chayes, Invaded cluster algorithm for Potts models, *Phys. Rev. E* **54**, 1332 (1996).

MacKay, G., and N. Jan, Forest fires as critical phenomena, *J. Phys. A* **17**, L757 (1984).

Madden, T.R., Microcrack connectivity in rocks: A renormalization group approach to the critical phenomena of conductivity and failure in crystalline rocks, *J. Geophys. Res.* **88**, 585 (1983).

Mahadevan, S., and A. Giridhar, On the chemical and mechanical thresholds of some chalcogenide glasses, *J. Non-Cryst. Solids* **110**, 118 (1989).

Maier, R., and W.G. Laidlaw, Fluid percolation in bond-site size-correlated three-dimensional networks, *Transp. Porous Media* **5**, 421 (1990).

Maier, R., and W.G. Laidlaw, Invariants for the critical points in network models of flow in porous media, *J. Stat. Phys.* **62**, 269 (1991a).

Maier, R., and W.G. Laidlaw, The statistical properties of immiscible fluid displacement in bond/site correlated network models of porous media, *Math. Geol.* **23**, 87 (1991b).

Majmudar, T.S., M. Sperl, S. Luding, and R.P. Behringer, Jamming transition in granular systems, *Phys. Rev. Lett.* **98**, 058001 (2007).

Makse, H.A., J.S. Andrade, and H.R. Stanley, Tracer dispersion in a percolation network with spatial correlations, *Phys. Rev. E* **61**, 583 (2000).

Makse, H.A., S. Buldyrev, H. Leschhorn, and H.E. Stanley, The pinning paths of an elastic interface, *Europhys. Lett.* **41**, 251 (1998).

Malek, K., and M. Sahimi, Molecular dynamics simulations of adsorption and diffusion of gases in silicon-carbide nanotubes, *J. Chem. Phys.* **132**, 014310 (2010).

Maliepaard, M.C., J.H. Page, J.P. Harrison, and R.J. Stubbs, Ultrasonic study of the vibrational modes of sintered metal powders, *Phys. Rev. B* **32**, 6261 (1985).

Malin, M.C., and K.S. Edgett, Oceans or seas in the Martian northern lowlands: High resolution imaging tests of proposed coastlines, *Geophys. Res. Lett.* **26**, 3049 (1999).

Mall, S., and W.R. Russel, Effective medium approximation for an elastic network model of flocculated suspensions, *J. Rheology* **31**, 651 (1987).

Mallamace, F., P. Gambadauro, N. Micali, P. Tartaglia, C. Liao, and S.H. Chen, Kinetic glass transition in a micellar system with short–range attractive interaction, *Phys. Rev. Lett.* **84**, 5431 (2000)

Malliaris, A., and D.T. Turner, Influence of particle size on the electrical resistivity of compacted mixtures of polymeric and metallic powders, *J. Appl. Phys.* **42**, 614 (1971).

Mallinson, J.B., S. Shirai, S.K. Acharya, S.K. Bose, E. Galli, and S.A. Brown, Avalanches and criticality in self-organized nanoscale networks, *Sci. Adv.* **5**, eaaw8438 (2019).

Malamud, B.D., G. Morein, and D.L. Turcotte, Forest fires: An example of self-organized critical behavior, *Science* **281**, 1840 (1998).

Maloufi, N., A. Audouard, M. Piecuch, M. Vergant, G. Marchal, and M. Gerl, Experimental study of the dc conductivity mechanism in amorphous Si_xSn_{1-x} alloys, *Phys. Rev. B* **37**, 8867 (1988).

Mamunya, Y.P., V.A. Davydenko, P. Pissis, and E.V. Lebedev, Electrical and thermal conductivity of polymers filled with metal powders, *Europ. Polymer J.* **38**, 1887 (2002).

Mandal, P., A. Neumann, A.G. Jansen, P. Wyder, and R. Deltour, Temperature and magnetic-field dependence of the resistivity of carbon-black polymer composites, *Phys. Rev. B* **55**, 452 (1997).

Mandelbrot, B.B., *The Fractal Geometry of Nature* (Freeman, San Francisco, 1983).

Manshour, P., S. Saberi, M. Sahimi, J. Peinke, A.F. Pacheco, and M.R. Rahimi Tabar, Turbulencelike behavior of seismic Time series, *Phys. Rev. Lett.* **102**, 014101 (2009).

Manshour, P., F. Ghasemi, T. Matsumoto, J. Gómez, M. Sahimi, J. Peinke, A.F. Pacheco, and M.R. Rahimi Tabar, Anomalous fluctuations of vertical velocity of Earth and their possible implications for earthquakes, *Phys. Rev. E* **82**, 036105 (2010).

Mao, X., O. Stenull, and T.C. Lubensky, Effective-medium theory of a filamentous triangular lattice, *Phys. Rev. E* **87**, 042601 (2013a).

Mao, X., O. Stenull, and T.C. Lubensky, Elasticity of a filamentous kagome lattice, *Phys. Rev. E* **87**, 042602 (2013b).

Marcq, P., H. Chaté, and P. Manneville, Universal critical behavior in two-dimensional coupled map lattices, *Phys. Rev. Lett.* **77**, 4003 (1996).

Marsden, A.J., D.G. Papageorgiou, C. Vallés, A. Liscio, V. Palermo, M.A. Bissett, R.J. Young, and I.A. Kinloch, Electrical percolation in graphene-polymer composites, **2D Mater. 5**, 032003 (2018).

Maritan, M., F. Colaiori, A. Flammini, M. Cieplak, and J.R. Banavar, Universality classes of optimal channel networks, *Science* **272**, 984 (1996).
Martin, J.E., D. Adolf, and J.P. Wilcoxon, Viscoelasticity of near-critical gels, *Phys. Rev. Lett.* **61**, 2620 (1988).
Martin, J.E., and J.P. Wilcoxon, Critical dynamics of the sol-gel transition, *Phys. Rev. Lett.* **61**, 373 (1988).
Martys, N., and E. Garboczi, Length scales relating the quid permeability and electrical conductivity in random two-dimensional model porous media, *Phys. Rev. B* **46**, 6080 (1992).
Mason, G., The effect of pore space connectivity on the hysteresis of capillary condensation in adsorption: desorption isotherms, *J. Colloid Interface Sci.* **88**, 36 (1982).
Mason, G., Model of adsorption-desorption hysteresis in which hysteresis is primarily developed by the interconnections in a network of pores, *Proc. Roy. Soc. Lond.* **A390**, 47 (1983).
Mason, G., Determination of the pore-size distributions and pore-space interconnectivity of Vycor porous glass from adsorption-desorption hysteresis capillary condensation isotherms, *Proc. Roy. Soc. Lond.* **A415**, 453 (1988).
Mason, T.G., J. Bibette, and D.A. Weitz, Elasticity of compressed emulsions, *Phys. Rev. Lett.* **75**, 2051 (1995).
Mastrogiannis, D., S.I. Andreopoulos, S.M. Potirakis, I. Stavrakas, D. Triantis, and S.K. Kourkoulis, Investigation of acoustic emissions and pressure stimulated currents detected during bending of restored marble epistyles within the frame of log-periodic power-law models, *Procedia Structural Integrity* **10**, 319 (2018).
Mata, V.G., J.C.B. Lopes, and M.M. Dias, Porous media characterization using mercury porosimetry simulation. 1. Description of the simulator and its sensitivity to model parameters, *Ind. Eng. Chem. Res.* **40**, 3511 (2001).
Matyka, M., A. Khalili, and Z. Koza, Tortuosity-porosity relation in porous media flow, *Phys. Rev. E* **78**, 026306 (2008).
McLachlan, D.S., An equation for the conductivity of binary mixtures with 763 anisotropic grain structures, *J. Phys. C* **20**, 865 (1987).
McLachlan, D.S., Measurement and analysis of a model dual-conductivity medium using a generalised effective-medium theory, *J. Phys. C* **21**, 1521 (1988).
Mclachlan, D.S., M. Blaszkiewicz, and R.E. Newnham, Electrical resistivity of composites, *J. Am. Ceram. Soc.* **73**, 2187 (1990).
Mead, C., Neuromorphic electronic systems, *Proc. IEEE* **78** (10), 1629 (1990).
Meakin, P., Formation of fractal clusters and networks by irreversible diffusion-limited aggregation, *Phys. Rev. Lett.* **51**, 1119 (1983)
Meakin, P., Simple models for heterogeneous catalysis with a poisoning transition, *J. Chem. Phys.* **93**, 2903 (1990).
Meakin, P., Invasion percolation on substrates with correlated disorder, *Physica A* **173**, 305 (1991).
Meakin, P., *Fractals, Scaling and Growth Far From Equilibrium* (Cambridge University Press, Cambridge, 1998).
Meakin, P., J. Feder, V. Frette, and T. Jøossang, Invasion percolation in a destabilizing gradient, *Phys. Rev. A* **46**, 3357 (1992).
Meakin, P., and H.E. Stanley, Novel dimension-independent behavior for diffusive annihilation on percolation fractals, *J. Phys. A* **17**, L173 (1984).
Medini, D., A. Covacci, and C. Donati, Protein homology network families reveal step-wise diversification of type III and type IV secretion systems, *PLoS Comput Biol.* **2**, 12 (2006).
Meester, R., and R. Roy, *Continuum Percolation* (Cambridge University Press, Cambridge, 1996).
Meester, R., R. Roy, and A. Sarkar, Nonuniversality and continuity of the critical covered volume fraction in continuum percolation, *J. Stat. Phys.* **75**, 123 (1994).
Mehbod, M., P. Wyder, R. Deltour, C. Pierre, and G. Geuskens, Temperature dependence of the resistivity in polymer-conducting-carbon-black composites, *Phys. Rev. B* **36**, 7627 (1987).
Meir, Y., Universal crossover between Efros-Shklovskii and Mott variable-range-hopping regimes, *Phys. Rev. Lett.* **77**, 5265 (1996).

Meir, Y., R. Blumenfeld, A. Aharony, and A.B. Harris, Series analysis of randomly diluted nonlinear resistor networks, *Phys. Rev. B* **34**, 3424 (1986).

Melkote, R.R., and K.F. Jensen, Models for catalytic pore plugging: application to hydrodemetallation, *Chem. Eng. Sci.* **44**, 649 (1989).

Melrose, J.C., and C.F. Brandner, Role of capillary forces in determining microscopic displacement efficiency for oil recovery by waterflooding, *J. Can. Pet. Tech.* **13**, 54 (1974).

Michalland, S., M. Rabaud, and Y. Couder, Transition to chaos by spatio-temporal intermittency in directional viscous fingering, *Europhys. Lett.* **22**, 17 (1993).

Michels, M.A.J., J.C.M. Brokken-Zijp, W.M. Groenewoud, and A. Knoester, Systematic study of percolative network formation and effective electric response in low-concentration-carbon-black/polymer composites, *Physica A* **157**, 529 (1989).

Miller, A., and E. Abrahams, Electro-optic kerr effect and polarization reversal in Deuterium doped Rochelle salt, *Phys. Rev.* **120**, 745 (1960).

Mines, G.R., On dynamic equilibrium in the heart, *J. Physiol.* **46**, 349 (1913).

Miri, R., R. van Noort, P. Aagaard, and H. Hellevang, New insights on the physics of salt precipitation during injection of CO_2 into saline aquifers, *Int. J. Greenhouse Gas Control* **43**, 10 (2015).

Mitzenmacher, M., The power of two choices in randomized load balancing, *Parallel Distrib. Syst.* **12**, 1094 (2001).

Mizuno, D., C. Tardin, C.F. Schmidt, and F.C. MacKintosh, Nonequilibrium mechanics of active cytoskeletal networks, *Science* **315**, 370 (2007).

Moatamed, F., M. Sahimi, and F. Naeim, Fractal dimensions in the bone marrows with metastatic lesions, *Human Pathology* **29**, 1299 (1998).

Modaresi Rad, A., B. Ghahraman, and M. Sadegh, Revising tortuosity and multi-fractal assumptions of unsaturated hydraulic conductivity from critical path analysis of percolation theory, *Geoderma* **352**, 213 (2019).

Mohammadi, H., E. Nedaaee Oskoee, M. Afsharchi, N. Yazdani, and M. Sahimi, A percolation model of mobile ad-hoc networks, *Int. J. Mod. Phys. C* **20**, 1871 (2009).

Mohanty, K.K., H.T. Davis, and L.E. Scriven, Physics of oil entrapment in water-wet rock, SPE Paper 9406 (1980).

Mohanty, K.K., J.M. Ottino, and H.T. Davis, Reaction and transport in disordered composite media: Introduction of percolation concepts, *Chem. Eng. Sci.* **37**, 905 (1982).

Moha-Ouchane, M., J. Peyrelasse, and C. Boned, Percolation transition in microemulsions: effect of water-surfactant ratio, temperature, and salinity, *Phys. Rev. A* **35**, 3027 (1987).

Moldrup, P., T. Oleson, T. Komatsu, P. Schjoning, and D.E. Rolston, Tortuosity, diffusivity, and permeability in the soil liquid and gaseous phases, *Soil Sci. Soc. Am. J.* **65**, 613 (2001).

Moldrup, P., T. Olesen, P. Schjnning, T. Yamaguchi, and D.E. Rolston, Predicting the gas diffusion coefficient in undisturbed soil from soil water characteristics, *Soil Sci. Soc. Am. J.* **64**, 94 (2000a).

Moldrup, P., T. Olesen, J. Gamst, P. Schjnning, T. Yamaguchi, and D.E. Rolston, Predicting the gas diffusion coefficient in repacked soil: Water-induced linear reduction model, *Soil Sci. Soc. Am. J.* **64**, 1588 (2000b).

Moldrup, P., T. Olesen, S. Yoshikawa, T. Komatsu, and D.E. Rolston, Three-porosity model for predicting the gas diffusion coefficient in undisturbed soil, *Soil Sci. Soc. Am. J.* **68**, 750 (2004).

Molloy, M., and B. Reed, A critical point for random graphs with a given degree sequence, *Random Struct. Algorithms* **6**, 161 (1995).

Molloy, M., and B. Reed, The size of the giant component of a random graph with a given degree sequence, *Combinatorics Probab. Comput.* **7**, 295 (1998).

Monceau, P.J.-M., and J.-C.S. Levy, Monodimensional effects on elastic and vibrational properties of lacunary networks, *Phys. Rev. B* **49**, 1026 (1994).

Monin, A.S., and A.U. Yaglom, *Statistical Fluid Mechanics*, vols. 1 & 2 (MIT Press, Cambridge, 1975).

Montroll, E.W., and G.H. Weiss, Random walks on lattices. II, *J. Math. Phys.* **6**, 167 (1965).

Mora, A., P. Verma, and S. Kumar, Electrical conductivity of CNT/polymer composites: 3D printing, measurements and modeling, *Compos. B Eng.* **183**, 107600 (2020).

Mora, T., and W. Bialek, Are biological systems poised at criticality? *J. Stat. Phys.* **144**, 268 (2011).

Morone, F., and H.A. Makse, Influence maximization in complex networks through optimal percolation, *Nature* **524**, 65 (2015).

Mort, J., Perspective: the applicability of percolation theory to innovation, *J. Product Innovation Management* **8**, 32 (1991).

Mort, J., Xerography: a study in innovation and economic competitiveness, *Phys. Today* **47** (4), 32 (1994).

Moss de Oliveira, S., P.M.C. de Oliveira, and D. Stauffer., *Evolution, Money, War and Computers* (Springer, Berlin, 1999).

Mott, N.F., On the transition to metallic conduction in semiconductors, *Can. J. Phys.* **34**, 1356 (1956).

Mott, N.F., Conduction in glasses containing transition metal ions, *J. Non-Cryst. Solids* **1**, 1 (1968).

Moukarzel, C., and P.M. Duxbury, Stressed backbone and elasticity of random central-force systems, *Phys. Rev. Lett.* **75**, 4055 (1995).

Moukarzel, C., P.M. Duxbury, and P.L. Leath, First-order rigidity on Cayley trees *Phys. Rev. E* **55**, 5800 (1997a).

Moukarzel, C., P.M. Duxbury, and P.L. Leath, Infinite-cluster geometry in central-force networks, *Phys. Rev. Lett.* **78**, 1480 (1997b).

Mourzenko, V.V., J.-F. Thovert, and P.M. Adler, Percolation of three-dimensional fracture networks with power-law distributions, *Phys. Rev. E* **72**, 036103 (2005).

Mousseau, N., and M.F. Thorpe, Structural model for crystalline and amorphous Si-Ge alloys, *Phys. Rev. B* **48**, 5172 (1993).

Movaghar, B., B. Pohlmann, and W. Schirmacher, Theory of AC and DC conductivity in disordered hopping systems, *Phys. Stat. Sol. (b)* **97**, 533 (1980a).

Movaghar, B., B. Pohlmann, and W. Schirmacher, Random walk in disordered hopping systems, *Solid State Commun.* **34**, 451 (1980b).

Movaghar, B., and W. Schirmacher, On the theory of hopping conductivity in disordered systems, *J. Phys. C* **14**, 859 (1981).

Mukhopadhyay, S., and M. Sahimi, Scaling behavior of permeability and conductivity anisotropy near the percolation threshold, *J. Statist. Phys.* **74**, 1301 (1994).

Mukhopadhyay, S., and M. Sahimi, Calculation of the effective permeabilities of field-scale porous media, *Chem. Eng. Sci.* **55**, 4495 (2000).

Mundt, C.C., and K.E. Sackett, Spatial scaling relationships for spread of disease caused by a wind–dispersed plant pathogen, *Ecosphere* **3**, art24 (2012).

Muñoz, M.A., R. Dickman, A. Vespignani, and S. Zapperi, Avalanche and spreading exponents in systems with absorbing states, *Phys. Rev. E* **59**, 6175 (1999).

Murray, K.L., N.A. Seaton, and M.A. Day, Use of mercury intrusion data, combined with nitrogen adsorption measurements, as a probe of pore network connectivity, *Langmuir* **15**, 8155 (1999).

Murty, G.S., Enumeration of rooted trees and its applications to earthquake and acoustic-emission amplitudes, *Phys. Earth Plan. Int.* **32**, 160 (1983).

Mustonen, V., and M. Lassig, From fitness landscapes to seascapes: non-equilibrium dynamics of selection and adaptation, *Trends Genet.* **25**, 111 (2009).

Nadim, F., G.E. Hoag, S. Liu, R.J. Carley, and P. Zack, Detection and remediation of soil and aquifer systems contaminated with petroleum products: an overview, *J. Pet. Sci. Eng.* **26**, 169 (2000).

Naeim, F., F. Moatamed, and M. Sahimi, Morphogenesis of the bone marrow: Fractal structures and diffusion-limited growth, *Blood* **87**, 5027 (1996).

Nagata, K., H. Iwabuki, and H. Nigo, Effect of particle size of graphites on electrical conductivity of graphite/polymer composite, *Compos. Interf.* **6**, 483 (1999).

Nagumo, J.-I., S. Arimoto, and S. Yoshizawa, An active pulse transmission line simulating nerve axon, *Proc. IRE.* **50**, 2061 (1962).

Nakanishi, H., M. Sahimi, M.C. Robertson, C.G. Sammis, and M.D. Rintoul, Fractal properties of distribution of earthquake hypcenters, *J. Phys. France I* **3**, 773 (1993).

Nakaya, S., T. Yoshida, and N. Shioiri, Percolation conditions in binary fractal fracture networks: Applications to rock fractures and active and seismogenic faults, *J. Geophys. Res. B: Solid Earth* **108**, 2348 (2003).

Nakayama, T., K. Yakubo, and R. Orbach, Dynamical properties of fractal networks: Scaling, numerical simulations, and physical realizations, *Rev. Mod. Phys.* **66**, 381 (1994).

Nam, I.S., and G.F. Froment, Catalyst deactivation by site coverage through multi-site reaction mechanisms, *J. Catal.* **108**, 271 (1987).

Nan, C.-W., Y. Shen, and J. Ma, Physical properties of composites near percolation, *Annu. Rev. Mater. Sci.* **40**, 131 (2010).

Nandan, S., S. Kumar Ram, G. Ouillon, and D. Sornette, Is seismicity operating at a critical point? *Phys. Rev. Lett.* **126**, 128501 (2021).

Nastrom, J.C., W.H. Jasperson, and K.S. Gage, Horizontal spectra of atmospheric tracers measured during global atmospheric sampling program, *J. Geophys. Res. D* **91**, 13201 (1986).

Néda, Z., R. Florian, and Y. Brechet, Reconsideration of continuum percolation of isotropically oriented sticks in three dimensions, *Phys. Rev. E* **59**, 3717 (1999).

Nedaaee Oskoee, E., and M. Sahimi, Electric currents in networks of interconnected memristors, *Phys. Rev. E* **83**, 031105 (2011).

Neumann, A.U., and G. Weisbuch, Dynamics and topology of idiotypic networks, *Bull. Math. Biol.* **54**, 699 (1992).

Newman, M.E.J., Structure and function of complex networks, *SIAM Rev.* **45**, 167 (2003).

Newman, M.E.J., and R.M. Ziff, Fast Monte Carlo algorithm for site or bond percolation, *Phys. Rev. E* **64**, 016706 (2001).

Newman, W.I., and D.L. Turcotte, Cascade model for fluvial geomorphology, *Geophys. J. Int.* **100**, 433 (1990).

Newman, W.I., and D.L. Turcotte, A simple model for the earthquake cycle combining self-organized complexity with critical point behavior, *Nonlin. Proces. Geophys.* **9**, 453 (2002).

Ngai, K.L., and C.T. Moynihan, The dynamics of mobile ions in ionically conducting glasses and other materials, *Mater. Res. Soc. Bull.* **23** (11), 51 (1998).

Nguyen, T.P., Z. Qu, and J. N. Weiss, Cardiac fibrosis and arrhythmogenesis: The road to repair is paved with perils, *J. Mol. Cell. Cardiol.* **70**, 83 (2014).

Nickel, B., and D. Wilkinson, Invasion percolation on the Cayley tree: Exact solution of a modified percolation model, *Phys. Rev. Lett.* **51**, 71 (1983).

Niemann, J.D., R.L. Bras, D. Veneziano, and A. Rinaldo, Impacts of surface elevation on the growth and scaling properties of simulated river networks, *Geomorphology* **40**, 37 (2001).

Niklasson, G.A., and C.G. Granqvist, Optical properties and solar selectivity of coevaporated Co-Al_2O_3 composite films, *J. Appl. Phys.* **55**, 3382 (1984).

Noble, B., and J.W. Daniel, *Applied Linear Algebra*, 2nd ed. (Prentice-Hall, New Jersey, 1977).

Nolen-Hoeksema, R.C., and R.B. Gordon, Optical detection of crack patterns in the opening-mode fracture of marble, *Int. J. Rock Mech. Min. Sci. Geomech. Abstr.* **24**, 135 (1987).

Nyame, B.K., and J.M. Ilbston, Capillary pore structure and permeability of hardened cement paste, in *7th International Symposium on Chemistry of Cement Paste*, Paris, **3:VI**, pp. 181 (1980)

Oberlin, A., M. Endo, and T. Koyama, Filamentous growth of carbon through benzene decomposition, *J. Cryst. Growth* **32**, 335 (1976).

Obukhov, A.M., The structure of the temperature field in a turbulent flow, *Izv. Akad. Nauk. SSSR, Ser. Geogr. Geophys.* **13**, 58 (1949).

Obukhov, S.P., The problem of directed percolation, *Physica A* **101**, 145 (1980).

Obukhov, S.P., First order rigidity transition in random rod networks, *Phys. Rev. Lett.* **74**, 4472 (1995).

Octavio, M., A. Octavio, J. Aponte, R. Medina, and C.J. Lobb, Nonuniversal critical behavior in the critical current of superconducting composites, *Phys. Rev. B* **37**, 9292 (1988).

Odagaki, T., and M. Lax, ac Hopping conductivity of a one-dimensional bond-percolation model, *Phys. Rev. Lett.* **45**, 847 (1980).

Odagaki, T., and M. Lax, Coherent-medium approximation in the stochastic transport theory of random media, *Phys. Rev. B* **24**, 5284 (1981).
Odagaki, T., M. Lax, and A. Puri, Hopping conduction in the d-dimensional lattice bond-percolation problem, *Phys. Rev. B* **28**, 2755 (1983).
Ódor, G., Universality classes in nonequilibrium lattice systems, *Rev. Mod. Phys.* **76**, 663 (2004).
O'Hern, C.S., L.E. Silbert, A.J. Liu, and S.R. Nagel, Jamming at zero temperature and zero applied stress: The epitome of disorder, *Phys. Rev. E* **68**, 011306 (2003).
Ohtsuki, T., and T. Keyes, Biased percolation: forest fires with wind, *J. Phys. A* **19**, L281 (1986).
Oliveira, C.L.N., N.A.M. Araújo, J.S. Andrade, and H.J. Herrmann, Explosive electric breakdown due to conducting–particle deposition on an insulating substrate, *Phys. Rev. Lett.* 113, 155701 (2014).
Oliveira, E.A., R.S. Pires, R.S. Oliveira, V. Furtado, H.J. Herrmann, and J.S. Andrade Jr., A universal approach for drainage basins, *Sci. Rep.* **9**, 9845 (2019).
Oliverira, J.G., and A.-L. Barabási, Darwin and Einstein correspondence patterns, *Nature* **437**, 1251 (2005).
Ono, N., and T. Kasai, Surface layer salinity of young sea ice, *Annal. Glaciol.* **6**, 298 (1985)
Onsager, L., Statistical hydrodynamics, *Nuovo Cimento* (Suppl.) **6**, 279 (1949).
Or, D., P. Lehmann, E. Shahraeeni, and N. Shokri, Advances in soil evaporation physics - A review, *Vadose Zone J.* **12**, vzj2012.0163 (2013).
Ortuno, M., and M. Pollak, Hopping transport in a-Ge and a-Si, *Philos. Mag.* **47**, L93 (1983).
Ota, R., T. Yamate, N. Soga, and M. Kunugi, Elastic properties of Ge-Se glass under pressure, *J. Non-Cryst. Solids* **29**, 67 (1978).
Otsuka, M., A simulation of earthquake occurrence, *Zishin* **24**, 215 (1971).
Otsuka, M., A chain-reaction type source model as a tool to interpret the magnitude-frequency relation of earthquakes, *J. Phys. Earth* (Tokyo) **20**, 35 (1972).
Ottavi, H., J.P. Clerc, G. Giraud, J. Roussenq, E. Guyon, and C.D. Mitescu, Electrical conductivity of a mixture of conducting and insulating spheres: an application of some percolation concepts, *J. Phys. C* **11**, 1311 (1978).
Pan, R.K., M. Kivela, J. Saramäki, Kimmo Kaski, and Kertész, Using explosive percolation in analysis of real-world networks, *Phys. Rev. E* **83**, 046112 (2011).
Page, J.H., W.J.L. Buyers, G. Dolling, P. Gerlach, and J.P. Harrison, Neutron inelastic scattering from fumed silica, *Phys. Rev. B* **39**, 6180 (1989).
Palevski, A., M.L. Rappaport, A. Kapitulnik, A. Fried, and G. Deutscher, Hall coefficient and conduction in a 2D percolation system, *J. Physique Lett.* **45**, L367 (1984).
Pang, H., T. Chen, G. Zhang, B. Zeng, and Z.-M. Li, An electrically conducting polymer/graphene composite with a very low percolation threshold, *Mater. Lett.* **64**, 226 (2010).
Papandreou, N., and P. Ne'dellec, Interplay between microscopic and macroscopic disorder in percolating Pd films, *J. Physique.* I France **2**, 707 (1992).
Parisi, G., and N. Sourlas, Critical behavior of branched polymers in the Lee-Yang edge singularity, *Phys. Rev. Lett.* **46**, 891 (1981).
Parlar, M., and Y.C. Yortsos, Percolation theory of vapor adsorption-desorption processes in porous materials, *J. Colloid Interface Sci.* **124**, 162 (1988);
Parlar, M., and Y.C. Yortsos, Nucleation and pore geometry effects in capillary desorption processes in porous media, *J. Colloid Interface Sci.* **132**, 425 (1989).
Pascual, M., and F. Guichard, Criticality and disturbance in spatial ecological systems, *Trends in Ecology & Evolution* **20**, 88 (2005).
Pashakhanloo, F., and A.V. Panfilov, Minimal functional clusters predict the probability of reentry in cardiac fibrotic tissue, *Phys. Rev. Lett.* **127**, 098101 (2021).
Pastén, D., M. Muñoz, A. Armando, J. Rogan, and J.A. Valdivia, Monofractal and multifractal analysis of the spatial distribution of earthquakes in the central zone of Chile, *Phys. Rev. E* **84**, 066123 (2011).
Pastor–Satorras, R., and C. Castellano, The localization of non–backtracking centrality in networks and its physical consequences, *Sci. Rep.* **10**, 21539 (2020).

Pastor-Satorras, R., and A. Vespignani, *Evolution and Structure of the Internet: A Statistical Physics Approach* (Cambridge University Press, Cambridge, 2003)

Patton, E., J.A. Wesson, M. Rubinstein, J.E. Wilson, and L.E. Oppenheimer, Scaling properties of branched polyesters, *Macromolecules* **22**, 1946 (1989).

Payandeh, B., A block cluster approach to percolation, *Rivista del Nuovo Cimento* **3**, 1 (1980).

Payatakes, A.C., K.M. Ng, and R.W. Flumerfelt, Oil ganglion dynamics during immiscible displacement: Model formulation, *AIChE J.* **26**, 430 (1980).

Pazhoohesh, E., H. Hamzehpour, and M. Sahimi, Numerical simulation of ac conduction in three-dimensional heterogeneous materials, *Phys. Rev. B* **73**, 174206 (2006).

Pelc, M., M. Vučković, M.S. Grbić, M. Požek, G. Yu, T. Sasagawa, M. Greven, and N. Barišić, Emergence of superconductivity in the cuprates via a universal percolation process, *Nat. Commun.* **9**, 4327 (2018).

Pelous, J. R. Vacher, T. Woignier, J.L. Sauvajol, and E. Courtens, Scaling phonon-fracton dispersion laws in fractal aerogels, *Philos. Mag. B* **59**, 65 (1989).

Peng, X., J. He, and A.J. Niemi, Clustering and percolation in protein loop structure, *BMC Structural Biology*, **15**, 22 (2015).

Penney, M.D., Y. Yargic, L. Smolin, E.W. Thommes, M. Anand, and C.T. Bauch, Hot-spotting to improve vaccine allocation by harnessing digital contact tracing technology: An application of percolation theory, *PLoS ONE* **16**, e0256889 (2021).

Perelson, A.S., and G.F. Oster, Theoretical studies of clonal selection: minimal antibody repertoire size and reliability of self-non-self discrimination, *J. Theor. Biol.* **81**, 645 (1979).

Perelson, A.S., and G. Weisbuch, Immunology for physicists, *Rev. Mod. Phys.* **69**, 1219 (1997).

Perron, J.T., J.X., Mitrovica, M. Manga, I. Matsuyama, and M.A. Richards, Evidence for an ancient martian ocean in the topography of deformed shorelines, *Nature* **447**, 840 (2007).

Pervago, P., A. Mousato, E. Kazatchenko, and M. Markov, Computation of continuum percolation threshold for pore systems composed of vugs and fractures, *Computer. Geosci.* **116**, 53 (2018).

Petersen, E.E., Reaction of porous solids, *AIChE J.* **3**, 443 (1957).

Petridou, N.I., B. Corominas-Murtra C.–P. Heisenberg, and E. Hannezo, Rigidity percolation uncovers a structural basis for embryonic tissue phase transitions, *Cell* **184**, 1914 (2021).

Peyrelasse, J., M. Moha-Ouchane, and C. Boned, Dielectric relaxation and percolation phenomena in ternary microemulsions, *Phys. Rev. A* **38**, 904 (1988).

Pfeuty, P., and G. Toulouse, *Introduction to the Renormalization Group and to Critical Phenomena* (Wiley, New York, 1977).

Phillips, J.C., Topology of covalent non-crystalline solids I: short-range order in chalcogenide alloys, *J. Non-Crystalline Solids* **34**, 153 (1979).

Phillips, J.C., Topology of covalent non-crystalline solids II: medium-range order in chalcogenide alloys and A-Si(Ge), *J. Non-Cryst. Solids* **43**, 37 (1981).

Phillips, J.C., Constraint theory, stiffness percolation and the rigidity transition in network glasses, in *Rigidity Theory and Applications*, edited by M.F. Thorpe and P.M. Duxbury (Kluwer Academic, New York, 1999), p. 155.

Phillips, J.C., and M.F. Thorpe, Constraint theory, vector percolation, and glass formation, *Solid State Commun.* **53**, 699 (1985).

Piggot, A.R., Fractal relations for the diameter and trace length of disc-shaped fractures, *J. Geophys. Res. B* **102**, 18121 (1997).

Pike, G.E., and C.H. Seager, Percolation and conductivity: A computer study, *Phys. Rev. B* **10**, 1421 (1974).

Pike, M.D., S.K. Bose, J.B. Mallinson, S.K. Acharya, S. Shirai, E. Galli, S.J. Weddell, P.J. Bones, M.D. Arnold, and S.A. Brown, Atomic scale dynamics drive brain-like avalanches in percolating nanostructured networks, *Nano Lett.* **20**, 3935 (2020).

Piri, M., and M.J. Blunt, Three-dimensional mixed-wet random pore-scale network modeling of two-and three-phase flow in porous media. I. Model description, *Phys. Rev. E* **71**, 026301 (2005a).

Piri, M., and M.J. Blunt, Three-dimensional mixed-wet random pore-scale network modeling of two-and three-phase flow in porous media. II. Results, *Phys. Rev. E* **71**, 026302 (2005b).

Pitt, G., The kinetics of the evolution of volatile products from coal, *Fuel* **41**, 267 (1962).

Pittel, B., J. Spencer, and N. Wornald, Sudden emergence of a giant k–core on a random graph, *J. Comb. Theory B* **67**, 111 (1996).

Pla, O., R. Garcia-Molina, F. Guinea, and E. Louis, Properties of elastic percolating networks in isotropic media with arbitrary elastic constants, *Phys. Rev. B* **41**, 11449 (1990).

Plischke, M., and B. Joós, Entropic elasticity of diluted central force networks, *Phys. Rev. Lett.* **80**, 4907 (1998).

Plischke, M., D.C. Vernon, B. Joós, and Z. Zhou, Entropic rigidity of randomly diluted two- and three-dimensional networks, *Phys. Rev. E* **60**, 3129 (1999).

Poglazov, B.F., S.N. Borhsenius, N.A. Belozerska, and E.M. Belantseva, Study of aggregation of the tail sheaths of phage T2, *Virology* **30**, 36 (1967).

Pokhrel, S., B. Waters, S. Felton, Z.-F. Huang, and B. Nadgorny, Percolation in metal-insulator composites of randomly packed spherocylindrical nanoparticles, *Phys. Rev. B* **103**, 134110 (2021).

Pollak, M., A percolation treatment of dc hopping conduction, *J. Non-Cryst. Solids* **11**, 1 (1972).

Pollak, M., in *The Metal-Non-metal Transition in Disordered Systems*, edited by L.R. Friedman and D.P. Tunstall (Scottish Universities Summer School in Physics, St. Andrews, 1978), p. 95.

Pollak, M., and C.J. Adkins, Conduction in granular metals, *Phil. Mag. B* **65**, 855 (1992).

Pollak, M., and I. Riess, Application of percolation theory to 2D-3D Heisenberg ferromagnets, *Phys. Status Solidi (b)* **69**, K15 (1975).

Pollard, D.D., On the form and stability of open hydraulic fractures in the earth's crust, *Geophys. Res. Lett.* **3**, 513 (1976).

Pomeau, Y., Frontal motion, Metastability and subcritical bifurcations in hydrodynamics, *Physica D* **23**, 3 (1986).

Ponton, A., T.K. Bose, and G. Delbos, Dielectric study of percolation in an oil-continuous microemulsion, *J. Chem. Phys.* **94**, 6879 (1991).

Porto, M., S. Havlin, S. Schwarzer, and A. Bunde, Optimal path in strong disorder and shortest path in invasion percolation with trapping, *Phys. Rev. Lett.* **79**, 4060 (1997).

Portsmouth, R.L., and L.F. Gladden, Determination of pore connectivity by mercury porosimetry, *Chem. Eng. Sci.* **46**, 3023 (1991).

Posselt, D., J.K. Kjems, A. Bernasconi, T. Sleator, and H.R. Ott, The thermal conductivity of silica aerogel in the phonon, the fracton and the particle-mode regime, *Europhys. Lett.* **16**, 59 (1991).

Postma, H.W.Che., I. Kozinsky, A. Husain, and M.L. Roukes, Dynamic range of nanotube- and nanowire-based electromechanical systems, *Appl. Phys. Lett.* **86**, 223105 (2005).

Prakash, S., S. Havlin, M. Schwartz, and H.E. Stanley, Structural and dynamical properties of long-range correlated percolation, *Phys. Rev. A* **46**, R1724 (1992).

Prasher, R.S., X.J. Hu, Y. Chalopin, N. Mingo, K. Lofgreen, S. Volz, F. Cleri, and P. Keblinski, Turning carbon nanotubes from exceptional heat conductors into insulators, *Phys. Rev. Lett.* **102**, 105901 (2009).

Prat, M., Isothermal drying on non-hygroscopic capillary-porous materials as an invasion percolation process, *Int. J. Multiphase Flow* **21**, 875 (1995).

Prat, M., Recent advances in pore-scale models for drying of porous media, *Chem. Eng. J.* **86**, 153 (2002).

Prat, M., Pore network models of drying, contact angle, and film flows, *Chem. Eng. Technol.* **34**, 1029 (2011).

V. Priesemann, M. Wibral, M. Valderrama, R. Pröpper, M. Le Van Quyen, T. Geisel, J. Triesch, D. Nikolić, and M.H.J. Munk, Spike avalanches *in vivo* suggest a driven, slightly subcritical brain state, *Front. Syst. Neurosci.* **8**, 108 (2014).

Priest, S.D., and J.A. Hudson, Discontinuity spacings in rock, *Int. J. Rock Mech. Min. Sci. Geomech. Abstr.* **13**, 135 (1976).

Priest, S.D., and J.A. Hudson, Estimating of discontinuous spacing and trace length using scanline surveys, *Int. J. Rock. Mech. Min. Sci. Geomech. Abstr.* **18**, 183 (1981)

Prim, R.C., Shortest connection networks and some generalizations, *Bell Syst. Tech. J.* **36**, 1389 (1957).

Queipo, N.V., R.T. Haftka, W. Shyy, T. Goel, R. Vaidyanathan, and P.K. Tucker, Surrogate-based analysis and optimization, *Prog. Aerospace Sci.* **41**, 1 (2005).
Quintanilla, J., S. Torquato, and R.M. Ziff, Efficient measurement of the percolation threshold for fully penetrable disks, *J. Phys. A* **33**, L399 (2000).
Rabinovitch, R., Y. Biton, D. Braunstein, I. Aviram, R. Thieberger, and A. Rabinovitch, Percolation and tortuosity in heart–like cells, *Sci. Rep.* **11**, 11441 (2021).
Rabinovich, S., P. Grinchuk, B.B. Khina, and A.V. Belyaev, Percolation combustion: is it possible in SHS? *Int. J. Self-Propagating High-Temperature Synthesis* **11**, 257 (2002).
Radicchi, F., and S. Fortunato, Explosive percolation in scale-free networks, *Phys. Rev. Lett.* **103**, 168701 (2009).
Radicchi, F., and S. Fortunato, Explosive percolation: A numerical analysis, *Phys. Rev. E* **81**, 036110 (2010).
Rahimi, A., T. Metzger, A. Kharaghani, and E. Tsotsas, Interaction of droplets with porous structures: Pore network simulation of wetting and drying, *Drying Technol.* **34**, 1129 (2016).
Rahimi Tabar, M.R., M. Sahimi, F. Ghasemi, K. Kaviani, M. Allamehzadeh, J. Peinke, M. Mokhtari, M. Vesaghi, M.D. Niry, A. Bahraminasab, S. Tabatabai, S. Fayazbakhsh, and M. Akbari, Short-term prediction of medium- and large-size earthquakes based on Markov and extended self-similarity analysis of seismic data, in, *Modelling Critical and Catastrophic Phenomena in Geoscience*, edited by P. Bhattacharyya and B.K. Chakrabarti (Springer, Berlin, 2006), p. 281.
Raj, A., and A. van Oudenaarden, Nature, nurture, or chance: Stochastic gene expression and its consequences, *Cell* **135**, 216 (2008).
Ramakrishnan, T.S., and D.T. Wasan, The relative permeability function for two-phase flow in porous media: Effect of capillary number, SPE Paper 12693 (1984).
Ramírez, J.E., E. Molina-Gayosso, J. Lozada-Lechuga, L.M. Flores-Rojas, M.I. Martínez, and A. Fernández Téllez, Percolation strategy to improve the production of plants with high pathogen susceptibility, *Phys. Rev. E* **98**, 062409 (2018).
Rammal, R., T.C. Lubensky, and G. Toulouse, Superconducting networks in a magnetic field, *Phys. Rev. B* **27**, 2820 (1983).
Rammal, R., and G. Toulouse, Random walks on fractal structures and percolation clusters, *J. Physique Lett.* **44**, L13 (1983).
Ransohoff, T.C., and C.J. Radke, Laminar flow of a wetting liquid along the corners of a predominantly gas-occupied noncircular pore, *J. Colloid Interface Sci.*. **121**, 392 (1988).
Rappoport, N., S. Karsenty, A. Stern, N. Linial, and M. Linial, ProtoNet 6.0: Organizing 10 million protein sequences in a compact hierarchical family tree, *Nucleic Acids Res.* **40**, 313 (2012).
Rapoport, A., A contribution to the theory of random and biased nets, **Bull. Math. Biophys. 19**, 257 (1957).
Rassamdana, H., B. Dabir, M. Nematy, M. Farhani, and M. Sahimi, Asphalt flocculation and deposition. I. The onset of precipitation, *AIChE J.* **42**, 10 (1996).
Rassamdana, H., and M. Sahimi, Asphalt flocculation and deposition: II. Formation and growth of fractal aggregates, *AIChE J.* **42**, 3318 (1996).
Recski, A., Application of combinatorics to statics - a second survery, *Disc. Math.* **108**, 183 (1992).
Redfield, D., Observation of $\log \sigma \sim T^{-1/2}$ in three-dimensional energy-band tails, *Phys. Rev. Lett.* **30**, 1319 (1973).
Redner, S., Directed and diode percolation, *Phys. Rev. B* **25**, 3242 (1982a).
Redner, S., Conductivity of random resistor-diode networks, *Phys. Rev. B* **25**, 5646 (1982b).
Redner, S., Directionality effects in percolation, in *The Mathematics and Physics of Disordered Media*, edited by B.D. Hughes and B.W. Ninham, Lecture Notes in Mathematics, **1035** (Springer, Berlin, 1983).
Reichenauer, G., J. Fricke, and U. Buchenau, Neutron scattering study of low-frequency vibrations in silica aerogels, *Europhys. Lett.* **8**, 415 (1989).
Reis, A.H., Constructal view of scaling laws of river basins, *Geomorphology* **78**, 201 (2006).
Renshaw, C.E., Sampling bias and the scaling of hydraulic conductivity in fractured rock, *Geophys. Res. Lett.* **25**, 121 (1998).

Resnick, D.J., J.C. Garland, and R. Newrock, Galvanomagnetic properties of randomly inhomogeneous superconductors, *Phys. Rev. Lett.* **43**, 1192 (1979).
Reyes, S., and K.F. Jensen, Percolation concepts in modelling of gas-solid reactions - I. Application to char gasification in the kinetic regime, *Chem. Eng. Sci.* **41**, 333 (1986).
Reyes, S., and K.F. Jensen, Percolation concepts in modelling of gas-solid reactions - III. Application to sulphation of calcined limestone, *Chem. Eng. Sci.* **42**, 565 (1987).
Reynolds, O., An experimental investigation of the circumstances which determine whether the motion of water shall be direct or sinuous, and of the law of resistance in parallel channels, *Philos. Trans. R. Soc., Ser. A* **174**, 935 (1883).
Reynolds, P.J., H.E. Stanley, and W. Klein, Large-cell Monte Carlo renormalization group for percolation, *Phys. Rev. B* **21**, 1223 (1980).
Ribeiro, T.L., M. Copelli, F. Caixeta, H. Belchior, D.R. Chialvo, M.A.L. Nicolelis, and S. Ribeiro, Spike avalanches exhibit universal dynamics across the sleep—wake cycle, *PLoS ONE* **5**, e14129 (2010).
Richesson, S., and M. Sahimi, Flow and transport properties of deforming porous media. I. Permeability, *Transp. Porous Media* **138**, 577 (2021a).
Richesson, S., and M. Sahimi, Flow and transport properties of deforming porous media. II. Electrical conductivity, *Transp. Porous Media* **138**, 611 (2021b).
Richesson, S., and M. Sahimi, Flow and transport properties of deforming porous media. II. Electrical conductivity, *Transp. Porous Media* **138**, 611 (2021).
Richter, D., and L. Passell, Neutron scattering as a probe of small-particle dynamics in hydroxylated amorphous silica, *Phys. Rev. Lett.* **44**, 1593 (1980).
Rieu, M., and G. Sposito, Fractal fragmentation, soil porosity, and soil water properties: II. Applications, *Soil Sci. Soc. Am. J.* **55**, 1231 (1991).
Rinaldoa, A., R. Rigon, J.R. Banavar, A. Maritan, and I. Rodriguez–Iturbe, Evolution and selection of river networks: Statics, dynamics, and complexity, *Proc. Natl. Acad. Sci. U.S.A.* **111**, 2417 (2014).
Rinaldo, A., I. Rodríguez-Iturbe, R. Rigon, E. Ijjasz–Vasquez, and R. L. Bras, Self-organized fractal river networks, *Phys. Rev. Lett.* **70**, 822 (1993).
Rintoul, M.D., and S. Torquato (1997), Precise determination of the critical threshold and exponents in a three-dimensional continuum percolation model, *J. Phys. A* **30**, L585L590 (1997).
Riordan, O., and L. Warnke, Explosive percolation is continuous, *Science* **333**, 322 (2011).
Roberts, J.J., and C. Lin, Electrical properties of partially saturated Topopah Spring tuff. Water distribution as a function of saturation, *Water Resour. Res.* **33**, 577 (1997).
Roberts, J.N., and L.M. Schwartz, Grain consolidation and electrical conductivity in porous media, *Phys. Rev. B* **31**, 5990 (1985).
Robertson, M.C., C.G. Sammis, M. Sahimi, and A.J. Martin, Fractal analysis of three-dimensional spatial distributions of earthquakes with a percolation interpretation, *J. Geophys. Res. B* **100**, 609 (1995).
Robinson, J.W., and J.E. Gale, A laboratory and numerical investigation of solute transport in discontinuous fracture systems, *Ground Water* **28**, 25 (1990).
Robinson, P.C., Connectivity of fracture systems: A percolation theory approach, *J. Phys. A* **16**, 605 (1983).
Robinson, P.C., Connectivity, flow and transport in network models of f ractured media, Ph.D. Thesis, St. Catherine's College, Oxford University (1984a).
Robinson, P.C., Numerical calculations of critical densities for lines and planes, *J. Phys. A* **17**, 2823 (1984b).
Rohde, M., and H. Micklitz, Critical behavior of the Hall conductivity near the percolation threshold in granular Sn:Ar mixtures, *Phys. Rev. B* 36, 7289(R) (1987).
Rolf, J., T. Bohr, and M.H. Jensen, Directed percolation universality in asynchronous evolution of spatiotemporal intermittency, *Phys. Rev. E* **57**, R2503 (1998).
Roling, B., Scaling properties of the conductivity spectra of glasses and supercooled melts, *Solid State Ionics* **105**, 185 (1998).

Ronceray, P., C.P. Broederszb, and M. Lenz, Fiber networks amplify active stress, *Proc. Natl. Acad. Sci. U.S.A.* **113**, 2827 (2016).

Roshko, A. Experiments on the flow past a circular cylinder at very high Reynolds number. *J. Fluid Mech.* **10**, 345 (1961).

Roux, J.-N., and D. Wilkinson, Resistance jumps in mercury injection in porous media, *Phys. Rev. A* **37**, 3921 (1988).

Roux, S., and E. Guyon, Mechanical percolation: a small beam lattice study, *J. Physique Lett.* **46**, L999 (1985).

Roux, S., and E. Guyon, Transport exponents in percolation, in *On Growth and Form*, edited by H.E. Stanley and N. Ostrowsky (Martin Nijhoff, Dordrecht, 1986), p. 273.

Roux, S., and E. Guyon, Temporal development of invasion percolation, *J. Phys. A* **22**, 3693 (1989).

Roux, S., and A. Hansen, Transfer-matrix study of the elastic properties of central-force percolation, *Europhys. Lett.* **6**, 301 (1988).

Royer, E., C. Beniot, and G. Poussigue, Dynamics of percolating networks, *J. Phys.: Condens. Matter* **4**, 561 (1992).

Rozanski, S.A., G. Kermer, P. Köberle, and A. Laschewsky, Relaxation and charge transport in mixtures of zwitterionic polymers and inorganic salts, *Macromol. Chem. Phys.* **196**, 877 (1995).

Rozenfeld, H.D., L.K. Gallos, and H.A. Makse, Explosive percolation in the human protein homology network, *Eur. Phys. J. B* **75**, 305 (2010).

Rubio, M.A., C.A. Edwards, A. Dougherty, and J.P. Gollub, Self-affine fractal interfaces from immiscible displacement in porous media, *Phys. Rev. Lett.* **63**, 1685 (1989).

Rudman, D.A., J.J. Calabrese, and J.C. Garland, Noise spectra of three-dimensional random metal-insulator composites, *Phys. Rev. B* **32**, 1456 (1986).

Rundle, J.B., R. Ortez, J. Knigslieb, and D.L. Turcotte, Constrained invasion percolation model: growth via leath bursts and the origin of seismic *b*-value, *Phys. Rev. Lett.* **124**, 068501 (2020).

Rupley, J.A. , I. Siemenkowski, G. Careri, and F. Bruni, Two-dimensional protonic percolation on lightly hydrated purple membrane, *Proc. Natl. Acad. Sci. U.S.A.* **85**, 9022 (1988).

Rupp, P., R. Richter, and I. Rehberg, Critical exponents of directed percolation measured in spatiotemporal intermittency, *Phys. Rev. E* **67**, 036209 (2003).

Saberi, A.A., Percolation description of the global topography of Earth and Moon, *Phys. Rev. Lett.* **110**, 178501 (2013).

Saberi, A.A., Recent advances in percolation theory and its applications, *Phys. Rep.* **578**, 1 (2015).

Saberi, A.A., Evidence for an ancient sea level on mars, *Astrophysical J. Lett.* **876**, L25 (2020).

Saberi, A.A., Application of percolation theory to statistical topographies, in *Complex Media and Percolation Theory*, edited by M. Sahimi and A.G. Hunt (Springer, New York, 2021),

Sadakata, M., A. Yoshino, T. Sakai, and H. Matsuoka, Formation of submicron unburnt carbon and NO_x from pulverized coal combustion System, in *Twentieth Symposium (International) on Combustion*, the Combustion Institute, Ann Arbor, Michigan (1984), p. 913.

Saeger, R.B., L.E. Scriven, and H.T. Davis, Flow, conduction, and a characteristic length in periodic bicontinuous porous media, *Phys. Rev. A* **44**, 5087 (1991).

Saffman, P.G., A theory of dispersion in a porous medium, *J. Fluid Mech.* **6**, 321 (1959).

Sahimi, M., Exact solutin of the master equation in a Bethe lattice, unpublished notes, University of Minnesota, Minneapolis (1980).

Sahimi, M., Effective-medium approximation for density of states and the spectral dimension of percolation networks, *J. Phys. C* **17**, 3957 (1984).

Sahimi, M., Fractal dimension in a percolation model of fluid displacement, *Phys. Rev. Lett.* **55**, 1698 (1985).

Sahimi, M., Dynamic percolation and diffusion in disordered systems, *J. Phys. C* **19**, 1311 (1986a).

Sahimi, M., Relation between the critical exponent of elastic percolation networks and the dynamical and geometrical exponents, *J. Phys. C* **19**, L79 (1986b).

Sahimi, M., Hydrodynamic dispersion near the percolation threshold: scaling and probability densities, *J. Phys. A* **20**, L1293 (1987).

Sahimi, M., On the determination of transport properties of disordered systems, *Chem. Eng. Commun.* **64**, 179 (1988).

Sahimi, M., Transport, reaction and fragmentation in evolving porous media, *Phys. Rev. A* **43**, 5367 (1991).

Sahimi, M., Nonlinear transport processes in disordered media, *AIChE J.* **39**, 369 (1993a).

Sahimi, M., Flow phenomena in rocks: From continuum models to fractals, percolation, cellular automata and simulated annealing, *Rev. Mod. Phys.* **65**, 1395 (1993b).

Sahimi, M., Long-range correlated percolation and flow and transport in heterogeneous porous media, *J. Phys. I France* **4**, 1263 (1994).

Sahimi, M., Effect of long-range correlations on transport phenomena in disordered media, *AIChE J.* **41**, 229 (1995).

Sahimi, M., Non-linear and non-local transport in heterogeneous media: From long-range correlated percolation to fracture and materials breakdown, *Phys. Rep.* **306**, 213 (1998).

Sahimi, M., *Heterogeneous Materials I. Linear Transport and Optical Properties* (Springer, New York, 2003a).

Sahimi, M., *Heterogeneous Materials I. Nonlinear and Breakdown Properties and Atomistic Modeling* (Springer, New York, 2003b).

Sahimi, M., *Flow and Transport in Porous Media and Fractured Rock*, 2nd ed. (Wiley-VCH, Weinheim, 2011).

Sahimi M., Dispersion in porous media, continuous-time random walk, and percolation, *Phys. Rev. E* **85**, 016316 (2012).

Sahimi, M., AC hopping conduction at extreme disorder takes place on the bond invasion percolation cluster, arXiv:2208.00770v1 [cond-mat.dis-nn] (archived 8 July 2022) (2022a).

Sahimi, M., Universal frequency-dependent permeability of heterogeneous porous media: Effective-medium approximation and critical-path analysis, *Transp. Porous Media* **144**, 759 (2022b).

Sahimi, M., and S. Arbabi, Force distribution, multiscaling and fluctuations in disordered elastic media, *Phys. Rev. B* **40**, 4975 (1989).

Sahimi, M., and S. Arbabi, On correction to scaling for two- and three-dimensional scalar and vector percolation, *J. Stat. Phys.* **62**, 453 (1991).

Sahimi, M., and S. Arbabi, Percolation and fracture in disordered solids and granular media: Approach to a fixed point, *Phys. Rev. Lett.* **68**, 608 (1992).

Sahimi, M., and S. Arbabi, Mechanics of disordered solids. II. Percolation on elastic networks with bond-bending forces, *Phys. Rev. B* **47**, 703 (1993).

Sahimi, M., and S. Arbabi, Scaling laws for fracture of heterogeneous materials and rock, *Phys. Rev. Lett.* **77**, 3689 (1996).

Sahimi, M., H.T. Davis, and L.E. Scriven, Dispersion in disordered porous media, *Chem. Eng. Commun.* **23**, 329 (1983a).

Sahimi, M., G.R. Gavalas, and T.T. Tsotsis, Statistical and continuum models of fluid-solid reactions in porous media, *Chem. Eng. Sci.* **45**, 1443 (1990).

Sahimi, M., and J.D. Goddard, Superelastic percolation networks and the viscosity of gels, *Phys. Rev. B* **32**, 1869 (1985).

Sahimi, M., and J.D. Goddard, Elastic percolation models for cohesive mechanical failure in heterogeneous systems, *Phys. Rev. B* **33**, 7848 (1986).

Sahimi, M., M. Hashemi, and J. Ghassemzadeh, Site-bond invasion percolation with fluid trapping, *Physica A* **260**, 231 (1998).

Sahimi, M., A.A. Heiba, H.T. Davis, and L.E. Scriven, Dispersion in flow through porous media: II. Two-phase flow, *Chem. Eng. Sci.* **41**, 2123 (1986a).

Sahimi, M., A.A. Heiba, B.D. Hughes, H.T. Davis, and L.E. Scriven, Dispersion in flow through porous media, SPE Paper 10969 (1982).

Sahimi, M., B.D. Hughes, L.E. Scriven, and H.T. Davis, Stochastic transport in disordered systems, *J. Chem. Phys.* **78**, 6849 (1983b).

Sahimi, M., B.D. Hughes, L.E. Scriven, and H.T. Davis, Real-space renormalization and effective-medium approximation to the percolation conduction problem, *Phys. Rev. B* **28**, 307 (1983c).

Sahimi, M., B.D. Hughes, L.E. Scriven, and H.T. Davis, Dispersion in flow through porous media: I. One-phase flow, *Chem. Eng. Sci.* **41**, 2103 (1986b).
Sahimi, M., and A.G. Hunt (eds.), *Complex Media and Percolation Theory* (Springer, New York, 2021).
Sahimi, M., and A.O. Imdakm, The effect of morphological disorder on hydrodynamic dispersion in flow through porous media, *J. Phys. A* **21**, 3833 (1988).
Sahimi, M., and A.O. Imdakm, Hydrodynamics of particulate motion in porous media, *Phys. Rev. Lett.* **66**, 1169 (1991).
Sahimi, M., and S. Mukhopadhyay, Scaling properties of a percolation model with long-range correlations, *Phys. Rev. E* **54**, 3870 (1996).
Sahimi, M., M. Naderian, and F. Ebrahimi, Efficient simulation of ac conduction in heterogeneous materials at low temperatures, *Phys. Rev. B* **71**, 094208 (2005).
Sahimi, M., and H. Rassamdana, On position-space renormalization group approach to percolation, *J. Stat. Phys.* **78**, 1157 (1995).
Sahimi, M., and T.S. Ray, Transport through bootstrap percolation clusters, *J. Phys. I* **1**, 685 (1991).
Sahimi, M., M.C. Robertson, and C.G. Sammis, Fractal distribution of earthquake hypocenters and its relation with fault patterns and percolation, *Phys. Rev. Lett.* **70**, 2186 (1993).
Sahimi, M., L.E. Scriven, and H.T. Davis, On the improvement of the effective-medium approximation to the percolation conductivity problem, *J. Phys. C* **17**, 1941 (1984)
Sahimi, M., and D. Stauffer, Ising model above the upper critical dimension: An application to biology, *Phys. Rev. Lett.* **71**, 4271 (1993).
Sahimi, M., and P. Tahmasebi, Reconstruction, optimization, and design of heterogeneous materials and media: Basic principles, computational algorithms, and applications, *Phys. Rep.* **939**, 1 (2021).
Sahimi, M., and S.E. Tajer, Self-affine distributions of the bulk density, elastic moduli, and seismic wave velocities of rock. *Phys. Rev. E* **71**, 046301 (2005).
Sahimi, M., and T.T. Tsotsis, A percolation model of catalyst deactivation by site coverage and pore blockage, *J. Catal.* **96**, 552 (1985).
Sahimi, M., and T.T. Tsotsis, Dynamic scaling for the fragmentation of reactive porous media, *Phys. Rev. Lett.* **59**, 888 (1987).
Sahimi, M., and T.T. Tsotsis, Statistical modeling of gas-solid reactions with Pore volume growth: Kinetic regime, *Chem. Eng. Sci.* **43**, 113 (1988).
Sahimi, M., and T.T. Tsotsis, Transient diffusion and conduction in heterogeneous media: Beyond the classical effective-medium approximation. *Ind. Eng. Chem. Res.* **36**, 3043 (1997).
Sahni, A., J. Burger, and M.J. Blunt, Measurement of three phase relative permeability during gravity drainage using CT scanning, SPE paper 39655 (1998).
Saito, M., M. Kikushi, and K. Kudo, Analytical solution of 'Go-game model of earthquakes', *Zishin* **26**, 19 (1973).
Saleur, H., and B. Duplantier, Exact determination of the percolation hull exponent in two dimensions, *Phys. Rev. Lett.* **58**, 2325 (1987).
Salimi, H., J. Bruining, and V. Joekar-Niasar, Comparison of modified effective-medium approximation to pore-network theory for relative permeabilities, *J. Pet. Sci. Eng.* **184**, 106594 (2020).
Sammis, C.G., and D. Sornette, Positive feedback, memory, and the predictability of earthquakes, *Proc. Natl. Acad. Sci. USA* **99**, 2501 (2002).
Sampaio Filho1, C.I.N., J.S. Andrade, H.J. Herrmann, and A.A. Moreira, Elastic backbone defines a new transition in the percolation model, *Phys. Rev. Lett.* **120**, 175701 (2018).
Sangare, D., and P.M. Adler, Continuum percolation of isotropically oriented circular cylinders, *Phys. Rev. E* **79**, 052101 (2009).
Sano, M., and K. Tamai, A universal transition to turbulence in channel flow, *Nat. Phys.* **12**, 249 (2016).
Sapoval, B., M. Rosso, and J.F. Gouyet, The fractal nature of a diffusion front and the relation to percolation, *J. Physique Lett.* **46**, L149 (1985).
Sastry, S., P.G. Debenedetti, F. Sciortino, and H.E. Stanley, Singularity–free interpretation of the thermodynamics of supercooled water, *Phys. Rev. E* **53**, 6144 (1996).

Sayles, R.S., and T.R. Thomas, Surface topography as a non-stationary random process, *Nature* **271**, 431 (1978).
Schaefer, D.W., C.J. Brinker, D. Richter, B. Farago, and B. Frick, Dynamics of weakly connected solids: silica aerogels, *Phys. Rev. Lett.* **64**, 2316 (1990).
Schaefer, D.W., and K.D. Keefer, Structure of random porous materials: silica aerogel, *Phys. Rev. Lett.* **56**, 2199 (1986).
Scheidegger, A.E., A stochastic model for drainage patterns into an intramontane trench, *Bull. Assoc. Sci. Hydrol.* **12**, 15 (1967).
Scher, H., and R. Zallen, Critical density in percolation processes, *J. Chem. Phys.* **53**, 3759 (1970).
Scherer, G.W., Theory of drying, *J. Am. Ceramic Soc.* **73**, 3 (1990).
Schertzer, D., and S. Lovejoy, Physical modeling and analysis of rain and clouds by anisotropic scaling multiplicative processes, *J. Geophys. Res.* **92**, 9693 (1987).
Schiff, J.L., *Cellular Automata: A Discrete View of the World* (Wiley, New York, 2011).
Schiulaz, M., C.R. Laumann, A.V. Balatsky, and B.Z. Spivak, Theory of combustion in disordered media, *Phys. Rev. E* **97**, 062133 (2018).
Schmidt, M., and W. Burchard, Critical exponents in polymers: A sol-gel study of anionically prepared styrene-divinilbenzene copolymers, *Macromolecules* **14**, 370 (1981).
Schneider, C.M., T. Mihaljev, S. Havlin, and H.J. Herrmann, Suppressing epidemics with a limited amount of immunization units, *Phys. Rev. E* **84**, 061911 (2011).
Schneider, C.M., T. Mihaljev, and H.J. Herrmann, Inverse targeting: An effective immunization strategy, *Europhys. Lett.* **98**, 46002 (2012).
Schneider, C.M., N. Yazdani, N.A.M. Araújo, S. Havlin, and H.J. Herrmann, Towards designing robust coupled networks, *Sci. Rep.* **3**, 1969 (2013).
Scholz, C.H., The frequency-magnitude relation of microfracturing in rock and its relation to earthquakes, *Bull. Seism. Soc. Am.* **58**, 399 (1968).
Scholz, C.H., *The Mechanics of Eathquakes and Faulting* (Cambridge University Press, Cambridge, 1990).
Schrenk, K.J., A. Felder, S. Deflorin, N. A. M. Araújo, R.M. D'Souza, and H.J. Herrmann, Bohman-Frieze-Wormald model on the lattice, yielding a discontinuous percolation transition, *Phys. Rev. E* **85** 031103 (2012).
Schröder, M., S.H. Ebrahimnazhad Rahbari, and J. Nagler, Crackling noise in fractional percolation, *Nature Commun.* **4**, Article number 2222 (2013).
Schröder, M., J. Nagler, M. Timme, and D. Witthaut, Hysteretic Percolation from locally optimal individual decisions, *Phys. Rev. Lett.* **120**, 248302 (2018).
Schrøder, T.B., and J.C. Dyre, Scaling and universality of ac conduction in disordered solids, *Phys. Rev. Lett.* **101**, 025901 (2008).
Schulman, L.S., and P.E. Seiden, Percolation and galaxies, *Science* **233**, 425 (1986).
Schwartz, F.W., L. Smith, and A.S. Crowe, Stochastic analysis of macroscopic dispersion in fractured media, *Water Resour. Res.* **19**, 1253 (1983).
Schwartz, L.M., J.R. Banavar, and B.I. Halperin, Biased-diffusion calculation of electrical transport in inhomogeneous continuum systems, *Phys. Rev. B* **40**, 9155 (1989).
Schwartz, L.M., S. Feng, M.F. Thorpe, and P.N. Sen, Behavior of depleted elastic networks: Comparison of effective-medium and numerical calculations, *Phys. Rev. B* **32**, 4607 (1985).
Schwartz, L.M., D.L. Johnson, and S. Feng, Vibrational modes in granular materials, *Phys. Rev. Lett.* **52**, 831 (1984).
Seager, C.H., and G.E. Pike, Percolation and conductivity: A computer study II, *Phys. Rev. B* **10**, 1435 (1974).
Seaton, N.A., Determination of the connectivity of porous solids from nitrogen sorption measurements, *Chem. Eng. Sci.* **46**, 1895 (1991).
Seiden, P.E., and L.S. Schulman, Percolation model of galactic structure, *Adv. Phys.* **39**, 1 (1990).
Serio, M.A., D.G. Hamblen, J.R. Markham, and P.R. Solomon, Kinetics of volatile product evolution in coal pyrolysis: experiment and theory, *Energy Fuels* **1**, 138 (1987).

Seyed-allaei, H., G. Bianconi, and M. Marsili, Scale-free networks with an exponent less than two, *Phys. Rev. E* **73**, 046113 (2006).
Seymour, J.D., and P.T. Callaghan, Generalized approach to NMR analysis of flow and dispersion in porous media, *AIChE J.* **43**, 2096 (1997).
Shah, N., and J.M. Ottino, Transport and reaction in evolving, disordered composites - I. Gasification of porous solids, *Chem. Eng. Sci.* **42**, 63 (1987a).
Shah, N., and J.M. Ottino, Transport and reaction in evolving, disordered composites - II. Coke deposition in a catalytic pellet, *Chem. Eng. Sci.* **42**, 73 (1987b).
Shante, V.K.S., Hopping conduction in quasi-one-dimensional disordered compounds, *Phys. Rev. B* **16**, 2597 (1977).
Shante, V.K.S., and S. Kirkpatrick, An introduction to percolation theory, *Adv. Phys.* **20**, 325 (1971).
Shao, J., S. Havlin, and H.E. Stanley, Dynamic opinion model and invasion percolation, *Phys. Rev. Lett.* **103** (2009) 018701.
Sharma, M.M., A. Garrouch, and H.F. Dunlap, Effects of wettability, pore geometry, and stress on electrical conduction in fluid-saturated rocks, *Log Analyst* **32**, 511 (1991).
Shaw, M.W., Modeling stochastic processes in plant pathology, *Annu. Rev. Phytopathol.* **32**, 523 (1994).
Shaw, T.M., Drying as an immiscible displacement process with fluid counterflow, *Phys. Rev. Lett.* **59**, 1671 (1987).
Sheffield, R.E., and A.B. Metzner, Flow of nonlinear fluids through porous media, *AIChE J.* **22**, 736 (1976).
Sheinman, M., C.P. Broedersz, and F.C. MacKintosh, Actively stressed marginal networks, *Phys. Rev. Lett.* **109**, 238101 (2012).
Sheinman, M., A. Sharma, J. Alvarado, G.H. Koenderink, and F.C. MacKintosh, Anomalous discontinuity at the percolation critical Point of active gels, *Phys. Rev. Lett.* **114**, 098104 (2015).
Sheng, P., and J. Klafter, Hopping conductivity in granular disordered systems, *Phys. Rev. B* **27**, 2583 (1983).
Sheppard, A.P., M.A. Knackstedt, W.V. Pinczewski, and M. Sahimi, Invasion percolation: new algorithms and universality classes, *J. Phys. A* **32**, L521 (1999).
Sherman, S.M., Thalamocortical interactions, *Curr. Opin. Neurobiol.* **22**, 575 (2012).
Shew, W.L., W.P., Clawson, J. Pobst, Y. Karimipanah, N.C. Wright, and R. Wessel, Adaptation to sensory input tunes visual cortex to criticality, *Nat. Phys.* **11**, 659 (2015).
Shih, H.-Y., T.-L. Hsieh, and N. Goldenfeld, Ecological collapse and the emergence of travelling waves at the onset of shear turbulence, *Nat. Phys.* **12**, 245 (2016).
Shimo, M., and J.C.S. Long, in *Flow and Transport Through Unsaturated Rock*, AGU Geophysics Monographs, **43**, 121 (1987).
Shin, D., J. Lee, J.-R. Gong, and K.-H. Cho, Percolation transition of cooperative mutational effects in colorectal tumorigenesis, *Nat. Commun.* **8**, 1270 (2017).
Shirazi, A., G.R. Jafari, J. Davoudi, J. Peinke, M.R. Rahimi Tabar, and M. Sahimi, Mapping stochastic processes onto complex networks, *J. Stat. Mech.*, P07046 (2009).
Shklovskii, B.I., Anisotropy of percolation conduction, *Phys. Stat. Sol. B* **85**, K111 (1978).
Shklovskii, B.I., and A.L. Efros, Impurity band and conductivity of compensated semiconductors, *Zh. Eksp. Teor. Fiz.* **60**, 867 (1971) [*Sov. Phys. JETP* **33**, 468 (1971)].
Shklovskii, B.I., and A.L. Efros, Percolation theory and conductivity of strongly inhomogeneous media, *Soviet Phys. Uspekhi* **18**, 845 (1975).
Shklovskii, B.I., and Efros, A.L., *Electronic Properties of Doped Semiconductors* (Springer, Berlin, 1984).
Shokri, N., P. Lehmann, and D. Or, Effects of hydrophobic layers on evaporation from porous media, *Geophys. Res. Lett.* **35**, L19407 (2008a).
Shokri, N., P. Lehmann, P. Vontobel, and D. Or, Drying front and water content dynamics during evaporation from sand delineated by neutron radiography, *Water Resour. Res.* **44**, 1944 (2008b).
Shokri, N., P. Lehmann, and D. Or, Evaporation from layered porous media, *J. Geophys. Res: Solid Earth* **115**, B06204 (2010).

Shtein, M., R. Nadiv, M. Buzaglo, and O. Regev, Graphene-based hybrid composites for efficient thermal management of electronic devices, *ACS Appl. Mater. Interfaces* **7**, 23725 (2015a).
Shtein, M., R. Nadiv, M. Buzaglo, K. Kahil, and O. Regev, Thermally conductive graphene-polymer composites: size, percolation, and synergy effects, *Chem. Mater.* **27**, 2100 (2015b).
Sidebottom, D.L., Universal approach for scaling the ac conductivity in ionic glasses, *Phys. Rev. Lett.* **82**, 3653 (1999).
Sieh, K., M. Stuiver, and D. Brillinger, A more precise chronology of earthquakes produced by the San Andreas Fault in southern California, *J. Geophys. Res. B* **94**, 603 (1989).
Silva, K.P.T., T.I. Yusufaly, P. Chellamuthu, and J.Q. Boedicker, Disruption of microbial communication yields a two-dimensional percolation transition, *Phys. Rev. E* **99**, 042409 (2019).
Silverberg, G., and B. Verspagen, A percolation model of innovation in complex technology spaces, *J. Econ. Dynamics Control* **29**, 225 (2005).
Sing, K.S.W., D.H. Everett, R.A.W. Haul, L. Moscou, R.A. Pierotti, J. Rouquèrol, and T. Siemieniwska, Reporting physisorption data for gas/solid systems with special reference to the determination of surface area and porosity, *Pure Appl. Chem.* **57**, 603 (1985).
Singhal, A.K., and W.H. Somerton, Quantitative modeling of immiscible displacement in porous media. A network approach, *J. Inst. France Pet.* **32**, 897 (1977).
Sipos, M., and N. Goldenfeld, Directed percolation describes lifetime and growth of turbulent puffs and slugs, *Phys. Rev. E* **84**, 035304(R) (2011).
Skaggs, T.H., Assessment of critical path analyses of the relationship between permeability and electrical conductivity of pore networks, *Adv. Water Resour.* **34**, 1335 (2011).
Smith, J.M., Natural selection and the concept of a protein space, *Nature* **225**, 563 (1970).
Smith, L.N., and C.J. Lobb, Percolation in two-dimensional conductor-insulator networks with controllable anisotropy, *Phys. Rev. B* **20**, 3653 (1979).
Smith, T.F., and M.S. Waterman, Identification of common molecular subsequences, *J. Mol. Biol.* **147**, 195 (1981).
Smoluchowski, M., Drei Vorträge über diffusion, Brownsche molekularbewegung und koagulation von kolloidteilchen, *Z. Phys.* **17**, 557 (1916).
Snider, E., N. Dasenbrock-Gammon, R. McBride, M. Debessai, H. Vindana, K. Vencatasamy, K.V. Lawler, A. Salamat, R.P. Dias, Room-temperature superconductivity in a carbonaceous sulfur hydride, *Nature* **586**, 373 (2020).
Sobolev, G.A., and Y.S. Tyupkin, Analysis of energy release process during main rupture formation in laboratory studies of rock fracture and before strong earthquakes, *Izv. Phys. Solid Earth* **36**, 138 (2000).
Sofo, J., J. Lorenzana, and E.N. Martinez, Critical behavior of Young's modulus for two-dimensional randomly holed metallized Mylar, *Phys. Rev. B* **36**, 3960 (1987).
Solomon, P.R., and H.H. King, Tar evolution from coal and model polymers: theory and experiments, *Fuel* **63**, 1302 (1984).
Solomon, S., G. Weisbuch, L. de Arcangelis, N. Jan, and D. Stauffer, Social percolation models, *Physica A* **277**, 239 (2000).
Soltanian, M.R., M.A. Amooie, D.R. Cole, D.E. Graham, S.A. Hosseini, S. Hovorka, S.M. Pfiffner, T.J. Phelps, and J. Moortgat, Simulating the Cranfield geological carbon sequestration project with high-resolution static models and an accurate equation of state, *Int. J. Greenhouse Gas Control* **54**, 282 (2016).
Song, Y., T.W. Noh, S.-I. Lee, and J.R. Gaines, Experimental study of the three dimensional ac conductivity and dielectric constant of a conductor-insulator composite near the percolation threshold, *Phys. Rev. B* **33**, 904 (1986).
Sornette, D., *Critical Phenomena in Natural Sciences*, 2nd ed. (Springer, Berlin, 2004).
Sornette, D., and C.G. Sammis, Complex critical exponents Fom renormalization group theory of earthquakes: Implications for earthquake predictions, *J. Phys. I France* **5**, 607 (1995).
Sotirchos, S.V., and S. Zarkanitis, A distributed pore size and length model for porous media reacting with diminishing porosity, *Chem. Eng. Sci.* **48**, 1487 (1993).

Sporns, O., Network attributes for segregation and integration in the human brain, *Curr. Opin. Neurobiol.* **23**, 162 (2013).
Sporns, O., D.R. Chialvo, M. Kaiser, and C.C. Hilgetag, Organization, development and function of complex brain networks, *Trends Cognitive Sci.* **8**, 418 (2004).
Sreeram, A.N., A.K. Varshneya, and D.R. Swiler, Molar volume and elastic properties of multicomponent chalcogenide glasses, *J. Non-Cryst. Solids* **128**, 294 (1991).
Squire, K.R., P.R. Solomon, R.M. Carangelo, and M.B. DiTarento, Tar evolution from coal and model polymers. 2. The effects of aromatic ring sizes and donatable hydrogens, *Fuel* **65**, 833 (1986).
Srinivasan, A., K.N. Madhusoodanan, and E.S.R. Gopal, Thermal diffusivity of IV-V-VI glasses - An evidence for the existence of a mechanical threshold, *Solid State Commun.* **83**, 163 (1992).
Sreenivasan, K.R., and R. Ramshankar, Transition intermittency in open flows, and intermittency routes to chaos, *Physica D* **23**, 246 (1986).
Stachowiak, H., Effective electric conductivity tensor of polycrystalline metals in high magnetic fields, *Physica* **45**, 481 (1970).
Stanley, H.E., A polychromatic correlated–site percolation problem with possible relevance to the unusual behaviour of supercooled H_2O and D_2O, *J. Phys. A* **12**, L329 (1979).
Stanley, H.E., and J. Texeira, Are concepts of percolation and gelation of possible relevance to the behaviour of water at very low temperatures? *Ferroelectrics* **30**, 213 (1980).
Stark, C.P., An invasion percolation model of drainage network evolution, *Nature* **352**, 423 (1991).
Stark, C.P., and J.A. Stark, Seismic fluids and percolation theory, *J. Geophys. Res.* **B96**, 8417 (1991).
Stauffer, D., Gelation in concentrated critically branched polymersolutions. Percolation scaling theory of intramolecular bond cycles, *J. Chem. Soc. Faraday Trans. II* **72**, 1354 (1976).
Stauffer, D., J. Adler, and A. Aharony, Universality at the three-dimensional percolation threshold, *J. Phys. A* **27**, L475 (1994).
Stauffer, D. and Aharony, A., *Introduction to Percolation Theory*, 2nd ed. (Taylor & Francis, 1994).
Stauffer, D., A. Coniglio, and M. Adam, Gelation and critical phenomena, *Adv. Polymer Sci.* **44**, 105 (1982).
Stauffer, D., and L. de Arcangelis, Dynamics and strong size effects of a bootstrap percolation problem, *Int. J. Mod. Phys. C* **7**, 739 (1996).
Stauffer, D., and M. Sahimi, High-dimensional simulation and very large cellular automata for immunological shape space, *Int. J. Mod. Phys. C* **4**, 401 (1993).
Stauffer, D., and M. Sahimi, High-dimensional simulation of simple immunological models, *J. Theor. Biol.* **166**, 289 (1994).
Stauffer, D., and M. Sahimi, Diffusion in scale-free networks with annealed disorder, *Phys. Rev. E* **72**, 046128 (2005).
Stauffer, D., and M. Sahimi, Discrete simulation of the dynamics of spread of extreme opinions in a society, *Physica A* **364**, 537 (2006).
Stauffer, D., and M. Sahimi, Can a few fanatics influence the opinion of a large segment of a society? *Euro. Phys. J. B* **57**, 147 (2007).
Stauffer, D., and G. Weisbuch, High-dimensional simulation of the shape-space model for the immune system, *Physica A* **180**, 42 (1992).
Stewart, J., and F.J. Varela, Morphogenesis in shape-space. Elementary meta-dynamics in a model of the immune network, *J. Theor. Biol.* **153**, 477 (1991).
Steyn-Ross, D.A., M.L. Steyn-Ross, L.C. Wilcocks, and J.W. Sleigh, Toward a theory of the general-anestheticinduced phase transition of the cerebral cortex. II. Numerical simulations, spectral entropy, and correlation times, *Phys. Rev. E* **64**, 011918 (2001).
Stinchcombe, R.B., The branching model for percolation theory and electrical conductivity, *J. Phys. C* **6**, L1 (1973).
Stinchcombe, R.B., Conductivity and spin-wave stiffness in disordered systems-an exactly soluble model, *J. Phys. C* **7**, 179 (1974).
Stinchcombe, R.B., and B.P. Watson, Renormalization group approach for percolation conductivity, *J. Phys. C* **9**, 3221 (1976).

Stockmayer, W.H., Theory of molecular size distribution and gel formation in branched-chain polymers, *J. Chem. Phys.* **11**, 45 (1943).

Stokes, J.P., D.A. Weitz, J.P. Gollub, A. Dougherty, M.O. Robbins, P.M. Chaikin, and H.M. Lindsay, Interfacial stability of immiscible displacement in a porous medium, *Phys. Rev. Lett.* **57**, 1718 (1986).

Straley, C., A. Mateson, S. Feng, L.M. Schwartz, W.E. Kenyon and J.R. Banavar, Magnetic resonance, digital image analysis, and permeability of porous media, *Appl. Phys. Lett.* **51**, 1146 (1987).

Straley, J.P., Critical phenomena in resistor networks, *J. Phys. C* **9** 783 (1976).

Straley, J.P., Critical exponents for the conductivity of random resistor lattices, *Phys. Rev. B* **15**, 5733 (1977).

Straley, J.P., Non-universal threshold behaviour of random resistor networks with anomalous distribution of conductances, *J. Phys. C* **15**, 2343 (1982).

Strelniker, Y.M., A. Frydman, and S. Havlin, Percolation model for the superconductor-insulator transition in granular films, *Phys. Rev. B* **76**, 224528 (2007).

Strocka, B., J.A.M.S. Duarte, and M. Schreckenberg, The self-organized critical forest-fire model with trees and bushes, *J. Phys. I France* **5**, 1233 (1995).

Stroud, D., Generalized effective-medium approach to the conductivity of an inhomogeneous material, *Phys. Rev. B* **12**, 3368 (1975).

Stroud, D., Classical theory of the magnetoresistance and Hall coefficient of normal-superconducting composites, *Phys. Rev. Lett.* **44**, 1708 (1980).

Strukov, D.B., G.S. Snider, D.R. Stewart, and R.S. Williams, The missing memristor found, *Nature* **453**, 80 (2008).

Suarez-Martinez, I., N. Grobert, and C. Ewels, Nomenclature of sp^2 carbon nanoforms, *Carbon* **50**, 741 (2011).

Sumino, Y., K.H. Nagai, Y. Shitaka, D. Tanaka, K. Yoshikawa, H. Chaté, and K. Oiwa, Largescale vortex lattice emerging from collectively moving microtubules, *Nature* **483**, 448 (2012).

Sundback, C.A., J.M. Beér, and A.F. Sarofim, Fragmentation behavior of single coal particles in a fluidized bed, in *Twentieth Symposium (International) on Combustion*, the Combustion Institute, Ann Arbor, Michigan (1984), p. 1495.

Surasani, V.K., T. Metzger, and E. Tsotsas, Consideration of heat transfer in pore network modelling of convective drying, *Int. J. Heat Mass Transfer* **51**, 2506 (2008).

Swanson, B.F., A simple correlation between permeabilities and mercury capillary pressures, *J. Pet. Tech.* **33**, 2498 (1981).

Suzuki, M., A.C. hopping conductivity in Mn-Co-Ni-Cu complex oxide semiconductors with spinel structure, *J. Phys. Chem. Solids* **41**, 1253 (1980).

Sykes, L.R., and S. Jaumé, Seismic activity on neighboring faults as a long term precursor to large earthquakes in the San Francisco bay area, *Nature* **348**, 595 (1990).

Symmons, O., and A. Raj, What's luck got to do with it: Single cells, multiple fates, and biological nondeterminism, *Mol. Cell* **62**, 788 (2016).

Sznajd-Weron, K., and J. Sznajd, Opinion evolution in closed community, *Int. J. Mod. Phys. C* **11**, 1157 (2000).

Szpilka, A.M., and P. Viščor, D.C. electrical conductivity of evaporated amorphous germanium - the low temperature limit, *Philos. Mag.* **45**, 485 (1982).

Tafti, T.A., M. Sahimi, F. Aminzadeh, and C. G. Sammis, Use of microseismicity for determining the structure of the fracture network of large-scale porous media, *Phys. Rev. E* **87**, 032152 (2013).

Takagi, H., S.-i. Uchida, K. Kitazawa, and S. Tanaka, High$-T_c$ superconductivity of La–Ba–Cu oxides. II. - Specification of the superconducting phase, *Japn. J. Appl. Phys.* **26**, L123 (1987).

Takahashi, M., K. Yokoyama, and T. Masuda, Dynamic viscoelasticity and critical exponents in sol-gel transition of an end-linking polymer, *J. Chem. Phys.* **101**, 798 (1994).

Takeuchi, K.A., M. Kuroda, H. Chaté, and M. Sano, Directed percolation criticality in turbulent liquid crystals, *Phys. Rev. Lett.* **99**, 234503 (2007).

Takeuchi, K.A., M. Kuroda, H. Chaté, and M. Sano, Experimental realization of directed percolation criticality in turbulent liquid crystals. *Phys. Rev. E* **80**, 051116 (2009).

Takeuchi, K.A., and M. Sano, Universal fluctuations of growing interfaces: Evidence in turbulent liquid crystals, *Phys. Rev. Lett.* **104**, 230601 (2010).

Tanaka, K., Structural phase transitions in chalcogenide glasses, *Phys. Rev. B* **39**, 1270 (1989).

Tanaka, K., T. Nakagawa, and A. Odajima, Photodarkening in glassy chalcogenide systems, *Philos. Mag. Lett. B* **54**, L3 (1986).

Tanaka, T., G. Swislow, and I. Ohmine, Phase separation and gelation in gelatin gels, *Phys. Rev. Lett.* **42**, 1557 (1979).

Tang, A.M., Y.-J. Cui, G. Richard, and P. Défossez, A study on the air permeability as affected by compression of three French soils, *Geoderma* **162**, 171 (2011).

Tang, L.H., and H. Leschhorn, Pinning by directed percolation, *Phys. Rev. A* **45**, R8309 (1992).

Tarboton, D.G., R.L. Bras, and I. Rodriguez-Iturbe, The fractal nature of river networks, *Water Resour. Res.* **24**, 1317 (1988).

Tatsumisago, M., B.L. Halfpap, J.L. Green, S.M. Lindsay, and C.A. Angell, Fragility of Ge-As-Se glass-forming liquids in relations to rigidity percolation, and the Kauzmann paradox, *Phys. Rev. Lett.* **64**, 1549 (1990).

Taubert, F., R. Fischer, J. Groeneveld, S. Lehmann, M.S. Müller, E. Rödig, T. Wiegand, and A. Huth, Global patterns of tropical forest fragmentation, *Nature* **554**, 519 (2018).

Taylor, G.I., Dispersion of soluble matter in solvent slowly flowing through a tube, *Proc. Roy. Soc. Lond.* **A219**, 186 (1953).

Taylor, H.E., The dielectric relaxation spectrum of glass, *Trans. Faraday Soc.* **52**, 873 (1956).

Terabe, K., T. Hasegawa, T. Nakayama, and M. Aono, Quantized conductance atomic switch, *Nature* **433**, 47 (2005).

Tessier, J.J., and K.J. Packer, The characterization of multiphase fluid transport in a porous solid by pulsed gradient stimulated echo nuclear magnetic resonance, *Phys. Fluids* **10**, 75 (1998).

Tessler, L.R., and G. Deutscher, Experimental investigation of two-dimensional In-Ge thin films, *Physica A* **157**, 154 (1989).

Thampi, S.P., R. Golestanian, and J.M. Yeomans, Instabilities and topological defects in active nematics, *Europhys. Lett.* **105**, 18001 (2014).

Thess, A., R. Lee, P. Nikolaev, H. Dai, P. Petit, J. Robert, C. Xu, Y.H. Lee, S.G. Kim, A.G. Rinzler, D.T. Colbert, G.E. Scuseria, J.E. Fischer, and R.E. Smalley, Crystalline popes of metallic carbon nanotubes, *Science* **273**, 843 (1996).

Thompson, A.H., A.J. Katz, and C.E. Krohn, The microgeometry and transport properties of sedimentary rock, *Adv. Phys.* **36**, 625 (1987a).

Thompson, A.H., A.J. Katz, and R.A. Rashke, Mercury injection in porous media: A resistance devil's staircase with percolation geometry, *Phys. Rev. Lett.* **58**, 29 (1987b).

Thongruang, W., R.J. Spontak, and C.M. Balik, Bridged double percolation in conductive polymer composites: an electrical conductivity, morphology and mechanical property study, *Polymer* **43**, 3717 (2002).

Thornton, J.W., Resurrecting ancient genes: experimental analysis of extinct molecules, *Nat. Rev. Genet.* **5**, 366 (2004).

Thorpe, M.F., Continuous deformation in random networks, *J. Non-Cryst. Solids* **57**, 355 (1983).

Thorpe, M.F., and E.J. Garboczi, Site percolation on central-force elastic networks, *Phys. Rev. B* **35**, 8579 (1987).

Thorpe, M.F., D.J. Jacobs, M.V. Chubynsky, and J.C. Phillips, Self-organization in network glases, *J. Non-Cryst. Solids* **266-269**, 859 (2000).

Thovert, J.-F., V.V. Mourzenko, and P.M. Adler, Percolation in three-dimensional fracture networks for arbitrary size and shape distributions, *Phys. Rev. E* **95**, 042112 (2017).

Tien, C., and A.C. Payatakes, Advances in deep bed filtration, *AIChE J.* **25**, 737 (1979)

Tian, L., and D.-N. Shi, The nature of explosive percolation phase transition, *Phys. Lett. A* **376**, 286 (2012).

Todd, A.C., J.E. Somerville, and G. Scott, The Application of depth of formation damage measurements in predicting water injectivity decline, SPE paper 12498-MS (1984).

Toffoli, T., and N. Margolus, *Cellular Automata Machines: A New Environment for Modeling* (MIT Press, Cambridge, 1987).

Toker, D., D. Azulay, N. Shimoni, I. Balberg, and O. Millo, Tunneling and percolation in metal-insulator composite materials, *Phys. Rev. B* **68**, 041403(R) (2003).

Tokita, M., R. Niki, and K. Hikichi, Percolation theory and elastic modulus of gel, *J. Phys. Soc. Jpn.* **53**, 480 (1984).

Tokunaga, T., and J. Wan, Water film flow along fracture surfaces of porous rock, *Water Resour. Res.* **33**, 1287 (1997).

Torelli, L., and A.E. Scheidegger, Three-dimensional branching-type models of flow through porous media, *J. Hydrol.* **15**, 23 (1972).

Torquato, S., *Random Heterogeneous Materials* (Springer, New York, 2002).

Touboul, J., and A. Destexhe, Power-law statistics and universal scaling in the absence of criticality, *Phys. Rev. E* **95**, 012413 (2017).

Trappe, V., W. Richtering, and W. Burchard, Critical behavior of anhydride cured expoxies, *J. Physique II* **2**, 1453 (1992).

Trifu, C.-I., and M. Radulian, Asperity distribution and percolation as fundamentals of an earthquake cycle, *Phys. Earth Plant. Interiors* **58**, 277 (1989).

Troadec, J.P., D., Bideau, and E. Guyon, Transport properties of conducting and semiconducting anisotropic mixtures, *J. Phys. C* **14**, 4807 (1981).

Trugman, S.A., Localization, percolation, and the quantum Hall effect, *Phys. Rev. B* **27**, 7539 (1983).

Tsakiroglou, C.D., and A.C. Payatakes, A new simulator of mercury porosimetry for the characterization of porous materials, *J. Colloid Interface Sci.* **137**, 315 (1990).

Tsimpanogiannis, I.N., Y.C. Yortsos, S. Poulou, N. Kanellopoulos, and A. K. Stubos, Scaling theory of drying in porous media, *Phys. Rev. E* **59**, 4353 (1999).

Turcotte, D.L., Earthquake prediction, *Annu. Rev. Earth Planet. Sci.* **19**, 263 (1991).

Turcotte, D. L., *Fractals and Chaos in Geology and Geophysics*, 2nd ed. (Cambridge University Press, New York, 1997).

TyČ, S., and B.I. Halperin, Random resistor network with an exponentially wide distribution of bond conductances, *Phys. Rev. B* **39**, 877 (1989).

Uebbing, B., and A.J. Sievers, Influence of the glass network on the vibrational energy transfer from H_2O inpurities in the Ge-As-Se glass series, *J. Non-Cryst. Solids* **203**, 159 (1996).

Vacher, R., E. Courtens, G. Coddens, A. Heidemann, Y. Tsujimi, J. Pelous, and M. Foret, "Crossovers in the density of states of fractal silica aerogels," *Phys. Rev. Lett.* **65**, 1008 (1990).

Vacher, R., T. Woignier, J. Pelous, G. Coddens, and E. Courtens, "The density of vibrational states of silica aerogels," *Europhys. Lett.* **8**, 161 (1989).

Vacher, R., T. Woignier, J. Pelous, and E. Courtens, Structure and self-similarity of silica aerogels, *Phys. Rev. B* **37**, 6500 (1988).

Vallette, D.P., G. Jacobs, and J.P. Gollub, Oscillations and spatiotemporal chaos of one-dimensional fluid fronts, *Phys. Rev. E* **55**, 4274 (1997).

Valvatne, P.H., and M.J. Blunt, Predictive pore-scale modeling of two-phase flow in mixed wet media, *Water Resour. Res.* **40**, W07406 (2004).

Vanunu, O., O. Magger, E. Ruppin, T. Shlomi, and R. Sharan, Associating genes and protein complexes with disease via network propagation, *PLoS Comput. Biol.* **6**, e1000641 (2010).

van der Marck, S.C., Network approach to void percolation in a pack of unequal spheres, *Phys. Rev. Lett.* **77**, 1785 (1996).

van der Putten, D., J.T., Moonen, H.B. Brom, J.C.M. Brokken-Zijp, and M.A.J. Michels, Evidence for superlocalization on a fractal network in conductive carbon-black-polymer composites, *Phys. Rev. Lett.* **69**, 494 (1992).

van Dijk, M.A., Dielectric study of percolation phenomena in a microemulsion, *Phys. Rev. Lett.* **55**, 1003 (1985).

van Dijk, M.A., G. Castelejin, J.G.H. Joosten, and Y.K. Levine, Percolation in oil-continuous microemulsions. A dielectric study of aerosol OT/water/isooctane, *J. Chem. Phys.* **85**, 626 (1986).
van Staveren, M.P.J., H.B. Brom, and L.J. de Jongh, Metal-cluster compounds and universal features of the hopping conductivity of solids, *Phys. Rep.* **208**, 1 (1991).
Vázquez, A., Exact results for the Barabási model of human dynamics, *Phys. Rev. Lett.* **95**, 248701 (2005).
Vaźquez, A., and O.S. Costa, Self-organized criticality and directed percolation, *J. Phys. A* **32**, 2633 (1999).
Velde, B., J. Dubois, D. Moore, and G. Touchard, Fractal patterns of fractures in granites, *Earth Planet. Sci. Lett.* **104**, 25 (1991).
Vening Meinesz, F.A., A remarkable feature of the Earth's topography, origin of continents and oceans, *Proc. Koninkl. Ned. Akad. Wetensch. ser. B* **55**, 212 (1951).
Vere-Jones, D., A branching model for crack propagation, *Pure Appl. Geophys.* **114**, 711 (1976).
Vere-Jones, D., Statistical theories of crack propagation, *Math. Geol.* **9**, 455 (1977).
Vigmond, E., A. Pashaei, S. Amraoui, H. Cochet, and M. Hassaguerre, Percolation as a mechanism to explain atrial fractionated electrograms and reentry in a fibrosis model based on imaging data, *Heart Rhythm* **13**, 1536 (2016).
Visčor, P., and A.D. Yoffe, Amorphous germanium prepared in ultra high vacuum and measured in situ: The DC electrical conductivity, *J. Non-Cryst. Solids* **35/36**, 409 (1982).
Viswanathan, R., and M.B. Heaney, Direct imaging of the percolation network in a three-dimensional disordered conductor-insulator composite, *Phys. Rev. Lett.* **75**, 4433 (1995).
Voigt, C.A., and R.M. Ziff, Epidemic analysis of the second–order transition in the Ziff–Gulari–Barshad surface–reaction model, *Phys. Rev. E* **56**, R6241 (1997).
Voitalov, I., P. van der Hoorn, R. van der Hofstad, and D. Krioukov, Scale-free networks well done, *Phys. Rev. Res.* **1**, 033034 (2019).
Volger, J., Note on the Hall potential across an inhomogeneous conductor, *Phys. Rev.* **79**, 1023 (1950).
von Niessen, W., and A. Blumen, Dynamics of forest fires as a directed percolation model, *J. Phys. A* **19**, L289 (1986).
Vorhauer, N., E. Tsotsas, and M. Prat, Drying of thin porous disks from pore network simulations, *Drying Technol.* **36**, 651 (2017).
Voss, R.F., R.B. Laibowitz, and E.I. Allessandrini, Fractal (scaling) clusters in thin gold films near the percolation threshold, *Phys. Rev. Lett.* **44**, 1441 (1982).
Wagie, H.E., and P. Geissinger, Hole-burning spectroscopy as a probe of nano-environments and processes in biomolecules: A review, *Applied Spectroscopy* **66**, 609 (2012).
Wagner, T., and S.O. Kasap, Glass transformation, heat capacity and structure of AsxSe1-x glasses studied by modulated temperature differential scanning calorimetry experiments, *Philos. Mag. B* **74**, 667 (1996).
Wall, G.C., and R.J.C. Brown, The determination of pore-size distributions from sorption isotherms and mercury penetration in interconnected pores: The application of percolation theory, *J. Colloid Interface Sci.* **82**, 141 (1981).
Walter, U., D.L. Price, S. Susman, and K.J. Volin, Network dynamics of chalcogenide glasses. I. Germanium diselenide, *Phys. Rev. B* **37**, 4232 (1988).
Wang, J., The bond-bending model in three dimensions, *J. Phys. A* **22**, L291 (1989).
Wang, K., M.L. Steyn-Ross, D.A. Steyn-Ross, M.T. Wilson, and J.W. Sleigh, EEG slow-wave coherence changes in propofol-induced general anesthesia: experiment and theory, *Front. Syst. Neurosci.* **8**, 215 (2014).
Wang, S., and P.G. Wolynes, Active contractility in actomyosin networks, *Proc. Natl. Acad. Sci. U.S.A.* **109**, 6446 (2012).
Wang, Y., M. Nakamura, O. Matsuda, and K. Murase, Raman-spectroscopy studies on rigidity percolation and fragility on Ge-(S,Se) glasses, *J. Non-Cryst. Solids* **266-269**, 872 (2000).
Wanliss, J., V. Muñoz, D. Pastén, B. Toledo, and J.A. Valdivia, Critical behavior in earthquake energy dissipation, *Europ. Phys. J. B* **90**, Article 167 (2017).

Warburton, P.M., A stereological interpretationof joint trace data, *Int. J. Rock Mech. Min. Sci. Geomech. Abstr.* **17**, 181 (1980).

Wardlaw, N.C., Y. Li, and D. Forbes, Pore-throat size correlation from capillary pressure curves, *Transp. Porous Media* **2**, 597 (1987).

Wardlaw, N.C., and M. McKellar, Mercury porosimetry and the interpretation of pore geometry in sedimentary rocks and artificial models, *Powder Technol.* **29**, 127 (1981).

Wardlaw, N.C., and R.P. Taylor, Mercury capillary pressure curves and the interpretation of pore structure and capillary behavior in reservoir rocks, *Bull. Can. Pet. Geol.* **24**, 225 (1976).

Warren, B.E., X-ray diffraction of vireous silica, *J. Am. Ceram. Soc.* **17**, 249 (1934).

Washburn, E.W., Note on a method of determining the distribution of pore sizes in a porous, materia *Proc. Nat. Acad. Sci. USA* **7**, 115 (1921).

Watanabe, K., and H. Takahashi, Fractal geometry characterization of geothermal reservoir fracture networks, *J. Geophys. Res.* **100B**, 521 (1995).

Watts, D.J., and S.H. Strogatz, collective dynamics of "small world" networks, *Nature* **393**, 440 (1998).

Weaire, D., and S. Hutzler, *The Physics of Foams* (Oxford University Press, New York, 1999).

Weber, J.K., and V.S. Pande, Percolation-like phase transitions in network models of protein dynamics, *J. Chem. Phys.* **142**, 215105 (2015).

Webman, I., and G. Grest, Dynamical behavior of fractal structures, *Phys. Rev. B* **31**, 1689 (1985).

Webman, I., J. Jortner, and M.H. Cohen, Numerical simulation of electrical conductivity in microscopically inhomogeneous materials, *Phys. Rev. B* **11**, 2885 (1975).

Weinrib, A., Long-range correlated percolation, *Phys. Rev. B* **29**, 387 (1984).

Weinrib, A., and B.I. Halperin, Critical phenomena in systems with long-range-correlated quenched disorder, *Phys. Rev. B* **27**, 413 (1983).

Weisbuch, G., and G. Boudjema, Dynamical aspects in the adoption of agri-environmental measures, *Adv. Complex Systems* **2**, 11 (1999).

West, G.B., and J.H. Brown, The origin of allometric scaling laws in biology from genomes to ecosystems: towards a quantitative unifying theory of biological structure and organization, *J. Exp. Biolog.* **208**, 1575 (2005).

West, G.B., J.H. Brown, and B.J. Enquist, A general model for the origin of allometric scaling laws in biology, *Science* **276**, 122 (1997).

White, A.F., and S.L. Brantley, The effect of time on the weathering rates of silicate minerals. Why do weathering rates differ in the lab and in the field? *Chem. Geol.* **202**, 479 (2003).

Wierman, J.C., Exact percolation threshold, in *Complex Media and Percolation Theory*, edited by M. Sahimi and A.G. Hunt (Springer, New York, 2021).

Wilhelm, J., and E. Frey, Elasticity of stiff polymer networks, *Phys. Rev. Lett.* **91**, 108103 (2003).

Wilke, S., E. Guyon, and G. de Marsily, Water penetration through fractured rock: Test of a tridimensional percolation description, *Math. Geol.* **17**, 17 (1985).

Wilkinson, D., Percolation model of immiscible displacement in the presence of buoyancy forces, *Phys. Rev. A* **30**, 520 (1984).

Wilkinson, D., Percolation effects in immiscible displacement, *Phys. Rev. A* **34**, 1380 (1986).

Wilkinson, D., and J.F. Willemsen, Invasion percolation: a new form of percolation theory, *J. Phys. A* **16**, 3365 (1983).

Williams, M.L., and H.J. Maris, Numerical study of phonon localization in disordered systems, *Phys. Rev. B* **31**, 4508 (1985).

Wilting, J., and V. Priesemann, Between perfectly critical and fully irregular: a reverberating model captures and predicts cortical spike propagation, *Cereb. Cortex* **29**, 2759 (2019).

Winterfeld, P.H., L.E. Scriven, and H.T. Davis, Percolation and conductivity of random two-dimensional composites, *J. Phys. C* **14**, 2361 (1981).

Wioland, H., E. Lushi, and R.E. Goldstein, Directed collective motion of bacteria under channel confinement, *New J. Phys.* **18**, 075002 (2016).

Witten, T.A., and L.M. Sander, Diffusion–limited aggregation, a kinetic critical phenomenon, *Phys. Rev. Lett.* **47**, 1400 (1981).

Woignier, T., J. Phalippou, R. Sempere, and J. Pelons, Analysis of the elastic behaviour of silica aerogel taken as a percolating system, *J. Phys. France* **49**, 289 (1988).

Woignier, T., J. Phalippou, R. Vacher, J. Pelous, and E. Courtens, Different kind of fractal strucures in silica aerogels, *J. Non-Cryst. Solids* **121**, 198 (1990).

Wolfram, S., Statistical mechanics of cellular automata, *Rev. Mod. Phys.* **55**, 601 (1983).

Wong, P.-Z., J. Koplik, and J.P. Tomanic, conductivity and permeability of rocks, *Phys. Rev. B* **30**, 6606 (1984).

Worst, E.G., P. Zimmer, E. Wollrab, K. Kruse, and A. Ott, Unbounded growth patterns of reproducing, competing polymers' similarities to biological evolution, *New J. Phys.* **18**, 103003 (2016).

Wu, F.Y., The Potts model, *Rev. Mod. Phys.* **54**, 235 (1982).

Wu, R., T. Zhang, C. Ye, C.Y. Zhao, E. Tsotsas, and A. Kharaghani, Pore network model of evaporation in porous media with continuous and discontinuous corner films, *Phys. Rev. Fluids* **5**, 014307 (2020).

Wu, Z.L., Implications of a percolation model for earthquake, *Geophys. J. Int.* **133**, 104 (1998).

Wygnanski, I., and F.H. Champagne, On transition in a pipe. Part 1. The origin of puffs and slugs and the flow in a turbulent slug, *J. Fluid Mech.* **59**, 281 (1973).

Xu, B., Y.C. Yortsos, and D. Salin, Invasion percolation with viscous forces, *Phys. Rev. E* **57**, 739 (1998).

Xu, W., Y. Jiang, and Y. Jiao, Continuum percolation of complex polyhedral pores/fractures (preprint, 2022).

Xu, W., Z. Zhu, Y. Jiang, and Y. Jiao, Continuum percolation of congruent overlapping polyhedral particles: Finite-size-scaling analysis and renormalization-group method, *Phys. Rev. E* **99**, 032107 (2019).

Yakubo, K., E. Courtens, and T. Nakayama, Missing modes in the density of states of fractal networks, *Phys. Rev. B* **42**, 1078 (1990a).

Yakubo, K., and T. Nakayama, Fracton dynamics of percolating elastic networks: energy spectrum and localized nature, *Phys. Rev. B* **40**, 517 (1989).

Yakubo, K., T. Nakayama, and H.J. Maris, Analysis of a new method for finding eigenmodes of very large lattice systems, *J. Phys. Soc. Jpn.* **60**, 3249 (1991).

Yakubo, K., K. Takasugi, and T. Nakayama, Crossover behavior from vector to scalar fractons in the density of states of percolating networks, *J. Phys. Soc. Jpn.* **59**, 1909 (1990b).

Yang, J., and M.T. Balhoff, Pore-network modeling of particle retention in porous media, *AIChE J.* **63**, 3118 (2017).

Yang, X.S., Small-world networks in geophysics, *Geophys. Res. Lett.* **28**, 2549 (2001).

Yazdi, A., H. Hamzehpour, and M. Sahimi, Permeability, Porosity, and Percolation Properties of Two-Dimensional Disordered Fracture Networks, *Phys. Rev. E* **84**, 046317 (2011).

Ye, G., H. Wang, X. Duan, Z. Sui, X. Zhou, M.-O. Coppens, and W. Yuan, Pore network modeling of catalyst deactivation by coking, from single site to particle, during propane dehydrogenation, *AIChE J.* **65**, 141 (2019).

Ye, G., X. Zhou, W. Yuan, and M.-O. Coppens, Probing pore blocking effects on multiphase reactions within porous catalyst particles using a discrete model, *AIChE J.* **62**, 451 (2016).

Yiotis, A.G., A.G. Boudouvis, A.K. Stubos, I.N. Tsimpanogiannis, and Y.C. Yortsos, Effect of liquid films on the drying of porous media, *AIChE J.* **50**, 2721 (2004).

Yiotis, A.G., D. Salin, and Y.C. Yortsos, Pore network modeling of drying processes in macroporous materials: effects of gravity, mass moundary layer and pore microstructure, *Transp. Porous Media* **110**, 175 (2015).

Yiotis, A.G., A.K. Stubos, A.G. Boudouvis, and Y.C. Yortsos, A 2-D pore-network model of the drying of single-component liquids in porous media, *Adv. Water Resour.* **24**, 439 (2001).

Yonezawa, F., and K. Morigaki, Coherent potential approximation: Basic concepts and applications, *Prog. Theor. Phys. Supplement* **53**, 1 (1973).

Yoonessi, M., J.R. Gaier, M. Sahimi, T.L. Daulton, R.B. Kaner, and M.A. Meador, Fabrication of graphene-polyimide nanocomposites with superior electrical conductivity, *ACS Appl. Mater. Interfaces* **9**, 49 (2017).

Yortsos, Y.C., and M.M. Sharma, Application of percolation theory to noncatalytic gas-solid reactions, *AIChE J.* **32**, 46 (1986).
Yortsos, Y.C., B. Xu, and D. Salin, Phase diagram of fully developed drainage in porous media, *Phys. Rev. Lett.* **79**, 4581 (1997).
Young, A.P., and R.B. Stinchcombe, A renormalization group theory for percolation problems, *J. Phys. C* **8**, L535 (1975).
Young, T., *Miscellaneous Works*, Vol. 1, edited by G. Peacock (J. Murray, London, 1855), p. 418.
Yu, A., M.E. Itkis, E. Bekyarova, and R.C. Haddon, Effect of single-walled carbon nanotube purity on the thermal conductivity of carbon nanotube-based composites, *Appl. Phys. Lett.* **89**, 133102 (2006).
Yu, A., P. Ramesh, M.E. Itkis, E. Bekyarova, and R.C. Haddon, Graphite nanoplatelet-epoxy composite thermal interface materials, *J. Phys. Chem. C* **111**, 7565 (2007).
Yu, A., P. Ramesh, X. Sun, E. Bekyarova, M.E. Itkis, and R.C. Haddon, enhanced thermal conductivity in a hybrid graphite nanoplatelet - carbon nanotube filler for epoxy composites, *Adv. Mater.* **20**, 4740 (2008).
Yu, F., and A.G. Hunt, An examination of the steady-state assumption in certain soil production models, *Earth Surface Processes and Landforms* **42**, 2599 (2017a).
Yu, F., and A.G. Hunt, Damköhler number input to transport–limited chemical weathering and soil production calculations, *ACS Earth and Space Chemistry* **1**, 30 (2017b).
Yu, F., A. Hunt, M. Egli, and G. Raab, Comparison and contrast in soil depth evolution for steady state and stochastic erosion processes: Possible implications for landslide prediction, *Geochem. Geophys. Geosyst.* **20**, 2886 (2019).
Yu, H.-C., and S.V. Sotirchos, A generalized pore model for gas-solid reactions exhibiting pore closure, *AIChE J.* **33**, 382 (1987).
Yun, S.S., H. Li, R.L. Cappelletti, R.N. Enzweiler, and P. Boolchand, Onset of rigidity in $Se_{1-x}Ge_x$ glasses: Ultrasonic elastic moduli, *Phys. Rev. B* **39**, 8702 (1989).
Zachariasen, W.H., The atomic arrangement in glass, *J. Am. Chem. Soc.* **54**, 3841 (1932).
Zallen, R., Polychromatic percolation: Coexistence of percolating species in highly connected lattices, *Phys. Rev. B* **16** 1426 (1977).
Zallen, R., *The Physics of Amorphous Solids* (Wiley, New York, 1983).
Zdeborová, P. Zhang, and H.-J. Zhou, Fast and simple decycling and dismantling of networks, *Sci. Rep.* **6**, Article 37954 (2016).
Zeimetz, B., B.A. Glowacki, and J.E. Evetts, Application of percolation theory to current transfer in granular superconductors, *Europ. Phys. J. B* **29**, 359 (2002).
Zemlyanov, M.G., V.K. Malinovskii, V.N. Novikov, P.P. Parshin, and A.P. Sokolov, A study of fractons in polymers, *Zh. Eskp. Teor. Fix.* **101**, 284 (1992) [*Sov. Phys. JETP* **74**, 151 (1992)].
Zenklusen, R., Connectivity interdiction, *Operations Res. Lett.* **42**, :450 (2014).
Zeppini, P., and K. Frenken, Networks, percolation, and consumer demand, *J. Artificial Societies and Social Simul.* **21**, 1 (2018).
Zhang, L., J.H. Seong, and M. Bucci, Percolation scale-free behavior in boiling crisis, *Phys. Rev. Lett.* **122**, 134501 (2019).
Zhang, Y., T. Kogure, S. Chiyonobu, X. Lei, and Z. Xue, Influence of heterogeneity on relative permeability for CO_2/brine: CT observations and numerical modeling, *Energy Procedia* **37**, 4647 (2013).
Zhao, N., Z. Sun, X. Yang, and W. Xu, Explosive death of conjugate coupled Van der Pol oscillators on networks, *Phys. Rev. E* **97**, 062203 (2018)
Zhdanov, V.P., V.B. Fenelonov, and D.K. Efremov, Determination of pore-size distribution from sorption isotherms: Application of percolation theory, *J. Colloid Interface Sci.* **120**, 218 (1987).
Zhigalov, A., G. Arnulfo, L. Nobili, S. Palva, and J.M. Palva, Relationship of fast- and slow–timescale neuronal dynamics in human MEG and SEEG, *J. Neurosci.* **35**, 5385 (2015).
Zhou, D.W., D.D. Mowrey, P. Tang, Y. Xu, Percolation model of sensory transmission and loss of consciousness under general anesthesia, *Phys. Rev. Lett.* **115**, 108103 (2015).

Zhu, X., J. Song, J. Meng, and J. Liu, Information spreading in social network through explosive percolation theory, in *Lecture Notes in Computer Science*, edited by X. Chen, A. Sen, W.W. Li, and M.T. Tahi, **11280**, 487 (Springer Nature, Switzerland, 2018).

Ziff, R.M., Test of scaling exponents for percolation–cluster perimeters, *Phys. Rev. Lett.* **56**, 545 (1986).

Ziff, R.M., Spanning probability in 2D percolation, *Phys. Rev. Lett.* **69**, 2670 (1992).

Ziff, R.M., Explosive growth in biased dynamic percolation on two-dimensional regular lattice networks, *Phys. Rev. Lett.* **103**, 045701 (2009).

Ziff, R.M., Scaling behavior of explosive percolation on the square lattice, *Phys. Rev. E* **82**, 051105 (2010).

Ziff, R.M., Getting the jump on explosive percolation, *Science* **339**, 1159 (2013).

Ziff, R.M., E. Gulari, and Y. Barshad, Kinetic phase transitions in an irreversible surface–reaction model, *Phys. Rev. Lett.* **56**, 2553 (1986).

Ziman, J.M., The localization of electrons in ordered and disordered systems. I. Percolation of classical particles, *J. Phys. C* **1**, 1532 (1968).

Zimmer, P., K. Kruse, and J. Nagler, Anomalous percolation features in molecular evolution, *Phys. Rev. E* **98**, 022408 (2018).

Zhang, Y., P. Dai, M. Levy, and M.P. Sarachik, Probing the Coulomb gap in insulating n-type CdSe, *Phys. Rev. Lett.* **64**, 2687 (1990).

Zöller, G., S. Hainzl, and J. Kurths, Observation of growing correlation length as an indicator for critical point behavior prior to large earthquakes, *J. Geophys. Res. B* **106**, 2167 (2001).

Zuber, M.T., The crust and mantle of Mars, *Nature* **412**, 220 (2001).

Zuber, M.T., D. Smith, S. Solomon, D.O. Muhleman, J.W. Head, J.B. Garvin, J.B. Abshire, and J.L. Bufton, The Mars Observer laser altimeter investigation, *J. Geophys. Res.* **97**, 7781 (1992).

Index

A

AC conductivity
  effect of percolation, 62, 163
  scaling properties, 23, 43, 193, 257, 275, 357, 570
  universality, 415, 416
Activation energy, 77, 210, 219, 337, 389, 393, 401, 415, 416
Actomyosin networks
  myosin-induced dynamics, 461
  percolation model, 461, 462
Additive polymerization, 343
Aerogels, 38, 376–380, 386
Alexander-Orbach conjecture, 236, 368, 370–372
Anderson localization, 413
Anisotropic materials, 126, 309
Anomalous diffusion, 122, 164, 165, 300, 326, 370
Antigen–antibody reaction, 444–446, 448
Archie's law, 59, 120, 135, 149
Average molecular weight, 214, 345, 352

B

Backbone
  critical exponent, 18, 165, 283
  fractal dimension, 22, 27, 61, 89, 107, 110, 111, 169, 171, 258, 265, 421
Beam model, 273, 275
Bethe lattice, 8, 9, 11, 12, 21, 49, 74, 113, 179, 214, 215, 220, 221, 227, 228, 235, 265, 332, 342, 343, 349, 396, 397, 433, 441, 444–447, 483, 521
Biological systems
  percolation conductivity, 449, 451
  power law, 607
Bohr radius, 391
Boiling (continuum percolation), 42
Boltzmann's equation, 2
Bond-bending forces, 261, 262, 336, 382, 385
Bond-bending models
  force distribution, 263
  Kirkwood-Keating model, 262, 318, 336
Bond-bond Green function, 398
Bootstrap percolation, 48, 49, 517
Born model, 247, 248, 269–272, 276, 357, 358
Bose-Einstein distribution, 379
Brain
  collective influence, 467
  essential nodes, 465, 466
  neuronal avalanche, *see*Directed percolation
  optimal percolation model, 465, 466
Branched polymers
  fractal dimension, 345, 349, 350
  in good solvent, 348
  in swollen state, 348
  lattice animal model, 348, 350
  Parisi-Sourlas-Family relation, 349
  small-angle neutron scattering, 331, 350, 351

C

Capillary pressure, 50, 56, 67–76, 130, 175–177, 180, 181, 183, 185–188, 190–193, 195, 199–202, 204, 417, 532
Carbon black, 5, 286, 288, 291, 327, 411, 414
Cardiac fibrosis
  arrhythmia, 470

M. Sahimi, *Applications of Percolation Theory*, Applied Mathematical Sciences 213,
https://doi.org/10.1007/978-3-031-20386-2

cellular network, 470
correlation length, 472
critical wavelength, 472
governing equation, 470
percolation transition, 474
reentry formation, 470, 473
Catalyst deactivation
continuum model, 230, 232
diffusion-limited, 230
percolation model, 227, 228
pore-network model, 232–234
reaction-limited, 228
reactor scale, 234
Cayley tree, 432, 433, 483
Central-force percolation, *see* Rigidity percolation
Central forces, 314, 316, 342, 384
Chemical dimension, 441
Chemical gels, 332, 343, 351, 357, 358
Chemical path, 413, 501
Chemical weathering, 489–492, 494–497, 504
Clogging
experimental data, 218, 227, 239, 242
percolation aspects, 220, 227, 239
pore-network model, 234, 239, 241
precipitation, 238, 239, 241
size-exclusion, 241
Coal gasification
chemical reactions, 214, 224, 491
continuum models, 220
percolation statistics, 217
with pore blocking, 224, 234
Coherent-potential approximation, 399
Colorectal tumorigenesis
giant cluster, 463–465
percolation transition, 462, 465
protein-protein interaction network, 463–465
Complex networks
degree distribution, 434, 440, 441, 466
Erdös - Rényi graph, 436, 437
giant cluster, 437, 441
percolation threshold, 12, 24
scale-free graphs, 437
small-world networks, 437
Composite materials
metal-insulators, 279
Conductivity, *see also* Random resistor networks
anisotropy near percolation, 283, 284
effective-medium approximation, 127, 134, 135, 150
Connectivity, 2, 3, 7, 8, 10, 11, 15, 17, 20, 21, 25, 32, 42, 45, 46, 50, 55, 58, 62, 76, 77, 79, 81–83, 86, 87, 91–93, 95, 96, 98, 99, 117–119, 127, 144, 157, 175, 176, 203, 221, 234, 239, 251, 253, 266, 280, 289, 297, 333, 339, 340, 344, 345, 435, 437, 443, 460, 462, 464, 466, 478–480, 485, 493, 511, 517, 518, 520, 533, 539, 542–546, 585–588, 590, 607
Constraint-counting method
for Kirkwood-Keating model, 341
for rigidity percolation, 259
Coordination number, 11, 18, 29, 31, 37, 38, 40, 45, 48, 70, 71, 74, 81, 82, 87, 92, 113, 126, 130, 179, 186, 199, 214, 222, 227, 228, 250, 260, 262, 263, 272, 273, 276, 282, 333–341, 349, 387, 396–398, 400, 401, 435, 441, 445, 478
Corrections to scaling, 24, 105, 256–258
Correlated percolation
behavior of water, 555
site-bond percolation, 81
with long-range correlation, 43, 44
with short-range correlations, 43
Correlation function, 37, 43, 44, 108, 198, 349, 378, 403, 423, 517, 551, 564, 576
Covalent bonding, 335
Covid-19
heirarchical mobility, 474
percolation transition, 474, 475
temporal network, 475
Critical exponents. *see* percolation
Critical-path analysis
for dispersion, 164
for electrical conductivity, 127, 128, 131, 136–138, 145
for hopping conductivity, 411
for permeability, 118, 125, 127, 129–131, 133, 136
Critical point, 5, 23, 52, 103–105, 115, 266, 332, 377, 429, 443, 527, 563, 565, 568, 572, 574, 600, 609
Critical subnetwork, 404, 406, 408
Critical volume fraction, 133, 146, 171, 211, 229, 288, 312, 318, 327, 378
Crossover frequency, 367–369, 372, 374, 375, 379, 381, 385, 387
Crossover length, 576

D
Dead-end bonds, 552, 563
Debye process, 355
Degree of polymerization, 346–348
Devil's staircase, 195
Diffusion-limited annihilation
  experiental data, 236
  percolation model, 236
  reaction kinetics, 235
Directed percolation
  depinning of an interface, 575
  flowing sand, 579
  general model, 558
  meandering exponent, 557
  neuronal avalanche in brain, 580
  opening angle, 578
  relation with ecological collapse, 573
  relation with turbulent flow, 102, 549, 552, 555, 559, 560, 572
  satellite networks, 582
Dispersion
  Anomalous, 163, 171
  classification of, 160
  convective–diffusion equation, 157
  critical-path analysis, 169
  first-passage time distribution, 156–158, 164, 166, 168
  Gaussian, 157, 164, 171
  mechanisms of, 156, 157
  percolation effect, 163
  pore-network model, 162, 163
  scaling laws, 164
  Taylor–Aris dispersion, 159, 160, 163
Drying
  in porous media, 202, 203
  pore-network model, 203
  scaling theory, 206
Duality relation
  for resistor networks, 20
Dynamic-mechanical experiments, 352, 353
Dynamic percolation, 51, 306
Dynamic scaling, 223, 423, 574

E
Earthquakes
  as a critical phenomenon, 103
  b-values, 110
  fractal dimensions, 106, 107
  Gutenberg-richter law, 110
  log-periodic corrections, 105
  Markov model, 101
  microseismicity, 108
  seismicity, 103, 104
  structure function, 102
Effective-medium approximation
  for anisotropic materials, 309
  for bond-bending models, 273
  for Hall conductivity, 308, 310
  for rigidity percolation, 269
  for the Born model, 270, 272
  for vibrational density of states, 368
Einstein relation, 20, 394
Elastic energy, 105, 246–249, 253, 255, 256, 261, 263, 271, 275, 332, 357, 358, 382, 383, 385
Elastic moduli, 17, 18, 20, 41, 86, 248, 249, 254–258, 261, 263, 265, 267, 269–273, 275, 277, 287, 290, 316–318, 336–338, 340, 341, 344, 345, 351, 352, 356–358, 360, 362, 363, 378, 382, 387
Elastic network, *see* Rigidity percolation; Bondbending models
  fixed points, 277
  universality of Poisson's ratio, 277
Elastic spectral dimension, 384
Electrical conductivity, *see also* Conductivity; Random resistor networks
  critical exponent, 138
  critical-path analysis, 136, 137
  duality relation, 138
  effective-medium approximation, 134
  effective-medium approximation and percolation, 135
  percolation and critical-path analysis, 127, 132
  power-law for, 135
  renormalization group approach, 138, 143
  renormalized effective-medium approximation, 143, 144
  scaling properties, 146
Entropic elasticity, 358
Erdös and Réyni graph, 436, 467
Evolution
  multidimensional fitness landscape, 483
  percolation transition, 483, 484
Explosive death, 546, *see also* optimal dismantling
Explosive immunization
  infected cluster, 543
  number of leaves, 544
  percolation threshold, 543
  scale-free model, 544
Explosive percolation

Achlioptas process, 518, 519
Bohman–Frieze–Wormald model, 521
bundle of nanotubes formation, 531
cluster aggregation, 527
crackling noise, 533, *see also* fractional percolation
emergence of molecular life, 535
evolution of protein networks, 537, 538
n-edge model, 522
powder keg, 523, 524
product rule, 519, 531, 547
reversible and irreversible transition, 527
spanning cluster-avoiding, 522
sum rule, 522, 547

F

Faults (earthquakes), 85, 87, 101, 104, 107, 108, 111, 113
Fermi distribution, 390
Filling factor of a lattice, 36, 282
Finite-size scaling, 23, 24, 62, 82, 140, 195, 256, 257, 283, 292, 371, 542, 564
Fixed point, 140, 277, 562
Flexibility of thought
complex network model, 472
percolation analysis, 469
percolation integral, 469, 470
semantic networks, 468
Flexural dimension, 274, 382
Floppy state
geometrical, 251, 252
macroscopic, 251
microscopic, 251
Fluctuation-dissipation theorem, 394
Foams
mechanical properties, 313
relation to rigidity percolation, 315
Forrest fires
percolation scaling of burnt trees, 597, 599
propagation time, 599
trapping effect, 596
Fractal diffusion, 23, 107, 121, 300, 326
Fractal dimension, 21, 22, 26–28, 30, 48, 61, 62, 87–89, 93, 98, 106, 107, 109–111, 121, 166, 169–171, 180, 202, 206, 236, 238, 239, 258, 265, 267, 297, 345–347, 349, 350, 369–371, 378, 382, 413, 414, 417, 421, 424–427, 432, 434, 441, 482, 494, 499, 501, 502, 505, 507, 514, 523, 524, 526, 534, 551, 553–555, 595, 599
Fractals, 22, 23, 26–28, 38, 43, 62, 87, 89, 93, 107, 109, 110, 131, 163, 170, 182, 198, 206, 237, 238, 265, 297, 304, 345, 369, 372, 378, 380–383, 386, 413, 414, 417, 424, 426, 428, 431, 434, 441, 454, 459, 517, 529, 530, 534, 552
Fractional percolation, 533
Fractons
computer simulation, 384
crossover to phonons, 372
dispersion relation, 372
effect on sound velocity, 381
effect on thermal transport, 366
ensemble-averaged properties, 372
fracton dimension, 369–371, 417
localization of, 370, 414
mode patterns of, 372, 374, 375, 379, 380
phonon assisted hopping, 414, 417
Fractures
excluded surface, 89, 90
excluded volume, 89, 95–97
experimental data, 87, 93, 94, 122
network models, 85, 87, 88
percolation threshold, 91, 93, 95, 96, 98, 99, 107, 110
relevance of percolation, 86–89
self-similarity, 88
Frequency-dependent conductivity, 30, 394, 402
Fungi
invasion model, 511
percolation transition, 510–512
spreading in soil, 511

G

Gelation
Debye process, 355
determination of transition point, 479
dynamic-mechanical experiments, 352, 353
dynamic superelastic network model, 335
elastic moduli, 336, 337, 340, 341, 344, 345, 351, 352, 356
gel fraction, 345, 348
percolation models, 344
relaxation time spectrum, 354, 356
retardation time spectrum, 356
scaling properties, 357
viscosity, 351, 360–362
Gels

chemical, 332, 343, 351, 357, 358
complex shear compliance, 355
dynamic shear compliance, 355
dynamic tensile modulus, 352
enthalpic, 358
entropic, 358, 359
fluctuation algorithm for, 359
loss modulus, 353, 357
physical, 360
relaxation strength, 355
storage modulus, 353, 357
Geochemistry
mean distance for transport, 494
percolation model, 490
scaling theory, 492
Glasses
ageing, 335
Cohen-Grest model, 334
constraint-counting argument, 340
effect of high coordination number, 339
effect of onefold-coordinated atoms, 341
free volume, 334
mean-coordination number, 339
Phillips-Thorpe theory, 387
Raman spectra, 338
stress-free versus stressed, 342
structure of network, 335, 337
transition temperature, 334, 337
Global trade
Nash equilibrium, 540
percolation model, 539, 540
Gradient percolation, 196, 197, 423
Granular model, 239, 253, 275, 294, 295, 312, 319, 320, 324, 326, 406, 407, 607
Granular superconductors, 320
Green functions
for conduction, 269
for the Born model, 269–272

H

Hack's law (stream networks), 425–427, 501, 502
Hall conductivity
effective-medium approximation, 308
governing equations, 307
scaling properties, 310
Hopping conductivity
critical-path analysis, 389, 404, 416
effect of Coulomb interactions, 407
effect of density of states, 406
effect of fractal morphology, 413
exact solution for Bethe lattices, 396
exact solution in 1D, 394
Miller-Abrahams network, 390
Mott's formula, 401
network reduction, 393
symmetric hopping model, 393, 394, 396
universal scaling, 415
Hydraulic conductivity
critical-path analysis, 119, 121, 130, 132, 133, 144, 145
percolation model, 119, 121, 130, 133
Hydrology
hydrogeology, 489
infiltration, 489

I

Immunological systems
damage spreading, 449
Hamming distance, 449
localization, 448
percolation, 447, 448
Incoherent inelastic scattering, 379
Information spreading, 9, 51, 547, 606, 607, 609
Invasion percolation
efficient simulation, 185
experimental data, 185
fractal dimensions, 185, 202
gradient invasion percolation, 422
in cancerous vascular networks, 428
in queuing problem, 432
in social dynamics, 433
in stream networks, 425, 427
loopless model, 422
optimization problems, 430, 431
relation with self-organized criticality, 420, 423, 610
self-avoiding model, 422
simulation of spin systems, 429
site-bond model, 70, 198, 202
thermal invasion, 423, 424
without trapping, 193
with trapping, 171, 186, 432, 461, 498
with two defenders, 189
Inverted Swiss-cheese model, 38, 40, 41, 320, 325

K

$k-$core percolation, 50

L

Laplace equation, 65
Lattice animal, 348–350
Lattice structure factor, 294
Leath–Alexandrowicz algorithm, 13
Life cycle
  consumer demand, 603, 606
  industrial products, 603
  percolation model, 604, 605
  sale agents, 605
Linear chain, 394, 408
Localization, 319, 370, 413, 414, 448
London penetration depth, 321, 322
Loss modulus, 353, 354, 358
Loss angle, 301, 304–306, 353, 354
Loss of consciousness, 458–460
Lymphocyte membranes, 443, 446

M

Magnetoconductivity, 307, 308, 320
Mars, *see also* topography
  ancient sea level, 595
  isoheight lines, 595, 596
  percolation scaling, 595
Master equation, 366, 393–395
Maxwell equations, 321
Mean-field approximation, 235, 349
Mercury porosimetry
  percolation models, 56, 70, 71
Metal-insulator films, 279
Metal-insulator transition, 295, 296, 312
Microbial communications
  autoinducers, 484, 486
  firing cells, 485, 486
  percolation model, 465, 485–487
  quorum sensing, 482, 484, 487
Microemulsions, 34, 47, 51, 122, 304–306
Miller-Abrahams network, 390
Mobile Ad-hoc Networks (MANETs)
  coverage, 588
  critical transmission range, 589
  giant cluster, 589
  percolation, 589
Molecular motors
  active gels, 477
  cytoskeletal filamentous actin, 477
  effective-medium approximation, 478, 479
  elastic network model, 478
  elastic properties, 478
  extracellular matrix, 477
  percolation transition, 479
  shera modulus, 479
Molecular weight, 205, 210, 212, 226, 231
Mott hopping model, 401–403
Mott's formula, 389, 401, 402, 406

N

Network reduction (hopping conductivity), 393, 408
Neuromorphic Computing
  avalanche size distribution, 457
  conductance, 456
  memristors, 452, 453, 458
  percolation aspects, 454, 457
Newman–Ziff algorithm, 14, 15
Non-catalystic reactions
  diffusion-limited, 209, 211
  percolation models, 209, 214
  reaction-limited, 209, 210, 213
  with fragmentation, 211, 213
$n-$vector model, 331

O

Optimal dismantling
  C-dismantling set of a graph, 546
  interdiction problem, 546
  van der Pol oscillator, 546
Ornstein–Zernike equation, 36

P

Packing fraction, 266, 283
Percolation
  accessible fraction, 15, 57, 83, 165, 172, 226, 257
  backbone, 61, 107, 110, 165, 482, 494
  bond model, 43, 600
  cluster number, 16
  clusters, 14, 15, 22, 29, 56, 62, 71, 86, 106–112, 114, 169, 186, 206, 267, 283, 300, 304, 306, 348, 349, 358, 373, 386, 403, 414, 429, 434, 463, 473, 480, 482, 483, 524, 531, 552, 555, 583, 593, 599
  conductor–insulator composites, 15, 19, 23, 24, 28, 29
  conductor–superconductor composites, 19, 23, 322
  correlation length, 15, 25, 28, 43, 49, 61, 107, 121, 132, 138, 140, 141, 143, 163, 164, 170, 196, 197, 236, 292, 324, 325, 348, 369, 372, 374, 383, 413, 479
  critical exponents, 254, 441
  definition, 60, 61, 215, 386, 598

dimensional invariance, 6
flow and transport properties, 15, 17, 19, 24, 32
fractal dimensions, 21, 27, 29, 61, 107, 206, 482, 494, 505
galactic structure, 51
historical origin, 8
in large-scale systems, 585
Leath–Alexandrowicz algorithm, 13
mean cluster size, 16, 37, 47, 487, 517, 520, 594
measurement of its properties, 24
Newman–Ziff algorithm, 14, 15
percolation probability, 15, 61, 62, 119–121, 151, 279, 345, 369, 441, 487
rigid–soft composites, 19, 23
rigid–superrigid composites, 19, 23
scale dependence, 21
scaling properties, 23, 43, 44, 193, 357
site model, 44, 62, 600
spreading of technological innovations, 606
thresholds, 5, 11–13, 15, 18, 24–27, 29–31, 33–38, 40, 43–45, 48, 49, 57, 59, 61–64, 74, 76, 81, 83, 86, 87, 91, 93, 95, 96, 98, 99, 104, 107, 110, 112–116, 118–121, 124, 127, 133, 135, 137, 138, 140, 143, 144, 147, 149–151, 162, 163, 168, 169, 179, 182, 185, 194, 196, 198, 214, 215, 222, 232, 233, 236, 237, 248, 253, 254, 256–263, 271–273, 275, 277, 280–297, 302, 310, 312, 313, 315–318, 320–323, 326–328, 334–336, 344, 354, 357, 358, 368, 373, 374, 378, 381, 384, 385, 397, 399–401, 405, 414, 416, 417, 420, 424, 429–431, 440, 441, 445, 447, 449, 450, 454, 466, 470, 472–474, 479, 484–487, 494, 508, 511, 512, 514, 517, 519–521, 524, 525, 527, 532, 539, 542, 552, 553, 556, 558, 560, 561, 567, 593, 594, 598–600, 602, 603, 606, 608–610
universal scaling laws, 153
with trapping, 192, 193
Permeability
critical-path analysis, 127, 128, 130, 132
effective-medium approximation, 125–127
for power-law fluids, 132
link with electrical conductivity, 120, 126, 127, 131
with deformation, 126
Phase transition, 5, 23, 48–52, 115, 196, 221, 264, 265, 279, 332, 333, 345, 434, 458, 474, 475, 484, 517, 520, 522, 526–529, 532, 534, 535, 540, 559, 581, 582, 591, 594, 596, 600, 606
Phonon-assisted hopping, 389
Phonons, 367, 369, 370, 372–375, 379–381, 383, 385, 391
Physical gels, 343, 351, 360
Plague susceptibility
effective percolation threshold, 508
lattice model, 508
percolation transition, 510
Plants
percolation model, 507
Poisson's ratio
universality of, 277
Polycondensation, 343, 357, 358
Polycrystalline materials, 415
Polymer chains, 331, 343, 358
Polymer composites
effect of thickness, 292
elastic moduli, 287
electrical conductivity, 281
with graphene, 289, 290
with nanotubes, 290, 291
Polymeric materials, 344
Polymerization, 5, 8, 332, 343, 345, 347, 348, 350, 356, 445, 534–536
Poor man's percolation, *see*Effective-medium approximation
Position-space renormalization, 139
Powders, 202, 253, 281, 282, 284–286, 303, 304, 316, 449, 450, 602, 603
Power laws, 18–20, 23, 24, 28, 29, 40, 42, 43, 48, 61, 63, 87, 103, 105, 107, 109, 112, 117–121, 124, 133–136, 147, 149–152, 165, 166, 169, 179, 182, 183, 254, 257, 258, 265, 281, 287, 291, 293, 296, 297, 300, 303, 312, 316–318, 324, 326, 328, 345, 346, 349, 352, 356, 362, 369, 381, 409, 415, 427, 438–440, 442, 457, 486, 494, 495, 499, 507, 520, 521, 524, 525, 545, 561, 563, 568, 570, 572, 575, 580, 588–590, 603, 606
Protein structure
connected networks, 476
percolation transition, 477
power-law distribution, 477, 478

Q

Quenched disorder, 51, 570
Queuing problem, 432

R

Radial distribution function
  Ornstein-Zernike formulation, 36
Raman scattering, 338
Random resistor network, *see also*Effectivemedium approximation; Percolation; Electrical conductivity
  exact solution for Bethe lattices, 179, 396
  Green function formulation, 269, 270, 272, 400
Rayleigh-Bénard convection, 568
Red bonds, 18, 22
Relative permeability
  Bethe lattice model, 179
  effective-medium approximation, 180
  random-bond correlated site model, 185
  random percolation model, 176, 179
Relaxation time spectrum, 356
Renormalization-group transformation
  for conductivity, 138, 141, 142
  for percolation, 138, 141, 142
Renormalized effective-medium approximation
  electrical conductivity, 143
Rigidity percolation
  constraint-counting method, 259, 260
  continuum model, 245
  correlation length, 256
  critical-path analysis, 276
  effective-medium approximation, 269, 270
  filamentous networks, 275
  finite-size scaling, 255–257
  floppy state, 251, 253, 276
  force distribution, 257
  nature of phase transition, 264
  numerical simulation, 255
  pebble game, 260, 261
  percolation threshold, 273, 315
  scaling properties, 267
  shortcomings of, 115
  static and dynamic rigidity, 251
  superelasticity, 253
Rigidity percolation
  length of a river, 501–503
  percolation scaling, 501
Rotational invariance (Born model), 252, 359

S

Sample-spanning cluster, 10–15, 17, 18, 20–27, 30, 34, 38, 42–44, 48, 52, 57, 61, 62, 81, 83, 87, 89, 118, 121, 122, 129, 130, 138, 140, 149, 163–166, 169, 171, 201, 202, 227, 229, 236, 245, 254, 257, 263, 265, 282, 286–289, 292, 293, 297, 323–325, 327, 334, 358, 359, 369, 382, 386, 393, 404, 417, 420, 430, 434, 485, 509, 517, 522, 523, 526, 531, 538, 539, 556, 609, 610
Scale-free networks, 434, 439, 477, 520, 547, 606
Scaling function, 19, 22, 23, 168, 194, 223, 268, 298, 299, 311, 345, 353, 356, 412, 415, 563, 564, 570, 576, 589, 599
Scaling theory, 28, 193, 256, 310, 311, 370, 375, 381, 412
Scattering intensity, 350, 379
Sea ice
  permeability, 600, 601
Self-avoiding walks, 9
Self-organized criticality, 103, 104, 106, 420, 423, 461, 600, 610, *see also* Eathquakes; Forrest fires; Directed percolation
Self-similarity, 88, 378, 427
Semiconductors, 118, 127, 301, 333, 389, 393, 402, 407, 411, 413–415, 452, 588
Sensory transmission
  complex network model, 459
  neural activity, 459
  percolation aspects, 458–460
Shear deformation, 335
Shear modulus, 18, 247, 250, 252, 254, 255, 264–266, 268, 273, 274, 276, 277, 313, 315, 316, 339, 358, 360–362, 479
Silica aerogels, 343, 360, 376–380, 386, 387
Singly-connected bonds, 358
Site-site Green function, 398
Small-angle scattering, 378
Small-world networks, 12, 438
Social dynamics, 433
Social networks, 439, 541, 547, 604–606
Sol, 332, 344, 345, 347, 351, 352, 360, 362
Sol-gel transition, 34, 332, 343, 344
Sorption
  determining connectivity, 82
  hysteresis loop, 78
  isotherm types, 65, 77
  percolation model, 79
Spectral density, 370, 379

Spectral dimension
  diffusion/conduction, 236
  elastic systems (flexural dimension), 382
Statistical self-similarity, 21
Stream networks, 425, 427
Structure factor
  numerical computation, 403
  scaling analysis, 383
Superconductivity
  Bardeen-Cooper-Schrieffer theory, 323
  coherence length, 326
  critical current, 324
  critical field, 320, 325
  diamagnetic susceptibility, 326
  differential magnetic susceptibility, 326
  London penetration depth, 321, 322
  specific heat, 323, 324
  Type I, 320, 325
  Type II, 320, 325, 326
  upper critical field, 320, 326
Superelastic percolation networks, 254, 257, 267, 335, 360–363
Superlocalization, 413, 415
Superrigid composites, 18, 19, 23, 254
Swiss-cheese model, 38–41, 316, 320
Symmetric hopping model, 393, 394, 396

T
Taylor-Isard scaling constant, 415, 416
Tessellation models, 34, 220, 280
Thermal percolation
  effective-medium approximation, 150
  effective-medium approximation and percolation, 150
  percolation model, 327
Thin metal films, 24, 25, 293, 294
Thin wetting films, 186, 188, 190, 192
Topography
  Earth, 590, 595
  fractal dimension, 26, 593
  Mars, 595, 596
  mean island size, 593, 594
  Moon, 590, 594
  power spectrum, 590, 591
  relief network, 593, 594
Tortuosity
  link with continuum percolation, 62
  percolation model for saturated and unsaturated media, 63
Transition rates (hopping model), 366, 368, 389, 394, 395, 397, 398, 408
Tropical deforestation
  forest fragmentation, 513, 514
  fragment-size distribution, 514
  power law, 514
Tunneling, 291, 297, 298, 389, 454–457
Turbulent flow, *see* Directed percolation

U
Universality class, 51, 115, 266, 296, 384, 431, 432, 434, 501, 520, 524, 542, 562, 568, 573–575, 577, 579, 580, 600
Universal scaling laws, 153, 280, 416
Upper critical dimension, 265, 349, 526, 556

V
Variable-range hopping, 402, 403, 406
Vector percolation, *see*Rigidity percolation
Vegetation growth
  percolation scaling, 499
  root radial extent, 499
  tree heights, 499, 500
Vibrational density of states
  crossover between scalar and vector models, 383
  effective-medium approximation, 368
  fracton regime, 384
  numerical computations, 367
  phonon regime, 369, 370, 380, 381, 383
  scalar approximation, 366–368, 370, 373
  scaling properties, 370, 381
  vector percolation model, 381, 384
Viscoelasticity, *see* Gelation; Gels
Voronoi network, 34, 38, 40, 235, 435
Vulcanization, 343

W
Water balance
  net primary productivity, 505
  percolation model, 504
Water cycle, 489–491, 513
Weathering rind, 498

X
X-ray scattering, 387

Y
Young-Laplace equation, 65, 67, *see also*